Analysis of Multivariate and High-Dimensional Data

'Big data' poses challenges that require both classical multivariate methods and contemporary techniques from machine learning and engineering. This modern text integrates the two strands into a coherent treatment, drawing together theory, data, computation and recent research.

The theoretical framework includes formal definitions, theorems and proofs which clearly set out the guaranteed 'safe operating zone' for the methods and allow users to assess whether data are in or near the zone. Extensive examples showcase the strengths and limitations of different methods in a range of cases: small classical data; data from medicine, biology, marketing and finance; high-dimensional data from bioinformatics; functional data from proteomics; and simulated data. High-dimension low sample size data get special attention. Several data sets are revisited repeatedly to allow comparison of methods. Generous use of colour, algorithms, MATLAB code and problem sets completes the package. The text is suitable for graduate students in statistics and researchers in data-rich disciplines.

INGE KOCH is Associate Professor of Statistics at the University of Adelaide, Australia.

CAMBRIDGE SERIES IN STATISTICAL AND PROBABILISTIC MATHEMATICS

Editorial Board
Z. Ghahramani (Department of Engineering, University of Cambridge)
R. Gill (Mathematical Institute, Leiden University)
F. P. Kelly (Department of Pure Mathematics and Mathematical Statistics, University of Cambridge)
B. D. Ripley (Department of Statistics, University of Oxford)
S. Ross (Department of Industrial and Systems Engineering, University of Southern California)
M. Stein (Department of Statistics, University of Chicago)

This series of high-quality upper-division textbooks and expository monographs covers all aspects of stochastic applicable mathematics. The topics range from pure and applied statistics to probability theory, operations research, optimization, and mathematical programming. The books contain clear presentations of new developments in the field and also of the state of the art in classical methods. While emphasizing rigorous treatment of theoretical methods, the books also contain applications and discussions of new techniques made possible by advances in computational practice.

A complete list of books in the series can be found at www.cambridge.org/statistics.
Recent titles include the following:

11. *Statistical Models,* by A. C. Davison
12. *Semiparametric Regression,* by David Ruppert, M. P. Wand and R. J. Carroll
13. *Exercises in Probability,* by Loïc Chaumont and Marc Yor
14. *Statistical Analysis of Stochastic Processes in Time,* by J. K. Lindsey
15. *Measure Theory and Filtering,* by Lakhdar Aggoun and Robert Elliott
16. *Essentials of Statistical Inference,* by G. A. Young and R. L. Smith
17. *Elements of Distribution Theory,* by Thomas A. Severini
18. *Statistical Mechanics of Disordered Systems,* by Anton Bovier
19. *The Coordinate-Free Approach to Linear Models,* by Michael J. Wichura
20. *Random Graph Dynamics,* by Rick Durrett
21. *Networks,* by Peter Whittle
22. *Saddlepoint Approximations with Applications,* by Ronald W. Butler
23. *Applied Asymptotics,* by A. R. Brazzale, A. C. Davison and N. Reid
24. *Random Networks for Communication,* by Massimo Franceschetti and Ronald Meester
25. *Design of Comparative Experiments,* by R. A. Bailey
26. *Symmetry Studies,* by Marlos A. G. Viana
27. *Model Selection and Model Averaging,* by Gerda Claeskens and Nils Lid Hjort
28. *Bayesian Nonparametrics,* edited by Nils Lid Hjort *et al.*
29. *From Finite Sample to Asymptotic Methods in Statistics,* by Pranab K. Sen, Julio M. Singer and Antonio C. Pedrosa de Lima
30. *Brownian Motion,* by Peter Mörters and Yuval Peres
31. *Probability (Fourth Edition),* by Rick Durrett
33. *Stochastic Processes,* by Richard F. Bass
34. *Regression for Categorical Data,* by Gerhard Tutz
35. *Exercises in Probability (Second Edition),* by Loïc Chaumont and Marc Yor
36. *Statistical Principles for the Design of Experiments,* by R. Mead, S. G. Gilmour and A. Mead

Analysis of Multivariate and High-Dimensional Data

Inge Koch
University of Adelaide, Australia

CAMBRIDGE
UNIVERSITY PRESS

CAMBRIDGE
UNIVERSITY PRESS

University Printing House, Cambridge CB2 8BS, United Kingdom

One Liberty Plaza, 20th Floor, New York, NY 10006, USA

477 Williamstown Road, Port Melbourne, VIC 3207, Australia

314–321, 3rd Floor, Plot 3, Splendor Forum, Jasola District Centre,
New Delhi – 110025, India

79 Anson Road, #06–04/06, Singapore 079906

Cambridge University Press is part of the University of Cambridge.

It furthers the University's mission by disseminating knowledge in the pursuit of education, learning and research at the highest international levels of excellence.

www.cambridge.org
Information on this title: www.cambridge.org/9780521887939

First published 2014
Reprinted 2020

Printed in the United Kingdom by TJ International Ltd. Padstow, Cornwall

A catalog record for this publication is available from the British Library.

Library of Congress Cataloging in Publication Data
Koch, Inge, 1952–
Analysis of multivariate and high-dimensional data / Inge Koch.
pages cm
ISBN 978-0-521-88793-9 (hardback)
1. Multivariate analysis. 2. Big data. I. Title.
QA278.K5935 2013
519.5′35–dc23 2013013351

ISBN 978-0-521-88793-9 Hardback

Additional resources for this publication at www.cambridge.org/9780521887939

To Alun, Graeme and Reiner

Contents

List of Algorithms

Notation

$\mathbf{a} = [a_1 \cdots a_d]^\top$	Column vector in $\mathbb{R}^d$
$A,\ A^\top,\ B$	Matrices, with $A^\top$ the transpose of A
$A_{p\times q},\ B_{r\times s}$	Matrices A and B of size $p \times q$ and $r \times s$
$A = (a_{ij})$	Matrix A with entries a_{ij}
$A = \begin{bmatrix}\mathbf{a}_1 \cdots \mathbf{a}_p\end{bmatrix} = \begin{bmatrix}\mathbf{a}_{\bullet 1}\\ \vdots \\ \mathbf{a}_{\bullet q}\end{bmatrix}$	Matrix A of size $p \times q$ with columns $\mathbf{a}_i$, rows $\mathbf{a}_{\bullet j}$
A_{diag}	Diagonal matrix consisting of the diagonal entries of A
$\mathbf{0}_{k\times\ell}$	$k \times \ell$ matrix with all entries 0
$\mathbf{1}_k$	Column vector with all entries 1
$\mathbf{I}_{d\times d}$	$d \times d$ identity matrix
$\mathbf{I}_{k\times\ell}$	$\begin{pmatrix}\mathbf{I}_{k\times k} & \mathbf{0}_{k\times(\ell-k)}\end{pmatrix}$ if $k \le \ell$, and $\begin{pmatrix}\mathbf{I}_{\ell\times\ell}\\ \mathbf{0}_{(k-\ell)\times\ell}\end{pmatrix}$ if $k \ge \ell$
$\mathbf{X},\ \mathbf{Y}$	d-dimensional random vectors
$\mathbf{X}\vert\mathbf{Y}$	Conditional random vector $\mathbf{X}$ given $\mathbf{Y}$
$\mathbf{X} = \begin{bmatrix}X_1 \ldots X_d\end{bmatrix}^\top$	d-dimensional random vector with entries (or variables) X_j
$\mathbb{X} = \begin{bmatrix}\mathbf{X}_1\ \mathbf{X}_2 \cdots \mathbf{X}_n\end{bmatrix}$	$d \times n$ data matrix of random vectors $\mathbf{X}_i$, $i \le n$
$\mathbf{X}_i = \begin{bmatrix}X_{i1} \cdots X_{id}\end{bmatrix}^\top$	Random vector from $\mathbb{X}$ with entries X_{ij}
$\mathbf{X}_{\bullet j} = \begin{bmatrix}X_{1j} \cdots X_{nj}\end{bmatrix}$	$1 \times n$ row vector of the jth variable of $\mathbb{X}$
$\boldsymbol{\mu} = \mathbb{E}\mathbf{X}$	Expectation of a random vector $\mathbf{X}$, also denoted by $\boldsymbol{\mu}_X$
$\overline{\mathbf{X}}$	Sample mean
$\overline{\overline{\mathbf{X}}}$	Average sample class mean
$\sigma^2 = \text{var}(X)$	Variance of random variable X
$\Sigma = \text{var}(\mathbf{X})$	Covariance matrix of $\mathbf{X}$ with entries σ_{jk} and $\sigma_{jj} = \sigma_j^2$
S	Sample covariance matrix of $\mathbb{X}$ with entries s_{ij} and $s_{jj} = s_j^2$
$R = (\rho_{ij}),\ R_S = (\widehat{\rho}_{ij})$	Matrix of correlation coefficients for the population and sample
$Q_{\langle n\rangle},\ Q^{\langle d\rangle}$	Dual matrices $Q_{\langle n\rangle} = \mathbb{X}\mathbb{X}^\top$ and $Q^{\langle d\rangle} = \mathbb{X}^\top\mathbb{X}$
Σ_{diag}	Diagonal matrix with entries σ_j^2 obtained from Σ
S_{diag}	Diagonal matrix with entries s_j^2 obtained from S
$\Sigma = \Gamma\Lambda\Gamma^\top$	Spectral decomposition of Σ

$\Gamma_k = \begin{bmatrix}\boldsymbol{\eta}_1 \cdots \boldsymbol{\eta}_k\end{bmatrix}$	$d \times k$ matrix of (orthogonal) eigenvectors of Σ, $k \le d$
$\Lambda_k = \mathrm{diag}(\lambda_1, \ldots, \lambda_k)$	Diagonal $k \times k$ matrix with diagonal entries the eigenvalues of Σ, $k \le d$
$S = \widehat{\Gamma}\widehat{\Lambda}\widehat{\Gamma}^{\mathsf{T}}$	Spectral decomposition of S with eigenvalues $\widehat{\lambda}_j$ and eigenvectors $\widehat{\boldsymbol{\eta}}_j$
$\beta_3(\mathbf{X})$, $b_3(\mathbb{X})$	Multivariate skewness of $\mathbf{X}$ and sample skewness of $\mathbb{X}$
$\beta_4(\mathbf{X})$, $b_4(\mathbb{X})$	Multivariate kurtosis of $\mathbf{X}$ and sample kurtosis of $\mathbb{X}$
f, F	Multivariate probability density and distribution functions
ϕ, Φ	Standard normal probability density and distribution functions
f, f_G	Multivariate probability density functions; f and f_G have the same mean and covariance matrix, and f_G is Gaussian
$L(\theta)$ or $L(\theta \mid \mathbb{X})$	Likelihood function of the parameter θ, given $\mathbb{X}$
$\mathbf{X} \sim (\boldsymbol{\mu}, \Sigma)$	Random vector with mean $\boldsymbol{\mu}$ and covariance matrix Σ
$\mathbf{X} \sim \mathcal{N}(\boldsymbol{\mu}, \Sigma)$	Random vector from the multivariate normal distribution with mean $\boldsymbol{\mu}$ and covariance matrix Σ
$\mathbb{X} \sim \mathsf{Sam}(\overline{\mathbf{X}}, S)$	Data – with sample mean $\overline{\mathbf{X}}$ and sample covariance matrix S
$\mathbb{X}_{\text{cent}}$	Centred data $\begin{bmatrix}\mathbf{X}_1 - \overline{\mathbf{X}} \cdots \mathbf{X}_n - \overline{\mathbf{X}}\end{bmatrix}$, also written as $\mathbb{X} - \overline{\mathbf{X}}$
$\mathbf{X}_\Sigma$, $\mathbb{X}_S$	Sphered vector and data $\Sigma^{-1/2}(\mathbf{X} - \boldsymbol{\mu})$, $S^{-1/2}(\mathbb{X} - \overline{\mathbf{X}})$
$\mathbf{X}_{\text{scale}}$, $\mathbb{X}_{\text{scale}}$	Scaled vector and data $\Sigma_{\text{diag}}^{-1/2}(\mathbf{X} - \boldsymbol{\mu})$, $S_{\text{diag}}^{-1/2}(\mathbb{X} - \overline{\mathbf{X}})$
$\mathbf{X}^{\diamond}$, $\mathbb{X}^{\diamond}$	(Spatially) whitened random vector and data
$\mathbf{W}^{(k)} = \begin{bmatrix}W_1 \cdots \mathbf{W}_k\end{bmatrix}^{\mathsf{T}}$	Vector of first k principal component scores
$\mathbb{W}^{(k)} = \begin{bmatrix}\mathbf{W}_{\bullet 1} \ldots \mathbf{W}_{\bullet k}\end{bmatrix}^{\mathsf{T}}$	$k \times n$ matrix of first k principal component scores
$\mathbf{P}_k = W_k \boldsymbol{\eta}_k$	Principal component projection vector
$\mathbb{P}_{\bullet k} = \widehat{\boldsymbol{\eta}}_k \mathbf{W}_{\bullet k}$	$d \times n$ matrix of principal component projections
$\mathbf{F}$, $\mathbb{F}$	Common factor for population and data in k-factor model
$\mathbf{S}$, $\mathbb{S}$	Source for population and data in Independent Component Analysis
$\mathcal{O} = \{O_1, \ldots, O_n\}$	Set of objects corresponding to data $\mathbb{X} = \begin{bmatrix}\mathbf{X}_1\ \mathbf{X}_2 \ldots \mathbf{X}_n\end{bmatrix}$
$\{\mathcal{O}, \varrho\}$	Observed data, consisting of objects O_i and dissimilarities ϱ_{ik} between pairs of objects
$\mathfrak{f}$, $\mathfrak{f}(\mathbf{X})$, $\mathfrak{f}(\mathbb{X})$	Feature map, feature vector and feature data
$\mathrm{cov}(\mathbf{X}, \mathbf{T})$	$d_X \times d_T$ (between) covariance matrix of $\mathbf{X}$ and $\mathbf{T}$
$\Sigma_{12} = \mathrm{cov}(\mathbf{X}^{[1]}, \mathbf{X}^{[2]})$	$d_1 \times d_2$ (between) covariance matrix of $\mathbf{X}^{[1]}$ and $\mathbf{X}^{[2]}$
$S_{12} = \mathrm{cov}(\mathbb{X}^{[1]}, \mathbb{X}^{[2]})$	$d_1 \times d_2$ sample (between) covariance matrix of $d_\ell \times n$ data $\mathbb{X}^{[\ell]}$, for $\ell = 1, 2$
$C = \Sigma_1^{-1/2} \Sigma_{12} \Sigma_2^{-1/2}$	Canonical correlation matrix of $\mathbf{X}^{[\ell]} \sim (\boldsymbol{\mu}_\ell, \Sigma_\ell)$, for $\ell = 1, 2$
$C = P \Upsilon Q^{\mathsf{T}}$	Singular value decomposition of C with singular values υ_j and eigenvectors $\mathbf{p}_j, \mathbf{q}_j$
$\widehat{C} = \widehat{P}\widehat{\Upsilon}\widehat{Q}^{\mathsf{T}}$	Sample canonical correlation matrix and its singular value decomposition
$R^{[C,1]} = CC^{\mathsf{T}}$, $R^{[C,2]} = C^{\mathsf{T}}C$	Matrices of multivariate coefficients of determination, with C the canonical correlation matrix

$\mathbf{U}^{(k)}, \mathbf{V}^{(k)}$	Pair of vectors of k-dimensional canonical correlations
$\boldsymbol{\varphi}_k, \boldsymbol{\psi}_k$	kth pair of canonical (correlation) transforms
$\mathbb{U}^{(k)}, \mathbb{V}^{(k)}$	$k \times n$ matrices of k-dimensional canonical correlation data
$\mathcal{C}_\nu$	νth class (or cluster)
$\mathfrak{r}$	Discriminant rule or classifier
$h,\ h_\beta$	Decision function for a discriminant rule $\mathfrak{r}$ (h_β depends on β)
$\mathfrak{b}, \mathfrak{w}$	Between-class and within-class variability
$\mathcal{E}$	(Classification) error
$\mathcal{P}(\mathbb{X}, k)$	k-cluster arrangement of $\mathbb{X}$
$A \circ B$	Hadamard or Schur product of matrices A and B
$\operatorname{tr}(A)$	Trace of a matrix A
$\det(A)$	Determinant of a matrix A
$\operatorname{dir}(\mathbf{X})$	Direction (vector) $\mathbf{X}/\|\mathbf{X}\|$ of $\mathbf{X}$
$\|\cdot\|,\ \|\cdot\|_p,$	(Euclidean) norm, p-norm of a vector or matrix
$\|\mathbf{X}\|_{tr}$	Trace norm of $\mathbf{X}$ given by $[\operatorname{tr}(\Sigma)]^{1/2}$
$\|A\|_{\text{Frob}}$	Frobenius norm of a matrix A given by $[\operatorname{tr}(AA^{\mathsf{T}})]^{1/2}$
$\Delta(\mathbf{X}, \mathbf{Y})$	Distance between vectors $\mathbf{X}$ and $\mathbf{Y}$
$\varrho(\mathbf{X}, \mathbf{Y})$	Dissimilarity of vectors $\mathbf{X}$ and $\mathbf{Y}$
$\mathfrak{a}(\boldsymbol{\alpha}, \boldsymbol{\beta})$	Angle between directions $\boldsymbol{\alpha}$ and $\boldsymbol{\beta}$: $\arccos(\boldsymbol{\alpha}^{\mathsf{T}}\boldsymbol{\beta})$
$\mathcal{H},\ \mathcal{I},\ \mathcal{J},\ \mathcal{K}$	Entropy, mutual information, negentropy and Kullback-Leibler divergence
$\mathcal{Q}$	Projection index
$\mathbf{n} \succ \mathbf{d}$	Asymptotic domain, d fixed and $n \to \infty$
$\mathbf{n} \succeq \mathbf{d}$	Asymptotic domain, $d, n \to \infty$ and $d = O(n)$
$\mathbf{n} \preceq \mathbf{d}$	Asymptotic domain, $d, n \to \infty$ and $n = o(d)$
$\mathbf{n} \prec \mathbf{d}$	Asymptotic domain, n fixed and $d \to \infty$

Preface

This book is about data in many – and sometimes very many – variables and about analysing such data. The book attempts to integrate classical multivariate methods with contemporary methods suitable for high-dimensional data and to present them in a coherent and transparent framework. Writing about ideas that emerged more than a hundred years ago and that have become increasingly relevant again in the last few decades is exciting and challenging. With hindsight, we can reflect on the achievements of those who paved the way, whose methods we apply to ever bigger and more complex data and who will continue to influence our ideas and guide our research. Renewed interest in the classical methods and their extension has led to analyses that give new insight into data and apply to bigger and more complex problems.

There are two players in this book: *Theory* and *Data*. *Theory* advertises its wares to lure *Data* into revealing its secrets, but *Data* has its own ideas. *Theory* wants to provide elegant solutions which answer many but not all of *Data*'s demands, but these lead *Data* to pose new challenges to *Theory*. Statistics thrives on interactions between theory and data, and we develop better theory when we 'listen' to data. Statisticians often work with experts in other fields and analyse data from many different areas. We, the statisticians, need and benefit from the expertise of our colleagues in the analysis of their data and interpretation of the results of our analysis. At times, existing methods are not adequate, and new methods need to be developed.

This book attempts to combine theoretical ideas and advances with their application to data, in particular, to interesting and real data. I do not shy away from stating theorems as they are an integral part of the ideas and methods. Theorems are important because they summarise what we know and the conditions under which we know it. They tell us when methods may work with particular data; the hypotheses may not always be satisfied exactly, but a method may work nevertheless. The precise details do matter sometimes, and theorems capture this information in a concise way.

Yet a balance between theoretical ideas and data analysis is vital. An important aspect of any data analysis is its interpretation, and one might ask questions like: What does the analysis tell us about the data? What new insights have we gained from a particular analysis? How suitable is my method for my data? What are the limitations of a particular method, and what other methods would produce more appropriate analyses? In my attempts to answer such questions, I endeavour to be objective and emphasise the strengths and weaknesses of different approaches.

Who Should Read This Book?

This book is suitable for readers with various backgrounds and interests and can be read at different levels. It is appropriate as a graduate-level course – two course outlines are suggested in the section ‘Teaching from This Book’. A second or more advanced course could make use of the more advanced sections in the early chapters and include some of the later chapters. The book is equally appropriate for working statisticians who need to find and apply a relevant method for analysis of their multivariate or high-dimensional data and who want to understand how the chosen method deals with the data, what its limitations might be and what alternatives are worth considering.

Depending on the expectation and aims of the reader, different types of backgrounds are needed. Experience in the analysis of data combined with some basic knowledge of statistics and statistical inference will suffice if the main aim involves applying the methods of this book. To understand the underlying theoretical ideas, the reader should have a solid background in the theory and application of statistical inference and multivariate regression methods and should be able to apply confidently ideas from linear algebra and real analysis.

Readers interested in **statistical ideas and their application to data** may benefit from the theorems and their illustrations in the examples. These readers may, in a first journey through the book, want to focus on the basic ideas and properties of each method and leave out the last few more advanced sections of each chapter. For possible paths, see the models for a one-semester course later in this preface.

Researchers and graduate students with a good background in statistics and mathematics who are primarily interested in the **theoretical developments of the different topics** will benefit from the formal setting of definitions, theorems and proofs and the careful distinction of the population and the sample case. This setting makes it easy to understand what each method requires and which ideas can be adapted. Some of these readers may want to refer to the recent literature and the references I provide for theorems that I do not prove.

Yet another broad group of readers may want to focus on **applying the methods of this book to particular data**, with an emphasis on the results of the data analysis and the new insight they gain into their data. For these readers, the interpretation of their results is of prime interest, and they can benefit from the many examples and discussions of the analysis for the different data sets. These readers could concentrate on the descriptive parts of each method and the interpretative remarks which follow many theorems and need not delve into the theorem/proof framework of the book.

Outline

This book consists of **three parts**. Typically, each method corresponds to a single chapter, and because the methods have different origins and varied aims, it is convenient to group the chapters into parts. The methods focus on two main themes:

1. *Component Analysis*, which aims to simplify the data by summarising them in a smaller number of more relevant or more interesting components

2. *Statistical Learning*, which aims to group, classify or regress the (component) data by partitioning the data appropriately or by constructing rules and applying these rules to new data.

The two themes are related, and each method I describe addresses at least one of the themes.

The first chapter in each part presents notation and summarises results required in the following chapters. I give references to background material and to proofs of results, which may help readers not acquainted with some topics. Readers who are familiar with the topics of the three first chapters in their part may only want to refer to the notation. Properties or theorems in these three chapters are called *Results* and are stated without proof. Each of the main chapters in the three parts is dedicated to a specific method or topic and illustrates its ideas on data.

Part I deals with the classical methods *Principal Component Analysis, Canonical Correlation Analysis* and *Discriminant Analysis*, which are 'musts' in multivariate analysis as they capture essential aspects of analysing multivariate data. The later sections of each of these chapters contain more advanced or more recent ideas and results, such as Principal Component Analysis for high-dimension low sample size data and Principal Component Regression. These sections can be left out in a first reading of the book without greatly affecting the understanding of the rest of Parts I and II.

Part II complements Part I and is still classical in its origin: *Cluster Analysis* is similar to Discriminant Analysis and partitions data but without the advantage of known classes. *Factor Analysis* and Principal Component Analysis enrich and complement each other, yet the two methods pursue distinct goals and differ in important ways. Classical *Multidimensional Scaling* may seem to be different from Principal Component Analysis, but Multidimensional Scaling, which ventures into non-linear component analysis, can be regarded as a generalisation of Principal Component Analysis. The three methods, Principal Component Analysis, Factor Analysis and Multidimensional Scaling, paved the way for non-Gaussian component analysis and in particular for Independent Component Analysis and Projection Pursuit.

Part III gives an overview of more recent and current ideas and developments in component analysis methods and links these to statistical learning ideas and research directions for high-dimensional data. A natural starting point are the twins, *Independent Component Analysis* and *Projection Pursuit*, which stem from the signal-processing and statistics communities, respectively. Because of their similarities as well as their resulting analysis of data, we may regard both as non-Gaussian component analysis methods. Since the early 1980s, when Independent Component Analysis and Projection Pursuit emerged, the concept of independence has been explored by many authors. Chapter 12 showcases *Independent Component Methods* which have been developed since about 2000. There are many different approaches; I have chosen some, including *Kernel Independent Component Analysis*, which have a more statistical rather than heuristic basis. The final chapter returns to the beginning – *Principal Component Analysis* – but focuses on current ideas and research directions: feature selection, component analysis of high-dimension low sample size data, decision rules for such data, asymptotics and consistency results when the dimension increases faster than the sample size. This last chapter includes *inconsistency* results and concludes with a new and general asymptotic framework for Principal Component Analysis which covers the different asymptotic domains of sample size and dimension of the data.

Data and Examples

This book uses many contrasting data sets: small classical data sets such as Fisher's four-dimensional iris data; data sets of moderate dimension (up to about thirty) from medical, biological, marketing and financial areas; and big and complex data sets. The data sets vary in the number of observations from fewer than fifty to about one million. We will also generate data and work with simulated data because such data can demonstrate the performance of a particular method, and the strengths and weaknesses of particular approaches more clearly. We will meet high-dimensional data with more than 1,000 variables and high-dimension low sample size (HDLSS) data, including data from genomics with dimensions in the tens of thousands, and typically fewer than 100 observations. In addition, we will encounter functional data from proteomics, for which each observation is a curve or profile.

Visualising data is important and is typically part of a first exploratory step in data analysis. If appropriate, I show the results of an analysis in graphical form.

I describe the analysis of data in *Examples* which illustrate the different tools and methodologies. In the examples I provide relevant information about the data, describe each analysis and give an interpretation of the outcomes of the analysis. As we travel through the book, we frequently return to data we previously met. The *Data Index* shows, for each data set in this book, which chapter contains examples pertaining to these data. Continuing with the same data throughout the book gives a more comprehensive picture of how we can study a particular data set and what methods and analyses are suitable for specific aims and data sets. Typically, the data sets I use are available on the Cambridge University Press website www.cambridge.org/9780521887939.

Use of Software and Algorithms

I use MATLAB for most of the examples and make generic MATLAB code available on the Cambridge University Press website. Readers and data analysts who prefer R could use that software instead of MATLAB. There are, however, some differences in implementation between MATLAB and R, in particular, in the Independent Component Analysis algorithms. I have included some comments about these differences in Section 11.4.

Many of the methods in Parts I and II have a standard one-line implementation in MATLAB and R. For example, to carry out a Principal Component Analysis or a likelihood-based Factor Analysis, all that is needed is a single command which includes the data to be analysed. These stand-alone routines in MATLAB and R avoid the need for writing one's own code. The MATLAB code I provide typically includes the initial visualisation of the data and, where appropriate, code for a graphical presentation of the results.

This book contains algorithms, that is, descriptions of the mathematical or computational steps that are needed to carry out particular analyses. Algorithm 4.2, for example, details the steps that are required to carry out classification for principal component data. A list of all algorithms in this book follows the Contents.

Theoretical Framework

I have chosen the conventional format with *Definitions, Theorems, Propositions* and *Corollaries*. This framework allows me to state assumptions and conclusions precisely and in an easily recognised form. If the assumptions of a theorem deviate substantially from properties of the data, the reader should recognise that care is required when applying a particular result or method to the data.

In the first part of the book I present proofs of many of the theorems and propositions. For the later parts, the emphasis changes, and in Part III in particular, I typically present theorems without proof and refer the reader to the relevant literature. This is because most of the theorems in Part III refer to recent results. Their proofs can be highly technical and complex without necessarily increasing the understanding of the theorem for the general readership of this book.

Many of the methods I describe do not require knowledge of the underlying distribution. For this reason, my treatment is mostly distribution-free. At times, stronger properties can be derived when we know the distribution of the data. (In practice, this means that the data come from the Gaussian distribution.) In such cases, and in particular in non-asymptotic situations I will explicitly point out what extra mileage knowledge of the Gaussian distribution can gain us. The development of asymptotic theory typically requires the data to come from the Gaussian distribution, and in these theorems I will state the necessary distributional assumptions. Asymptotic theory provides a sound theoretical framework for a method, and in my opinion, it is important to detail the asymptotic results, including the precise conditions under which these results hold.

The formal framework and proofs of theorems should not deter readers who are more interested in an analysis of their data; the theory guides us regarding a method's appropriateness for specific data. However, the methods of this book can be used without dipping into a single proof, and most of the theorems are followed by remarks which explain aspects or implications of the theorems. These remarks can be understood without a deep knowledge of the assumptions and mathematical details of the theorems.

Teaching from This Book

This book is suitable as a graduate-level textbook and can be taught as a one-semester course with an optional advanced second semester. Graduate students who use this book as their textbook should have a solid general knowledge of statistics, including statistical inference and multivariate regression methods, and a good background in real analysis and linear algebra and, in particular, matrix theory. In addition, the graduate student should be interested in analysing real data and have some experience with MATLAB or R.

Each part of this book ends with a set of problems for the preceding chapters. These problems represent a mixture of theoretical exercises and data analysis and may be suitable as exercises for students.

Some models for a one-semester graduate course are

- A 'classical' course which focuses on the following sections of Chapters 2 to 7

  Chapter 2, Principal Component Analysis: up to Section 2.7.2
  Chapter 3, Canonical Correlation Analysis: up to Section 3.7
  Chapter 4, Discriminant Analysis: up to Section 4.8
  Chapter 6, Cluster Analysis: up to Section 6.6.2
  Chapter 7, Factor Analysis: up to Sections 7.6.2 and 7.6.5.

- A mixed 'classical' and 'modern' course which includes an introduction to Independent Component Analysis, based on Sections 10.1 to 10.3 and some subsections from Section 10.6. Under this model I would focus on a subset of the classical model, and leave out some or all of the following sections

  Section 4.7.2 to end of Chapter 4
  Section 6.6.
  Section 7.6.

Depending on the choice of the first one-semester course, a more advanced one-semester graduate course could focus on the extension sections in Parts I and II and then, depending on the interest of teacher and students, pick and choose material from Chapters 8 and 10 to 13.

Choice of Topics and How This Book Differs from Other Books on Multivariate Analysis

This book deviates from classical and newer books on multivariate analysis (MVA) in a number of ways.

1. To begin with, I have only a single chapter on background material, which includes summaries of pertinent material on the multivariate normal distribution and relevant results from matrix theory.
2. My emphasis is on the interplay of theory and data. I develop theoretical ideas and apply them to data. In this I differ from many books on MVA, which focus on either data analysis or treat theory.
3. I include analysis of high-dimension low sample size data and describe new and very recent developments in Principal Component Analysis which allow the dimension to grow faster than the sample size. These approaches apply to gene expression data and data from proteomics, for example.

Many of the methods I discuss fall into the category of *component analysis*, including the newer methods in Part III. As a consequence, I made choices to leave out some topics which are often treated in classical multivariate analysis, most notably ANOVA and MANOVA. The former is often taught in undergraduate statistics courses, and the latter is an interesting extension of ANOVA which does not fit so naturally into our framework but is easily accessible to the interested reader.

Another more recent topic, *Support Vector Machines*, is also absent. Again, it does not fit naturally into our component analysis framework. On the other hand, there is a growing number of books which focus exclusively on Support Vector Machines and which remedy my omission. I do, however, make the connection to Support Vector Machines and cite relevant papers and books, in particular in Sections 4.7 and 12.1.

And Finally ...

It gives me great pleasure to thank the people who helped while I was working on this monograph. First and foremost, my deepest thanks to my husband, Alun Pope, and sons, Graeme and Reiner Pope, for your love, help, encouragement and belief in me throughout the years it took to write this book. Thanks go to friends, colleagues and students who provided valuable feedback. I specially want to mention and thank two friends: Steve Marron, who introduced me to Independent Component Analysis and who continues to inspire me with his ideas, and Kanta Naito, who loves to argue with me until we find *good* answers to the problems we are working on. Finally, I thank Diana Gillooly from Cambridge University Press for her encouragement and help throughout this journey.

Part I

Classical Methods

1

Multidimensional Data

Denken ist interessanter als Wissen, aber nicht als Anschauen (Johann Wolfgang von Goethe, Werke – Hamburger Ausgabe Bd. 12, Maximen und Reflexionen, 1749–1832).
Thinking is more interesting than knowing, but not more interesting than looking at.

1.1 Multivariate and High-Dimensional Problems

Early in the twentieth century, scientists such as Pearson (1901), Hotelling (1933) and Fisher (1936) developed methods for analysing multivariate data in order to

- understand the structure in the data and summarise it in simpler ways;
- understand the relationship of one part of the data to another part; and
- make decisions and inferences based on the data.

The early methods these scientists developed are *linear*; their conceptual simplicity and elegance still strike us today as natural and surprisingly powerful. Principal Component Analysis deals with the first topic in the preceding list, Canonical Correlation Analysis with the second and Discriminant Analysis with the third. As time moved on, more complex methods were developed, often arising in areas such as psychology, biology or economics, but these linear methods have not lost their appeal. Indeed, as we have become more able to collect and handle very large and high-dimensional data, renewed requirements for linear methods have arisen. In these data sets essential structure can often be obscured by noise, and it becomes vital to

> reduce the original data in such a way that informative and interesting structure in the data is preserved while noisy, irrelevant or purely random variables, dimensions or features are removed, as these can adversely affect the analysis.

Principal Component Analysis, in particular, has become indispensable as a dimension-reduction tool and is often used as a first step in a more comprehensive analysis.

The data we encounter in this book range from two-dimensional samples to samples that have thousands of dimensions or consist of continuous functions. Traditionally one assumes that the dimension d is small compared to the sample size n, and for the asymptotic theory, n increases while the dimension remains constant. Many recent data sets do not fit into this framework; we encounter

- data whose dimension is comparable to the sample size, and both are large;
- high-dimension low sample size (HDLSS) data whose dimension d vastly exceeds the sample size n, so $d \gg n$; and
- functional data whose observations are functions.

High-dimensional and functional data pose special challenges, and their theoretical and asymptotic treatment is an active area of research. Gaussian assumptions will often not be useful for high-dimensional data. A deviation from normality does not affect the applicability of Principal Component Analysis or Canonical Correlation Analysis; however, we need to exercise care when making inferences based on Gaussian assumptions or when we want to exploit the normal asymptotic theory.

The remainder of this chapter deals with a number of topics that are needed in subsequent chapters. Section 1.2 looks at different ways of displaying or visualising data, Section 1.3 introduces notation for random vectors and data and Section 1.4 discusses Gaussian random vectors and summarises results pertaining to such vectors and data. Finally, in Section 1.5 we find results from linear algebra, which deal with properties of matrices, including the spectral decomposition. In this chapter, I state results without proof; the references I provide for each topic contain proofs and more detail.

1.2 Visualisation

Before we analyse a set of data, it is important to *look* at it. Often we get useful clues such as skewness, bi- or multi-modality, outliers, or distinct groupings; these influence or direct our analysis. Graphical displays are exploratory data-analysis tools, which, if appropriately used, can enhance our understanding of data. The insight obtained from graphical displays is more *subjective* than quantitative; for most of us, however, visual cues are easier to understand and interpret than numbers alone, and the knowledge gained from graphical displays can complement more quantitative answers.

Throughout this book we use graphic displays extensively and typically in the examples. In addition, in the introduction to non-Gaussian analysis, Section 9.1 of Part III, I illustrate with the simple graphical displays of Figure 9.1 the difference between *interesting* and purely random or *non-informative* data.

1.2.1 Three-Dimensional Visualisation

Two-dimensional scatterplots are a natural – though limited – way of looking at data with three or more variables. As the number of variables, and therefore the dimension, increases, sequences of two-dimensional scatterplots become less feasible to interpret. We can, of course, still display three of the d dimensions in scatterplots, but it is less clear how one can look at more than three dimensions in a single plot.

We start with visualising three data dimensions. These arise as three-dimensional data or as three specified variables of higher-dimensional data. Commonly the data are displayed in a *default* view, but rotating the data can better reveal the structure of the data. The scatterplots in Figure 1.1 display the 10,000 observations and the three variables *CD3*, *CD8* and *CD4* of the five-dimensional HIV$^+$ and HIV$^-$ data sets, which contain measurements of blood cells relevant to HIV. The left panel shows the HIV$^+$ data and the right panel the

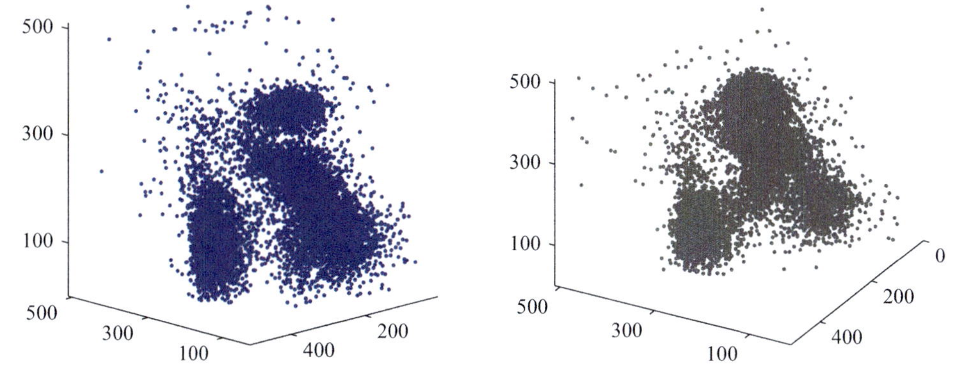

Figure 1.1 HIV$^+$data (*left*) and HIV$^-$data (*right*) of Example 2.4 with variables *CD3*, *CD8* and *CD4*.

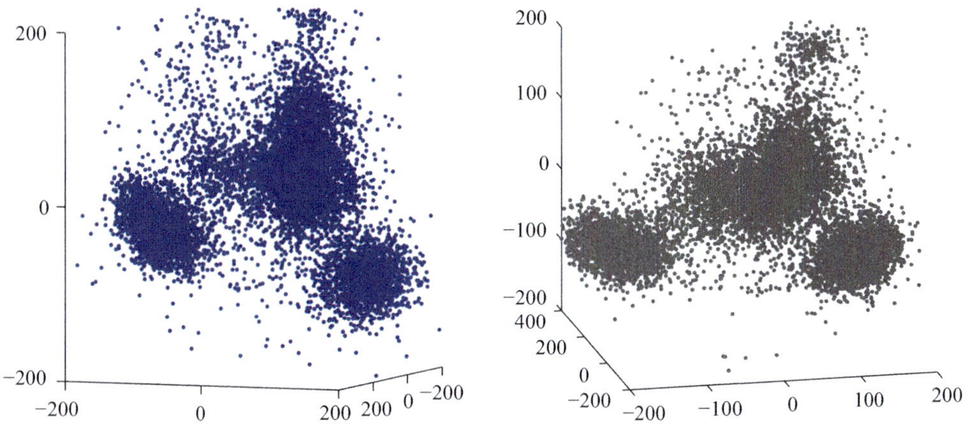

Figure 1.2 Orthogonal projections of the five-dimensional HIV$^+$data (*left*) and the HIV$^-$data (*right*) of Example 2.4.

HIV$^-$ data. There are differences between the point clouds in the two figures, and an important task in the analysis of such data is to exhibit and quantify the differences. The data are described in Example 2.4 of Section 2.3.

It may be helpful to present the data in the form of movies or combine a series of different views of the same data. Other possibilities include projecting the five-dimensional data onto a smaller number of orthogonal directions and displaying the lower-dimensional projected data as in Figure 1.2. These figures, again with HIV$^+$ in the left panel and HIV$^-$ in the right panel, highlight the cluster structure of the data in Figure 1.1. We can see a smaller fourth cluster in the top right corner of the HIV$^-$ data, which seems to have almost disappeared in the HIV$^+$ data in the left panel.

We return to these figures in Section 2.4, where I explain how to find *informative* projections. Many of the methods we explore use projections: Principal Component Analysis, Factor Analysis, Multidimensional Scaling, Independent Component Analysis and Projection Pursuit. In each case the projections focus on different aspects and properties of the data.

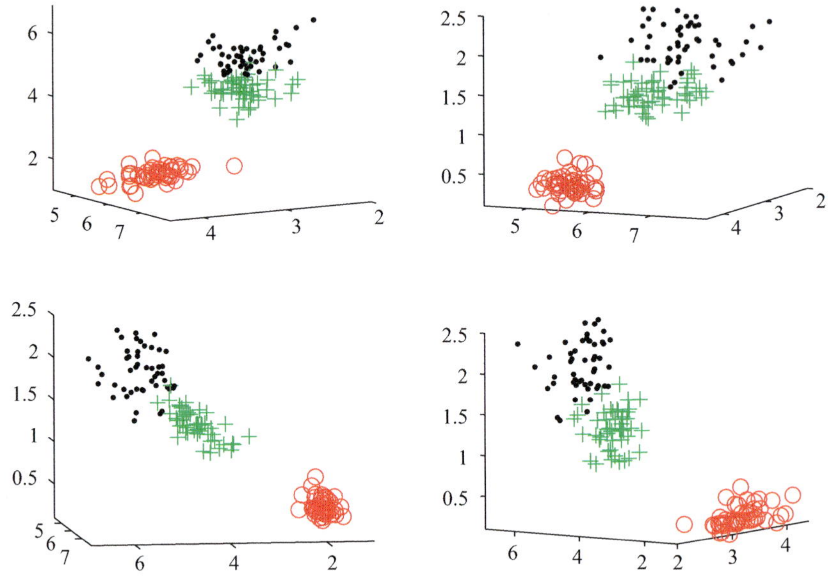

Figure 1.3 Three species of the iris data: dimensions 1, 2 and 3 (*top left*), dimensions 1, 2 and 4 (*top right*), dimensions 1, 3 and 4 (*bottom left*) and dimensions 2, 3 and 4 (*bottom right*).

Another way of representing low-dimensional data is in a number of three-dimensional scatterplots – as seen in Figure 1.3 – which make use of colour and different plotting symbols to enhance interpretation. We display the four variables of Fisher's *iris* data – *sepal length, sepal width, petal length* and *petal width* – in a sequence of three-dimensional scatterplots. The data consist of three species: red refers to *Setosa*, green to *Versicolor* and black to *Virginica*. We can see that the red observations are well separated from the other two species for all combinations of variables, whereas the green and black species are not as easily separable. I describe the data in more detail in Example 4.1 of Section 4.3.2.

1.2.2 Parallel Coordinate Plots

As the dimension grows, three-dimensional scatterplots become less relevant, unless we know that only some variables are important. An alternative, which allows us to *see* all variables at once, is to follow Inselberg (1985) and to present the data in the form of **parallel coordinate plots**. The idea is to present the data as two-dimensional graphs. Two different versions of parallel coordinate plots are common. The main difference is an interchange of the axes. In **vertical parallel coordinate plots** – see Figure 1.4 – the variable numbers are represented as values on the y-axis. For a vector $\mathbf{X} = [X_1, \ldots, X_d]^{\mathsf{T}}$ we represent the first variable X_1 by the point $(X_1, 1)$ and the jth variable X_j by (X_j, j). Finally, we connect the d points by a line which goes from $(X_1, 1)$ to $(X_2, 2)$ and so on to (X_d, d). We apply the same rule to the next d-dimensional datum. Figure 1.4 shows a vertical parallel coordinate plot for Fisher's *iris* data. For easier visualisation, I have used the same colours for the three species as in Figure 1.3, so red refers to the observations of species 1, green to those of species 2 and black to those of species 3. The parallel coordinate plot of the iris data shows that the data fall into two distinct groups – as we have also seen in Figure 1.3, but unlike the previous figure it tells us that dimension 3 separates the two groups most strongly.

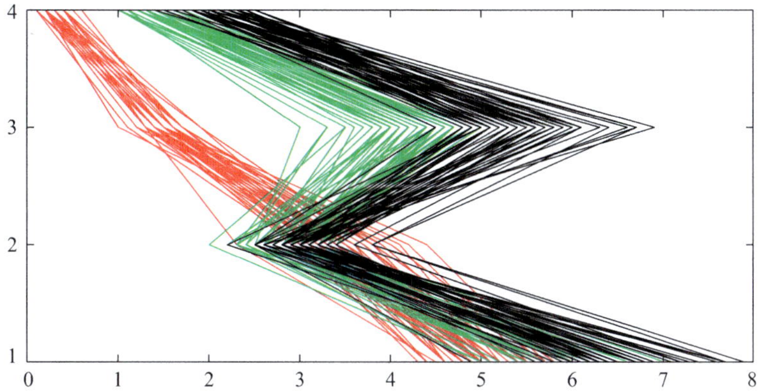

Figure 1.4 Iris data with variables represented on the y-axis and separate colours for the three species as in Figure 1.3.

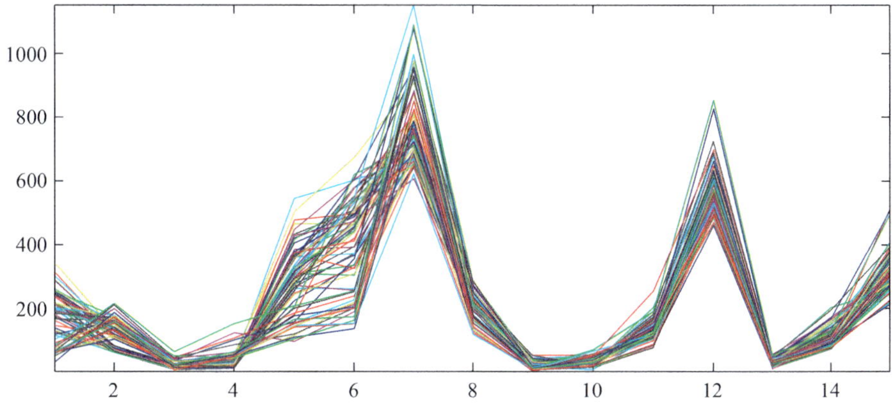

Figure 1.5 Parallel coordinate view of the illicit drug market data of Example 2.14.

Instead of the three colours shown in the plot of the iris data, different colours can be used for each observation, as in Figure 1.5. In a **horizontal parallel coordinate plot**, the x-axis represents the variable numbers $1,\ldots,d$. For a datum $\mathbf{X} = [X_1 \cdots X_d]^{\mathsf{T}}$, the first variable gives rise to the point $(1, X_1)$ and the jth variable X_j to (j, X_j). The d points are connected by a line, starting with $(1, X_1)$, then $(2, X_2)$, until we reach (d, X_d). Because we typically identify variables with the x-axis, we will more often use horizontal parallel coordinate plots. The differently coloured lines make it easier to trace particular observations.

Figure 1.5 shows the 66 monthly observations on 15 features or variables of the *illicit drug market* data which I describe in Example 2.14 in Section 2.6.2. Each observation (month) is displayed in a different colour. I have excluded two variables, as these have much higher values and would obscure the values of the remaining variables. Looking at variable 5, heroin overdose, the question arises whether there could be two groups of observations corresponding to the high and low values of this variable. The analyses of these data throughout the book will allow us to look at this question in different ways and provide answers.

Interactive graphical displays and movies are also valuable visualisation tools. They are beyond the scope of this book; I refer the interested reader to Wegman (1991) or Cook and Swayne (2007).

1.3 Multivariate Random Vectors and Data

Random vectors are vector-valued functions defined on a sample space. We consider a single random vector and collections of random vectors. For a single random vector we assume that there is a model such as the first few moments or the distribution, or we might assume that the random vector satisfies a 'signal plus noise' model. We are then interested in deriving properties of the random vector under the model. This scenario is called the **population** case.

For a collection of random vectors, we assume the vectors to be independent and identically distributed and to come from the same model, for example, have the same first and second moments. Typically we do not know the true moments. We use the collection to construct estimators for the moments, and we derive properties of the estimators. Such properties may include how 'good' an estimator is as the number of vectors in the collection grows, or we may want to draw inferences about the appropriateness of the model. This scenario is called the **sample** case, and we will refer to the collection of random vectors as the **data** or the **(random) sample**.

In applications, specific values are measured for each of the random vectors in the collection. We call these values the **realised** or **observed values** of the data or simply the **observed data**. The observed values are no longer random, and in this book we deal with the observed data in examples only. If no ambiguity exists, I will often refer to the observed data as data throughout an example.

Generally, I will treat the population case and the sample case separately and start with the population. The distinction between the two scenarios is important, as we typically have to switch from the population parameters, such as the mean, to the sample parameters, in this case the sample mean. As a consequence, the definitions for the population and the data are similar but not the same.

1.3.1 The Population Case

Let

$$\mathbf{X} = \begin{bmatrix} X_1 \\ \vdots \\ X_d \end{bmatrix}$$

be a random vector from a distribution $F\colon \mathbb{R}^d \to [0,1]$. The individual X_j, with $j \leq d$, are **random variables**, also called the **variables**, **components** or **entries** of $\mathbf{X}$, and $\mathbf{X}$ is ***d*-dimensional** or ***d*-variate**. We assume that $\mathbf{X}$ has a finite d-dimensional mean or expected value $\mathbb{E}\mathbf{X}$ and a finite $d \times d$ covariance matrix $\operatorname{var}(\mathbf{X})$. We write

$$\boldsymbol{\mu} = \mathbb{E}\mathbf{X} \qquad \text{and} \qquad \Sigma = \operatorname{var}(\mathbf{X}) = \mathbb{E}\left[(\mathbf{X}-\boldsymbol{\mu})(\mathbf{X}-\boldsymbol{\mu})^{\mathsf{T}}\right].$$

The entries of $\boldsymbol{\mu}$ and Σ are

$$\boldsymbol{\mu} = \begin{bmatrix} \mu_1 \\ \vdots \\ \mu_d \end{bmatrix} \quad \text{and} \quad \Sigma = \begin{pmatrix} \sigma_1^2 & \sigma_{12} & \cdots & \sigma_{1d} \\ \sigma_{21} & \sigma_2^2 & \cdots & \sigma_{2d} \\ \vdots & \vdots & \ddots & \vdots \\ \sigma_{d1} & \sigma_{d2} & \cdots & \sigma_d^2 \end{pmatrix}, \tag{1.1}$$

where $\sigma_j^2 = \mathrm{var}(X_j)$ and $\sigma_{jk} = \mathrm{cov}(X_j, X_k)$. More rarely, I write σ_{jj} for the diagonal elements σ_j^2 of Σ.

Of primary interest in this book are the mean and covariance matrix of a random vector rather than the underlying distribution F. We write

$$\mathbf{X} \sim (\boldsymbol{\mu}, \Sigma) \tag{1.2}$$

as shorthand for a random vector $\mathbf{X}$ which has mean $\boldsymbol{\mu}$ and covariance matrix Σ.

If $\mathbf{X}$ is a d-dimensional random vector and A is a $d \times k$ matrix, for some $k \geq 1$, then $A^{\mathsf{T}}\mathbf{X}$ is a k-dimensional random vector. Result 1.1 lists properties of $A^{\mathsf{T}}\mathbf{X}$.

Result 1.1 *Let $\mathbf{X} \sim (\boldsymbol{\mu}, \Sigma)$ be a d-variate random vector. Let A and B be matrices of size $d \times k$ and $d \times \ell$, respectively.*

1. *The mean and covariance matrix of the k-variate random vector $A^{\mathsf{T}}\mathbf{X}$ are*

$$A^{\mathsf{T}}\mathbf{X} \sim \left(A^{\mathsf{T}}\boldsymbol{\mu}, A^{\mathsf{T}}\Sigma A\right). \tag{1.3}$$

2. *The random vectors $A^{\mathsf{T}}\mathbf{X}$ and $B^{\mathsf{T}}\mathbf{X}$ are uncorrelated if and only if $A^{\mathsf{T}}\Sigma B = \mathbf{0}_{k\times\ell}$, where $\mathbf{0}_{k\times\ell}$ is the $k \times n$ matrix all of whose entries are* 0.

Both these results can be strengthened when $\mathbf{X}$ is Gaussian, as we shall see in Corollary 1.6.

1.3.2 The Random Sample Case

Let $\mathbf{X}_1, \ldots, \mathbf{X}_n$ be d-dimensional random vectors. Unless otherwise stated, we assume that the $\mathbf{X}_i$ are independent and from the same distribution $F\colon \mathbb{R}^d \to [0,1]$ with finite mean $\boldsymbol{\mu}$ and covariance matrix Σ. We omit reference to F when knowledge of the distribution is not required.

In statistics one often identifies a random vector with its observed values and writes $\mathbf{X}_i = \mathbf{x}_i$. We explore properties of random samples but only encounter observed values of random vectors in the examples. For this reason, I will typically write

$$\mathbb{X} = \left[\mathbf{X}_1 \; \mathbf{X}_2 \cdots \mathbf{X}_n\right], \tag{1.4}$$

for the sample of independent random vectors $\mathbf{X}_i$ and call this collection a **random sample** or **data**. I will use the same notation in the examples as it will be clear from the context whether I refer to the random vectors or their observed values. If a clarification is necessary, I will provide it. We also write

$$\mathbb{X} = \begin{bmatrix} X_{11} & X_{21} & \cdots & X_{n1} \\ X_{12} & X_{22} & \cdots & X_{n2} \\ \vdots & \vdots & \ddots & \vdots \\ X_{1d} & X_{2d} & \cdots & X_{nd} \end{bmatrix} = \begin{bmatrix} \mathbf{X}_{\bullet 1} \\ \mathbf{X}_{\bullet 2} \\ \vdots \\ \mathbf{X}_{\bullet d} \end{bmatrix}. \tag{1.5}$$

The ith column of $\mathbb{X}$ is the ith random vector $\mathbf{X}_i$, and the jth row $\mathbf{X}_{\bullet j}$ is the jth variable across all n random vectors. Throughout this book the first subscript i in X_{ij} refers to the ith vector $\mathbf{X}_i$, and the second subscript j refers to the jth variable.

A Word of Caution. The data $\mathbb{X}$ are $d \times n$ matrices. This notation differs from that of some authors who write data as an $n \times d$ matrix. I have chosen the $d \times n$ notation for one main reason: The population random vector is regarded as a column vector, and it is therefore more natural to regard the random vectors of the sample as column vectors. An important consequence of this notation is the fact that the population and sample cases can be treated the same way, and no additional transposes are required.[1]

For data, the mean $\boldsymbol{\mu}$ and covariance matrix Σ are usually not known; instead, we work with the sample mean $\overline{\mathbf{X}}$ and the sample covariance matrix S and sometimes write

$$\mathbb{X} \sim \mathsf{Sam}(\overline{\mathbf{X}}, S) \tag{1.6}$$

in order to emphasise that we refer to the sample quantities, where

$$\overline{\mathbf{X}} = \frac{1}{n} \sum_{i=1}^{n} \mathbf{X}_i \qquad \text{and} \tag{1.7}$$

$$S = \frac{1}{n-1} \sum_{i=1}^{n} (\mathbf{X}_i - \overline{\mathbf{X}})(\mathbf{X}_i - \overline{\mathbf{X}})^{\mathsf{T}}. \tag{1.8}$$

As before, $\mathbb{X} \sim (\boldsymbol{\mu}, \Sigma)$ refers to the data with population mean $\boldsymbol{\mu}$ and covariance matrix Σ.

The sample mean and sample covariance matrix depend on the sample size n. If the dependence on n is important, for example, in the asymptotic developments, I will write S_n instead of S, but normally I will omit n for simplicity.

Definitions of the sample covariance matrix use n^{-1} or $(n-1)^{-1}$ in the literature. I use the $(n-1)^{-1}$ version. This notation has the added advantage of being compatible with software environments such as MATLAB and R.

Data are often centred. We write $\mathbb{X}_{\text{cent}}$ for the centred data and adopt the (*unconventional*) notation

$$\mathbb{X}_{\text{cent}} \equiv \mathbb{X} - \overline{\mathbf{X}} = \begin{bmatrix} \mathbf{X}_1 - \overline{\mathbf{X}} & \ldots & \mathbf{X}_n - \overline{\mathbf{X}} \end{bmatrix}. \tag{1.9}$$

The centred data are of size $d \times n$. Using this notation, the $d \times d$ sample covariance matrix S becomes

$$S = \frac{1}{n-1} \left(\mathbb{X} - \overline{\mathbf{X}}\right) \left(\mathbb{X} - \overline{\mathbf{X}}\right)^{\mathsf{T}}. \tag{1.10}$$

In analogy with (1.1), the entries of the sample covariance matrix S are s_{jk}, and

$$s_{jk} = \frac{1}{n-1} \sum_{i=1}^{n} (X_{ij} - m_j)(X_{ik} - m_k), \tag{1.11}$$

with $\overline{\mathbf{X}} = [m_1, \ldots, m_d]^{\mathsf{T}}$, and m_j is the sample mean of the jth variable. As for the population, we write s_j^2 or s_{jj} for the diagonal elements of S.

[1] Consider $\mathbf{a} \in \mathbb{R}^d$; then the projection of $\mathbf{X}$ onto $\mathbf{a}$ is $\mathbf{a}^{\mathsf{T}}\mathbf{X}$. Similarly, the projection of the matrix $\mathbb{X}$ onto $\mathbf{a}$ is done elementwise for each random vector $\mathbf{X}_i$ and results in the $1 \times n$ vector $\mathbf{a}^{\mathsf{T}}\mathbb{X}$. For the $n \times d$ matrix notation, the projection of $\mathbb{X}$ onto $\mathbf{a}$ is $\mathbb{X}^{\mathsf{T}}\mathbf{a}$, and this notation differs from the population case.

1.4 Gaussian Random Vectors

1.4.1 The Multivariate Normal Distribution and the Maximum Likelihood Estimator

Many of the techniques we consider do not require knowledge of the underlying distribution. However, the more we know about the data, the more we can exploit this knowledge in the derivation of rules or in decision-making processes. Of special interest is the Gaussian distribution, not least because of the Central Limit Theorem. Much of the asymptotic theory in multivariate statistics relies on normality assumptions. In this preliminary chapter, I present results without proof. Readers interested in digging a little deeper might find Mardia, Kent, and Bibby (1992) and Anderson (2003) helpful.

We write

$$\mathbf{X} \sim \mathcal{N}(\boldsymbol{\mu}, \Sigma) \tag{1.12}$$

for a random vector from the **multivariate normal distribution** with mean $\boldsymbol{\mu}$ and covariance matrix Σ. We begin with the population case and consider transformations of $\mathbf{X}$ and relevant distributional properties.

Result 1.2 *Let $\mathbf{X} \sim \mathcal{N}(\boldsymbol{\mu}, \Sigma)$ be d-variate, and assume that Σ^{-1} exists.*

1. *Let $\mathbf{X}_\Sigma = \Sigma^{-1/2}(\mathbf{X} - \boldsymbol{\mu})$; then $\mathbf{X}_\Sigma \sim \mathcal{N}(\mathbf{0}, \mathbf{I}_{d\times d})$, where $\mathbf{I}_{d\times d}$ is the $d \times d$ identity matrix.*
2. *Let $X^2 = (\mathbf{X} - \boldsymbol{\mu})^{\mathsf{T}} \Sigma^{-1} (\mathbf{X} - \boldsymbol{\mu})$; then $X^2 \sim \chi^2_d$, the χ^2 distribution in d degrees of freedom.*

The first part of the result states the multivariate analogue of standardising a normal random variable. The result is a multivariate random vector with independent variables. The second part generalises the square of a *standard normal* random variable. The quantity X^2 is a scalar random variable which has, as in the one-dimensional case, a χ^2-distribution, but this time in d degrees of freedom.

From the population we move to the sample. Fix a dimension $d \geq 1$. Let $\mathbf{X}_i \sim \mathcal{N}(\boldsymbol{\mu}, \Sigma)$ be independent d-dimensional random vectors for $i = 1, \ldots, n$ with sample mean $\overline{\mathbf{X}}$ and sample covariance matrix S. We define **Hotelling's** T^2 by

$$T^2 = n\left(\overline{\mathbf{X}} - \boldsymbol{\mu}\right)^{\mathsf{T}} S^{-1} \left(\overline{\mathbf{X}} - \boldsymbol{\mu}\right). \tag{1.13}$$

Further let $\mathbf{Z}_j \sim \mathcal{N}(\mathbf{0}, \Sigma)$ for $j = 1, \ldots, m$ be independent d-dimensional random vectors, and let

$$W = \sum_{j=1}^{m} \mathbf{Z}_j \mathbf{Z}_j^{\mathsf{T}} \tag{1.14}$$

be the $d \times d$ random matrix generated by the $\mathbf{Z}_j$; then W has the **Wishart** distribution $W(m, \Sigma)$ with m degrees of freedom and covariance matrix Σ, where m is the number of summands and Σ is the common $d \times d$ covariance matrix. Result 1.3 lists properties of these sample quantities.

Result 1.3 *Let $\mathbf{X}_i \sim \mathcal{N}(\boldsymbol{\mu}, \Sigma)$ be d-dimensional random vectors for $i = 1, \ldots, n$. Let S be the sample covariance matrix, and assume that S is invertible.*

1. *The sample mean* $\overline{\mathbf{X}}$ *satisfies* $\overline{\mathbf{X}} \sim \mathcal{N}(\boldsymbol{\mu}, \Sigma/n)$.
2. *Assume that* $n > d$. *Let* T^2 *be given by (1.13). It follows that*

$$\frac{n-d}{(n-1)d}T^2 \sim F_{d,n-d},$$

the F distribution in d and $n-d$ *degrees of freedom.*
3. *For n observations* $\mathbf{X}_i$ *and their sample covariance matrix S there exist* $n-1$ *independent random vectors* $\mathbf{Z}_j \sim \mathcal{N}(\mathbf{0}, \Sigma)$ *such that*

$$S = \frac{1}{n-1}\sum_{j=1}^{n-1} \mathbf{Z}_j \mathbf{Z}_j^{\mathsf{T}},$$

and $(n-1)S$ *has a* $W((n-1), \Sigma)$ *Wishart distribution.*

From the distributional properties we turn to estimation of the parameters. For this we require the likelihood function.

Let $\mathbf{X} \sim \mathcal{N}(\boldsymbol{\mu}, \Sigma)$ be d-dimensional. The **multivariate normal probability density function** f is

$$f(\cdot) = (2\pi)^{-d/2} \det(\Sigma)^{-1/2} \exp\left[-\frac{1}{2}(\cdot - \boldsymbol{\mu})^{\mathsf{T}} \Sigma^{-1} (\cdot - \boldsymbol{\mu})\right], \tag{1.15}$$

where $\det(\Sigma)$ is the determinant of Σ. For a sample $\mathbb{X} = \begin{bmatrix}\mathbf{X}_1\ \mathbf{X}_2 \cdots \mathbf{X}_n\end{bmatrix}$ of independent random vectors from the normal distribution with the same mean and covariance matrix, the joint probability density function is the product of the functions (1.15).

If attention is focused on a parameter θ of the distribution, which we want to estimate, we define the **normal** or **Gaussian likelihood (function)** L as a function of the parameter θ of interest conditional on the data

$$L(\theta|\mathbb{X}) = (2\pi)^{-nd/2} \det(\Sigma)^{-n/2} \exp\left[-\frac{1}{2}\sum_{i=1}^{n} (\mathbf{X}_i - \boldsymbol{\mu})^{\mathsf{T}} \Sigma^{-1} (\mathbf{X}_i - \boldsymbol{\mu})\right]. \tag{1.16}$$

Assume that the parameter of interest is the mean and the covariance matrix, so $\theta = (\boldsymbol{\mu}, \Sigma)$; then, for the normal likelihood, the **maximum likelihood estimator (MLE)** of θ, denoted by $\widehat{\theta}$, is

$$\widehat{\theta} = (\widehat{\boldsymbol{\mu}}, \widehat{\Sigma}),$$

where

$$\widehat{\boldsymbol{\mu}} = \frac{1}{n}\sum_{i=1}^{n} \mathbf{X}_i = \overline{\mathbf{X}} \qquad \text{and}$$

$$\widehat{\Sigma} = \frac{1}{n}\sum_{i=1}^{n} \left(\mathbf{X}_i - \overline{\mathbf{X}}\right)\left(\mathbf{X}_i - \overline{\mathbf{X}}\right)^{\mathsf{T}} = \frac{n-1}{n} S. \tag{1.17}$$

Remark. In this book we distinguish between $\widehat{\Sigma}$ and S, the sample covariance matrix defined in (1.8). For details and properties of MLEs, see chapter 7 of Casella and Berger (2001).

1.4.2 Marginal and Conditional Normal Distributions

Consider a normal random vector $\mathbf{X} = [X_1, X_2, \ldots, X_d]^{\mathsf{T}}$. Let $\mathbf{X}^{[1]}$ be a vector consisting of the first d_1 entries of $\mathbf{X}$, and let $\mathbf{X}^{[2]}$ be the vector consisting of the remaining d_2 entries; then

$$\mathbf{X} = \begin{bmatrix} \mathbf{X}^{[1]} \\ \mathbf{X}^{[2]} \end{bmatrix}. \tag{1.18}$$

For $\iota = 1, 2$ we let $\boldsymbol{\mu}_\iota$ be the mean of $\mathbf{X}^{[\iota]}$ and Σ_ι its covariance matrix.

Result 1.4 *Assume that* $\mathbf{X}^{[1]}$, $\mathbf{X}^{[2]}$ *and* $\mathbf{X}$ *are given by (1.18) for some* $d_1, d_2 < d$ *such that* $d_1 + d_2 = d$. *Assume also that* $\mathbf{X} \sim \mathcal{N}(\boldsymbol{\mu}, \Sigma)$. *The following hold:*

1. *For* $j = 1, \ldots, d$, *the* j*th variable* X_j *of* $\mathbf{X}$ *has the distribution* $\mathcal{N}(\mu_j, \sigma_j^2)$.
2. *For* $\iota = 1, 2$, $\mathbf{X}^{[\iota]}$ *has the distribution* $\mathcal{N}(\boldsymbol{\mu}_\iota, \Sigma_\iota)$.
3. *The (between) covariance matrix* $\operatorname{cov}(\mathbf{X}^{[1]}, \mathbf{X}^{[2]})$ *of* $\mathbf{X}^{[1]}$ *and* $\mathbf{X}^{[2]}$ *is the* $d_1 \times d_2$ *submatrix* Σ_{12} *of*

$$\Sigma = \begin{pmatrix} \Sigma_1 & \Sigma_{12} \\ \Sigma_{12}^{\mathsf{T}} & \Sigma_2 \end{pmatrix}.$$

Result 1.4 tells us that the marginal distributions of normal random vectors are normal with means and covariance matrices extracted from those of the original random vector. The next result considers the relationship between different subvectors of $\mathbf{X}$ further.

Result 1.5 *Assume that* $\mathbf{X}^{[1]}$, $\mathbf{X}^{[2]}$ *and* $\mathbf{X}$ *are given by (1.18) for some* $d_1, d_2 < d$ *such that* $d_1 + d_2 = d$. *Assume also that* $\mathbf{X} \sim \mathcal{N}(\boldsymbol{\mu}, \Sigma)$ *and that* Σ_1 *and* Σ_2 *are invertible.*

1. *The covariance matrix* Σ_{12} *of* $\mathbf{X}^{[1]}$ *and* $\mathbf{X}^{[2]}$ *satisfies*

$$\Sigma_{12} = \mathbf{0}_{d_1 \times d_2} \qquad \text{if and only if } \mathbf{X}^{[1]} \text{ and } \mathbf{X}^{[2]} \text{ are independent.}$$

2. *Assume that* $\Sigma_{12} \neq \mathbf{0}_{d_1 \times d_2}$. *Put* $\mathbf{X}_{2\backslash 1} = \mathbf{X}_2 - \Sigma_{12}^{\mathsf{T}} \Sigma_1^{-1} \mathbf{X}_1$. *Then* $\mathbf{X}_{2\backslash 1}$ *is a* d_2*-dimensional random vector which is independent of* $\mathbf{X}_1$ *and* $\mathbf{X}_{2\backslash 1} \sim \mathcal{N}(\boldsymbol{\mu}_{2\backslash 1}, \Sigma_{2\backslash 1})$ *with*

$$\boldsymbol{\mu}_{2\backslash 1} = \boldsymbol{\mu}_2 - \Sigma_{12}^{\mathsf{T}} \Sigma_1^{-1} \boldsymbol{\mu}_1 \qquad \text{and} \qquad \Sigma_{2\backslash 1} = \Sigma_2 - \Sigma_{12}^{\mathsf{T}} \Sigma_1^{-1} \Sigma_{12}.$$

3. *Let* $\left(\mathbf{X}^{[1]} \mid \mathbf{X}^{[2]}\right)$ *be the conditional random vector* $\mathbf{X}^{[1]}$ *given* $\mathbf{X}^{[2]}$. *Then*

$$\left(\mathbf{X}^{[1]} \mid \mathbf{X}^{[2]}\right) \sim \mathcal{N}\left(\boldsymbol{\mu}_{X_1|X_2}, \Sigma_{X_1|X_2}\right),$$

where

$$\boldsymbol{\mu}_{X_1|X_2} = \boldsymbol{\mu}_1 + \Sigma_{12} \Sigma_2^{-1} (\mathbf{X}^{[2]} - \boldsymbol{\mu}_2)$$

and

$$\Sigma_{X_1|X_2} = \Sigma_1 - \Sigma_{12} \Sigma_2^{-1} \Sigma_{12}^{\mathsf{T}}.$$

The first property is specific to the normal distribution: independence always implies uncorrelatedness, and for the normal distribution the converse holds, too. The second part shows how one can uncorrelate the vectors $\mathbf{X}^{[1]}$ and $\mathbf{X}^{[2]}$, and the last part details the adjustments that are needed when the subvectors have a non-zero covariance matrix.

A combination of Result 1.1 and the results of this section leads to

Corollary 1.6 *Let* $\mathbf{X} \sim \mathcal{N}(\boldsymbol{\mu}, \Sigma)$. *Let* A *and* B *be matrices of size* $d \times k$ *and* $d \times \ell$, *respectively.*

1. *The k-variate random vector* $A^{\mathsf{T}}\mathbf{X}$ *satisfies* $A^{\mathsf{T}}\mathbf{X} \sim \mathcal{N}\big(A^{\mathsf{T}}\boldsymbol{\mu}, A^{\mathsf{T}}\Sigma A\big)$.
2. *The random vectors* $A^{\mathsf{T}}\mathbf{X}$ *and* $B^{\mathsf{T}}\mathbf{X}$ *are independent if and only if* $A^{\mathsf{T}}\Sigma B = \mathbf{0}_{k\times\ell}$.

1.5 Similarity, Spectral and Singular Value Decomposition

There are a number of results from linear algebra that we require in many of the following chapters. This section provides a list of relevant results. Readers familiar with linear algebra will only want to refer to the notation I establish. For readers who want more detail and access to proofs, Strang (2005), Harville (1997) and Searle (1982) are useful resources; in addition, Gentle (2007) has an extensive collection of matrix results for the working statistician.

1.5.1 Similar Matrices

Definition 1.7 Let $A = (a_{ij})$ and $B = (b_{ij})$ be square matrices, both of size $d \times d$.

1. The **trace** of A is the sum of its diagonal elements

$$\operatorname{tr}(A) = \sum_{i=1}^{d} a_{ii}.$$

2. The matrices A and B are **similar** if there exists an invertible matrix E such that

$$A = EBE^{-1}.$$

□

The next result lists properties of similar matrices and traces that we require throughout this book.

Result 1.8 *Let A and B be similar matrices of size* $d \times d$, *and assume that A is invertible and has d distinct eigenvalues* $\lambda_1, \ldots, \lambda_d$. *Let* $\det(A)$ *be the determinant of A. The following hold:*

1. *The sum of the eigenvalues of A equals the sum of the diagonal elements of A, and hence*

$$\operatorname{tr}(A) = \sum_{i=1}^{d} a_{ii} = \sum_{i=1}^{d} \lambda_i.$$

2. *The matrices A and B have the same eigenvalues and hence the same trace.*
3. *For any* $d \times d$ *matrix C, the trace and the determinant of AC satisfy*

$$\operatorname{tr}(AC) = \operatorname{tr}(CA) \qquad \textit{and} \qquad \det(AC) = \det(A)\det(C).$$

1.5.2 Spectral Decomposition for the Population Case

Result 1.9 *Let* Σ *be a* $d \times d$ *matrix, and assume that* Σ *is symmetric and positive definite. Then* Σ *can be expressed in the form*

$$\Sigma = \Gamma \Lambda \Gamma^{\mathsf{T}}, \tag{1.19}$$

where Λ is the diagonal matrix which consists of the d non-zero eigenvalues of Σ

$$\Lambda = diag(\lambda_1,\ldots,\lambda_d) = \begin{pmatrix} \lambda_1 & 0 & \cdots & 0 \\ 0 & \lambda_2 & \cdots & 0 \\ \vdots & \vdots & \ddots & \vdots \\ 0 & \cdots & 0 & \lambda_d \end{pmatrix} \tag{1.20}$$

arranged in decreasing order: $\lambda_1 \geq \lambda_2 \geq \cdots \geq \lambda_d > 0$. The columns of the matrix Γ are eigenvectors $\boldsymbol{\eta}_k$ of Σ corresponding to the eigenvalues λ_k such that

$$\Sigma \boldsymbol{\eta}_k = \lambda_k \boldsymbol{\eta}_k \qquad \textit{and} \qquad \left\| \boldsymbol{\eta}_k \right\| = 1 \qquad \textit{for } k = 1,\ldots,d.$$

The eigenvalue–eigenvector decomposition (1.19) of Σ is called the **spectral decomposition**. Unless otherwise stated, I will assume that the eigenvalues of Σ are distinct, non-zero and given in decreasing order.

We often encounter submatrices of Σ. For $1 \leq q \leq d$, we write

$$\Lambda_q = \mathrm{diag}(\lambda_1,\ldots,\lambda_q) \qquad \text{and} \qquad \Gamma_q = \Gamma_{d\times q} = \left[\boldsymbol{\eta}_1\ \boldsymbol{\eta}_2 \cdots \boldsymbol{\eta}_q\right], \tag{1.21}$$

so Λ_q is the $q \times q$ submatrix of Λ which consists of the *top left corner* of Λ, and Γ_q is the $d \times q$ matrix consisting of the first q eigenvectors of Σ.

Result 1.10 *Let Σ be a $d \times d$ matrix with rank d and spectral decomposition $\Gamma\Lambda\Gamma^{\mathsf{T}}$.*

1. *The columns of Γ are linearly independent, and Γ is orthogonal and therefore satisfies*

$$\Gamma\Gamma^{\mathsf{T}} = \Gamma^{\mathsf{T}}\Gamma = \mathbf{I}_{d\times d} \qquad \textit{or equivalently} \qquad \Gamma^{\mathsf{T}} = \Gamma^{-1}, \tag{1.22}$$

where $\mathbf{I}_{d\times d}$ is the $d \times d$ identity matrix.

2. *For $p < d$,*

$$\Gamma_p^{\mathsf{T}}\Gamma_p = \mathbf{I}_{p\times p} \qquad \textit{but} \qquad \Gamma_p\Gamma_p^{\mathsf{T}} \neq \mathbf{I}_{d\times d}. \tag{1.23}$$

3. *For $q < p \leq d$,*

$$\Gamma_q^{\mathsf{T}}\Gamma_p = \mathbf{I}_{q\times p} \qquad \textit{and} \qquad \Gamma_p^{\mathsf{T}}\Gamma_q = \mathbf{I}_{p\times q}, \tag{1.24}$$

where $\mathbf{I}_{q\times p} = \begin{pmatrix}\mathbf{I}_{q\times q} & \mathbf{0}_{q\times(p-q)}\end{pmatrix}$ and $\mathbf{I}_{p\times q} = \begin{pmatrix}\mathbf{I}_{q\times q} \\ \mathbf{0}_{(p-q)\times q}\end{pmatrix}$.

4. *The eigenvectors of Σ satisfy*

$$\sum_{k=1}^{d} \eta_{jk}^2 = \sum_{\ell=1}^{d} \eta_{\ell k}^2 = 1 \qquad \textit{and} \qquad \sum_{k=1}^{d} \eta_{jk}\eta_{\ell k} = \sum_{j=1}^{d} \eta_{jk}\eta_{jl} = 0,$$

for $j,\ell = 1,\ldots,d$ and $j \neq \ell$ in the second set of equalities.

If the rank r of Σ is strictly smaller than the dimension d, then Σ has the spectral decomposition $\Sigma = \Gamma_r\Lambda_r\Gamma_r^{\mathsf{T}}$, where Γ_r is of size $d \times r$, and Λ_r is of size $r \times r$. Further, Γ_r is not orthogonal, and (1.22) is replaced with the weaker relationship

$$\Gamma_r^{\mathsf{T}}\Gamma_r = \mathbf{I}_{r\times r}. \tag{1.25}$$

We call such a matrix Γ_r ***r*-orthogonal** and note that $\Gamma_r\Gamma_r^{\mathsf{T}} \neq \mathbf{I}_{d\times d}$.

I will often omit the subscripts r in the spectral decomposition of Σ even if the rank is smaller than the dimension.

Results 1.9 and 1.10 apply to a much larger class of matrices than the covariance matrices used here.

Result 1.11 *Let Σ be a $d \times d$ matrix with rank d and spectral decomposition $\Gamma\Lambda\Gamma^{\mathsf{T}}$ with eigenvalues $\lambda_1, \ldots, \lambda_d$ and corresponding eigenvectors $\boldsymbol{\eta}_1, \ldots, \boldsymbol{\eta}_d$.*

1. *The matrix Σ is given by*

$$\Sigma = \sum_{j=1}^{d} \lambda_j \boldsymbol{\eta}_j \boldsymbol{\eta}_j^{\mathsf{T}}.$$

2. *For any $\mathbf{a} \in \mathbb{R}^d$, $\mathbf{a}^{\mathsf{T}}\Sigma\mathbf{a} \geq 0$.*
3. *There exists a matrix Θ such that $\Sigma = \Theta^{\mathsf{T}}\Theta$.*
4. *For any integers k and m,*

$$\Sigma^{k/m} = \Gamma\Lambda^{k/m}\Gamma^{\mathsf{T}}.$$

The matrix Θ of part 3 of the result does not need to be symmetric, nor is it unique. An example of Θ is the matrix $\Theta = \Lambda^{1/2}\Gamma^{\mathsf{T}}$. Part 4 of the result is of particular interest for $k = -1$ and $m = 2$.

1.5.3 Decompositions for the Sample Case

For data $\mathbb{X} = \begin{bmatrix}\mathbf{X}_1 \ \mathbf{X}_2 \ldots \mathbf{X}_n\end{bmatrix}$ with sample mean $\overline{\mathbf{X}}$ and sample covariance matrix S, the **spectral decomposition** of S is

$$S = \widehat{\Gamma}\widehat{\Lambda}\widehat{\Gamma}^{\mathsf{T}}. \tag{1.26}$$

The 'hat' notation reminds us that $\widehat{\Gamma}$ and $\widehat{\Lambda}$ are estimators of Γ and Λ, respectively. The eigenvalues of S are $\widehat{\lambda}_1 \geq \widehat{\lambda}_2 \geq \cdots \geq \widehat{\lambda}_r > 0$, where r is the rank of S. The columns of the matrix $\widehat{\Gamma}$ are the eigenvectors $\widehat{\boldsymbol{\eta}}_k$ of S corresponding to the eigenvalues $\widehat{\lambda}_k$ so that

$$S\widehat{\boldsymbol{\eta}}_k = \widehat{\lambda}_k \widehat{\boldsymbol{\eta}}_k \qquad \text{for } k = 1, \ldots, r.$$

Although the sample mean and sample covariance matrix are estimators of the respective population quantities, mean and covariance matrix, we use the common notation $\overline{\mathbf{X}}$ and S. The sample covariance matrix is positive definite (or positive semidefinite if it is rank-deficient), and properties corresponding to those listed in Results 1.10 apply to S.

In addition to the spectral decomposition of S, the data $\mathbb{X}$ have a decomposition which is related to the spectral decomposition of S.

Definition 1.12 Let $\mathbb{X}$ be data of size $d \times n$ with sample mean $\overline{\mathbf{X}}$ and sample covariance matrix S, and let $r \leq d$ be the rank of S. The **singular value decomposition** of $\mathbb{X}$ is

$$\mathbb{X} = UDV^{\mathsf{T}}, \tag{1.27}$$

where D is an $r \times r$ diagonal matrix with diagonal entries $d_1 \geq d_2 \geq \cdots \geq d_r > 0$, the **singular values** of $\mathbb{X}$.

The matrices U and V are of size $d \times r$ and $n \times r$, respectively, and their columns are the **left** (and respectively **right**) **eigenvectors** of $\mathbb{X}$. The left eigenvectors $\mathbf{u}_j$ and the right eigenvectors $\mathbf{v}_j$ of $\mathbb{X}$ are unit vectors and satisfy

$$\mathbb{X}^{\mathsf{T}}\mathbf{u}_j = \left(\mathbf{u}_j^{\mathsf{T}}\mathbb{X}\right)^{\mathsf{T}} = d_j\mathbf{v}_j \qquad \text{and} \qquad \mathbb{X}\mathbf{v}_j = d_j\mathbf{u}_j \qquad \text{for } j = 1, \ldots, r.$$

□

The notation of 'left' and 'right' eigenvectors of $\mathbb{X}$ is natural and extends the concept of an eigenvector for symmetric matrices. The decompositions of $\mathbb{X}$ and S are functions of the sample size n. For fixed n, the singular value decomposition of the data and the spectral decomposition of the sample covariance matrix are related.

Result 1.13 *Let* $\mathbb{X} = \begin{bmatrix}\mathbf{X}_1\ \mathbf{X}_2 \ldots \mathbf{X}_n\end{bmatrix}$ *be a random sample with mean* $\overline{\mathbf{X}}$ *and sample covariance matrix* S*. For the centred data* $\mathbb{X}_{cent} = \mathbb{X} - \overline{\mathbf{X}}$*, let*

$$\mathbb{X}_{cent} = UDV^{\mathsf{T}}$$

be the singular value decomposition, where U *and* V *are the matrices of left and right eigenvectors, and* D *is the diagonal matrix of singular values of* $\mathbb{X}_{cent}$*. Further let*

$$S = \widehat{\Gamma}\widehat{\Lambda}\widehat{\Gamma}^{\mathsf{T}}$$

be the spectral decomposition of S*. The two decompositions are related by*

$$U = \widehat{\Gamma} \qquad \textit{and} \qquad \frac{1}{n-1}D^2 = \widehat{\Lambda}.$$

The interplay between the spectral decomposition of S and the singular value decomposition of $\mathbb{X}$ becomes of special interest when the dimension exceeds the sample size. We will return to these ideas in more detail in Part II, initially in Section 5.5 but then more fully in Chapter 8. In Canonical Correlation Analysis (Chapter 3) we meet the singular value decomposition of the matrix of canonical correlations C and exploit its relationship to the two different square matrices CC^{T} and $C^{\mathsf{T}}C$.

Notation and Terminology. The notion of a projection map has a precise mathematical meaning. In this book I use the word **projection** in the following ways.

1. Let $\mathbf{X}$ be a d-dimensional random vector. For $k \le d$, let E be a $d \times k$ matrix whose columns are orthonormal, so the columns $\mathbf{e}_i$ satisfy $\mathbf{e}_i^{\mathsf{T}}\mathbf{e}_j = \delta_{ij}$, where δ is the Kronecker delta function with $\delta_{ii} = 1$ and $\delta_{ij} = 0$ if $i \ne j$. The columns $\mathbf{e}_i$ of E are called **directions** or **direction vectors**. The **projection** of $\mathbf{X}$ onto E, or in the direction of E, is the k-dimensional vector $E^{\mathsf{T}}\mathbf{X}$. The eigenvectors $\boldsymbol{\eta}_k$ of the covariance matrix Σ of $\mathbf{X}$ will be regarded as directions in Chapter 2; a projection of $\mathbf{X}$ onto $\boldsymbol{\eta}_k$ is the scalar $\boldsymbol{\eta}_k^{\mathsf{T}}\mathbf{X}$.
2. A **projection (vector)** $\mathbf{b} \in \mathbb{R}^d$ is a linear transform of a direction given by

$$\mathbf{b} = b\mathbf{e} \qquad \text{or} \qquad \mathbf{b} = B\mathbf{e},$$

   where $\mathbf{e} \in \mathbb{R}^d$ is a direction, $b \ne 0$ is a scalar, and B is a $d \times d$ matrix. In Section 2.2, I define the principal component projections $\alpha\boldsymbol{\eta}_k$, with scalars α obtained in Definition 2.2; in contrast, the canonical projections of Section 3.2 are of the form $B\mathbf{e}$.

2

Principal Component Analysis

Mathematics, rightly viewed, possesses not only truth, but supreme beauty (Bertrand Russell, Philosophical Essays No. 4, 1910).

2.1 Introduction

One of the aims in multivariate data analysis is to summarise the data in fewer than the original number of dimensions without losing essential information. More than a century ago, Pearson (1901) considered this problem, and Hotelling (1933) proposed a solution to it: instead of treating each variable separately, he considered combinations of the variables. Clearly, the average of all variables is such a combination, but many others exist. Two fundamental questions arise:

1. How should one choose these combinations?
2. How many such combinations should one choose?

There is no single strategy that always gives the right answer. This book will describe many ways of tackling at least the first problem.

Hotelling's proposal consisted in finding those linear combinations of the variables which best explain the variability of the data. Linear combinations are relatively easy to compute and interpret. Also, linear combinations have nice mathematical properties. Later methods, such as Multidimensional Scaling, broaden the types of combinations, but this is done at a cost: The mathematical treatment becomes more difficult, and the practical calculations will be more complex. The complexity increases with the size of the data, and it is one of the major reasons why Multidimensional Scaling has taken rather longer to regain popularity.

The second question is of a different nature, and its answer depends on the solution to the first. In particular cases, one can take into account prior information or the accuracy of the available data. Possible solutions range from using visual cues to adopting objective or data-driven solutions; the latter are still actively researched. In this chapter we will only touch on this second question, and we then return to it in later chapters.

Traditionally, the number of combinations represents a compromise between accuracy and efficiency. The more combinations we use, the closer we can get to the original data, and the greater the computational effort becomes. Another approach is to regard the observations as *signal plus noise*. The signal is the part which we want to preserve and could therefore be thought of as contained in the combinations we keep, whereas the noise can be discarded. The aim then becomes that of separating the signal from the noise. There is no single best method, nor is there one that always works. Low-dimensional multivariate data will need to be treated differently from high-dimensional data, and for practical applications, the size of the sample will often play a crucial role.

In this chapter we explore how Principal Component Analysis combines the original variables into a smaller number of variables which lead to a simpler description of the data. Section 2.2 describes the approach for the population, Section 2.3 deals with the sample, and Section 2.4 explores ways of visualising principal components. In Section 2.5 we derive properties of these new sets of variables. Section 2.6 looks at special classes of observations: those for which the variables have different ranges and scales, high-dimension low sample size and functional data. Section 2.7 details asymptotic results for the classical case with fixed dimension, as well as the case when the dimension increases. Section 2.8 explores two extensions of Principal Component Analysis: a likelihood-based approach for selecting the number of combinations and Principal Component Regression, an approach to variable selection in linear regression which is anchored in Principal Component Analysis. Problems pertaining to the material of this chapter are listed at the end of Part I. Throughout this chapter, examples demonstrate how Principal Component Analysis works and when it works well. In some of our examples, Principal Component Analysis does not lead to a good answer. Such examples are equally important as they improve our understanding of the limitations of the method.

2.2 Population Principal Components

We begin with properties of linear combinations of random variables, and then I define principal components for a random vector with known mean and covariance matrix.

Proposition 2.1 *Let* $\mathbf{a} = [a_1 \ldots a_d]^{\mathsf{T}}$ *and* $\mathbf{b} = [b_1 \ldots b_d]^{\mathsf{T}}$ *be d-dimensional vectors with real entries. Let* $\mathbf{X} \sim (\boldsymbol{\mu}, \Sigma)$, *and put*

$$V_{\mathbf{a}} = \mathbf{a}^{\mathsf{T}}\mathbf{X} = \sum_{j=1}^{d} a_j X_j \qquad \textit{and} \qquad V_{\mathbf{b}} = \mathbf{b}^{\mathsf{T}}\mathbf{X} = \sum_{j=1}^{d} b_j X_j.$$

1. *The expected value and variance of* $V_{\mathbf{a}}$ *are*

$$\mathbb{E}(V_{\mathbf{a}}) = \mathbf{a}^{\mathsf{T}}\boldsymbol{\mu} = \sum_{j=1}^{d} a_j \mu_j$$

and

$$\operatorname{var}(V_{\mathbf{a}}) = \mathbf{a}^{\mathsf{T}}\Sigma\mathbf{a}.$$

2. *For the vectors* **a** *and* **b**, *the covariance of* $V_{\mathbf{a}}$ *and* $V_{\mathbf{b}}$ *is*

$$\operatorname{cov}(V_{\mathbf{a}}, V_{\mathbf{b}}) = \mathbf{a}^{\mathsf{T}}\Sigma\mathbf{b}.$$

Proof The expectation result in part 1 of the proposition follows by linearity. To calculate the variance, note that

$$\begin{aligned}\operatorname{var}(V_{\mathbf{a}}) &= \mathbb{E}\left\{\left[\mathbf{a}^{\mathsf{T}}\mathbf{X} - \mathbb{E}(\mathbf{a}^{\mathsf{T}}\mathbf{X})\right]\left[\mathbf{X}^{\mathsf{T}}\mathbf{a} - \mathbb{E}(\mathbf{X}^{\mathsf{T}}\mathbf{a})\right]\right\}\\ &= \mathbb{E}\left[\mathbf{a}^{\mathsf{T}}(\mathbf{X} - \boldsymbol{\mu})(\mathbf{X}^{\mathsf{T}} - \boldsymbol{\mu}^{\mathsf{T}})\mathbf{a}\right]\\ &= \mathbf{a}^{\mathsf{T}}\mathbb{E}\left[(\mathbf{X} - \boldsymbol{\mu})(\mathbf{X}^{\mathsf{T}} - \boldsymbol{\mu}^{\mathsf{T}})\right]\mathbf{a} = \mathbf{a}^{\mathsf{T}}\Sigma\mathbf{a}. \end{aligned} \tag{2.1}$$

The proof of part 2 is deferred to the Problems at the end of Part I. ∎

The proposition relates the first- and second-moment properties of random vectors projected onto fixed vectors. We now start with the covariance matrix, and we find *interesting* projections. As in (1.19) of Section 1.5.2, the spectral decomposition of Σ is

$$\Sigma = \Gamma\Lambda\Gamma^{\mathsf{T}} \qquad \text{with } \Sigma\boldsymbol{\eta}_k = \lambda_k\boldsymbol{\eta}_k \qquad \text{for } k = 1,\ldots,d,$$

where $\Lambda = \mathrm{diag}(\lambda_1,\ldots,\lambda_d)$, with $\lambda_1 > \lambda_2 > \cdots > \lambda_d > 0$, and Γ is the orthogonal matrix with columns $\boldsymbol{\eta}_k$, the eigenvectors of Σ corresponding to the eigenvalues λ_k. If the rank r of Σ is less than the dimension d, we use the submatrix notation Λ_r and Γ_r of (1.21) and recall that Γ_r is r-orthogonal and satisfies (1.25).

As I said in Section 1.5.2, I assume that the eigenvalues of Σ are distinct, non-zero, and listed in decreasing order, unless otherwise stated.

Definition 2.2 Consider $\mathbf{X} \sim (\boldsymbol{\mu}, \Sigma)$, with $\Sigma = \Gamma\Lambda\Gamma^{\mathsf{T}}$. Let r be the rank of Σ. Let $k = 1,\ldots,r$.

1. The kth **principal component score** is the scalar

$$W_k = \boldsymbol{\eta}_k^{\mathsf{T}}(\mathbf{X} - \boldsymbol{\mu});$$

2. the k-dimensional **principal component vector** is

$$\mathbf{W}^{(k)} = \begin{bmatrix} W_1 \\ \vdots \\ W_k \end{bmatrix} = \Gamma_k^{\mathsf{T}}(\mathbf{X} - \boldsymbol{\mu}); \tag{2.2}$$

3. and the kth **principal component projection (vector)** is

$$\mathbf{P}_k = \boldsymbol{\eta}_k\boldsymbol{\eta}_k^{\mathsf{T}}(\mathbf{X} - \boldsymbol{\mu}) = W_k\boldsymbol{\eta}_k. \tag{2.3}$$

□

We use the letter W for the principal component scores, indicating that they are *weighted* random variables. Informally, we refer to the *PCs* of $\mathbf{X}$, meaning any one or all of the principal component scores, vectors or projections, and in particular, we will use the notation PC_k for the kth principal component score. When a distinction is necessary, I will refer to the individual objects by their precise names. For brevity we call the principal component scores *PC scores*, or just *PCs*.

The kth principal component score W_k represents the contribution of $\mathbf{X}$ in the direction $\boldsymbol{\eta}_k$. The vectors $\boldsymbol{\eta}_k$ are sometimes called the **loadings** as they represent the 'load' or 'weight' each variable is accorded in the projection. Mathematically, W_k is obtained by projecting the centred $\mathbf{X}$ onto $\boldsymbol{\eta}_k$. Collecting the contributions of the first k scores into one object leads to the principal component vector $\mathbf{W}^{(k)}$, a k-dimensional random vector which summarises the contributions of $\mathbf{X}$ along the first k eigen directions, the eigenvectors of Σ.

The principal component projection $\mathbf{P}_k$ is a d-dimensional random vector which points in the same – or opposite – direction as the kth eigenvector $\boldsymbol{\eta}_k$, and the length or Euclidean norm of $\mathbf{P}_k$ is the absolute value of W_k. In the sample case, the scores of different observations $\mathbf{X}_i$ will vary, and similarly, the PC projections arising from different $\mathbf{X}_i$ will vary. We make use of these different contributions of the $\mathbf{X}_i$ in an analysis of the data.

Example 2.1 Let $\mathbf{X} = [X_1\ X_2]^\top$ be a two-dimensional **random vector** with mean $\boldsymbol{\mu}$ and covariance matrix Σ given by

$$\boldsymbol{\mu} = [0, -1]^\top \quad \text{and} \quad \Sigma = \begin{pmatrix} 2.4 & -0.5 \\ -0.5 & 1 \end{pmatrix}. \tag{2.4}$$

The eigenvalues and eigenvectors of Σ are

$$(\lambda_1, \boldsymbol{\eta}_1) = \left(2.5602, \begin{bmatrix} 0.9523 \\ -0.3052 \end{bmatrix}\right) \quad \text{and} \quad (\lambda_2, \boldsymbol{\eta}_2) = \left(0.8398, \begin{bmatrix} 0.3052 \\ 0.9523 \end{bmatrix}\right).$$

The eigenvectors show the axes along which the data vary most, with the first vector $\sqrt{\lambda_1}\boldsymbol{\eta}_1$ pointing along the direction in which the data have the largest variance. The vectors $\sqrt{\lambda_j}\boldsymbol{\eta}_j$ with $j = 1, 2$ are given in Figure 2.1. In this figure I have drawn them so that the x-values of the vectors are positive. I could instead have shown the direction vectors $\boldsymbol{\eta}_1' = (-0.9523, 0.3052)^\top$ and $\boldsymbol{\eta}_2' = (-0.3052, -0.9523)^\top$ because they are also eigenvectors of Σ.

The first eigenvalue is considerably bigger than the second, so much more variability exists along this direction. Theorem 2.5 reveals the relationship between the eigenvalues and the variance of $\mathbf{X}$.

The two principal component scores are

$$W_1 = 0.953 X_1 - 0.305(X_2 + 1) \quad \text{and} \quad W_2 = 0.305 X_1 + 0.953(X_2 + 1).$$

The first PC score is heavily weighted in the direction of the first variable, implying that the first variable contributes more strongly than the second to the variance of $\mathbf{X}$. The reverse holds for the second component. The first and second PC projections are

$$\mathbf{P}_1 = W_1 \begin{bmatrix} 0.9523 \\ -0.3052 \end{bmatrix} \quad \text{and} \quad \mathbf{P}_2 = W_2 \begin{bmatrix} 0.3052 \\ 0.9523 \end{bmatrix},$$

with W_1 and W_2 as earlier. ■

The major data axes do not generally coincide with the (x, y)-coordinate system, and typically the variables of the random vector $\mathbf{X}$ are correlated. In the following section we will

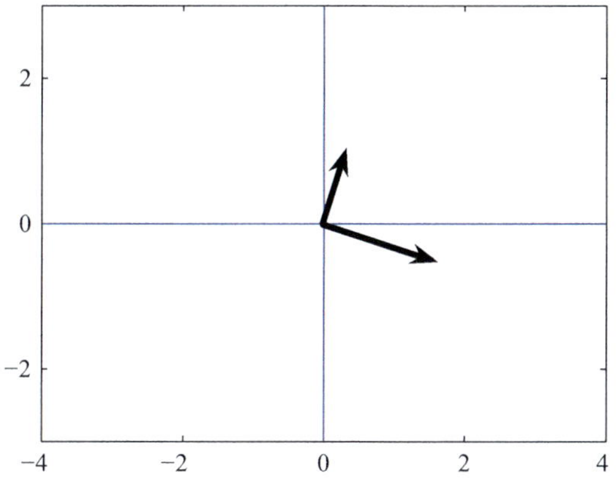

Figure 2.1 Major axes for Example 2.1, given by $\sqrt{\lambda_1}\boldsymbol{\eta}_1$ and $\sqrt{\lambda_2}\boldsymbol{\eta}_2$.

Table 2.1 *Eigenvalues and eigenvectors for* Σ *of (2.5)*

	1	2	3	4	5	6
λ	3.0003	0.9356	0.2434	0.1947	0.0852	0.0355
η	−0.0438	−0.0107	0.3263	0.5617	−0.7526	−0.0981
	0.1122	−0.0714	0.2590	0.4555	0.3468	0.7665
	0.1392	−0.0663	0.3447	0.4153	0.5347	−0.6317
	0.7683	0.5631	0.2180	−0.1861	−0.1000	0.0222
	0.2018	−0.6593	0.5567	−0.4507	−0.1019	0.0349
	−0.5789	0.4885	0.5918	−0.2584	0.0845	0.0457

see that a principal component analysis rotates the data. Figure 2.2 illustrates this rotation for a random sample with true mean and covariance matrix as in Example 2.1.

Example 2.2 Consider a six-dimensional **random vector** with covariance matrix

$$\Sigma = \begin{pmatrix} 0.1418 & 0.0314 & 0.0231 & -0.1032 & -0.0185 & 0.0843 \\ 0.0314 & 0.1303 & 0.1084 & 0.2158 & 0.1050 & -0.2093 \\ 0.0231 & 0.1084 & 0.1633 & 0.2841 & 0.1300 & -0.2405 \\ -0.1032 & 0.2158 & 0.2841 & 2.0869 & 0.1645 & -1.0370 \\ -0.0185 & 0.1050 & 0.1300 & 0.1645 & 0.6447 & -0.5496 \\ 0.0843 & -0.2093 & -0.2405 & -1.0370 & -0.5496 & 1.3277 \end{pmatrix}. \tag{2.5}$$

The eigenvalues and eigenvectors of Σ are given in Table 2.1, starting with the first (and largest) eigenvalue. The entries for each eigenvector show the contribution or weight of each variable: $\boldsymbol{\eta}_2$ has the entry of 0.5631 for the fourth variable X_4.

The eigenvalues decrease quickly: the second is less than one-third of the first, and the last two eigenvalues are about 3 and 1 per cent of the first and therefore seem to be negligible.

An inspection of the eigenvectors shows that the first eigenvector has highest absolute weights for variables X_4 and X_6, but these two weights have opposite signs. The second eigenvector points most strongly in the direction of X_5 and also has large weights for X_4 and X_6. The third eigenvector, again, singles out variables X_5 and X_6, and the remaining three eigenvectors have large weights for the variables X_1, X_2 and X_3. Because the last three eigenvalues are considerably smaller than the first two, we conclude that variables X_4 to X_6 contribute more to the variance than the other three. ■

2.3 Sample Principal Components

In this section we consider definitions for samples of independent random vectors. Let $\mathbb{X} = [\mathbf{X}_1\ \mathbf{X}_2 \cdots \mathbf{X}_n]$ be $d \times n$ data. In general, we do not know the mean and covariance structure of $\mathbb{X}$, so we will, instead, use the sample mean $\overline{\mathbf{X}}$, the centred data $\mathbb{X}_{\text{cent}}$, the sample covariance matrix S, and the notation $\mathbb{X} \sim \mathsf{Sam}(\overline{\mathbf{X}}, S)$ as in (1.6) through (1.10) of Section 1.3.2.

Let $r \leq d$ be the rank of S, and let

$$S = \widehat{\Gamma}\widehat{\Lambda}\widehat{\Gamma}^{\mathsf{T}}$$

be the spectral decomposition of S, as in (1.26) of Section 1.5.3, with eigenvalue–eigenvector pairs $(\widehat{\lambda}_j, \widehat{\boldsymbol{\eta}}_j)$. For $q \leq d$, we use the submatrix notation $\widehat{\Lambda}_q$ and $\widehat{\Gamma}_q$, similar to the population case. For details, see (1.21) in Section 1.5.2.

Definition 2.3 Consider the random sample $\mathbb{X} = [\mathbf{X}_1\ \mathbf{X}_2 \cdots \mathbf{X}_n] \sim \mathrm{Sam}(\overline{\mathbf{X}}, S)$ with sample mean $\overline{\mathbf{X}}$ and sample covariance matrix S of rank r. Let $S = \widehat{\Gamma}\widehat{\Lambda}\widehat{\Gamma}^\top$ be the spectral decomposition of S. Consider $k = 1, \ldots, r$.

1. The kth **principal component score** of $\mathbb{X}$ is the row vector
$$\mathbf{W}_{\bullet k} = \widehat{\boldsymbol{\eta}}_k^\top (\mathbb{X} - \overline{\mathbf{X}});$$

2. the **principal component data** $\mathbb{W}^{(k)}$ consist of the first k principal component vectors $\mathbf{W}_{\bullet j}$, with $j = 1, \ldots, k$, and
$$\mathbb{W}^{(k)} = \begin{bmatrix} \mathbf{W}_{\bullet 1} \\ \vdots \\ \mathbf{W}_{\bullet k} \end{bmatrix} = \widehat{\Gamma}_k^\top (\mathbb{X} - \overline{\mathbf{X}}); \tag{2.6}$$

3. and the $d \times n$ matrix of the kth **principal component projections** $\mathbb{P}_{\bullet k}$ is
$$\mathbb{P}_{\bullet k} = \widehat{\boldsymbol{\eta}}_k \widehat{\boldsymbol{\eta}}_k^\top (\mathbb{X} - \overline{\mathbf{X}}) = \widehat{\boldsymbol{\eta}}_k \mathbf{W}_{\bullet k}. \tag{2.7}$$

□

The row vector
$$\mathbf{W}_{\bullet k} = \begin{bmatrix} W_{1k}\ W_{2k} \cdots W_{nk} \end{bmatrix} \tag{2.8}$$

has n entries: the kth scores of all n observations. The first subscript of $\mathbf{W}_{\bullet k}$, here written as $\bullet$, runs over all n observations, and the second subscript, k, refers to the kth dimension or component. Because $\mathbf{W}_{\bullet k}$ is the vector of scores of all observations, we write it as a row vector. The $k \times n$ matrix $\mathbb{W}^{(k)}$ follows the same convention as the data $\mathbb{X}$: the rows correspond to the variables or dimensions, and each column corresponds to an observation.

Next, we consider $\mathbb{P}_{\bullet k}$. For each k, $\mathbb{P}_{\bullet k} = \widehat{\boldsymbol{\eta}}_k \mathbf{W}_{\bullet k}$ is a $d \times n$ matrix. The columns of $\mathbb{P}_{\bullet k}$ are the kth principal component projections or projection vectors, one column for each observation. The n columns share the same direction – $\widehat{\boldsymbol{\eta}}_k$ – however, the values of their entries differ as they reflect each observation's contribution to a particular direction.

Before we move on, we compare the population case and the random sample case. Table 2.2 summarises related quantities for a single random vector and a sample of size n. We can think of the population case as the *ideal*, where truth is known. In this case, we establish properties pertaining to the single random vector and its distribution. For data, we generally do not know the truth, and the best we have are estimators such as the sample mean and sample covariance, which are derived from the available data. From the strong law of large numbers, we know that the sample mean converges to the true mean, and the sample covariance matrix converges to the true covariance matrix as the sample size increases. In Section 2.7 we examine how the behaviour of the sample mean and covariance matrix affects the convergence of the eigenvalues, eigenvectors and principal components.

The relationship between the population and sample quantities in the table suggests that the population random vector could be regarded as one of the columns of the sample – a reason for writing the data as a $d \times n$ matrix.

Table 2.2 *Relationships of population and sample principal components*

	Population		Random Sample	
Random vectors	$\mathbf{X}$	$d\times 1$	$\mathbb{X}$	$d\times n$
kth PC score	W_k	1×1	$\mathbf{W}_{\bullet k}$	$1\times n$
PC vector/data	$\mathbf{W}^{(k)}$	$k\times 1$	$\mathbb{W}^{(k)}$	$k\times n$
kth PC projection	$\mathbf{P}_k$	$d\times 1$	$\mathbb{P}_{\bullet k}$	$d\times n$

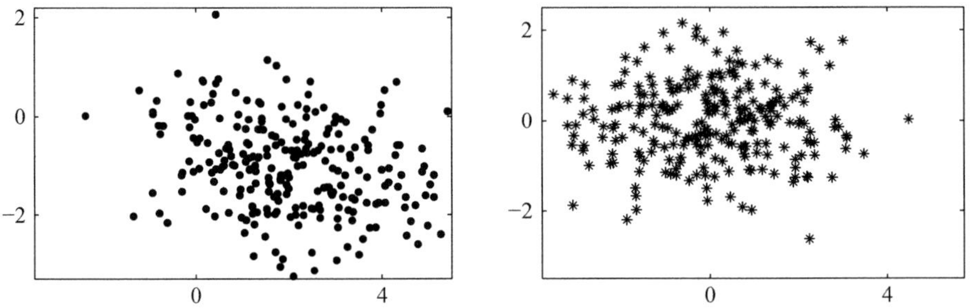

Figure 2.2 Two-dimensional simulated data of Example 2.3 (*left panel*) and PC data (*right panel*).

The following two examples show sample PCs for a randomly generated sample with mean and covariance matrix as in Example 2.1 and for five-dimensional *flow cytometric* measurements.

Example 2.3 The two-dimensional **simulated data** consist of 250 vectors $\mathbf{X}_i$ from the bivariate normal distribution with mean $[2,-1]^{\mathsf{T}}$ and covariance matrix Σ as in (2.4). For these data, the sample covariance matrix

$$S=\begin{pmatrix}2.2790 & -0.3661\\ -0.3661 & 0.8402\end{pmatrix}.$$

The two sample eigenvalue–eigenvector pairs of S are

$$(\hat{\lambda}_1,\hat{\eta}_1)=\left(2.3668,\begin{bmatrix}-0.9724\\ +0.2332\end{bmatrix}\right)\qquad\text{and}\qquad(\hat{\lambda}_2,\hat{\eta}_2)=\left(0.7524,\begin{bmatrix}-0.2332\\ -0.9724\end{bmatrix}\right).$$

The left panel of Figure 2.2 shows the data, and the right panel shows the PC data $\mathbb{W}^{(2)}$, with $\mathbf{W}_{\bullet 1}$ on the x-axis and $\mathbf{W}_{\bullet 2}$ on the y-axis. We can see that the PC data are centred and rotated, so their first major axis is the x-axis, and the second axis agrees with the y-axis.

The data are simulations based on the population case of Example 2.1. The calculations show that the sample eigenvalues and eigenvectors are close to the population quantities. ■

Example 2.4 The **HIV flow cytometry** data of Rossini, Wan, and Moodie (2005) consist of fourteen subjects: five are HIV$^+$, and the remainder are HIV$^-$. Multiparameter flow cytometry allows the analysis of cell surface markers on white blood cells with the aim of finding cell subpopulations with similar combinations of markers that may be used for diagnostic

Table 2.3 *Eigenvalues and eigenvectors of* HIV$^+$ *and* HIV$^-$ *data from Example 2.4*

		HIV$^+$				
$\widehat{\lambda}$		12,118	8,818	4,760	1,326	786
$\widehat{\eta}$	*FS*	0.1511	0.3689	0.7518	−0.0952	−0.5165
	SS	0.1233	0.1448	0.4886	−0.0041	0.8515
	CD3	0.0223	0.6119	−0.3376	−0.7101	0.0830
	CD8	−0.7173	0.5278	−0.0332	0.4523	0.0353
	CD4	0.6685	0.4358	−0.2845	0.5312	−0.0051
		HIV$^-$				
$\widehat{\lambda}$		13,429	7,114	4,887	1,612	598
$\widehat{\eta}$	*FS*	0.1456	0.5765	−0.6512	0.1522	0.4464
	SS	0.0860	0.2336	−0.3848	−0.0069	−0.8888
	CD3	0.0798	0.4219	0.4961	0.7477	−0.1021
	CD8	−0.6479	0.5770	0.2539	−0.4273	−0.0177
	CD4	0.7384	0.3197	0.3424	−0.4849	0.0110

purposes. Typically, five to twenty quantities – based on the markers – are measured on tens of thousands of blood cells. For an introduction and background, see Givan (2001).

Each new marker potentially leads to a split of a subpopulation into parts, and the discovery of these new parts may lead to a link between markers and diseases. Of special interest are the number of modes and associated clusters, the location of the modes and the relative size of the clusters. New technologies allow, and will continue to allow, the collection of more *parameters*, and thus flow cytometry measurements provide a rich source of multidimensional data.

We consider the first and second subjects, who are HIV$^+$ and HIV$^-$, respectively. These two subjects have five measurements on 10,000 blood cells: *forward scatter (FS)*, *side scatter (SS)*, and the three intensity measurements *CD4*, *CD8* and *CD3*, called *colours* or *parameters*, which arise from different antibodies and markers. The colours *CD4* and *CD8* are particularly important for differentiating between HIV$^+$ and HIV$^-$ subjects because it is known that the level of *CD8* increases and that of *CD4* decreases with the onset of HIV$^+$. Figure 1.1 of Section 1.2 shows plots of the three colours for these two subjects. The two plots look different, but it is difficult to quantify these differences from the plots.

A principal component analysis of these data leads to the eigenvalues and eigenvectors of the two sample covariance matrices in Table 2.3. The second column in the table shows the variable names, and I list the eigenvectors in the same column as their corresponding eigenvalues. The eigenvalues decrease quickly and at about the same rate for the HIV$^+$ and HIV$^-$ data. The first eigenvectors of the HIV$^+$ and HIV$^-$ data have large weights of opposite signs in the variables *CD4* and CD8. For HIV$^+$, the largest contribution to the first principal component is *CD8*, whereas for HIV$^-$, the largest contribution is CD4. This change reflects the shift from *CD4* to *CD8* with the onset of HIV$^+$. For PC_2 and HIV$^+$, *CD3* becomes important, whereas FS and *CD8* have about the same high weights for the HIV$^-$ data.

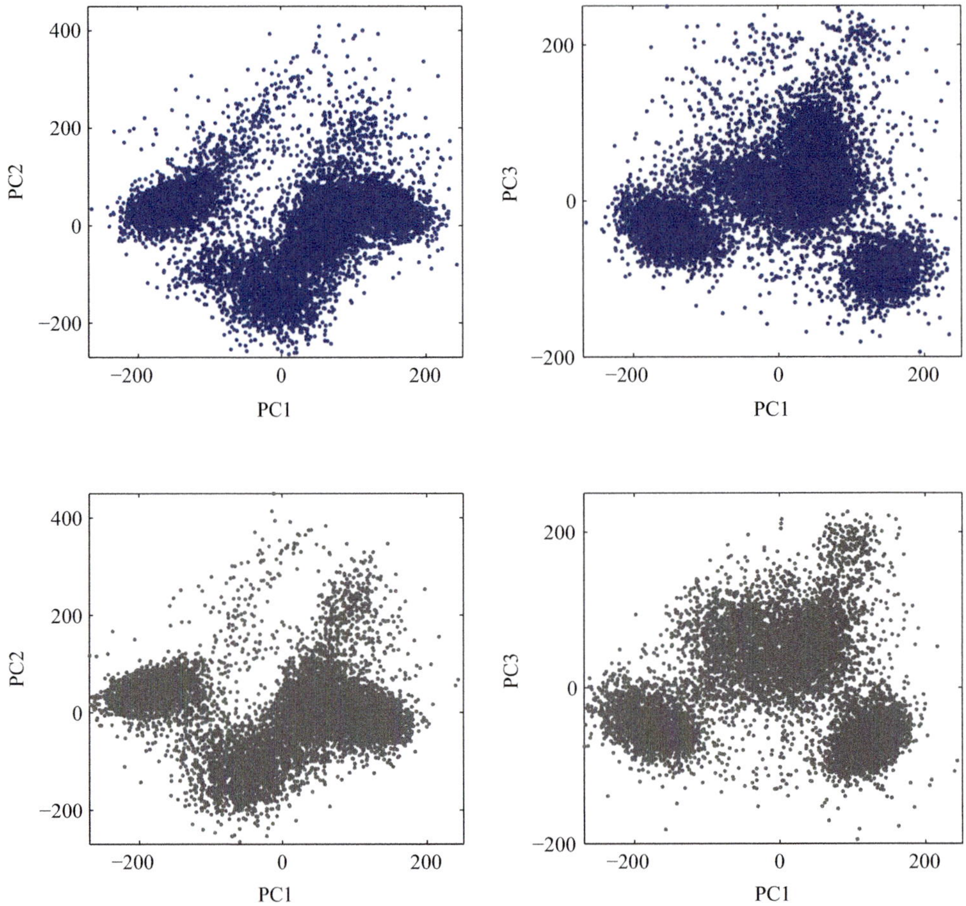

Figure 2.3 Principal component scores for HIV^+ data (*top row*) and HIV^- data (*bottom row*) of Example 2.4.

Figure 2.3 shows plots of the principal component scores for both data sets. The top row displays plots in blue which relate to the HIV^+ data: PC_1 on the x-axis against PC_2 on the left and against PC_3 on the right. The bottom row shows similar plots, but in grey, for the HIV^- data. The patterns in the two PC_1/PC_2 plots are similar; however, the 'grey' PC_1/PC_3 plot exhibits a small fourth cluster in the top right corner which has almost disappeared in the corresponding 'blue' HIV^+ plot. The PC_1/PC_3 plots suggest that the cluster configurations of the HIV^+ and HIV^- data could be different.

A comparison with Figure 1.2 of Section 1.2, which depicts the three-dimensional score plots PC_1, PC_2 and PC_3, shows that the information contained in both sets of plots is similar. In the current figures we can see more easily which principal components are responsible for the extra cluster in the HIV^- data, and we also note that the orientation of the main cluster in the PC_1/PC_3 plot differs between the two subjects. In contrast, the three-dimensional views of Figure 1.2 avail themselves more readily to a spatial interpretation of the clusters. ■

The example allows an interesting interpretation of the first eigenvectors of the HIV^+ and HIV^- data: the largest weight of $\widehat{\boldsymbol{\eta}}_1$ of the HIV^- data, associated with *CD4*, has decreased to second largest for the HIV$^+$ data, whereas the second-largest weight (in absolute value) of the HIV$^-$ data, which is associated with *CD8*, has become the largest weight for the HIV$^+$ data. This shift reflects the increase of *CD8* and decrease of *CD4* that occurs with the onset of the disease and which I mentioned earlier in this example. A more comprehensive analysis involving a number of subjects of each type and different analytical methods is necessary, however, to understand and quantify the differences between HIV$^+$ and HIV$^-$ data.

2.4 Visualising Principal Components

Visual inspection of the principal components helps to see what is going on. Example 2.4 exhibits scatterplots of principal component scores, which display some of the differences between the two data sets. We may obtain additional information by considering

1. eigenvalue, variance and scree plots;
2. plots of two- and three-dimensional principal component scores; and
3. projection plots and estimates of the density of the scores.

I explain each idea briefly and then illustrate with data. Section 2.5, which deals with properties of principal components, will complement the visual information we glean from the figures of this section.

2.4.1 Scree, Eigenvalue and Variance Plots

We begin with summary statistics that are available from an analysis of the covariance matrix.

Definition 2.4 Let $\mathbf{X} \sim (\boldsymbol{\mu}, \Sigma)$. Let r be the rank of Σ, and for $k \le r$, let λ_k be the eigenvalues of Σ. For $\kappa \le r$, let $\mathbf{W}^{(\kappa)}$ be the κth principal component vector. The **proportion of total variance** or the **contribution to total variance** explained by the kth principal component score W_k is

$$\frac{\lambda_k}{\sum_{j=1}^r \lambda_j} = \frac{\lambda_k}{\operatorname{tr}(\Sigma)}.$$

The **cumulative contribution to total variance** of the κ-dimensional principal component vector $\mathbf{W}^{(\kappa)}$ is

$$\frac{\sum_{k=1}^{\kappa} \lambda_k}{\sum_{j=1}^r \lambda_j} = \frac{\sum_{k=1}^{\kappa} \lambda_k}{\operatorname{tr}(\Sigma)}.$$

A **scree** plot is a plot of the eigenvalues λ_k against the **index** k. □

For data, the **(sample) proportion of** and **contribution to total variance** are defined analogously using the sample covariance matrix S and its eigenvalues $\widehat{\lambda}_k$.

It may be surprising to use the term *variance* in connection with the eigenvalues of Σ or S. Theorems 2.5 and 2.6 establish the relationship between the eigenvalues λ_k and the variance of the scores W_k and thereby justify this terminology.

Scree is the accumulation of rock fragments at the foot of a cliff or hillside and is derived from the Old Norse word *skrītha*, meaning a landslide, to slip or to slide – see Partridge

Table 2.4 *Variables of the Swiss bank notes data from Example 2.5*

1	Length of the bank notes
2	Height of the bank notes, measured on the left
3	Height of the bank notes, measured on the right
4	Distance of inner frame to the lower border
5	Distance of inner frame to the upper border
6	Length of the diagonal

(1982). Scree plots, or plots of the eigenvalues $\widehat{\lambda}_k$ against their index k, tell us about the distribution of the eigenvalues and, in light of Theorem 2.5, about the decrease in variance of the scores. Of particular interest is the ratio of the first eigenvalue to the trace of Σ or S. The actual size of the eigenvalues may not be important, so the proportion of total variance provides a convenient standardisation of the eigenvalues.

Scree plots may exhibit an *elbow* or a *kink*. Folklore has it that the index κ at which an elbow appears is the number of principal components that adequately represent the data, and this κ is interpreted as the dimension of the reduced or principal component data. However, the existence of elbows is not guaranteed. Indeed, as the dimension increases, elbows do not usually appear. Even if an elbow is visible, there is no real justification for using its index as the dimension of the PC data. The words *knee* or *kink* also appear in the literature instead of elbow.

Example 2.5 The **Swiss bank notes** data of Flury and Riedwyl (1988) contain six variables measured on 100 genuine and 100 counterfeit old Swiss 1,000-franc bank notes. The variables are shown in Table 2.4.

A first inspection of the data (which I do not show here) reveals that the values of the largest variable are 213 to 217 mm, whereas the smallest two variables (4 and 5) have values between 7 and 12 mm. Thus the largest variable is about twenty times bigger than the smallest.

Table 2.1 shows the eigenvalues and eigenvectors of the sample covariance matrix which is given in (2.5). The left panel of Figure 2.4 shows the size of the eigenvalues on the y-axis against their index on the x-axis. We note that the first eigenvalue is large compared with the second and later ones.

The lower curve in the right panel shows, on the y-axis, the contribution to total variance, that is, the standardised eigenvalues, and the upper curve shows the cumulative contribution to total variance – both as percentages – against the index on the x-axis. The largest eigenvalue contributes well over 60 per cent of the total variance, and this percentage may be more useful than the actual size of $\widehat{\lambda}_1$. In applications, I recommend using a combination of both curves as done here.

For these data, an *elbow* at the third eigenvalue is visible, which may lead to the conclusion that three PCs are required to represent the data. This elbow is visible in the lower curve in the right subplot but not in the cumulative upper curve. ■

Our second example looks at two thirty-dimensional data sets of very different origins.

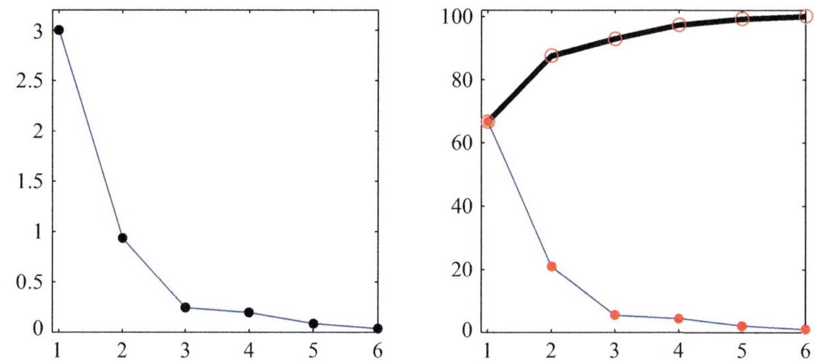

Figure 2.4 Swiss bank notes of Example 2.5; eigenvalues (*left*) and simple and cumulative contributions to variance (*right*) against the number of PCs, given as percentages.

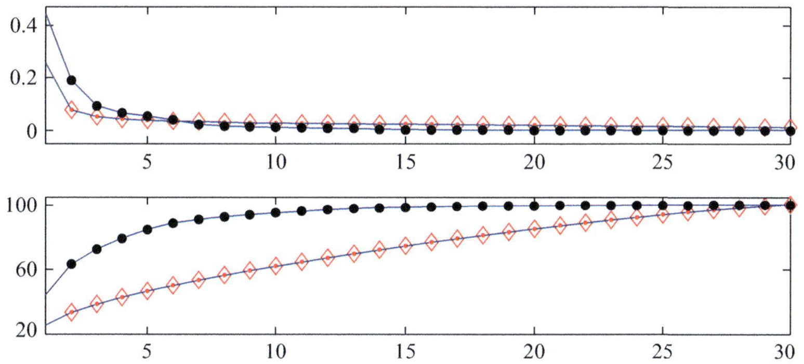

Figure 2.5 Scree plots (*top*) and cumulative contributions to total variance (*bottom*) for the breast cancer data (*black dots*) and the Dow Jones returns (*red diamonds*) of Example 2.6 – in both cases against the index on the x-axis.

Example 2.6 The **breast cancer** data of Blake and Merz (1998) consist of 569 records and thirty variables. The **Dow Jones returns** consist of thirty stocks on 2,528 days over the period from January 1991 to January 2001. Of these, twenty-two stocks are still in the 2012 Dow Jones 30 Index.

The breast cancer data arise from two groups: 212 malignant and 357 benign cases. And for each record, this status is known. We are not interested in this status here but focus on the sample covariance matrix.

The Dow Jones observations are the 'daily returns', the differences of log prices taken on consecutive days.

Figure 2.5 shows the contributions to variance in the top panel and the cumulative contributions to variance in the lower panel against the index of the PCs on the x-axis. The curves with black dots correspond to the breast cancer data, and those with red diamonds correspond to the Dow Jones returns.

The eigenvalues of the breast cancer data decrease more quickly than those of the Dow Jones returns. For the breast cancer data, the first PC accounts for 44 per cent, and the second for 19 per cent. For $k = 10$, the total contribution to variance amounts to more than 95 per cent, and at $k = 17$ to just over 99 per cent. This rapid increase suggests that principal components 18 and above may be negligible. For the Dow Jones returns, the first PC accounts for 25.5 per cent of variance, the second for 8 per cent, the first ten PCs account for about 60 per cent of the total contribution to variance, and to achieve 95 per cent of variance, the first twenty-six PCs are required. ■

The two data sets have the same number of variables and share the lack of an elbow in their scree plots. The absence of an elbow is more common than its presence. Researchers and practitioners have used many different schemes for choosing the number of PCs to represent their data. We will explore two dimension-selection approaches in Sections 2.8.1 and 10.8, respectively, which are more objective than some of the available ad hoc methods for choosing the number of principal components.

2.4.2 Two- and Three-Dimensional PC Score Plots

In this section we consider two- and three-dimensional scatterplots of principal component scores which I refer to as **(principal component) score plots** or **PC score plots**. Score plots summarise the data and can exhibit pattern in the data such as clusters that may not be apparent in the original data.

We could consider score plots of all PC scores, but principal component scores corresponding to relatively small eigenvalues are of lesser interest. I recommend looking at the first four PCs, but often fewer than four PCs exhibit the structure of the data. In such cases, a smaller number of PCs suffices. For these first few PCs, we consider pairwise (two-dimensional) and also three-dimensional (3D) score plots.

For PC data $\mathbb{W}^{(3)}$, which consist of the first three PC scores, 3D score plots are scatterplots of the the first three PC scores. As we shall see in Theorem 2.5, the variance of the first few PC scores is larger than that of later scores, and one therefore hopes that structure or pattern becomes visible in these plots. A series of six score plots as shown in the next example is a convenient way of displaying the information of the pairwise score plots, and colour may be effective in enhancing the structure. The examples illustrate that interesting structure may not always appear in a plot of the first two or three scores and more or different scores may be required to find the pattern.

Example 2.7 Figure 2.6 shows pairwise score plots of the **Swiss bank notes** data of Example 2.5. In each of the plots, the 100 genuine bank notes are shown in blue and the 100 counterfeits in black.

The plots in the top row show the PC_1 scores on the x-axis; on the y-axis we have the PC_2 scores (*left panel*), the PC_3 scores (*middle*) and the PC_4 scores (*right panel*). The plots in the lower row show from left to right: PC_2 scores on the x-axis against PC_3 scores and PC_4 scores, respectively, and finally, PC_3 scores on the x-axis against PC_4 scores.

The score plots involving PC_1 clearly show the two parts of the data, and the colour confirms that the data split into a genuine part (in *blue*) and a counterfeit part (in *black*). The separation is clearest in the scores of the first two PCs. The remaining score plots also show outliers in the data, which are apparent in the plots of the upper row. ■

For the *Swiss bank notes* data, the pairwise score plots involving PC_1 brought out the grouping of the data into separate parts. For these data, 3D score plots do not add any further information. The opposite is the case in our next example.

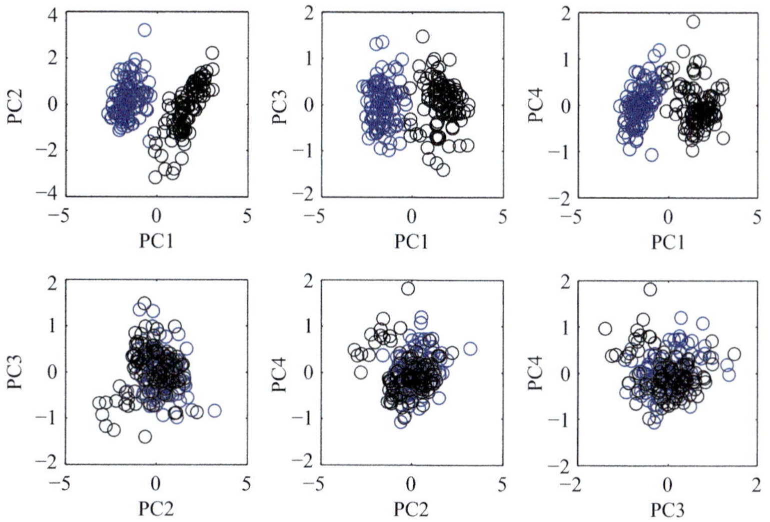

Figure 2.6 Score plots of the Swiss bank notes of Example 2.7. (*Top row*): PC_1 scores (x-axis) against PC_2–PC_4 scores. (*Bottom row*): PC_2 scores against PC_3 and PC_4 scores (*left* and *middle*), and PC_3 scores against PC_4 scores (*right*).

Example 2.8 The **wine recognition** data of Forina et al (see Aeberhard, Coomans, and de Vel, 1992) are obtained as a result of a chemical analysis of three types of wine grown in the same region in Italy but derived from three different cultivars. The analysis resulted in measurements of thirteen variables, called the *constituents*. From the 178 observations, 59 belong to the first cultivar, 71 to the second, and the remaining 48 to the third. In Example 4.5 in Section 4.4.2 we explore rules for dividing these data into the three cultivars. In the current example, we examine the PC scores of the data.

For an easier visual inspection, I plot the scores of the three cultivars in different colours: black for the first, red for the second, and blue for the third cultivar. Two-dimensional score plots, as shown in Figure 2.6, do not exhibit a separation of the cultivars, and for this reason, I have not shown them here. The left subplot of Figure 2.7 shows the PC data $\mathbb{W}^{(3)}$ and so scores of the first three principal components. For these data, the red and blue observations overlap almost completely. The right subplot shows PC_1, PC_3 and PC_4 scores, and in contrast to the configuration in the left panel, here we obtain a reasonable – though not perfect – separation of the three cultivars. ■

Although one might expect the 'most interesting' structure to be visible in the first two or three PC scores, this is not always the case. The principal component directions exhibit variability in the data, and the directions which exhibit the clusters may differ from the former.

2.4.3 Projection Plots and Estimates of the Density of the Scores

The principal component projections $\mathbb{P}_{\bullet k}$ are $d \times n$ matrices. For each index $k \le d$, the ith column of $\mathbb{P}_{\bullet k}$ represents the contribution of the ith observation in the direction of $\widehat{\eta}_k$. It is convenient to display $\mathbb{P}_{\bullet k}$, separately for each k, in the form of parallel coordinate plots with

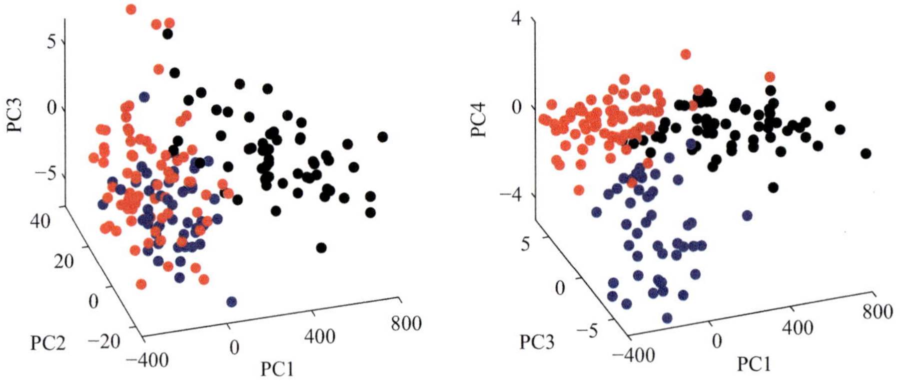

Figure 2.7 3D score plots from the wine recognition data of Example 2.8 with different colours for the three cultivars. (*left*): $\mathbb{W}^{(k)}$; (*right*): PC_1, PC_3 and PC_4.

the variable number on the x-axis. As we shall see in Theorem 2.12 and Corollary 2.14, the principal component projections 'make up the data' in the sense that we can reconstruct the data arbitrarily closely from these projections.

The kth principal component projections show how the eigenvector $\widehat{\boldsymbol{\eta}}_k$ has been modified by the kth scores $\mathbf{W}_{\bullet k}$, and it is therefore natural to look at the distribution of these scores, here in the form of density estimates. The shape of the density provides valuable information about the distribution of the scores. I use the MATLAB software *curvdatSM* of Marron (2008), which calculates non-parametric density estimates based on Gaussian kernels with suitably chosen bandwidths.

Example 2.9 We continue with our parallel analysis of the **breast cancer** data and the **Dow Jones returns**. Both data sets have thirty variables, but the Dow Jones returns have about five times as many observations as the breast cancer data. We now explore parallel coordinate plots and estimates of density of the first and second principal component projections for both data sets.

The top two rows in Figure 2.8 refer to the breast cancer data, and the bottom two rows refer to the Dow Jones returns. The left column of the plots shows the principal component projections, and the right column shows density estimates of the scores. Rows one and three refer to PC_1, and rows two and four refer to PC_2.

In both data sets, all entries of the first eigenvector have the same sign; we can verify this in the projection plots in the first and third rows, where each observation is either positive for each variable or remains negative for all variables. This behaviour is unusual and could be exploited in a later analysis: it allows us to split the data into two groups, the positives and the negatives. No single variable stands out; the largest weight for both data sets is about 0.26. Example 6.9 of Section 6.5.1 looks at splits of the first PC scores for the breast cancer data. For the Dow Jones returns, a split into 'positive' and 'negative' days does not appear to lead to anything noteworthy.

The projection plots of the second eigenvectors show the more common pattern, with positive and negative entries, which show the opposite effects of the variables. A closer inspection of the second eigenvector of the Dow Jones returns shows that the variables 3, 13, 16, 17 and 23 have negative weights. These variables correspond to the five information

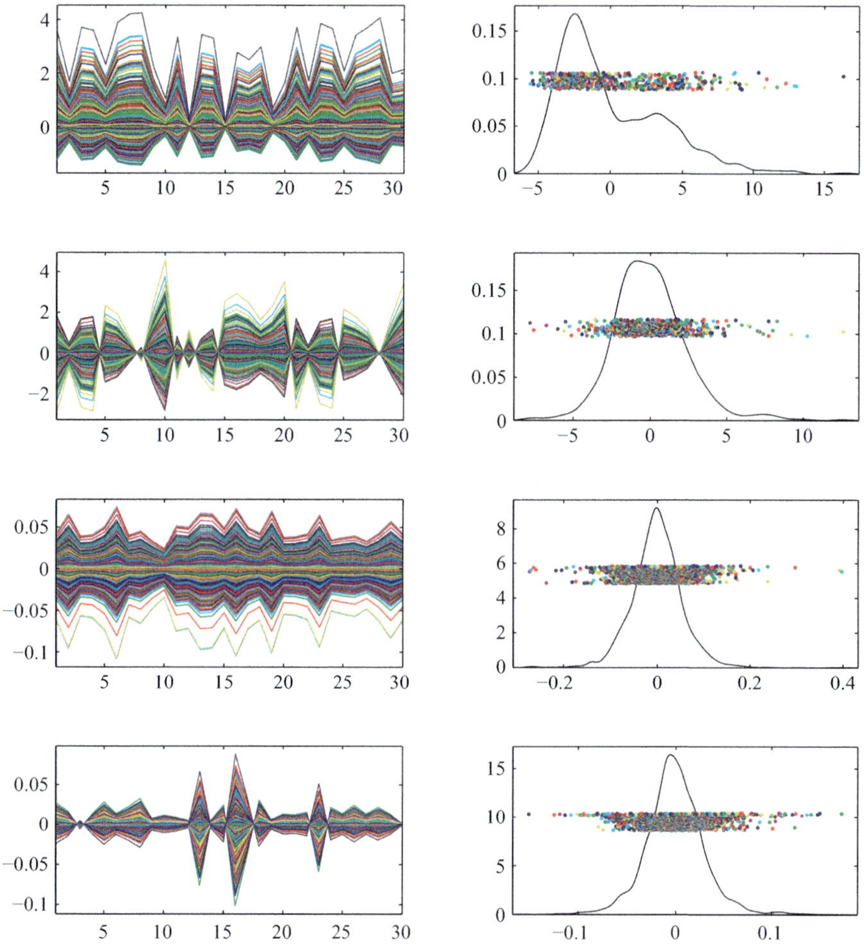

Figure 2.8 Projection plots (*left*) and density estimates (*right*) of PC_1 scores (rows 1 and 3) and PC_2 scores (rows 2 and 4) of the breast cancer data (*top two rows*) and the Dow Jones returns (*bottom two rows*) from Example 2.9.

technology (IT) companies in the list of stocks, namely, *AT&T, Hewlett-Packard, Intel Corporation, IBM* and *Microsoft*. With the exception of AT&T, the IT companies have the largest four weights (in absolute value). Thus PC_2 clearly separates the IT stocks from all others.

It is interesting to see the much larger spread of scores of the breast cancer data, both for PC_1 and for PC_2: y-values in the projection plots and the range of x-values in the density plots. In Example 2.6 we have seen that the first two PCs of the breast cancer data contribute more than 60 per cent to the variance, whereas the corresponding PCs of the Dow Jones returns only make up about 33 per cent of variance. Parallel coordinate views of subsequent PC projections may sometimes contain extra useful information. In these two data sets they do not.

The plots in the right column of Figure 2.8 show the scores and their non-parametric density estimates, which I produced with the *curvdatSM* software of Marron (2008). The score of each observation is given by its value on the x-axis; for easier visual inspection, the actual values of the scores are displayed at random heights y as coloured dots, and each observation is represented by the same colour in the two plots in one row: the outlier at the right end in the PC_1 breast cancer density plot corresponds to the most positive curve in the corresponding projection plot.

An inspection of the density estimates shows that the PC_1 scores of the breast cancer data deviate substantially from the normal density; the bimodal and right skewed shape of this density could reflect the fact that the data consist of benign and malignant observations, and we can infer that the distribution of the first scores is *not* Gaussian. The other three density plots look symmetric and reasonably normal. For good accounts of non-parametric density estimation, the interested reader is referred to Scott (1992) and Wand and Jones (1995). ■

As mentioned at the beginning of Section 2.4, visual inspections of the principal components help to see what is going on, and we have seen that suitable graphical representations of the PC data may lead to new insight. Uncovering clusters, finding outliers or deducing that the data may not be Gaussian, all these properties aid our understanding and inform subsequent analyses.

The next section complements the visual displays of this section with theoretical properties of principal components.

2.5 Properties of Principal Components

2.5.1 Correlation Structure of **X** *and Its PCs*

The variables of the random vector **X** are commonly correlated, and the covariance matrix of **X** is not diagonal. Principal Component Analysis 'untangles' the dependent components and results in uncorrelated combinations of random variables or components. In this section we explore properties of the PCs and examine the correlation structure that **X** and its PCs share.

Theorem 2.5 *Let* $\mathbf{X} \sim (\boldsymbol{\mu}, \Sigma)$*, and let* r *be the rank of* Σ*. For* $\kappa \le r$*, put* $\mathbf{W}^{(\kappa)} = \begin{bmatrix} W_1 \cdots W_\kappa \end{bmatrix}^{\mathsf{T}}$ *as in (2.2).*

1. *The mean and covariance matrix of* $\mathbf{W}^{(\kappa)}$ *are*

$$\mathbb{E}\mathbf{W}^{(\kappa)} = \mathbf{0} \qquad \textit{and} \qquad \operatorname{var}\left[\mathbf{W}^{(\kappa)}\right] = \Lambda_\kappa,$$

   where $\Lambda_\kappa = \operatorname{diag}(\lambda_1, \ldots, \lambda_\kappa)$*. Compare (1.21) of Section 1.5.2.*

2. *Variance and covariance properties of the individual principal component scores* W_k *are*

$$\operatorname{var}(W_k) = \lambda_k \qquad \textit{for } k = 1, \ldots, \kappa;$$
$$\operatorname{cov}(W_k, W_\ell) = \boldsymbol{\eta}_k^{\mathsf{T}} \Sigma \boldsymbol{\eta}_\ell = \lambda_k \delta_{k\ell} \qquad \textit{for } k, \ell = 1, \ldots, \kappa,$$

   where δ *is the Kronecker delta function with* $\delta_{kk} = 1$ *and* $\delta_{k\ell} = 0$ *if* $k \ne \ell$*; and*

$$\operatorname{var}(W_1) \ge \operatorname{var}(W_2) \ge \cdots \ge \operatorname{var}(W_\kappa) > 0.$$

The theorem presents moment properties of the PCs: The PC vectors are centred and uncorrelated. If the random vector is Gaussian, then the PCs are also independent. Non-Gaussian uncorrelated random vectors, however, are in general not independent, and other approaches are required to find independent components for such random vectors. In Chapters 10 to 12 we will meet approaches which address this task.

Proof The result for the mean follows by linearity from the definition of the principal component vector. For the variance calculations in part 1 of the theorem, we use Result 1.10, part 3, of Section 1.5.2, so for $k \le r$,

$$\Gamma_k^{\mathsf{T}}\Gamma = \mathbf{I}_{k\times d} \qquad \text{and} \qquad \Gamma^{\mathsf{T}}\Gamma_k = \mathbf{I}_{d\times k},$$

with $\mathbf{I}_{j\times m}$ as defined in (1.24) of Section 1.5.2. Using these relationships, for $\kappa \le r$, we have

$$\begin{aligned}\operatorname{var}\left[\mathbf{W}^{(\kappa)}\right] &= \mathbb{E}\left[\Gamma_\kappa^{\mathsf{T}}(\mathbf{X}-\boldsymbol{\mu})(\mathbf{X}-\boldsymbol{\mu})^{\mathsf{T}}\Gamma_\kappa\right]\\ &= \Gamma_\kappa^{\mathsf{T}}\mathbb{E}\left[(\mathbf{X}-\boldsymbol{\mu})(\mathbf{X}-\boldsymbol{\mu})^{\mathsf{T}}\right]\Gamma_\kappa\\ &= \Gamma_\kappa^{\mathsf{T}}\Sigma\Gamma_\kappa = \Gamma_\kappa^{\mathsf{T}}\Gamma\Lambda\Gamma^{\mathsf{T}}\Gamma_\kappa = \Lambda_\kappa.\end{aligned}$$

For part 2, note that the first two statements follow immediately from part 1 because Λ_κ is diagonal. The last result in part 2 is a consequence of the ordering of the eigenvalues. ■

I illustrate the theorem with the random vector of Example 2.1.

Example 2.10 We continue with the 2D **simulated data** and calculate the variance for the scores W_1 and W_2:

$$\begin{aligned}\operatorname{var}(W_1) &= \operatorname{var}\left(\boldsymbol{\eta}_1^{\mathsf{T}}\mathbf{X}\right) = \operatorname{var}(\eta_{11}X_1 + \eta_{12}X_2)\\ &= \eta_{11}^2\operatorname{var}(X_1) + \eta_{12}^2\operatorname{var}(X_2) + 2\eta_{11}\eta_{12}\operatorname{cov}(X_1, X_2)\\ &= 0.9523^2 \times 2.4 + 0.3052^2 \times 1 + 2\times 0.9523 \times (-0.3052)\times(-0.5) = 2.5602\end{aligned}$$

and similarly

$$\operatorname{var}(W_2) = \operatorname{var}\left(\boldsymbol{\eta}_2^{\mathsf{T}}\mathbf{X}\right) = 0.8398.$$

A comparison with the eigenvalues listed in Example 2.1 shows that these variances agree with the eigenvalues.

Because there are two variables, we only need to calculate one covariance:

$$\begin{aligned}\operatorname{cov}(W_1, W_2) &= \operatorname{cov}\left(\boldsymbol{\eta}_1^{\mathsf{T}}\mathbf{X}, \boldsymbol{\eta}_2^{\mathsf{T}}\mathbf{X}\right)\\ &= \operatorname{cov}(0.9523X_1 - 0.3052X_2, 0.3052X_1 + 0.9523X_2) = 0.\end{aligned}$$

■

The eigenvalues play an important role, as we shall see in the next result.

Theorem 2.6 *If* $\mathbf{X} \sim (\boldsymbol{\mu}, \Sigma)$, *and* Σ *has eigenvalues* λ_j, *then*

$$\sum_{j=1}^{d}\operatorname{var}(X_j) = \sum_{j=1}^{d}\operatorname{var}(W_j),$$

or equivalently,

$$\sum_{j=1}^{d}\sigma_j^2 = \sum_{j=1}^{d}\lambda_j.$$

Proof By definition, $\mathrm{var}(X_j) = \sigma_j^2$. From part 2 of Theorem 2.5, $\mathrm{var}(W_j) = \lambda_j$. Because Σ and Λ are similar matrices, Result 1.8 of Section 1.5.1 leads to

$$\sum_{j=1}^{d} \mathrm{var}(X_j) = \sum_{j=1}^{d} \sigma_j^2 = \sum_{j=1}^{d} \lambda_j = \sum_{j=1}^{d} \mathrm{var}(W_j).$$

∎

The next corollary states a data version of Theorem 2.6.

Corollary 2.7 *If data* $\mathbb{X} \sim \mathsf{Sam}(\overline{\mathbf{X}}, S)$ *and* S *has positive eigenvalues* $\widehat{\lambda}_j$*, then*

$$\sum_{j=1}^{d} s_j^2 = \sum_{j=1}^{d} \widehat{\lambda}_j,$$

where s_j^2 *is the sample variance of the* j*th variable.*

The theorem and corollary assert the equality of the cumulative contributions to variance of all variables and the total variance of the PC scores. For the population, the total variance of the PC scores is the trace of Σ or, equivalently, the sum of the eigenvalues of Σ, and S plays a similar role for data. The first PC makes the largest contribution to variance, and if there are many variables, we can approximate the total variance as closely as required by considering the most important contributions to variance. Theorem 2.12 will provide a rigorous foundation for this idea.

Looking back at Definition 2.4, we can now appreciate the notion 'contribution to total variance', and the two data sets of Example 2.6 illustrate that the variance terms decrease as the number of PCs increases.

As we have seen in Theorem 2.5, the principal component scores are uncorrelated, but they are correlated with the original random vector. Propositions 2.8 and 2.9 make these relationships explicit.

Proposition 2.8 *Let* $\mathbf{X} \sim (\boldsymbol{\mu}, \Sigma)$*, and assume that* Σ *has rank* d*. For* $k \leq d$*, let* $\mathbf{W}^{(k)}$ *be the* k*th principal component vector of* $\mathbf{X}$*. If* Σ *has spectral decomposition* $\Sigma = \Gamma \Lambda \Gamma^{\mathsf{T}}$*, then*

$$\mathrm{cov}\left[\mathbf{X}, \mathbf{W}^{(k)}\right] = \Gamma \Lambda \mathbf{I}_{d \times k}.$$

In particular, the covariance of the j*th variable* X_j *of* $\mathbf{X}$ *and the* ℓ*th score* W_ℓ *of* $\mathbf{W}^{(k)}$ *is given by*

$$\mathrm{cov}(X_j, W_\ell) = \lambda_\ell \eta_{\ell j},$$

where $\eta_{\ell j}$ *is the* j*th entry of the* ℓ*th eigenvector of* Σ*.*

Proof For notational simplicity, we consider random variables $\mathbf{X}$ with mean zero. From the definition of the principal component vector given in (2.2), we find that

$$\mathrm{cov}\left[\mathbf{X}, \mathbf{W}^{(k)}\right] = \mathbb{E}\left(\mathbf{X}\mathbf{X}^{\mathsf{T}}\Gamma_k\right) = \Sigma\Gamma_k = \Gamma\Lambda\mathbf{I}_{d\times k}$$

because $\Gamma^{\mathsf{T}}\Gamma_k = \mathbf{I}_{d\times k}$. If $\mathbf{X}$ has non-zero mean $\boldsymbol{\mu}$, then

$$\mathrm{cov}\left[\mathbf{X}, \mathbf{W}^{(k)}\right] = \mathbb{E}\left[(\mathbf{X} - \boldsymbol{\mu})\mathbf{W}^{(k)^{\mathsf{T}}}\right] = \mathbb{E}\left[\mathbf{X}\mathbf{W}^{(k)^{\mathsf{T}}}\right]$$

because $\mathbb{E}\mathbf{W}^{(k)} = \mathbf{0}$. From these calculations, the desired result follows. ∎

Next, we turn to properties of the PC projection $\mathbf{P}_k$. We recall that $\mathbf{P}_k$ is a scalar multiple of the eigenvector $\boldsymbol{\eta}_k$, and the norm of $\mathbf{P}_k$ is $|W_k|$. The definition of $\mathbf{P}_k$ involves the matrix $\boldsymbol{\eta}_k\boldsymbol{\eta}_k^{\mathsf{T}}$. Put

$$H_k = \boldsymbol{\eta}_k\boldsymbol{\eta}_k^{\mathsf{T}} = \begin{pmatrix} \eta_{k1}^2 & \eta_{k1}\eta_{k2} & \eta_{k1}\eta_{k3} & \cdots & \eta_{k1}\eta_{kd} \\ \eta_{k2}\eta_{k1} & \eta_{k2}^2 & \eta_{k2}\eta_{k3} & \cdots & \eta_{k2}\eta_{kd} \\ \eta_{k3}\eta_{k1} & \eta_{k3}\eta_{k2} & \eta_{k3}^2 & \cdots & \eta_{k3}\eta_{kd} \\ \vdots & \vdots & \vdots & \ddots & \vdots \\ \eta_{kd}\eta_{k1} & \eta_{kd}\eta_{k2} & \eta_{kd}\eta_{k3} & \cdots & \eta_{kd}^2 \end{pmatrix}. \tag{2.9}$$

For $\mathbf{X} \sim (\boldsymbol{\mu}, \Sigma)$ and $\Sigma = \Gamma\Lambda\Gamma^{\mathsf{T}}$, the matrices H_k with $k \le d$ enjoy the following properties:

- H_k is positive semidefinite and $\operatorname{tr}(H_k) = 1$;
- H_k is idempotent, that is, $H_k H_k = H_k$,
- $H_k H_\ell = \mathbf{0}_{d\times d}$, for $k \neq \ell$; and
- $\Sigma = \sum_{k\ge 1} \lambda_k H_k$, where the λ_k are the diagonal entries of Λ.

Some authors refer to the equality $\Sigma = \sum \lambda_k H_k$ as the spectral decomposition of Σ.

Proposition 2.9 *Let $\mathbf{X} \sim (\boldsymbol{\mu}, \Sigma)$, and let r be the rank of Σ. For $k \le r$, the principal component projection $\mathbf{P}_k$ of $\mathbf{X}$ satisfies*

$$\operatorname{var}(\mathbf{P}_k) = \lambda_k H_k \qquad \text{and} \qquad \operatorname{cov}(\mathbf{X}, \mathbf{P}_k) = \Sigma H_k,$$

with H_k as in (2.9).

Proof Because $\mathbb{E}\mathbf{P}_k = \mathbf{0}$, the covariance matrix of $\mathbf{P}_k$ is

$$\operatorname{var}(\mathbf{P}_k) = \mathbb{E}\left(\mathbf{P}_k\mathbf{P}_k^{\mathsf{T}}\right) = \mathbb{E}\left(W_k\boldsymbol{\eta}_k\boldsymbol{\eta}_k^{\mathsf{T}}W_k\right) = \boldsymbol{\eta}_k\mathbb{E}(W_kW_k)\boldsymbol{\eta}_k^{\mathsf{T}} = \lambda_k H_k.$$

Here we have used the fact that $\mathbb{E}(W_kW_k) = \lambda_k$, which is shown in part 2 of Theorem 2.5.

Next, we turn to the covariance of $\mathbf{X}$ and $\mathbf{P}_k$. Because $\mathbb{E}\mathbf{P}_k = 0$, we have

$$\begin{aligned} \operatorname{cov}(\mathbf{X}, \mathbf{P}_k) &= \mathbb{E}\left[(\mathbf{X} - \boldsymbol{\mu})\mathbf{P}_k^{\mathsf{T}}\right]) - \mathbb{E}(\mathbf{X} - \boldsymbol{\mu})\mathbb{E}\mathbf{P}_k \\ &= \mathbb{E}\left[(\mathbf{X} - \boldsymbol{\mu})\mathbf{P}_k^{\mathsf{T}}\right] \\ &= \mathbb{E}\left[(\mathbf{X} - \boldsymbol{\mu})(\mathbf{X} - \boldsymbol{\mu})^{\mathsf{T}}\boldsymbol{\eta}_k\boldsymbol{\eta}_k^{\mathsf{T}}\right] \\ &= \Sigma\boldsymbol{\eta}_k\boldsymbol{\eta}_k^{\mathsf{T}} = \Sigma H_k. \end{aligned}$$

∎

2.5.2 Optimality Properties of PCs

In addition to yielding uncorrelated projections, principal components have a number of uniqueness properties which make them valuable and useful.

Theorem 2.10 *Let $\mathbf{X} \sim (\boldsymbol{\mu}, \Sigma)$. For $j \le d$, let λ_j be the eigenvalues of Σ. For any unit vector $\mathbf{u} \in \mathbb{R}^d$, put $v_{\mathbf{u}} = \operatorname{var}(\mathbf{u}^{\mathsf{T}}\mathbf{X})$.*

1. *If $\mathbf{u}^*$ maximises $v_{\mathbf{u}}$ over unit vectors $\mathbf{u}$, then*

$$\mathbf{u}^* = \pm\boldsymbol{\eta}_1 \qquad \text{and} \qquad v_{\mathbf{u}^*} = \lambda_1,$$

where $(\lambda_1, \boldsymbol{\eta}_1)$ is the first eigenvalue–eigenvector pair of Σ.

2. *If* $\mathbf{u}^*$ *maximises* $v_{\mathbf{u}}$ *over unit vectors* $\mathbf{u}$ *such that* $\mathbf{u}^{\mathsf{T}}\mathbf{X}$ *and* $\boldsymbol{\eta}_1^{\mathsf{T}}\mathbf{X}$ *are uncorrelated, then*

$$\mathbf{u}^* = \pm\boldsymbol{\eta}_2 \qquad \textit{and} \qquad v_{\mathbf{u}^*} = \lambda_2.$$

3. *Consider* $k = 2,\ldots,d$. *If* $\mathbf{u}^*$ *maximises* $v_{\mathbf{u}}$ *over unit vectors* $\mathbf{u}$ *such that* $\mathbf{u}^{\mathsf{T}}\mathbf{X}$ *and* $\boldsymbol{\eta}_\ell^{\mathsf{T}}\mathbf{X}$ *are uncorrelated for* $\ell < k$, *then*

$$\mathbf{u}^* = \pm\boldsymbol{\eta}_k \qquad \textit{and} \qquad v_{\mathbf{u}^*} = \lambda_k.$$

This theorem states that the first principal component score results in the largest variance. Furthermore, the second principal component score has the largest contribution to variance amongst all projections $\mathbf{u}^{\mathsf{T}}\mathbf{X}$ of $\mathbf{X}$ that are uncorrelated with the first principal component score. For the third and later principal component scores, similar results hold: the kth score is optimal in that it maximises variance while being uncorrelated with the first $k-1$ scores.

Theorem 2.10 holds under more general assumptions than stated earlier. If the rank r of Σ is smaller than d, then at most r eigenvalues can be found in part 3 of the theorem. Further, if the eigenvalues are not distinct, say λ_1 has multiplicity $j > 1$, then the maximiser in part 1 is not unique.

In terms of the combinations of the original variables, referred to at the beginning of this chapter, Principal Component Analysis furnishes us with combinations which

- are uncorrelated; and
- explain the variability inherent in $\mathbf{X}$.

Proof Because Σ has full rank, the d eigenvectors of Σ form a basis, so we can write

$$\mathbf{u} = \sum_{j=1}^{d} c_j \boldsymbol{\eta}_j \qquad \text{with constants } c_j \text{ satisfying} \qquad \sum c_j^2 = 1.$$

For $U = \mathbf{u}^{\mathsf{T}}\mathbf{X}$, the variance of U is

$$\begin{aligned}
\operatorname{var}(U) &= \mathbf{u}^{\mathsf{T}}\Sigma\mathbf{u} && \text{by part 1 of Proposition 2.1}\\
&= \sum_{j,k} c_j c_k \boldsymbol{\eta}_j^{\mathsf{T}}\Sigma\boldsymbol{\eta}_k\\
&= \sum_j c_j^2 \boldsymbol{\eta}_j^{\mathsf{T}}\Sigma\boldsymbol{\eta}_j && \text{by part 2 of Theorem 2.5}\\
&= \sum_j c_j^2 \lambda_j.
\end{aligned} \tag{2.10}$$

The last equality holds because the λ_j are the eigenvalues of Σ. By assumption, $\mathbf{u}$ is a unit vector, so $\sum c_j^2 = 1$. It follows that

$$\operatorname{var}(U) = \sum_j c_j^2\lambda_j \le \sum_j c_j^2 \lambda_1 = \lambda_1 \tag{2.11}$$

because λ_1 is the largest eigenvalue. We obtain an equality in (2.11) if and only if $c_1 = \pm 1$ and $c_j = 0$ for $j > 1$. This shows that $\mathbf{u} = \pm\boldsymbol{\eta}_1$.

For the second part, observe that $\mathbf{u}^{\mathsf{T}}\mathbf{X}$ is uncorrelated with $\boldsymbol{\eta}_1^{\mathsf{T}}\mathbf{X}$. This implies that $\mathbf{u}^{\mathsf{T}}\boldsymbol{\eta}_1 = \mathbf{0}$, and hence $\mathbf{u}$ is a linear combination of the remaining basis vectors $\boldsymbol{\eta}_j$ for

$j \geq 2$. The eigenvalues of Σ are ordered, with λ_2 the second largest. An inspection of the calculation of $\operatorname{var}(U)$ shows that

$$\operatorname{var}(U) = \sum_{j>1} c_j^2 \lambda_j \tag{2.12}$$

because $\mathbf{u}$ and $\boldsymbol{\eta}_1$ are uncorrelated. The result follows as earlier by replacing the variance term in (2.11) by the corresponding expression in (2.12). The third part follows by a similar argument. ■

In Section 1.5.1 we defined the trace of a square matrix and listed some properties. Here we use the trace to define a norm.

Definition 2.11 Let $\mathbf{X} \sim (\boldsymbol{\mu}, \Sigma)$. The **trace norm** of $\mathbf{X}$ is

$$\|\mathbf{X}\|_{tr} = \left[\operatorname{tr}(\Sigma)\right]^{1/2}. \tag{2.13}$$

Let $\mathbb{X} \sim \mathsf{Sam}(\overline{\mathbf{X}}, S)$. The **(sample) trace norm** of $\mathbf{X}_i$ from $\mathbb{X}$ is

$$\|\mathbf{X}_i\|_{tr} = \left[\operatorname{tr}(S)\right]^{1/2}. \tag{2.14}$$

□

A Word of Caution. The trace norm is not a norm. It is defined on the (sample) covariance matrix of the random vector or data of interest. As we shall see in Theorem 2.12, it is the right concept for measuring the error between a random vector $\mathbf{X}$ and its PC-based approximations. As a consequence of Theorem 2.6, the trace norm of $\mathbf{X}$ is related to the marginal variances and the eigenvalues of the covariance matrix, namely,

$$\|\mathbf{X}\|_{tr} = \left(\sum \sigma_j^2\right)^{1/2} = \left(\sum \lambda_j\right)^{1/2}.$$

Theorem 2.12 *Let $\mathbf{X} \sim (\boldsymbol{\mu}, \Sigma)$, and assume that Σ has rank d. For $k \leq d$, let $(\lambda_k, \boldsymbol{\eta}_k)$ be the eigenvalue–eigenvector pairs of Σ. For $j \leq d$, put*

$$\mathbf{p}_j = \boldsymbol{\eta}_j \boldsymbol{\eta}_j^{\mathsf{T}} \mathbf{X}.$$

Then

1.

$$\left\| \mathbf{X} - \mathbf{p}_j \right\|_{tr}^2 = \sum_{k\geq 1,\, k\neq j} \lambda_k,$$

2. *and for $\kappa \leq d$,*

$$\sum_{j=1}^{\kappa} \mathbf{p}_j = \Gamma_\kappa \Gamma_\kappa^{\mathsf{T}} \mathbf{X} \qquad \textit{with} \quad \left\| \mathbf{X} - \sum_{j=1}^{\kappa} \mathbf{p}_j \right\|_{tr}^2 = \sum_{k>\kappa} \lambda_k.$$

The vectors $\mathbf{p}_j$ are closely related to the PC projections $\mathbf{P}_j$ of (2.3): the PC projections arise from centred vectors, whereas the $\mathbf{p}_j$ use $\mathbf{X}$ directly, so for $j \leq d$,

$$\mathbf{P}_j = \mathbf{p}_j - \boldsymbol{\eta}_j \boldsymbol{\eta}_j^{\mathsf{T}} \boldsymbol{\mu}.$$

Because of this relationship between $\mathbf{P}_j$ and $\mathbf{p}_j$, we refer to the $\mathbf{p}_j$ as the **uncentred PC projections**.

The theorem quantifies the error between each $\mathbf{p}_j$ and $\mathbf{X}$, or equivalently, it shows how close each PC projection is to $\mathbf{X}-\boldsymbol{\mu}$, where the distance is measured by the trace norm. The second part of Theorem 2.12 exhibits the size of the error when $\mathbf{X}$ is approximated by the first κ of the uncentred PC projections. The size of the error can be used to determine the number of PCs: if we want to guarantee that the error is not bigger than α per cent, then we determine the index κ such that the first κ uncentred PC projections approximate $\mathbf{X}$ to within α per cent.

Corollary 2.13 *Let $\mathbf{X}$ satisfy the assumptions of Theorem 2.12. Then $\mathbf{X}$ can be reconstructed from its uncentred principal component projections*

$$\mathbf{X} \approx \sum_{j=1}^{\kappa} \mathbf{p}_j = \sum_{j=1}^{\kappa} \boldsymbol{\eta}_j \boldsymbol{\eta}_j^{\mathsf{T}} \mathbf{X} = \Gamma_\kappa \Gamma_\kappa^{\mathsf{T}} \mathbf{X},$$

with $\kappa \le d$, and equality holds for $\kappa = d$.

I have stated Theorem 2.12 and Corollary 2.13 for the random vector $\mathbf{X}$. In applications, this vector is replaced by the data $\mathbb{X}$. As these error bounds are important in a data analysis, I will explicitly state an analogous result for the sample case. In Theorem 2.12 we assumed that the rank of the covariance matrix is d. The approximation and the error bounds given in the theorem remain the same if the covariance matrix does not have full rank; however, the proof is more involved.

Corollary 2.14 *Let data $\mathbb{X} = \left[\mathbf{X}_1 \cdots \mathbf{X}_n\right] \sim \mathsf{Sam}(\overline{\mathbf{X}}, S)$, with r the rank of S. For $k \le r$, let $(\widehat{\lambda}_k, \widehat{\boldsymbol{\eta}}_k)$ be the eigenvalue–eigenvector pairs of S. For $i \le n$ and $j \le r$, put*

$$\widehat{\mathbf{P}}_{ij} = \widehat{\boldsymbol{\eta}}_j \widehat{\boldsymbol{\eta}}_j^{\mathsf{T}} \mathbf{X}_i \qquad \text{and} \qquad \mathbb{P}_{\bullet j} = \left[\widehat{\mathbf{P}}_{1j} \ldots \widehat{\mathbf{P}}_{nj}\right].$$

Then the sample trace norm of $\mathbf{X}_i - \sum \widehat{\mathbf{P}}_{ij}$ is

$$\left\| \mathbf{X}_i - \sum_{j=1}^{\kappa} \widehat{\mathbf{P}}_{ij} \right\|_{tr}^2 = \sum_{k>\kappa} \widehat{\lambda}_k,$$

and the data $\mathbb{X}$ are reconstructed from the uncentred principal component projections

$$\mathbb{X} \approx \sum_{j=1}^{\kappa} \mathbb{P}_{\bullet j} = \sum_{j=1}^{\kappa} \widehat{\boldsymbol{\eta}}_j \widehat{\boldsymbol{\eta}}_j^{\mathsf{T}} \mathbb{X} = \widehat{\Gamma}_\kappa \widehat{\Gamma}_\kappa^{\mathsf{T}} \mathbb{X},$$

for $\kappa \le r$, with equality when $\kappa = r$.

We call $\mathbb{P}_{\bullet j}$ the matrix of **uncentred *j*th principal component projections**. The corollary states that the same error bound holds for all observations.

Proof of Theorem 2.12. For part 1, fix $j \le d$. Put

$$\mathbf{p} = \boldsymbol{\eta}_j \boldsymbol{\eta}_j^{\mathsf{T}} \mathbf{X} = H_j \mathbf{X} \qquad \text{and} \qquad \boldsymbol{\Delta} = \mathbf{X} - \mathbf{p},$$

with H_j as in (2.9). The expected value of Δ is $\mathbb{E}\Delta = (\mathbf{I} - H_j)\boldsymbol{\mu}$, where $\mathbf{I} = \mathbf{I}_{d\times d}$, and

$$\begin{aligned}
\operatorname{var}(\boldsymbol{\Delta}) &= \mathbb{E}\left[(\mathbf{I} - H_j)(\mathbf{X}-\boldsymbol{\mu})(\mathbf{X}-\boldsymbol{\mu})^{\mathsf{T}}(\mathbf{I} - H_j)\right] \\
&= (\mathbf{I} - H_j)\operatorname{var}(\mathbf{X})(\mathbf{I} - H_j) \\
&= \Sigma - H_j\Sigma - \Sigma H_j + H_j \Sigma H_j \\
&= \Sigma - 2\lambda_j H_j + \lambda_j H_j H_j. \qquad (2.15)
\end{aligned}$$

To appreciate why the last equality in (2.15) holds, recall that η_j is the jth eigenvector of Σ, and so $\eta_j\eta_j^{\top}\Sigma = \eta_j\lambda_j\eta_j^{\top}$. Similarly, one can show, and we do so in the Problems at the end of Part I, that $\Sigma H_j = \lambda_j H_j$ and $H_j\Sigma H_j = \lambda_j H_j H_j$.

Next, for $d \times d$ matrices A and B, it follows that $\operatorname{tr}(A+B) = \operatorname{tr}(A) + \operatorname{tr}(B)$, and therefore,

$$\operatorname{tr}\{\operatorname{var}(\Delta)\} = \sum_k \lambda_k - 2\lambda_j + \lambda_j = \sum_{k\neq j} \lambda_k$$

because $\operatorname{tr}(H_j H_j) = 1$, from which the desired result follows. The proof of part 2 uses similar arguments. ■

I illustrate some of the theoretical results with the HIV$^+$ data.

Example 2.11 We continue with the five-dimensional **HIV flow cytometry** data. For easier visualisation, we use a small subset of the cell population rather than all 10,000 blood cells of the HIV$^+$ data that we considered in Example 2.4.

The sample covariance matrix of the first 100 blood cells is

$$S_{\mathrm{HIV}^+} = 10^3 \times \begin{pmatrix} 4.7701 & 2.0473 & 1.1205 & 0.0285 & 1.3026 \\ 2.0473 & 1.8389 & 0.0060 & -0.4693 & 0.6515 \\ 1.1205 & 0.0060 & 4.8717 & 3.0141 & 1.4854 \\ 0.0285 & -0.4693 & 3.0141 & 9.3740 & -4.2233 \\ 1.3026 & 0.6515 & 1.4854 & -4.2233 & 8.0923 \end{pmatrix},$$

and the corresponding matrix of eigenvalues is

$$\widehat{\Lambda}_{\mathrm{HIV}^+} = 10^3 \times \begin{pmatrix} 13.354 & 0 & 0 & 0 & 0 \\ 0 & 8.504 & 0 & 0 & 0 \\ 0 & 0 & 4.859 & 0 & 0 \\ 0 & 0 & 0 & 1.513 & 0 \\ 0 & 0 & 0 & 0 & 0.717 \end{pmatrix}.$$

The traces of the two matrices both equal 28,947, thus confirming Theorem 2.6.

Figure 2.9 and Table 2.5 illustrate the reconstruction results of Theorem 2.12 and Corollary 2.14. Figure 2.9 shows parallel coordinate plots of the first 100 blood cells. In the top row, the uncentred PC projections $\mathbb{P}_{\bullet 1} \cdots \mathbb{P}_{\bullet 5}$ are shown. The variability decreases as the index increases. In the $\mathbb{P}_{\bullet 1}$ plot, variable four (*CD8*) shows the largest variation because the first eigenvector has the largest entry in variable four. In Example 2.4, I mentioned the increase in *CD8* and decrease in *CD4* with the onset of HIV$^+$. Here we notice that the increasing level of *CD8* is responsible for the variability. The $\mathbb{P}_{\bullet 2}$ plot picks up the variability in the third and fifth variables, which are also known to be associated with HIV$^+$.

The bottom-left subplot shows the parallel coordinate plot of the data $[\mathbf{X}_1 \cdots \mathbf{X}_{100}]$. The remaining panels show the partial sums $\sum_{j\leq k} \mathbb{P}_{\bullet j}$ in the kth panel from the left and starting with $k = 2$. This sequence of figures shows how the data in the left-most panel gradually emerge from the projections. As expected, the last plot, bottom right, agrees with the left-most plot.

Table 2.5 shows the improvement in accuracy of the reconstruction as an increasing number of principal component projections is used. The table focuses on the first of the 100

Table 2.5 *Reconstructing* $\mathbf{X}_1$ *of the* HIV$^+$ *data from Example 2.11*

	FS	*SS*	*CD3*	*CD8*	*CD4*	$\| \|_2$	$\| \|_{tr}$
$\mathbf{X}$	191	250	175	114	26	378.52	170.14
$\mathfrak{e}_1 = \mathbf{X} - \widehat{\mathbf{p}}_1$	196.50	255.05	165.33	66.17	63.65	373.39	124.87
$\mathfrak{e}_2 = \mathbf{X} - \sum_{j\leq 2} \widehat{\mathbf{p}}_j$	63.36	206.64	−17.39	−32.37	−82.57	234.27	84.20
$\mathfrak{e}_3 = \mathbf{X} - \sum_{j\leq 3} \widehat{\mathbf{p}}_j$	−69.67	128.14	38.53	−15.44	−16.74	152.57	47.22
$\mathfrak{e}_4 = \mathbf{X} - \sum_{j\leq 4} \widehat{\mathbf{p}}_j$	−70.81	131.03	24.22	−4.19	−6.35	151.09	26.78
$\mathfrak{e}_5 = \mathbf{X} - \sum_{j\leq 5} \widehat{\mathbf{p}}_j$	0.00	0.00	0.00	0.00	0.00	0.00	0

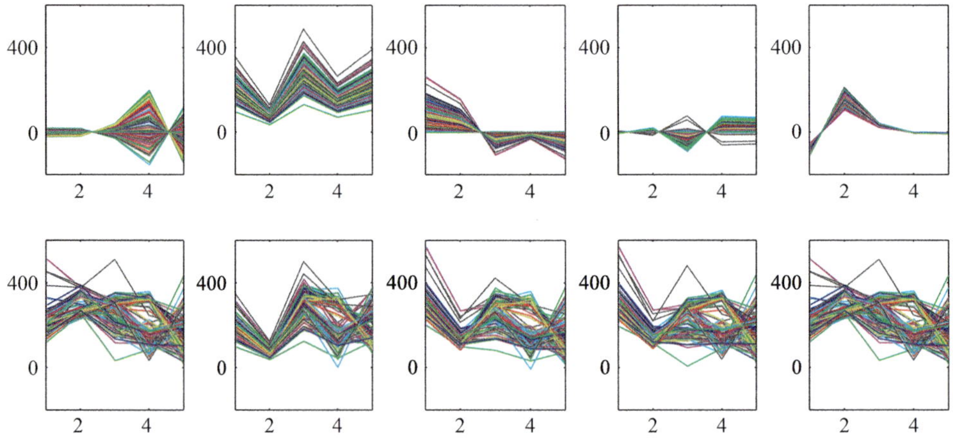

Figure 2.9 Reconstructing $\mathbb{X}$ of Example 2.11 from PC projections; $\mathbb{P}_{\bullet 1} \cdots \mathbb{P}_{\bullet 5}$ (*top*) and their partial sums $\sum_{j\leq k} \mathbb{P}_{\bullet j}$ (*columns two to five bottom*), with the original data bottom left.

blood cells, and $\mathbf{X} = \mathbf{X}_1$; the eigenvectors are those of the covariance matrix S_{HIV^+} of the first 100 blood cells. The rows of Table 2.5 show the values of the entries *FS*, *SS*, *CD3*, *CD8* and *CD4* at each stage of the approximation. For $\mathfrak{e}_3$, the value of *SS* is 128.14, which is a substantial decrease from the initial value of 250. Some entries increase from one approximation to the next for a specific variable, but the total error decreases. At $\mathfrak{e}_5$, the observation $\mathbf{X}$ is completely recovered, and all entries of $\mathfrak{e}_5$ are zero.

The last two columns of the table measure the error in the $\mathfrak{e}_k$ for $k \leq 5$ with the Euclidean norm $\| \|_2$ and the trace norm $\| \|_{tr}$ of (2.13). The Euclidean norm varies with the observation number – here I use $\mathbf{X} = \mathbf{X}_1$, and the 'errors' $\mathfrak{e}_k$ relate to the first observation $\mathbf{X}_1$. The matrix $\widehat{\Lambda}_{\text{HIV}^+}$ is given at the beginning of the example. It is therefore easy to verify that the square of the trace norm of $\mathfrak{e}_k$ equals the sum of last few eigenvalues, starting with the kth one. This equality is stated in Corollary 2.14. ■

2.6 Standardised Data and High-Dimensional Data

2.6.1 Scaled and Sphered Data

Example 2.6 shows the variance contributions of the *breast cancer* data. I did not mention in Examples 2.6 and 2.9 that the PCs of the *breast cancer* data are not based on the raw data but on transformed data.

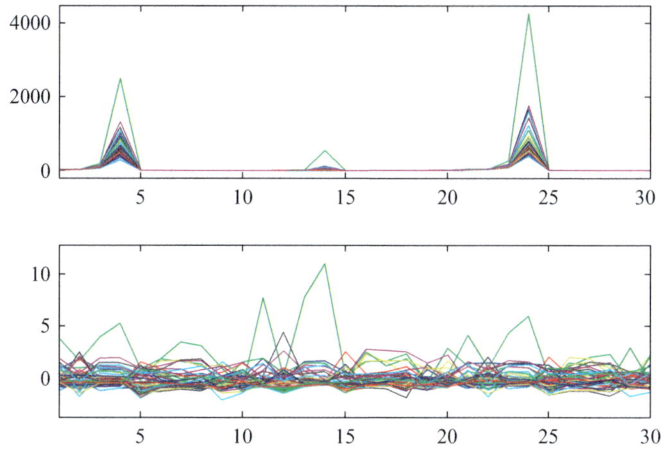

Figure 2.10 Parallel coordinate views of samples 461 to 500 from the breast cancer data of Example 2.6. (*Top panel*) Raw data; (*bottom panel*) scaled data.

Figure 2.10 shows parallel coordinate plots of a small subset of the data, namely, observations 461 to 500. In the top plot, which shows the raw data, we observe three variables which are several orders of magnitudes bigger than the rest: variables 24 and 4 have values above 2,000, and variable 14 has values around 400, whereas all other variables have values in the hundreds, tens, or smaller. One observation, shown in green, has particularly large values for these three variables. The bottom panel shows the same observations after transformation. The transformation I use here is the same as in Examples 2.6 and 2.9. I will give details of this transformation, called *scaling*, in Definition 2.15. Because the sequence of colours is the same in both panels of Figure 2.10, we can see that the green observation has a big value at variable 14 in the scaled data, but the large values at variables 4 and 24 of the top panel have decreased as a result of scaling.

The results of a principal component analysis of the raw data provide some insight into why we may not want to use the raw data directly. The results reveal that

1. the first eigenvector is concentrated almost exclusively on the largest two variables, with weights of 0.85 (for variable 24) and 0.517 (for variable 4);
2. the second eigenvector is concentrated on the same two variables, but the loadings are interchanged;
3. the third eigenvector has a weight of 0.99 along the third-largest variable; and
4. the contribution to total variance of the first PC is 98.2 per cent, and that of the first two PCs is 99.8 per cent.

These facts speak for themselves: if some variables are orders of magnitude bigger than the rest, then these *big* variables dominate the principal component analysis to the almost total exclusion of the other variables. Some of the *smaller* variables may contain pertinent information regarding *breast cancer* which is therefore lost. If the three variables 24, 4 and 14 are particularly informative for breast cancer, then the preceding analysis may be exactly what is required. If, on the other hand, these three variables happen to be measured on a different scale and thus result in numerical values that are much larger than the other variables, then they should not dominate the analysis merely because they are measured on a different scale.

Definition 2.15 Let $\mathbf{X} \sim (\boldsymbol{\mu}, \Sigma)$, and put

$$\Sigma_{\text{diag}} = \begin{pmatrix} \sigma_1^2 & 0 & 0 & \cdots & 0 \\ 0 & \sigma_2^2 & 0 & \cdots & 0 \\ 0 & 0 & \sigma_3^2 & \cdots & 0 \\ \vdots & \vdots & \vdots & \ddots & \vdots \\ 0 & 0 & 0 & \cdots & \sigma_d^2 \end{pmatrix}. \tag{2.16}$$

The **scaled** or **standardised** random vector

$$\mathbf{X}_{\text{scale}} = \Sigma_{\text{diag}}^{-1/2} (\mathbf{X} - \boldsymbol{\mu}). \tag{2.17}$$

Similarly for data $\mathbb{X} = \left[\mathbf{X}_1\ \mathbf{X}_2 \cdots \mathbf{X}_n\right] \sim \mathsf{Sam}(\overline{\mathbf{X}}, S)$ and analogously defined diagonal matrix S_{diag}, the **scaled** or **standardised** data

$$\mathbb{X}_{\text{scale}} = S_{\text{diag}}^{-1/2} \left(\mathbb{X} - \overline{\mathbf{X}}\right). \tag{2.18}$$

□

In the literature, *standardised data* sometimes refers to standardising with the full covariance matrix. To avoid ambiguities between these two concepts, I prefer to use the expression *scaled*, which standardises each variable separately.

Definition 2.16 Let $\mathbf{X} \sim (\boldsymbol{\mu}, \Sigma)$, and assume that Σ is invertible. The **sphered** random vector

$$\mathbf{X}_{\Sigma} = \Sigma^{-1/2} (\mathbf{X} - \boldsymbol{\mu}).$$

For data $\mathbb{X} = \left[\mathbf{X}_1\ \mathbf{X}_2 \cdots \mathbf{X}_n\right] \sim \mathsf{Sam}(\overline{\mathbf{X}}, S)$ and non-singular S, the **sphered** data

$$\mathbb{X}_S = S^{-1/2} \left(\mathbb{X} - \overline{\mathbf{X}}\right).$$

□

If Σ is singular with rank r, then we may want to replace Σ by $\Gamma_r \Lambda_r \Gamma_r^{\mathsf{T}}$ and $\Sigma^{-1/2}$ by $\Gamma_r \Lambda_r^{-1/2} \Gamma_r^{\mathsf{T}}$, and similarly for the sample covariance matrix.

The name *sphered* is used in appreciation of what happens for Gaussian data. If the covariance matrix of the resulting vector or data is the identity matrix, then the multivariate ellipsoidal shape becomes spherical. For arbitrary random vectors, sphering makes the variables of the random vector $\mathbf{X}$ uncorrelated, as the following calculation shows.

$$\operatorname{var}\left[\Sigma^{-1/2} (\mathbf{X} - \boldsymbol{\mu})\right] = \mathbb{E}\left(\Sigma^{-1/2}(\mathbf{X} - \boldsymbol{\mu})(\mathbf{X} - \boldsymbol{\mu})^{\mathsf{T}} \Sigma^{-1/2}\right) = \Sigma^{-1/2} \Sigma \Sigma^{-1/2} = \mathbf{I}_{d \times d}. \tag{2.19}$$

If $\mathbf{X}$ has the identity covariance matrix, then Principal Component Analysis does not produce any new results, and consequently, sphering should **not** be used prior to a principal component analysis.

How does scaling affect the variance? Scaling is applied when the variables are measured on different scales – hence the name. It does not result in the identity covariance matrix but merely makes the variables more comparable.

Theorem 2.17 *Let* $\mathbf{X} \sim (\boldsymbol{\mu}, \Sigma)$, *and assume that* Σ *has rank* d. *Let* Σ_{diag} *be the diagonal matrix given in (2.16), and put* $\mathbf{X}_{scale} = \Sigma_{diag}^{-1/2}(\mathbf{X} - \boldsymbol{\mu})$.

1. *The covariance matrix of* $\mathbf{X}_{scale}$ *is the matrix of correlation coefficients* R, *that is,*

$$\operatorname{var}(\mathbf{X}_{scale}) = \Sigma_{diag}^{-1/2}\, \Sigma\, \Sigma_{diag}^{-1/2} = R.$$

2. *The covariance of the* j*th and* k*th entries of* $\mathbf{X}_{scale}$ *is*

$$\rho_{jk} = \frac{\operatorname{cov}(X_j, X_k)}{[\operatorname{var}(X_j)\operatorname{var}(X_k)]^{1/2}}.$$

The proof of this theorem is deferred to the Problems at the end of Part I.

For data $\mathbb{X}$, we obtain analogous results by using the sample quantities instead of the population quantities. For convenience, I state the result corresponding to Theorem 2.17.

Corollary 2.18 *Let* $\mathbb{X} \sim \mathsf{Sam}(\overline{\mathbf{X}}, S)$, *and let* S_{diag} *be the diagonal matrix obtained from* S. *Then the sample covariance matrix of the scaled data* $\mathbb{X}_{scale}$ *is*

$$R_S = S_{diag}^{-1/2}\, S\, S_{diag}^{-1/2}, \tag{2.20}$$

and the (j,k)*th entry of this matrix is*

$$\widehat{\rho}_{jk} = \frac{s_{jk}}{s_j s_k}.$$

The covariance matrices of the data and the scaled data satisfy (2.20), but the eigenvectors of the data and scaled data are not related by simple relationships. However, the eigenvalues of the covariance matrix of the scaled data have some special features which are a consequence of the trace properties given in Theorem 2.6.

Corollary 2.19 *The following hold.*

1. *Let* $\mathbf{X} \sim (\boldsymbol{\mu}, \Sigma)$, *and let* r *be the rank of* Σ. *For* R *as in part 1 of Theorem 2.17,*

$$\operatorname{tr}(R) = \sum_{j=1}^{r} \lambda_j^{(scale)} = r,$$

where $\lambda_j^{(scale)}$ *is the* j*th eigenvalue of* R.

2. *Let* $\mathbb{X}$ *be the data as in Corollary 2.18, and let* r *be the rank of* S. *Then*

$$\operatorname{tr}(R_S) = \sum_{j=1}^{r} \widehat{\lambda}_j^{(scale)} = r,$$

where $\widehat{\lambda}_j^{(scale)}$ *is the* j*th eigenvalue of the sample correlation matrix of* $\mathbb{X}_{scale}$.

A natural question arising from the discussion and the results in this section is: When should we use the raw data, and when should we analyse the scaled data instead?

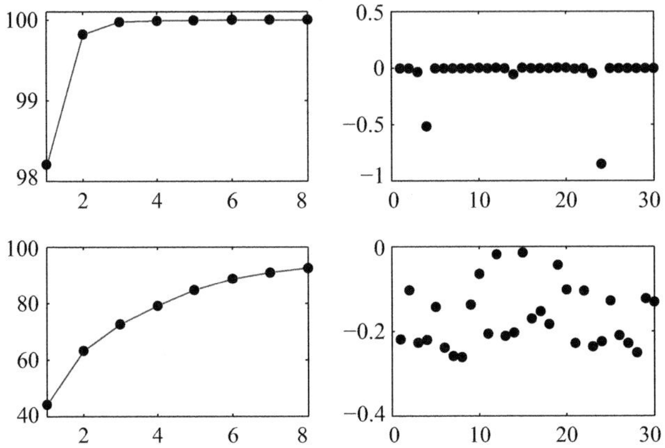

Figure 2.11 Comparison of raw (*top*) and scaled (*bottom*) data from Example 2.12. (*Left*) Cumulative contribution to variance of the first eight components; (*right*) weights of the first eigenvector.

The theory is very similar for both types of analyses, so it does not help us answer this question. And, indeed, there is no definitive answer. If the variables are comparable, then no scaling is required, and it is preferable to work with the raw data. If some of the variables are one or more magnitudes larger than others, the large variables dominate the analysis. If the *large* variables are of no special interest, then scaling is advisable, and I recommend carrying out parallel analyses for raw and scaled data and to compare the results. This is the approach I take with the *breast cancer* data.

Example 2.12 We continue with the **breast cancer** data and compare principal component analyses for the raw and scaled data. The raw data have two variables that dominate the analysis and explain most of the variance. In contrast, for the scaled data, there no longer exist any dominating variables. Plots of samples 461 to 500 of both the raw data and the scaled data, are displayed in Figure 2.10. The contributions to total variance of the first few scaled PCs are much smaller than the corresponding variance contributions of the raw data: the two left panels in Figure 2.11 show the contribution to total variance of the first eight PCs. The 'kink' in the top panel at PC_2 marks the effect of the two very large variables. No such behaviour is noticeable in the bottom panel of the figure.

The two right panels of Figure 2.11 display the entries of the first eigenvector along the thirty variables shown on the x-axis. We notice the two large negative values in the top panel, whereas all other variables have negligible weights. The situation in the bottom-right panel is very different: about half the entries of the scaled first eigenvector lie around the -0.2 level, about seven or eight lie at about -0.1, and the rest are smaller in absolute value. This analysis is more appropriate in extracting valuable information from the principal components because it is not dominated by the two large variables. ■

The next example explores Principal Component Analysis for highly correlated variables.

Example 2.13 The **abalone** data of Blake and Merz (1998) from the Marine Research Laboratories in Taroona, Tasmania, in Australia, contain 4,177 samples in eight variables. The

Table 2.6 *First eigenvector $\widehat{\eta}_1$ for raw and scaled data from Example 2.13*

	Length	*Diameter*	*Height*	*Whole Weight*	*Shucked Weight*	*Viscera Weight*	*Dried-Shell Weight*
Raw $\widehat{\eta}_1$	0.1932	0.1596	0.0593	0.8426	0.3720	0.1823	0.2283
Scaled $\widehat{\eta}_1$	0.3833	0.3836	0.3481	0.3907	0.3782	0.3815	0.3789

last variable, *age*, is to be estimated from the remaining seven variables. In this analysis, I use the first seven variables only. We will return to an estimation of the eighth variable in Section 2.8.2. Meanwhile, we compare principal component analyses of the raw and scaled data.

The seven variables used in this analysis are *length*, *diameter* and *height* (all in mm); *whole weight*, *shucked weight*, *viscera weight* (the shell weight after bleeding) and the *dried-shell weight*. The last four are given in grams.

Table 2.6 shows the entries of the first eigenvector for the raw and scaled data. We note that for the raw data, the entry for 'whole weight' is about twice as big as that of the next-largest variable, 'shucked weight'. The cumulative contribution to total variance of the first component is above 97 per cent, so almost all variance is explained by this component. In comparison, the cumulative variance contribution of the first scaled PC is about 90 per cent.

The correlation between the variables is very high; it ranges from 0.833 between variables 6 and 7 to above 0.98 for variables 1 and 2. In regression analysis, *collinearity* refers to a linear relationship between predictor variables. Table 2.6 shows that the first eigenvector of the scaled data has very similar weights across all seven variables, with practically identical weights for variables 1 and 2. For highly correlated predictor variables, it may be advisable to replace the variables with a smaller number of less correlated variables. I will not do so here.

An inspection of the second eigenvector of the scaled data shows that the weights are more varied. However, the second eigenvalue is less than 3 per cent of the first and thus almost negligible. ■

In the analysis of the *abalone* data, the variables exhibit a high positive correlation which could be a consequence of multicollinearity of the variables. The analysis shows that the first principal component of the scaled data does not select particular variables – having almost equal weights for all variables. An elimination of some variables might be appropriate for these data.

2.6.2 High-Dimensional Data

So far we considered data with a relatively small number of variables – at most thirty for the *breast cancer* data and the *Dow Jones returns* – and in each case the number of variables was considerably smaller than the number of observations. These data sets belong to the *classical* domain, for which the sample size is much larger than the dimension. Classical limit theorems apply, and we understand the theory for the $n > d$ case.

In high dimension, the space becomes emptier as the dimension increases. The simple example in Figure 2.12 tries to give an idea how data 'spread out' when the number of dimensions increases, here from two to three. We consider 100 independent and identically

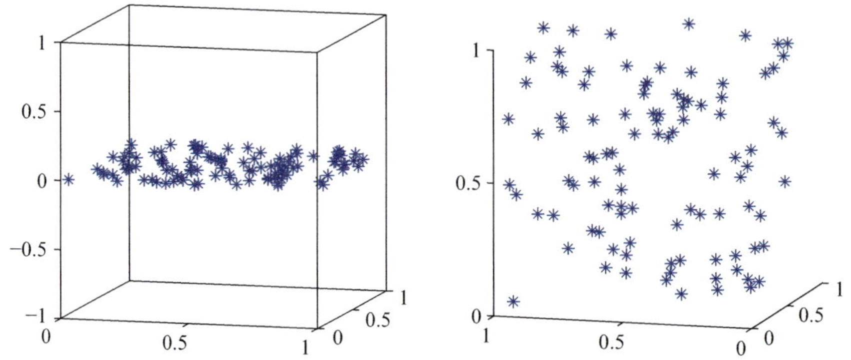

Figure 2.12 Distribution of 100 points in 2D and 3D unit space.

distributed points from the uniform distribution in the unit cube. The projection of these points onto the (x, y)-plane is shown in the left panel of Figure 2.12. The points seem to cover the unit square quite well. In the right panel we see the same 100 points with their third dimension; many *bare* patches have now appeared between the points. If we generated 100 points within the four-dimensional unit volume, the empty regions would further increase. As the dimension increases to *very many*, the space becomes very thinly populated, point clouds tend to disappear and generally it is quite 'lonely' in high-dimensional spaces.

The term *high-dimensional* is not clearly defined. In some applications, anything beyond a *handful* of dimensions is regarded as high-dimensional. Generally, we will think of high-dimensional as much higher, and the thirty dimensions we have met so far I regard as a moderate number of dimensions. Of course, the relationship between d and n plays a crucial role. Data with a large number of dimensions are called **high-dimensional data (HDD)**. We distinguish different groups of HDD which are characterised by

1. d is large but smaller than n;
2. d is large and larger than n: the **high-dimension low sample size** data (**HDLSS**); and
3. the data are functions of a continuous variable d: the **functional data**.

Our applications involve all these types of data. In functional data, the observations are curves rather than consisting of individual variables. Example 2.16 deals with functional data, where the curves are mass spectrometry measurements.

Theoretical advances for HDD focus on large n and large d. In the research reported in Johnstone (2001), both n and d grow, with d growing as a function of n. In contrast, Hall, Marron, and Neeman (2005) and Jung and Marron (2009) focus on the non-traditional case of a fixed sample size n and let $d \to \infty$. We look at some of these results in Section 2.7 and return to more asymptotic results in Section 13.5.

A Word of Caution. Principal Component Analysis is an obvious candidate for summarising HDD into a smaller number of components. However, care needs to be taken when the dimension is large, especially when $d > n$, because the rank r of the covariance matrix S satisfies

$$r \leq \min\{d, n\}.$$

For HDLSS data, this statement implies that one cannot obtain more than n principal components. The rank serves as an upper bound for the number of derived variables that can be constructed with Principal Component Analysis; variables with large variance are 'in', whereas those with small variance are 'not in'. If this criterion is not suitable for particular HDLSS data, then either Principal Component Analysis needs to be adjusted, or other methods such as Independent Component Analysis or Projection Pursuit could be used.

We look at two HDLSS data sets and start with the smaller one.

Example 2.14 The Australian **illicit drug market** data of Gilmour et al. (2006) contain monthly counts of events recorded by key health, law enforcement and drug treatment agencies in New South Wales, Australia. These data were collected across different areas of the three major stakeholders. The combined count or *indicator* data consist of seventeen separate data series collected over sixty-six months between January 1997 and June 2002. The series are listed in Table 3.2, Example 3.3, in Section 3.3, as the split of the data into the two groups fits more naturally into the topic of Chapter 3. In the current analysis, this partition is not relevant. Heroin, cocaine and amphetamine are the quantities of main interest in this data set. The relationship between these drugs over a period of more than five years has given rise to many analyses, some of which are used to inform policy decisions. Figure 1.5 in Section 1.2.2 shows a parallel coordinate plot of the data with the series numbers on the x-axis.

In this analysis we consider each of the seventeen series as an observation, and the sixty-six months represent the variables. An initial principal component analysis shows that the two series *break and enter dwelling* and *steal from motor vehicles* are on a much larger scale than the others and dominate the first and second PCs. The scaling of Definition 2.15 is not appropriate for these data because the mean and covariance matrix naturally pertain to the months. For this reason, we scale each series and call the new data the *scaled* (*indicator*) *data*.

Figure 2.13 shows the raw data (*top*) and the scaled data (*bottom*). For the raw data I have excluded the two observations, *break and enter dwelling* and *steal from motor vehicles*, because they are on a much bigger scale and therefore obscure the remaining observations. For the scaled data we observe that the spread of the last ten to twelve months is larger than that of the early months.

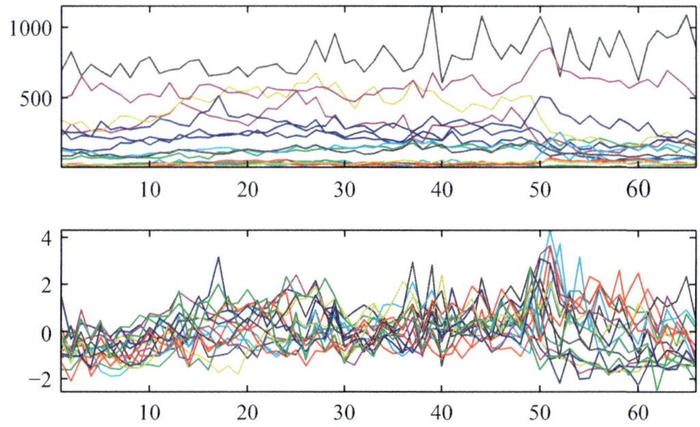

Figure 2.13 Illicit drug market data of Example 2.14 with months as variables: (*top*) raw data; (*bottom*) scaled data.

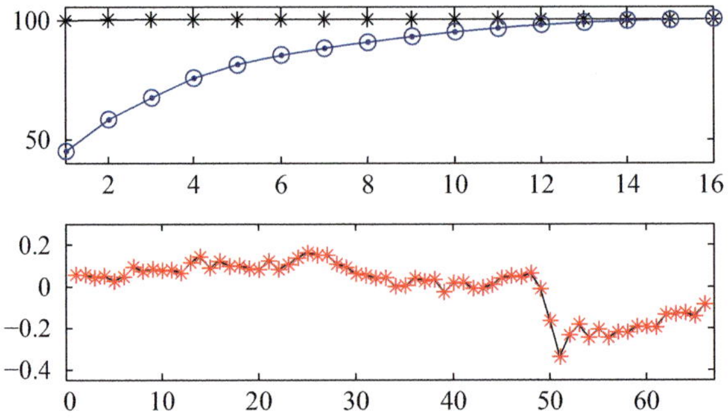

Figure 2.14 (*Top*): Cumulative contributions to variance of the raw (*black*) and scaled (*blue*) illicit drug market data of Example 2.14 (*bottom*): Weights of the first eigenvector of the scaled data with the variable number on the x-axis.

Because $d = 66$ is much larger than $n = 17$, there are at most seventeen PCs. The analysis shows that the rank of S is 16, so $r < n < d$. For the raw data, the first PC scores account for 99.45 per cent of total variance, and the first eigenvalue is more than 200 times larger than the second. Furthermore, the weights of the first eigenvector are almost uniformly distributed over all sixty-six dimensions, so they do not offer much insight into the structure of the data. For this reason, we analyse the scaled data.

Figure 2.14 displays the cumulative contribution to total variance of the raw and scaled data in the top plot. For the scaled data, the first PC scores account for less than half the total variance, and the first ten PCs account for about 95 per cent of total variance. The first eigenvalue is about three times larger than the second; the first eigenvector shows an interesting pattern which is displayed in the lower part of Figure 2.14: For the first forty-eight months the weights have small positive values, whereas at month forty-nine the sign is reversed, and all later months have negative weights. This pattern is closely linked to the Australian heroin shortage in early 2001, which is analysed in Gilmour et al. (2006). It is interesting that the first eigenvector shows this phenomenon so clearly. ■

Our next HDLSS example is much bigger, in terms of both variables and samples.

Example 2.15 The **breast tumour (gene expression)** data of van't Veer et al. (2002) consist of seventy-eight observations and 4,751 gene expressions. Typically, gene expression data contain intensity levels or *expression indices* of genes which are measured for a large number of genes. In bioinformatics, the results of pre-processed gene microarray experiments are organised in an $N_c \times N_g$ 'expression index' matrix which consists of $n = N_c$ chips or slides and $d = N_g$ genes or probesets. The number of genes may vary from a few hundred to many thousands, whereas the number of chips ranges from below 100 to maybe 200 to 400. The chips are the observations, and the genes are the variables. The data are often accompanied by survival times or binary responses. The latter show whether the individual has survived beyond a certain time. Gene expression data belong to the class of high-dimension low sample size data.

Genes are often grouped into subgroups, and within these subgroups one wants to find genes that are 'differentially expressed' and those which are non-responding. Because of the very large number of genes, a first step in many analyses is dimension reduction. Later

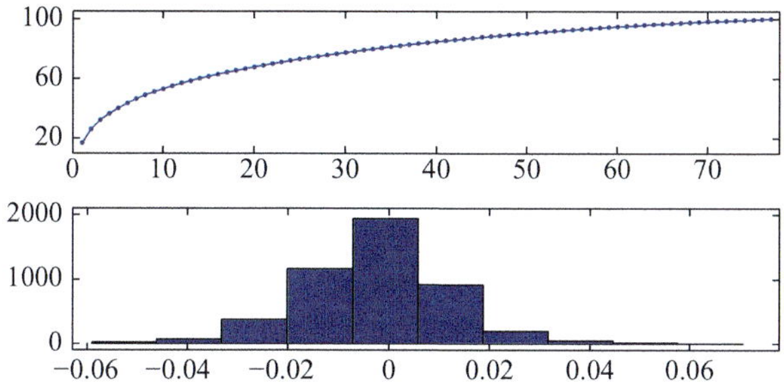

Figure 2.15 The breast tumour data of Example 2.15: cumulative contributions to variance against the index (*top*) and histogram of weights of the first eigenvector (*bottom*).

steps in the analysis are concerned with finding genes that are responsible for particular diseases or are good predictors of survival times.

The breast tumour data of van't Veer et al. (2002) are given as $\log_{10}$ transformations. The data contain survival times in months as well as binary responses regarding survival. Patients who left the study or metastasised before the end of five years were grouped into the first class, and those who survived five years formed the second class. Of the seventy-eight patients, forty-four survived the critical five years.

The top panel of Figure 2.15 shows the cumulative contribution to variance. The rank of the covariance matrix is 77, so smaller than n. The contributions to variance increase slowly, starting with the largest single variance contribution of 16.99 per cent. The first fifty PCs contribute about 90 per cent to total variance. The lower panel of Figure 2.15 shows the weights of the first eigenvector in the form of a histogram. In a principal component analysis, all eigenvector weights are non-zero. For the 4,751 genes, the weights of the first eigenvector range from -0.0591 to 0.0706, and as we can see in the histogram, most are very close to zero.

A comparison of the first four eigenvectors based on the fifty variables with the highest absolute weights for each vector shows that PC_1 has no 'high-weight' variables in common with PC_2 or PC_3, whereas PC_2 and PC_3 share three such variables. All three eigenvectors share some 'large' variables with the fourth. Figure 2.16 also deals with the first four principal components in the form of 2D score plots. The blue scores correspond to the forty-four patients who survived five years, and the black scores correspond to the other group. The blue and black point clouds overlap in these score plots, indicating that the first four PCs cannot separate the two groups. However, it is interesting that there is an outlier, observation 54 marked in red, which is clearly separate from the other points in all but the last plot.

The principal component analysis has reduced the total number of variables from 4,751 to 77 PCs. This reduction is merely a consequence of the HDLSS property of the data, and a further reduction in the number of variables may be advisable. In a later analysis we will examine how many of these seventy-seven PCs are required for reliable prediction of the time to metastasis. ■

A little care may be needed to distinguish the gene expression breast tumour data from the thirty-dimensional breast cancer data which we revisited earlier in this section. Both data sets deal with breast cancer, but they are very different in content and size. We refer to the

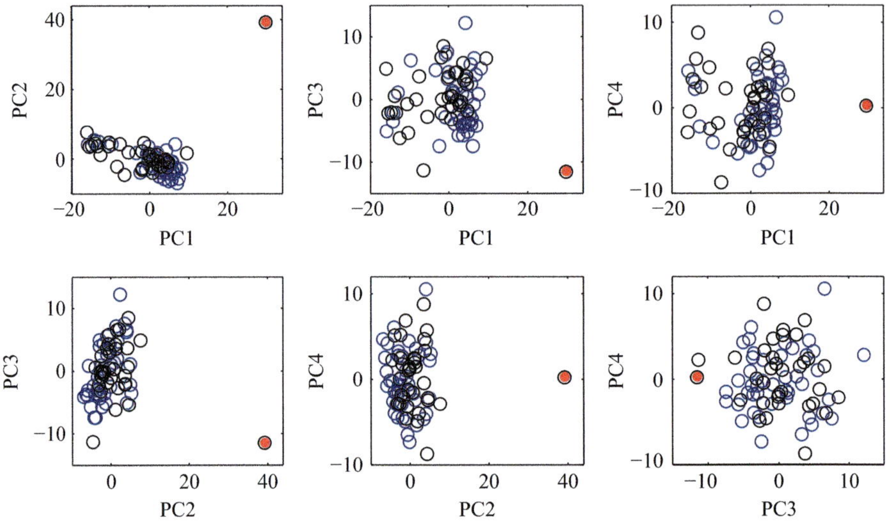

Figure 2.16 Score plots of the breast tumour data of Example 2.15: (*top row*): PC_1 scores (x-axis) against PC_2–PC_4 scores; (*bottom row*) PC_2 scores against PC_3 and PC_4 scores (*left* and *middle*) and PC_3 scores against PC_4 scores (*right*).

smaller one simply as the *breast cancer* data and call the HDLSS data the *breast tumour (gene expression)* data.

Our third example deals with functional data from bioinformatics. In this case, the data are measurements on proteins or, more precisely, on the simpler peptides rather than genes.

Example 2.16 The **ovarian cancer proteomics** data of Gustafsson (2011)[1] are mass spectrometry profiles or curves from a tissue sample of a patient with high-grade serous ovarian cancer. Figure 2.17 shows an image of the tissue sample stained with haematoxylin and eosin, with the high-grade cancer regions marked. The preparation and analysis of this and similar samples are described in chapter 6 of Gustafsson (2011), and Gustafsson et al. (2011) describe matrix-assisted laser desorption–ionisation imaging mass spectrometry (MALDI-IMS), which allows acquisition of mass data for proteins or the simpler peptides used here. For an introduction to mass spectrometry–based proteomics, see Aebersold and Mann (2003).

The observations are the profiles, which are measured at 14,053 regularly spaced points given by their (x, y)-coordinates across the tissue sample. At each grid point, the counts – detections of peptide ion species at instrument-defined mass-to-charge m/z intervals – are recorded. There are 1,331 such intervals, and their midpoints are the variables of these data. Because the ion counts are recorded in adjoining intervals, the profiles may be regarded as discretisations of curves. For MALDI-IMS, the charge z is one and thus could be ignored. However, in proteomics, it is customary to use the notation m/z, and despite the simplification $z = 1$, we use the mass-to-charge terminology.

Figure 2.18 shows two small subsets of the data, with the m/z values on the x-axis. The top panel shows the twenty-one observations or profiles indexed 400 to 420, and the middle

[1] For these data, contact the author directly.

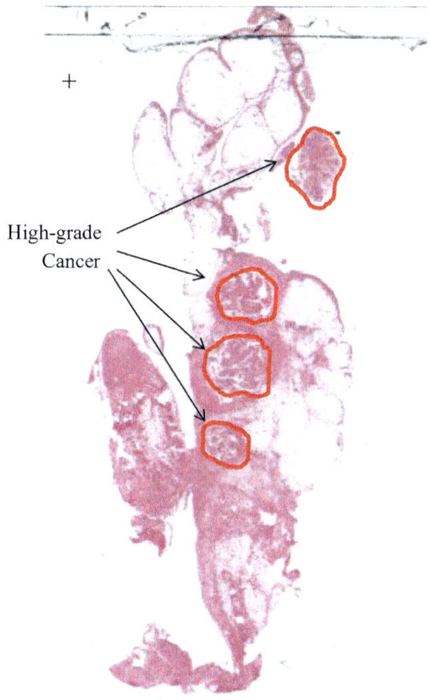

Figure 2.17 Image of tissue sample with regions of ovarian cancer from Example 2.16.

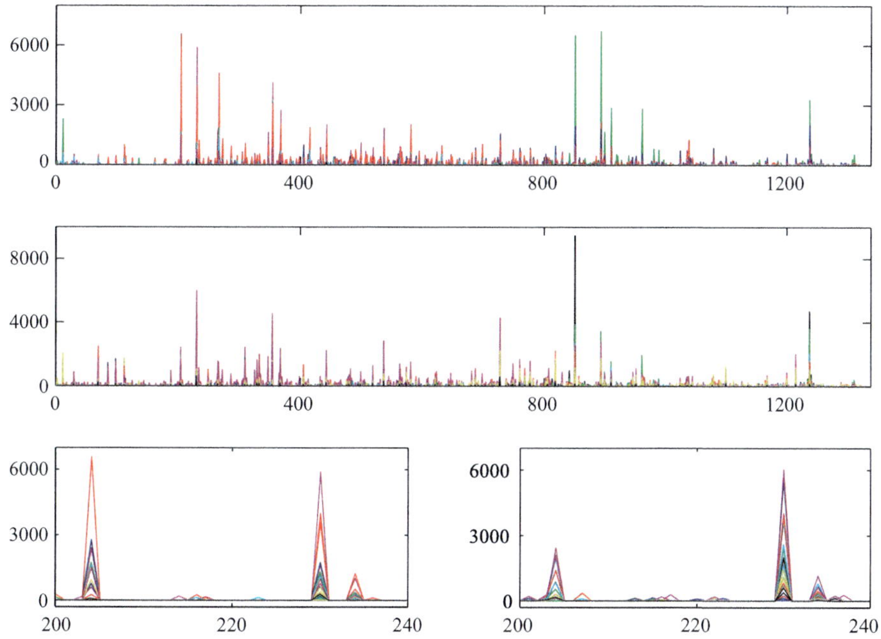

Figure 2.18 Mass-spectrometry profiles from the ovarian cancer data of Example 2.16 with m/z values on the x-axis and counts on the y-axis. Observations 400 to 420 (*top*) with their zoom-ins (*bottom left*) and observations 1,000 to 1,020 (*middle*) with their zoom-ins (*bottom right*).

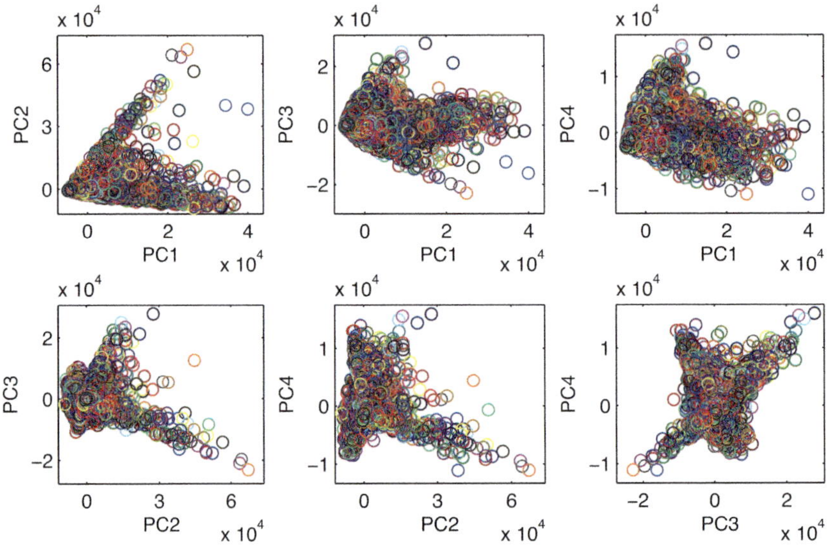

Figure 2.19 Score plots of the ovarian cancer data of Example 2.16: (*top row*) PC_1 scores (x-axis) against PC_2–PC_4 scores; (*bottom row*) PC_2 scores against PC_3 and PC_4 scores (*left* and *middle*) and PC_3 scores against PC_4 scores (*right*).

panel shows the profiles indexed 1,000 to 1,020 of 14,053. The plots show that the number and position of the peaks vary across the tissue. The plots in the bottom row are 'zoom-ins' of the two figures above for m/z values in the range 200 to 240. The left panel shows the zoom-ins of observations 400 to 420, and the right panel corresponds to observations 1,000 to 1,020. The peaks differ in size, and some of the smaller peaks that are visible in the right panel are absent in the left panel. There are many m/z values which have zero counts.

The rank of the covariance matrix of all observations agrees with $d = 1,331$. The eigenvalues decrease quickly; the tenth is about 2 per cent of the size of the first, and the one-hundredth is about 0.03 per cent of the first. The first principal component score contributes 44.8 per cent to total variance, the first four contribute 79 per cent, the first ten result in 89.9 per cent, and the first twenty-five contribute just over 95 per cent. It is clear from these numbers that the first few PCs contain most of the variability of the data.

Figure 2.19 shows 2D score plots of the first four principal component scores of the raw data. As noted earlier, these four PCs contribute about 80 per cent to total variance. PC_1 is shown on the x-axis against PC_2 to PC_4 in the top row, and the remaining combinations of score plots are shown in the lower panels.

The score plots exhibit interesting shapes which deviate strongly from Gaussian shapes. In particular, the PC_1 and PC_2 data are highly skewed. Because the PC data are centred, the figures show that a large proportion of the observations have very small negative values for both the PC_1 and PC_2 scores. The score plots appear as connected regions. From these figures, it is not clear whether the data split into clusters, and if so, how many. We return to these data in Section 6.5.3, where we examine this question further and find partial answers. ■

In an analysis of high-dimensional data, important steps are a reduction of the number of variables and a separation of the *signal* from the *noise*. The distinction between signal and noise is not well defined, and the related questions of how many variables we want to preserve and how many are negligible do not have clear answers. For HDLSS data, a natural upper bound for the number of variables is the rank of the covariance matrix. But often this rank is still considerably larger than the number of 'useful' variables, and further dimension-reduction is necessary. Throughout this book we will encounter dimension reduction aspects, and at times I will suggest ways of dealing with these problems and give partial solutions.

2.7 Asymptotic Results

2.7.1 Classical Theory: Fixed Dimension d

The law of large numbers tells us that the sample mean converges to the expectation and that similar results hold for higher-order moments. These results are important as they provide theoretical justification for the use of sample moments instead of the population parameters.

Do similar results hold for the spectral decomposition obtained from Principal Component Analysis? To examine this question rigorously, we would need to derive the sampling distributions of the eigenvalues and their corresponding eigenvectors. This is beyond the scope of this book but is covered in Anderson (2003). I will merely present the approximate distributions of the sample eigenvalues and sample eigenvectors. The asymptotic results are important in their own right. In addition, they enable us to construct approximate confidence intervals.

For Gaussian data with fixed dimension d, Anderson (1963) contains a proof of the following Central Limit Theorem.

Theorem 2.20 *Let $\mathbb{X} = \begin{bmatrix}\mathbf{X}_1\ \mathbf{X}_2 \cdots \mathbf{X}_n\end{bmatrix}$ be a sample of independent random vectors from the multivariate normal distribution $\mathcal{N}(\boldsymbol{\mu}, \Sigma)$. Assume that Σ has rank d and spectral decomposition $\Sigma = \Gamma\Lambda\Gamma^{\mathsf{T}}$. Let S be the sample covariance matrix of $\mathbb{X}$.*

1. *Let $\boldsymbol{\lambda}$ and $\widehat{\boldsymbol{\lambda}}$ be the vectors of eigenvalues of Σ and S, respectively, ordered in decreasing size. Then, as $n \to \infty$, the approximate distribution of $\widehat{\boldsymbol{\lambda}}$ is*

$$\widehat{\boldsymbol{\lambda}} \sim \mathcal{N}\left(\boldsymbol{\lambda}, 2\Lambda^2/n\right).$$

2. *For $k \le d$, let $\boldsymbol{\eta}_k$ be the kth eigenvector of Σ and $\widehat{\boldsymbol{\eta}}_k$ the corresponding sample eigenvector of S. Put*

$$\Theta_k = \sum_{j \neq k} \frac{\lambda_k \lambda_j}{(\lambda_k - \lambda_j)^2} \boldsymbol{\eta}_j \boldsymbol{\eta}_j^{\mathsf{T}}.$$

Then, as $n \to \infty$, the approximate distribution of $\widehat{\boldsymbol{\eta}}_k$ is

$$\widehat{\boldsymbol{\eta}}_k \sim \mathcal{N}\left(\boldsymbol{\eta}_k, \Theta_k/n\right).$$

3. *Asymptotically, the entries $\widehat{\lambda}_j$ of $\widehat{\boldsymbol{\lambda}}$ and the $\widehat{\boldsymbol{\eta}}_k$ are independent for $j, k = 1, \ldots, d$.*

The theorem states that asymptotically, the eigenvalues and eigenvectors converge to the *right* quantities; the sample eigenvalues and eigenvectors are asymptotically independent, unbiased and normally distributed. If the rank of Σ is smaller than the dimension, the theorem still holds with the obvious replacement of d by r in all three parts.

The results in the theorem imply that the sample quantities are consistent estimators for the true parameters, provided that the dimension d is fixed and is smaller than the sample size. If the dimension grows, then these results may no longer hold, as we shall see in Section 2.7.2.

Based on the asymptotic distributional results for the eigenvalues, one could consider hypothesis tests for the rank of Σ, namely,

$$H_0\colon \lambda_j = 0 \qquad \text{versus} \qquad H_1\colon \lambda_j > 0 \qquad \text{for some } j,$$

or

$$H_0\colon \lambda_p = \lambda_{p+1} = \cdots = \lambda_d = 0 \qquad \text{versus} \qquad H_1\colon \lambda_p > 0 \qquad \text{for some } p \geq 1.$$

Asymptotically, the eigenvalues are independent, and we could carry out a series of tests starting with the smallest non-zero sample eigenvalue and appropriate test statistics derived from Theorem 2.20. We will not pursue these tests; instead, we look at confidence intervals for the eigenvalues.

Corollary 2.21 *Let* $\mathbb{X} = \begin{bmatrix}\mathbf{X}_1\ \mathbf{X}_2 \cdots \mathbf{X}_n\end{bmatrix}$ *satisfy the assumptions of Theorem 2.20. Let* z_α *be the critical value of the normal distribution* $\mathcal{N}(0,1)$ *such that* $P\{Z \geq z_\alpha\} = 1-\alpha$. *For* $j \leq d$, *approximate two-sided and lower one-sided confidence intervals for* λ_j, *with confidence level* $(1-\alpha)$, *are*

$$\left[\frac{\widehat{\lambda}_j}{1+z_{\alpha/2}\sqrt{2/n}},\ \frac{\widehat{\lambda}_j}{1-z_{\alpha/2}\sqrt{2/n}}\right] \quad \textit{and} \quad \left[\frac{\widehat{\lambda}_j}{1+z_\alpha\sqrt{2/n}},\ \infty\right). \tag{2.21}$$

The lower bound of the confidence intervals is positive. The length of the two-sided jth interval is proportional to $\widehat{\lambda}_j$, and this length decreases because the $\widehat{\lambda}_j$ values decrease as the index increases. The confidence intervals differ from what we might expect: Because they do not include zero, in the classical sense, they are not suitable for testing whether λ_j could be zero. However, they assist in an assessment of which eigenvalues should be deemed negligible.

Proof We use the asymptotic distribution established in Theorem 2.20. For $j \leq d$, let λ_j and $\widehat{\lambda}_j$ be the jth eigenvalue of Σ and S, respectively. Because we work with only one eigenvalue, I drop the subscript in this proof. Fix $\alpha > 0$. Then

$$\begin{aligned}
1-\alpha &= \mathbb{P}\left\{|\widehat{\lambda} - \lambda| \leq z_{\alpha/2}\lambda\sqrt{2/n}\right\} \\
&= \mathbb{P}\left\{1 - z_{\alpha/2}\sqrt{2/n} \leq \frac{\widehat{\lambda}}{\lambda} \leq 1 + z_{\alpha/2}\sqrt{2/n}\right\} \\
&= \mathbb{P}\left\{\frac{\widehat{\lambda}}{1+z_{\alpha/2}\sqrt{2/n}} \leq \lambda \leq \frac{\widehat{\lambda}}{1-z_{\alpha/2}\sqrt{2/n}}\right\}.
\end{aligned}$$

The proof for the one-sided intervals is similar. ∎

In Theorem 2.20 and Corollary 2.21, we assume that the data are normal. If we relax the assumptions of multivariate normality, then these results may no longer hold. In practice, we want to assess whether particular multivariate data are normal. A good indication of the Gaussianity or deviation from Gaussianity is the distribution of the first few principal component scores. If this distribution is strongly non-Gaussian, then this fact is evidence against the Gaussianity of the data. Many of the data sets we encountered so far deviate from the Gaussian distribution because their PC scores

- are bimodal as in the *Swiss bank notes* data – see the $\mathbb{W}^{(2)}$ data in Figure 2.6 of Example 2.7;
- have a number of clusters, as in the *HIV* data – see Figure 2.3 of Example 2.4; or
- are strongly skewed, as in the *ovarian cancer proteomics* data – see the $\mathbb{W}^{(2)}$ data in Figure 2.19 of Example 2.16.

For any of these and similar data, we can still calculate the principal component data because these calculations are 'just' based on linear algebra; however, some caution is appropriate in the interpretation of these quantities. In particular, the sample quantities may not converge to the population quantities if the data deviate strongly from the Gaussian distribution.

2.7.2 Asymptotic Results when d Grows

This section attempts to give a flavour of theoretical developments when the dimension grows. In the classical framework, where the dimension d is fixed and smaller than the sample size n, the d-dimensional random vector has a covariance matrix Σ. If we let the dimension grow, then we no longer have a single fixed mean and covariance matrix. For the population, we consider instead a sequence of random vectors indexed by d:

$$\mathbf{X}_d \sim (\boldsymbol{\mu}_d, \Sigma_d), \tag{2.22}$$

where $\boldsymbol{\mu}_d \in \mathbb{R}^d$, Σ_d is a $d \times d$ matrix, and $d = 1, 2, \ldots$.

I will often drop the subscript d and write $\mathbf{X}$ instead of $\mathbf{X}_d$ or Σ instead of Σ_d and refer to the covariance matrix rather than a sequence of covariance matrices. For each d, Σ has a spectral decomposition and gives rise to principal component scores and vectors. In general, however, there is no simple relationship between the scores pertaining to d and $d+1$.

The setting for the sample is more complicated as it involves the relationship between d and n. We distinguish between the four scenarios which I define now and which I refer to by the relationship between n and d shown symbolically in bold at the beginning of each line.

Definition 2.22 We consider the following asymptotic domains for the relationships of n and d.

n≻d refers to classical asymptotics, where d is fixed, and $n \to \infty$;

n⪰d refers to the random matrix theory domain, where d and n grow, and $d = O(n)$, that is, d grows at a rate not faster than that of n;

n⪯d refers to the random matrix theory domain, where d and n grow, and $n = o(d)$, that is, d grows at a faster rate than n; and

n≺d refers to the HDLSS domain, where $d \to \infty$, and n remains fixed. □

The first case, **n≻d**, is the classical set-up, and asymptotic results are well established for normal random vectors. The cases **n⪰d** and **n⪯d** are sometimes combined into one case. Asymptotic results refer to n and d because both grow. The last case, **n≺d**, refers to HDLSS data, a clear departure from classical asymptotics. In this section we consider asymptotic results for $\mathbf{n} \succeq \mathbf{d}$ and $\mathbf{n} \prec \mathbf{d}$. In Section 13.5 we will return to $\mathbf{n} \prec \mathbf{d}$ and explore $\mathbf{n} \succeq \mathbf{d}$ and $\mathbf{n} \preceq \mathbf{d}$.

For data $\mathbb{X} = \mathbb{X}_d$, we define a sequence of sample means $\overline{\mathbf{X}}_d$ and sample covariance matrices S_d as d grows. For $d > n$, S_d is singular, even though the corresponding population matrix Σ_d may be non-singular. As a result, the eigenvalues and eigenvectors of S may fail to converge to the population parameters.

As in the classical asymptotic setting, for $\mathbf{n} \succeq \mathbf{d}$, we assume that the data $\mathbb{X}$ consist of n independent random vectors from the multivariate normal distribution in d dimensions, and d and n satisfy

$$n/d \to \gamma \geq 1.$$

Unlike the classical case, the sample eigenvalues become more spread out than the population values as the dimension grows. This phenomenon follows from Marčenko and Pastur (1967), who show that the empirical distribution function G_d of all eigenvalues of the sample covariance matrix S converges almost surely:

$$G_d(t) = \frac{1}{d}\#\{\lambda_j : \lambda_j \leq nt\} \to G(t) \qquad \text{as } n/d \to \gamma \geq 1,$$

where the probability density function g of the limiting distribution G is

$$g(t) = \frac{\gamma}{2\pi t}\sqrt{(b-t)(t-a)} \qquad \text{for } a \leq t \leq b,$$

with $a = (1-\gamma^{-1/2})^2$ and $b = (1+\gamma^{-1/2})^2$. Direct computations show that decreasing the ratio n/d increases the spread of the eigenvalues.

Of special interest is the largest eigenvalue of the sample covariance matrix. Tracy and Widom (1996) use techniques from random matrix theory to show that for a symmetric $n \times n$ matrix with Gaussian entries, as $n \to \infty$, the limiting distribution of the largest eigenvalue follows the *Tracy-Widom law of order 1*, F_1, defined by

$$F_1(t) = \exp\left\{-\frac{1}{2}\int_t^\infty q(x) + (x-t)q^2(x)\, dx\right\} \qquad \text{for } s \in \mathbb{R}, \tag{2.23}$$

where q is the solution of a non-linear Painlevé II differential equation. Tracy and Widom (1996) have details on the theory, and Tracy and Widom (2000) contain numerical results on F_1. Johnstone (2001) extends their results to the sample covariance matrix of $\mathbb{X}$.

Theorem 2.23 [Johnstone (2001)] *Let $\mathbb{X}$ be $d \times n$ data, and assume that $X_{ij} \sim \mathcal{N}(0,1)$. Assume that the dimension $d = d(n)$ grows with the sample size n. Let $\widehat{\lambda}_1$ be the largest eigenvalue of $\mathbb{X}\mathbb{X}^{\mathsf{T}}$, which depends on n and d. Put $k_{nd} = (\sqrt{n-1} + \sqrt{d})$,*

$$\mu_{nd} = k_{nd}^2, \qquad \textit{and} \qquad \sigma_{nd} = k_{nd}\left(\frac{1}{\sqrt{n-1}} + \frac{1}{\sqrt{d}}\right)^{1/3}.$$

If $n/d \to \gamma \geq 1$, then

$$\frac{\widehat{\lambda}_1 - \mu_{nd}}{\sigma_{nd}} \sim F_1, \tag{2.24}$$

where F_1 as in (2.23) is the Tracy-Widom law of order 1.

Theorem 2.23 is just the beginning, and extensions and new results for high dimensions are emerging (see Baik and Silverstein 2006 and Paul 2007). The theorem shows a departure from the conclusions in Theorem 2.20, the Central Limit Theorem for fixed dimension d: if the dimension grows with the sample size, then even for normal data, asymptotically, the first eigenvalue no longer has a normal distribution. In the proof, Johnstone makes use of random matrix theory and shows how the arguments of Tracy and Widom (1996) can be extended to his case.

In the proof of Theorem 2.23, the author refers to $d \leq n$; however, the theorem applies equally to the case $n < d$ by reversing the roles of n and d in the definitions of μ_{nd} and σ_{nd}. Johnstone (2001) illustrates that the theorem holds approximately for very small values of n and d such as $n = d = 5$. For such small values of n, the classical result of Theorem 2.20 may not apply.

In practice, one or more eigenvalues of the covariance matrix can be much larger than the rest of the eigenvalues: in the *breast cancer* data of Example 2.12, the first two eigenvalues contribute well over 99 per cent of total variance. To capture settings with a small number of distinct, that is, large eigenvalues, Johnstone (2001) introduces the term **spiked covariance** model, which assumes that the first κ eigenvalues of the population covariance matrix are greater than 1 and the remaining $d - \kappa$ are 1. Of particular interest is the case $\kappa = 1$. The spiked covariance models have lead to new results on convergence of eigenvalues and eigenvectors when d grows, as we see in the next result.

For HDLSS data, case $\mathbf{n \prec d}$, I follow Jung and Marron (2009). There is a small difference between their notation and ours: their sample covariance matrix used the scalar $(1/n)$. This difference does not affect the ideas or the results. We begin with a special case of their theorem 2 and explore the subtleties that can occur in more detail in Section 13.5.2.

For high-dimensional data, a natural measure of closeness between two unit vectors $\boldsymbol{\alpha}$, $\boldsymbol{\beta} \in \mathbb{R}^d$ is the angle $\angle(\boldsymbol{\alpha}, \boldsymbol{\beta})$ between these vectors. We also write $\mathfrak{a}(\alpha, \beta)$ for this angle, where

$$\mathfrak{a}(\boldsymbol{\alpha}, \boldsymbol{\beta}) = \angle(\boldsymbol{\alpha}, \boldsymbol{\beta}) = \arccos(\boldsymbol{\alpha}^{\top} \boldsymbol{\beta}). \tag{2.25}$$

Definition 2.24 For fixed n, let $d \geq n$. Let $\mathbb{X} = \mathbb{X}_d \sim (\mathbf{0}_d, \Sigma_d)$ be a sequence of $d \times n$ data with sample covariance matrices S_d. Let

$$\Sigma_d = \Gamma_d \Lambda_d \Gamma_d \qquad \text{and} \qquad S_d = \widehat{\Gamma}_d \widehat{\Lambda}_d \widehat{\Gamma}_d$$

be the spectral decompositions of Σ_d and S_d, respectively. The jth eigenvector $\widehat{\boldsymbol{\eta}}_j$ of S_d is HDLSS **consistent** with its population counterpart $\boldsymbol{\eta}_j$ if the angle $\mathfrak{a}$ between the vectors satisfies

$$\mathfrak{a}(\boldsymbol{\eta}_j, \widehat{\boldsymbol{\eta}}_j) \xrightarrow{p} 0 \qquad \text{as } d \to \infty,$$

where $\xrightarrow{p}$ refers to convergence in probability. Let λ_j be the jth eigenvalue of Σ_d. For $k = 1, 2, \ldots$, put

$$\varepsilon_k(d) = \frac{\left(\sum_{j=k}^d \lambda_j\right)^2}{d \sum_{j=k}^d \lambda_j^2}; \tag{2.26}$$

then ε_k defines a measure of **spikiness** or **sphericity** for Σ_d, and Σ_d satisfies the **ε_k-condition** $\varepsilon_k(d) \gg 1/d$ if

$$\left[d\varepsilon_k(d)\right]^{-1} \to 0 \qquad \text{as } d \to \infty. \tag{2.27}$$

□

For notational convenience, I omit the subscript d when referring to the eigenvalues and eigenvectors of Σ_d or S_d. We note that $\varepsilon_1(d) = \left[\operatorname{tr}(\Sigma_d)\right]^2/[d\operatorname{tr}(\Sigma_d^2)]$, and thus $d^{-1} \leq \varepsilon_1(d) \leq 1$. If all eigenvalues of Σ_d are 1, then $\varepsilon_1(d) = 1$. The sphericity idea was proposed by John (1971, 1972). Ahn et al. (2007) introduced the concept of HDLSS consistency and showed that if $\varepsilon_1(d) \gg 1/d$, then the eigenvalues of the sample covariance matrix S tend to be identical in probability as $d \to \infty$.

To gain some intuition into spiked covariance models, we consider a simple scenario which is given as example 4.1 in Jung and Marron (2009). Let F_d be a symmetric $d \times d$ matrix with diagonal entries 1 and off-diagonal entries $\rho_d \in (0,1)$. Let $\Sigma_d = F_d F_d^{\mathsf{T}}$ be a sequence of covariance matrices indexed by d. The eigenvalues of Σ_d are

$$\lambda_1 = (d\rho_d + 1 - \rho_d)^2 \qquad \text{and} \qquad \lambda_2 = \cdots = \lambda_d = (1 - \rho_d)^2. \tag{2.28}$$

The first eigenvector of Σ_d is $\boldsymbol{\eta}_1 = [1, \ldots, 1]^{\mathsf{T}}/\sqrt{d}$. Further, $\sum_{j \geq 2} \lambda_j = (d-1)(1-\rho_d)^2$, and

$$\varepsilon_2(d) = \frac{(d-1)^2(1-\rho_d)^4}{d(d-1)(1-\rho_d)^4} = \frac{d-1}{d},$$

so Σ_d satisfies the ε_2 condition $\varepsilon_2(d) \gg 1/d$.

In the classical asymptotics of Theorem 2.20 and in theorem 2.23 of Johnstone (2001), the entries of $\mathbb{X}$ are Gaussian. Jung and Marron (2009) weaken the Gaussian assumption and only require the ρ-mixing condition of the variables given next. This condition suffices to develop appropriate laws of large numbers as it leads to an asymptotic independence of the variables, possibly after rearranging the variables.

For random variables X_i with $J \leq i \leq L$ and $-\infty \leq J \leq L \leq \infty$, let $\mathcal{F}_J^L$ be the σ-field generated by the X_i, and let $L_2(\mathcal{F}_J^L)$ be the space of square integrable $\mathcal{F}_J^L$-measurable random variables. For $m \geq 1$, put

$$\rho(m) = \sup_{f,g,j \in \mathbb{Z}} |\operatorname{cor}(f,g)|, \tag{2.29}$$

where $f \in L_2(\mathcal{F}_{-\infty}^j)$ and $g \in L_2(\mathcal{F}_{j+m}^{\infty})$. The sequence $\{X_i\}$ is **ρ-mixing** if $\rho(m) \to 0$ as $m \to \infty$.

In the next theorem, the columns of $\mathbb{X}$ are not required to be ρ-mixing; instead, the principal component data satisfy this condition. Whereas the order of the entries in $\mathbb{X}$ may be important, the same is not true for the PCs, and an appropriate permutation of the PC entries may be found which leads to ρ-mixing sequences.

Theorem 2.25 [Jung and Marron (2009)] *For fixed n and $d \geq n$, let $\mathbb{X}_d \sim (\mathbf{0}_d, \Sigma_d)$ be a sequence of $d \times n$ data, with Σ_d, S_d, their eigenvalues and eigenvectors as in Definition 2.24. Put $\mathbb{W}_{\Lambda,d} = \Lambda_d^{-1/2}\Gamma_d^{\mathsf{T}}\mathbb{X}_d$, and assume that $\mathbb{W}_{\Lambda,d}$ have uniformly bounded fourth moments and are ρ-mixing for some permutation of the rows. Assume further that for $\alpha > 1$, $\lambda_1/d^\alpha \to c$ for some $c > 0$ and that the Σ_d satisfy $\varepsilon_2(d) \gg 1/d$ and $\sum_{j=2}^d \lambda_j = O(d)$. Then the following hold:*

1. *The first eigenvector* $\widehat{\eta}_1$ *of* S_d *is consistent in the sense that*

$$\mathfrak{a}(\widehat{\eta}_1, \eta_1) \xrightarrow{p} 0 \qquad \text{as } d \to \infty. \tag{2.30}$$

2. *The first eigenvalue* $\widehat{\lambda}_1$ *of* S_d *converges in distribution*

$$\frac{\widehat{\lambda}_1}{d^\alpha} \xrightarrow{D} \nu_1 \qquad \text{as } d \to \infty \qquad \text{and} \qquad \nu_1 = \frac{c}{n\lambda_1}\, \eta_1^{\mathsf{T}} \mathbb{X}\mathbb{X}^{\mathsf{T}} \eta_1.$$

3. *If the* $\mathbb{X}_d$ *are also normal, then as* $d \to \infty$, *the asymptotic distribution of* $n\widehat{\lambda}_1/\lambda_1$ *is* χ^2_n.

Theorem 2.25 shows the HDLSS consistency of the first sample eigenvector and the convergence in distribution of $\widehat{\lambda}_1$ to a random variable ν_1, which has a χ^2_n distribution if the data are normal. In contrast to Theorem 2.23, which holds for arbitrary covariance matrices Σ, in the HDLSS context, the sample eigenvalues and eigenvectors only converge for suitably spiked matrices Σ. This is the price we pay when $d \to \infty$.

Theorem 2.25 is a special case of theorem 2 of Jung and Marron (2009) and is more specifically based on their lemma 1, corollary 3, and parts of proposition 1. In the notation of their theorem 2, our Theorem 2.25 treats the case $p = 1$ and $k_1 = 1$ and deals with consistency results only. We will return to the paper of Jung and Marron in Section 13.5 and then consider their general setting, including some inconsistency results, which I present in Theorems 13.15 and 13.16.

In their proof, Jung and Marron (2009) exploit the fact that the $d \times d$ matrix $(n-1)S = \mathbb{X}\mathbb{X}^{\mathsf{T}}$ and the $n \times n$ matrix $Q_{(n)} = \mathbb{X}^{\mathsf{T}}\mathbb{X}$ have the same eigenvalues. We derive this property in Proposition 3.1 of Section 3.2. In Multidimensional Scaling, Gower (1966) uses the equality of the eigenvalues of $\mathbb{X}\mathbb{X}^{\mathsf{T}}$ and $\mathbb{X}^{\mathsf{T}}\mathbb{X}$ in the formulation of Principal Coordinate Analysis – see Theorem 8.3 in Section 8.2.1. Since the early days of Multidimensional Scaling, switching between $\mathbb{X}\mathbb{X}^{\mathsf{T}}$ and $\mathbb{X}^{\mathsf{T}}\mathbb{X}$ has become common practice. Section 5.5 contains background for the duality of $\mathbb{X}$ and $\mathbb{X}^{\mathsf{T}}$, and Section 8.5.1 highlights some of the advantages of working with the dual matrix $Q_{(n)}$.

Note that $\mathbb{W}_{\Lambda,d}$ in Theorem 2.25 are not the PC data because Jung and Marron apply the population eigenvalues and eigenvectors to the data $\mathbb{X}$, which makes sense in their theoretical development. A consequence of this definition is that $Q = \mathbb{W}^{\mathsf{T}}_{\Lambda,d}\Lambda_d\mathbb{W}_{\Lambda,d}$, and this Q bridges the gap between Σ and S. Jung and Marron show (in their corollary 1) that the eigenvectors in (2.30) converge almost surely if the vectors of $\mathbb{W}_{\Lambda,d}$ are independent and have uniformly bounded eighth moments.

For covariance matrices Σ_d with eigenvalues (2.28) and normal data $\mathbb{X} \sim \mathcal{N}(\mathbf{0}_d, \Sigma_d)$ with fixed n and $d = n+1, n+2, \ldots$, the assumptions of Theorem 2.25 are satisfied if ρ_d of (2.28) is constant or if ρ_d decreases to 0 so slowly that $\rho_d \gg d^{-1/2}$. In this case, the first sample eigenvector is a consistent estimator of η_1 as $d \to \infty$. If ρ_d decreases to 0 too quickly, then there is no $\alpha > 1$ for which $\lambda_1/d^\alpha \to c$, and then the first sample eigenvector $\widehat{\eta}_1$ does not converge to η_1. We meet precise conditions for the behaviour of $\widehat{\eta}_1$ in Theorem 13.15 of Section 13.5.

Our results for high-dimensional settings, Theorems 2.23 and 2.25, show that the $d \to \infty$ asymptotics differ considerably from the classical $n \to \infty$ asymptotics. Even for normal data, the sample eigenvalues and eigenvectors may fail to converge to their population parameters, and if they converge, the limit distribution is not normal. For HDLSS problems, we can still calculate the sample PCs and reduce the dimension of the original data, but the estimates may not converge to their population PCs. In some applications this matters, and then other avenues need to be explored. In Section 13.5 we return to asymptotic properties of high-dimensional data and explore conditions on the covariance matrices which determine the behaviour and, in particular, the consistency of the eigenvectors.

2.8 Principal Component Analysis, the Number of Components and Regression

Regression and prediction are at the heart of data analysis, and we therefore *have to* investigate the connection between principal components and regression. I give an overview and an example in Section 2.8.2. First, though, we explore a method for choosing the number of components.

2.8.1 Number of Principal Components Based on the Likelihood

As we have seen in the discussion about scree plots and their use in selecting the number of components, more often than not there is no 'elbow'. Another popular way of choosing the number of PCs is the so-called 95 per cent rule. The idea is to pick the index κ such that the first κ PC scores contribute 95 per cent to total variance. Common to both ways of choosing the number of PCs is that they are easy, but both are ad hoc, and neither has a mathematical foundation.

In this section I explain the **Probabilistic Principal Component Analysis** of Tipping and Bishop (1999), which selects the number of components in a natural way, namely, by maximising the likelihood.

Tipping and Bishop (1999) consider a model-based framework for the population in which the d-dimensional random vector $\mathbf{X}$ satisfies

$$\mathbf{X} = A\mathbf{Z} + \boldsymbol{\mu} + \boldsymbol{\epsilon}, \tag{2.31}$$

where $\mathbf{Z}$ is a p-dimensional random vector with $p \le d$, A is a $d \times p$ matrix, $\boldsymbol{\mu} \in \mathbb{R}^d$ is the mean of $\mathbf{X}$, and $\boldsymbol{\epsilon}$ is d-dimensional random noise. Further, $\mathbf{X}$, $\mathbf{Z}$ and $\boldsymbol{\epsilon}$ satisfy G1 to G4.

G1: The vector $\boldsymbol{\epsilon}$ is independent Gaussian random noise with $\boldsymbol{\epsilon} \sim \mathcal{N}(\mathbf{0}, \sigma^2 \mathbf{I}_{d\times d})$.

G2: The vector $\mathbf{Z}$ is independent Gaussian with $\mathbf{Z} \sim \mathcal{N}(\mathbf{0}, \mathbf{I}_{p\times p})$.

G3: The vector $\mathbf{X}$ is multivariate normal with $\mathbf{X} \sim \mathcal{N}(\boldsymbol{\mu}, AA^{\mathrm{T}} + \sigma^2 \mathbf{I}_{d\times d})$.

G4: The conditional random vectors $\mathbf{X}|\mathbf{Z}$ and $\mathbf{Z}|\mathbf{X}$ are multivariate normal with means and covariance matrices derived from the means and covariance matrices of $\mathbf{X}$, $\mathbf{Z}$ and $\boldsymbol{\epsilon}$.

In this framework, A, p and $\mathbf{Z}$ are unknown. Without further assumptions, we cannot find all these unknowns. From a PC perspective, model (2.31) is interpreted as the approximation of the centred vector $\mathbf{X} - \boldsymbol{\mu}$ by the first p principal components, and $\boldsymbol{\epsilon}$ is the error arising from this approximation. The vector $\mathbf{Z}$ is referred to as the **unknown source** or the vector of **latent variables**. The latent variables are not of interest in their own right for this analysis; instead, the goal is to determine the dimension p of $\mathbf{Z}$. The term *latent variables* is used in Factor Analysis. We return to the results of Tipping and Bishop in Section 7.4.2 and in particular in Theorem 7.6.

Principal Component Analysis requires no assumptions about the distribution of the random vector $\mathbf{X}$ other than finite first and second moments. If Gaussian assumptions are made or known to hold, they invite the use of powerful techniques such as likelihood methods. For data $\mathbb{X}$ satisfying (2.31) and G1 to G4, Tipping and Bishop (1999) show the connection between the likelihood of the data and the principal component projections. The multivariate normal likelihood function is defined in (1.16) in Section 1.4. Tipping and Bishop's key idea is that the dimension of the latent variables is the value p which maximises the likelihood. For details and properties of the likelihood function, see chapter 7 of Casella and Berger (2001).

Theorem 2.26 [Tipping and Bishop (1999)] *Assume that the data $\mathbb{X} = \begin{bmatrix}\mathbf{X}_1 \ \mathbf{X}_2 \cdots \mathbf{X}_n\end{bmatrix}$ satisfy (2.31) and G1 to G4. Let S be the sample covariance matrix of $\mathbb{X}$, and let $\widehat{\Gamma}\widehat{\Lambda}\widehat{\Gamma}^{\mathsf{T}}$ be the spectral decomposition of S. Let $\widehat{\Lambda}_p$ be the diagonal $p \times p$ matrix of the first p eigenvalues $\widehat{\lambda}_j$ of S, and let $\widehat{\Gamma}_p$ be the $d \times p$ matrix of the first p eigenvectors of S.*

Put $\theta = (A, \boldsymbol{\mu}, \sigma)$. Then the log-likelihood function of θ given $\mathbb{X}$ is

$$\log L(\theta|\mathbb{X}) = -\frac{n}{2}\left\{d\ \log(2\pi) + \log[\det(C)] + \operatorname{tr}(C^{-1}S)\right\},$$

where

$$C = AA^{\mathsf{T}} + \sigma^2 \mathbf{I}_{d\times d},$$

and L is maximised by

$$\widehat{A}_{(ML,p)} = \widehat{\Gamma}_p(\widehat{\Lambda}_p - \widehat{\sigma}^2_{(ML,p)}\mathbf{I}_{p\times p})^{1/2}, \quad \widehat{\sigma}^2_{(ML,p)} = \frac{1}{d-p}\sum_{j>p}\widehat{\lambda}_j \quad \text{and} \quad \widehat{\boldsymbol{\mu}} = \overline{\mathbf{X}}. \tag{2.32}$$

To find the number of latent variables, Tipping and Bishop propose to use

$$p^* = \operatorname{argmax}\ \log L(p).$$

In practice, Tipping and Bishop find the maximiser p^* of the likelihood as follows: Determine the principal components of $\mathbb{X}$. For $p < d$, use the first p principal components, and calculate $\widehat{\sigma}^2_{(\mathrm{ML},p)}$ and $\widehat{A}_{(\mathrm{ML},p)}$ as in (2.32). Put

$$\widehat{C}_p = \widehat{A}_{(\mathrm{ML},p)}\widehat{A}^{\mathsf{T}}_{(\mathrm{ML},p)} + \widehat{\sigma}^2_{(\mathrm{ML},p)}\mathbf{I}_{d\times d}, \tag{2.33}$$

and use $\widehat{C}_p$ instead of C in the calculation of the log likelihood. Once the log likelihood has been calculated for all values of p, find its maximiser p^*.

How does the likelihood approach relate to PCs? The matrix A in the model (2.31) is not specified. For fixed $p \le d$, the choice based on principal components is $A = \widehat{\Lambda}_p^{-1/2}\widehat{\Gamma}^{\mathsf{T}}_p$. With this A and ignoring the error term $\boldsymbol{\epsilon}$, the observations $\mathbf{X}_i$ and the p-dimensional vectors $\mathbf{Z}_i$ are related by

$$\mathbf{Z}_i = \widehat{\Lambda}_p^{-1/2}\widehat{\Gamma}^{\mathsf{T}}_p\mathbf{X}_i \qquad \text{for } i = 1, \ldots, n.$$

From $\mathbf{W}^{(p)}_i = \widehat{\Gamma}^{\mathsf{T}}_p(\mathbf{X}_i - \overline{\mathbf{X}})$, it follows that

$$\mathbf{Z}_i = \widehat{\Lambda}_p^{-1/2}\mathbf{W}^{(p)}_i + \widehat{\Lambda}_p^{-1/2}\widehat{\Gamma}^{\mathsf{T}}_p\overline{\mathbf{X}},$$

so the $\mathbf{Z}_i$ are closely related to the PC scores. The extra term $\widehat{\Lambda}_p^{-1/2}\widehat{\Gamma}^{\mathsf{T}}_p\overline{\mathbf{X}}$ accounts for the lack of centring in Tipping and Bishop's approach. The dimension which maximises the likelihood is therefore a natural candidate for the dimension of the latent variables.

Our next result, which follows from Theorem 2.26, explicitly shows the relationship between $\widehat{C}_p$ of (2.33) and S.

Corollary 2.27 *Assume* $\mathbb{X} = \begin{bmatrix}\mathbf{X}_1\ \mathbf{X}_2 \cdots \mathbf{X}_n\end{bmatrix}$ *and S satisfy the assumptions of Theorem 2.26. Let* $\widehat{A}_{(ML,p)}$ *and* $\widehat{\sigma}_{(ML,p)}$ *be the maximisers of the likelihood as in (2.32). Then, for* $p \le d$,

$$\widehat{C}_p = \widehat{\Gamma}\widehat{\Lambda}_p\widehat{\Gamma}^{\mathsf{T}} + \widehat{\sigma}^2_{(ML,p)} \begin{pmatrix} \mathbf{0}_{p\times p} & \mathbf{0}_{p\times(d-p)} \\ \mathbf{0}_{(d-p)\times p} & \mathbf{I}_{(d-p)\times(d-p)} \end{pmatrix}. \tag{2.34}$$

The proof of the corollary is deferred to the Problems at the end of Part I. The corollary states that S and $\widehat{C}_p$ agree in their upper $p \times p$ submatrices. In the lower right part of $\widehat{C}_p$, the eigenvalues of S have been replaced by sums, thus $\widehat{C}_p$ could be regarded as a stabilised or robust form of S.

A similar framework to that of Tipping and Bishop (1999) has been employed by Minka (2000), who integrates Bayesian ideas and derives a rigorous expression for an approximate likelihood function.

We consider two data sets and calculate the number of principal components with the method of Tipping and Bishop (1999). One of the data sets could be approximately Gaussian, whereas the other is not.

Example 2.17 We continue with the seven-dimensional **abalone** data. The first eigenvalue contributes more than 97 per cent to total variance, and the PC_1 data (not shown) look reasonably normal, so Tipping and Bishop's assumptions apply.

In the analysis I show results for the raw data only; however, I use all 4,177 observations, as well as subsets of the data. In each case I calculate the log likelihood as a function of the index p at the maximiser $\widehat{\theta}_p = (\widehat{A}_{(\mathrm{ML},p)}, \widehat{\boldsymbol{\mu}}, \widehat{\sigma}_{(\mathrm{ML},p)})$ as in Theorem 2.26 and (2.33). Figure 2.20 shows the results with $p \le 6$ on the x-axis. Going from the left to the right panels, we look at the results of all 4,177 observations, then at the first 1,000, then at the first 100, and finally at observations 1,001 to 1,100. The range of the log likelihood – on the y-axes – varies as a consequence of the different values of n and actual observations.

In all four plots, the maximiser $p^* = 1$. One might think that this is a consequence of the high contribution to variance of the first PC. A similar analysis based on the scaled data also leads to $p^* = 1$ as the optimal number of principal components. These results suggest that because the variables are highly correlated, and because PC_1 contributes 90 per cent (for the scaled data) or more to variance, PC_1 captures the essence of the likelihood, and it therefore suffices to represent the data by their first principal component scores. ■

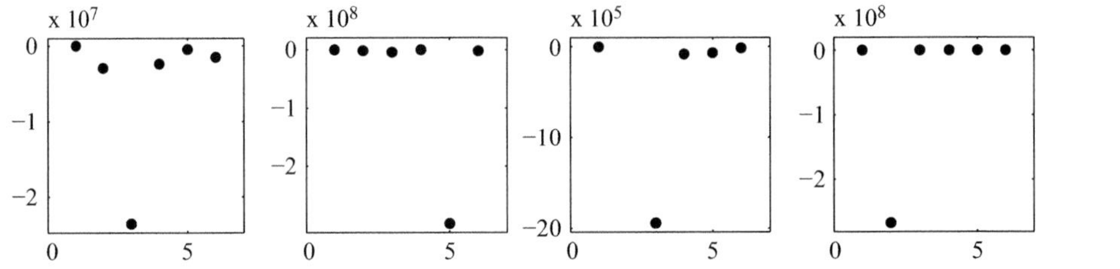

Figure 2.20 Maximum log likelihood versus index of PCs on the x-axis for the abalone data of Example 2.17. All 4,177 observations are used in the left panel, the first 1,000 are used in the second panel, and 100 observations each are used in the two right panels.

Table 2.7 *Indices p^* for the Number of PCs of the Raw and Scaled Breast Cancer Data from Example 2.18*

	Raw Data			Scaled Data		
Observations	1:569	1:300	270:569	1:569	1:300	270:569
p^*	29	29	29	1	1	3

The *abalone* data set has a relatively small number of dimensions and is approximately normal. The method of Tipping and Bishop (1999) has produced appropriate results here. In the next example we return to the thirty-dimensional *breast cancer* data.

Example 2.18 We calculate the dimension p^* for the **breast cancer** data. The top right panel of Figure 2.8 shows that the density estimate of the scaled PC_1 scores has one large and one small mode and is right-skewed, so the PC_1 data deviate considerably from the normal distribution. The scree plot of Figure 2.5 does not provide any information about a suitable number of PCs. Figure 2.10, which compares the raw and scaled data, shows the three large variables in the raw data.

We calculate the maximum log likelihood for $1 \le p \le 29$, as described in Example 2.17, for the scaled and raw data and for the whole sample as well as two subsets of the sample, namely, the first 300 observations and the last 300 observations, respectively. Table 2.7 shows the index p^* for each set of observations.

Unlike the preceding example, the raw data select $p^* = 29$ for the full sample and the two subsamples, whereas the scaled data select $p^* = 1$ for the whole data and the first half and $p^* = 3$ for the second half of the data. Neither choice is convincing, which strongly suggests that Tipping and Bishop's method is not appropriate here. ■

In the last example we used non-normal data. In this case, the raw and scaled data lead to completely different best dimensions – the largest and smallest possible value of p. Although it is not surprising to obtain different values for the 'best' dimensions, these extremes could be due to the strong deviation from normality of the data. The results also indicate that other methods of dimension selection need to be explored which are less dependent on the distribution of the data.

2.8.2 Principal Component Regression

So far we considered the structure within the data and attempted to simplify the data by summarising the original variables in a smaller number of uncorrelated variables. In multivariate Linear Regression we attempt to reduce the number of variables, the 'predictors' rather than combining them; we either keep a variable or discard it, unlike the principal components approach, which looks for the optimal way of combining all variables. It is natural to apply the PC strategy to Linear Regression.

For the population, we model the scalar response variable Y as

$$Y = \beta_0 + \boldsymbol{\beta}^{\mathsf{T}}\mathbf{X} + \epsilon, \tag{2.35}$$

where $\beta_0 \in \mathbb{R}$, $\boldsymbol{\beta} \in \mathbb{R}^d$, $\mathbf{X} \sim (\boldsymbol{\mu}, \Sigma)$ is a d-dimensional random vector, $\epsilon \sim (0, \sigma^2)$ is the measurement error for some $\sigma^2 > 0$ and $\mathbf{X}$ and ϵ are independent of each other. The corresponding model for the sample is

$$\mathbf{Y} = \beta_0 \mathbf{I}_{1\times n} + \boldsymbol{\beta}^{\mathsf{T}}\mathbb{X} + \boldsymbol{\epsilon}, \tag{2.36}$$

where $\mathbf{Y} = \begin{bmatrix} Y_1\, Y_2 \cdots Y_n \end{bmatrix}$ is a $1 \times n$ row vector, $\mathbb{X} = \begin{bmatrix} \mathbf{X}_1\, \mathbf{X}_2 \cdots \mathbf{X}_n \end{bmatrix}$ are $d \times n$ and the $\boldsymbol{\epsilon} \sim (\mathbf{0}, \sigma^2 \mathbf{I}_{n\times n})$ consists of n identically and independently distributed random variables. We can think of normal errors, but we do not require this distributional assumption in our present discussion. The $\mathbb{X}$ consist of random vectors, and we will tacitly assume that $\mathbb{X}$ and $\boldsymbol{\epsilon}$ are independent of each other.

A Word of Caution. I use the notation (2.36) throughout this book. This notation differs from many regression settings in that our response vector $\mathbf{Y}$ is a row vector. The results are unaffected by this change in notation.

For notational convenience, we assume that $\beta_0 = 0$ and that the columns of $\mathbb{X}$ have mean zero. In practice, we can consider centred response variables, which eliminates the need for the intercept term β_0.

From observations $\begin{bmatrix} \mathbb{X} \\ \mathbf{Y} \end{bmatrix}$ one finds an estimator $\widehat{\boldsymbol{\beta}}$ for $\boldsymbol{\beta}$ in (2.36). The **least squares estimator** $\widehat{\boldsymbol{\beta}}_{\mathrm{LS}}^{\mathsf{T}}$ and the fitted $\widehat{\mathbf{Y}}$ are

$$\widehat{\boldsymbol{\beta}}_{\mathrm{LS}}^{\mathsf{T}} = \mathbf{Y}\mathbb{X}^{\mathsf{T}}(\mathbb{X}\mathbb{X}^{\mathsf{T}})^{-1}$$

and

$$\widehat{\mathbf{Y}} = \widehat{\boldsymbol{\beta}}_{\mathrm{LS}}^{\mathsf{T}}\mathbb{X} = \mathbf{Y}\mathbb{X}^{\mathsf{T}}(\mathbb{X}\mathbb{X}^{\mathsf{T}})^{-1}\mathbb{X}. \tag{2.37}$$

To appreciate these definitions, assume first that $\mathbf{Y} = \boldsymbol{\beta}^{\mathsf{T}}\mathbb{X}$ and that $\mathbb{X}\mathbb{X}^{\mathsf{T}}$ is invertible. The following steps 1 to 3 show how to obtain $\widehat{\mathbf{Y}}$, but they are not a proof that the resulting $\widehat{\mathbf{Y}}$ is a least squares estimator.

1. Because $\mathbb{X}$ is not a square matrix, post-multiply both sides by $\mathbb{X}^{\mathsf{T}}$: $\mathbf{Y}\mathbb{X}^{\mathsf{T}} = \boldsymbol{\beta}^{\mathsf{T}}\mathbb{X}\mathbb{X}^{\mathsf{T}}$.
2. Apply the $d \times d$ matrix inverse $(\mathbb{X}\mathbb{X}^{\mathsf{T}})^{-1}$ to both sides $\mathbf{Y}\mathbb{X}^{\mathsf{T}}\left(\mathbb{X}\mathbb{X}^{\mathsf{T}}\right)^{-1} = \boldsymbol{\beta}^{\mathsf{T}}\mathbb{X}\mathbb{X}^{\mathsf{T}}\left(\mathbb{X}\mathbb{X}^{\mathsf{T}}\right)^{-1}$.
3. Put

$$\widehat{\boldsymbol{\beta}}^{\mathsf{T}} = \mathbf{Y}\mathbb{X}^{\mathsf{T}}\left(\mathbb{X}\mathbb{X}^{\mathsf{T}}\right)^{-1} \quad \text{and} \quad \widehat{\mathbf{Y}} = \mathbf{Y}\mathbb{X}^{\mathsf{T}}\left(\mathbb{X}\mathbb{X}^{\mathsf{T}}\right)^{-1}\mathbb{X}. \tag{2.38}$$

For measurements $\mathbf{Y}$ as in (2.36), we calculate the least squares estimator $\widehat{\boldsymbol{\beta}}$ and the 'fitted' $\widehat{\mathbf{Y}}$ directly from the $\mathbb{X}$, as described in step 3.

I have introduced the least squares solution because it is a basic solution which will allow us to understand how principal components fit into this framework. Another, and more stable, estimator is the *ridge* estimator $\widehat{\boldsymbol{\beta}}_{\mathrm{ridge}}$ which is related to $\widehat{\boldsymbol{\beta}} = \widehat{\boldsymbol{\beta}}_{\mathrm{LS}}$, namely,

$$\widehat{\boldsymbol{\beta}}_{\mathrm{ridge}}^{\mathsf{T}} = \mathbf{Y}\mathbb{X}^{\mathsf{T}}(\mathbb{X}\mathbb{X}^{\mathsf{T}} + \zeta \mathbf{I}_{d\times d})^{-1}, \tag{2.39}$$

where $\zeta \geq 0$ is a penalty parameter which controls the solution. For details of the ridge solution, see section 3.4.3 in Hastie, Tibshirani, and Friedman (2001). Other regression estimators, and in particular, the LASSO estimator of Tibshirani (1996), could be used instead of the least squares estimator. We briefly consider the LASSO in Section 13.4; I will explain then how the LASSO can be used to reduce the number of non-zero weights in Principal Component Analysis.

Remark. In regression, one makes the distinction between estimation and prediction: $\widehat{\boldsymbol{\beta}}$ is an estimator of $\boldsymbol{\beta}$, and this estimator is used in two ways:

- to calculate the fitted $\widehat{\mathbf{Y}}$ from $\mathbb{X}$ as in (2.38); and
- to predict a new response $\widehat{Y}_{\text{new}}$ from a new datum $\mathbf{X}_{\text{new}}$.

In prediction a new response $\widehat{Y}_{\text{new}}$ is obtained from the current $\widehat{\boldsymbol{\beta}}^{\mathsf{T}}$ of (2.38), and the new $\mathbf{X}_{\text{new}}$

$$\widehat{Y}_{\text{new}} = \widehat{\boldsymbol{\beta}}^{\mathsf{T}} \mathbf{X}_{\text{new}} = \mathbf{Y}\mathbb{X}^{\mathsf{T}} \left(\mathbb{X}\mathbb{X}^{\mathsf{T}}\right)^{-1} \mathbf{X}_{\text{new}} \tag{2.40}$$

and $\mathbb{X}$ are the original $d \times n$ data that are used to derive $\widehat{\boldsymbol{\beta}}$.

To find a parsimonious solution $\widehat{\boldsymbol{\beta}}$, we require a good technique for selecting the variables which are important for prediction. The least squares solution in its basic form does not include a selection of variables – unlike the ridge regression solution or the LASSO which *shrink* the number of variables – see section 3.4.3 in Hastie, Tibshirani, and Friedman (2001).

To include variable selection into the basic regression solution, we apply ideas from Principal Component Analysis. Assume that the data $\mathbb{X}$ and $\mathbf{Y}$ are centred. Let $S = \widehat{\Gamma}\widehat{\Lambda}\widehat{\Gamma}^{\mathsf{T}}$ be the sample covariance matrix of $\mathbb{X}$. Fix $\kappa \le r$, where r is the rank of S. Consider the principal component data $\mathbb{W}^{(\kappa)} = \widehat{\Gamma}_{\kappa}^{\mathsf{T}}\mathbb{X}$, which we call **derived variables** or **derived predictors**, and put $\mathbb{P}^{(\kappa)} = \widehat{\Gamma}_{\kappa}\widehat{\Gamma}_{\kappa}^{\mathsf{T}}\mathbb{X} = \widehat{\Gamma}_{\kappa}\mathbb{W}^{(\kappa)}$. By Corollary 2.13, $\mathbb{P}^{(\kappa)}$ approximate $\mathbb{X}$.

There are two natural ways of selecting variables, as illustrated in the following diagram. The first downward arrows in the diagram point to the new predictor variables, and the second downward arrows give expressions for $\widehat{\mathbf{Y}}$ in terms of the new predictors. For a fixed κ, the two branches in the diagram result in the same estimator $\widehat{\mathbf{Y}}$, and I therefore only discuss one branch.

$$\begin{array}{ccc} & \mathbf{Y} = \boldsymbol{\beta}^{\mathsf{T}}\mathbb{X} + \boldsymbol{\epsilon} & \\ \swarrow & & \searrow \\ \mathbb{X} \to \mathbb{W}^{(\kappa)} = \widehat{\Gamma}_{\kappa}^{\mathsf{T}}\mathbb{X} & & \mathbb{X} \approx \mathbb{P}^{(\kappa)} = \widehat{\Gamma}_{\kappa}\widehat{\Gamma}_{\kappa}^{\mathsf{T}}\mathbb{X} \\ \downarrow & & \downarrow \\ \widehat{\mathbf{Y}} = \widehat{\boldsymbol{\beta}}_W^{\mathsf{T}}\mathbb{W}^{(\kappa)} & & \widehat{\mathbf{Y}} = \widehat{\boldsymbol{\beta}}_P^{\mathsf{T}}\mathbb{P}^{(\kappa)} \end{array}$$

We follow the left branch and leave similar calculations along the right branch for the Problems at the end of Part I. We replace $\mathbb{X}$ by $\mathbb{W}^{(\kappa)}$ and then calculate the least squares estimator directly from $\mathbb{W}^{(\kappa)}$. From (2.38), we get

$$\widehat{\boldsymbol{\beta}}_W = \mathbf{Y}\mathbb{W}^{(\kappa)^{\mathsf{T}}}\left[\mathbb{W}^{(\kappa)}\mathbb{W}^{(\kappa)^{\mathsf{T}}}\right]^{-1}. \tag{2.41}$$

Observe that

$$\begin{aligned} \mathbb{W}^{(\kappa)}\mathbb{W}^{(\kappa)^{\mathsf{T}}} &= \widehat{\Gamma}_{\kappa}^{\mathsf{T}}\mathbb{X}\mathbb{X}^{\mathsf{T}}\widehat{\Gamma}_{\kappa} = (n-1)\widehat{\Gamma}_{\kappa}^{\mathsf{T}}S\widehat{\Gamma}_{\kappa} \\ &= (n-1)\widehat{\Gamma}_{\kappa}^{\mathsf{T}}\widehat{\Gamma}\widehat{\Lambda}\widehat{\Gamma}^{\mathsf{T}}\widehat{\Gamma}_{\kappa} = (n-1)\widehat{\Lambda}_{\kappa} \end{aligned} \tag{2.42}$$

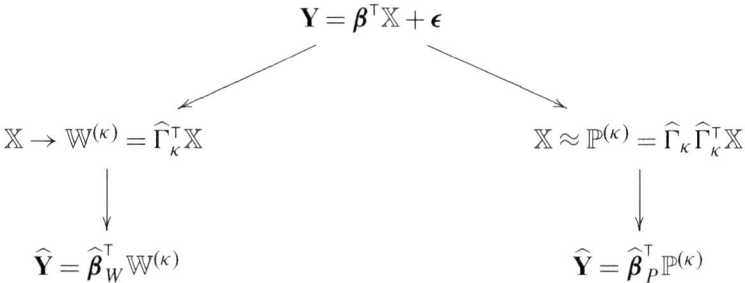

because $\widehat{\Gamma}_\kappa^{\mathsf{T}}\widehat{\Gamma} = \mathbf{I}_{\kappa\times d}$. Substituting (2.42) into (2.41) leads to

$$\widehat{\boldsymbol{\beta}}_W^{\mathsf{T}} = (n-1)^{-1}\mathbf{Y}\mathbb{X}^{\mathsf{T}}\widehat{\Gamma}_\kappa\widehat{\Lambda}_\kappa^{-1}$$

and

$$\begin{aligned}\widehat{\mathbf{Y}} = \widehat{\boldsymbol{\beta}}_W^{\mathsf{T}}\mathbb{W}^{(\kappa)} &= (n-1)^{-1}\mathbf{Y}\mathbb{W}^{(\kappa)^{\mathsf{T}}}\widehat{\Lambda}_\kappa^{-1}\mathbb{W}^{(\kappa)} \\ &= (n-1)^{-1}\mathbf{Y}\mathbb{W}_\Lambda^{(\kappa)^{\mathsf{T}}}\mathbb{W}_\Lambda^{(\kappa)},\end{aligned} \tag{2.43}$$

where $\mathbb{W}_\Lambda^{(\kappa)} = \widehat{\Lambda}_\kappa^{-1/2}\mathbb{W}^{(\kappa)}$. For $j = 1, \ldots, \kappa$, the jth entry of $\widehat{\boldsymbol{\beta}}_W$ is

$$\widehat{\beta}_{W,j} = \frac{\left(\sum_{i=1}^n Y_i W_{ij}\right)}{(n-1)\widehat{\lambda}_j},$$

with W_{ij} the ijth entry of $\mathbb{W}^{(\kappa)}$.

We call the estimation of $\widehat{\boldsymbol{\beta}}_W^{\mathsf{T}}$ and $\widehat{\mathbf{Y}}$ from $\mathbb{W}^{(\kappa)}$ as in (2.43) **Principal Component Regression**. Both $\widehat{\boldsymbol{\beta}}_W^{\mathsf{T}}$ and $\widehat{\mathbf{Y}}$ vary with κ, and we therefore need to specify a value for κ. For some applications, $\kappa = 1$ suffices, but in general, the choice of κ is not uniquely defined; the problem of finding the 'best' dimension in a principal component analysis reappears here in a different guise. A few comments are in order about the type of solution and the number of variables.

1. The predictors $\mathbb{W}^{(\kappa)}$ are uncorrelated – this follows from Theorem 2.5 – whereas the original variables $\mathbb{X}$ are correlated.
2. Typically, $\kappa < r$, so the principal component approach leads to a reduction in the number of variables.
3. If the contribution to variance of PC_1 is sufficiently large compared with the total variance, then (2.43) with $\kappa = 1$ reduces to

$$\widehat{\mathbf{Y}} = \widehat{\beta}_W^{\mathsf{T}}\mathbb{W}^{(1)} = \frac{1}{(n-1)\widehat{\lambda}_1}\mathbf{Y}\mathbf{W}_{\bullet 1}^{\mathsf{T}}\mathbf{W}_{\bullet 1}. \tag{2.44}$$

4. If some of the eigenvalues are close to zero but need to be retained in the regression solution, a more stable solution similar to (2.39) can be integrated into the expression for $\widehat{\beta}_W$.
5. For $\kappa = r$, the PC solution is the least squares solution (2.37) as a comparison of (2.37) and (2.43) shows.

The expression (2.44) is of special interest. It is often used in applications because PC_1 is a weighted sum of all variables.

Example 2.19 We consider all eight variables of the **abalone** data. The eighth variable, which we previously ignored, contains the number of rings, and these are regarded as a measure of the age of abalone. If abalone are fished when they are too young, this may lead to a later shortage.

The age of abalone is determined by cutting the shell through the cone, staining it, and counting the number of rings through a microscope – a tedious and time-consuming task. Predicting the age accurately from the other measurements is therefore a sensible approach.

I use the raw data for this analysis with PC_1 as the derived predictor because it contributes more than 97 per cent to total variance. The choice of one derived predictor is supported by the likelihood results in Figure 2.20. Put

$$\widehat{\mathbf{Y}} = \widehat{\beta}_W^{\mathsf{T}} \mathbf{W}_{\bullet 1}.$$

Table 2.6 of Example 2.13 shows that the fourth variable, *whole weight*, has the largest eigenvector weight, namely, 0.8426. The other variables have much smaller absolute weights for PC_1. We could calculate the error between $\mathbf{Y}$ and the fitted $\widehat{\mathbf{Y}}$, and I will do so in the continuation of this example in Section 3.7, when I compare the performance of Principal Component Regression with that of other techniques. Suffice it to say that the error between $\mathbf{Y}$ and $\widehat{\mathbf{Y}}$ decreases if we use more than one derived predictor, and for $\kappa = 3$, the error is smaller than that obtained from the single best least squares predictor. ■

It may seem surprising that Principal Component Regression with PC_1 performs worse than Linear Regression with its best single predictor. We will examine this behaviour further in Section 3.7 and more fully in Section 13.3.

With the growing interest in high-dimensional data, the principal component approach to regression has gained renewed interest. An inspection of (2.38) shows why we need to consider other methods when the dimension becomes larger than the sample size: for $d > n$, $\mathbb{X}\mathbb{X}^{\mathsf{T}}$ is singular. Bair et al. (2006) and references therein have used a principal component approach for latent variable models. For $\begin{bmatrix} \mathbb{X} \\ \mathbf{Y} \end{bmatrix}$ data, Bair et al. (2006) assume that the responses $\mathbf{Y}$ are related to $\mathbb{X}$ through an underlying model of the form

$$\mathbf{Y} = \beta_0 + \boldsymbol{\beta}^{\mathsf{T}} \mathbb{F} + \boldsymbol{\epsilon}, \tag{2.45}$$

where $\mathbb{F}$ is a matrix of latent variables which are linked to a subset of the variables in $\mathbb{X}$ via

$$\mathbf{X}_{\bullet j} = \alpha_{0j} + \boldsymbol{\alpha}_{1j} \mathbb{F} + \boldsymbol{\varepsilon} \qquad \text{for some } j. \tag{2.46}$$

A goal is to identify the variables of $\mathbb{X}$ which contribute to (2.46) and to determine the number of latent variables for the prediction model (2.45). We return to this topic in Section 13.3.

3

Canonical Correlation Analysis

Alles Gescheite ist schon gedacht worden, man muß nur versuchen, es noch einmal zu denken (Johann Wolfgang von Goethe, Wilhelm Meisters Wanderjahre, 1749–1832.)
Every clever thought has been thought before, we can only try to recreate these thoughts.

3.1 Introduction

In Chapter 2 we represented a random vector as a linear combination of uncorrelated vectors. From one random vector we progress to two vectors, but now we look for correlation between the variables of the first and second vectors, and in particular, we want to find out which variables are correlated and how strong this relationship is.

In medical diagnostics, for example, we may meet multivariate measurements obtained from tissue and plasma samples of patients, and the tissue and plasma variables typically differ. A natural question is: What is the relationship between the tissue measurements and the plasma measurements? A strong relationship between a combination of tissue variables and a combination of plasma variables typically indicates that either set of measurements could be used for a particular diagnosis. A very weak relationship between the plasma and tissue variables tells us that the sets of variables are not equally appropriate for a particular diagnosis.

On the share market, one might want to compare changes in the price of industrial shares and mining shares over a period of time. The time points are the observations, and for each time point, we have two sets of variables: those arising from industrial shares and those arising from mining shares. We may want to know whether the industrial and mining shares show a similar growth pattern over time. In these scenarios, two sets of observations exist, and the data fall into two parts; each part consists of the same n objects, such as people, time points, or locations, but the variables in the two parts or sets differ, and the number of variables is typically not the same in the two parts.

To find high correlation between two sets of variables, we consider linear combinations of variables and then ask the questions:

1. How should one choose these combinations?
2. How many such combinations should one consider?

We do not ask, 'Is there a relationship?' Instead, we want to find the combinations of variables separately for each of the two sets of variables which exhibit the strongest correlation between the two sets of variables. The second question can now be re-phrased: How many such combinations are correlated enough?

If one of the two sets of variables consists of a single variable, and if the single variable is linearly related to the variables in the other set, then Multivariate Linear Regression can be used to determine the nature and strength of the relationship. The two methods, Canonical Correlation Analysis and Linear Regression, agree for this special setting. The two methods are closely related but differ in a number of aspects:

- Canonical Correlation Analysis exhibits a symmetric relationship between the two sets of variables, whereas Linear Regression focuses on predictor variables and response variables, and the roles of predictors and responses cannot be reversed in general.
- Canonical Correlation Analysis determines *optimal* combinations of variables simultaneously for both sets of variables and finds the strongest overall relationship. In Linear Regression with vector-valued responses, relationships between each scalar response variable and combinations of the predictor variables are determined separately for each component of the response.

The pioneering work of (Hotelling 1935, 1936) paved the way in this field. Hotelling published his seminal work on Principal Component Analysis in 1933, and only two years later his next big advance in multivariate analysis followed!

In this chapter we consider different subsets of the variables, and *parts* of the data will refer to the subsets of variables. The chapter describes how Canonical Correlation Analysis finds combinations of variables of two vectors or two parts of data that are more strongly correlated than the original variables. We begin with the population case in Section 3.2 and consider the sample case in Section 3.3. We derive properties of canonical correlations in Section 3.4. Transformations of variables are frequently encountered in practice. We investigate correlation properties of transformed variables in Section 3.5 and distinguish between transformations resulting from singular and non-singular matrices. Maximum Covariance Analysis is also mentioned in this section. In Section 3.6 we briefly touch on asymptotic results for multivariate normal random vectors and then consider hypothesis tests of the strength of the correlation. Section 3.7 compares Canonical Correlation Analysis with Linear Regression and Partial Least Squares and shows that these three approaches are special instances of the generalised eigenvalue problems. Problems pertaining to the material of this chapter are listed at the end of Part I.

A Word of Caution. The definitions I give in this chapter use the centred (rather than the raw) data and thus differ from the usual treatment. Centring the data is a natural first step in many analyses. Further, working with the centred data will make the theoretical underpinnings of this topic more obviously compatible with those of Principal Component Analysis. The basic building stones are covariance matrices, and for this reason, my approach does not differ much from that of other authors. In particular, properties relating to canonical correlations are not affected by using centred vectors and centred data.

3.2 Population Canonical Correlations

The main goal of this section is to define the key ideas of Canonical Correlation Analysis for random vectors. Because we are dealing with two random vectors rather than a single vector, as in Principal Component Analysis, the matrices which link the two vectors are important. We begin with properties of matrices based on Definition 1.12 and Result 1.13 of

Section 1.5.3 and exploit an important link between the singular value decomposition and the spectral decomposition.

Proposition 3.1 *Let A be a $p \times q$ matrix with rank r and singular value decomposition $A = E\Lambda F^{\top}$. Put $B = AA^{\top}$ and $K = A^{\top}A$. Then*

1. *the matrices B and K have rank r;*
2. *the spectral decompositions of B and K are*
$$B = EDE^{\top} \qquad \text{and} \qquad K = FDF^{\top},$$
where $D = \Lambda^2$; and
3. *the eigenvectors $\mathbf{e}_k$ of B and $\mathbf{f}_k$ of K satisfy*
$$A\mathbf{f}_k = \lambda_k \mathbf{e}_k \qquad \text{and} \qquad A^{\top}\mathbf{e}_k = \lambda_k \mathbf{f}_k,$$
where the λ_k are the diagonal elements of Λ, for $k = 1, \ldots, r$.

Proof Because r is the rank of A, the $p \times r$ matrix E consists of the left eigenvectors of A, and the $q \times r$ matrix F consists of the right eigenvectors of A. The diagonal matrix Λ is of size $r \times r$. From this and the definition of B, it follows that

$$B = AA^{\top} = E\Lambda F^{\top} F \Lambda E^{\top} = E\Lambda^2 E^{\top}$$

because $\Lambda = \Lambda^{\top}$ and $F^{\top}F = \mathbf{I}_{r\times r}$. By the uniqueness of the spectral decomposition, it follows that E is the matrix of the first r eigenvectors of B, and $D = \Lambda^2$ is the matrix of eigenvalues λ_k^2 of B, for $k \leq r$. A similar argument applies to K. It now follows that B and K have the same rank and the same eigenvalues. The proof of part 3 is considered in the Problems at the end of Part I. ■

We consider two random vectors $\mathbf{X}^{[1]}$ and $\mathbf{X}^{[2]}$ such that $\mathbf{X}^{[\rho]} \sim (\boldsymbol{\mu}_\rho, \Sigma_\rho)$, for $\rho = 1, 2$. Throughout this chapter, $\mathbf{X}^{[\rho]}$ will be a d_ρ-dimensional random vector, and $d_\rho \geq 1$. Unless otherwise stated, we assume that the covariance matrix Σ_ρ has rank d_ρ. In the random-sample case, which I describe in the next section, the observations $\mathbf{X}_i^{[\rho]}$ (for $i = 1, \ldots, n$) are d_ρ-dimensional random vectors, and their sample covariance matrix has rank d_ρ. We write

$$\mathbf{X} = \begin{bmatrix} \mathbf{X}^{[1]} \\ \mathbf{X}^{[2]} \end{bmatrix} \sim (\boldsymbol{\mu}, \Sigma),$$

for the d-dimensional random vector $\mathbf{X}$, with $d = d_1 + d_2$, mean $\boldsymbol{\mu} = \begin{bmatrix} \boldsymbol{\mu}_1 \\ \boldsymbol{\mu}_2 \end{bmatrix}$ and covariance matrix

$$\Sigma = \begin{pmatrix} \Sigma_1 & \Sigma_{12} \\ \Sigma_{12}^{\top} & \Sigma_2 \end{pmatrix}. \tag{3.1}$$

To distinguish between the different submatrices of Σ in (3.1), we call the $d_1 \times d_2$ matrix Σ_{12} the **between covariance matrix** of the vectors $\mathbf{X}^{[1]}$ and $\mathbf{X}^{[2]}$. Let r be the rank of Σ_{12}, so $r \leq \min(d_1, d_2)$. In general, $d_1 \neq d_2$, as the dimensions reflect specific measurements, and there is no reason why the number of measurements in $\mathbf{X}^{[1]}$ and $\mathbf{X}^{[2]}$ should be the same.

We extend the notion of projections onto direction vectors from Principal Component Analysis to Canonical Correlation Analysis, which has pairs of vectors. For $\rho = 1,2$, let $\mathbf{X}_\Sigma^{[\rho]} = \Sigma_\rho^{-1/2}\left(\mathbf{X}^{[\rho]} - \boldsymbol{\mu}_\rho\right)$ be the sphered vectors. For $k = 1,\ldots,r$, let $\mathbf{a}_k \in \mathbb{R}^{d_1}$ and $\mathbf{b}_k \in \mathbb{R}^{d_2}$ be unit vectors, and put

$$U_k = \mathbf{a}_k^{\top}\mathbf{X}_\Sigma^{[1]} \qquad \text{and} \qquad V_k = \mathbf{b}_k^{\top}\mathbf{X}_\Sigma^{[2]}. \tag{3.2}$$

The aim is to find direction vectors $\mathbf{a}_k$ and $\mathbf{b}_k$ such that the dependence between U_k and V_k is strongest for the pair (U_1, V_1) and decreases with increasing index k. We will measure the strength of the relationship by the absolute value of the covariance, so we require that

$$|\operatorname{cov}(U_1, V_1)| \geq |\operatorname{cov}(U_2, V_2)| \geq \cdots \geq |\operatorname{cov}(U_r, V_r)| > 0.$$

The between covariance matrix Σ_{12} of (3.1) is the link between the vectors $\mathbf{X}^{[1]}$ and $\mathbf{X}^{[2]}$. It turns out to be more useful to consider a *standardised* version of this matrix rather than the matrix itself.

Definition 3.2 Let $\mathbf{X}^{[1]} \sim (\boldsymbol{\mu}_1, \Sigma_1)$ and $\mathbf{X}^{[2]} \sim (\boldsymbol{\mu}_2, \Sigma_2)$, and assume that Σ_1 and Σ_2 are invertible. Let Σ_{12} be the between covariance matrix of $\mathbf{X}^{[1]}$ and $\mathbf{X}^{[2]}$. The **matrix of canonical correlations** or the **canonical correlation matrix** is

$$C = \Sigma_1^{-1/2}\Sigma_{12}\Sigma_2^{-1/2}, \tag{3.3}$$

and the **matrices of multivariate coefficients of determination** are

$$R^{[C,1]} = CC^{\top} \qquad \text{and} \qquad R^{[C,2]} = C^{\top}C. \tag{3.4}$$

□

In Definition 3.2 we assume that the covariance matrices Σ_1 and Σ_2 are invertible. If Σ_1 is singular with rank r, then we may want to replace Σ_1 by its spectral decomposition $\Gamma_{1,r}\Lambda_{1,r}\Gamma_{1,r}^{\top}$ and $\Sigma_1^{-1/2}$ by $\Gamma_{1,r}\Lambda_{1,r}^{-1/2}\Gamma_{1,r}^{\top}$ and similarly for Σ_2.

A little reflection reveals that C is a generalisation of the univariate correlation coefficient to multivariate random vectors. Theorem 2.17 of Section 2.6.1 concerns the matrix of correlation coefficients R, which is the covariance matrix of the scaled vector $\mathbf{X}_{\text{scale}}$. The entries of this $d \times d$ matrix R are the correlation coefficients arising from pairs of variables of $\mathbf{X}$. The canonical correlation matrix C compares each variable of $\mathbf{X}^{[1]}$ with each variable of $\mathbf{X}^{[2]}$ because the focus is on the between covariance or the correlation between $\mathbf{X}^{[1]}$ and $\mathbf{X}^{[2]}$. So C has $d_1 \times d_2$ entries.

We may interpret the matrices $R^{[C,1]}$ and $R^{[C,2]}$ as two natural generalisations of the coefficient of determination in Linear Regression; $R^{[C,1]} = \Sigma_1^{-1/2}\Sigma_{12}\Sigma_2^{-1}\Sigma_{12}^{\top}\Sigma_1^{-1/2}$ is of size $d_1 \times d_1$, whereas $R^{[C,2]} = \Sigma_2^{-1/2}\Sigma_{12}^{\top}\Sigma_1^{-1}\Sigma_{12}\Sigma_2^{-1/2}$ is of size $d_2 \times d_2$.

The matrix C is the object which connects the two vectors $\mathbf{X}^{[1]}$ and $\mathbf{X}^{[2]}$. Because the dimensions of $\mathbf{X}^{[1]}$ and $\mathbf{X}^{[2]}$ differ, C is not a square matrix. Let

$$C = P\Upsilon Q^{\top}$$

be its singular value decomposition. By Proposition 3.1, $R^{[C,1]}$ and $R^{[C,2]}$ have the spectral decompositions

$$R^{[C,1]} = P\Upsilon^2 P^{\top} \qquad \text{and} \qquad R^{[C,2]} = Q\Upsilon^2 Q^{\top},$$

where Υ^2 is diagonal with diagonal entries $\upsilon_1^2 \ge \upsilon_2^2 \ge \cdots \ge \upsilon_r^2 > 0$. For $k \le r$, we write $\mathbf{p}_k$ for the eigenvectors of $R^{[C,1]}$ and $\mathbf{q}_k$ for the eigenvectors of $R^{[C,2]}$, so

$$P = \begin{bmatrix}\mathbf{p}_1\ \mathbf{p}_2 \cdots \mathbf{p}_r\end{bmatrix} \qquad \text{and} \qquad Q = \begin{bmatrix}\mathbf{q}_1\ \mathbf{q}_2 \cdots \mathbf{q}_r\end{bmatrix}.$$

The eigenvectors $\mathbf{p}_k$ and $\mathbf{q}_k$ satisfy

$$C\mathbf{q}_k = \upsilon_k \mathbf{p}_k \qquad \text{and} \qquad C^{\mathsf{T}}\mathbf{p}_k = \upsilon_k \mathbf{q}_k, \tag{3.5}$$

and because of this relationship, we call them the **left** and **right eigenvectors of** C. See Definition 1.12 in Section 1.5.3.

Throughout this chapter I use the submatrix notation (1.21) of Section 1.5.2 and so will write Q_m for the $m \times r$ submatrix of Q, where $m \le r$, and similarly for other submatrices.

We are now equipped to define the canonical correlations.

Definition 3.3 Let $\mathbf{X}^{[1]} \sim (\boldsymbol{\mu}_1, \Sigma_1)$ and $\mathbf{X}^{[2]} \sim (\boldsymbol{\mu}_2, \Sigma_2)$. For $\rho = 1, 2$, let $\mathbf{X}_\Sigma^{[\rho]}$ be the sphered vector

$$\mathbf{X}_\Sigma^{[\rho]} = \Sigma_\rho^{-1/2}\left(\mathbf{X}^{[\rho]} - \boldsymbol{\mu}_\rho\right).$$

Let Σ_{12} be the between covariance matrix of $\mathbf{X}^{[1]}$ and $\mathbf{X}^{[2]}$. Let C be the matrix of canonical correlations of $\mathbf{X}^{[1]}$ and $\mathbf{X}^{[2]}$, and write $C = P\Upsilon Q^{\mathsf{T}}$ for its singular value decomposition. Consider $k = 1, \ldots, r$.

1. The kth **pair of canonical correlation scores** or **canonical variates** is

$$U_k = \mathbf{p}_k^{\mathsf{T}}\mathbf{X}_\Sigma^{[1]} \qquad \text{and} \qquad V_k = \mathbf{q}_k^{\mathsf{T}}\mathbf{X}_\Sigma^{[2]}; \tag{3.6}$$

2. the k-dimensional **pair of vectors of canonical correlations** or **vectors of canonical variates** is

$$\mathbf{U}^{(k)} = \begin{bmatrix} U_1 \\ \vdots \\ U_k \end{bmatrix} \qquad \text{and} \qquad \mathbf{V}^{(k)} = \begin{bmatrix} V_1 \\ \vdots \\ V_k \end{bmatrix}; \tag{3.7}$$

3. and the kth **pair of canonical (correlation) transforms** is

$$\boldsymbol{\varphi}_k = \Sigma_1^{-1/2}\mathbf{p}_k \qquad \text{and} \qquad \boldsymbol{\psi}_k = \Sigma_2^{-1/2}\mathbf{q}_k. \tag{3.8}$$

□

For brevity, I sometimes refer to the pair of canonical correlations scores or vectors as *CC scores*, or *vectors of CC scores* or simply as *canonical correlations*.

I remind the reader that our definitions of the canonical correlation scores and vectors use the centred data, unlike other treatments of this topic, which use uncentred vectors. It is worth noting that the canonical correlation transforms of (3.8) are, in general, **not** unit vectors because they are linear transforms of unit vectors, namely, the eigenvectors $\mathbf{p}_k$ and $\mathbf{q}_k$.

The canonical correlation scores are also called the **canonical (correlation) variables**. Mardia, Kent, and Bibby (1992) use the term *canonical correlation vectors* for the $\boldsymbol{\varphi}_k$ and $\boldsymbol{\psi}_k$ of (3.8). To distinguish the vectors of canonical correlations $\mathbf{U}^{(k)}$ and $\mathbf{V}^{(k)}$ of part 2 of the definition from the φ_k and ψ_k, I prefer the term *transforms* for the φ_k and ψ_k because these vectors are transformations of the directions $\mathbf{p}_k$ and $\mathbf{q}_k$ and result in the pair of scores

$$U_k = \varphi_k^{\mathsf{T}}(\mathbf{X}^{[1]} - \mu_1) \qquad \text{and} \qquad V_k = \psi_k^{\mathsf{T}}(\mathbf{X}^{[2]} - \mu_2) \qquad \text{for } k = 1,\ldots,r. \tag{3.9}$$

At times, I refer to a specific pair of transforms, but typically we are interested in the first p pairs with $p \le r$. We write

$$\Phi = \begin{bmatrix}\varphi_1 \; \varphi_2 \cdots \varphi_r\end{bmatrix} \qquad \text{and} \qquad \Psi = \begin{bmatrix}\psi_1 \; \psi_2 \cdots \psi_r\end{bmatrix}$$

for the matrices of canonical correlation transforms. The entries of the vectors φ_k and ψ_k are the weights of the variables of $\mathbf{X}^{[1]}$ and $\mathbf{X}^{[2]}$ and so show which variables contribute strongly to correlation and which might be negligible. Some authors, including Mardia, Kent, and Bibby (1992), define the CC scores as in (3.9). Naively, one might think of the vectors φ_k and ψ_k as sphered versions of the eigenvectors $\mathbf{p}_k$ and $\mathbf{q}_k$, but this is incorrect; Σ_1 is the covariance matrix of $\mathbf{X}^{[1]}$, and $\mathbf{p}_k$ is the kth eigenvector of the non-random CC^{T}.

I prefer the definition (3.6) to (3.9) for reasons which are primarily concerned with the interpretation of the results, namely,

1. the vectors $\mathbf{p}_k$ and $\mathbf{q}_k$ are unit vectors, and their entries are therefore easy to interpret, and
2. the scores are given as linear combinations of uncorrelated random vectors, the sphered vectors $\mathbf{X}_\Sigma^{[1]}$ and $\mathbf{X}_\Sigma^{[2]}$. Uncorrelated variables are more amenable to an interpretation of the contribution of each variable to the correlation between the pairs U_k and V_k.

Being eigenvectors, the $\mathbf{p}_k$ and $\mathbf{q}_k$ values play a natural role as directions, and they are some of the key quantities when dealing with correlation for transformed random vectors in Section 3.5.2 as well as in the variable ranking based on the correlation matrix which I describe in Section 13.3.

The canonical correlation scores (3.6) and vectors (3.7) play a role similar to the PC scores and vectors in Principal Component Analysis, and the vectors $\mathbf{p}_k$ and $\mathbf{q}_k$ remind us of the eigenvector η_k. However, there is a difference: Principal Component Analysis is based on the raw or scaled data, whereas the vectors $\mathbf{p}_k$ and $\mathbf{q}_k$ relate to the sphered data. This difference is exhibited in (3.6) but is less apparent in (3.9). The explicit nature of (3.6) is one of the reasons why I prefer (3.6) as the definition of the scores.

Before we leave the population case, we compare the between covariance matrix Σ_{12} and the canonical correlation matrix in a specific case.

Example 3.1 The **car** data is a subset of the 1983 ASA Data Exposition of Ramos and Donoho (1983). We use their five continuous variables: *displacement*, *horsepower*, *weight*, *acceleration* and *miles per gallon* (*mpg*). The first three variables correspond to physical properties of the cars, whereas the remaining two are performance-related. We combine the first three variables into one part and the remaining two into the second part. The random vectors $\mathbf{X}_i^{[1]}$ have the variables *displacement*, *horsepower*, and *weight*, and the random vectors $\mathbf{X}_i^{[2]}$ have the variables *acceleration* and *mpg*. We consider the sample variance and covariance matrix of the $\mathbf{X}^{[\rho]}$ in lieu of the respective population quantities. We obtain the 3×2 matrices

$$\Sigma_{12} = \begin{pmatrix} -157.0 & -657.6 \\ -73.2 & -233.9 \\ -976.8 & -5517.4 \end{pmatrix}$$

and

$$C = \begin{pmatrix} -0.3598 & -0.1131 \\ -0.5992 & -0.0657 \\ -0.1036 & -0.8095 \end{pmatrix}. \tag{3.10}$$

Both matrices have negative entries, but the entries are very different: the between covariance matrix Σ_{12} has entries of arbitrary size. In contrast, the entries of C are correlation coefficients and are in the interval $[-1, 0]$ in this case and, more generally, in $[-1, 1]$. As a consequence, C explicitly reports the strength of the relationship between the variables. *Weight* and *mpg* are most strongly correlated with an entry of $-5{,}517.4$ in the covariance matrix, and -0.8095 in the correlation matrix. Although $-5{,}517.4$ is a large negative value, it does not lead to a natural interpretation of the strength of the relationship between the variables.

An inspection of C shows that the strongest absolute correlation exists between the variables *weight* and *mpg*. In Section 3.4 we examine whether a combination of variables will lead to a stronger correlation. ∎

3.3 Sample Canonical Correlations

In Example 3.1, I calculate the covariance matrix from data because we do not know the true population covariance structure. In this section, I define canonical correlation concepts for data. At the end of this section, we return to Example 3.1 and calculate the CC scores. The sample definitions are similar to those of the preceding section, but because we are dealing with a sample and do not know the true means and covariances, there are important differences. Table 3.1 summarises the key quantities for both the population and the sample.

We begin with some notation for the sample. For $\rho = 1, 2$, let

$$\mathbb{X}^{[\rho]} = \left[\mathbf{X}_1^{[\rho]}\, \mathbf{X}_2^{[\rho]} \cdots \mathbf{X}_n^{[\rho]}\right]$$

be $d_\rho \times n$ data which consist of n independent d_ρ-dimensional random vectors $\mathbf{X}_i^{[\rho]}$. The data $\mathbb{X}^{[1]}$ and $\mathbb{X}^{[2]}$ usually have a different number of variables, but measurements on the *same n* objects are carried out for $\mathbb{X}^{[1]}$ and $\mathbb{X}^{[2]}$. This fact is essential for the type of comparison we want to make.

We assume that the $\mathbf{X}_i^{[\rho]}$ have sample mean $\overline{\mathbf{X}}_\rho$ and sample covariance matrix S_ρ. Sometimes it will be convenient to consider the combined data. We write

$$\mathbb{X} = \begin{bmatrix} \mathbb{X}^{[1]} \\ \mathbb{X}^{[2]} \end{bmatrix} \sim \mathsf{Sam}(\overline{\mathbf{X}}, S),$$

so $\mathbb{X}$ is a $d \times n$ matrix with $d = d_1 + d_2$, sample mean $\overline{\mathbf{X}} = \begin{bmatrix} \overline{\mathbf{X}}_1 \\ \overline{\mathbf{X}}_2 \end{bmatrix}$ and sample covariance matrix

$$S = \begin{pmatrix} S_1 & S_{12} \\ S_{12}^{\mathsf{T}} & S_2 \end{pmatrix}.$$

Here S_{12} is the $d_1 \times d_2$ **(sample) between covariance matrix** of $\mathbb{X}^{[1]}$ and $\mathbb{X}^{[2]}$ defined by

$$S_{12} = \frac{1}{n-1} \sum_{i=1}^{n} (\mathbf{X}_i^{[1]} - \overline{\mathbf{X}}_1)(\mathbf{X}_i^{[2]} - \overline{\mathbf{X}}_2)^{\mathsf{T}}. \tag{3.11}$$

Unless otherwise stated, in this chapter, S_ρ has rank d_ρ for $\rho = 1, 2$, and $r \le \min(d_1, d_2)$ is the rank of S_{12}.

Definition 3.4 Let $\mathbb{X}^{[1]} \sim \mathsf{Sam}\big(\overline{\mathbf{X}}_1, S_1\big)$ and $\mathbb{X}^{[2]} \sim \mathsf{Sam}\big(\overline{\mathbf{X}}_2, S_2\big)$, and let S_{12} be their sample between covariance matrix. Assume that S_1 and S_2 are non-singular. The **matrix of sample canonical correlations** or the **sample canonical correlation matrix** is

$$\widehat{C} = S_1^{-1/2} S_{12} S_2^{-1/2},$$

and the **pair of matrices of sample multivariate coefficients of determination** are

$$\widehat{R}^{[C,1]} = \widehat{C}\widehat{C}^{\mathsf{T}} \qquad \text{and} \qquad \widehat{R}^{[C,2]} = \widehat{C}^{\mathsf{T}}\widehat{C}. \tag{3.12}$$

□

As in (1.9) of Section 1.3.2, we use the subscript *cent* to refer to the centred data. With this notation, the $d_1 \times d_2$ matrix

$$\widehat{C} = \left(\mathbb{X}_{\text{cent}}^{[1]} \mathbb{X}_{\text{cent}}^{[1]\,\mathsf{T}}\right)^{-1/2} \left(\mathbb{X}_{\text{cent}}^{[1]} \mathbb{X}_{\text{cent}}^{[2]\,\mathsf{T}}\right) \left(\mathbb{X}_{\text{cent}}^{[2]} \mathbb{X}_{\text{cent}}^{[2]\,\mathsf{T}}\right)^{-1/2}, \tag{3.13}$$

and $\widehat{C}$ has the singular value decomposition

$$\widehat{C} = \widehat{P}\widehat{\Upsilon}\widehat{Q}^{\mathsf{T}},$$

where

$$\widehat{P} = \begin{bmatrix} \widehat{\mathbf{p}}_1 \; \widehat{\mathbf{p}}_2 \cdots \widehat{\mathbf{p}}_r \end{bmatrix} \qquad \text{and} \qquad \widehat{Q} = \begin{bmatrix} \widehat{\mathbf{q}}_1 \; \widehat{\mathbf{q}}_2 \cdots \widehat{\mathbf{q}}_r \end{bmatrix},$$

and r is the rank of S_{12} and hence also of $\widehat{C}$. In the population case I mentioned that we may want to replace a singular $\Sigma_1^{-1/2}$, with rank $r < d_1$, by its $\Gamma_r \Lambda_r^{-1/2} \Gamma_r^{\mathsf{T}}$. We may want to make an analogous replacement in the sample case.

In the population case, we define the canonical correlation scores U_k and V_k of the vectors $\mathbf{X}^{[1]}$ and $\mathbf{X}^{[2]}$. The sample CC scores will be vectors of size n – similar to the PC sample scores – with one value for each observation.

Definition 3.5 Let $\mathbb{X}^{[1]} \sim \mathsf{Sam}\big(\overline{\mathbf{X}}_1, S_1\big)$ and $\mathbb{X}^{[2]} \sim \mathsf{Sam}\big(\overline{\mathbf{X}}_2, S_2\big)$. For $\rho = 1, 2$, let $\mathbb{X}_S^{[\rho]}$ be the sphered data. Let S_{12} be the sample between covariance matrix and $\widehat{C}$ the sample canonical correlation matrix of $\mathbb{X}^{[1]}$ and $\mathbb{X}^{[2]}$. Write $\widehat{P}\widehat{\Upsilon}\widehat{Q}^{\mathsf{T}}$ for the singular value decomposition of $\widehat{C}$. Consider $k = 1, \ldots, r$, with r the rank of S_{12}.

1. The kth **pair of canonical correlation scores** or **canonical variates** is

$$\mathbf{U}_{\bullet k} = \widehat{\mathbf{p}}_k^{\mathsf{T}} \mathbb{X}_S^{[1]} \qquad \text{and} \qquad \mathbf{V}_{\bullet k} = \widehat{\mathbf{q}}_k^{\mathsf{T}} \mathbb{X}_S^{[2]};$$

2. the k-dimensional **canonical correlation data** or **data of canonical variates** consist of the first k pairs of canonical correlation scores:

$$\mathbb{U}^{(k)} = \begin{bmatrix} \mathbf{U}_{\bullet 1}, \\ \vdots \\ \mathbf{U}_{\bullet k} \end{bmatrix} \qquad \text{and} \qquad \mathbb{V}^{(k)} = \begin{bmatrix} \mathbf{V}_{\bullet 1}, \\ \vdots \\ \mathbf{V}_{\bullet k} \end{bmatrix};$$

3. and the kth **pair of canonical (correlation) transforms** consists of

$$\widehat{\boldsymbol{\varphi}}_k = S_1^{-1/2} \widehat{\mathbf{p}}_k \qquad \text{and} \qquad \widehat{\boldsymbol{\psi}}_k = S_2^{-1/2} \widehat{\mathbf{q}}_k.$$

□

When we go from the population to the sample case, the size of some of the objects changes: we go from a d-dimensional random vector to data of size $d \times n$. For the parameters of the random variables, such as the mean or the covariance matrix, no such change in dimension occurs because the sample parameters are estimators of the true parameters. Similarly, the eigenvectors of the sample canonical correlation matrix $\widehat{C}$ are estimators of the corresponding population eigenvectors and have the same dimension as the population eigenvectors. However, the scores are defined for each observation, and we therefore obtain n pairs of scores for data compared with a single pair of scores for the population.

For a row vector of scores of n observations, we write

$$\mathbf{V}_{\bullet k} = \begin{bmatrix} V_{1k} \; V_{2k} \cdots V_{nk} \end{bmatrix},$$

which is similar to the notation for the PC vector of scores in (2.8) of Section 2.3. As in the PC case, $\mathbf{V}_{\bullet k}$ is a vector whose first subscript, $\bullet$, runs over all n observations and thus contains the kth CC score of all n observations of $\mathbb{X}^{[2]}$. The second subscript, k, refers to the index or numbering of the scores. The kth canonical correlation matrices $\mathbb{U}^k$ and $\mathbb{V}^k$ have dimensions $k \times n$; that is, the jth row summarises the contributions of the jth CC scores for all n observations. Using parts 1 and 2 of Definition 3.5, $\mathbb{U}^k$ and $\mathbb{V}^k$ are

$$\mathbb{U}^{(k)} = \widehat{P}_k^{\mathsf{T}} \, \mathbb{X}_S^{[1]} \qquad \text{and} \qquad \mathbb{V}^{(k)} = \widehat{Q}_k^{\mathsf{T}} \, \mathbb{X}_S^{[2]}, \tag{3.14}$$

where $\widehat{P}_k$ contains the first k vectors of the matrix $\widehat{P}$, and $\widehat{Q}_k$ contains the first k vectors of the matrix $\widehat{Q}$.

The relationship between pairs of canonical correlation scores and canonical transforms for the sample is similar to that of the population – see (3.9). For the centred data $\mathbb{X}_{\text{cent}}^{[\rho]}$ and $k \le r$, we have

$$\mathbf{U}_{\bullet k} = \widehat{\boldsymbol{\varphi}}_k^{\mathsf{T}} \mathbb{X}_{\text{cent}}^{[1]} \qquad \text{and} \qquad \mathbf{V}_{\bullet k} = \widehat{\boldsymbol{\psi}}_k^{\mathsf{T}} \mathbb{X}_{\text{cent}}^{[2]}. \tag{3.15}$$

Table 3.1 summarises the population and sample quantities for the CC framework.

Small examples can be useful in seeing how a method works. We start with the five-dimensional *car* data and then consider the canonical correlation scores for data with more variables.

Table 3.1 *Relationships of population and sample canonical correlations: $d_\rho = d_1$ for the first vector/data and $d_\rho = d_2$ for the second vector/data.*

	Population		Random Sample	
Random variables	$\mathbf{X}^{[1]}, \mathbf{X}^{[2]}$	$d_\rho \times 1$	$\mathbb{X}^{[1]}, \mathbb{X}^{[2]}$	$d_\rho \times n$
kth CC scores	U_k, V_k	1×1	$\mathbf{U}_{\bullet k}, \mathbf{V}_{\bullet k}$	$1 \times n$
CC vector/data	$\mathbf{U}^{(k)}, \mathbf{V}^{(k)}$	$k \times 1$	$\mathbb{U}^{(k)}, \mathbb{V}^{(k)}$	$k \times n$

Example 3.2 We continue with the 392 observations of the **car** data, and as in Example 3.1, $\mathbb{X}^{[1]}$ consist of the first three variables and $\mathbb{X}^{[2]}$ the remaining two. The sample between covariance matrix and the sample canonical correlation matrix are shown in (3.10), and we now refer to them as S and $\widehat{C}$, respectively. The singular values of $\widehat{C}$ are 0.8782 and 0.6328, and the left and right eigenvectors of $\widehat{C}$ are

$$\widehat{P} = \begin{bmatrix} \widehat{\mathbf{p}}_1 \; \widehat{\mathbf{p}}_2 \; \widehat{\mathbf{p}}_3 \end{bmatrix} = \begin{bmatrix} 0.3218 & 0.3948 & -0.8606 \\ 0.4163 & 0.7574 & 0.5031 \\ 0.8504 & -0.5202 & 0.0794 \end{bmatrix}$$

and

$$\widehat{Q} = \begin{bmatrix} \widehat{\mathbf{q}}_1 \; \widehat{\mathbf{q}}_2 \end{bmatrix} = \begin{bmatrix} -0.5162 & -0.8565 \\ -0.8565 & 0.5162 \end{bmatrix}.$$

The sample canonical transforms, which are obtained from the eigenvectors, are

$$\widehat{\Phi} = \begin{bmatrix} \widehat{\boldsymbol{\phi}}_1 \; \widehat{\boldsymbol{\phi}}_2 \; \widehat{\boldsymbol{\phi}}_3 \end{bmatrix} = \begin{bmatrix} 0.0025 & 0.0048 & -0.0302 \\ 0.0202 & 0.0409 & 0.0386 \\ 0.0000 & -0.0027 & 0.0020 \end{bmatrix}$$

and

$$\widehat{\Psi} = \begin{bmatrix} \widehat{\boldsymbol{\psi}}_1 \; \widehat{\boldsymbol{\psi}}_2 \end{bmatrix} = \begin{bmatrix} -0.1666 & -0.3637 \\ -0.0916 & 0.1078 \end{bmatrix}. \tag{3.16}$$

The rank of $\widehat{C}$ is two. Because $\mathbb{X}^{[1]}$ has three variables, MATLAB gives a third eigenvector, which – together with the other two eigenvectors – forms a basis for $\mathbb{R}^3$. Typically, we do not require this additional vector because the rank determines the number of pairs of scores we consider.

The signs of the vectors are identical for $\widehat{P}$ and $\widehat{\Phi}$ (and $\widehat{Q}$ and $\widehat{\Psi}$, respectively), but the entries of $\widehat{P}$ and $\widehat{\Phi}$ (and similarly of $\widehat{Q}$ and $\widehat{\Psi}$) differ considerably. In this case the entries of $\widehat{\Phi}$ and $\widehat{\Psi}$ are much smaller than the corresponding entries of the eigenvector matrices.

The $\mathbb{X}^{[2]}$ data are two-dimensional, so we obtain two pairs of vectors and CC scores. Figure 3.1 shows two-dimensional scatterplots of the scores, with the first scores in the left panel and the second scores in the right panel. The x-axis displays the scores for $\mathbb{X}^{[1]}$, and the y-axis shows the scores for $\mathbb{X}^{[2]}$. Both scatterplots have positive relationships. This might appear contrary to the negative entries in the between covariance matrix and the canonical correlation matrix of (3.10). Depending on the sign of the eigenvectors, the CC scores could have a positive or negative relationship. It is customary to show them with a positive

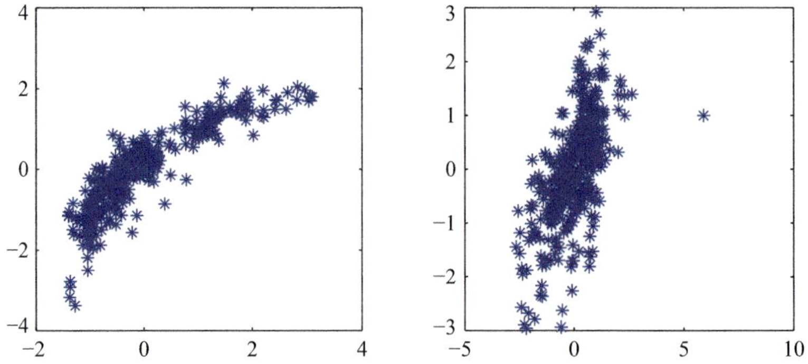

Figure 3.1 Canonical correlation scores of Example 3.2: (*left panel*) first scores; (*right panel*) second scores with $\mathbb{X}^{[1]}$ values on the x-axis, and $\mathbb{X}^{[2]}$ values on the y-axis.

relationship. By doing so, comparisons are easier, but we have lost some information: we are no longer able to tell from these scatterplots whether the original data have a positive or negative relationship.

We see from Figure 3.1 that the first scores in the left panel form a tight curve, whereas the second scores are more spread out and so are less correlated. The sample correlation coefficients for first and second scores are 0.8782 and 0.6328, respectively. The first value, 0.8782, is considerably higher than the largest entry, 0.8095, of the canonical correlation matrix (3.10). Here the *best* combinations are

$$0.0025 * (displacement - 194.4) + 0.0202 * (horsepower - 104.5) + 0.000025 *$$
$$(weight - 2977.6) - 0.1666 * (acceleration - 15.54) - 0.0916 * (mpg - 23.45).$$

The coefficients are those obtained from the canonical transforms in (3.16). Although the last entry in the first column of the matrix $\widehat{\Phi}$ is zero to the first four decimal places, the actual value is 0.000025.

The analysis shows that strong correlation exists between the two data sets. By considering linear combinations of the variables in each data set, the strength of the correlation between the two parts can be further increased, which shows that the combined physical properties of cars are very strongly correlated with the combined performance-based properties. ■

The next example, the Australian *illicit drug market* data, naturally split into two parts and are thus suitable for a canonical correlation analysis.

Example 3.3 We continue with the **illicit drug market** data for which seventeen different series have been measured over sixty-six months. Gilmour and Koch (2006) show that the data split into two distinct groups, which they call the *direct measures* and the *indirect measures* of the illicit drug market. The two groups are listed in separate columns in Table 3.2. The series numbers are given in the first and third columns of the table.

The direct measures are less likely to be affected by external forces such as health or law enforcement policy and economic factors but are more vulnerable to direct effects on the markets, such as successful intervention in the supply reduction and changes in the

Table 3.2 *Direct and indirect measures of the illicit drug market*

Series	Direct measures $\mathbb{X}^{[1]}$	Series	Indirect measures $\mathbb{X}^{[2]}$
1	Heroin possession offences	4	Prostitution offences
2	Amphetamine possession offences	7	PSB reregistrations
3	Cocaine possession offences	8	PSB new registrations
5	Heroin overdoses (ambulance)	12	Robbery 1
6	ADIS heroin	17	Steal from motor vehicles
9	Heroin deaths		
10	ADIS cocaine		
11	ADIS amphetamines		
13	Amphetamine overdoses		
14	Drug psychoses		
15	Robbery 2		
16	Break and enter dwelling		

Note: From Example 3.3. ADIS refers to the Alcohol and Drug Information Service, and ADIS heroin/cocaine/amphetamine refers to the number of calls to ADIS by individuals concerned about their own or another's use of the stated drug. PSB registrations refers to the number of individuals registering for pharmacotherapy.

availability or purity of the drugs. The twelve variables of $\mathbb{X}^{[1]}$ are the direct measures of the market, and the remaining five variables are the indirect measures and make up $\mathbb{X}^{[2]}$.

In this analysis we use the raw data. A calculation of the correlation coefficients between all pairs of variables in $\mathbb{X}^{[1]}$ and $\mathbb{X}^{[2]}$ yields the highest single correlation of 0.7640 between *amphetamine possession offences* (series 2) and *steal from motor vehicles* (series 17). The next largest coefficient of 0.6888 is obtained for *ADIS heroin* (series 6) and *PSB new registrations* (series 8). Does the overall correlation between the two data sets increase when we consider combinations of variables within $\mathbb{X}^{[1]}$ and $\mathbb{X}^{[2]}$? The result of a canonical correlation analysis yields five pairs of CC scores with correlation coefficients

$$0.9543 \quad 0.8004 \quad 0.6771 \quad 0.5302 \quad 0.3709.$$

The first two of these coefficients are larger than the correlation between *amphetamine possession offences* and *steal from motor vehicles*. The first CC score is based almost equally on *ADIS heroin* and *amphetamine possession offences* from among the $\mathbb{X}^{[1]}$ variables. The $\mathbb{X}^{[2]}$ variables with the largest weights for the first CC score are, in descending order, *PSB new registrations*, *steal from motor vehicles* and *robbery 1*. The other variables have much smaller weights. As in the previous example, the weights are those of the canonical transforms $\widehat{\boldsymbol{\varphi}}_1$ and $\widehat{\boldsymbol{\psi}}_1$.

It is interesting to note that the two variables *amphetamine possession offences* and *steal from motor vehicles*, which have the strongest single correlation, do not have the highest absolute weights in the first CC score. Instead, the pair of variables *ADIS heroin* and *PSB new registrations*, which have the second-largest single correlation coefficient, are the strongest contributors to the first CC scores. Overall, the correlation increases from 0.7640 to 0.9543 if the other variables from both data sets are taken into account.

The scatterplots corresponding to the CC data $\mathbb{U}^5$ and $\mathbb{V}^5$ are shown in Figure 3.2 starting with the most strongly correlated pair $(\mathbf{U}_{\bullet 1}, \mathbf{V}_{\bullet 1})$ in the top-left panel. The last subplot in the bottom-right panel shows the scatterplot of *amphetamine possession offences* versus

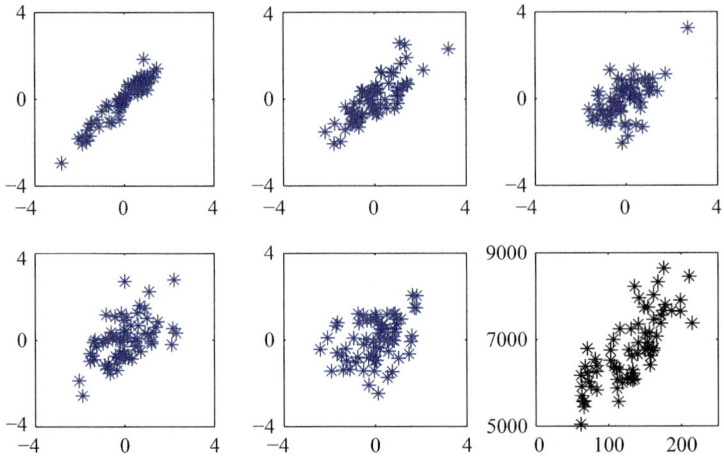

Figure 3.2 Canonical correlation scores of Example 3.3. CC scores 1 to 5 and best single variables in the bottom right panel. $\mathbb{X}^{[1]}$ values are shown on the x-axis and $\mathbb{X}^{[2]}$ values on the y-axis.

steal from motor vehicles. The variables corresponding to $\mathbb{X}^{[1]}$ are displayed on the x-axis, and the $\mathbb{X}^{[2]}$ variables are shown on the y-axis. The progression of scatterplots shows the decreasing strength of the correlation between the combinations as we go from the first to the fifth pair.

The analysis shows that there is a very strong, almost linear relationship between the direct and indirect measures of the illicit drug market, which is not expected from an inspection of the correlation plots of pairs of variables (not shown here). This relationship far exceeds that of the 'best' individual variables, *amphetamine possession offences* and *steal from motor vehicles*, and shows that the two parts of the data are very strongly correlated. ■

Remark. The singular values of C and $\widehat{C}$, being positive square roots of their respective multivariate coefficients of determination, reflect the strength of the correlation between the scores. If we want to know whether the combinations of variables are positively or negatively correlated, we have to calculate the actual correlation coefficient between the two linear combinations.

3.4 Properties of Canonical Correlations

In this section we consider properties of the CC scores and vectors which include optimality results for the CCs. Our first result relates the correlation coefficients, such as those calculated in Example 3.3, to properties of the matrix of canonical correlations, C.

Theorem 3.6 *Let* $\mathbf{X}^{[1]} \sim (\boldsymbol{\mu}_1, \Sigma_1)$ *and* $\mathbf{X}^{[2]} \sim (\boldsymbol{\mu}_2, \Sigma_2)$. *Let* Σ_{12} *be the between covariance matrix of* $\mathbf{X}^{[1]}$ *and* $\mathbf{X}^{[2]}$, *and let* r *be its rank. Let* C *be the matrix of canonical correlations with singular value decomposition* $C = P \Upsilon Q^{\top}$. *For* $k, \ell = 1, \ldots, r$, *let* $\mathbf{U}^{(k)}$ *and* $\mathbf{V}^{(\ell)}$ *be the canonical correlation vectors of* $\mathbf{X}^{[1]}$ *and* $\mathbf{X}^{[2]}$, *respectively.*

1. *The mean and covariance matrix of* $\begin{bmatrix}\mathbf{U}^{(k)}\\ \mathbf{V}^{(\ell)}\end{bmatrix}$ *are*

$$\mathbb{E}\begin{bmatrix}\mathbf{U}^{(k)}\\ \mathbf{V}^{(\ell)}\end{bmatrix} = \begin{bmatrix}\mathbf{0}_k\\ \mathbf{0}_\ell\end{bmatrix} \qquad \textit{and} \qquad \operatorname{var}\left(\begin{bmatrix}\mathbf{U}^{(k)}\\ \mathbf{V}^{(\ell)}\end{bmatrix}\right) = \begin{pmatrix}\mathbf{I}_{k\times k} & \Upsilon_{k\times\ell}\\ \Upsilon_{k\times\ell}^{\top} & \mathbf{I}_{\ell\times\ell},\end{pmatrix},$$

where $\Upsilon_{k\times\ell}$ *is the* $k\times\ell$ *submatrix of* Υ *which consists of the 'top-left corner' of* Υ.

2. *The variances and covariances of the canonical correlation scores* U_k *and* V_ℓ *are*

$$\operatorname{var}(U_k) = \operatorname{var}(V_\ell) = 1 \qquad \textit{and} \qquad \operatorname{cov}(U_k, V_\ell) = \pm\upsilon_k\delta_{k\ell},$$

where υ_k *is the kth singular value of* Υ, *and* $\delta_{k\ell}$ *is the Kronecker delta function.*

This theorem shows that the object of interest is the submatrix $\Upsilon_{k\times\ell}$. We explore this matrix further in the following corollary.

Corollary 3.7 *Assume that for* $\rho = 1, 2$, *the* $\mathbf{X}^{[\rho]}$ *satisfy the conditions of Theorem 3.6. Then the covariance matrix of* $\mathbf{U}^{(k)}$ *and* $\mathbf{V}^{(\ell)}$ *is*

$$\operatorname{cov}\left(\mathbf{U}^{(k)}, \mathbf{V}^{(\ell)}\right) = \operatorname{cor}\left(\mathbf{U}^{(k)}, \mathbf{V}^{(\ell)}\right),$$

where $\operatorname{cor}\left(\mathbf{U}^{(k)}, \mathbf{V}^{(\ell)}\right)$ *is the matrix of correlation coefficients of* $\mathbf{U}^{(k)}$ *and* $\mathbf{V}^{(\ell)}$.

Proof of Theorem 3.6 For $k, \ell = 1, \ldots, r$, let P_k be the submatrix of P which consists of the first k left eigenvectors of C, and let Q_ℓ be the submatrix of Q which consists of the first ℓ right eigenvectors of C.

For part 1 of the theorem and k, recall that $\mathbf{U}^{(k)} = P_k^{\top}\Sigma_1^{-1/2}(\mathbf{X}^{[1]} - \boldsymbol{\mu}_1)$, so

$$\mathbb{E}\mathbf{U}^{(k)} = P_k^{\top}\Sigma_1^{-1/2}\mathbb{E}(\mathbf{X}^{[1]} - \boldsymbol{\mu}_1) = \mathbf{0}_k,$$

and similarly, $\mathbb{E}\mathbf{V}^{(\ell)} = \mathbf{0}_\ell$.

The calculations for the covariance matrix consist of two parts: the separate variance calculations for $\mathbf{U}^{(k)}$ and $\mathbf{V}^{(\ell)}$ and the between covariance matrix $\operatorname{cov}\left(\mathbf{U}^{(k)}, \mathbf{V}^{(\ell)}\right)$. All vectors have zero means, which simplifies the variance calculations. Consider $\mathbf{V}^{(\ell)}$. We have

$$\begin{aligned}\operatorname{var}(\mathbf{V}^{(\ell)}) &= \mathbb{E}\left(\mathbf{V}^{(\ell)}\mathbf{V}^{(\ell)^{\top}}\right)\\ &= \mathbb{E}\left[Q_\ell^{\top}\Sigma_2^{-1/2}\left(\mathbf{X}^{[2]} - \boldsymbol{\mu}_2\right)\left(\mathbf{X}^{[2]} - \boldsymbol{\mu}_2\right)^{\top}\Sigma_2^{-1/2}Q_\ell\right]\\ &= Q_\ell^{\top}\Sigma_2^{-1/2}\Sigma_2\Sigma_2^{-1/2}Q_\ell = \mathbf{I}_{\ell\times\ell}.\end{aligned}$$

In these equalities we have used the fact that $\operatorname{var}(\mathbf{X}^{[2]}) = \mathbb{E}\left[\mathbf{X}^{[2]}(\mathbf{X}^{[2]})^{\top}\right] - \boldsymbol{\mu}_2\boldsymbol{\mu}_2^{\top} = \Sigma_2$ and that the matrix Q_ℓ consists of ℓ orthonormal vectors.

The between covariance matrix of $\mathbf{U}^{(k)}$ and $\mathbf{V}^{(\ell)}$ is

$$
\begin{aligned}
\operatorname{cov}(\mathbf{U}^{(k)},\mathbf{V}^{(\ell)}) &= \mathbb{E}\left(\mathbf{U}^{(k)}\mathbf{V}^{(\ell)^{\mathsf{T}}}\right)\\
&= \mathbb{E}\left[P_k^{\mathsf{T}}\Sigma_1^{-1/2}\left(\mathbf{X}^{[1]}-\boldsymbol{\mu}_1\right)\left(\mathbf{X}^{[2]}-\boldsymbol{\mu}_2\right)^{\mathsf{T}}\Sigma_2^{-1/2}Q_\ell\right]\\
&= P_k^{\mathsf{T}}\Sigma_1^{-1/2}\mathbb{E}\left[\left(\mathbf{X}^{[1]}-\boldsymbol{\mu}_1\right)\left(\mathbf{X}^{[2]}-\boldsymbol{\mu}_2\right)^{\mathsf{T}}\right]\Sigma_2^{-1/2}Q_\ell\\
&= P_k^{\mathsf{T}}\Sigma_1^{-1/2}\Sigma_{12}\Sigma_2^{-1/2}Q_\ell\\
&= P_k^{\mathsf{T}}CQ_\ell = P_k^{\mathsf{T}}P\Upsilon Q^{\mathsf{T}}Q_\ell\\
&= \mathbf{I}_{k\times r}\Upsilon\mathbf{I}_{r\times\ell} = \Upsilon_{k\times\ell}.
\end{aligned}
$$

In this sequence of equalities we have used the singular value decomposition $P\Upsilon Q^{\mathsf{T}}$ of C and the fact that $P_k^{\mathsf{T}}P = \mathbf{I}_{k\times r}$. A similar relationship is used in the proof of Theorem 2.6 in Section 2.5. Part 2 follows immediately from part 1 because Υ is a diagonal matrix with non-zero entries υ_k for $k = 1,\ldots,r$. ■

The theorem is stated for random vectors, but an analogous result applies to random data: the CC vectors become CC matrices, and the matrix $\Upsilon_{k\times\ell}$ is replaced by the sample covariance matrix $\widehat{\Upsilon}_{k\times\ell}$. This is an immediate consequence of dealing with a random sample. I illustrate Theorem 3.6 with an example.

Example 3.4 We continue with the **illicit drug market** data and use the direct and indirect measures of the market as in Example 3.3. The singular value decomposition $\widehat{C} = \widehat{P}\widehat{\Upsilon}\widehat{Q}^{\mathsf{T}}$ yields the five singular values

$$\widehat{\upsilon}_1 = 0.9543, \qquad \widehat{\upsilon}_2 = 0.8004, \qquad \widehat{\upsilon}_3 = 0.6771, \qquad \widehat{\upsilon}_4 = 0.5302, \qquad \widehat{\upsilon}_5 = 0.3709.$$

The five singular values of $\widehat{C}$ agree with (the moduli of) the correlation coefficients calculated in Example 3.3, thus confirming the result of Theorem 3.6, stated there for the population. The biggest singular value is very close to 1, which shows that there is a very strong relationship between the direct and indirect measures. ■

Theorem 3.6 and its corollary state properties of the CC scores. In the next proposition we examine the relationship between the $\mathbf{X}^{[\rho]}$ and the vectors $\mathbf{U}$ and $\mathbf{V}$.

Proposition 3.8 *Let* $\mathbf{X}^{[1]} \sim (\boldsymbol{\mu}_1,\Sigma_1)$, $\mathbf{X}^{[2]} \sim (\boldsymbol{\mu}_2,\Sigma_2)$. *Let* C *be the canonical correlation matrix with rank* r *and singular value decomposition* $P\Upsilon Q^{\mathsf{T}}$. *For* $k,\ell \le r$, *let* $\mathbf{U}^{(k)}$ *and* $\mathbf{V}^{(\ell)}$ *be the* k*- and* ℓ*-dimensional canonical correlation vectors of* $\mathbf{X}^{[1]}$ *and* $\mathbf{X}^{[2]}$. *The random vectors and their canonical correlation vectors satisfy*

$$\operatorname{cov}(\mathbf{X}^{[1]},\mathbf{U}^{(k)}) = \Sigma_1^{1/2}P_k \qquad \textit{and} \qquad \operatorname{cov}(\mathbf{X}^{[2]},\mathbf{V}^{(\ell)}) = \Sigma_2^{1/2}Q_\ell.$$

The proof of Proposition 3.8 is deferred to the Problems at the end of Part I.

In terms of the canonical transforms, by (3.8) the equalities stated in Proposition 3.8 are

$$\operatorname{cov}(\mathbf{X}^{[1]},\mathbf{U}^{(k)}) = \Sigma_1\Phi_k \qquad \text{and} \qquad \operatorname{cov}(\mathbf{X}^{[2]},\mathbf{V}^{(\ell)}) = \Sigma_2\Psi_\ell. \tag{3.17}$$

The equalities (3.17) look very similar to the covariance relationship between the random vector $\mathbf{X}$ and its principal component vector $\mathbf{W}^{(k)}$ which we considered in Proposition 2.8, namely,

$$\mathrm{cov}(\mathbf{X}, \mathbf{W}^{(k)}) = \Gamma \Lambda_k = \Sigma \Gamma_k, \tag{3.18}$$

with $\Sigma = \Gamma \Lambda \Gamma^{\mathsf{T}}$. In (3.17) and (3.18), the relationship between the random vector $\mathbf{X}^{[\rho]}$ or $\mathbf{X}$ and its CC or PC vector is described by a matrix, which is related to eigenvectors and the covariance matrix of the appropriate random vector. A difference between the two expressions is that the columns of Γ_k are the eigenvectors of Σ, whereas the columns of Φ_k and Ψ_ℓ are multiples of the left and right eigenvectors of the between covariance matrix Σ_{12} which satisfy

$$\Sigma_{12} \boldsymbol{\psi}_k = \upsilon_k \boldsymbol{\varphi}_k \quad \text{and} \quad \Sigma_{12}^{\mathsf{T}} \boldsymbol{\varphi}_k = \upsilon_k \boldsymbol{\psi}_k, \tag{3.19}$$

where υ_k is the kth singular value of C.

The next corollary looks at uncorrelated random vectors with covariance matrices $\Sigma_\rho = \mathbf{I}$ and non-trivial between covariance matrix Σ_{12}.

Corollary 3.9 *If* $\mathbf{X}^{[\rho]} \sim \left(\mathbf{0}_{d_\rho}, \mathbf{I}_{d_\rho \times d_\rho}\right)$, *for* $\rho = 1,2$, *the canonical correlation matrix* C *reduces to the covariance matrix* Σ_{12}, *and the matrices* P *and* Q *agree with the matrices of canonical transforms* $\mathbf{\Phi}$ *and* $\mathbf{\Psi}$ *respectively.*

This result follows from the variance properties of the random vectors because

$$C = \Sigma_1^{-1/2} \Sigma_{12} \Sigma_2^{-1/2} = \mathbf{I}_{d_1 \times d_1} \Sigma_{12} \mathbf{I}_{d_2 \times d_2} = \Sigma_{12}.$$

For random vectors with the identity covariance matrix, the corollary tells us that C and Σ_{12} agree, which is not the case in general. Working with the matrix C has the advantage that its entries are scaled and therefore more easily interpretable. In some areas or applications, including climate research and Partial Least Squares, the covariance matrix Σ_{12} is used directly to find the canonical transforms.

So far we have explored the covariance structure of pairs of random vectors and their CCs. The reason for constructing the CCs is to obtain combinations of $\mathbf{X}^{[1]}$ and $\mathbf{X}^{[2]}$ which are more strongly correlated than the individual variables taken separately. In our examples we have seen that correlation decreases for the first few CC scores. The next result provides a theoretical underpinning for these observations.

Theorem 3.10 *For* $\rho = 1,2$, *let* $\mathbf{X}^{[\rho]} \sim \left(\boldsymbol{\mu}_\rho, \Sigma_\rho\right)$, *with* d_ρ *the rank of* Σ_ρ. *Let* Σ_{12} *be the between covariance matrix and* C *the canonical correlation matrix of* $\mathbf{X}^{[1]}$ *and* $\mathbf{X}^{[2]}$. *Let* r *be the rank of* C, *and for* $j \le r$, *let* $(\mathbf{p}_j, \mathbf{q}_j)$ *be the left and right eigenvectors of* C. *Let* $\mathbf{X}_\Sigma^{[\rho]}$ *be the sphered vector derived from* $\mathbf{X}^{[\rho]}$. *For unit vectors* $\mathbf{u} \in \mathbb{R}^{d_1}$ *and* $\mathbf{v} \in \mathbb{R}^{d_2}$, *put*

$$c_{(\mathbf{u},\mathbf{v})} = \mathrm{cov}\left(\mathbf{u}^{\mathsf{T}} \mathbf{X}_\Sigma^{[1]}, \mathbf{v}^{\mathsf{T}} \mathbf{X}_\Sigma^{[2]}\right).$$

1. *It follows that* $c_{(\mathbf{u},\mathbf{v})} = \mathbf{u}^{\mathsf{T}} C \mathbf{v}$.
2. *If* $c_{(\mathbf{u}^*,\mathbf{v}^*)}$ *maximises the covariance* $c_{(\mathbf{u},\mathbf{v})}$ *over all unit vectors* $\mathbf{u} \in \mathbb{R}^{d_1}$ *and* $\mathbf{v} \in \mathbb{R}^{d_2}$, *then*

$$\mathbf{u}^* = \pm \mathbf{p}_1, \qquad \mathbf{v}^* = \pm \mathbf{q}_1 \qquad \textit{and} \qquad c_{(\mathbf{u}^*,\mathbf{v}^*)} = \upsilon_1,$$

where υ_1 *is the largest singular value of* C, *which corresponds to the eigenvectors* $\mathbf{p}_1$ *and* $\mathbf{q}_1$.

3. *Fix $1 < k \le r$. Consider unit vectors $\mathbf{u} \in \mathbb{R}^{d_1}$ and $\mathbf{v} \in \mathbb{R}^{d_2}$ such that*
   (a) $\mathbf{u}^{\mathsf{T}}\mathbf{X}_{\Sigma}^{[1]}$ is uncorrelated with $\mathbf{p}_j^{\mathsf{T}}\mathbf{X}_{\Sigma}^{[1]}$, and
   (b) $\mathbf{v}^{\mathsf{T}}\mathbf{X}_{\Sigma}^{[2]}$ is uncorrelated with $\mathbf{q}_j^{\mathsf{T}}\mathbf{X}_{\Sigma}^{[2]}$,
   for $j < k$. If $c_{(\mathbf{u}^*,\mathbf{v}^*)}$ maximise $c_{(\mathbf{u},\mathbf{v})}$ over all such unit vectors $\mathbf{u}$ and $\mathbf{v}$, then

$$\mathbf{u}^* = \pm\mathbf{p}_k, \qquad \mathbf{v}^* = \pm\mathbf{q}_k \qquad \textit{and} \qquad c_{(\mathbf{u}^*,\mathbf{v}^*)} = \upsilon_k.$$

Proof To show part 1, consider unit vectors $\mathbf{u}$ and $\mathbf{v}$ which satisfy the assumptions of the theorem. From the definition of the canonical correlation matrix, it follows that

$$\begin{aligned}
c_{(\mathbf{u},\mathbf{v})} &= \operatorname{cov}\left[\mathbf{u}^{\mathsf{T}}\Sigma_1^{-1/2}\left(\mathbf{X}^{[1]} - \boldsymbol{\mu}_1\right), \mathbf{v}^{\mathsf{T}}\Sigma_2^{-1/2}\left(\mathbf{X}^{[2]} - \boldsymbol{\mu}_2\right)\right] \\
&= \mathbb{E}\left[\mathbf{u}^{\mathsf{T}}\Sigma_1^{-1/2}\left(\mathbf{X}^{[1]} - \boldsymbol{\mu}_1\right)\left(\mathbf{X}^{[2]} - \boldsymbol{\mu}_2\right)^{\mathsf{T}}\Sigma_2^{-1/2}\mathbf{v}\right] \\
&= \mathbf{u}^{\mathsf{T}}\Sigma_1^{-1/2}\Sigma_{12}\Sigma_2^{-1/2}\mathbf{v} = \mathbf{u}^{\mathsf{T}}C\mathbf{v}.
\end{aligned}$$

To see why part 2 holds, consider unit vectors $\mathbf{u}$ and $\mathbf{v}$ as in part 1. For $j = 1, \ldots, d_1$, let $\mathbf{p}_j$ be the left eigenvectors of C. Because Σ_1 has full rank,

$$\mathbf{u} = \sum_j \alpha_j \mathbf{p}_j \qquad \text{with } \alpha_j \in \mathbb{R} \quad \text{and} \quad \sum_j \alpha_j^2 = 1,$$

and similarly,

$$\mathbf{v} = \sum_k \beta_k \mathbf{q}_k \qquad \text{with } \beta_k \in \mathbb{R} \quad \text{and} \quad \sum_k \beta_k^2 = 1,$$

where the $\mathbf{q}_k$ are the right eigenvectors of C. From part 1, it follows that

$$\begin{aligned}
\mathbf{u}^{\mathsf{T}}C\mathbf{v} &= \left(\sum_j \alpha_j \mathbf{p}_j^{\mathsf{T}}\right) C \left(\sum_k \beta_k \mathbf{q}_k\right) \\
&= \sum_{j,k} \alpha_j \beta_k \, \mathbf{p}_j^{\mathsf{T}} C \mathbf{q}_k \\
&= \sum_{j,k} \alpha_j \beta_k \, \mathbf{p}_j^{\mathsf{T}} \upsilon_k \mathbf{p}_k \qquad \text{by (3.5)} \\
&= \sum_j \alpha_j \beta_j \upsilon_j.
\end{aligned}$$

The last equality follows because the $\mathbf{p}_j$ are orthonormal, so $\mathbf{p}_j^{\mathsf{T}}\mathbf{p}_k = \delta_{jk}$, where δ_{jk} is the Kronecker delta function. Next, we use the fact that the singular values υ_j are positive and ordered, with υ_1 the largest. Observe that

$$\begin{aligned}
|\mathbf{u}^{\mathsf{T}}C\mathbf{v}| &= |\sum_j \alpha_j \beta_j \upsilon_j| \le \upsilon_1 |\sum_j \alpha_j \beta_j| \\
&\le \upsilon_1 \sum_j |\alpha_j \beta_j| \le \upsilon_1 \left(\sum_j |\alpha_j|^2\right)^{1/2} \left(\sum_j |\beta_j|^2\right)^{1/2} = \upsilon_1.
\end{aligned}$$

Here we have used Hölder's inequality, which links the sum $\sum_j |\alpha_j \beta_j|$ to the product of the norms $\left(\sum_j |\alpha_j|^2\right)^{1/2}$ and $\left(\sum_j |\beta_j|^2\right)^{1/2}$. For details, see, for example, theorem 5.4 in Pryce (1973). Because the $\mathbf{p}_j$ are unit vectors, the norms are one. The maximum is attained when

$$\alpha_1 = \pm 1, \qquad \beta_1 = \pm 1, \qquad \text{and} \qquad \alpha_j = \beta_j = 0 \qquad \text{for } j > 1.$$

Table 3.3 *Variables of the Boston housing data from Example 3.5, where* $^+$ *indicates that the quantity is calculated as a centred proportion*

Environmental and social measures $\mathbb{X}^{[1]}$	Individual measures $\mathbb{X}^{[2]}$
Per capita crime rate by town	Average number of rooms per dwelling
Proportion of non-retail business acres per town	Proportion of owner-occupied units built prior to 1940
Nitric oxide concentration (parts per 10 million)	Full-value property-tax rate per \$10,000
Weighted distances to Boston employment centres	Median value of owner-occupied homes in \$1000s
Index of accessibility to radial highways	
Pupil-teacher ratio by town	
Proportion of blacks by town$^+$	

This choice of coefficients implies that $\mathbf{u}^* = \pm\mathbf{p}_1$ and $\mathbf{v}^* = \pm\mathbf{q}_1$, as desired.

For a proof of part 3, one first shows the result for $k = 2$. The proof is a combination of the proof of part 2 and the arguments used in the proof of part 2 of Theorem 2.10 in Section 2.5.2. For $k > 2$, the proof works in almost the same way as for $k = 2$. ■

It is not possible to demonstrate the optimality of the CC scores in an example, but examples illustrate that the first pair of CC scores is larger than correlation coefficients of individual variables.

Example 3.5 The **Boston housing** data of Harrison and Rubinfeld (1978) consist of 506 observations with fourteen variables. The variables naturally fall into two categories: those containing information regarding the individual, such as house prices, and property tax; and those which deal with environmental or social factors. For this reason, the data are suitable for a canonical correlation analysis. Some variables are binary; I have omitted these in this analysis. The variables I use are listed in Table 3.3. (The variable names in the table are those of Harrison and Rubinfeld, rather than names I would have chosen.)

There are four variables in $\mathbb{X}^{[2]}$, so we obtain four CC scores. The scatterplots of the CC scores are displayed in Figure 3.3 starting with the first CC scores on the left. For each scatterplot, the environmental/social variables are displayed on the x-axis and the individual measures on the y-axis.

The four singular values of the canonical correlation matrix are

$$\widehat{\upsilon}_1 = 0.9451, \qquad \widehat{\upsilon}_2 = 0.6787, \qquad \widehat{\upsilon}_3 = 0.5714, \qquad \widehat{\upsilon}_4 = 0.2010.$$

These values decrease quickly. Of special interest are the first CC scores, which are shown in the left subplot of Figure 3.3. The first singular value is higher than any correlation coefficient calculated from the individual variables.

A singular value as high as 0.9451 expresses a very high correlation, which seems at first not consistent with the spread of the first scores. There is a reason for this high value: The first pair of scores consists of two distinct clusters, which behave like two points. So the large singular value reflects the linear relationship of the two clusters rather than a tight fit of the scores to a line. It is not clear without further analysis whether there is a strong positive correlation within each cluster. We also observe that the cluster structure is only present in the scatterplot of the first CC scores. ■

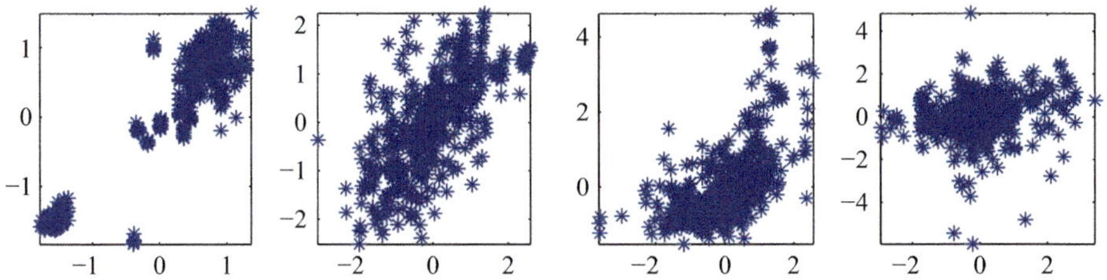

Figure 3.3 Canonical correlation scores of Example 3.5. CC scores 1 to 4 with environmental variables on the x-axis.

3.5 Canonical Correlations and Transformed Data

It is a well-known fact that random variables of the form $aX+b$ and $cY+d$ (with $ac \neq 0$) have the same absolute correlation coefficient as the original random variables X and Y. We examine whether similar relationships hold in the multivariate context. For this purpose, we derive the matrix of canonical correlations and the CCs for transformed random vectors. Such transformations could be the result of exchange rates over time on share prices or a reduction of the original data to a simpler form.

Transformations based on thresholds could result in binary data, and indeed, Canonical Correlation Analysis works for binary data. I will not pursue this direction but focus on linear transformations of the data and Canonical Correlation Analysis for such data.

3.5.1 Linear Transformations and Canonical Correlations

Let $\rho = 1,2$. Consider the random vectors $\mathbf{X}^{[\rho]} \sim \left(\boldsymbol{\mu}_{\boldsymbol{\rho}}, \Sigma_\rho\right)$. Let A_ρ be $\kappa_\rho \times d_\rho$ matrices with $\kappa_\rho \leq d_\rho$, and let $\mathbf{a}_\rho$ be fixed κ_ρ-dimensional vectors. We define the transformed random vector $\mathbf{T}$ by

$$\mathbf{T} = \begin{bmatrix} \mathbf{T}^{[1]} \\ \mathbf{T}^{[2]} \end{bmatrix} \qquad \text{with } \mathbf{T}^{[\rho]} = A_\rho \mathbf{X}^{[\rho]} + \mathbf{a}_\rho \quad \text{for } \rho = 1,2. \tag{3.20}$$

We begin with properties of transformed vectors.

Theorem 3.11 *Let $\rho = 1,2$ and $\mathbf{X}^{[\rho]} \sim \left(\boldsymbol{\mu}_{\boldsymbol{\rho}}, \Sigma_\rho\right)$. Let Σ_{12} be the between covariance matrix of $\mathbf{X}^{[1]}$ and $\mathbf{X}^{[2]}$. Let A_ρ be $\kappa_\rho \times d_\rho$ matrices with $\kappa_\rho \leq d_\rho$, and let $\mathbf{a}_\rho$ be fixed κ_ρ-dimensional vectors. Put $\mathbf{T}^{[\rho]} = A_\rho \mathbf{X}^{[\rho]} + \mathbf{a}_\rho$ and*

$$\mathbf{T} = \begin{bmatrix} \mathbf{T}^{[1]} \\ \mathbf{T}^{[2]} \end{bmatrix}.$$

1. *The mean of* $\mathbf{T}$ *is*

$$\mathbb{E}\mathbf{T} = \begin{bmatrix} A_1\boldsymbol{\mu}_1 + \mathbf{a}_1 \\ A_2\boldsymbol{\mu}_2 + \mathbf{a}_2 \end{bmatrix}.$$

2. *The covariance matrix of* $\mathbf{T}$ *is*

$$\operatorname{var}(\mathbf{T}) = \begin{pmatrix} A_1\Sigma_1 A_1^{\mathsf{T}} & A_1\Sigma_{12} A_2^{\mathsf{T}} \\ A_2\Sigma_{12}^{\mathsf{T}} A_1^{\mathsf{T}} & A_2\Sigma_2 A_2^{\mathsf{T}} \end{pmatrix}.$$

3. *If, for* $\rho = 1,2$, *the matrices* $A_\rho \Sigma_\rho A_\rho^{\mathsf{T}}$ *are non-singular, then the canonical correlation matrix* C_{T} *of* $\mathbf{T}^{[1]}$ *and* $\mathbf{T}^{[2]}$ *is*

$$C_{\mathrm{T}} = \left(A_1 \Sigma_1 A_1^{\mathsf{T}}\right)^{-1/2} \left(A_1 \Sigma_{12} A_2^{\mathsf{T}}\right) \left(A_2 \Sigma_2 A_2^{\mathsf{T}}\right)^{-1/2}.$$

Proof Part 1 follows by linearity, and the expressions for the covariance matrices follow from Result 1.1 in Section 1.3.1. The calculation of the covariance matrices of $\mathbf{T}^{[1]}$ and $\mathbf{T}^{[2]}$ and the canonical correlation matrix are deferred to the Problems at the end of Part I. ∎

As for the original vectors $\mathbf{X}^{[\rho]}$, we construct the canonical correlation scores from the matrix of canonical correlations; so for the transformed vectors we use C_{T}. Let $C_{\mathrm{T}} = P_{\mathrm{T}} \Upsilon_{\mathrm{T}} Q_{\mathrm{T}}$ be the singular value decomposition, and assume that the matrices $\Sigma_{T,\rho} = A_\rho \Sigma_\rho A_\rho^{\mathsf{T}}$ are invertible for $\rho = 1,2$. The canonical correlation scores are the projections of the sphered vectors onto the left and right eigenvectors $\mathbf{p}_{T,k}^{\mathsf{T}}$ and $\mathbf{q}_{T,k}^{\mathsf{T}}$ of C_{T}:

$$U_{T,k} = \mathbf{p}_{T,k}^{\mathsf{T}}\, \Sigma_{T,1}^{-1/2} (\mathbf{T}^{[1]} - \mathbb{E}\mathbf{T}^{[1]})$$

and

$$V_{T,k} = \mathbf{q}_{T,k}^{\mathsf{T}}\, \Sigma_{T,2}^{-1/2} (\mathbf{T}^{[2]} - \mathbb{E}\mathbf{T}^{[2]}). \tag{3.21}$$

Applying Theorem 3.6 to the transformed vectors, we find that

$$\operatorname{cov}(U_{T,k}, V_{T,\ell}) = \pm \upsilon_{T,k} \delta_{k\ell},$$

where $\upsilon_{T,k}$ is the kth singular value of Υ_{T}, and $k, \ell \le \min\{\kappa_1, \kappa_2\}$.

If the matrices $\Sigma_{T,\rho}$ are not invertible, the singular values of Υ and Υ_{T}, and similarly, the CC scores of the original and transformed vectors, can differ, as the next example shows.

Example 3.6 We continue with the direct and indirect measures of the **illicit drug market** data which have twelve and five variables in $\mathbb{X}^{[1]}$ and $\mathbb{X}^{[2]}$, respectively. Figure 3.2 of Example 3.3 shows the scores of all five pairs of CCs.

To illustrate Theorem 3.11, we use the first four principal components of $\mathbb{X}^{[1]}$ and $\mathbb{X}^{[2]}$ as our transformed data. So, for $\rho = 1,2$,

$$\mathbb{X}^{[\rho]} \longmapsto \mathbb{T}^{[\rho]} = \Gamma_{\rho,4}^{\mathsf{T}} (\mathbb{X}^{[\rho]} - \overline{\mathbf{X}}_\rho).$$

The scores of the four pairs of CCs of the $\mathbb{T}^{[\rho]}$ are shown in the top row of Figure 3.4 with the scores of $\mathbb{T}^{[1]}$ on the x-axis. The bottom row of the figure shows the first four CC pairs of the original data for comparison. It is interesting to see that all four sets of transformed scores are less correlated than their original counterparts.

Table 3.4 contains the correlation coefficients for the transformed and raw data for a more quantitative comparison. The table confirms the visual impression obtained from Figure 3.4: the correlation strength of the transformed CC scores is considerably lower than that of the original CC scores. Such a decrease does not *have to happen*, but it is worth reflecting on why it might happen.

The direction vectors in a principal component analysis of $\mathbb{X}^{[1]}$ and $\mathbb{X}^{[2]}$ are chosen to maximise the variability *within* these data. This means that the first and subsequent eigenvectors point in the direction with the largest variability. In this case, the variables *break and enter* (series 16) and *steal from motor vehicles* (series 17) have the highest PC_1 weights for $\mathbb{X}^{[1]}$ and $\mathbb{X}^{[2]}$, respectively, largely because these two series have much higher values than the remaining series, and a principal component analysis will therefore find

Table 3.4 *Correlation coefficients of the scores in Figure 3.4, from Example 3.6.*

	Canonical correlations				
Transformed	0.8562	0.7287	0.3894	0.2041	—
Original	0.9543	0.8004	0.6771	0.5302	0.3709

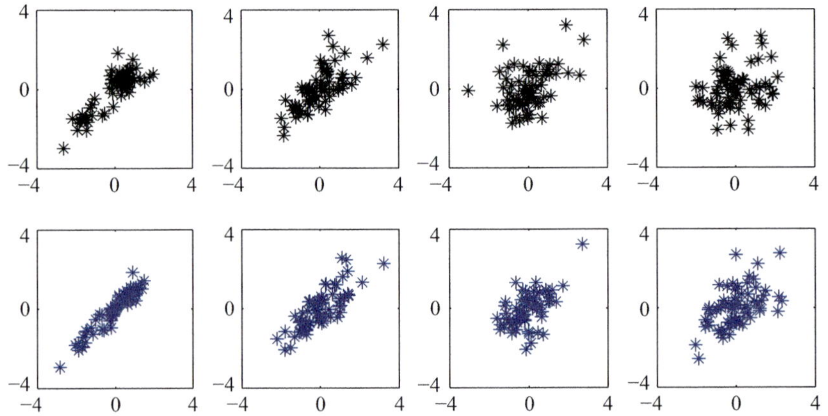

Figure 3.4 Canonical correlation scores of Example 3.6. CC scores 1 to 4 of transformed data (*top row*) and raw data (*bottom row*).

these variables first. In contrast, the canonical transforms maximise a different criterion: the between covariance matrix of $\mathbb{X}^{[1]}$ and $\mathbb{X}^{[2]}$. Because the criteria differ, the direction vectors differ too; the canonical transforms are best at exhibiting the strongest relationships *between* different parts of data. ■

The example shows the effect of transformations on the scores and the strength of their correlations. We have seen that the strength of the correlations decreases for the PC data. Principal Component Analysis effectively reduces the dimension, but by doing so, important structure in data may be lost or obscured. The highest correlation between the parts of the reduced data in the preceding example illustrates the loss in correlation strength between the parts of the transformed data that has occurred.

As we will see in later chapters, too, Principal Component Analysis is an effective dimension-reduction method, but structure may be obscured in the process. Whether this structure is relevant needs to be considered in each case.

3.5.2 Transforms with Non-Singular Matrices

The preceding section demonstrates that the CCs of the original data can differ from those of the transformed data for linear transformations with singular matrices. In this section we focus on non-singular matrices. Theorem 3.11 remains unchanged, but we are now able to explicitly compare the CCs of the original and the transformed vectors. The key properties of the transformed CCs are presented in Theorem 3.12. The results are useful in their own right but also show some interesting features of CCs, the associated eigenvectors and canonical transforms. Because Theorem 3.12 is of particular interest for data, I summarise the data results and point out relevant changes from the population case.

Theorem 3.12 *For $\rho = 1,2$, let $\mathbf{X}^{[\rho]} \sim (\boldsymbol{\mu}_\rho, \Sigma_\rho)$, and assume that the Σ_ρ are non-singular with rank d_ρ. Let C be the canonical correlation matrix of $\mathbf{X}^{[1]}$ and $\mathbf{X}^{[2]}$, and let r be the rank of C and $P\Upsilon Q^{\mathsf{T}}$ its singular value decomposition. Let A_ρ be non-singular matrices of size $d_\rho \times d_\rho$, and let $\mathbf{a}_\rho$ be d_ρ-dimensional vectors. Put*

$$\mathbf{T}^{[\rho]} = A_\rho \mathbf{X}^{[\rho]} + \mathbf{a}_\rho.$$

Let C_{T} be the canonical correlation matrix of $\mathbf{T}^{[1]}$ and $\mathbf{T}^{[2]}$, and write $C_{\mathrm{T}} = P_{\mathrm{T}}\Upsilon_{\mathrm{T}} Q_{\mathrm{T}}^{\mathsf{T}}$ for its singular value decomposition. The following hold:

1. *C_{T} and C have the same singular values, and hence $\Upsilon_{\mathrm{T}} = \Upsilon$.*
2. *For $k, \ell \le r$, the kth left and the ℓth right eigenvectors $\mathbf{p}_{T,k}$ and $\mathbf{q}_{T,\ell}$ of C_{T} and the corresponding canonical transforms $\boldsymbol{\varphi}_{T,k}$ and $\boldsymbol{\psi}_{T,\ell}$ of the $\mathbf{T}^{[\rho]}$ are related to the analogous quantities of the $\mathbf{X}^{[\rho]}$ by*

$$\mathbf{p}_{T,k} = \left(A_1\Sigma_1 A_1^{\mathsf{T}}\right)^{1/2} (A_1^{\mathsf{T}})^{-1} \Sigma_1^{-1/2}\mathbf{p}_k \quad \textit{and} \quad \mathbf{q}_{T,\ell} = \left(A_2\Sigma_2 A_2^{\mathsf{T}}\right)^{1/2} (A_2^{\mathsf{T}})^{-1} \Sigma_2^{-1/2}\mathbf{q}_\ell,$$

$$\boldsymbol{\varphi}_{T,k} = (A_1^{\mathsf{T}})^{-1} \Sigma_1^{-1/2}\boldsymbol{\varphi}_k \qquad \textit{and} \quad \boldsymbol{\psi}_{T,\ell} = (A_2^{\mathsf{T}})^{-1} \Sigma_2^{-1/2}\boldsymbol{\psi}_\ell.$$

3. *The kth and ℓth canonical correlation scores of $\mathbf{T}$ are*

$$U_{T,k} = \mathbf{p}_k^{\mathsf{T}} \Sigma_1^{-1/2} (\mathbf{X}^{[1]} - \boldsymbol{\mu}_1)$$

and

$$V_{T,\ell} = \mathbf{q}_\ell^{\mathsf{T}} \Sigma_2^{-1/2} (\mathbf{X}^{[2]} - \boldsymbol{\mu}_2),$$

and their covariance matrix is

$$\operatorname{var}\left(\begin{bmatrix} \mathbf{U}_{\mathrm{T}}^{(k)} \\ \mathbf{V}_{\mathrm{T}}^{(\ell)} \end{bmatrix}\right) = \begin{pmatrix} \mathbf{I}_{k\times k} & \boldsymbol{\Upsilon}_{k\times\ell} \\ \boldsymbol{\Upsilon}_{k\times\ell}^{\mathsf{T}} & \mathbf{I}_{\ell\times\ell} \end{pmatrix}.$$

The theorem states that the strength of the correlation is the same for the original and transformed data. The weights which combine the raw or transformed data may, however, differ. Thus the theorem establishes the invariance of canonical correlations under non-singular linear transformations, and it shows this invariance by comparing the singular values and CC scores of the original and transformed data. We find that

- the singular values of the canonical correlation matrices of the random vectors and the transformed vectors are the same,
- the canonical correlation scores of the random vectors and the transformed random vectors are identical (up to a sign), that is,

$$U_{T,k} = U_k \qquad \text{and} \qquad V_{T,\ell} = V_\ell,$$

  for $k, \ell = 1, \ldots, r$, and
- consequently, the covariance matrix of the CC scores remains the same, namely,

$$\operatorname{cov}(U_{T,k}, V_{T,\ell}) = \operatorname{cov}(U_k, V_\ell) = \upsilon_k \delta_{k\ell}.$$

Before we look at a proof of Theorem 3.12, we consider what changes occur when we deal with transformed data

$$\mathbb{T}^{[\rho]} = A_\rho \mathbb{X}^{[\rho]} + \mathbf{a}_\rho \qquad \text{for } \rho = 1, 2.$$

Going from the population to the sample, we replace the true parameters by their estimators. So the means are replaced by their sample means, the covariance matrices Σ by the sample covariance matrices S and the canonical correlation matrix C_T by $\widehat{C}_T$. The most noticeable difference is the change from the pairs of scalar canonical correlation scores to pairs of vectors of length n when we consider data.

I present the proof of Theorem 3.12, because it reveals important facts and relationships. To make the proof more transparent, I begin with some notation and then prove two lemmas. For $\mathbf{X}^{[\rho]}$, $\mathbf{T}^{[\rho]}$, C and C_T as in Theorem 3.12, put

$$\begin{aligned} R^{[C]} &= CC^\top \quad \text{and} \quad K = [\operatorname{var}(\mathbf{X}^{[1]})]^{-1/2} R^{[C]} [\operatorname{var}(\mathbf{X}^{[1]})]^{1/2}; \\ R_T^{[C]} &= C_T C_T^\top \quad \text{and} \quad K_T = [\operatorname{var}(\mathbf{T}^{[1]})]^{-1/2} R_T^{[C]} [\operatorname{var}(\mathbf{T}^{[1]})]^{1/2}. \end{aligned} \tag{3.22}$$

A comparison with (3.4) shows that I have omitted the second superscript '1' in $R^{[C]}$. In the current proof we refer to $CC^\top$ and so only make the distinction when necessary. The sequence

$$C_T \longleftrightarrow R_T^{[C]} \longleftrightarrow K_T \longleftrightarrow K \longleftrightarrow R^{[C]} \longleftrightarrow C \tag{3.23}$$

will be useful in the proof of Theorem 3.12 because the theorem makes statements about the endpoints C_T and C in the sequence. As we shall see in the proofs, relationships about K_T and K are the starting points because we can show that they are similar matrices.

Lemma 1 *Assume that the $\mathbf{X}^{[\rho]}$ satisfy the assumptions of Theorem 3.12. Let $\upsilon_1 > \upsilon_2 > \cdots > \upsilon_r$ be the singular values of C. The following hold.*

1. *The matrices $R^{[C]}$, K, $R_T^{[C]}$ and K_T as in (3.22) have the same eigenvalues*

$$\upsilon_1^2 > \upsilon_2^2 > \cdots > \upsilon_r^2.$$

2. *The singular values of C_T coincide with those of C.*

Proof To prove the statements about the eigenvalues and singular values, we will make repeated use of the fact that similar matrices have the same eigenvalues; see Result 1.8 of Section 1.5.1. So our proof needs to establish similarity relationships between matrices.

The aim is to relate the singular values of C_T and C. As it is not easy to do this directly, we travel along the path in (3.23) and exhibit relationships between the neighbours in (3.23).

By Proposition 3.1, $R^{[C]}$ has positive eigenvalues, and the singular values of C are the positive square roots of the eigenvalues of $R^{[C]}$. A similar relationship holds for $R_T^{[C]}$ and C_T. These relationships deal with the two ends of the sequence (3.23).

The definition of K implies that it is similar to $R^{[C]}$, so K and $R^{[C]}$ have the same eigenvalues. An analogous result holds for K_T and $R_T^{[C]}$. It remains to establish the similarity of K and K_T. This last similarity will establish that the singular values of C and C_T are identical.

We begin with K. We substitute the expression for $R^{[C]}$ and re-write K as follows:

$$\begin{aligned} K &= \Sigma_1^{-1/2} R^{[C]} \Sigma_1^{1/2} \\ &= \Sigma_1^{-1/2} \Sigma_1^{-1/2} \Sigma_{12} \Sigma_2^{-1/2} \Sigma_2^{-1/2} \Sigma_{12}^{\mathsf{T}} \Sigma_1^{-1/2} \Sigma_1^{1/2} \\ &= \Sigma_1^{-1} \Sigma_{12} \Sigma_2^{-1} \Sigma_{12}^{\mathsf{T}}. \end{aligned} \tag{3.24}$$

A similar expression holds for K_{T}. It remains to show that K and K_{T} are similar. To do this, we go back to the definition of K_{T} and use the fact that A_ρ and $\mathrm{var}(\mathbf{T})$ are invertible. Now

$$\begin{aligned} K_{\mathrm{T}} &= [\,\mathrm{var}(\mathbf{T}^{[1]})]^{-1} \mathrm{cov}(\mathbf{T}^{[1]}, \mathbf{T}^{[2]}) [\,\mathrm{var}(\mathbf{T}^{[2]})]^{-1} \mathrm{cov}(\mathbf{T}^{[1]}, \mathbf{T}^{[2]})^{\mathsf{T}} \\ &= (A_1 \Sigma_1 A_1^{\mathsf{T}})^{-1} (A_1 \Sigma_{12} A_2^{\mathsf{T}}) (A_2 \Sigma_2 A_2^{\mathsf{T}})^{-1} (A_2 \Sigma_{12}^{\mathsf{T}} A_1^{\mathsf{T}}) \\ &= (A_1^{\mathsf{T}})^{-1} \Sigma_1^{-1} A_1^{-1} A_1 \Sigma_{12} A_2^{\mathsf{T}} (A_2^{\mathsf{T}})^{-1} \Sigma_2^{-1} A_2^{-1} A_2 \Sigma_{12}^{\mathsf{T}} A_1^{\mathsf{T}} \\ &= (A_1^{\mathsf{T}})^{-1} \Sigma_1^{-1} \Sigma_{12} \Sigma_2^{-1} \Sigma_{12}^{\mathsf{T}} A_1^{\mathsf{T}} \\ &= (A_1^{\mathsf{T}})^{-1} K A_1^{\mathsf{T}}. \end{aligned} \tag{3.25}$$

The second equality in (3.25) uses the variance results of part 2 of Theorem 3.11. To show the last equality, use (3.24). The sequence of equalities establishes the similarity of the two matrices.

So far we have shown that the four terms in the middle of the sequence (3.23) are similar matrices, so have the same eigenvalues. This proves part 1 of the lemma. Because the singular values of C_{T} are the square roots of the eigenvalues of $R_{\mathrm{T}}^{[C]}$, C_{T} and C have the same singular values. ∎

Lemma 2 *Assume that the $\mathbf{X}^{[\rho]}$ satisfy the assumptions of Theorem 3.12. Let $\upsilon > 0$ be a singular value of C with corresponding left eigenvector $\mathbf{p}$. Define $R^{[C]}$, K and K_{T} as in (3.22).*

1. *If $\mathbf{r}$ is the eigenvector of $R^{[C]}$ corresponding to υ, then*

$$\mathbf{r} = \mathbf{p}.$$

2. *If $\mathbf{s}$ is the eigenvector of K corresponding to υ, then*

$$\mathbf{s} = \frac{\Sigma_1^{-1/2}\mathbf{p}}{\left\|\Sigma_1^{-1/2}\mathbf{p}\right\|} \quad \textit{and} \quad \mathbf{p} = \frac{\Sigma_1^{1/2}\mathbf{s}}{\left\|\Sigma_1^{1/2}\mathbf{s}\right\|}.$$

3. *If $\mathbf{s}_{\mathrm{T}}$ is the eigenvector of K_{T} corresponding to υ, then*

$$\mathbf{s}_{\mathrm{T}} = \frac{(A_1^{\mathsf{T}})^{-1}\mathbf{s}}{\left\|(A_1^{\mathsf{T}})^{-1}\mathbf{s}\right\|} \quad \textit{and} \quad \mathbf{s} = \frac{A_1^{\mathsf{T}}\mathbf{s}_{\mathrm{T}}}{\left\|A_1^{\mathsf{T}}\mathbf{s}_{\mathrm{T}}\right\|}.$$

Proof Part 1 follows directly from Proposition 3.1 because the left eigenvectors of C are the eigenvectors of $R^{[C]}$. To show part 2, we establish relationships between appropriate eigenvectors of objects in the sequence (3.23).

We first exhibit relationships between eigenvectors of similar matrices. For this purpose, let B and D be similar matrices which satisfy $B = EDE^{-1}$ for some matrix E. Let λ be an eigenvalue of D and hence also of B. Let $\mathbf{e}$ be the eigenvector of B which corresponds to λ. We have

$$B\mathbf{e} = \lambda\mathbf{e} = EDE^{-1}\mathbf{e}.$$

Pre-multiplying by the matrix E^{-1} leads to

$$E^{-1}B\mathbf{e} = \lambda E^{-1}\mathbf{e} = DE^{-1}\mathbf{e}.$$

Let $\boldsymbol{\eta}$ be the eigenvector of D which corresponds to λ. The uniqueness of the eigenvalue–eigenvector decomposition implies that $E^{-1}\mathbf{e}$ is a scalar multiple of the eigenvector $\boldsymbol{\eta}$ of D. This last fact leads to the relationships

$$\boldsymbol{\eta} = \frac{1}{c_1}E^{-1}\mathbf{e}$$

or equivalently,

$$\mathbf{e} = c_1 E\boldsymbol{\eta} \qquad \text{for some real } c_1, \tag{3.26}$$

and E therefore is the link between the eigenvectors. Unless E is an isometry, c_1 is required because eigenvectors in this book are vectors of norm 1.

We return to the matrices $R^{[C]}$ and K. Fix $k \le r$, the rank of C. Let υ be the kth eigenvalue of $R^{[C]}$ and hence also of K, and consider the eigenvector $\mathbf{p}$ of $R^{[C]}$ and $\mathbf{s}$ of K which correspond to υ. Because $K = [\operatorname{var}(\mathbf{X}^{[1]})]^{-1/2} R^{[C]} [\operatorname{var}(\mathbf{X}^{[1]})]^{1/2}$, (3.26) implies that

$$\mathbf{p} = c_2[\operatorname{var}(\mathbf{X}^{[1]})]^{1/2}\mathbf{s} = c_2\Sigma_1^{1/2}\mathbf{s},$$

for some real c_2. Now $\mathbf{p}$ has unit norm, so $c_2^{-1} = \left\|\Sigma_1^{1/2}\mathbf{s}\right\|$, and the results follows. A similar calculation leads to the results in part 3. ■

We return to Theorem 3.12 and prove it with the help of the two lemmas.

Proof of Theorem 3.12 Part 1 follows from Lemma 1. For part 2, we need to find relationships between the eigenvectors of C and C_{T}. We obtain this relationship via the sequence (3.23) and with the help of Lemma 2. By part 1 of Lemma 2 it suffices to consider the sequence

$$R_{\mathrm{T}}^{[C]} \longleftrightarrow K_{\mathrm{T}} \longleftrightarrow K \longleftrightarrow R^{[C]}.$$

We start with the eigenvectors of $R_{\mathrm{T}}^{[C]}$. Fix $k \le r$. Let υ^2 be the kth eigenvalue of $R_{\mathrm{T}}^{[C]}$ and hence also of K_{T}, K and $R^{[C]}$. Let $\mathbf{p}_{\mathrm{T}}$ and $\mathbf{p}$ be the corresponding eigenvectors of $R_{\mathrm{T}}^{[C]}$ and $R^{[C]}$, and $\mathbf{s}_{\mathrm{T}}$ and $\mathbf{s}$ those of K_{T} and K, respectively. We start with the pair $(\mathbf{p}_{\mathrm{T}}, \mathbf{s}_{\mathrm{T}})$. From the definitions (3.22), we obtain

$$\begin{aligned}
\mathbf{p}_{\mathrm{T}} &= c_1[\operatorname{var}(\mathbf{T}^{[1]})]^{1/2}\mathbf{s}_{\mathrm{T}}\\
&= c_1c_2[\operatorname{var}(\mathbf{T}^{[1]})]^{1/2}(A_1^{\mathsf{T}})^{-1}\mathbf{s}\\
&= c_1c_2c_3[\operatorname{var}(\mathbf{T}^{[1]})]^{1/2}(A_1^{\mathsf{T}})^{-1}\Sigma_1^{-1/2}\mathbf{p}
\end{aligned}$$

by parts 2 and 3 of Lemma 2, where the constants c_i are appropriately chosen. Put $c = c_1c_2c_3$. We find the value of c by calculating the norm of $\widetilde{\mathbf{p}} = [\operatorname{var}(\mathbf{T}^{[1]})]^{1/2}(A_1^{\mathsf{T}})^{-1}\Sigma_1^{-1/2}\mathbf{p}$. In the next calculation, I omit the subscript and superscript 1 in $\mathbf{T}$, A and Σ. Now,

$$\begin{aligned}\|\widetilde{\mathbf{p}}\|^2 &= \mathbf{p}^\top \Sigma^{-1/2} A^{-1} (A\Sigma A^\top)^{1/2} (A\Sigma A^\top)^{1/2} (A^\top)^{-1} \Sigma^{-1/2} \mathbf{p} \\ &= \mathbf{p}^\top \Sigma^{-1/2} A^{-1} (A\Sigma A^\top)(A^\top)^{-1} \Sigma^{-1/2} \mathbf{p} \\ &= \mathbf{p}^\top \Sigma^{-1/2} \Sigma \Sigma^{-1/2} \mathbf{p} = \|\mathbf{p}\|^2 = 1\end{aligned}$$

follows from the definition of $\operatorname{var}(\mathbf{T})$ and the fact that $(A\Sigma A^\top)^{1/2}(A\Sigma A^\top)^{1/2} = A\Sigma A^\top$. The calculations show that $c = \pm 1$, thus giving the desired result.

For the eigenvectors $\mathbf{q}_T$ and $\mathbf{q}$, we base the calculations on $R^{[C,2]} = C^\top C$ and recall that the eigenvectors of $C^\top C$ are the right eigenvectors of C. This establishes the relationship between $\mathbf{q}_T$ and $\mathbf{q}$. The results for canonical transforms follow from the preceding calculations and the definition of the canonical transforms in (3.8).

Part 3 is a consequence of the definitions and the results established in parts 1 and 2. I now derive the results for $\mathbf{T}^{[2]}$. Fix $k \le r$. I omit the indices k for the eigenvector and the superscript 2 in $\mathbf{T}^{[2]}$, $\mathbf{X}^{[2]}$ and the matrices A and Σ. From (3.6), we find that

$$V_T = \mathbf{q}_T^\top [\operatorname{var}(\mathbf{T})]^{-1/2} (\mathbf{T} - \mathbb{E}\mathbf{T}). \tag{3.27}$$

We substitute the expressions for the mean and covariance matrix, established in Theorem 3.11, and the expression for $\mathbf{q}$ from part 2 of the current theorem, into (3.27). It follows that

$$\begin{aligned}V_T &= \left[\mathbf{q}^\top \Sigma^{-1/2} A^{-1} \left(A\Sigma A^\top\right)^{1/2}\right] \left(A\Sigma A^\top\right)^{-1/2} (A\mathbf{X} + \mathbf{a} - A\boldsymbol{\mu} - \mathbf{a}) \\ &= \mathbf{q}^\top \Sigma^{-1/2} A^{-1} A (\mathbf{X} - \boldsymbol{\mu}) \\ &= \mathbf{q}^\top \Sigma^{-1/2} (\mathbf{X} - \boldsymbol{\mu}) = V,\end{aligned}$$

where V is the corresponding CC score of $\mathbf{X}$. Of course, $V_T = -\mathbf{q}^\top \Sigma^{-1/2}(\mathbf{X} - \boldsymbol{\mu})$ is also a solution because eigenvectors are unique only up to a sign. The remainder follows from Theorem 3.6 because the CC scores of the raw and transformed vectors are the same. ■

In the proof of Theorem 3.12 we explicitly use the fact that the transformations A_ρ are non-singular. If this assumption is violated, then the results may no longer hold. I illustrate Theorem 3.12 with an example.

Example 3.7 The **income** data are an extract from a survey in the San Francisco Bay Area based on more than 9,000 questionnaires. The aim of the survey is to derive a prediction of the annual household income from the other demographic attributes. The income data are also used in Hastie, Tibshirani, and Friedman (2001).

Some of the fourteen variables are not suitable for our purpose. We consider the nine variables listed in Table 3.5 and the first 1,000 records, excluding records with missing data. Some of these nine variables are categorical, but in the analysis I will not distinguish between the different types of variables. The purpose of this analysis is to illustrate the effect of transformations of the data, and we are not concerned here with interpretations or effect of individual variables. I have split the variables into two groups: $\mathbb{X}^{[1]}$ are the personal attributes, other than income, and $\mathbb{X}^{[2]}$ are the household attributes, with income as the first variable. The raw data are shown in the top panel of Figure 3.5, with the variables shown

Table 3.5 *Variables of the income data from Example 3.7*

Personal $\mathbb{X}^{[1]}$	Household $\mathbb{X}^{[2]}$
Marital status	Income
Age	No. in household
Level of education	No. under 18
Occupation	Householder status
	Type of home

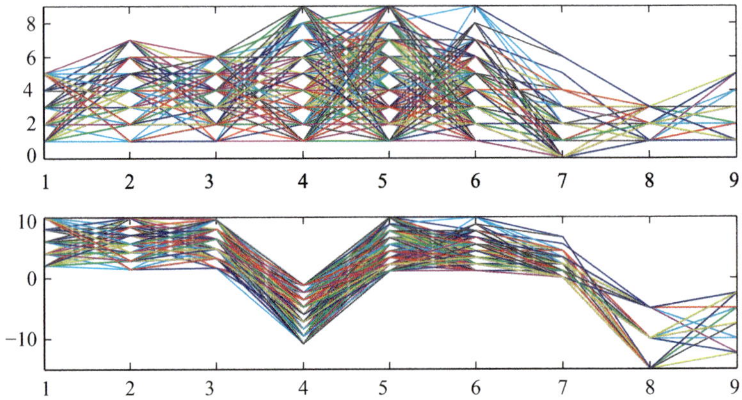

Figure 3.5 Income data from Example 3.7: (*top*): raw data; (*bottom*): transformed data.

on the x-axis, starting with the variables of $\mathbb{X}^{[1]}$, and followed by those of $\mathbb{X}^{[2]}$ in the order they are listed in the Table 3.5.

It is not easy to understand or interpret the parallel coordinate plot of the raw data. The lack of clarity is a result of the way the data are coded: large values for income represent a large income, whereas the variable occupation has a 'one' for 'professional', and its largest positive integer refers to 'unemployed'; hence occupation is negatively correlated with income. A consequence is the criss-crossing of the lines in the top panel.

We transform the data in order to disentangle this crossing over. Put $\mathbf{a} = \mathbf{0}$ and

$$A = \text{diag}\big(2.0 \quad 1.4 \quad 1.6 \quad -1.2 \quad 1.1 \quad 1.1 \quad 1.1 \quad -5.0 \quad -2.5\big).$$

The transformation $\mathbb{X} \to A\mathbb{X}$ scales the variables and changes the sign of variables such as *occupation*. The transformed data are displayed in the bottom panel of Figure 3.5. Variables 4, 8, and 9 have smaller values than the others, a consequence of the particular transformation I have chosen.

The matrix of canonical correlations has singular values 0.7762, 0.4526, 0.3312, and 0.1082, and these coincide with the singular values of the transformed canonical correlation matrix. The entries of the first normalised canonical transforms for both raw and transformed data are given in Table 3.6. The variable *age* has the highest weight for both the raw and transformed data, followed by *education*. *Occupation* has the smallest weight and opposite signs for the raw and transformed data. The change in sign is a consequence of the negative entry in A for *occupation*. *Householder status* has the highest weight among the $\mathbb{X}^{[2]}$ variables and so is most correlated with the $\mathbb{X}^{[1]}$ data. This is followed by the *income* variable.

Table 3.6 *First raw and transformed normalised canonical transforms from Example 3.7*

$\mathbb{X}^{[1]}$	$\widehat{\varphi}$ raw	$\widehat{\varphi}$ trans	$\mathbb{X}^{[2]}$	$\widehat{\psi}$ raw	$\widehat{\psi}$ trans
Marital status	0.4522	0.3461	Income	−0.1242	−0.4565
Age	−0.6862	−0.7502	No. in household	0.1035	0.3802
Education	−0.5441	−0.5205	No. under 18	0.0284	0.1045
Occupation	0.1690	−0.2155	Householder status	0.9864	−0.7974
			Type of home	−0.0105	0.0170

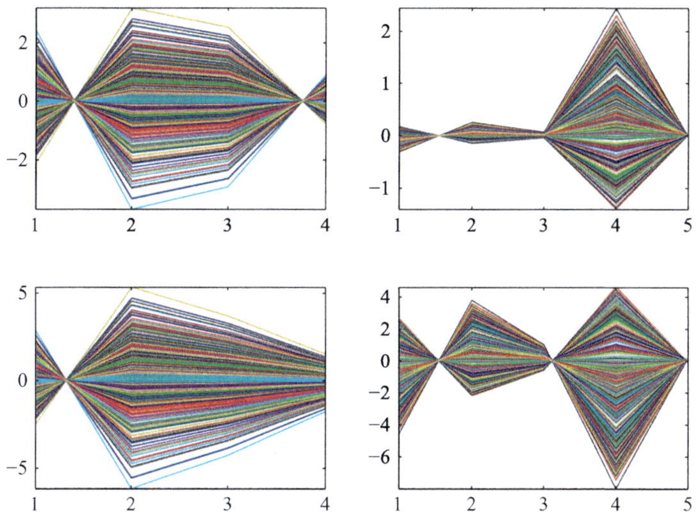

Figure 3.6 Contributions of CC scores along first canonical transforms from Example 3.7: (*top row*) raw data; (*bottom row*) transformed data. The $\mathbb{X}^{[1]}$ plots are shown in the left panels and the $\mathbb{X}^{[2]}$ plots on the right.

Again, we see that the signs of the weights change for negative entries of A, here for the variables *householder status* and *type of home*.

Figure 3.6 shows the information given in Table 3.6, and in particular highlights the change in sign of the weights of the canonical transforms. The figure shows the contributions of the first CC scores in the direction of the first canonical transforms, that is, parallel coordinate plots of $\widehat{\varphi}_1 \mathbf{U}_{\bullet 1}$ for $\mathbb{X}^{[1]}$ and $\widehat{\psi}_1 \mathbf{V}_{\bullet 1}$ for $\mathbb{X}^{[2]}$ with the variable numbers on the x-axis. The $\mathbb{X}^{[1]}$ plots are displayed in the left panels and the corresponding $\mathbb{X}^{[2]}$ plots in the right panels. The top row shows the raw data, and the bottom row shows the transformed data.

The plots show clearly where a change in sign occurs in the entries of the canonical transforms: the lines cross over. The sign change between variables 3 and 4 of $\widehat{\varphi}$ is apparent in the raw data but no longer exists in the transformed data. Similar sign changes exist for the $\mathbb{X}^{[2]}$ plots. Further, because of the larger weights of the first two variables of the $\mathbb{X}^{[2]}$ transformed data, these two variables have much more variability for the transformed data.

It is worth noting that the CC scores of the raw and transformed data agree because the matrices S_ρ and A are invertible. Hence, as stated in part 3 of Theorem 3.12, the CC scores are invariant under this transformation. ∎

For the *income* data, we applied a transformation to the data, but in other cases the data may only be available in transformed form. Example 3.7 shows the differences between the analysis of the raw and transformed data. If the desired result is the strength of the correlation between combinations of variables, then the transformation is not required. If a more detailed analysis is appropriate, then the raw and transformed data allow different insights into the data. The correlation analysis only shows the strength of the relationship and not the sign, and the decrease rather than an increase of a particular variable could be important.

The transformation of Example 3.6 is based on a singular matrix, and as we have seen there, the CCs are not invariant under the transformation. In Example 3.7, A is non-singular, and the CCs remain the same. Thus the simple univariate case does not carry across to the multivariate scenario in general, and care needs to be taken when working with transformed data.

3.5.3 Canonical Correlations for Scaled Data

Scaling of a random vector or data decreases the effect of variables whose scale is much larger than that of the other variables. In Principal Component Analysis, variables with large values dominate and can hide important information in the process. Scaling such variables prior to a principal component analysis is often advisable.

In this section we explore scaling prior to a Canonical Correlation Analysis. Scaling is a linear transformation, and Theorems 3.11 and 3.12 therefore apply to scaled data.

For $\rho = 1,2$, let $\mathbf{X}^{[\rho]} \sim (\boldsymbol{\mu}_\rho, \Sigma_\rho)$, and let $\Sigma_{\text{diag},\rho}$ be the diagonal matrix as in (2.16) of Section 2.6.1. Then

$$\mathbf{X}^{[\rho]}_{\text{scale}} = \Sigma_{\text{diag},\rho}^{-1/2}\left(\mathbf{X}^{[\rho]} - \boldsymbol{\mu}_\rho\right)$$

is the scaled vector. Similarly, for data $\mathbb{X}^{[\rho]} \sim \mathsf{Sam}(\overline{\mathbf{X}}_\rho, S_\rho)$ and diagonal matrix $S_{\text{diag},\rho}$, the scaled data are

$$\mathbb{X}^{[\rho]}_{\text{scale}} = S_{\text{diag},\rho}^{-1/2}\left(\mathbb{X}^{[\rho]} - \overline{\mathbf{X}}_\rho\right).$$

Using the transformation set-up, the scaled vector

$$\mathbf{T}^{[\rho]} = \Sigma_{\text{diag},\rho}^{-1/2}\left(\mathbf{X}^{[\rho]} - \boldsymbol{\mu}_\rho\right), \tag{3.28}$$

with

$$A_\rho = \Sigma_{\text{diag},\rho}^{-1/2} \qquad \text{and} \qquad \mathbf{a}_\rho = -\Sigma_{\text{diag},\rho}^{-1/2}\boldsymbol{\mu}_\rho.$$

If the covariance matrices Σ_ρ of $\mathbb{X}^{[\rho]}$ are invertible, then, by Theorem 3.12, the CC scores of the scaled vector are the same as those of the original vector, but the eigenvectors $\widehat{\mathbf{p}}_1$ of $\widehat{C}$ and $\widehat{\mathbf{p}}_{T,1}$ of $\widehat{C}_T$ differ, as we shall see in the next example. In the Problems at the end of Part I we derive an expression for the canonical correlation matrix of the scaled data and interpret it.

Example 3.8 We continue with the direct and indirect measures of the **illicit drug market** data and focus on the weights of the first CC vectors for the raw and scaled data. Table 3.7

Table 3.7 *First left and right eigenvectors of the canonical correlation matrix from Example 3.8*

Variable no.	1	2	3	4	5	6
Raw: $\widehat{\mathbf{p}}_1$	0.30	−0.58	−0.11	0.39	0.47	0.02
Scaled: $\widehat{\mathbf{p}}_{T,1}$	0.34	−0.54	−0.30	0.27	0.38	0.26
Variable no.	7	8	9	10	11	12
Raw: $\widehat{\mathbf{p}}_1$	−0.08	−0.31	−0.13	−0.10	−0.24	0.07
Scaled: $\widehat{\mathbf{p}}_{T,1}$	−0.19	−0.28	−0.26	0.01	−0.18	0.03
Variable no.	1	2	3	4	5	—
Raw: $\widehat{\mathbf{q}}_1$	−0.25	−0.22	0.49	−0.43	−0.68	—
Scaled: $\widehat{\mathbf{q}}_{T,1}$	−0.41	−0.28	0.50	−0.49	−0.52	—

lists the entries of the first left and right eigenvectors $\widehat{\mathbf{p}}_1$ and $\widehat{\mathbf{q}}_1$ of the original data and $\widehat{\mathbf{p}}_{T,1}$ and $\widehat{\mathbf{q}}_{T,1}$ of the scaled data. The variables in the table are numbered 1 to 12 for the direct measures $\mathbb{X}^{[1]}$ and 1 to 5 for the indirect measures $\mathbb{X}^{[2]}$. The variable names are given in Table 3.2.

For $\mathbb{X}^{[2]}$, the signs of the eigenvector weights and their ranking (in terms of absolute value) are the same for the raw and scaled data. This is not the case for $\mathbb{X}^{[1]}$.

The two pairs of variables with the largest absolute weights deserve further comment. For the $\mathbb{X}^{[1]}$ data, variable 2 (*amphetamine possession offences*) has the largest absolute weight, and variable 5 (*ADIS heroin*) has the second-largest weight, and this order is the same for the raw and scaled data. The two largest absolute weights for the $\mathbb{X}^{[2]}$ data belong to variable 5 (*steal from motor vehicles*) and variable 3 (*PSB new registrations*). These four variables stand out in our previous analysis in Example 3.3: *amphetamine possession offences* and *steal from motor vehicles* have the highest single correlation coefficient, and *ADIS heroin* and *PSB new registrations* have the second highest. Further, the highest contributors to the canonical transforms $\widehat{\boldsymbol{\varphi}}_1$ and $\widehat{\boldsymbol{\psi}}_1$ are also these four variables, but in this case in opposite order, as we noted in Example 3.3. These observations suggest that these four variables are jointly responsible for the correlation behaviour of the data.

A comparison with the CC scores obtained in Example 3.6 leads to interesting observations. The first four PC transformations of Example 3.6 result in different CC scores and different correlation coefficients from those obtained in the preceding analysis. Further, the two sets of CC scores obtained from the four-dimensional PC data differ depending on whether the PC transformations are applied to the raw or scaled data. If, on the other hand, a canonical correlation analysis is applied to all d_ρ PCs, then the derived CC scores are related to the sphered PC vectors by an orthogonal transformation. We derive this orthogonal matrix E in in the Problems at the end of Part I. ■

In light of Theorem 3.12 and the analysis in Example 3.8, it is worth reflecting on the circumstances under which a canonical correlation analysis of PCs is advisable. If the main focus of the analysis is the examination of the relationship between two parts of the data, then a prior partial principal component analysis could decrease the effect of variables which do not contribute strongly to variance but which might be strongly related to the other part of the data. On the other hand, if the original variables are ranked as described in Section 13.3, then a correlation analysis of the PCs can decrease the effect of *noise* variables in the analysis.

3.5.4 Maximum Covariance Analysis

In geophysics and climatology, patterns of spatial dependence between different types of geophysical measurements are the objects of interest. The observations are measured on a number of quantities from which one wants to extract the most coherent patterns. This type of problem fits naturally into the framework of Canonical Correlation Analysis. Traditionally, however, the geophysics and related communities have followed a slightly different path, known as Maximum Covariance Analysis.

We take a brief look at Maximum Covariance Analysis without going into details. As in Canonical Correlation Analysis, in Maximum Covariance Analysis one deals with two distinct parts of a vector or data and aims at finding the strongest relationship between the two parts. The fundamental object is the between covariance matrix, which is analysed directly.

For $\rho = 1, 2$, let $\mathbf{X}^{[\rho]}$ be d_ρ-variate random vectors. Let Σ_{12} be the between covariance matrix of the two vectors, with singular value decomposition $\Sigma_{12} = EDF^{\mathsf{T}}$ and rank r. We define r-variate coefficient vectors $\mathbf{A}$ and $\mathbf{B}$ by

$$\mathbf{A} = E^{\mathsf{T}}\mathbf{X}^{[1]} \qquad \text{and} \qquad \mathbf{B} = F^{\mathsf{T}}\mathbf{X}^{[2]}$$

for suitable matrices E and F. The vectors $\mathbf{A}$ and $\mathbf{B}$ are analogous to the canonical correlation vectors and so could be thought of as 'covariance scores'. The pair (A_1, B_1), the first entries of $\mathbf{A}$ and $\mathbf{B}$, are most strongly correlated. Often the coefficient vectors $\mathbf{A}$ and $\mathbf{B}$ are normalised. The normalised pairs of coefficient vectors are further analysed and used to derive patterns of spatial dependence.

For data $\mathbb{X}^{[1]}$ and $\mathbb{X}^{[2]}$, the sample between covariance matrix S_{12} replaces the between covariance matrix Σ_{12}, and the coefficient vectors become coefficient matrices whose columns correspond to the n observations.

The basic difference between Maximum Covariance Analysis and Canonical Correlation Analysis lies in the matrix which drives the analysis: Σ_{12} is the central object in Maximum Covariance Analysis, whereas $C = \Sigma_1^{-1/2}\Sigma_{12}\Sigma_2^{-1/2}$ is central to Canonical Correlation Analysis. The between covariance matrix contains the raw quantities, whereas the matrix C has an easier statistical interpretation in terms of the strengths of the relationships. For more information on and interpretation of Maximum Covariance Analysis in the physical and climate sciences, see von Storch and Zwiers (1999).

3.6 Asymptotic Considerations and Tests for Correlation

An asymptotic theory for Canonical Correlation Analysis is more complex than that for Principal Component Analysis, even for normally distributed data. The main reason for the added complexity is the fact that Canonical Correlation Analysis involves the singular values and pairs of eigenvectors of the matrix of canonical correlations C. In the sample case, the matrix $\widehat{C}$ is the product of functions of the covariance matrices S_1 and S_2 and the between covariance matrix S_{12}. In a Gaussian setting, the matrices S_1 and S_2, as well as the combined covariance matrix

$$S = \begin{pmatrix} S_1 & S_{12} \\ S_{12}^{\mathsf{T}} & S_2 \end{pmatrix},$$

converge to the corresponding true covariance matrices. Convergence of $\widehat{C}$ to C further requires the convergence of inverses of the covariance matrices and their products. We will not pursue these convergence ideas. Instead, I only mention that under the assumptions of normality of the random samples, the singular values and the eigenvectors of $\widehat{C}$ converge to the corresponding true population parameters for large enough sample sizes. For details, see Kshirsagar (1972, pp. 261ff).

The goal of Canonical Correlation Analysis is to determine the relationship between two sets of variables. It is therefore relevant to examine whether such a relationship actually exists, that is, we want to know whether the correlation coefficients *differ significantly* from 0. We consider two scenarios:

$$H_0 \colon \Sigma_{12} = \mathbf{0} \qquad \text{versus} \qquad H_1 \colon \Sigma_{12} \neq \mathbf{0} \tag{3.29}$$

and

$$H_0 \colon \upsilon_j = \upsilon_{j+1} = \cdots = \upsilon_r = 0, \qquad \text{versus} \qquad H_1 \colon \upsilon_j > 0, \tag{3.30}$$

for some j, where the υ_j are the singular values of $C = \Sigma_1^{-1/2}\Sigma_{12}\Sigma_2^{-1/2}$, and the υ_j are listed in decreasing order. Because the singular values appear in decreasing order, the second scenario is described by a sequence of tests, one for each j:

$$H_0^j \colon \upsilon_{j+1} = 0 \qquad \text{versus} \qquad H_1^j \colon \upsilon_{j+1} > 0.$$

Assuming that the covariance matrices Σ_1 and Σ_2 are invertible, then the scenario (3.29) can be cast in terms of the matrix of canonical correlations C instead of Σ_{12}. In either case, there is no dependence relationship under the null hypothesis, whereas in the tests of (3.30), non-zero correlation exists, and one tests *how many* of the correlations differ significantly from zero. The following theorem, given without proof, addresses both test scenarios (3.29) and (3.30). Early results and proofs, which are based on normal data and on approximations of the likelihood ratio, are given in Bartlett (1938) for part 1 and Bartlett (1939) for part 2, and Kshirsagar (1972) contains a comprehensive proof of part 1 of Theorem 3.13.

To test the hypotheses (3.29) and (3.30), we use the likelihood ratio test statistic. Let L be the likelihood of the data $\mathbb{X}$, and let θ be the parameter of interest. We consider the likelihood ratio test statistic Λ for testing H_0 against H_1:

$$\Lambda(\mathbb{X}) = \frac{\sup_{\theta \in H_0} L(\theta|\mathbb{X})}{\sup_\theta L(\theta|\mathbb{X})}.$$

For details of the likelihood ratio test statistic Λ, see chapter 8 of Casella and Berger (2001).

Theorem 3.13 *Let $\rho = 1, 2$, and let $\mathbb{X}^{[\rho]} = \left[\mathbf{X}_1^{[\rho]} \cdots \mathbf{X}_n^{[\rho]}\right]$ be samples of independent d_ρ-dimensional random vectors such that*

$$\mathbb{X} = \begin{bmatrix} \mathbb{X}^{[1]} \\ \mathbb{X}^{[2]} \end{bmatrix} \sim \mathcal{N}(\boldsymbol{\mu}, \Sigma) \qquad \textit{with} \quad \Sigma = \begin{pmatrix} \Sigma_1 & \Sigma_{12} \\ \Sigma_{12}^{\mathsf{T}} & \Sigma_2 \end{pmatrix}.$$

Let r be the rank of Σ_{12}. Let S and S_ℓ be the sample covariance matrices corresponding to Σ and Σ_ℓ, for $\ell = 1,2$ and 12, and assume that Σ_1 and Σ_2 are invertible. Let $\widehat{\boldsymbol{\upsilon}} = \left[\widehat{\upsilon}_1 \cdots \widehat{\upsilon}_r\right]$ be the singular values of $\widehat{C}$ listed in decreasing order.

1. *Let Λ_1 be the likelihood ratio statistic for testing*

$$H_0\colon C = \mathbf{0} \qquad \text{versus} \qquad H_1\colon C \neq \mathbf{0}.$$

Then the following hold.

(a)

$$-2\log\Lambda_1(\mathbb{X}) = n\log\left[\frac{\det(S_1)\,\det(S_2)}{\det(S)}\right],$$

(b)

$$-2\log\Lambda_1(\mathbb{X}) \approx -\left[n - \frac{1}{2}(d_1 + d_2 + 3)\right]\log\prod_{j=1}^{r}\left(1 - \widehat{\upsilon}_j^2\right). \tag{3.31}$$

(c) *Further, the distribution of $-2\log\Lambda_1(\mathbb{X})$ converges to a χ^2distribution in $d_1 \times d_2$ degrees of freedom as $n \to \infty$.*

2. *Fix $k \le r$. Let $\Lambda_{2,k}$ be the likelihood ratio statistic for testing*

$$H_0^k\colon \upsilon_1 \neq 0, \ldots, \upsilon_k \neq 0 \qquad \text{and} \qquad \upsilon_{k+1} = \cdots = \upsilon_r = 0$$

versus

$$H_1^k\colon \upsilon_j \neq 0 \quad \text{for some } j \ge k+1.$$

Then the following hold.

(a)

$$-2\log\Lambda_{2,k}(\mathbb{X}) \approx -\left[n - \frac{1}{2}(d_1 + d_2 + 3)\right]\log\prod_{j=k+1}^{r}\left(1 - \widehat{\upsilon}_j^2\right), \tag{3.32}$$

(b) *$-2\log\Lambda_{2,k}(\mathbb{X})$ has an approximate χ^2distribution in $(d_1 - k) \times (d_2 - k)$ degrees of freedom as $n \to \infty$.*

In practice, the tests of part 2 of the theorem are more common and typically they are also applied to non-Gaussian data. In the latter case, care may be required in the interpretation of the results. If $C = \mathbf{0}$ is rejected, then at least the largest singular value is non-zero. Obvious starting points for the individual tests are therefore either the second-largest singular value or the smallest. Depending on the decision of the test H_0^1 and, respectively, H_0^{r-1}, one may continue with further tests. Because the tests reveal the number of non-zero singular values, the tests can be employed for estimating the rank of C.

Example 3.9 We continue with the **income** data and test for non-zero canonical correlations. Example 3.7 finds the singular values

$$0.7762, \qquad 0.4526, \qquad 0.3312, \qquad 0.1082$$

for the first 1,000 records. These values express the strength of the correlation between pairs of CC scores.

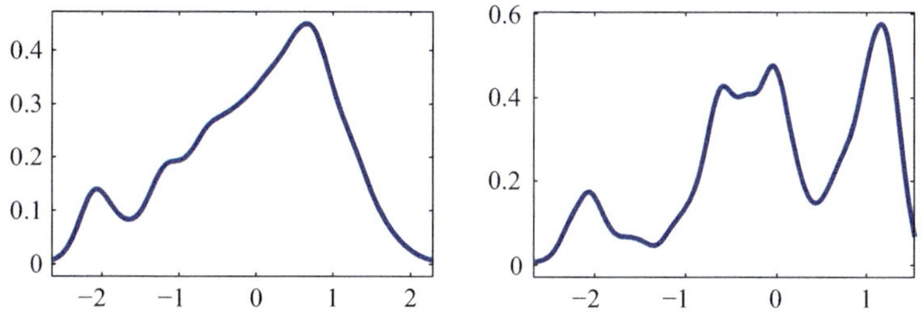

Figure 3.7 Kernel density estimates of the CC scores of Example 3.9 with $\mathbf{U}_{\bullet 1}$ in the left panel and $\mathbf{V}_{\bullet 1}$ in the right panel.

An inspection of the CC scores $\mathbf{U}_{\bullet 1}$ and $\mathbf{V}_{\bullet 1}$, in the form of kernel density estimates in Figure 3.7, shows that these density estimates deviate considerably from the normal density. Because the CC scores are linear combinations of the variables of $\mathbb{X}^{[1]}$ and $\mathbb{X}^{[2]}$, respectively, normality of the $\mathbb{X}^{[1]}$ and $\mathbb{X}^{[2]}$ leads to normal CC scores. Thus, strictly speaking, the hypothesis tests of Theorem 3.13 do not apply because these tests are based on Gaussian assumptions. We apply them to the data, but the interpretation of the test results should be treated with caution.

In the tests, I use the significance level of 1 per cent. We begin with the test of part 1 of Theorem 3.13. The value of the approximate likelihood ratio statistic, calculated as in (3.31), is 1,271.9, which greatly exceeds the critical value 37.57 of the χ^2 distribution in 20 degrees of freedom. So we reject the null hypothesis and conclude that the data are consistent with a non-zero matrix C.

We next test whether the smallest singular value could be zero; to do this, we apply the test of part 2 of the theorem with the null hypothesis H_0^3, which states that the first three singular values are non-zero, and the last equals zero. The value of the approximate test statistic is 11.85, which still exceeds the critical value 9.21 of the χ^2 distribution in 2 degrees of freedom. Consequently, we conclude that υ_4 could be non-zero, and so all pairs of CC scores could be correlated.

As we know, the outcome of a test depends on the sample size. In the initial tests, I considered all 1,000 observations. If one considers a smaller sample, the null hypothesis is less likely to be rejected. For the *income* data, we now consider separately the first and second half of the first 1,000 records. Table 3.8 contains the singular values of the complete data (the 1,000 records) as well as those of the two parts. The singular values of the two parts differ from each other and from the corresponding value of the complete data. The smallest singular value of the first 500 observations in particular has decreased considerably.

The test for $C = \mathbf{0}$ is rejected for both subsets, so we look at tests for individual singular values. Both parts of the data convincingly reject the null hypothesis H_0^2 with test statistics above 50 and a corresponding critical value of 16.81. However, in contrast to the test on all 1000 records, the null hypothesis H_0^3 is accepted by both parts at the 1 per cent level, with a test statistic of 1.22 for the first part and 8.17 for the second part. The discrepancy in the decisions of the test between the 1000 observations and the two parts is a consequence of (3.32), which explicitly depends on n. For these data, $n = 500$ is not large enough to reject the null hypothesis. ■

Table 3.8 *Singular values of the matrix $\widehat{C}$ from Example 3.9 for the complete data and parts subsets of the data.*

Records	$\widehat{\upsilon}_1$	$\widehat{\upsilon}_2$	$\widehat{\upsilon}_3$	$\widehat{\upsilon}_4$
1–1,000	0.7762	0.4526	0.3312	0.1082
1–500	0.8061	0.4827	0.3238	0.0496
501–1,000	0.7363	0.4419	0.2982	0.1281

As we have seen in the example, the two subsets of the data can result in different test decisions from that obtained for the combined sample. There are a number of reasons why this can happen.

- The test statistic is approximately proportional to the sample size n.
- For non-normal data, a larger value of n is required before the test statistic is approximately χ^2.
- Large values of n are more likely to reject a null hypothesis.

What we have seen in Example 3.9 may be a combination of all of these.

3.7 Canonical Correlations and Regression

In Canonical Correlation Analysis, the two random vectors $\mathbf{X}^{[1]}$ and $\mathbf{X}^{[2]}$ or the two data sets $\mathbb{X}^{[1]}$ and $\mathbb{X}^{[2]}$ play a symmetric role. This feature does not apply in a regression setting. Indeed, one of the two data sets, say $\mathbb{X}^{[1]}$, plays the role of the explanatory or predictor variables, whereas the second, $\mathbb{X}^{[2]}$, acquires the role of the response variables. Instead of finding the strongest relationship between the two parts, in regression, we want to predict the responses from the predictor variables, and the roles are not usually reversible.

In Section 2.8.2, I explain how ideas from Principal Component Analysis are adapted to a Linear Regression setting: A principal component analysis reduces the *predictor* variables, and the lower-dimensional PC data are used as the derived predictor variables. The dimension-reduction step is carried out entirely among the predictors without reference to the response variables. Like Principal Component Analysis, Canonical Correlation Analysis is related to Linear Regression, and the goal of this section is to understand this relationship better.

We deviate from the CC setting of two symmetric objects and return to the notation of Section 2.8.2: we use $\mathbf{X}$ instead of $\mathbf{X}^{[1]}$ for the predictor variables and $\mathbf{Y}$ instead of $\mathbf{X}^{[2]}$ for the responses. This notation carries over to the covariance matrices. For data, we let $\mathbb{X} = \begin{bmatrix}\mathbf{X}_1 \ \mathbf{X}_2 \cdots \mathbf{X}_n\end{bmatrix}$ be the d-variate predictor variables and $\mathbb{Y} = \begin{bmatrix}\mathbf{Y}_1 \ \mathbf{Y}_2 \cdots \mathbf{Y}_n\end{bmatrix}$ the q-variate responses with $q \geq 1$. In Linear Regression, one assumes that $q \leq d$, but this restriction is not necessary in our setting. We assume a linear relationship of the form

$$\mathbf{Y} = \boldsymbol{\beta}_0 + B^{\mathsf{T}}\mathbf{X}, \tag{3.33}$$

where $\boldsymbol{\beta}_0 \in \mathbb{R}^q$, and B is a $d \times q$ matrix. If $q = 1$, B reduces to the vector $\boldsymbol{\beta}$ of (2.35) in Section 2.8.2. Because the focus of this section centres on the estimation of B, unless otherwise stated, in the remainder of Section 3.7 we assume that the predictors $\mathbf{X}$ and responses

$\mathbf{Y}$ are centred, so

$$\mathbf{X} \in \mathbb{R}^d, \qquad \mathbf{Y} \in \mathbb{R}^q, \qquad \mathbf{X} \sim (\mathbf{0}, \Sigma_X) \qquad \text{and} \qquad \mathbf{Y} \sim (\mathbf{0}, \Sigma_Y).$$

3.7.1 The Canonical Correlation Matrix in Regression

We begin with the univariate relationship $Y = \beta_0 + \beta_1 X$, and take $\beta_0 = 0$. For the standardised variables Y_Σ and X_Σ, the correlation coefficient ϱ measures the strength of the correlation:

$$Y_\Sigma = \varrho X_\Sigma. \tag{3.34}$$

The matrix of canonical correlations generalises the univariate correlation coefficient to a multivariate setting. If the covariance matrices of $\mathbf{X}$ and $\mathbf{Y}$ are invertible, then it is natural to explore the multivariate relationship

$$\mathbf{Y}_\Sigma = C^{\mathsf{T}} \mathbf{X}_\Sigma \tag{3.35}$$

between the random vectors $\mathbf{X}$ and $\mathbf{Y}$. Because $\mathbf{X}$ and $\mathbf{Y}$ are centred, (3.35) is equivalent to

$$\mathbf{Y} = \Sigma_{XY}^{\mathsf{T}} \Sigma_X^{-1} \mathbf{X}. \tag{3.36}$$

In the Problems at the end of Part I, we consider the estimation of $\widehat{\mathbb{Y}}$ by the data version of (3.36), which yields

$$\widehat{\mathbb{Y}} = \mathbb{Y}\mathbb{X}^{\mathsf{T}} \left(\mathbb{X}\mathbb{X}^{\mathsf{T}}\right)^{-1} \mathbb{X}. \tag{3.37}$$

The variables of $\mathbf{X}$ that have low correlation with $\mathbf{Y}$ do not contribute much to the relationship (3.35) and may therefore be omitted or weighted down. To separate the highly correlated combinations of variables from those with low correlation, we consider approximations of C. Let $C = Q \Upsilon P^{\mathsf{T}}$ be the singular value decomposition, and let r be its rank. For $\kappa \le r$, we use the submatrix notation (1.21) of Section 1.5.2; thus P_κ and Q_κ are the $\kappa \times r$ submatrices of P and Q, Υ_κ is the $\kappa \times \kappa$ diagonal submatrix of Υ, which consists of the first κ diagonal elements of Υ, and

$$C^{\mathsf{T}} \approx Q_\kappa \Upsilon_\kappa P_\kappa^{\mathsf{T}}.$$

Substituting the last approximation into (3.35), we obtain the equivalent expressions

$$\mathbf{Y}_\Sigma \approx Q_\kappa \Upsilon_\kappa P_\kappa^{\mathsf{T}} \mathbf{X}_\Sigma \qquad \text{and} \qquad \mathbf{Y} \approx \Sigma_Y^{1/2} Q_\kappa \Upsilon_\kappa P_\kappa^{\mathsf{T}} \Sigma_X^{-1/2} \mathbf{X} = \Sigma_Y \Psi_\kappa \Upsilon_\kappa \Phi_\kappa^{\mathsf{T}} \mathbf{X},$$

where we have used the relationship (3.8) between the eigenvectors $\mathbf{p}$ and $\mathbf{q}$ of C and the canonical transforms $\boldsymbol{\varphi}$ and $\boldsymbol{\psi}$, respectively. Similarly, the estimator $\widehat{\mathbb{Y}}$ for data, based on $\kappa \le r$ predictors, is

$$\begin{aligned} \widehat{\mathbb{Y}} &= S_Y \widehat{\Psi}_\kappa \widehat{\Upsilon}_\kappa \widehat{\Phi}_\kappa^{\mathsf{T}} \mathbb{X} \\ &= \frac{1}{n-1} \mathbb{Y}\mathbb{Y}^{\mathsf{T}} \widehat{\Psi}_\kappa \widehat{\Upsilon}_\kappa \widehat{\Phi}_\kappa^{\mathsf{T}} \mathbb{X} \\ &= \frac{1}{n-1} \mathbb{Y}\mathbb{V}^{(\kappa)^{\mathsf{T}}} \widehat{\Upsilon}_\kappa \mathbb{U}^{(\kappa)}, \end{aligned} \tag{3.38}$$

where the sample canonical correlation scores $\mathbb{U}^{(\kappa)} = \widehat{\Phi}_\kappa^{\mathsf{T}}\mathbb{X}$ and $\mathbb{V}^{(\kappa)} = \widehat{\Psi}_\kappa^{\mathsf{T}}\mathbb{Y}$ are derived from (3.15). For the special case $\kappa = 1$, (3.38) reduces to a form similar to (2.44), namely,

$$\widehat{\mathbb{Y}} = \frac{\widehat{\upsilon}_1}{n-1}\mathbb{Y}\mathbb{V}^{(1)^{\mathsf{T}}}\mathbb{U}^{(1)}. \tag{3.39}$$

This last equality gives an expression of an estimator for $\mathbb{Y}$ which is derived from the first canonical correlations scores alone. Clearly, this estimator will differ from an estimator derived for $k > 1$. However, if the first canonical correlation is very high, this estimator may convey most of the relevant information.

The next example explores the relationship (3.38) for different combinations of predictors and responses.

Example 3.10 We continue with the direct and indirect measures of the **illicit drug market** data in a linear regression framework. We regard the twelve direct measures as the predictors and consider three different responses from among the indirect measures: *PSB new registrations, robbery 1*, and *steal from motor vehicles*. I have chosen *PSB new registrations* because an accurate prediction of this variable is important for planning purposes and policy decisions. Regarding *robbery 1*, commonsense tells us that it should depend on many of the direct measures. *Steal from motor vehicles* and the direct measure *amphetamine possession offences* exhibit the strongest single correlation, as we have seen in Example 3.3, and it is therefore interesting to consider *steal from motor vehicles* as a response.

All calculations are based on the scaled data, and *correlation coefficient* in this example means the absolute value of the correlation coefficient, as is common in Canonical Correlation Analysis. For each of the response variables, I calculate the correlation coefficient based on the derived predictors

1. the first PC score $\mathbb{W}^{(1)} = \widehat{\boldsymbol{\eta}}_1^{\mathsf{T}}(\mathbb{X}^{[1]} - \overline{\mathbf{X}}^{[1]})$ of $\mathbb{X}^{[1]}$, and
2. the first CC score $\mathbb{U}^{(1)} = \widehat{\mathbf{p}}_1^{\mathsf{T}}\mathbb{X}_S^{[1]}$ of $\mathbb{X}^{[1]}$,

where $\widehat{\boldsymbol{\eta}}_1$ is the first eigenvector of the sample covariance matrix S_1 of $\mathbb{X}^{[1]}$, and $\widehat{\mathbf{p}}_1$ is the first left eigenvector of the canonical correlation matrix $\widehat{C}$. Table 3.9 and Figure 3.8 show the results. Because it is interesting to know which variables contribute strongly to $\mathbb{W}^{(1)}$ and $\mathbb{U}^{(1)}$, I list the variables and their weights for which the entries of $\widehat{\boldsymbol{\eta}}_1$ and $\widehat{\mathbf{p}}_1$ exceed 0.4. The correlation coefficient, however, is calculated from $\mathbb{W}^{(1)}$ or $\mathbb{U}^{(1)}$ as appropriate. The last column of the table refers to Figure 3.8, which shows scatterplots of $\mathbb{W}^{(1)}$ and $\mathbb{U}^{(1)}$ on the x-axis and the response on the y-axis.

The PC scores $\mathbb{W}^{(1)}$ are calculated without reference to the response and thus lead to the same combination (weights) of the predictor variables for the three responses. There are four variables, all heroin-related, with absolute weights between 0.42 and 0.45, and all other weights are much smaller. The correlation coefficient of $\mathbb{W}^{(1)}$ with *PSB new registration* is 0.7104, which is slightly higher than 0.6888, the correlation coefficient of *PSB new registration* and *ADIS heroin*. *Robbery 1* has its strongest correlation of 0.58 with *cocaine possession offences*, but this variable has the much lower weight of 0.25 in $\mathbb{W}^{(1)}$. As a result the correlation coefficient of *robbery 1* and $\mathbb{W}^{(1)}$ has decreased compared with the single correlation of 0.58 owing to the low weight assigned to cocaine possession offences in the linear combination $\mathbb{W}^{(1)}$. A similar remark applies to *steal from motor vehicles*; the correlation with

Table 3.9 *Strength of correlation for the illicit drug market data from Example 3.10*

Response	Method	Predictor variables	Eigenvector weights	Corr. Coeff.	Figure position
PSB new reg.	PC	Heroin poss. off.	−0.4415	0.7104	Top left
		ADIS heroin	−0.4319		
		Heroin o/d	−0.4264		
		Heroin deaths	−0.4237		
PSB new reg.	CC	ADIS heroin	−0.5796	0.8181	Bottom left
		Drug psych.	−0.4011		
Robbery 1	PC	Heroin poss. off.	−0.4415	0.4186	Top middle
		ADIS heroin	−0.4319		
		Heroin o/d	−0.4264		
		Heroin deaths	−0.4237		
Robbery 1	CC	Robbery 2	0.4647	0.8359	Bottom middle
		Cocaine poss. off.	0.4001		
Steal m/vehicles	PC	Heroin poss. off.	−0.4415	0.3420	Top right
		ADIS heroin	−0.4319		
		Heroin o/d	−0.4264		
		Heroin deaths	−0.4237		
Steal m/vehicles	CC	Amphet. poss. off.	0.7340	0.8545	Bottom right

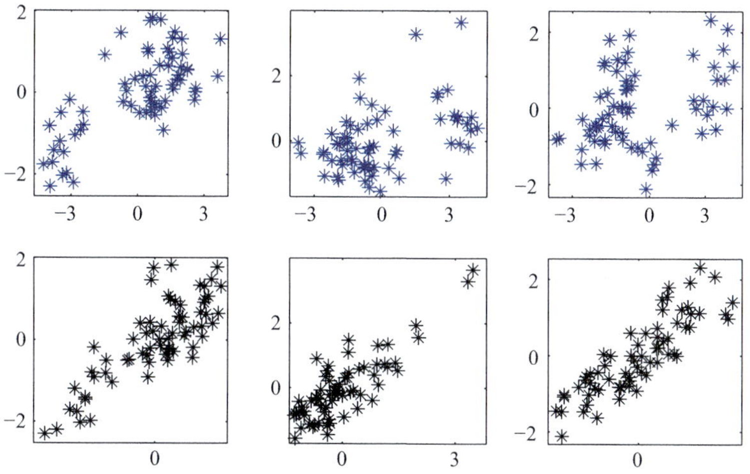

Figure 3.8 Scatterplots for Example 3.10. The x-axis shows the PC predictors $\mathbb{W}^{(1)}$ in the top row and the CC predictors $\mathbb{U}^{(1)}$ in the bottom row against the responses *PSB new registrations* (*left*), *robbery 1* (*middle*) and *steal from motor vehicles* (*right*) on the y-axis.

$\mathbb{W}^{(1)}$ has decreased from the best single of 0.764 to 0.342, a consequence of the low weight 0.2 for *amphetamine possession offences*.

Unlike the PC scores, the CC scores depend on the response variable. The weights of the relevant variables are higher than the highest weights of PC scores, and the correlation coefficient for each response is higher than that obtained with the PC scores as predictors. This difference is particularly marked for *steal from motor vehicles*.

The scatterplots of Figure 3.8 confirm the results shown in the table, and the table together with the plots show the following

1. Linear Regression based on the first CC scores exploits the relationship between the response and the original data.
2. In Linear Regression with the PC_1 predictors, the relationship between response and original predictor variables may not have been represented appropriately.

The analysis shows that we may lose valuable information when using PCs as predictors in Linear Regression. ■

For the three response variables in Example 3.10, the CC-based predictors result in much stronger relationships. It is natural to ask: Which approach is better, and why? The two approaches maximise different criteria and hence solve different problems: in the PC approach, the variance of the predictors is maximised, whereas in the CC approach, the correlation between the variables is maximised. If we want to find the best linear predictor, the CC scores are more appropriate than the first PC scores. In Section 13.3 we explore how one can combine PC and CC scores to obtain better regression predictors.

3.7.2 Canonical Correlation Regression

In Section 3.7.1 we explored Linear Regression based on the first pair of CC scores as a single predictor. In this section we combine the ideas of the preceding section with the more traditional estimation of regression coefficients (2.38) in Section 2.8.2. I refer to this approach as *Canonical Correlation Regression* by analogy with Principal Component Regression.

Principal Component Regression applies to multivariate predictor variables and univariate responses. In Canonical Correlation Regression it is natural to allow multivariate responses. Because we will be using derived predictors instead of the original predictor variables, we adopt the notation of Koch and Naito (2010) and define estimators $\widehat{B}$ for the coefficient matrix B of (3.33) in terms of derived data $\widetilde{\mathbb{X}}$. We consider specific forms of $\widetilde{\mathbb{X}}$ below and require that $(\widetilde{\mathbb{X}}\widetilde{\mathbb{X}}^{\mathsf{T}})^{-1}$ exists. Put

$$\widehat{B} = (\widetilde{\mathbb{X}}\widetilde{\mathbb{X}}^{\mathsf{T}})^{-1}\widetilde{\mathbb{X}}\mathbb{Y}^{\mathsf{T}} \quad \text{and} \quad \widehat{\mathbb{Y}} = \widehat{B}^{\mathsf{T}}\widetilde{\mathbb{X}} = \mathbb{Y}\widetilde{\mathbb{X}}^{\mathsf{T}}(\widetilde{\mathbb{X}}\widetilde{\mathbb{X}}^{\mathsf{T}})^{-1}\widetilde{\mathbb{X}}. \tag{3.40}$$

The dimension of $\widetilde{\mathbb{X}}$ is generally smaller than that of $\mathbb{X}$, and consequently, the dimension of $\widehat{B}$ is decreased, too.

As in Principal Component Regression, we replace the original d-dimensional data $\mathbb{X}$ by lower-dimensional data. Let r be the rank of $\widehat{C}$. We project $\mathbb{X}$ onto the left canonical transforms, so for $\kappa \le r$,

$$\mathbb{X} \longmapsto \widetilde{\mathbb{X}} = \mathbb{U}^{(\kappa)} = \widehat{P}_{\kappa}^{\mathsf{T}}\mathbb{X}_S = \widehat{\Phi}_{\kappa}^{\mathsf{T}}\mathbb{X}. \tag{3.41}$$

The derived data $\widetilde{\mathbb{X}}$ are the CC data in κ variables. By Theorem 3.6, the covariance matrix of the population canonical variates is the identity, and the CC data satisfy

$$\mathbb{U}^{(\kappa)}\mathbb{U}^{(\kappa)^{\mathsf{T}}} = (n-1)\mathbf{I}_{\kappa\times\kappa}.$$

Substituting $\widetilde{\mathbb{X}} = \mathbb{U}^{(\kappa)}$ into (3.40) leads to

$$\widehat{B}_U = \left[\mathbb{U}^{(\kappa)}\mathbb{U}^{(\kappa)^{\mathsf{T}}}\right]^{-1}\mathbb{U}^{(\kappa)}\mathbb{Y}^{\mathsf{T}} = \frac{1}{n-1}\widehat{P}_\kappa^{\mathsf{T}}\mathbb{X}_S\mathbb{Y}^{\mathsf{T}} = \frac{1}{n-1}\widehat{\Phi}_\kappa^{\mathsf{T}}\mathbb{X}\mathbb{Y}^{\mathsf{T}}$$

and

$$\widehat{\mathbb{Y}} = \widehat{B}_U^{\mathsf{T}}\mathbb{U}^{(\kappa)} = \frac{1}{n-1}\mathbb{Y}\mathbb{X}_S^{\mathsf{T}}\widehat{P}_\kappa\widehat{P}_\kappa^{\mathsf{T}}\mathbb{X}_S = \frac{1}{n-1}\mathbb{Y}\mathbb{U}^{(\kappa)^{\mathsf{T}}}\mathbb{U}^{(\kappa)}. \tag{3.42}$$

The expression (3.42) is applied to the prediction of $\widehat{\mathbf{Y}}_{\mathrm{new}}$ from a new datum $\mathbf{X}_{\mathrm{new}}$ by putting

$$\widehat{\mathbf{Y}}_{\mathrm{new}} = \frac{1}{n-1}\mathbb{Y}\mathbb{U}^{(\kappa)^{\mathsf{T}}}\widehat{P}_\kappa^{\mathsf{T}}S^{-1/2}\mathbf{X}_{\mathrm{new}}.$$

In the expression for $\widehat{\mathbf{Y}}_{\mathrm{new}}$, I assume that $\mathbf{X}_{\mathrm{new}}$ is centred. In Section 9.5 we focus on training and testing. I will give an explicit expression, (9.16), for the predicted Y-value of new datum.

The two expressions (3.38) and (3.42) *look* different, but they agree for fixed κ. This is easy to verify for $\kappa = 1$ and is trivially true for $\kappa = r$. The general case follows from Theorem 3.6.

A comparison of (3.42) with (2.43) in Section 2.8.2 shows the similarities and subtle differences between Canonical Correlation Regression and Principal Component Regression. In (3.42), the data $\mathbb{X}$ are projected onto directions which take into account the between covariance matrix S_{XY} of $\mathbb{X}$ and $\mathbb{Y}$, whereas (2.43) projects the data onto directions purely based on the covariance matrix of $\mathbb{X}$. As a consequence, variables with low variance contributions will be down-weighted in the regression (2.43). A disadvantage of (3.42) compared with (2.43) is, however, that the number of components is limited by the dimension of the response variables. In particular, for univariate responses, $\kappa = 1$ in the prediction based on (3.42).

I illustrate the ideas of this section with an example at the end of the next section and then also include Partial Least Squares in the data analysis.

3.7.3 Partial Least Squares

In classical multivariate regression $d < n$ and $\mathbb{X}\mathbb{X}^{\mathsf{T}}$ is invertible. If $d > n$, then $\mathbb{X}\mathbb{X}^{\mathsf{T}}$ is singular, and (3.37) does not apply. Wold (1966) developed an approach to regression which circumvents the singularity of $\mathbb{X}\mathbb{X}^{\mathsf{T}}$ in the $d > n$ case. Wold's approach, which can be regarded as reversing the roles of n and d, is called *Partial Least Squares* or *Partial Least Squares Regression*. Partial Least Squares was motivated by the requirement to extend multivariate regression to the $d > n$ case but is not restricted to this case.

The problem statement is similar to that of Linear Regression: For predictors $\mathbf{X}$ with a singular covariance matrix Σ_X and responses $\mathbf{Y}$ which satisfy $\mathbf{Y} = B^{\mathsf{T}}\mathbf{X}$ for some matrix B, construct an estimator $\widehat{B}$ of B. Wold (1966) proposed an iterative approach to constructing the estimator $\widehat{B}$. We consider two such approaches, Helland (1988) and Rosipal and Trejo (2001), which are modifications of the original proposal of Wold (1966). The key idea is to exploit the covariance relationship between the predictors and responses. Partial Least Squares enjoys popularity in the social sciences, marketing and in chemometrics; see Rosipal and Trejo (2001).

The population model of Wold (1966) consists of a d-dimensional predictor $\mathbf{X}$, a q-dimensional response $\mathbf{Y}$, a κ-dimensional unobserved $\mathbf{T}$ with $\kappa < d$ and unknown linear transformations A_X and A_Y of size $d \times \kappa$ and $q \times \kappa$, respectively, such that

$$\mathbf{X} = A_X\mathbf{T} \qquad \text{and} \qquad \mathbf{Y} = A_Y\mathbf{T}. \tag{3.43}$$

The aim is to estimate $\mathbf{T}$ and $\widehat{\mathbf{Y}} = \mathbf{Y}G(\mathbf{T})$, where G is a function of the unknown $\mathbf{T}$.

For the sample, we keep the assumption that the $\mathbb{X}$ and $\mathbb{Y}$ are centred, and we put

$$\mathbb{X} = A_X\mathbb{T} \qquad \text{and} \qquad \mathbb{Y} = A_Y\mathbb{T}$$

for some $\kappa \times n$ matrix $\mathbb{T}$ and A_X and A_Y as in (3.43).

The approaches of Helland (1988) and Rosipal and Trejo (2001) differ in the way they construct the row vectors $\mathbf{t}_1, \ldots, \mathbf{t}_\kappa$ of $\mathbb{T}$. Algorithm 3.1 outlines the general idea for constructing a partial least squares solution which is common to both approaches. So, for given $\mathbb{X}$ and $\mathbb{Y}$, the algorithm constructs $\mathbb{T}$ and the transformations A_X and A_Y. Helland (1988) proposes two solution paths and shows the relationship between the two solutions, and Helland (1990) presents some population results for the set-up we consider below, which contains a comparison with Principal Component Analysis. Helland (1988, 1990) deal with univariate responses only. I restrict attention to the first solution in Helland (1988) and then move on to the approach of Rosipal and Trejo (2001).

Algorithm 3.1 ***Partial Least Squares Solution***

Construct κ row vectors $\mathbf{t}_1, \ldots, \mathbf{t}_\kappa$ of size $1 \times n$ iteratively starting from

$$\mathbb{X}_0 = \mathbb{X}, \quad \mathbb{Y}_0 = \mathbb{Y} \qquad \text{and some } 1 \times n \text{ vector } \mathbf{t}_0. \tag{3.44}$$

- In the kth step, obtain the triplet $(\mathbf{t}_k, \mathbb{X}_k, \mathbb{Y}_k)$ from $(\mathbf{t}_{k-1}, \mathbb{X}_{k-1}, \mathbb{Y}_{k-1})$ as follows:
  1. Construct the row vector $\mathbf{t}_k$, add it to the collection $T = \left[\mathbf{t}_1, \ldots, \mathbf{t}_{k-1}\right]^{\mathsf{T}}$.
  2. Update

$$\mathbb{X}_k = \mathbb{X}_{k-1}\left(\mathbf{I}_{n\times n} - \mathbf{t}_k^{\mathsf{T}}\mathbf{t}_k\right),$$

$$\mathbb{Y}_k = \mathbb{Y}_{k-1}\left(\mathbf{I}_{n\times n} - \mathbf{t}_k^{\mathsf{T}}\mathbf{t}_k\right). \tag{3.45}$$

- When $T = \left[\mathbf{t}_1, \ldots \mathbf{t}_\kappa\right]^{\mathsf{T}}$, put

$$\widehat{\mathbb{Y}} = \mathbb{Y}G(T) \qquad \text{for some function } G. \tag{3.46}$$

■

The construction of the row vector $\mathbf{t}_k$ in each step and the definition of the $n \times n$ matrix of coefficients G distinguish the different approaches.

Helland (1988): $\mathbf{t}_k$ for univariate responses $\mathbb{Y}$. Assume that for $1 < k \le \kappa$, we have constructed $(\mathbf{t}_{k-1}, \mathbb{X}_{k-1}, \mathbb{Y}_{k-1})$.

H1. Put $\mathbf{t}_{k-1,0} = \mathbb{Y}_{k-1}/\left\|\mathbb{Y}_{k-1}\right\|$.

H2. For $\ell = 1, 2, \ldots,$ calculate

$$\mathbf{w}_{k-1,\ell} = \mathbb{X}_{k-1}\mathbf{t}_{k-1,\ell-1}^{\mathsf{T}},$$

$$\mathbf{t}'_{k-1,\ell} = \mathbf{w}_{k-1,\ell}^{\mathsf{T}}\mathbb{X}_{k-1} \text{ and } \mathbf{t}_{k-1,\ell} = \mathbf{t}'_{k-1,\ell}/\left\|\mathbf{t}'_{k-1,\ell}\right\|. \tag{3.47}$$

H3. Repeat the calculations of step H2 until the sequence $\{\mathbf{t}_{k-1,\ell}\colon \ell = 1,2,\ldots\}$ has converged. Put

$$\mathbf{t}_k = \lim_{\ell} \mathbf{t}_{k-1,\ell},$$

and use $\mathbf{t}_k$ in (3.45).

H4. After the κth step of Algorithm 3.1, put

$$G(T) = G(\mathbf{t}_1,\ldots,\mathbf{t}_\kappa) = \sum_{k=1}^{\kappa}\left(\mathbf{t}_k\mathbf{t}_k^{\mathsf{T}}\right)^{-1}\mathbf{t}_k^{\mathsf{T}}\mathbf{t}_k$$

and

$$\widehat{\mathbb{Y}} = \mathbb{Y}G(T) = \mathbb{Y}\sum_{k=1}^{\kappa}\left(\mathbf{t}_k\mathbf{t}_k^{\mathsf{T}}\right)^{-1}\mathbf{t}_k^{\mathsf{T}}\mathbf{t}_k. \tag{3.48}$$

Rosipal and Trejo (2001): $\mathbf{t}_k$ for q-variate responses $\mathbb{Y}$. In addition to the $\mathbf{t}_k$, vectors $\mathbf{u}_k$ are constructed and updated in this solution, starting with $\mathbf{t}_0$ and $\mathbf{u}_0$. Assume that for $1 < k \le \kappa$, we have constructed $(\mathbf{t}_{k-1}, \mathbf{u}_{k-1}, \mathbb{X}_{k-1}, \mathbb{Y}_{k-1})$.

RT1. Let $\mathbf{u}_{k-1,0}$ be a random vector of size $1\times n$.

RT2. For $\ell = 1,2,\ldots$, calculate

$$\mathbf{w}_{\ell} = \mathbb{X}_{k-1}\mathbf{u}_{k-1,\ell-1}^{\mathsf{T}},$$

$$\mathbf{t}'_{k-1,\ell} = \mathbf{w}_{\ell}^{\mathsf{T}}\mathbb{X}_{k-1} \quad \text{and} \quad \mathbf{t}_{k-1,\ell} = \mathbf{t}'_{k-1,\ell}/\left\|\mathbf{t}'_{k-1,\ell}\right\|,$$

$$\mathbf{v}_{\ell} = \mathbb{Y}_{k-1}\mathbf{t}_{k-1,\ell}^{\mathsf{T}},$$

$$\mathbf{u}'_{k-1,\ell} = \mathbf{v}_{\ell}^{\mathsf{T}}\mathbb{Y}_{k-1} \quad \text{and} \quad \mathbf{u}_{k-1,\ell} = \mathbf{u}'_{k-1,\ell}/\left\|\mathbf{u}'_{k-1,\ell}\right\|. \tag{3.49}$$

RT3. Repeat the calculations of step RT2 until the sequences of vectors $\mathbf{t}_{k-1,\ell}$ and $\mathbf{u}_{k-1,\ell}$ have converged. Put

$$\mathbf{t}_k = \lim_{\ell}\mathbf{t}_{k-1,\ell} \quad \text{and} \quad \mathbf{u}_k = \lim_{\ell}\mathbf{t}_{k-1,\ell},$$

and use $\mathbf{t}_k$ in (3.45).

RT4. After the κth step of Algorithm 3.1, define the $\kappa\times n$ matrices

$$\mathbb{T} = \begin{bmatrix}\mathbf{t}_1\\ \vdots\\ \mathbf{t}_\kappa\end{bmatrix} \quad \text{and} \quad \mathbb{U} = \begin{bmatrix}\mathbf{u}_1\\ \vdots\\ \mathbf{u}_\kappa\end{bmatrix},$$

and put

$$G(\mathbb{T}) = G(\mathbf{t}_1,\ldots,\mathbf{t}_\kappa) = \mathbb{T}^{\mathsf{T}}\left(\mathbb{U}\mathbb{X}^{\mathsf{T}}\mathbb{X}\mathbb{T}^{\mathsf{T}}\right)^{-1}\mathbb{U}\mathbb{X}^{\mathsf{T}}\mathbb{X},$$

$$\widehat{B}^{\mathsf{T}} = \mathbb{Y}\mathbb{T}^{\mathsf{T}}\left(\mathbb{U}\mathbb{X}^{\mathsf{T}}\mathbb{X}\mathbb{T}^{\mathsf{T}}\right)^{-1}\mathbb{U}\mathbb{X}^{\mathsf{T}} \quad \text{and}$$

$$\widehat{\mathbb{Y}} = \widehat{B}^{\mathsf{T}}\mathbb{X} = \mathbb{Y}G(\mathbb{T}). \tag{3.50}$$

A comparison of (3.48) and (3.50) shows that both solutions use of the covariance relationship between $\mathbb{X}$ and $\mathbb{Y}$ (and of the updates $\mathbb{X}_{k-1}$ and $\mathbb{Y}_{k-1}$) in the calculation of $\mathbf{w}$ in (3.47) and of $\mathbf{v}_\ell$ in (3.49). The second algorithm calculates two sets of vectors $\mathbf{t}_k$ and $\mathbf{u}_k$, whereas Helland's solution only requires the $\mathbf{t}_k$. The second algorithm applies to multivariate responses; Helland's solution has no obvious extension to $q > 1$.

For multiple regression with univariate responses and $\kappa = 1$, write $\widehat{\mathbf{Y}}_{\mathrm{H}}$ and $\widehat{\mathbf{Y}}_{\mathrm{RT}}$ for the estimators defined in (3.48) and (3.50), respectively. Then

$$\widehat{\mathbf{Y}}_{\mathrm{H}} = \mathbf{Y}\mathbf{t}_1^{\mathsf{T}}\mathbf{t}_1/\|\mathbf{t}_1\|^2 \qquad \text{and} \qquad \widehat{\mathbf{Y}}_{\mathrm{RT}} = \mathbf{Y}\mathbf{t}_1^{\mathsf{T}}\left(\mathbf{u}_1\mathbb{X}^{\mathsf{T}}\mathbb{X}\mathbf{t}_1^{\mathsf{T}}\right)^{-1}\mathbf{u}_1\mathbb{X}^{\mathsf{T}}\mathbb{X}. \tag{3.51}$$

The more complex expression for $\widehat{\mathbf{Y}}_{\mathrm{RT}}$ could be interpreted as the cost paid for starting with a random vector $\mathbf{u}$. There is no clear winner among these two solutions; for univariate responses Helland's algorithm is clearly the simpler, whereas the second algorithm answers the needs of multivariate responses.

Partial Least Squares methods are based on all variables rather than on a reduced number of derived variables, as done in Principal Component Analysis and Canonical Correlation Analysis. The iterative process, which leads to the κ components $\mathbf{t}_j$ and $\mathbf{u}_j$ (with $j \le \kappa$) stops when the updated matrix $\mathbb{X}_k = \mathbb{X}_{k-1}\left(\mathbf{I}_{n\times n} - \mathbf{t}_k^{\mathsf{T}}\mathbf{t}_k\right)$ is the zero matrix.

In the next example I compare the two Partial Least Squares solutions to Principal Component Analysis, Canonical Correlation Analysis and Linear Regression. In Section 3.7.4, I indicate how these four approaches fit into the framework of generalised eigenvalue problems.

Example 3.11 For the **abalone** data, the number of rings allows the experts to estimate the age of the abalone. In Example 2.19 in Section 2.8.2, we explored Linear Regression with PC_1 as the derived predictor of the number of rings. Here we apply a number of approaches to the abalone data. To assess the performance of each approach, I use the mean sum of squared errors

$$\mathrm{MSSE} = \frac{1}{n}\sum_{i=1}^{n}\left\|\mathbf{Y}_i - \widehat{\mathbf{Y}}_i\right\|^2. \tag{3.52}$$

I will use only the first 100 observations in the calculations and present the results in Tables 3.10 and 3.11. The comparisons include classical Linear Regression (LR), Principal Component Regression (PCR), Canonical Correlation Regression (CCR) and Partial Least Squares (PLS). Table 3.10 shows the relative importance of the seven predictor variables, here given in decreasing order and listed by the variable number, and Table 3.11 gives the MSSE as the number of terms or components increases.

For LR, I order the variables by their significance obtained from the traditional p-values. Because the data are not normal, the p-values are approximate. The *dried shell weight*, variable 7, is the only variable that is significant, with a p-value of 0.0432. For PCR and CCR, Table 3.10 lists the ordering of variables induced by the weights of the first direction vector. For PCR, this vector is the first eigenvector $\widehat{\eta}_1$ of the sample covariance matrix S of $\mathbb{X}$, and for CCR, it is the first canonical transform $\widehat{\varphi}_1$. LR and CCR pick the same variable as most important, whereas PCR selects variable 4, *whole weight*. This difference is not surprising because variable 4 is chosen merely because of its large effect on the covariance matrix of the predictor data $\mathbb{X}$. The PLS components are calculated iteratively and are based

Table 3.10 *Relative importance of variables for prediction by method for the abalone data from Example 3.11.*

Method	Order of variables						
LR	7	5	1	4	6	2	3
PCR	4	5	7	1	6	2	3
CCR	7	1	5	2	4	6	3

Table 3.11 *MSSE for different prediction approaches and number of components for the abalone data from Example 3.11*

	Number of variables or components						
Method	1	2	3	4	5	6	7
LR	5.6885	5.5333	5.5260	5.5081	5.4079	5.3934	5.3934
PCR	5.9099	5.7981	5.6255	5.5470	5.4234	5.4070	5.3934
CCR	5.3934	—	—	—	—	—	—
PLS_H	5.9099	6.0699	6.4771	6.9854	7.8350	8.2402	8.5384
PLS_RT	5.9029	5.5774	5.5024	5.4265	5.3980	5.3936	5.3934

on all variables, so they cannot be compared conveniently in this way. For this reason, I do not include PLS in Table 3.10.

The MSSE results for each approach are given in Table 3.11. The column headed '1' shows the MSSE for one variable or one derived variable, and later columns show the MSSE for the number of (derived) variables shown in the top row. For CCR, there is only one derived variable because the response is univariate. The PLS solutions are based on all variables rather than on subsets, and the number of components therefore has a different interpretation. For simplicity, I calculate the MSSE based on the first component, first two components, and so on and include their MSSE in the table.

A comparison of the LR and PCR errors shows that all errors – except the last – are higher for PCR. When all seven variables are used, the two methods agree. The MSSE of CCR is the same as the smallest error for LR and PCR, and the CCR solution has the same weights as the LR solution with all variables. For the abalone data, PLS_H performs poorly compared with the other methods, and the MSSE increases if more components than the first are used, whereas the performance of PLS_RT is similar to that of LR. ■

The example shows that in the classical scenario of a single response and many more observations than predictor variables, Linear Regression does at least as well as the competitors I included. However, Linear Regression has limitations, in particular, $d \ll n$ is required, and it is therefore important to have methods that apply when these conditions are no longer satisfied. In Section 13.3.2 we return to regression and consider HDLSS data which require more sophisticated approaches.

3.7.4 The Generalised Eigenvalue Problem

I conclude this chapter with an introduction to *Generalised Eigenvalue Problems*, a topic which includes Principal Component Analysis, Canonical Correlation Analysis, Partial

Least Squares (PLS) and Multiple Linear Regression (MLR) as special cases. The ideas I describe are presented in Borga, Landelius, and Knutsson (1997) and extended in De Bie, Christianini, and Rosipal (2005).

Definition 3.14 Let A and B be square matrices of the same size, and assume that B is invertible. The task of the **generalised eigen(value) problem** is to find eigenvalue–eigenvector solutions $(\lambda, \mathbf{e})$ to the equation

$$A\mathbf{e} = \lambda B\mathbf{e} \qquad \text{or equivalently} \qquad B^{-1}A\mathbf{e} = \lambda\mathbf{e}. \tag{3.53}$$

□

Problems of this type, which involve two matrices, arise in physics and the engineering sciences. For A, B and a vector $\mathbf{e}$, (3.53) is related to the Rayleigh quotient, named after the physicist Rayleigh, which is the solution of defined by

$$\frac{\mathbf{e}^{\mathsf{T}}A\mathbf{e}}{\mathbf{e}^{\mathsf{T}}B\mathbf{e}}.$$

We restrict attention to those special cases of the generalised eigenvalue problem we have met so far. Each method is characterised by the *role* the eigenvectors play and by the choice of the two matrices. In each case the eigenvectors optimise specific criteria. Table 3.12 gives explicit expressions for the matrices A and B and the criteria the eigenvectors optimise. For details, see Borga, Knutsson, and Landelius (1997).

The setting of Principal Component Analysis is self-explanatory: the eigenvalues and eigenvectors of the generalised eigenvalue problem are those of the covariance matrix Σ of the random vector $\mathbf{X}$, and by Theorem 2.10 in Section 2.5.2, the eigenvectors $\boldsymbol{\eta}$ of Σ maximise $\mathbf{e}^{\mathsf{T}}\Sigma\mathbf{e}$.

For Partial Least Squares, two random vectors $\mathbf{X}^{[1]}$ and $\mathbf{X}^{[2]}$ and their covariance matrix Σ_{12} are the objects of interest. In this case, the singular values and the left and right eigenvectors of Σ_{12} solve the problem. Maximum Covariance Analysis (MCA), described in Section 3.5.4, shares these properties with PLS.

For Canonical Correlation Analysis, we remain with the pair of random vectors $\mathbf{X}^{[1]}$ and $\mathbf{X}^{[2]}$ but replace the covariance matrix of Partial Least Squares by the matrix of canonical correlations C. The vectors listed in Table 3.12 are the normalised canonical transforms of (3.8). To see why these vectors are appropriated, observe that the generalised eigen problem arising from A and B is the following

$$\begin{pmatrix} \Sigma_{12}\mathbf{e}_2 \\ \Sigma_{12}^{\mathsf{T}}\mathbf{e}_1 \end{pmatrix} = \lambda \begin{pmatrix} \Sigma_1\mathbf{e}_1 \\ \Sigma_2\mathbf{e}_2 \end{pmatrix}, \tag{3.54}$$

which yields

$$\Sigma_1^{-1}\Sigma_{12}\Sigma_2^{-1}\Sigma_{12}^{\mathsf{T}}\mathbf{e}_1 = \lambda^2\mathbf{e}_1. \tag{3.55}$$

A comparison of (3.55) and (3.24) reveals some interesting facts: the matrix on the left-hand side of (3.55) is the matrix K, which is similar to $R = CC^{\mathsf{T}}$. The matrix similarity implies that the eigenvalues of K are squares of the singular values υ of C. Further, the eigenvector $\mathbf{e}_1$ of K is related to the corresponding eigenvector $\mathbf{p}$ of R by

$$\mathbf{e}_1 = c\,\Sigma_1^{-1/2}\mathbf{p}, \tag{3.56}$$

Table 3.12 *Special cases of the generalised eigenvalue problem*

Method	A	B	Eigenvectors	Comments
PCA	Σ	$\mathbf{I}$	Maximise Σ	See Section 2.2
PLS/MCA	$\begin{pmatrix} \mathbf{0} & \Sigma_{12} \\ \Sigma_{12}^{\top} & \mathbf{0} \end{pmatrix}$	$\begin{pmatrix} \mathbf{I} & \mathbf{0} \\ \mathbf{0} & \mathbf{I} \end{pmatrix}$	Maximise Σ_{12}	See Section 3.7.3
CCA	$\begin{pmatrix} \mathbf{0} & \Sigma_{12} \\ \Sigma_{12}^{\top} & \mathbf{0} \end{pmatrix}$	$\begin{pmatrix} \Sigma_1 & \mathbf{0} \\ \mathbf{0} & \Sigma_2 \end{pmatrix}$	Maximise C	See Section 3.2
LR	$\begin{pmatrix} \mathbf{0} & \Sigma_{12} \\ \Sigma_{12}^{\top} & \mathbf{0} \end{pmatrix}$	$\begin{pmatrix} \Sigma_1 & \mathbf{0} \\ \mathbf{0} & \mathbf{I} \end{pmatrix}$	Minimise LSE	—

for some $c > 0$. The vector $\mathbf{p}$ is a left eigenvector of C, and so the eigenvector $\mathbf{e}_1$ of K is nothing but the normalised canonical transform $\boldsymbol{\varphi}/||\boldsymbol{\varphi}||$ because the eigenvectors have norm 1. A similar argument, based on the matrix $C^{\top}C$ instead of $CC^{\top}$, establishes that the second eigenvector equals $\boldsymbol{\psi}/||\boldsymbol{\psi}||$.

Linear Regression treats the two random vectors asymmetrically. This can be seen in the expression for B. We take $\mathbf{X}^{[1]}$ to be the predictor vector and $\mathbf{X}^{[2]}$ the response vector. The generalised eigen equations amount to

$$\begin{pmatrix} \Sigma_{12}\mathbf{e}_2 \\ \Sigma_{12}^{\top}\mathbf{e}_1 \end{pmatrix} = \lambda \begin{pmatrix} \Sigma_1\mathbf{e}_1 \\ \mathbf{e}_2 \end{pmatrix}, \tag{3.57}$$

and hence one needs to solve the equations

$$\Sigma_1^{-1}\Sigma_{12}\Sigma_{12}^{\top}\mathbf{e}_1 = \lambda^2\mathbf{e}_1 \qquad \text{and} \qquad \Sigma_{12}^{\top}\Sigma_1^{-1}\Sigma_{12}\mathbf{e}_2 = \lambda^2\mathbf{e}_2.$$

The matrix $\Sigma_1^{-1}\Sigma_{12}\Sigma_{12}^{\top}$ is not symmetric, so it has a singular value decomposition with left and right eigenvectors, whereas $\Sigma_{12}^{\top}\Sigma_1^{-1}\Sigma_{12}$ has a spectral decomposition with a unique set of eigenvectors.

Generalised eigenvalue problems are, of course, not restricted to these cases. In Section 4.3 we meet Fisher's discriminant function, which is the solution of another generalised eigenvalue problem, and in Section 12.4 we discuss approaches based on two scatter matrices which also fit into this framework.

4

Discriminant Analysis

'That's not a regular rule: you invented it just now.' 'It's the oldest rule in the book,' said the King. 'Then it ought to be Number One,' said Alice (Lewis Carroll, *Alice's Adventures in Wonderland*, 1865).

4.1 Introduction

To discriminate means to single out, to recognise and understand differences and to distinguish. Of special interest is discrimination in two-class problems: A tumour is benign or malignant, and the correct diagnosis needs to be obtained. In the finance and credit-risk area, one wants to assess whether a company is likely to go bankrupt in the next few years or whether a client will default on mortgage repayments. To be able to make decisions in these situations, one needs to understand what distinguishes a 'good' client from one who is likely to default or go bankrupt.

Discriminant Analysis starts with data for which the classes are known and finds characteristics of the observations that accurately predict each observation's class. One then combines this information into a *rule* which leads to a partitioning of the observations into disjoint classes. When using Discriminant Analysis for tumour diagnosis, for example, the first step is to determine the variables which best characterise the difference between the benign and malignant groups – based on data for tumours whose status (benign or malignant) is known – and to construct a decision rule based on these variables. In the second step one wants to apply this rule to tumours whose status is yet to be determined.

One of the best-known classical examples, Fisher's *iris* data, consists of three different species, and for each species, four measurements are made. Figures 1.3 and 1.4 in Section 1.2 show these data in four three-dimensional (3D) scatterplots and a parallel coordinate view, with different colours corresponding to the different species. We see that the red species, *Setosa*, is well separated from the other two species, whereas the second species, *Versicolor*, and the third species, *Virginica*, shown in green and black, are not as easily distinguishable. The aim of the analysis is to use the four variables and the species information to group the data into the three species. In this case, variables 3 and 4, the two petal properties, suffice to separate the red species from the other two, but more complicated techniques are required to separate the green and black species.

For discrimination and the construction of a suitable rule, Fisher (1936) suggested making use of the fact that

vectors in one class behave differently from vectors in the other classes; and
the variance within the classes differs maximally from that between the classes.

Exploiting Fisher's insight poses an important task in Discriminant Analysis: the construction of *(discriminant) rules* which best capture differences between the classes. The application of a rule is often called *Classification*, the *making of classes*, that is, the partitioning of the data into different classes based on a particular rule. There are no clear boundaries between Discriminant Analysis and Classification. It might be helpful to think of Discriminant Analysis as the analysis step, the construction of the rule, whereas Classification focuses more on the outcome and the interpretation of the class allocation, but mostly the two terms are used interchangeably. As in regression analysis, the use of rules for prediction, that is, the assigning of new observations to classes, is an important aspect of Discriminant Analysis.

Many discrimination methods and rules have been proposed, with new ones appearing all the time. Some rules are very general and perform well in a wide range of applications, whereas others are designed for a particular data set or exploit particular properties of data. Traditionally, Gaussian random vectors have played a special role in Discriminant Analysis. Gaussian assumptions are still highly relevant today as they form the framework for theoretical and probabilistic error calculations, in particular for high-dimensional low sample size (HDLSS) problems. This chapter focusses on ideas, and starts from the basic methods. In Section 13.3 we return to discrimination and look at newer developments for HDLSS data, including theoretical advances in Discriminant Analysis. For a classical theoretical treatment of Discriminant Analysis, see Devroye, Györfi, and Lugosi (1996).

Despite a fast growing body of research into more advanced classification tools, Hand (2006) points out that the 'apparent superiority of more sophisticated methods may be an illusion' and that 'simple methods typically yield performances almost as good as more sophisticated methods'. Of course, there is a trade-off between accuracy and efficiency, but his arguments pay homage to the classical and linear methods and re-acknowledge them as valuable discrimination tools.

To gain some insight into the way Discriminant Analysis relates to Principal Component Analysis and Canonical Correlation Analysis, we consider the following schematic diagram which starts with $d \times n$ data $\mathbb{X}$ and a PC setting:

$$\mathbb{X}_{pca}$$

same n (left) — same d (right)

$$\mathbb{X}_{cca} = \begin{bmatrix} \mathbb{X}^{[1]} \\ \hline \mathbb{X}^{[2]} \end{bmatrix} \qquad\qquad \mathbb{X}_{da} = \left[\mathbb{X}^{[1]} \,|\, \mathbb{X}^{[2]} \,|\, \cdots \,|\, \mathbb{X}^{[\kappa]} \right]$$

$$\downarrow \qquad\qquad\qquad\qquad \downarrow$$

$$\mathbb{X}_{cca}^{[\rho]} \text{ is } d_\rho \times n \qquad\qquad \mathbb{X}_{da}^{[\nu]} \text{ is } d \times n_\nu$$

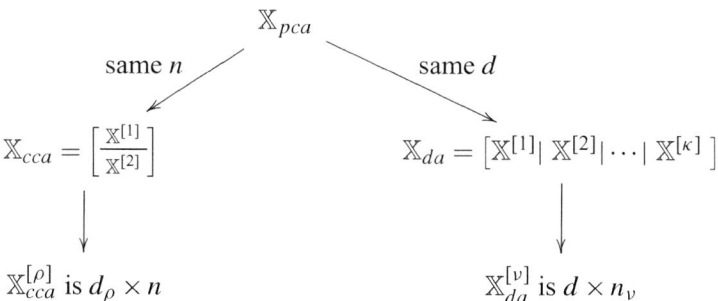

In Principal Component Analysis we regard the data $\mathbb{X}$ as an entity, and the analysis applies to all observations and all variables simultaneously. In Canonical Correlation Analysis and Discriminant Analysis we divide the data; in Canonical Correlation Analysis we split the *variables* into two parts – shown schematically by the horizontal bar which separates the first group of variables from the second group. The aim in Canonical Correlation Analysis is to find the combination of variables within $\mathbb{X}^{[1]}$ and within $\mathbb{X}^{[2]}$ that are most strongly correlated. In Discriminant Analysis we divide the *observations* into κ groups – indicated by the vertical bars. In this case, all observations have the same variables, irrespective of the group or class to which they belong. In Discriminant Analysis we want to find properties which distinguish the observations in one group from those in another. We know which group each observation belongs to, and the aim is to find the variables which best characterise the split into the different groups.

The terms *Statistical Learning*, *Machine Learning*, *Supervised Learning* and *Unsupervised Learning* have evolved in the statistical and computer science communities in recent years. Statistical Learning and Machine Learning are basically synonymous terms which started in different disciplines. Following Hastie et al. (2001, 9), 'statistical learning is the prediction of outputs from inputs'. The inputs refer to multivariate predictors, and the outputs are the response variables in regression or the classes in Discriminant Analysis.

Supervised Learning includes Discriminant Analysis, Classification and Regression, and involves a *training* and a *testing* component. In the training step, a part of the data is used to derive a discriminant rule or the regression coefficients and estimates. The performance of the rule is assessed or 'tested' on the remaining data. For regression, this second part corresponds to prediction of responses for new variables. This chapter focuses primarily on methods of Discriminant Analysis and less so on training and testing issues, which I cover in Section 9.5. In contrast to Supervised Learning, Unsupervised Learning refers to Cluster Analysis, which we consider in Chapter 6.

In this chapter we explore the basic and some more complex approaches and learn when to apply the different rules. Section 4.2 introduces notation and definitions. Section 4.3 looks at linear discriminant rules, starting with Fisher's linear discriminant rule, which is treated separately for the population and the sample, and then moving on to linear rules for Gaussian populations. Section 4.4 describes how to evaluate and compare rules and introduces the probability of misclassification. We return to a Gaussian setting in Section 4.5, which extends the linear approach of Section 4.3 to κ classes and quadratic discriminant rules. We also explore the range of applicability of these rules when the Gaussian assumptions are violated. Section 4.6 considers discrimination in a Bayesian framework. In Section 4.7, I describe nearest-neighbour discrimination, the logistic regression approach to discrimination, Friedman's regularised discriminant rule, and also mention Support Vector Machines. Our final Section 4.8 explores the relationship between Discriminant Analysis and Linear Regression and then looks at dimension-reduction ideas in discrimination, with a focus on Principal Component Discriminant Analysis, variable ranking and variable selection. Problems pertaining to the material of this chapter are listed at the end of Part I.

4.2 Classes, Labels, Rules and Decision Functions

The classes are the fundamental objects in Discriminant Analysis. Classes are distinct subsets of the space of d-dimensional random vectors. The classes can correspond to species of plants or animals, or they may be characterised by the status of a disease, such as present or absent. More abstractly, classes can refer to families of distributions which differ in some of their population parameters. Of special interest among the latter are the classes $\mathcal{N}(\boldsymbol{\mu}_\nu, \Sigma_\nu)$, which differ in their means and covariance matrices.

Throughout this chapter, the integer $\kappa \geq 2$ denotes the number of classes. In Discriminant Analysis, this number is fixed and known. We write $\mathcal{C}_1, \ldots, \mathcal{C}_\kappa$ for the κ classes.

This section establishes notation we require throughout this chapter. The reader may want to ignore some of the notation, for example, Definition 4.2, in a first perusal of Discriminant Analysis or treat this notation section more like the background chapters at the beginning of each part of the book.

Definition 4.1 Consider classes $\mathcal{C}_1, \ldots, \mathcal{C}_\kappa$. A d-dimensional random vector $\mathbf{X}$ **belongs to class** $\mathcal{C}_\nu$ or $\mathbf{X}$ is **a member of class** $\mathcal{C}_\nu$ for some $\nu \le \kappa$, if $\mathbf{X}$ satisfies the properties that characterise $\mathcal{C}_\nu$. To show the class membership of $\mathbf{X}$, we write

$$\mathbf{X} \in \mathcal{C}_\nu \qquad \text{or} \qquad \mathbf{X}^{[\nu]}. \tag{4.1}$$

For random vectors $\mathbf{X}$ from κ classes, the **label** Y of $\mathbf{X}$ is a random variable which takes discrete values $1, \ldots, \kappa$ such that

$$Y = \nu \qquad \text{if } \mathbf{X} \in \mathcal{C}_\nu. \tag{4.2}$$

We regard $\begin{bmatrix} \mathbf{X} \\ Y \end{bmatrix}$ as a $d+1$-dimensional random vector called the labelled random vector. If Y is the label for $\mathbf{X}$, then we also refer to $\mathbf{X}$ as the labelled random vector. □

The classes are sometimes called the **populations**; we will use both notions interchangeably. If the κ classes are characterised by their means $\boldsymbol{\mu}_\nu$ and covariance matrices Σ_ν so that $\mathcal{C}_\nu \equiv (\boldsymbol{\mu}_\nu, \Sigma_\nu)$, we write

$$\mathbf{X} \sim (\boldsymbol{\mu}_\nu, \Sigma_\nu) \qquad \text{or} \qquad \mathbf{X} \in \mathcal{C}_\nu.$$

Unless otherwise stated, random vectors $\mathbf{X}_1, \ldots, \mathbf{X}_n$ belonging to the classes $\mathcal{C}_1, \ldots, \mathcal{C}_\kappa$ have the same dimension d. If the dimensions of the random vectors were unequal, one could classify the vectors in terms of their dimension, and more sophisticated ideas would not be required.

Instead of the scalar-valued labels Y of (4.2) with values $1 \le \nu \le \kappa$, we also consider equivalent vector-valued labels. These labels are used in Section 13.3.

Definition 4.2 Let $\mathcal{C}_1, \ldots, \mathcal{C}_\kappa$ be κ classes. Let $\mathbf{X}$ be a random vector which belongs to class $\mathcal{C}_\nu$ for some $\nu \le \kappa$. A **(vector-valued) label** $\mathbf{Y}$ of $\mathbf{X}$ is a κ-dimensional random vector with entries

$$\mathbf{Y} = \begin{bmatrix} Y_1 \\ \vdots \\ Y_\kappa \end{bmatrix} \tag{4.3}$$

such that

$$Y_\nu = \begin{cases} 1 & \text{since } \mathbf{X} \text{ belongs to the } \nu\text{th class} \\ 0 & \text{otherwise.} \end{cases}$$

We write $\begin{bmatrix} \mathbf{X} \\ \mathbf{Y} \end{bmatrix}$ for the $(d+\kappa)$-dimensional labelled random vector. □

From a single random vector $\mathbf{X}$, we move to data. Let $\mathbb{X} = \begin{bmatrix} \mathbf{X}_1 \cdots \mathbf{X}_n \end{bmatrix}$ be $d \times n$ data. We assume that each $\mathbf{X}_i$ belongs to one of the classes $\mathcal{C}_\nu$ for $\nu \le \kappa$. If the order of the $\mathbf{X}_i$ is not important, for notational convenience, we regroup the observations and write

$$\mathbb{X}^{[\nu]} = \left\{ \mathbf{X}_i^{[\nu]} \colon i = 1, \ldots, n_\nu \right\},$$

where $\mathbf{X}_i^{[\nu]}$ is the labelled random vector as in (4.1), which belongs to class $\mathcal{C}_\nu$, and – after regrouping – $\mathbb{X}$ becomes

$$\mathbb{X} = \left[\mathbb{X}^{[1]}\, \mathbb{X}^{[2]} \ldots \mathbb{X}^{[\kappa]}\right]. \tag{4.4}$$

The number of observations in the νth class is n_ν. The numbers vary from class to class, and $n = \sum n_\nu$.

Definition 4.3 Let $\mathcal{C}_1, \ldots, \mathcal{C}_\kappa$ be κ classes. Let $\mathbf{X}$ be a random vector which belongs to one of these classes. A **(discriminant) rule** or **classifier** $\mathfrak{r}$ for $\mathbf{X}$ is a map which assigns $\mathbf{X}$ a number $\ell \le \kappa$. We write

$$\mathfrak{r}(\mathbf{X}) = \ell \qquad \text{for } 1 \le \ell \le \kappa.$$

The rule $\mathfrak{r}$ **assigns X to the correct class** or **classifies X correctly** if

$$\mathfrak{r}(\mathbf{X}) = \nu \qquad \text{when } \mathbf{X} \in \mathcal{C}_\nu \tag{4.5}$$

and **misclassifies** or **incorrectly classifies X** otherwise. □

The term *classifier* has become more common in the Statistical Learning literature. In the machine-learning community, a classifier is also called a **learner**. This terminology is closely related to the training idea, which we discuss in more detail in Section 9.5. I will mostly use the term *rule*, which has a natural interpretation in linear regression, whereas the terms *classifier* and *learner* do not.

Of special interest are random vectors from two classes. For such random vectors, we interchangeably use a rule or its associated decision function.

Definition 4.4 Let $\mathbf{X}$ be a random vector which belongs to one of the two classes $\mathcal{C}_1$ or $\mathcal{C}_2$. Let $\mathfrak{r}$ be a discriminant rule for $\mathbf{X}$. A **decision function** for $\mathbf{X}$, associated with $\mathfrak{r}$, is a real-valued function h such that

$$h(\mathbf{X}) \begin{cases} > 0 & \text{if} \quad \mathfrak{r}(\mathbf{X}) = 1 \\ < 0 & \text{if} \quad \mathfrak{r}(\mathbf{X}) = 2. \end{cases} \tag{4.6}$$

□

A decision function corresponding to a rule is not unique; for example, any multiple of a decision function by a positive scalar is also a decision function for the same rule.

Ideally, the number assigned to $\mathbf{X}$ by the rule $\mathfrak{r}$ would be the same as the value of its label. In practice, this will not always occur, and one therefore considers assessment criteria which allow comparisons of the performance of different rules. Initially we will just count the number of observations that are misclassified and then look at assessment criteria and ways of evaluating rules in Section 4.4.2.

4.3 Linear Discriminant Rules

4.3.1 Fisher's Discriminant Rule for the Population

In the early part of the last century, Fisher (1936) developed a framework for discriminating between different populations. His key idea was to partition the data in such a way that the variability in each class is small and the variability between classes is large.

In the introduction to this chapter I quoted Fisher's observation that 'vectors in one class behave differently from vectors in the other classes; and the variance within the classes differs maximally from that between the classes.' We now explore Fisher's strategy for finding such classes. In Principal Component and Canonical Correlation Analysis we do not work with the multivariate random vectors directly but project them onto one-dimensional directions and then consider these much simpler one-dimensional variables. Fisher's (1936) strategy for Discriminant Analysis is similar: find the direction $\mathbf{e}$ which minimises the within-class variance and maximises the between-class variance of $\mathbf{e}^{\mathsf{T}}\mathbb{X}$, and work with these one-dimensional quantities. As in the previous analyses, the variability or, in this case, the difference in variability within and between classes drives the process.

Definition 4.5 Let $\mathbf{X}^{[\nu]} \sim (\boldsymbol{\mu}_\nu, \Sigma_\nu)$ be labelled random vectors with $\nu \le \kappa$. Let $\bar{\boldsymbol{\mu}} = (\sum \boldsymbol{\mu}_\nu)/\kappa$ be the average of the means. Let $\mathbf{e}$ be a d-dimensional vector of unit length.

1. The **between-class variability** $\mathfrak{b}$ is

$$\mathfrak{b}(\mathbf{e}) = \sum_{\nu=1}^{\kappa} |\mathbf{e}^{\mathsf{T}}(\boldsymbol{\mu}_\nu - \bar{\boldsymbol{\mu}})|^2.$$

2. The **within-class variability** $\mathfrak{w}$ is

$$\mathfrak{w}(\mathbf{e}) = \sum_{\nu=1}^{\kappa} \operatorname{var}(\mathbf{e}^{\mathsf{T}}\mathbf{X}^{[\nu]}).$$

3. For $\mathfrak{q}(\mathbf{e}) = \mathfrak{b}(\mathbf{e})/\mathfrak{w}(\mathbf{e})$, Fisher's **discriminant** $\mathfrak{d}$ is

$$\mathfrak{d} = \max_{\{\mathbf{e}:\ \|\mathbf{e}\|=1\}} \mathfrak{q}(\mathbf{e}) = \max_{\{\mathbf{e}:\ \|\mathbf{e}\|=1\}} \frac{\mathfrak{b}(\mathbf{e})}{\mathfrak{w}(\mathbf{e})}.$$

□

The between-class variability, the within-class variability and the quotient $\mathfrak{q}$ are functions of the unit vector $\mathbf{e}$, whereas Fisher's discriminant $\mathfrak{d}$, which he called 'discriminant function', is *not* a function but the maximum $\mathfrak{q}$ attains. For this reason, I drop the word *function* when referring to $\mathfrak{d}$. Immediate questions are: How do we find the direction $\mathbf{e}$ which maximises $\mathfrak{q}$, and what is this maximum?

Theorem 4.6 *Let* $\mathbf{X}^{[\nu]} \sim (\boldsymbol{\mu}_\nu, \Sigma_\nu)$ *be labelled random vectors, and let* $\bar{\boldsymbol{\mu}} = (\sum \boldsymbol{\mu}_\nu)/\kappa$ *for* $\nu \le \kappa$. *Put*

$$B = \sum_{\nu=1}^{\kappa} (\boldsymbol{\mu}_\nu - \bar{\boldsymbol{\mu}})(\boldsymbol{\mu}_\nu - \bar{\boldsymbol{\mu}})^{\mathsf{T}} \qquad \text{and} \qquad W = \sum_{\nu=1}^{\kappa} \Sigma_\nu, \tag{4.7}$$

and assume that W *is invertible. Let* $\mathbf{e}$ *be a* d*-dimensional unit vector. The following hold:*

1. *The between-class variability* $\mathfrak{b}$ *is related to* B *by*

$$\mathfrak{b}(\mathbf{e}) = \mathbf{e}^{\mathsf{T}} B \mathbf{e}.$$

2. *The within-class variability* $\mathfrak{w}$ *is related to* W *by*

$$\mathfrak{w}(\mathbf{e}) = \mathbf{e}^{\mathsf{T}} W \mathbf{e}.$$

3. *The largest eigenvalue of* $W^{-1}B$ *is Fisher's discriminant* $\mathfrak{d}$.

4. *If η is a maximiser of the quotient $\mathfrak{q}(\mathbf{e})$ over all unit vectors $\mathbf{e}$, then η is the eigenvector of $W^{-1}B$ which corresponds to $\mathfrak{d}$.*

Proof Let $\mathbf{e}$ be a unit vector. Put $\mathbf{a} = (\boldsymbol{\mu}_\nu - \bar{\boldsymbol{\mu}})$ and $A = \mathbf{a}\mathbf{a}^{\mathsf{T}}$. Because $(\mathbf{e}^{\mathsf{T}}\mathbf{a})^2 = \mathbf{e}^{\mathsf{T}} A\mathbf{e}$, part 1 of the theorem follows. Part 2 follows from Proposition 2.1 in Section 2.2.

To see parts 3 and 4, let $\mathfrak{q}(\mathbf{e}) = (\mathbf{e}^{\mathsf{T}} B\mathbf{e})/(\mathbf{e}^{\mathsf{T}} W\mathbf{e})$. A calculation of $d\mathfrak{q}/d\mathbf{e}$ shows that

$$\frac{d\mathfrak{q}}{d\mathbf{e}} = \frac{2}{\mathbf{e}^{\mathsf{T}} W\mathbf{e}}\left[B\mathbf{e} - \mathfrak{q}(\mathbf{e})W\mathbf{e}\right].$$

The maximum value of $\mathfrak{q}$ is $\mathfrak{q}(\eta) = \mathfrak{d}$, and because W is invertible,

$$B\eta = \mathfrak{d}W\eta \qquad \text{or, equivalently,} \qquad W^{-1}B\eta = \mathfrak{d}\eta. \tag{4.8}$$

The last equality shows that η is the eigenvector of $W^{-1}B$ which corresponds to $\mathfrak{d}$ as desired. ∎

We recognise that $\mathfrak{q}(\mathbf{e})$ represents the Rayleigh quotient of the generalised eigenvalue problem $B\mathbf{e} = \mathfrak{d}W\mathbf{e}$ (see Definition 3.53 in Section 3.7.4).

We are ready to define our first discriminant rule.

Definition 4.7 For $\nu \le \kappa$, let $\mathcal{C}_\nu$ be classes characterised by $(\boldsymbol{\mu}_\nu, \Sigma_\nu)$. Let $\mathbf{X}$ be a random vector from one of the classes $\mathcal{C}_\nu$. Define the matrices B and W as in (4.7) in Theorem 4.6, and assume that W is invertible. Let η be the eigenvector of $W^{-1}B$ corresponding to the discriminant $\mathfrak{d}$, and call η the **discriminant direction**.

Fisher's (linear) discriminant rule or **Fisher's rule** $\mathfrak{r}_F$ is defined by

$$\mathfrak{r}_F(\mathbf{X}) = \ell \qquad \text{if } |\eta^{\mathsf{T}}\mathbf{X} - \eta^{\mathsf{T}}\boldsymbol{\mu}_\ell| < |\eta^{\mathsf{T}}\mathbf{X} - \eta^{\mathsf{T}}\boldsymbol{\mu}_\nu| \qquad \text{for all } \nu \ne \ell. \tag{4.9}$$

□

Fisher's rule assigns $\mathbf{X}$ the number ℓ if the scalar $\eta^{\mathsf{T}}\mathbf{X}$ is closest to the scalar mean $\eta^{\mathsf{T}}\boldsymbol{\mu}_\ell$. Thus, instead of looking for the true mean $\boldsymbol{\mu}_\ell$ which is closest to $\mathbf{X}$, we pick the simpler scalar quantity $\eta^{\mathsf{T}}\boldsymbol{\mu}_\ell$ which is closest to the weighted average $\eta^{\mathsf{T}}\mathbf{X}$. Using the scalar $\eta^{\mathsf{T}}\mathbf{X}$ has advantages over considering the d-dimensional vector $\mathbf{X}$:

- It reduces and simplifies the multivariate comparisons to univariate comparisons.
- It can give more weight to *important* variables of $\mathbf{X}$ while reducing the effect of variables that do not contribute much to $W^{-1}B$.

For two classes with means $\boldsymbol{\mu}_1$ and $\boldsymbol{\mu}_2$, we derive a decision function h for $\mathfrak{r}_F$ as in Definition 4.4. Starting from the squares of the inequality in (4.9) leads to the following sequence of equivalent statements and finally to a function which is linear in $\mathbf{X}$:

$$\begin{aligned}
& (\eta^{\mathsf{T}}\mathbf{X} - \eta^{\mathsf{T}}\boldsymbol{\mu}_1)^{\mathsf{T}}(\eta^{\mathsf{T}}\mathbf{X} - \eta^{\mathsf{T}}\boldsymbol{\mu}_1) < (\eta^{\mathsf{T}}\mathbf{X} - \eta^{\mathsf{T}}\boldsymbol{\mu}_2)^{\mathsf{T}}(\eta^{\mathsf{T}}\mathbf{X} - \eta^{\mathsf{T}}\boldsymbol{\mu}_2) \\
\iff\ & 2\mathbf{X}^{\mathsf{T}}\eta\eta^{\mathsf{T}}(\boldsymbol{\mu}_1 - \boldsymbol{\mu}_2) > (\boldsymbol{\mu}_1 + \boldsymbol{\mu}_2)^{\mathsf{T}}\eta\eta^{\mathsf{T}}(\boldsymbol{\mu}_1 - \boldsymbol{\mu}_2) \\
\iff\ & \left[\mathbf{X} - \frac{1}{2}(\boldsymbol{\mu}_1 + \boldsymbol{\mu}_2)\right]^{\mathsf{T}} \eta\eta^{\mathsf{T}}(\boldsymbol{\mu}_1 - \boldsymbol{\mu}_2) = c_\eta \left[\mathbf{X} - \frac{1}{2}(\boldsymbol{\mu}_1 + \boldsymbol{\mu}_2)\right]^{\mathsf{T}} \eta > 0,
\end{aligned} \tag{4.10}$$

where $c_\eta = \eta^{\mathsf{T}}(\boldsymbol{\mu}_1 - \boldsymbol{\mu}_2) > 0$ does not depend on $\mathbf{X}$. Define a decision function h by

$$h(\mathbf{X}) = \left[\mathbf{X} - \frac{1}{2}(\boldsymbol{\mu}_1 + \boldsymbol{\mu}_2)\right]^{\mathsf{T}} \eta. \tag{4.11}$$

To check that h is a decision function for $\mathfrak{r}_F$, note that $h(\mathbf{X}) > 0$ if and only if $\mathfrak{r}_F(\mathbf{X}) = 1$ and $h(\mathbf{X}) < 0$ if and only if $\mathfrak{r}_F(\mathbf{X}) = 2$.

4.3.2 Fisher's Discriminant Rule for the Sample

For data, we modify Fisher's discriminant rule by replacing the means and covariance matrices by their sample quantities. For $\nu \le \kappa$, the **sample mean of the νth class** is

$$\overline{\mathbf{X}}_\nu = \frac{1}{n_\nu}\sum_{i=1}^{n_\nu} \mathbf{X}_i^{[\nu]}. \tag{4.12}$$

By analogy with the notation for the population mean, I use a subscript to indicate the class for the sample mean. The **average of the sample class means** or the **average sample class mean** is

$$\overline{\overline{\mathbf{X}}} = \frac{1}{\kappa}\sum_{\nu=1}^{\kappa} \overline{\mathbf{X}}_\nu.$$

This average agrees with the sample mean $\overline{\mathbf{X}} = (\sum_{i=1}^{n} \mathbf{X}_i)/n$ if each class has the same number of observations, $n_\nu = n/\kappa$.

Definition 4.8 Let $\mathbb{X} = \left[\mathbb{X}^{[1]}\ \mathbb{X}^{[2]} \ldots \mathbb{X}^{[\kappa]}\right]$ be d-dimensional labelled data as in (4.4) which belong to classes $\mathcal{C}_\nu$ for $\nu \le \kappa$. Let $\mathbf{e}$ be a d-dimensional unit vector.

1. The **between-class sample variability** $\widehat{\mathfrak{b}}$ is given by

$$\widehat{\mathfrak{b}}(\mathbf{e}) = \sum_{\nu=1}^{\kappa} |\mathbf{e}^{\mathsf{T}}(\overline{\mathbf{X}}_\nu - \overline{\overline{\mathbf{X}}})|^2.$$

2. The **within-class sample variability** $\widehat{\mathfrak{w}}$ is given by

$$\widehat{\mathfrak{w}}(\mathbf{e}) = \sum_{\nu=1}^{\kappa} \frac{1}{n_\nu - 1}\sum_{i=1}^{n_\nu}\left[\mathbf{e}^{\mathsf{T}}(\mathbf{X}_i^{[\nu]} - \overline{\mathbf{X}}_\nu)\right]^2.$$

3. For $\widehat{\mathfrak{q}}(\mathbf{e}) = \widehat{\mathfrak{b}}(\mathbf{e})/\widehat{\mathfrak{w}}(\mathbf{e})$, Fisher's **sample discriminant** $\widehat{\mathfrak{d}}$ is

$$\widehat{\mathfrak{d}} = \max_{\{\mathbf{e}:\ \|\mathbf{e}\|=1\}} \widehat{\mathfrak{q}}(\mathbf{e}) = \max_{\{\mathbf{e}:\ \|\mathbf{e}\|=1\}} \frac{\widehat{\mathfrak{b}}(\mathbf{e})}{\widehat{\mathfrak{w}}(\mathbf{e})}.$$

□

I typically drop the word *sample* when referring to the discriminant in the sample case. However, the notation $\widehat{\mathfrak{d}}$ indicates that this quantity is an estimator of the population quantity, and hence the distinction between the population and the sample discriminant is, at times, necessary.

The next corollary is a sample analogue of Theorem 4.6.

Corollary 4.9 *Consider the data* $\mathbb{X} = \left[\mathbb{X}^{[1]}\ \mathbb{X}^{[2]} \ldots \mathbb{X}^{[\kappa]}\right]$ *from* κ *different classes which have sample class means* $\overline{\mathbf{X}}_\nu$ *and sample class covariance matrices* S_ν. *Put*

$$\widehat{B} = \sum_{\nu=1}^{\kappa}\left(\overline{\mathbf{X}}_\nu - \overline{\overline{\mathbf{X}}}\right)\left(\overline{\mathbf{X}}_\nu - \overline{\overline{\mathbf{X}}}\right)^{\mathsf{T}} \quad \text{and} \quad \widehat{W} = \sum_{\nu=1}^{\kappa} S_\nu.$$

Let $\mathbf{e}$ *be a unit vector. The following hold:*

1. *The between-class sample variability* $\widehat{\mathfrak{b}}$ *is related to* $\widehat{B}$ *by*

$$\widehat{\mathfrak{b}}(\mathbf{e}) = \mathbf{e}^{\mathsf{T}}\widehat{B}\mathbf{e}.$$

2. *The within-class sample variability* $\widehat{\mathfrak{w}}$ *is related to* $\widehat{W}$ *by*

$$\widehat{\mathfrak{w}}(\mathbf{e}) = \mathbf{e}^{\mathsf{T}}\widehat{W}\mathbf{e}.$$

3. *The largest eigenvalue of* $\widehat{W}^{-1}\widehat{B}$ *is Fisher's sample discriminant* $\widehat{\mathfrak{d}}$.
4. *If* $\widehat{\eta}$ *is the maximiser of the quotient* $\widehat{\mathfrak{q}}(\mathbf{e})$ *over all unit vectors* $\mathbf{e}$, *then* $\widehat{\eta}$ *is the eigenvector of* $\widehat{W}^{-1}\widehat{B}$ *which corresponds to* $\widehat{\mathfrak{d}}$.

The matrices $\widehat{B}$ and $\widehat{W}$ are not quite covariance matrices. The matrix $\widehat{B}$ is the covariance matrix of the sample class means modulo the factor $(\kappa - 1)$, and $\widehat{W}$ is the sum of the within-class sample covariance matrices.

The **(sample) discriminant direction** $\widehat{\eta}$ of Corollary 4.9 defines **Fisher's (linear) discriminant rule** $\mathfrak{r}_F$:

$$\mathfrak{r}_F(\mathbf{X}) = \ell \qquad \text{if } |\widehat{\eta}^{\mathsf{T}}\mathbf{X} - \widehat{\eta}^{\mathsf{T}}\overline{\mathbf{X}}_\ell| < |\widehat{\eta}^{\mathsf{T}}\mathbf{X} - \widehat{\eta}^{\mathsf{T}}\overline{\mathbf{X}}_\nu| \qquad \text{for all } \nu \neq \ell, \tag{4.13}$$

and as in Definition 4.3, we say that $\mathfrak{r}_F$ has **misclassified** an observation $\mathbf{X}$ if the true label and the label assigned to $\mathbf{X}$ by the rule are different. The interpretation of Fisher's rule is the same for the population and the sample apart from the obvious replacement of the true means by the sample means of the classes. There is generally no confusion between the population and sample versions, and I will therefore use the notation $\mathfrak{r}_F$ for the sample as well as the population.

We apply Fisher's rule to Fisher's *iris* data and to *simulated* data.

Example 4.1 Figures 1.3 and 1.4 in Section 1.2 show Fisher's **iris** data, which consist of three species of iris. The species are distinguished by their petals and sepals. For each species, fifty samples and four different measurements are available, resulting in a total of 150 observations in four variables. The data are labelled. Because the means and covariances matrices of the three classes are not known, we apply Fisher's sample discriminant rule (4.13).

Although the sample size and the dimension are small by today's standards, Fisher's iris data have remained popular in Discriminant Analysis and Cluster Analysis and often serve as an natural candidate for testing algorithms.

To gain some insight into the data and Fisher's discriminant rule, I use all four variables, as well as subsets of the variables. Figure 1.4 shows that the samples are better separated for variables 3 and 4 than the other two, and it will therefore be of interest to see how the exclusion of variables 1 or 4 affect the performance of Fisher's rule.

I calculate the discriminant $\widehat{\mathfrak{d}}$, the discriminant direction $\widehat{\eta}$ and the number of misclassified observations for all four variables and then separately for variables 1 to 3 and 2 to 4. Table 4.1 shows the results. The first column of the table refers to the variables used in the derivation of the rule. The last column, 'No. misclassified', tallies the number of misclassified samples.

Table 4.1 *Fisher's LDA for the Iris Data of Example 4.1*

Variables	$\widehat{\mathfrak{d}}$	Entries of $\widehat{\eta}^{\mathsf{T}}$	No. misclassified
1–4	31.6	$(0.20 \quad 0.39 \; -0.55 \; -0.71)$	2
1–3	24.7	$(0.34 \quad 0.26 \; -0.90)$	5
2–4	29.6	$(0.49 \; -0.38 \; -0.78)$	5

The calculated discriminants and discriminant directions differ. Because variables 3 and 4 separate the data better than the other two, they are more relevant in discrimination and hence have larger (absolute) entries for $\widehat{\eta}$.

Table 4.1 shows that the discriminant rule performs best when all variables are used. The two misclassified observations belong to classes 2 and 3, the green and black classes in the figures. The results tell us that leaving out one of the variables has a negative impact on the performance of Fisher's rule, contrary to the initial visual impression gleaned from Figure 1.4.

One could consider variables 3 and 4 only because the figures suggest that these two variables are the crucial ones. It turns out that the discriminant rule based on these variable performs more poorly than the rule based on variables 1 to 3, which shows that the best two variables are not enough and that care needs to be taken when leaving out variables in the design of the rule. ■

The example shows that the discrimination rule complements and improves on information obtained from visual representations of the data.

Example 4.2 We consider 3D **simulated data** arising from three classes $C_\nu \equiv (\boldsymbol{\mu}_\nu, \Sigma_\nu)$, with $\nu = 1, 2$ and 3. The number of classes is the same as in Fisher's iris data, but now we explore Fisher's rule when the classes become less well separated. We keep the three covariance matrices the same but allow the means to move closer together and consider four separate simulations which become increasingly harder as a classification task. The covariance matrices are

$$\Sigma_1 = \begin{pmatrix} 1/4 & 0 & 0 \\ 0 & 1/8 & 0 \\ 0 & 0 & 1/8 \end{pmatrix}, \quad \Sigma_2 = \begin{pmatrix} 1/4 & 0 & 0 \\ 0 & 1/4 & 0 \\ 0 & 0 & 1/4 \end{pmatrix}, \quad \Sigma_3 = \begin{pmatrix} 1/8 & 0 & 0 \\ 0 & 1/4 & 0 \\ 0 & 0 & 1/4 \end{pmatrix}.$$

I repeat each of the four simulations 1,000 times. For each simulation I generate 250 vectors $\mathbf{X}_i^{[1]} \sim \mathcal{N}(\boldsymbol{\mu}_1, \Sigma_1)$, 100 vectors $\mathbf{X}_i^{[2]} \sim \mathcal{N}(\boldsymbol{\mu}_2, \Sigma_2)$ and 150 vectors $\mathbf{X}_i^{[3]} \sim \mathcal{N}(\boldsymbol{\mu}_3, \Sigma_3)$. The class means for the four simulations are shown in Table 4.2. Typical randomly generated data are shown in Figure 4.1. The four subplots show the same 500 generated points, shifted by the means in all but the top-left plot. The points from class C_1 are shown in red, those from class C_2 in black and those from class C_3 in blue. The red class remains centred at the origin. The class means of the blue and black classes move closer to the mean of the red class in the later simulations. In the bottom-right subplot the class membership is only apparent

Table 4.2 *Class Means and Performance of Fisher's Discriminant Rule (4.9) for Four Simulations, Sim 1, ..., Sim 4, from Example 4.2*

	Class means			Misclassified points		
	μ_1	μ_2	μ_3	Average no.	Percentage	Std
Sim 1	(0, 0, 0)	(2, 1, -2)	(-1, 0, 2)	2.60	0.52	0.33
Sim 2	(0, 0, 0)	(1, 1, -2)	(-1, 0, 1)	24.69	4.94	0.98
Sim 3	(0, 0, 0)	(1, 0.5, 0)	(-1, 0, 1)	80.91	16.18	1.63
Sim 4	(0, 0, 0)	(1, 0.5, 0)	(0, 0, 1)	152.10	30.42	1.98

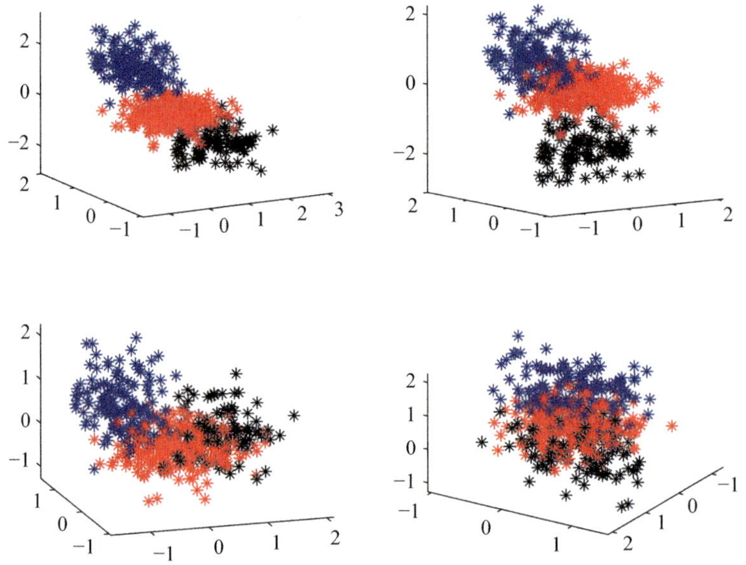

Figure 4.1 Three-dimensional simulated data from Example 4.2. The means of the three classes $\boldsymbol{\mu}_\nu$, given in Table 4.2, move closer together from the top-left to the bottom-right subplot.

because of the different colours. Each subfigure is shown from a different perspective in order to demonstrate the configuration of the classes best.

I calculate the value Fisher's rule (4.9) assigns to each random vector and find the number of misclassified points. Table 4.2 shows the average number of misclassified points in 1,000 runs for each of the four simulations, the percentage of misclassified points and the standard deviation (Std) – on the same scale as the percentage error. Because the class means $\boldsymbol{\mu}_\nu$ vary from one simulation to the next, the discriminant directions $\hat{\eta}$ also vary. In all cases the second coordinate of $\hat{\eta}$ is smallest and so least important, and for simulations 3 and 4, the third coordinate of $\hat{\eta}$ is dominant.

For simulation 1 (Sim 1 in Table 4.2), the classes are well separated, and on average, 2.6 of the 500 points are incorrectly classified. The discriminant rule performs well. In the top-right subplot we see more overlap of the classes, especially between the blue and red points. The discriminant rule misclassifies about 5 per cent of the observations, an almost 10-fold increase. When we get to the bottom-right subplot, Fisher's rule for simulation 4 misclassifies almost one-third of the points. The example illustrates that the classification becomes more difficult and more observations are incorrectly classified as the separation between the class means decreases. ■

So far we have discussed how to obtain Fisher's rule for the population and data. The prediction part of Discriminant Analysis is of great importance, similar to prediction in regression. If $\mathbf{X}_{\text{new}}$ is a new datum which we want to assign to one of the existing classes, we put

$$\mathfrak{r}_F(\mathbf{X}_{\text{new}}) = \ell \qquad \text{if } |\widehat{\eta}^{\mathsf{T}}\mathbf{X}_{\text{new}} - \widehat{\eta}^{\mathsf{T}}\overline{\mathbf{X}}_\ell| < |\widehat{\eta}^{\mathsf{T}}\mathbf{X}_{\text{new}} - \widehat{\eta}^{\mathsf{T}}\overline{\mathbf{X}}_\nu| \qquad \text{for all } \nu \neq \ell.$$

Thus, to determine the value the rule assigns to a new datum $\mathbf{X}_{\text{new}}$, we apply the rule with the direction $\widehat{\eta}$ and the sample means $\overline{\mathbf{X}}_\nu$, which have been calculated from the labelled data $\mathbb{X}$.

4.3.3 Linear Discrimination for Two Normal Populations or Classes

Better knowledge leads to better decisions. If we know, for example, that the data belong to the normal distribution, then we should incorporate this knowledge into the decision-making process. We now investigate how this can be done. We begin with univariate observations from the normal distribution and then consider discrimination rules for multivariate normal data.

Consider the two classes $C_1 = \mathcal{N}(\mu_1, \sigma^2)$ and $C_2 = \mathcal{N}(\mu_2, \sigma^2)$ which differ only in their means, and assume that $\mu_1 > \mu_2$. If a univariate random variable X belongs to one of the two classes, we design a rule which assigns X to class one if the observed X is more *likely* to come from C_1. This last statement suggests that one should take into account the likelihood, which is defined in (1.16) of Section 1.4. We define a discriminant rule

$$\mathfrak{r}(X) = 1 \qquad \text{if } L(\mu_1|X) > L(\mu_2|X),$$

where $L(\mu_\nu|X)$ is the likelihood function of $\mathcal{N}(\mu_\nu, \sigma^2)$. To exhibit a decision function for this rule, we begin with a comparison of the likelihood functions and work our way along the sequence of equivalent statements:

$$\begin{aligned}
& L(\mu_1|X) > L(\mu_2|X) \\
\Longleftrightarrow\ & |2\pi\sigma^2|^{-1/2}\exp\left[-\frac{(X-\mu_1)^2}{2\sigma^2}\right] > |2\pi\sigma^2|^{-1/2}\exp\left[-\frac{(X-\mu_2)^2}{2\sigma^2}\right] \\
\Longleftrightarrow\ & \exp\left\{-\frac{1}{2\sigma^2}\left[(X-\mu_1)^2 - (X-\mu_2)^2\right]\right\} > 1 \\
\Longleftrightarrow\ & (X-\mu_2)^2 > (X-\mu_1)^2 \\
\Longleftrightarrow\ & 2X(\mu_1-\mu_2) > (\mu_1^2 - \mu_2^2) \\
\Longleftrightarrow\ & X > \frac{1}{2}(\mu_1+\mu_2).
\end{aligned} \tag{4.14}$$

The third line combines terms, and the last equivalence follows from the assumption that $\mu_1 > \mu_2$. The calculations show that $L(\mu_1|X)$ is bigger than $L(\mu_2|X)$ precisely when X exceeds the average of the two means. Because $\mu_1 > \mu_2$, this conclusion makes sense. The one-dimensional result generalises naturally to random vectors.

Theorem 4.10 *Let* $\mathbf{X}$ *be a Gaussian random vector which belongs to one of the classes* $\mathcal{C}_\nu = \mathcal{N}(\boldsymbol{\mu}_\nu, \Sigma)$, *for* $\nu = 1, 2$, *and assume that* $\boldsymbol{\mu}_1 \neq \boldsymbol{\mu}_2$. *Let* $L(\boldsymbol{\mu}_\nu|\mathbf{X})$ *be the likelihood function of the* ν*th class. Let* $\mathfrak{r}_{norm}$ *be the rule which assigns* $\mathbf{X}$ *to class one if* $L(\boldsymbol{\mu}_1|\mathbf{X}) > L(\boldsymbol{\mu}_2|\mathbf{X})$. *If*

$$h(\mathbf{X}) = \left[\mathbf{X} - \frac{1}{2}(\boldsymbol{\mu}_1 + \boldsymbol{\mu}_2)\right]^{\mathsf{T}} \Sigma^{-1}(\boldsymbol{\mu}_1 - \boldsymbol{\mu}_2), \tag{4.15}$$

then h is a decision function for $\mathfrak{r}_{norm}$, *and* $h(\mathbf{X}) > 0$ *if and only if* $L(\boldsymbol{\mu}_1|\mathbf{X}) > L(\boldsymbol{\mu}_2|\mathbf{X})$.

We call the rule based on the decision function (4.15) the **normal (linear) discriminant rule**. This rule is based on normal classes which have the same covariance matrix Σ. The proof of Theorem 4.10 is similar to the one-dimensional argument earlier; hints are given in Problem 42 at the end of Part I of Theorem 3.13.

Often the rule and its decision function are regarded as one and the same thing. To define a sample version of the normal rule, we merely replace the means and the common covariance matrix by the respective sample quantities and obtain for $\mathbf{X}_i$ from data $\mathbb{X}$

$$\mathfrak{r}_{\rm norm}(\mathbf{X}_i) = \begin{cases} 1 & \text{if } \left[\mathbf{X}_i - \frac{1}{2}(\overline{\mathbf{X}}_1 + \overline{\mathbf{X}}_2)\right]^{\mathsf{T}} S^{-1}(\overline{\mathbf{X}}_1 - \overline{\mathbf{X}}_2) > 0, \\ 2 & \text{otherwise.} \end{cases}$$

For a new observation $\mathbf{X}_{\rm new}$, we put $\mathfrak{r}_{\rm norm}(\mathbf{X}_{\rm new}) = 1$ if

$$h(\mathbf{X}_{\rm new}) = \left[\mathbf{X}_{\rm new} - \frac{1}{2}(\overline{\mathbf{X}}_1 + \overline{\mathbf{X}}_2)\right]^{\mathsf{T}} S^{-1}(\overline{\mathbf{X}}_1 - \overline{\mathbf{X}}_2) > 0,$$

where the sample quantities $\overline{\mathbf{X}}_\nu$ and S are those calculated from the data $\mathbb{X}$ without reference to $\mathbf{X}_{\rm new}$.

We consider the performance of the normal discriminant rule in a simulation study.

Example 4.3 We consider 3D **simulated data** from two Gaussian classes which share the common covariance matrix Σ_1 of Example 4.2 and differ in their class means. The likelihood-based normal rule of Theorem 4.10 compares two classes, and we therefore have a simpler scenario than that considered in Example 4.2.

I simulate 300 samples from the first class and 200 from the second class, with means $\boldsymbol{\mu}_1 = (0, 0, 0)$ and $\boldsymbol{\mu}_2 = (1.25, 1, -0.5)$, and display them in red and blue, respectively, in the left subplot of Figure 4.2. There is a small overlap between the two samples. I apply the normal rule (4.15) to these data with the true class means and common covariance matrix. Normally we use the sample class means and a pooled covariance, since we do not know the true parameters. Six of the red samples are misclassified and two of the blue samples. The misclassification of these eight samples gives the low error rate of 1.6 per cent.

The data in the right subplot of Figure 4.2 pose a greater challenge. The red data are the same as in the left plot. The blue data are obtained from the blue data in the left panel by a shift of -1.25 in the first variable, so the class mean of the blue data is $\boldsymbol{\mu}_3 = (0, 1, -0.5)$. The two point clouds are not as well separated as before. The view in the right panel is obtained by a rotation of the axes to highlight the regions of overlap. As a result, the red data in the two panels no longer appear to be the same, although they are. The normal rule applied to the data in the right plot misclassifies eighteen of the red points and eight of the blue points. These misclassified points include most of the previously misclassified data from the right panel. The misclassification rate has increased to 5.2 per cent. ■

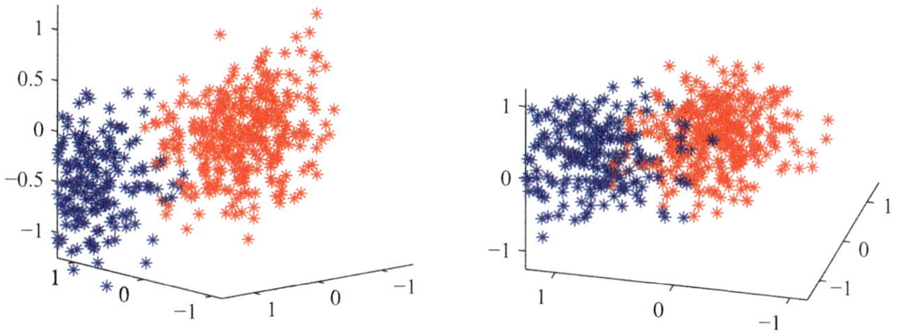

Figure 4.2 Three-dimensional simulated normal data belonging to two classes from Example 4.3. The two figues differ by a shift of the blue data towards the red in the right subplot.

Some authors refer to the normal linear rule as *Fisher's rule*. This implicitly makes an assumption about the underlying distribution of the data, something I have not done here, because Fisher's rule is a rule based on matrix and eigenvector calculations. However, there is a reason for identifying the two rules, at least for two class problems. Consider the *form* of the decision functions (4.11) for Fisher's rule and (4.15) for the normal rule. The two functions *look* similar, but there is a difference: the normal rule contains the matrix Σ^{-1}, whereas Fisher's rule uses the matrix $\eta\eta^{\top}$. Writing W for the within-class variances and applying Theorem 4.6, it follows that

$$B = (\mu_1 - \mu_2)(\mu_1 - \mu_2)^{\top}/2 \qquad \text{and} \qquad \eta = \frac{W^{-1}(\mu_1 - \mu_2)}{\left\|W^{-1}(\mu_1 - \mu_2)\right\|}. \tag{4.16}$$

If $W = 2\Sigma$, then the decision functions of Fisher's rule and the normal rule are equivalent. This is the reason why some authors regard the two rules as one and the same. I defer a proof of these statements to the Problems at the end of Part I.

The comparison of Fisher's rule and the normal rule motivates a general class of linear decision functions for two-class discrimination problems. Put

$$h_{\beta}(\mathbf{X}) = \left[\mathbf{X} - \frac{1}{2}(\mu_1 + \mu_2)\right]^{\top} \boldsymbol{\beta}, \tag{4.17}$$

where $\boldsymbol{\beta}$ is a suitably chosen vector: $\boldsymbol{\beta} = \eta$ for Fisher's rule, and $\boldsymbol{\beta} = \Sigma^{-1}(\mu_1 - \mu_2)$ for the normal linear rule. We will meet other choices for $\boldsymbol{\beta}$ in Sections 13.2.1 and 13.3.3.

4.4 Evaluation of Rules and Probability of Misclassification

4.4.1 Boundaries and Discriminant Regions

For two-class problems, a decision about the class membership can be based on a rule or a decision function which separates regions. Boundaries between regions are appealing when we deal with two classes or with low-dimensional data, where they give further visual insight.

Definition 4.11 Let $\mathbf{X}$ be a random vector which belongs to one of the two classes C_1 and C_2. Let $\mathfrak{r}$ be a discriminant rule for $\mathbf{X}$, and let h be a corresponding decision function.

1. The **decision boundary** B of the rule $\mathfrak{r}$ consists of all d-dimensional vectors $\mathbf{X}$ such that $h(\mathbf{X}) = 0$.
2. For $\nu \leq \kappa$, the **discriminant region** G_ν of the rule $\mathfrak{r}$ is defined by

$$G_\nu = \{\mathbf{X}\colon \mathfrak{r}(\mathbf{X}) = \nu\}. \tag{4.18}$$

□

When the rule is linear in $\mathbf{X}$, then the decision function is a line or a hyperplane. More complex boundaries arise for non-linear rules. I defined a decision boundary based on a rule; instead, one could start with a decision boundary and then derive a rule from the boundary. For data, the decision boundary and the discriminant regions are defined in an analogous way.

Decision boundaries and decision functions exist for more than two classes: for three classes and a linear rule, we obtain intersecting lines or hyperplanes as the decision functions. Although it is clear how to extend boundaries to κ classes mathematically, because of the increased complexity, decision functions and boundaries become less useful as the number of classes increases.

The discriminant regions G_ν are disjoint because each $\mathbf{X}$ is assigned one number $\nu \leq \kappa$ by the discriminant rule. We can interpret the discriminant regions as 'classes assigned by the rule', that is, what the rule determines the classes to be. Given disjoint regions, we define a discriminant rule by

$$\mathfrak{r}(\mathbf{X}) = \nu \qquad \text{if } \mathbf{X} \in G_\nu, \tag{4.19}$$

so the two concepts, regions and rule, define each other completely. If the rule were *perfect*, then for each ν, the class $\mathcal{C}_\nu$ and the region G_ν would agree. The decision boundary separates the discriminant regions. This is particularly easy to see in two-class problems.

I illustrate decision functions and boundaries for two-class problems with the normal linear rule and the decision function h of (4.15). For d-dimensional random vectors from two classes, the decision boundary is characterised by vectors $\mathbf{X}$ such that

$$B_h = \left\{\mathbf{X}\colon \left[\mathbf{X} - \frac{1}{2}(\boldsymbol{\mu}_1 + \boldsymbol{\mu}_2)\right]^{\mathsf{T}} \Sigma^{-1}(\boldsymbol{\mu}_1 - \boldsymbol{\mu}_2) = 0\right\}.$$

This boundary is a line in two dimensions and a hyperplane in d dimensions because its decision function is linear in $\mathbf{X}$.

Example 4.4 We consider two-class problems in two- and three-dimensional **simulated data**. For the two-dimensional case, let

$$\boldsymbol{\mu}_1 = \begin{bmatrix}0 & 0\end{bmatrix}^{\mathsf{T}} \qquad \text{and} \qquad \boldsymbol{\mu}_2 = \begin{bmatrix}1 & 1\end{bmatrix}^{\mathsf{T}}$$

be the class means of two classes from the normal distribution with common covariance matrix $\Sigma = 0.5\mathbf{I}_{2\times 2}$, where $\mathbf{I}_{2\times 2}$ is the 2×2 identity matrix. The decision function for the normal linear rule and $\mathbf{X} = [X_1\ X_2]^{\mathsf{T}}$ is

$$h(\mathbf{X}) = \left\{\mathbf{X} - 0.5 \times \begin{bmatrix}1\\1\end{bmatrix}\right\}^{\mathsf{T}} \begin{pmatrix}2 & 0\\0 & 2\end{pmatrix} \begin{bmatrix}-1\\-1\end{bmatrix} = 1 - X_1 - X_2.$$

It follows that

$$h(\mathbf{X}) = 0 \qquad \text{if and only if} \qquad X_1 + X_2 = 1.$$

The decision boundary is therefore

$$B_h = \{\mathbf{X}\colon X_2 = -X_1 + 1\}.$$

The left subplot of Figure 4.3 shows this boundary for 125 red points from the first class and 75 points from the second class. As the classes have some overlap, some of the blue points are on the red side of the boundary and vice versa.

In the 3D problem we consider two normal populations with class means and common covariance matrix given by

$$\boldsymbol{\mu}_1 = \begin{bmatrix}0 & 0 & 0\end{bmatrix}^{\mathsf{T}}, \qquad \boldsymbol{\mu}_2 = \begin{bmatrix}1.25 & 1 & -0.5\end{bmatrix}^{\mathsf{T}} \qquad \text{and} \qquad \Sigma = \frac{1}{8}\begin{pmatrix}2 & 0 & 0\\ 0 & 1 & 0\\ 0 & 0 & 1\end{pmatrix}.$$

For these parameters and $\mathbf{X} = [X_1\ X_2\ X_3]^{\mathsf{T}}$,

$$\begin{aligned} h(\mathbf{X}) &= \left\{\mathbf{X} - 0.5 \times \begin{bmatrix}1.25\\ 1\\ -0.5\end{bmatrix}\right\}^{\mathsf{T}} \begin{pmatrix}4 & 0 & 0\\ 0 & 8 & 0\\ 0 & 0 & 8\end{pmatrix} \begin{bmatrix}-1.25\\ -1\\ 0.5\end{bmatrix} \\ &= -5X_1 - 8X_2 + 4X_3 + 8.125. \end{aligned}$$

The solution $h(\mathbf{X}) = 0$ is a plane with

$$B_h = \{\mathbf{X}\colon X_3 = 1.25X_1 + 2X_2 - 65/32\}.$$

For 150 red points from the class $\mathcal{N}(\boldsymbol{\mu}_1, \Sigma)$ and 75 blue points from the class $\mathcal{N}(\boldsymbol{\mu}_2, \Sigma)$, I display the separating boundary in the right subplot of Figure 4.3. In this sample, three red points, marked with an additional blue circle, and two blue points, marked with a red circle, are on the 'wrong side of the fence'. As a closer inspection of the subplot shows, three of the incorrectly classified points are very close to the left vertical line of the separating hyperplane. I have rotated this figure to allow a clear view of the hyperplane. The line in the left plot and the plane in the right plot divide the space into two disjoint regions, the discriminant regions of the rule. ■

4.4.2 Evaluation of Discriminant Rules

There are a number of reasons why we want to assess the performance of a discriminant rule. If two or more rules are available, it is important to understand how well each rule performs, which rule performs best, and under what conditions a rule performs well. There is not a unique or even a best performance measure. Because of the various possible measures, it is good practice to refer explicitly to the performance measure that is used.

Suppose that the data $\mathbb{X} = \begin{bmatrix}\mathbf{X}_1\ \mathbf{X}_2 \cdots \mathbf{X}_n\end{bmatrix}$ belong to classes C_ν with $\nu \le \kappa$. Let $\mathfrak{r}$ be a discriminant rule for these data. As we have seen in Example 4.4, Fisher's rule does not assign all points to their correct class. Table 4.3 shows schematically what can happen to each observation. If the rule's assignment agrees with the value of the label, as happens along the diagonal of the table, then the rule has correctly classified the observation. The

Table 4.3 *Labels, Rules and Misclassifications*

		Rule's assignment			
		1	2	$\cdots$	κ
	1	(1, 1)	(1, 2)	$\cdots$	(1, κ)
Label's	2	(2, 1)	(2, 2)	$\cdots$	(2, κ)
value	$\vdots$	$\vdots$	$\vdots$	$\ddots$	$\vdots$
	κ	(κ, 1)	(κ, 2)	$\cdots$	(κ, κ)

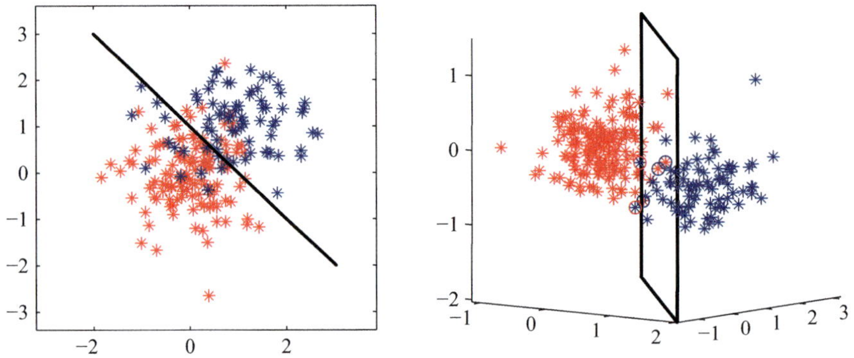

Figure 4.3 Linear boundaries for Example 4.4 with 2D data (*left*) and 3D data (*right*). Incorrectly classified points are circled in the right plot.

table shows that the more classes there are, the more 'mistakes' the rule can make. Indeed, for a two-class problem, random guessing gets the right answer 50 per cent of the time, whereas random guessing gets worse quickly when the number of classes increases.

In this section we consider two performance measures for assessing discriminant rules: a simple classification error and the *leave-one-out* error, which I introduce in Definition 4.13. In Section 9.5 we look at generalisations of the leave-one-out error and consider in more detail training and testing aspects in Supervised Learning. Our performance criteria are data-based. I briefly define and discuss error probabilities in Definition 4.14 and Proposition 4.15 at the end of this section and in Section 4.6. For more details see Devroye, Györfi, and Lugosi (1996).

Definition 4.12 Let $\mathbb{X} = \begin{bmatrix}\mathbf{X}_1\ \mathbf{X}_2 \cdots \mathbf{X}_n\end{bmatrix}$ be labelled data, and let $\mathfrak{r}$ be a rule. A **cost factor** $\mathbf{c} = [c_1, \ldots, c_n]$ associated with $\mathfrak{r}$ is defined by

$$c_i \begin{cases} = 0 & \text{if } \mathbf{X}_i \text{ has been correctly classified by } \mathfrak{r} \\ > 0 & \text{if } \mathbf{X}_i \text{ has been incorrectly classified by } \mathfrak{r}. \end{cases}$$

The **classification error** $\mathcal{E}_{\text{mis}}$ of the rule $\mathfrak{r}$ and cost factor $\mathbf{c}$ is the percentage error given by

$$\mathcal{E}_{\text{mis}} = \left(\frac{1}{n} \sum_{i=1}^{n} c_i \right) \times 100. \tag{4.20}$$

□

Remark. I use the term *classification error* for general error measures used in classification, including those in Section 9.5, but I reserve the symbol $\mathcal{E}_{\text{mis}}$ for the specific error (4.20).

If $\mathbf{X}_i$ has been incorrectly classified, we typically use $c_i = 1$. A classification error with cost factors 0 and 1 is simple and natural; it does not care how many classes there are, just counts the number of observations that are misclassified. Non-constant positive cost factors are useful

1. for two classes if we want to distinguish between 'false positives' and 'false negatives', and so distinguish between the entries $(2,1)$ and $(1,2)$ in Table 4.3 and
2. for more than two classes if the classes are not categorical but represent, for example, the degree of severity of a condition. One might choose higher cost factors if the rule gets it 'badly' wrong.

In addition, cost factors can include prior information or information based on conditional probabilities (see Section 4.6.1).

Performance measures play an important role when a rule is constructed from a part of the data and then applied to the remainder of the data. For $\mathbb{X} = [\mathbf{X}_1\ \mathbf{X}_2 \cdots \mathbf{X}_n]$, let $\mathbb{X}_0$ consist of $m < n$ of the original observations, and let $\mathbb{X}_p$ be the subsample containing the remaining $n - m$ observations. We first derive a rule $\mathfrak{r}_0$ from $\mathbb{X}_0$ (and without reference to $\mathbb{X}_p$), then predict the class membership of the observations in $\mathbb{X}_p$ using $\mathfrak{r}_0$, and finally calculate the classification error of the rule $\mathfrak{r}_0$ for the observations from $\mathbb{X}_p$. This process forms the basis of the training and testing idea in Supervised Learning, see Section 9.5. There are many ways we can choose the subset $\mathbb{X}_0$; here we focus on a simple case: $\mathbb{X}_0$ corresponds to all observations but one.

Definition 4.13 Consider labelled data $\mathbb{X} = [\mathbf{X}_1\ \mathbf{X}_2 \cdots \mathbf{X}_n]$ and a rule $\mathfrak{r}$. For $i \le n$, put

$$\begin{aligned}\mathbb{X}_{0,(-i)} &= [\mathbf{X}_1\ \mathbf{X}_2 \cdots \mathbf{X}_{i-1}\quad \mathbf{X}_{i+1} \cdots \mathbf{X}_n]\\ \mathbb{X}_{p,(-i)} &= \mathbf{X}_i,\end{aligned} \tag{4.21}$$

and call $\mathbb{X}_{0,(-i)}$ the **ith leave-one-out training set**. Let $\mathfrak{r}_{(-i)}$ be the rule which is constructed like $\mathfrak{r}$ but only based on $\mathbb{X}_{0,(-i)}$, and let $\mathfrak{r}_{(-i)}(\mathbf{X}_i)$ be the value the rule $\mathfrak{r}_{(-i)}$ assigns to the observation $\mathbf{X}_i$. A **cost factor** $\mathbf{k} = [k_{-1}, \ldots, k_{-n}]$ associated with the rules $\mathfrak{r}_{(-i)}$ is defined by

$$k_{-i} \begin{cases} = 0 & \text{if } \mathbf{X}_i \text{ has been correctly classified by } \mathfrak{r}_{(-i)},\\ > 0 & \text{if } \mathbf{X}_i \text{ has been incorrectly classified by } \mathfrak{r}_{(-i)}.\end{cases}$$

The **leave-one-out error** $\mathcal{E}_{loo}$ based on the n rules $\mathfrak{r}_{(-i)}$ with $i \le n$, the assigned values $\mathfrak{r}_{(-i)}(\mathbf{X}_i)$ and cost factor $\mathbf{k}$, is

$$\mathcal{E}_{\text{loo}} = \left(\frac{1}{n}\sum_{i=1}^{n} k_{-i}\right) \times 100. \tag{4.22}$$

The choice of these n leave-one-out training sets $\mathbb{X}_{0,(-i)}$, the derivation of the n rules $\mathfrak{r}_{(-i)}$ and the calculation of the leave-one-out error $\mathcal{E}_{\text{loo}}$ are called the **leave-one-out method**. □

Note that the ith leave-one-out training set $\mathbb{X}_{0,(-i)}$ omits precisely the ith observation. The singleton $\mathbb{X}_{p,(-i)} = \mathbf{X}_i$ is regarded as a 'new' observation which is to be classified by the rule $\mathfrak{r}_{(-i)}$. This process is repeated for each $i \le n$. Thus, each observation $\mathbf{X}_i$ is left out exactly once. The error $\mathcal{E}_{\text{loo}}$ collects contributions from those observations $\mathbf{X}_i$ for which $\mathfrak{r}_{(-i)}(\mathbf{X}_i)$ differs from the value of the label of $\mathbf{X}_i$. Typically we take cost factors 0 and 1 only, so $\mathcal{E}_{loo}$ simply counts the number of $\mathbf{X}_i$ that are misclassified by $\mathfrak{r}_{(-i)}$.

The key idea of the leave-one-out method is the prediction of the ith observation from all observations minus the ith. This motivates the notation $\mathfrak{r}_{(-i)}$. The rules $\mathfrak{r}_{(-i)}$ will differ from each other and from $\mathfrak{r}$. However, for a good discriminant rule, this difference should become negligible with increasing sample size.

The leave-one-out method is a special case of cross-validation which I describe in Definition 9.10 in Section 9.5. The leave-one-out method is also called **n-fold cross-validation**, and the error (4.22) is referred to as the **(n-fold) cross-validation error**.

A comparison of the classification error $\mathcal{E}_{\text{mis}}$ of (4.20) and the leave-one-out error $\mathcal{E}_{\text{loo}}$ of (4.22) suggests that $\mathcal{E}_{\text{mis}}$ should be smaller. The leave-one-out method, however, paves the way for prediction because it has an in-built prediction error through leaving out each point, one at a time.

Remark. Unless otherwise stated, we work with cost factors 0 and 1 for the classification error $\mathcal{E}_{\text{mis}}$ and the leave-one-out error $\mathcal{E}_{\text{loo}}$ in this chapter.

Example 4.5 We continue with the **wine recognition** data which arise from three different cultivars. There are a total of 178 observations, 59 belong to the first cultivar, 71 to the second, and the remaining 48 to the third, and for each observation there are thirteen measurements.

Two of the measurements, variable 6, *magnesium*, and variable 14, *proline*, are on a much larger scale than the other measurements and could dominate the analysis. For this reason, I apply Fisher's discriminant rule to the raw and scaled data and compare the performance of

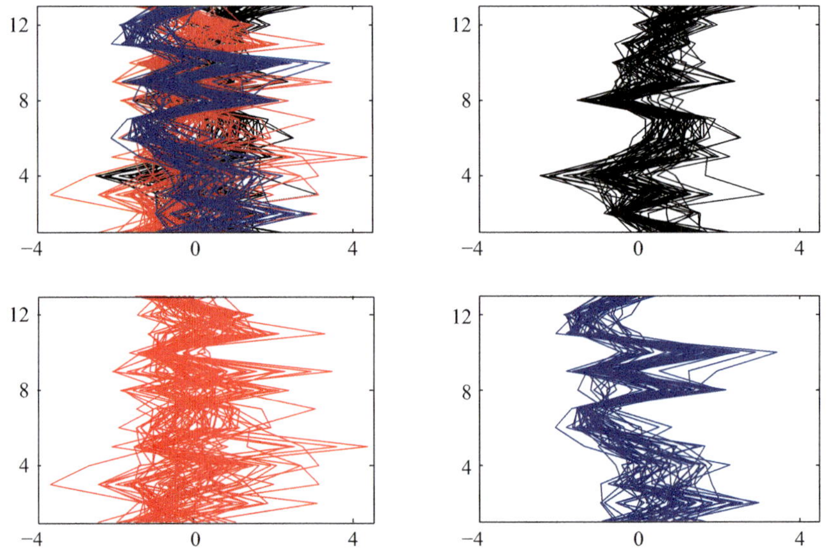

Figure 4.4 Parallel coordinate plots of the scaled wine data of Example 4.5. All data (*top left*), first cultivar (*top right*), second cultivar (*bottom left*) third cultivar (*bottom right*).

Table 4.4 *Classification Error $\mathcal{E}_{mis}$ and Leave-One-Out Error $\mathcal{E}_{loo}$ for the Wine Recognition Data of Example 4.5*

Totals	% error	Total, 178	Class 1, 59	Class 2, 71	Class 3, 48
$\mathcal{E}_{\text{mis}}$	6.74	12	3	9	0
$\mathcal{E}_{\text{loo}}$	8.43	15	3	12	0

Note: Columns 3 to 6 show numbers of misclassified observations.

the rule with $\mathcal{E}_{\text{mis}}$ and $\mathcal{E}_{\text{loo}}$. It will be interesting to see what, if any, effect scaling has. The weights of Fisher's direction vector $\widehat{\eta}$ are very similar for the raw and scaled data, and in particular, their signs agree for each variable.

Figure 4.4 shows vertical parallel coordinate plots of the scaled data. The top left panel contains all observations, with a different colour for each cultivar. The top right panel shows the observations from cultivar 1, the bottom left those from cultivar 2, and bottom right those from cultivar 3. A visual inspection of the scaled data tells us that cultivars 1 and 3 have sufficiently different measurements and so should not be too difficult to distinguish, whereas cultivar 2 may be harder to distinguish from the others.

The calculations show that the classification errors are identical for the raw and scaled data, so in this case scaling has no effect. Table 4.4 details the performance results for the classification error $\mathcal{E}_{\text{mis}}$ and the leave-one-out error $\mathcal{E}_{\text{loo}}$. In addition to the total number of misclassified observations, I have listed the number of misclassified observations from each class. We note that $\mathcal{E}_{\text{mis}}$ is overall smaller than $\mathcal{E}_{\text{loo}}$, but the two rules perform equally well on the observations from cultivars 1 and 3. The intuition gained from the visual inspection of the data is confirmed by the poorer performance of the two rules on the observations of cultivar 2. ■

For a theoretical assessment of rules, the error probability is an important concept. I introduce this idea for the special case of two-class problems.

Definition 4.14 Let $\mathbf{X}$ be a random vector which belongs to one of the two classes C_1 and C_2, and let Y be the label for $\mathbf{X}$. Let $\mathfrak{r}$ be a discriminant rule for $\mathbf{X}$. The **probability of misclassification** or the **error probability** of the rule $\mathfrak{r}$ is $\mathbb{P}\{\mathfrak{r}(\mathbf{X}) \neq Y\}$. □

Typically, we want the error probability to be as small as possible, and one is therefore interested in rules that minimise this error.

Proposition 4.15 *Let $\mathbf{X}$ be a random vector which belongs to one of the two classes C_1 and C_2, and let Y be the label for $\mathbf{X}$. Let*

$$p(\mathbf{X}) = \mathbb{P}\{Y = 1 \mid \mathbf{X}\}$$

be the conditional probability that $Y = 1$ given $\mathbf{X}$, and let $\mathfrak{r}^*$ be the discriminant rule defined by

$$\mathfrak{r}^*(\mathbf{X}) = \begin{cases} 1 & \text{if } p(\mathbf{X}) > 1/2, \\ 0 & \text{if } p(\mathbf{X}) < 1/2. \end{cases}$$

If $\mathfrak{r}$ is another discriminant rule for **X***, then*

$$\mathbb{P}\{\mathfrak{r}^*(\mathbf{X}) \neq Y\} \leq \mathbb{P}\{\mathfrak{r}(\mathbf{X}) \neq Y\}. \tag{4.23}$$

The inequality (4.23) shows the optimality of the rule $\mathfrak{r}^*$. We will return to this rule and associated optimality in Section 4.6.

Proof Consider a rule $\mathfrak{r}$ for $\mathbf{X}$. Then

$$\begin{aligned}\mathbb{P}\{\mathfrak{r}(\mathbf{X}) \neq Y \mid \mathbf{X}\} &= 1 - \mathbb{P}\{\mathfrak{r}(\mathbf{X}) = Y \mid \mathbf{X}\}\\ &= 1 - (\mathbb{P}\{Y = 1, \mathfrak{r}(\mathbf{X}) = 1 \mid \mathbf{X}\} + \mathbb{P}\{Y = 0, \mathfrak{r}(\mathbf{X}) = 0 \mid \mathbf{X}\})\\ &= 1 - (I_{\{\mathfrak{r}(\mathbf{X})=1\}}\mathbb{P}\{Y = 1 \mid \mathbf{X}\} + I_{\{\mathfrak{r}(\mathbf{X})=0\}}\mathbb{P}\{Y = 0 \mid \mathbf{X}\})\\ &= 1 - \{(I_{\{\mathfrak{r}(\mathbf{X})=1\}}\, p(\mathbf{X}) + I_{\{\mathfrak{r}(\mathbf{X})=0}[1 - p(\mathbf{X})]\},\end{aligned}$$

where I_G is the indicator function of a set G. Next, consider an arbitrary rule $\mathfrak{r}$ and the rule $\mathfrak{r}^*$. From the preceding calculations, it follows that

$$\begin{aligned}\delta^* &\equiv \mathbb{P}\{\mathfrak{r}(\mathbf{X}) \neq Y \mid \mathbf{X}\} - \mathbb{P}\{\mathfrak{r}^*(\mathbf{X}) \neq Y \mid \mathbf{X}\}\\ &= p(\mathbf{X})(I_{\{\mathfrak{r}^*(\mathbf{X})=1\}} - I_{\{\mathfrak{r}(\mathbf{X})=1\}}) + [1 - p(\mathbf{X})](I_{\{\mathfrak{r}^*(\mathbf{X})=0\}} - I_{\{\mathfrak{r}(\mathbf{X})=0\}})\\ &= [2p(\mathbf{X}) - 1]\,(I_{\{\mathfrak{r}^*(\mathbf{X})=1\}} - I_{\{\mathfrak{r}(\mathbf{X})=1\}}) \geq 0\end{aligned}$$

by the definition of $\mathfrak{r}^*$, and hence the result follows. ■

In Section 4.6.1 we extend the rule of Proposition 4.15 to more than two classes and consider prior probabilities for each class.

4.5 Discrimination under Gaussian Assumptions

4.5.1 Two and More Normal Classes

If the random vectors in a two-class problem are known to come from the normal distribution with the same covariance matrix, then the normal discriminant rule of Theorem 4.10 is appropriate. In many instances the distributional properties of the data differ from the assumptions of Theorem 4.10; for example:

- The covariance matrices of the two classes are not known or are not the same.
- The distributions of the random vectors are not known or are not normal.

If the true covariance matrices are not known but can be assumed to be the same, then *pooling* is an option. The **pooled sample covariance matrix** is

$$S_{\text{pool}} = \sum_{\nu=1}^{2} \frac{n_\nu - 1}{n - 2} S_\nu, \tag{4.24}$$

where S_ν is the sample covariance matrix, n_ν is the size of the νth class and $n = n_1 + n_2$. To justify the use of the pooled covariance matrix, one has to check the adequacy of pooling. In practice, this step is often omitted. If the class covariance matrices are clearly not the same, then, strictly speaking, the normal discriminant rule does not apply. In Example 4.8 we consider such cases.

If the distribution of the random vectors is not known, or if the random vectors are known to be non-normal, then the normal discriminant rule may not perform so well. However, for data that do not deviate 'too much' from the Gaussian, the normal discriminant rule still leads to good results. It is not an easy task to assess whether the data are 'normal enough' for the normal discriminant rule to yield reasonable results, nor do I want to specify what I mean by 'too much'. For data of moderate dimensionality, visual inspections or tests for normality can provide guidelines.

A sensible strategy is to apply more than one rule and to evaluate the performance of the rules. In practice, Fisher's rule and the normal rule do not always agree, as we shall see in the next example.

Example 4.6 We continue with the **breast cancer** data which come from two classes. In the previous analyses of these data – throughout Chapter 2 – we were interested in summarising these 30-dimensional data in a simpler form and comparing a principal component analysis of the raw and scaled data. We ignored the fact that the data belong to two classes – malignant and benign – and for each observation the status is recorded. Our attention now focuses on discrimination between the two classes.

We have 569 observations, 212 malignant and 357 benign. We consider two views of these data: parallel coordinate plots of the raw data (Figure 4.5) and principal component (PC) score plots (Figure 4.6). The top-left panel of Figure 4.5 displays the raw data. The two large variables, 4 and 24, account for most of the variance and obscure all other variables, and there is no indication of class membership visible. The top-right panel shows the same data, but now the variables have been centred. A naive hope might be that the positive and negative deviations are related to class membership. The subplots in the bottom row show separately, but on the same scale, the centred malignant data in the left panel and the centred benign data in the right panel. We glean from these figures that most of the three largest variables are positive in the malignant data and negative in the benign data.

Figure 4.6 shows principal component views of the raw data: PC score plots in the style of Section 2.4. In the PC score plots, PC_1 is shown on the y-axis against PC_2 on the left, PC_3 in the middle, and PC_4 on the right. The denser 'black' point clouds belong to the benign class, and the more spread out 'blue' point clouds correspond to the malignant data. The two sets of figures show that there is an overlap between the two classes. The bigger spread of the malignant data is visible in the bottom-left subplot of Figure 4.5 and becomes more apparent in the PC score plots.

We compare the performance of Fisher's rule with that of the normal rule. In Section 4.8, I explain how to include dimension-reduction methods prior to discrimination and then use Fisher's rule and the normal rule as benchmarks.

For Fisher's rule, 15 out of 569 observations are misclassified, so $\mathcal{E}_{\mathrm{mis}} = 2.63$ per cent. Three benign observations are classified as malignant, and twelve malignant observations are misclassified.

For the normal rule, I use the pooled covariance matrix, although the PC_1 data deviate from the normal distribution. We can see this in the PC score plots in Figure 4.6 and the density estimate of the PC_1 data in the top-right panel of Figure 2.8 in Section 2.4.3. The normal rule misclassifies eighteen observations and leads to $\mathcal{E}_{\mathrm{mis}} = 3.16$ per cent. This is slightly higher than that obtained with Fisher's rule. The normal rule does better on the

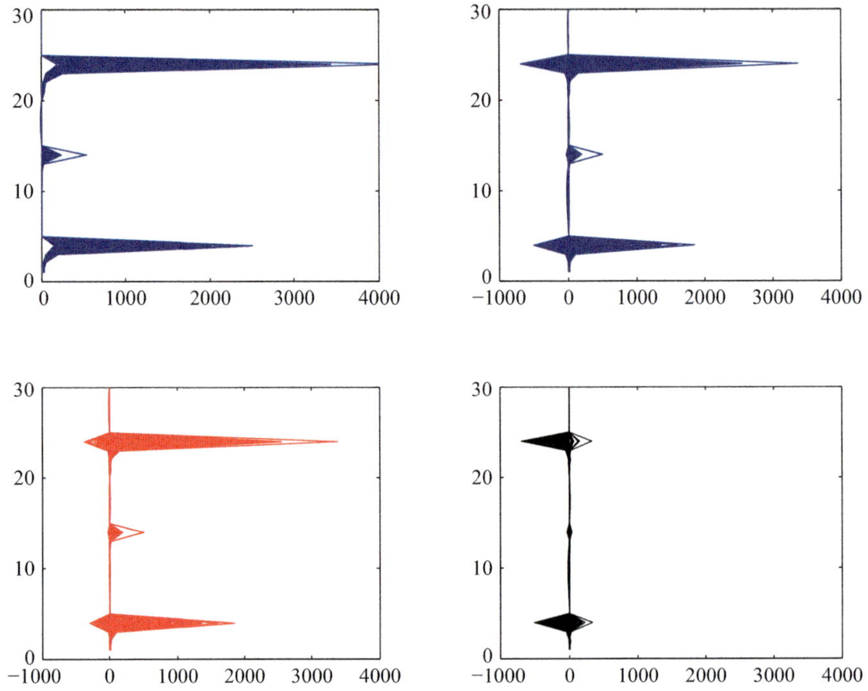

Figure 4.5 Breast cancer data of Example 4.6 . (*Top row*): Raw (*left*) and centred data (*right*). (*Bottom row*): Malignant centred data (*left*) and benign centred data (*right*).

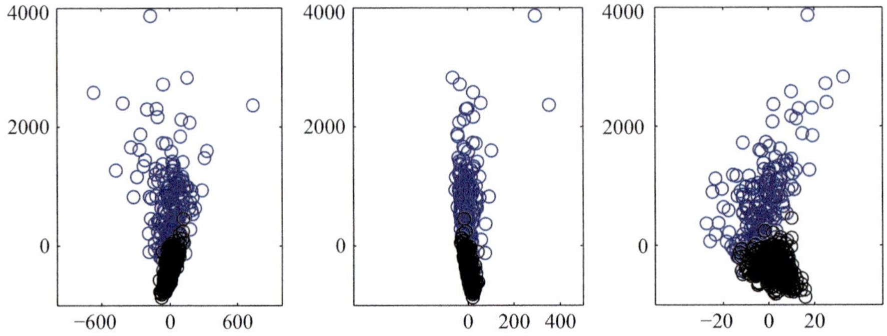

Figure 4.6 Breast cancer data of Example 4.6 . PC score plots showing benign observations in black and malignant in blue: PC_1 on the y-axis versus PC_2 on the left, PC_3 in the middle, and PC_4 on the right.

benign observations but worse on the malignant ones. Of the fifteen observations Fisher's rule misclassified, two are correctly classified by the normal rule.

The results show that the two rules produce similar results. Fisher's rule is slightly better than the normal rule, most likely because the data deviate considerably from the normal distribution. ■

Our next example consists of three classes. The normal rule is defined for two-class problems because it compares the likelihood of two parameters. There are natural extensions of the normal rule which I explain briefly before analysing the *wine recognition* data.

Assume that random vectors $\mathbf{X}$ belong to one of κ classes $\mathcal{C}_\nu = \mathcal{N}(\boldsymbol{\mu}_\nu, \Sigma)$, with $\nu \le \kappa$, which differ in their class means only. The aim is to find the parameter $\boldsymbol{\mu}_k$ which maximises the likelihood and then define a rule which assigns $\mathbf{X}$ to $\mathcal{C}_k$. To determine k, define **preferential decision functions**

$$h_{(\ell,\nu)}(\mathbf{X}) = \left[\mathbf{X} - \frac{1}{2}(\boldsymbol{\mu}_\ell + \boldsymbol{\mu}_\nu)\right]^{\mathsf{T}} \Sigma^{-1}(\boldsymbol{\mu}_\ell - \boldsymbol{\mu}_\nu), \qquad \text{for } \ell, \nu = 1, 2, \ldots, \kappa \text{ and } \ell \neq \nu,$$

and observe that the order of the subscripts is important because $h_{(\ell,\nu)}$ and $h_{(\nu,\ell)}$ have opposite signs. Next, put

$$h_{\text{norm}}(\mathbf{X}) = \max_{(\ell,\nu)=1,2,\ldots,\kappa} h_{(\ell,\nu)}(\mathbf{X}) \qquad \text{and} \qquad \mathfrak{r}_{\text{norm}}(\mathbf{X}) = k, \tag{4.25}$$

where k is the first of the pair of indices (ℓ, ν) which maximises $h_{\text{norm}}(\mathbf{X})$. Alternatively, we can consider the likelihoods $L(\boldsymbol{\mu}_\nu|\mathbf{X})$, for $\nu \le \kappa$, and define a rule

$$\mathfrak{r}_{\text{norm1}}(\mathbf{X}) = k \qquad \text{if } L(\boldsymbol{\mu}_k|\mathbf{X}) = \max_{1\le\nu\le\kappa} L(\boldsymbol{\mu}_\nu|\mathbf{X}). \tag{4.26}$$

For the population, the two rules assign $\mathbf{X}$ to the same class. For data, they may lead to different decisions depending on the variability within the classes and the choice of estimator S for Σ. In the next example I will explicitly state how S is calculated.

Example 4.7 We continue with the thirteen-dimensional **wine recognition** data which come from three classes. Previously, I applied Fisher's rule to classify these data. Figure 4.4 and Table 4.4 show that the second class is the hardest to classify correctly; indeed, from the seventy-one observations in this class, Fisher's rule misclassified nine.

To apply the normal rule to these data, I replace the class means by the sample means in each class and Σ by the pooled covariance matrix (4.24). With these choices, the rule (4.25) classifies all observations correctly and is thus an improvement over Fisher's rule, which resulted in a classification error of 6.74 per cent. The leave-one-out error $\mathcal{E}_{\text{loo}}$ is 1.12 per cent for the normal rule (4.25). It classifies two observations from the second class incorrectly: it assigns observation 97 to class 3 and observation 122 to class 1. These results are consistent with our previous results, shown in Table 4.4 in Example 4.5: The leave-one-out error is slightly higher than the classification error, as expected.

For these data, the classification results are the same when I calculate rule (4.26) with S, the covariance matrix of all observations. Further, as in the classification with Fisher's rule, the normal rule yields the same classification error for the raw and scaled data.

For the **wine recognition** data, the normal rule performs better than Fisher's rule. If the data are normal, such a result is not surprising. For the small number of observations – 178 spread across thirteen dimensions – it may not be possible to show that the data differ significantly from the normal distribution. The normal rule is therefore appropriate. ■

In the two examples of this section, the *wine recognition* data and the *breast cancer* data, neither rule – Fisher or normal – performs better both times. If the data are close to normal, as could be the case in the *wine recognition* data, then the normal rule performs *at par* or better. In the breast cancer data, Fisher's rule is superior. In general, it is advisable to apply more than one rule to the same data and to compare the performances of the rules, as I have done here. Similar comparisons can be made for the leave-one-out error $\mathcal{E}_{\text{loo}}$.

4.5.2 Gaussian Quadratic Discriminant Analysis

The two discriminant rules discussed so far, namely, Fisher's rule and the normal rule, are linear in the random vector $\mathbf{X}$; that is, their decision functions are of the form

$$h(\mathbf{X}) = \mathbf{a}^{\mathsf{T}}\mathbf{X} + c,$$

for some vector $\mathbf{a}$ and scalar c which do not depend on $\mathbf{X}$.

In the normal discriminant rule (4.15), the linearity is a consequence of the fact that the *same* covariance matrix is used across all classes. If the covariance matrices are known to be different – and are in fact sufficiently different that pooling is not appropriate – then the separate covariance matrices need to be taken into account. This leads to a non-linear decision function and, consequently, a non-linear rule.

For $\nu \le \kappa$, consider random vectors $\mathbf{X}$ from one of the classes $\mathcal{C}_\nu = \mathcal{N}(\boldsymbol{\mu}_\nu, \Sigma_\nu)$. We use the normal likelihoods L to define a discriminant rule. Put $\theta_\nu = (\boldsymbol{\mu}_\nu, \Sigma_\nu)$, so θ_ν is the parameter of interest for the likelihood function. Assign $\mathbf{X}$ to class ℓ if

$$L(\theta_\ell|\mathbf{X}) > L(\theta_\nu|\mathbf{X}) \qquad \text{for } \ell \ne \nu.$$

Theorem 4.16 *For $\nu \le \kappa$, consider the classes $\mathcal{C}_\nu = \mathcal{N}(\boldsymbol{\mu}_\nu, \Sigma_\nu)$, which differ in their means $\boldsymbol{\mu}_\nu$ and their covariance matrices Σ_ν. Let $\mathbf{X}$ be a random vector which belongs to one of these classes.*

The discriminant rule $\mathfrak{r}_{quad}$, which is based on the likelihood functions of these κ classes, assigns $\mathbf{X}$ to class $\mathcal{C}_\ell$ if

$$\left\|\mathbf{X}_{\Sigma_\ell}\right\|^2 + \log[\det(\Sigma_\ell)] = \min_{1\le\nu\le\kappa}\left\{\left\|\mathbf{X}_{\Sigma_\nu}\right\|^2 + \log[\det(\Sigma_\nu)]\right\}, \tag{4.27}$$

where $\mathbf{X}_\Sigma = \Sigma^{-1/2}(\mathbf{X} - \boldsymbol{\mu})$, the sphered version of $\mathbf{X}$.

We call the rule $\mathfrak{r}_{\text{quad}}$ the **(normal) quadratic discriminant rule** because it is quadratic in $\mathbf{X}$. Although I will often drop the word *normal* when referring to this rule, we need to keep in mind that the rule is derived for normal random vectors, and its performance for non-Gaussian data may not be optimal.

Proof Consider a random vector $\mathbf{X}$ which belongs to one of the κ classes $\mathcal{C}_\nu$. From the multivariate likelihood (1.16) in Section 1.4, we obtain

$$\begin{aligned}\log L(\theta|\mathbf{X}) &= -\frac{d}{2}\log(2\pi) - \frac{1}{2}\left[\log[\det(\Sigma)] + (\mathbf{X}-\boldsymbol{\mu})^{\mathsf{T}}\Sigma^{-1}(\mathbf{X}-\boldsymbol{\mu})\right]\\ &= -\frac{1}{2}\left[\|\mathbf{X}_\Sigma\|^2 + \log[\det(\Sigma)]\right] + c,\end{aligned} \tag{4.28}$$

where $c = -d\log(2\pi)/2$ is independent of the parameter $\theta = (\boldsymbol{\mu}, \Sigma)$. A Gaussian discriminant rule decides in favour of class $\mathcal{C}_\ell$ over $\mathcal{C}_\nu$ if $L(\theta_\ell|\mathbf{X}) > L(\theta_\nu|\mathbf{X})$ or, equivalently, if

$$\left\|\mathbf{X}_{\Sigma_\ell}\right\|^2 + \log[\det(\Sigma_\ell)] < \left\|\mathbf{X}_{\Sigma_\nu}\right\|^2 + \log[\det(\Sigma_\nu)].$$

The result follows from this last inequality. ■

For two classes, the quadratic rule admits a simpler explicit expression.

Corollary 4.17 *For $\kappa = 2$, consider the two classes $\mathcal{C}_\nu = \mathcal{N}(\boldsymbol{\mu}_\nu, \Sigma_\nu)$, with $\nu = 1, 2$, which differ in their means $\boldsymbol{\mu}_\nu$ and their covariance matrices Σ_ν. Let $\mathbf{X}$ be a random vector which belongs to one of the two classes. Assume that both covariance matrices have rank r. Let $\Sigma_\nu = \Gamma_\nu \Lambda_\nu \Gamma_\nu^{\mathsf{T}}$ be the spectral decomposition of Σ_ν, and let $\lambda_{\nu,j}$ be the jth eigenvalue of Λ_ν.*

The quadratic discriminant rule $\mathfrak{r}_{quad}$, based on the likelihood function of these two classes, assigns $\mathbf{X}$ to class $\mathcal{C}_1$ if

$$\left\|\Lambda_1^{-1/2}\Gamma_1^{\mathsf{T}}(\mathbf{X}-\boldsymbol{\mu}_1)\right\|^2 - \left\|\Lambda_2^{-1/2}\Gamma_2^{\mathsf{T}}(\mathbf{X}-\boldsymbol{\mu}_2)\right\|^2 + \sum_{j=1}^{r} \log\frac{\lambda_{1,j}}{\lambda_{2,j}} < 0.$$

In Example 4.3, I used the class means and the covariance matrix in the calculation of the normal rule. Because I generate the data, the sample means and the sample covariance matrices differ from the population parameters. If I use the pooled covariance matrix (4.24) and the sample means instead of the population parameters, I obtain the slightly lower classification error of 5 per cent for the second data. This performance is identical to that obtained with Fisher's discriminant rule, and the same observations are misclassified. In the next example we compare Fisher's rule with the linear and quadratic normal rules for different data sets. One of the purposes of the next example is to explore how well the normal rule performs when the classes have different covariance matrices or when they clearly deviate from the normal distribution.

Example 4.8 We continue with the 3D **simulated data** from two classes and also consider ten-dimensional simulated data. I apply Fisher's rule and the normal linear and quadratic rules to these data. As before, I refer to the normal linear rule simply as the normal rule.

Each data set consists of two classes; the first class contains 300 random samples; and the second class contains 200 random samples. This study is divided into pairs of simulations: $\mathcal{N}1$ and $\mathcal{N}2$, $\mathcal{N}3$ and $\mathcal{N}4$, $\mathcal{N}5$ and $\mathcal{N}6$ and $\mathcal{P}7$ and $\mathcal{P}8$, where $\mathcal{N}$ refers to data from the normal distribution, and $\mathcal{P}$ refers to Poisson-based data. With the exception of $\mathcal{N}5$ and $\mathcal{N}6$, which are ten-dimensional, all other simulations are three-dimensional.

The data sets in a pair have the same class means, but their covariance matrices differ. The means and covariance matrices are given in Table 4.5. In the odd-numbered simulations, $\mathcal{N}1$, $\mathcal{N}3$, $\mathcal{N}5$ and $\mathcal{P}7$, both classes have a common covariance matrix, which we call Σ_1. In the even-numbered simulations, the 300 samples from the first class have the covariance matrix called Σ_1 in the table, and the 200 samples from the second class have the covariance matrix Σ_2. We write $\mathcal{N}2$ for the simulation with data from the classes $\mathcal{N}(\boldsymbol{\mu}_1, \Sigma_1)$ and $\mathcal{N}(\boldsymbol{\mu}_2, \Sigma_2)$.

Typical representatives of the 3D data are displayed in Figure 4.7. The red and maroon points in this figure are from the first class, and the blue and black points are from the second class. For each data set in the top row, the covariance matrix is the same across the two classes, here called Σ_1. The data sets in the bottom row have different covariance matrices for the two classes. The top-left subplot is a representative of $\mathcal{N}1$, the middle of $\mathcal{N}3$, and the two subplots in the bottom row correspond to $\mathcal{N}2$ (*left*), and $\mathcal{N}4$ (*middle*), respectively.

Table 4.5 *Means and Covariance Matrices of Data from Example 4.8 and Shown in Figure 4.7 from the Gaussian ($\mathcal{N}$) and Poisson-Based ($\mathcal{P}$) Distributions*

	$\mathcal{N}1$ and $\mathcal{N}2$	$\mathcal{N}3$ and $\mathcal{N}4$	$\mathcal{P}7$	$\mathcal{P}8$
μ_1	$(0,0,0)$	$(0,0,0)$	$(10,10,10)$	$(10,10,10)$
μ_2	$(0,1,-0.5)$	$(1,1,-0.5)$	$(15,5,14)$	$(10,20,15)$
Σ_1	$\begin{pmatrix} 0.25 & 0 & 0 \\ 0 & 0.125 & 0 \\ 0 & 0 & 0.125 \end{pmatrix}$		$\begin{pmatrix} 10 & 0 & 0 \\ 0 & 10 & 0 \\ 0 & 0 & 10 \end{pmatrix}$	
Σ_2	$\begin{pmatrix} 0.125 & 0.05 & 0.02 \\ 0.05 & 0.25 & 0.0125 \\ 0.02 & 0.0125 & 0.2 \end{pmatrix}$			$\begin{pmatrix} 10 & 0 & 0 \\ 0 & 20 & 0 \\ 0 & 0 & 30 \end{pmatrix}$

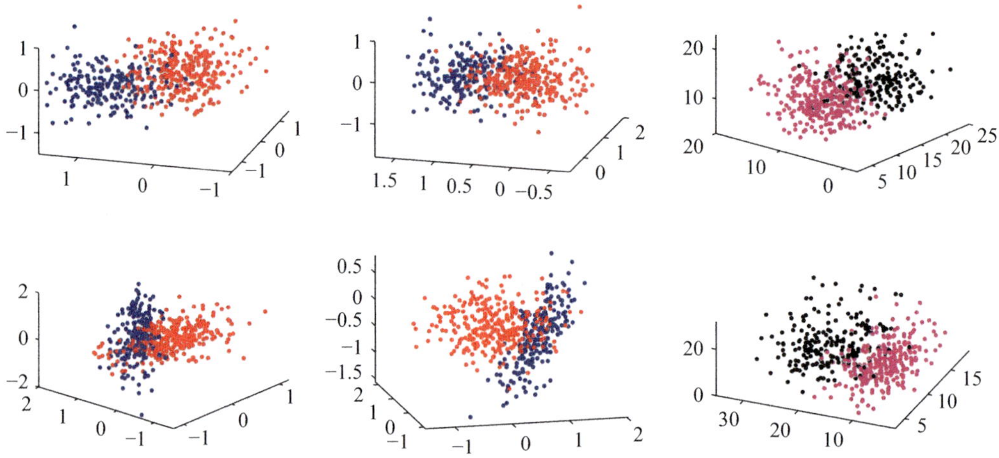

Figure 4.7 Three-dimensional simulated data from Example 4.8. The data $\mathcal{N}1$ (*left*), $\mathcal{N}3$ (*middle*) and $\mathcal{P}7$ (*right*) are shown in the top row and the data $\mathcal{N}2$ (*left*), $\mathcal{N}4$ (*middle*) and $\mathcal{P}8$ (*right*) in the bottom row, with means and covariance matrices given in Table 4.5.

The correlated covariance matrices of the blue data in the bottom row are noticeable by the non-spherical and rotated appearance of the cluster.

Columns $\mathcal{P}7$ and $\mathcal{P}8$ in Table 4.5 show the means and covariance matrices of the two right subplots in Figure 4.7. The random samples are generated from Poisson distributions but are transformed as described below. Column $\mathcal{P}7$ refers to the top-right subplot. The 300 maroon samples are independently generated from the univariate Poisson distribution with mean and variance 10. The 200 black samples are generated from the same distribution but are subsequently shifted by the vector $(5, -5, 4)$, so these 200 samples have a different class mean. The covariance matrix remains unchanged, but the data from the second class are no longer Poisson – because of the shift in the mean. This does not matter for our purpose.

Column $\mathcal{P}8$ in Table 4.5 relates to the bottom-right subplot in Figure 4.7. The maroon data are the same as in the top-right subplot, but the 200 black samples are generated from a Poisson distribution with increasing means $(10, 20, 30)$. Because the mean 30 for the third

Table 4.6 *Mean Classification Error and Standard Deviation (in parentheses) Based on Fisher's Rule and the Normal and the Quadratic Rules for Six Normal Data Sets and Two Poisson-Based Data Sets from Example 4.8*

	Data							
	$\mathcal{N}1$	$\mathcal{N}2$	$\mathcal{N}3$	$\mathcal{N}4$	$\mathcal{N}5$	$\mathcal{N}6$	$\mathcal{P}7$	$\mathcal{P}8$
Fisher	5.607	6.368	3.016	4.251	6.968	13.302	9.815	7.125
	(1.071)	(1.143)	(0.767)	(0.897)	(1.160)	(1.537)	(1.364)	(1.175)
Normal	5.570	6.337	3.015	4.205	6.958	13.276	9.816	7.105
	(1.068)	(1.138)	(0.766)	(0.894)	(1.154)	(1.526)	(1.374)	(1.174)
Quadratic	5.558	4.578	2.991	3.157	6.531	3.083	9.741	6.925
	(1.085)	(0.979)	(0.750)	(0.782)	(1.103)	(0.789)	(1.3565)	(1.142)

dimension separates the classes excessively, I have shifted the third dimension of each black observation by -15. As a result, the two classes overlap.

In addition, we consider ten-dimensional normal data, $\mathcal{N}5$ and $\mathcal{N}6$. The means of these data are

$$\boldsymbol{\mu}_1 = \mathbf{0} \qquad \text{and} \qquad \boldsymbol{\mu}_2 = (0, 0.25, 0, -0.5, 0, 0.5, 1, 0.25, 0.5, -0.5).$$

The variables are independent with $\sigma_1^2 = 0.25$ along the diagonal of the common covariance matrix and the much bigger variance of $\sigma_2^2 = 1$ for the second diagonal covariance matrix.

I present the performance of all eight data sets in Table 4.6. In each case I report the mean classification error over 1,000 repetitions and the standard deviation in parentheses. In the table, I refer to the normal linear discriminant rule as 'normal' and to the quadratic normal rule as 'quadratic'. All parameters, such as mean and covariance matrices, refer to the sample parameters. For the normal rule I use the pooled covariance matrix.

Table 4.6 shows that the quadratic rule performs better than the two linear rules. This difference in performance is not very big when the two classes have the same covariance matrix, but it increases for the data sets with different covariance matrices for the two classes. In the case of the ten-dimensional data with different covariance matrices ($\mathcal{N}6$), the quadratic rule strongly outperforms the other two, with a classification error of 3 per cent compared with more than 13 per cent for the linear methods. The two linear discriminant rules perform similarly for the normal data and the two Poisson-based data sets, and many of the same points are misclassified by both methods. ■

The example illustrates that the performance of the normal rule is similar to that of Fisher's rule. However, the quadratic rule outperforms the linear rules when the covariance matrices of the classes are clearly different, and in such cases, the quadratic rule is therefore preferable.

4.6 Bayesian Discrimination

4.6.1 Bayes Discriminant Rule

In a Bayesian framework we assume that we know the probabilities π_ℓ that an observation belongs to class C_ℓ. The probabilities $\pi_1, \ldots, \pi_\kappa$ are called the **prior probabilities**. The inclusion of the prior probabilities is advantageous when the prior probabilities differ substantially from one class to the next. Consider a two-class problem where two-thirds of the data are from the first class, and the remaining third is from the second class. A discriminant rule should be more likely to assign a new point to class 1. If such knowledge about the class membership is available, then it can enhance the performance of a discriminant rule, provided that it is integrated judiciously into the decision-making process.

Although the Bayesian setting does not depend on the assumption of Gaussianity of the random vectors, in practice, we often tacitly assume that the random vectors or the data are normal.

Definition 4.18 Let $\mathcal{C}_1,\ldots,\mathcal{C}_\kappa$ be classes which differ in their means and covariance matrices. Let $\mathbf{X}$ be a random vector which belongs to one of these classes. Let $\boldsymbol{\pi} = \left[\pi_1\ \pi_2 \cdots \pi_\kappa\right]$ be the prior probabilities for the classes $\mathcal{C}_\nu$. Let $\mathfrak{r}$ be a discriminant rule derived from regions G_ν as in (4.19).

1. The **conditional probability** that $\mathfrak{r}$ assigns $\mathbf{X}$ the value ν (or equivalently assigns $\mathbf{X}$ to G_ν) given that $\mathbf{X}$ belongs to class $\mathcal{C}_\ell$ is
$$p(\nu|\ell) = \mathbb{P}\left(\mathfrak{r}(\mathbf{X}) = \nu \mid \mathbf{X} \in \mathcal{C}_\ell\right).$$
2. The **posterior probability** that an observation $\mathbf{X}$ belongs to class $\mathcal{C}_\ell$ given that the rule $\mathfrak{r}$ assigns $\mathbf{X}$ the value ν is
$$\mathbb{P}\left(\mathbf{X} \in \mathcal{C}_\ell \mid \mathfrak{r}(\mathbf{X}) = \nu\right) = \frac{\mathbb{P}\left(\mathfrak{r}(\mathbf{X}) = \nu \mid \mathbf{X} \in \mathcal{C}_\ell\right)\pi_\ell}{\mathbb{P}\left(\mathfrak{r}(\mathbf{X}) = \nu\right)} = \frac{p(\nu|\ell)\pi_\ell}{\mathbb{P}\left(\mathfrak{r}(\mathbf{X}) = \nu\right)}. \tag{4.29}$$

□

Because the discriminant rule and the discriminant regions determine each other, it follows that the probability of assigning $\mathbf{X}$ to G_ν when it belongs to class $\mathcal{C}_\ell$ is
$$p(\nu|\ell) = \int I_{G_\nu} L_\ell, \tag{4.30}$$
where I_{G_ν} is the indicator function of G_ν, and L_ℓ is the likelihood function of the ℓth class $\mathcal{C}_\ell$. From (4.29), the posterior probability that $\mathbf{X}$ belongs to $\mathcal{C}_\nu$, given that $\mathfrak{r}$ has assigned it the value ν, is
$$\mathbb{P}\left(\mathbf{X} \in \mathcal{C}_\nu \mid \mathfrak{r}(\mathbf{X}) = \nu\right) = \frac{p(\nu|\nu)\pi_\nu}{\mathbb{P}\left(\mathfrak{r}(\mathbf{X}) = \nu\right)},$$
where $p(\nu|\nu)$ is the probability of correctly assigning an observation to class ν. Similarly, $1 - p(\nu|\nu)$ is the probability of not assigning an observation from class $\mathcal{C}_\nu$ to the correct class.

The next theorem defines the Bayesian discriminant rule and states some of its properties.

Theorem 4.19 *Let $\mathcal{C}_1,\ldots,\mathcal{C}_\kappa$ be classes with different means. Let $\boldsymbol{\pi} = \left[\pi_1\ \pi_2 \cdots \pi_\kappa\right]$ be the prior probabilities for the classes $\mathcal{C}_\nu$. Let $\mathbf{X}$ be a random vector which belongs to one of the κ classes, and write L_ν for the likelihood function of the νth class. For $\nu \le \kappa$, define regions*
$$G_\nu = \{\mathbf{X}\colon L_\nu(\mathbf{X})\pi_\nu = \max_{1\le\ell\le\kappa}\left[L_\ell(\mathbf{X})\pi_\ell\right]\},$$
and let $\mathfrak{r}_{Bayes}$ be the discriminant rule derived from the G_ν so that $\mathfrak{r}_{Bayes}(\mathbf{X}) = \nu$ if $\mathbf{X} \in G_\nu$.

1. *The rule* $\mathfrak{r}_{Bayes}$ *assigns an observation* $\mathbf{X}$ *to* G_ν *in preference to* G_ℓ *if*

$$\frac{L_\nu(\mathbf{X})}{L_\ell(\mathbf{X})} > \frac{\pi_\ell}{\pi_\nu}.$$

2. *If all classes are normally distributed and share the same covariance matrix* Σ, *then* $\mathfrak{r}_{Bayes}$ *assigns* $\mathbf{X}$ *to* G_ν *in preference to* G_ℓ *if*

$$\mathbf{X}^\mathsf{T}\Sigma^{-1}(\boldsymbol{\mu}_\nu - \boldsymbol{\mu}_\ell) > \frac{1}{2}(\boldsymbol{\mu}_\nu + \boldsymbol{\mu}_\ell)\Sigma^{-1}(\boldsymbol{\mu}_\nu - \boldsymbol{\mu}_\ell) + \log(\pi_\ell/\pi_\nu).$$

For the classes and distributions of part 2 of the theorem, we consider preferential decision functions

$$h_{(\nu,\ell)}(\mathbf{X}) = \left[\mathbf{X} - \frac{1}{2}(\boldsymbol{\mu}_\nu + \boldsymbol{\mu}_\ell)\right]^\mathsf{T} \Sigma^{-1}(\boldsymbol{\mu}_\nu - \boldsymbol{\mu}_\ell) - \log(\pi_\ell/\pi_\nu)$$

and note that $h_{(\nu,\ell)}(\mathbf{X}) > 0$ is equivalent to the rule $\mathfrak{r}_{\text{Bayes}}$, which assigns $\mathbf{X}$ to G_ν in preference to G_ℓ.

Theorem 4.19 follows directly from the definition of the rule $\mathfrak{r}_{\text{Bayes}}$. We call $\mathfrak{r}_{\text{Bayes}}$ the **Bayesian discriminant rule** or **Bayes (discriminant) rule**. Under this rule, the probability $\mathfrak{p}$ of assigning $\mathbf{X}$ to the correct class is

$$\mathfrak{p} = \sum_{\nu=1}^{\kappa} \mathbb{P}(\mathfrak{r}(\mathbf{X}) = \nu \mid \mathbf{X} \in \mathcal{C}_\nu)\,\pi_\nu = \sum_{\nu=1}^{\kappa} \int I_{G_\nu} L_\nu \pi_\nu,$$

and the indicator function I_{G_ν} of G_ν satisfies

$$I_{G_\nu}(\mathbf{X}) = \begin{cases} 1 & \text{if } L_\nu(\mathbf{X})\pi_\nu \geq L_\ell(\mathbf{X})\pi_\ell, \\ 0 & \text{otherwise.} \end{cases}$$

A comparison with Definition 4.14 and Proposition 4.15 shows that Bayes discriminant rule reduces to the rule $\mathfrak{r}^*$ in the two-class case.

The second part of the theorem invites a comparison with the normal rule $\mathfrak{r}_{\text{norm}}$: The only difference between the two rules is that the Bayes rule has the extra term $\log(\pi_\ell/\pi_\nu)$. If π_ν is smaller than π_ℓ, then $\log(\pi_\ell/\pi_\nu) > 0$, and it is therefore harder to assign $\mathbf{X}$ to class ν. Thus, the extra term incorporates the prior knowledge. When all prior probabilities are equal, then the Bayes rule reduces to the normal rule.

Example 4.9 We compare the performance of Bayes discriminant rule with that of the normal linear rule for the 3D **simulated data** from Example 4.3: two classes in three dimensions with an 'easy' and a 'hard' case which are distinguished by the distance of the two class means. The two classes have the common covariance matrix Σ_1 given in Example 4.2. Point clouds for the 'easy' and 'hard' cases are displayed in Figure 4.2 for 300 red points and 200 blue points.

In the simulations I vary the proportions and the overall sample sizes in each class, as shown in Table 4.7. I carry out 500 simulations for each combination of sample sizes and report the mean classification errors for the normal and Bayes discriminant rule in the table. Standard deviations are shown in parentheses.

The table shows that Bayes rule has a lower mean classification error than the normal linear rule and that the errors depend more on the proportion of the two samples than on

Table 4.7 *Mean Classification Error for Normal Linear and Bayes Discriminant Rules from Example 4.9 with Standard Deviations in parentheses. Easy and hard case refer to Figure 4.2*

Sample sizes		Easy case		Hard case	
n_1	n_2	*Mean* $\mathfrak{r}_{\text{norm}}$	*Mean* $\mathfrak{r}_{\text{Bayes}}$	*Mean* $\mathfrak{r}_{\text{norm}}$	*Mean* $\mathfrak{r}_{\text{Bayes}}$
30	10	2.2900	2.0500	5.6950	4.6950
		(2.4480)	(2.1695)	(3.5609)	(3.3229)
300	100	2.1985	1.8480	5.6525	4.6925
		(0.7251)	(0.6913)	(1.0853)	(0.9749)
20	30	2.1280	2.1040	5.8480	5.8240
		(2.0533)	(1.9547)	(3.2910)	(3.2751)
200	300	2.1412	2.0824	5.7076	5.5984
		(0.6593)	(0.6411)	(1.0581)	(1.0603)

the total sample size. (Recall that the misclassification error is given as a per cent error.) As the sample sizes in the two classes become more unequal, the advantage of Bayes rule increases. A comparison between the 'easy' and 'hard' cases shows that the Bayes rule performs relatively better in the harder cases.

To appreciate why Bayes rule performs better than the normal rule, let us assume that the first (red) class is bigger than the second (blue) class, with proportions π_ν of the two classes which satisfy $\pi_1 > \pi_2$. Put

$$h(\mathbf{X}) = \left[\mathbf{X} - \frac{1}{2}(\boldsymbol{\mu}_1 + \boldsymbol{\mu}_2)\right]^{\mathsf{T}} \Sigma^{-1}(\boldsymbol{\mu}_1 - \boldsymbol{\mu}_2).$$

The normal discriminant rule assigns $\mathbf{X}$ to 'red' if $h(\mathbf{X}) > 0$, whereas Bayes rule assigns $\mathbf{X}$ to 'red' if $h(\mathbf{X}) > \log(\pi_2/\pi_1)$. Observations $\mathbf{X}$ satisfying

$$\log(\pi_2/\pi_1) < h(\mathbf{X}) < 0$$

are classified differently by the two rules. Because the first (red) class is larger than the second class, on average, Bayes rule classifies more red observations correctly but possibly misclassifies more blue observations. An inspection of observations that are misclassified by both rules confirms this 'shift from blue to red' under Bayes rule. If the proportions are reversed, then the opposite shift occurs.

In Example 4.3, I reported classification errors of 1.6 and 5.2 for a single simulation of the 'easy' and 'hard' cases, respectively. These values are consistent with those in Table 4.7. Overall, the simulations show that the performance improves when we take the priors into account. ■

In the example we have assumed that the priors are correct. If the priors are wrong, then they could adversely affect the decisions.

4.6.2 Loss and Bayes Risk

In a Bayesian framework, loss and risk are commonly used to assess the performance of a method and to compare different methods. We consider these ideas from decision theory in the context of discriminant rules. For more detail, see Berger (1993).

For κ classes, a discriminant rule can result in the correct decision, it can lead to a 'not quite correct' answer, or it can lead to a vastly incorrect classification. The degree of incorrectness is captured by a loss function K which assigns a cost or loss to an incorrect decision.

Definition 4.20 Let $\mathfrak{C}$ be the collection of classes $C_1, \ldots, C_\kappa$, and let $\mathbf{X}$ be a random vector which belongs to one of these classes. Let $\mathfrak{r}$ be a discriminant rule for $\mathbf{X}$.

1. A **loss function** K is a map which assigns a non-negative number, called a **loss** or a **cost**, to $\mathbf{X}$ and the rule $\mathfrak{r}$. If $\mathbf{X} \in C_\ell$ and $\mathfrak{r}(\mathbf{X}) = \nu$, then the loss $c_{\ell,\nu}$ incurred by making the decision ν when the true class is ℓ is
$$\mathsf{K}(\mathbf{X}, \mathfrak{r}) = c_{\ell,\nu} \qquad \text{with} \quad \begin{cases} c_{\ell,\nu} = 0 & \text{if } \ell = \nu, \\ c_{\ell,\nu} > 0 & \text{otherwise.} \end{cases}$$
2. The **risk (function)** R is the expected loss that is incurred if the rule $\mathfrak{r}$ is used. We write
$$\mathsf{R}(\mathfrak{C}, \mathfrak{r}) = \mathbb{E}[\mathsf{K}(\cdot, \mathfrak{r})],$$
where the expectation is taken with respect to the distribution of $\mathbf{X}$.
3. Let $\boldsymbol{\pi} = \begin{bmatrix} \pi_1 \ \pi_2 \cdots \pi_\kappa \end{bmatrix}$ be the prior probabilities for the classes in $\mathfrak{C}$. **Bayes risk** B of a discriminant rule $\mathfrak{r}$ with respect to prior probabilities $\boldsymbol{\pi}$ is
$$\mathsf{B}(\boldsymbol{\pi}, \mathfrak{r}) = \mathbb{E}_{\boldsymbol{\pi}}[\mathsf{R}(\mathfrak{C}, \mathfrak{r})],$$
where the expectation is taken with respect to $\boldsymbol{\pi}$. □

The risk takes into account the loss across all classes for a specific rule. If we have a choice between rules, we select the rule with the smallest risk. In practice, it is hard, if not impossible, to find a rule that performs best across all classes, and additional knowledge, such as that obtained from the prior distribution, can aid the decision-making process.

I have used the symbol K for the loss function, which reminds us, at least phonetically, of 'cost' because we use the more common symbol L for the likelihood.

A common type of loss is the *zero-one* loss: $\mathsf{K}(\mathbf{X}, \mathfrak{r}) = 0$ if $\ell = \nu$, and $\mathsf{K}(\mathbf{X}, \mathfrak{r}) = 1$ otherwise. This loss reminds us of the cost factors 0 and 1 in the classification error $\mathcal{E}_{\text{mis}}$. We can 'grade' the degree of incorrectness: similar to the more general cost factors in Definition 4.12, $c_{\ell,\nu}$ and $c_{\nu,\ell}$ could differ, and both could take values other than 0 and 1. Such ideas are common in decision theory. For our purpose, it suffices to consider the simple zero-one loss. For the Bayesian discriminant rule, a zero-one loss function has an interpretation in terms of the posterior probability, which is a consequence of Theorem 4.19:

$$\mathsf{K}(\mathbf{X}, \mathfrak{r}) = \begin{cases} 0 & \text{if } L_\ell(\mathbf{X})\pi_\ell \geq L_\nu(\mathbf{X})\pi_\nu \quad \text{for } \ell \leq \kappa, \\ 1 & \text{otherwise.} \end{cases}$$

Definition 4.14 introduces the error probability of a rule, and Proposition 4.15 states the optimality of Bayes rule for two classes. We now include the loss function in the previous framework.

Theorem 4.21 *Let $\mathfrak{C}$ be the collection of classes $C_1, \ldots, C_\kappa$. Let $\boldsymbol{\pi} = \begin{bmatrix} \pi_1 \ \pi_2 \cdots \pi_\kappa \end{bmatrix}$ be the vector of prior probabilities for $\mathfrak{C}$. Let $\mathbf{X}$ be a random vector which belongs to one of the* *classes $\mathcal{C}_\nu$. Let $\mathfrak{r}_{Bayes}$ be Bayes discriminant rule, then $\mathfrak{r}_{Bayes}$ has the discriminant regions $G_\nu = \{\mathbf{X}\colon L_\nu(\mathbf{X})\pi_\nu = \max_\ell [L_\ell(\mathbf{X})\pi_\ell]\}$.*

For the zero-one loss function based on the discriminant regions G_ν, Bayes discriminant rule is optimal among all discriminant rules in the sense that it has

1. *the biggest probability of assigning* **X** *to the correct class, and*
2. *the smallest Bayes risk for the zero-one loss function.*

Proof The two parts of the theorem are closely related via the indicator functions I_{G_ν} because the loss function $\mathsf{K}(\mathbf{X},\mathfrak{r}) = 0$ if $I_{G_\nu}(\mathbf{X}) = 1$.

To show the optimality of Bayes discriminant rule, a proof by contradiction is a natural option. Let $\mathfrak{r}_{\text{Bayes}}$ be Bayes rule, and let $\mathfrak{r}^+$ be a rule based on the same loss function. Assume that the probability of assigning $\mathbf{X}$ to the correct class is larger for $\mathfrak{r}^+$ than for $\mathfrak{r}_{\text{Bayes}}$. Let $p^+(\nu|\nu)$ be the probability of correctly assigning an observation the value ν under $\mathfrak{r}^+$, and let G_ν^+ be the discriminant regions of $\mathfrak{r}^+$. If p^+ is the probability of assigning $\mathbf{X}$ correctly under $\mathfrak{r}^+$, then

$$\begin{aligned}
\mathrm{p}^+ &= \sum_{\nu=1}^{\kappa} p^+(\nu|\nu)\pi_\nu = \sum_{\nu=1}^{\kappa} \int I_{G_\nu^+} L_\nu \pi_\nu \\
&\le \sum_{\nu=1}^{\kappa} \int I_{G_\nu^+} \max_\nu\{L_\nu\pi_\nu\} = \int \max_\nu\{L_\nu\pi_\nu\} \\
&= \sum_{\nu=1}^{\kappa} \int I_{G_\nu} L_\nu \pi_\nu = \sum_{\nu=1}^{\kappa} p(\nu|\nu)\pi_\nu \\
&= \mathrm{p}_{\text{Bayes}}.
\end{aligned}$$

The calculations show that $\mathrm{p}^+ \le \mathrm{p}_{\text{Bayes}}$, in contradiction to the assumption; thus Bayes rule is optimal.

The second part follows similarly by making use of the relationship between the indicator functions of the discriminant regions and the loss function. The proof is left to the reader. ■

The optimality of Bayes rule is of great theoretical interest. It allows us to check how well a rule performs and whether a rule performs asymptotically as well as Bayes rule. For details on this topic, see Devroye, Györfi, and Lugosi (1996).

4.7 Non-Linear, Non-Parametric and Regularised Rules

A large body of knowledge and research exists on discrimination and classification methods – especially for discrimination problems with a low to moderate number of variables. A theoretical and probabilistic treatment of *Pattern Recognition* – yet another name for Discriminant Analysis – is the emphasis of the comprehensive book by Devroye, Györfi, and Lugosi (1996) which covers many of the methods I describe plus others. Areas that are closely related to Discriminant Analysis include *Neural Networks*, *Support Vector Machines* (*SVM*) and kernel density–based methods. It is not possible to do justice to these topics as part of a single chapter, but good accounts can be found in Ripley (1996), Cristianini and Shawe-Taylor (2000), Hastie, Tibshirani, and Friedman (2001) and Schölkopf and Smola (2002).

Some discrimination methods perform well for large samples but may not be so suitable for high dimensions. Kernel density methods appeal because of their non-parametric properties, but they become computationally inefficient with an increasing number of variables. Support Vector Machines work well for small to moderate dimensions but become less suitable as the dimension increases due to a phenomenon called *data piling* which I briefly return to in Section 4.7.4. The 'classical' or basic Support Vector Machines have a nice connection with Fisher's rule; for this reason, I will mention their main ideas at the end of this section. Support Vector Machines have become popular classification tools partly because they are accompanied by powerful optimisation algorithms which perform well in practice for a small to moderate number of variables.

4.7.1 Nearest-Neighbour Discrimination

An intuitive non-linear and non-parametric discrimination approach is based on properties of neighbours. I explain nearest-neighbour discrimination, which was initially proposed in Fix and Hodges (1951, 1952). We also look at an adaptive extension due to Hastie and Tibshirani (1996). Nearest-neighbour methods enjoy great popularity because they are easy to understand and have an intuitive interpretation.

Let

$$\begin{bmatrix} \mathbb{X} \\ \mathbf{Y} \end{bmatrix} = \begin{bmatrix} \mathbf{X}_1 & \mathbf{X}_2 & \cdots & \mathbf{X}_n \\ Y_1 & Y_2 & \cdots & Y_n \end{bmatrix} \tag{4.31}$$

be n labelled random vectors as in (4.2) which belong to κ classes. Let Δ be a distance between two vectors, such as the Euclidean norm, so

$$\Delta(\mathbf{X}_i, \mathbf{X}_j) = \rho \qquad \text{with} \begin{cases} \rho > 0 & \text{if } i \neq j, \\ \rho = 0 & \text{if } i = j. \end{cases}$$

Other distances, such as those summarised in Section 5.3.1, can be used instead of the Euclidean distance. The choice of distance (measure) might affect the performance of the method, but the key steps of the nearest-neighbour approach do not rely on the choice of distance.

For a fixed random vector $\mathbf{X}_0$ with the same distributional properties as the members of $\mathbb{X}$ in (4.31), put

$$\delta_i(\mathbf{X}_0) \equiv \Delta(\mathbf{X}_0, \mathbf{X}_i). \tag{4.32}$$

The δ_i are the distances from $\mathbf{X}_0$ to the vectors of $\mathbb{X}$. Write

$$\delta_{(1)}(\mathbf{X}_0) \leq \delta_{(2)}(\mathbf{X}_0) \leq \cdots \leq \delta_{(n)}(\mathbf{X}_0) \tag{4.33}$$

for the ordered distances between $\mathbf{X}_0$ and members of $\mathbb{X}$. If $\mathbf{X}_0$ belongs to $\mathbb{X}$, then we exclude the trivial distance of $\mathbf{X}_0$ to itself and start with the first non-zero distance.

Definition 4.22 Let $\begin{bmatrix} \mathbb{X} \\ \mathbf{Y} \end{bmatrix}$ be n labelled random vectors as in (4.31) which belong to κ distinct classes. Consider a random vector $\mathbf{X}$ from the same population as $\mathbb{X}$. Let Δ be a distance, and write $\{\delta_{(i)}(\mathbf{X})\}$ for the ordered distances from $\mathbf{X}$ to the vectors of $\mathbb{X}$ as in (4.32) and (4.33). Let $k \geq 1$ be an integer.

1. The **k-nearest neighbourhood** of $\mathbf{X}$ with respect to $\mathbb{X}$ is the set

$$N(\mathbf{X},k) = \{\mathbf{X}_i \in \mathbb{X}\colon \Delta(\mathbf{X},\mathbf{X}_i) \le \delta_{(k)}(\mathbf{X})\}, \tag{4.34}$$

   which contains the k vectors from $\mathbb{X}$ which are closest to $\mathbf{X}$ with respect to Δ and the ordering in (4.33).
2. Write

$$N_Y(\mathbf{X},k) = \{Y_i : Y_i \text{ is the label of } \mathbf{X}_i \text{ and } \mathbf{X}_i \in N(\mathbf{X},k)\}$$

   for the labels of the random vectors in $N(\mathbf{X},k)$. The **k-nearest neighbour discriminant rule** or the **kNN rule** $\mathfrak{r}_{\text{kNN}}$ is defined by

$$\mathfrak{r}_{\text{kNN}}(\mathbf{X}) = \ell, \quad \text{where } \ell \text{ is the mode of } N_Y(\mathbf{X},k).$$

□

The k-nearest neighbour discriminant rule assigns $\mathbf{X}$ to the class which is most common among the neighbouring points. If two classes occur equally often and more often than all others, then the decision is an arbitrary choice between the two candidates. The mode depends on k and the distance measure but does not depend on distributional properties of the data.

The choice of distance affects the shape of the neighbourhood, but the more important parameter is k, the number of nearest neighbours we consider. The parameter k is likened to the tuning or smoothing parameter in density estimation and non-parametric regression, and the distance takes the place of the kernel. As in non-parametric smoothing, the main question is: How big should k be? The value $k = 1$ is simplest, and Cover and Hart (1967) give an asymptotic expression for the error for $k = 1$, but larger values of k are of interest in practice as they can lead to smaller errors. It suffices to calculate the distances of vectors in a region around $\mathbf{X}$ because the discriminant rule is restricted to local information. Devroye, Györfi, and Lugosi (1996) established consistency results of the k-nearest neighbour rule. In their chapter 11 they show that if $k \to \infty$ and $k/n \to 0$ as $n \to \infty$, then the k-nearest neighbour rule converges to the Bayes rule. In our examples that follow, we take a practical approach regarding the choice of k: we consider a range of values for k and find the k which results in the minimum error.

Example 4.10 Fisher's four-dimensional **iris** data are a test case for many classification rules, so it is natural to apply the k-nearest neighbour rule to these data. There is no obvious value for k, the number of observations in each neighbourhood, and we therefore consider a range of values and calculate the number of misclassified observations for each k.

There are three classes with fifty observations each, so we use $k \le 49$. Figure 4.8 shows the results with k on the x-axis and the number of misclassified observations on the y-axis. The number of misclassified observations initially decreases until it reaches its minimum of three misclassified points for $k = 19, 20$ and 21 and then increases with k. The initial decrease is not strictly monotonic, but the global behaviour shows clearly that $k = 1$ is not the best choice.

Observations 84 and 107 are misclassified for each value of k. These are observations in the 'green' and 'black' classes of Figure 1.4 in Section 1.2.2. Observation 84 is obviously a hard iris to classify correctly; Fisher's rule also gets this one wrong! A further comparison with Example 4.1 shows that Fisher's rule does slightly better: it misclassifies two

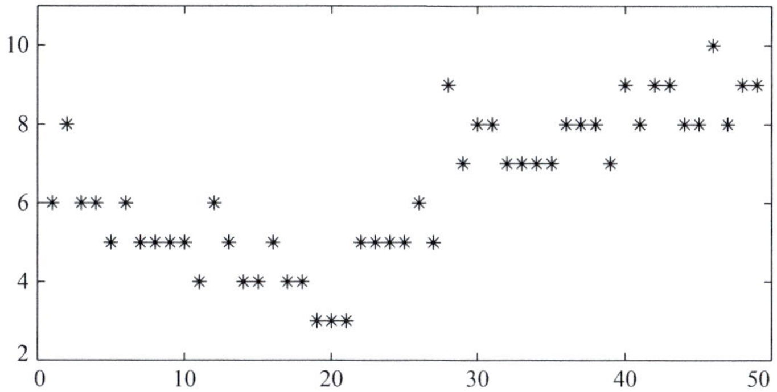

Figure 4.8 The kNN rule for the iris data from Example 4.10 with k on the x-axis and the number of misclassified observations on the y-axis.

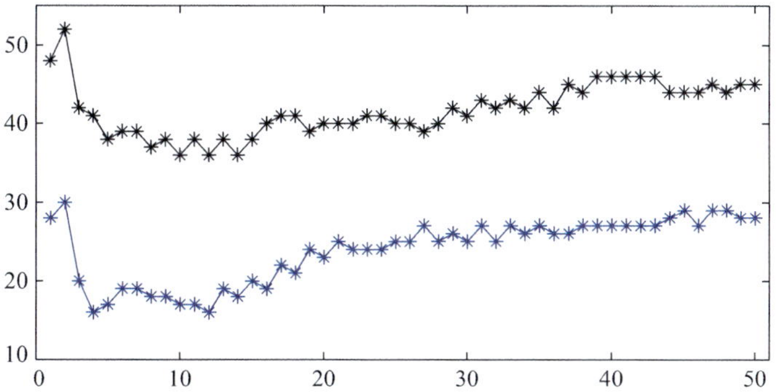

Figure 4.9 Number of misclassified observations against k (on the x-axis) with the kNN rule for the raw (*in black*) and the scaled (*in blue*) breast cancer data from Example 4.11.

observations from the 'green' class compared with the three misclassified irises for the kNN rule. ■

We next explore the performance of the kNN rule for a two-class problem with more observations and more dimensions.

Example 4.11 We continue with a classification of the **breast cancer** data. In Example 4.6, I applied the normal rule and Fisher's rule to these data. The normal rule resulted in sixteen misclassified observations and Fisher's rule in fifteen misclassified observations, and both rules have the same performance for the raw and scaled data.

For the kNN rule, the number of misclassified observations is shown on the y-axis in Figure 4.9, with k on the x-axis. The black curve shows the results for the raw data and the blue curve for the scaled data. In contrast to Fisher's rule, we note that the kNN rule performs considerably worse for the raw data than the for scaled data.

Both curves in Figure 4.9 show that neighbourhoods of more than one observation reduce the misclassification. For the raw data, the smallest number of misclassified observations is thirty-six, which is obtained for $k = 10, 12$ and 14. For the scaled data, a minimum of sixteen misclassified observations is obtained when $k=4$ and $k=12$, and thus the kNN rule performs as well as the normal rule. ■

Both examples show that $k=1$ is not the optimal choice for the kNN rule. We clearly do better when we consider larger neighbourhoods. For the *breast cancer* data we also observe that the scaled data result in a much smaller classification error than the raw data. Recall that three variables in these data are on a very different scale from the rest, so scaling the data has produced 'better' neighbourhoods in this case. In both examples, the best kNN rule performed slightly worse than Fisher's rule.

The kNN method has appeal because of its conceptual simplicity and because it is not restricted to Gaussian assumptions. The notion of distance is well defined in low and high dimensions alike, and the method therefore applies to high-dimensional settings, possibly in connection with dimension reduction. In supervised learning, the 'best' value for k can be determined by cross-validation, which I describe in Definition 9.10 in Section 9.5, but the simpler errors $\mathcal{E}_{\text{mis}}$ of (4.20) and $\mathcal{E}_{\text{loo}}$ of (4.22) are also useful for stepping through the values of k, as we have done here.

To predict the class of a new datum $\mathbf{X}_{\text{new}}$ with the kNN rule, we determine the optimal k, say k^*, as earlier (or by cross-validation). Next, we apply the rule with k^* to the new datum: we find the k^*-nearest neighbours of the new datum and then determine the mode of $N_Y(\mathbf{X}_{\text{new}}, k^*)$.

The basic idea of the kNN approach is used in more advanced methods such as the adaptive approach of Hastie and Tibshirani (1996), which I summarise from the simplified version given by the authors in section 13.4 of Hastie, Tibshirani, and Friedman (2001).

Let $\mathbb{X}$ be data from κ different classes. In the notation of Section 4.3.2, let

$$\widehat{B}_{\text{NN}} = \sum_{\nu=1}^{\kappa} \pi_\nu \left(\overline{\mathbf{X}}_\nu - \overline{\overline{\mathbf{X}}}\right)\left(\overline{\mathbf{X}}_\nu - \overline{\overline{\mathbf{X}}}\right)^{\mathsf{T}}$$

$$\widehat{W}_{\text{NN}} = \sum_{\nu=1}^{\kappa} \pi_\nu S_\nu, \tag{4.35}$$

be modified versions of Fisher's between-class variance and within-class variance, where the $\overline{\mathbf{X}}_\nu$ are the sample class means, $\overline{\overline{\mathbf{X}}}$ is the mean of these sample means and the π_ν are weights with values in $[0,1]$, such as class probabilities or weights associated with distances from a fixed point.

Hastie and Tibshirani (1996) consider the matrix

$$S_{\text{BW},\epsilon} = (\widehat{W}_{\text{NN}})^{-1/2}\widehat{B}^*_{\text{NN},\epsilon}(\widehat{W}_{\text{NN}})^{-1/2}, \tag{4.36}$$

with

$$\widehat{B}^*_{\text{NN},\epsilon} = (\widehat{W}_{\text{NN}})^{-1/2}\widehat{B}_{\text{NN}}(\widehat{W}_{\text{NN}})^{-1/2} + \epsilon\mathbf{I},$$

where $\epsilon \geq 0$. For observations $\mathbf{X}_i$ and $\mathbf{X}_j$, Hastie and Tibshirani (1996) define the **discriminant adaptive nearest-neighbour** (DANN) distance

$$\Delta_{\text{DANN}}(\mathbf{X}_i, \mathbf{X}_j) = \left\| S^{1/2}_{\text{BW},\epsilon}(\mathbf{X}_i - \mathbf{X}_j)\right\| \tag{4.37}$$

and recommend the following procedure as an adaptive kNN rule.

Algorithm 4.1 ***Discriminant Adaptive Nearest-Neighbour Rule***

Step 1. For $\mathbf{X}_0$, find a neighbourhood of about fifty points with the Euclidean distance.
Step 2. Calculate the matrices $\widehat{B}_{\text{NN}}$ and $\widehat{W}_{\text{NN}}$ of (4.35) and $S_{\text{BW},\epsilon}$ of (4.36) from points in the neighbourhood found in step 1.
Step 3. Combine steps 1 and 2 to calculate DANN distances (4.37) from $\mathbf{X}_0$ for $\mathbf{X}_i \in \mathbb{X}$:

$$\delta_{i,\text{DANN}}(\mathbf{X}_0) \equiv \Delta_{\text{DANN}}(\mathbf{X}_0, \mathbf{X}_i).$$

Step 4. Use the distances $\delta_{i,\text{DANN}}(\mathbf{X}_0)$ to define a kNN rule. ■

Hastie and Tibshirani suggest that the value $\epsilon = 1$ is appropriate in practice. Of course, a value for k is still required! Hastie and Tibshirani remark that a smaller number of neighbouring points is required in the final DANN rule than in the basic kNN rule, and the overall performance is improved when compared with the basic kNN rule. It is worth noting that the matrix $S_{\text{BW},\epsilon}$ of (4.37) is of the form $\widehat{W}^{-1}\widehat{B}\widehat{W}^{-1}$ rather than the form $\widehat{W}^{-1}\widehat{B}$ proposed by Fisher. The authors do not provide an explanation for their choice which includes the additional $\widehat{W}^{-1}$.

Other adaptive approaches have been proposed in the literature which are similar to that of Hastie and Tibshirani (1996) in that they rely on the basic nearest-neighbour ideas and adjust the distance measure. Some examples are Short and Fukunaga (1981) and Domeniconi, Peng, and Gunopulos (2002). Qiu and Wu (2006) replace Fisher's quotient $W^{-1}B$ by a difference of related matrices and then construct the eigenvectors of those matrices. Hinneburg, Aggarwal, and Keim (2000) propose adjusting the idea of nearest neighbours such that the dimensions are assigned different weights. This idea leads to the selection of a subset of dimensions, an approach which is closely related to dimension reduction prior to discrimination.

4.7.2 Logistic Regression and Discrimination

In logistic regression, one compares the probabilities $\mathbb{P}(\mathbf{X} \in C_1)$ and $\mathbb{P}(\mathbf{X} \in C_2)$ by modelling their log ratio as a linear relationship:

$$\log\left\{\frac{\mathbb{P}(\mathbf{X} \in C_1)}{\mathbb{P}(\mathbf{X} \in C_2)}\right\} = \beta_0 + \boldsymbol{\beta}^{\mathsf{T}}\mathbf{X}, \tag{4.38}$$

for some vector $\boldsymbol{\beta}$ and scalar β_0, and then derives the expressions

$$\mathbb{P}(\mathbf{X} \in C_1) = \frac{\exp(\beta_0 + \boldsymbol{\beta}^{\mathsf{T}}\mathbf{X})}{1 + \exp(\beta_0 + \boldsymbol{\beta}^{\mathsf{T}}\mathbf{X})} \quad \text{and} \quad \mathbb{P}(\mathbf{X} \in C_2) = \frac{1}{1 + \exp(\beta_0 + \boldsymbol{\beta}^{\mathsf{T}}\mathbf{X})}. \tag{4.39}$$

Instead of estimating the probabilities, one estimates the parameters β_0 and $\boldsymbol{\beta}$. The logistic regression model is a special case of Generalised Linear Models (see McCullagh and Nelder 1989).

How can we use these ideas in discrimination? For two-class problems, the formulation (4.39) is reminiscent of Bernoulli trials with success probability p. For a random vector $\mathbf{X}$, we put $Y = 1$ if $\mathbf{X} \in C_1$ and $Y = 2$ otherwise. For data $\mathbb{X} = \begin{bmatrix}\mathbf{X}_1 \cdots \mathbf{X}_n\end{bmatrix}$, we proceed similarly, and let $\mathbf{Y}$ be the $1 \times n$ row vector of labels. Write

$$\mathbf{p} = (p_i) \quad \text{with} \quad p_i = \mathbb{P}(\mathbf{X}_i \in C_1) = \frac{\exp(\beta_0 + \boldsymbol{\beta}^{\mathsf{T}}\mathbf{X}_i)}{1 + \exp(\beta_0 + \boldsymbol{\beta}^{\mathsf{T}}\mathbf{X}_i)}, \tag{4.40}$$

so $\mathbf{p}$ is a $1 \times n$ vector. Each observation $\mathbf{X}_i$ is regarded as a Bernoulli trial, which results in the likelihood function

$$L(\mathbf{p}|\, \mathbb{X}) = \prod_{i=1}^{n} \left[p_i^{Y_i} (1 - p_i)^{(1-Y_i)} \right].$$

Next, we augment the data by including the vector of ones, $\mathbf{1}_{1\times n} = [1 \cdots 1]$, and put

$$\mathbb{X}_{(+1)} = \left[\mathbf{X}_{1,(+1)} \cdots \mathbf{X}_{n,(+1)} \right] = \begin{bmatrix} \mathbf{1}_{1\times n} \\ \mathbb{X} \end{bmatrix} \qquad \text{and} \qquad B = \begin{bmatrix} \beta_0 \\ \boldsymbol{\beta} \end{bmatrix},$$

so (4.40) becomes

$$p_i = \frac{\exp(B^{\mathsf{T}} \mathbf{X}_{i,(+1)})}{1 + \exp(B^{\mathsf{T}} \mathbf{X}_{i,(+1)})}. \tag{4.41}$$

The probabilities p_i naturally lead to **logistic regression** discriminant rules: Fix $0 < \tau < 1$, and put

$$\mathfrak{r}_\tau(\mathbf{X}_i) = \begin{cases} 1 & \text{if } p_i \geq \tau, \\ 2 & \text{if } p_i < \tau. \end{cases} \tag{4.42}$$

For the common value $\tau = 1/2$, the rule assigns $\mathbf{X}_i$ to class 1 if $\beta_0 + \boldsymbol{\beta}^{\mathsf{T}} \mathbf{X}_i > 0$. Values for τ other than $\tau = 1/2$ can be chosen to minimise an error criterion such as the classification error $\mathcal{E}_{\text{mis}}$ or the leave-one-out error $\mathcal{E}_{\text{loo}}$. Alternatively, one may want to consider the p_i values for all observations $\mathbf{X}_i$ and look for a gap in the distribution of the p_i values as a suitable value for τ.

To classify a new datum $\mathbf{X}_{\text{new}}$ with a logistic regression discriminant rule, we derive estimates $\widehat{\beta}_0$ and $\widehat{\boldsymbol{\beta}}$ of the coefficients β_0 and $\boldsymbol{\beta}$ from the data $\mathbb{X}$ and determine a suitable value for τ. Next, we calculate

$$p_{\text{new}} = \frac{\exp(\widehat{\beta}_0 + \widehat{\boldsymbol{\beta}}^{\mathsf{T}} \mathbf{X}_{\text{new}})}{1 + \exp(\widehat{\beta}_0 + \widehat{\boldsymbol{\beta}}^{\mathsf{T}} \mathbf{X}_{\text{new}})}$$

and apply the logistic regression rule $\mathfrak{r}_\tau$ to $\mathbf{X}_{\text{new}}$ by comparing p_{new} with the threshold τ.

The rules (4.42) apply to two-class problems. In theory, they can by extended to more classes, but such a generalisation requires multiple pairwise comparisons, which become cumbersome as the number of classes increases. Alternatively, one can consider discrimination as a regression problem with discrete responses.

4.7.3 Regularised Discriminant Rules

In Section 4.5.2 we consider a quadratic discriminant rule for normal data. The accompanying simulation study, in Example 4.8, shows that the quadratic discriminant rule performs better than the normal linear rule and Fisher's rule. The simulations further indicate that the quadratic rule works well even for non-normal data.

The quadratic rule – like the normal linear rule – is based on the normal likelihood. If the covariance matrices of the κ classes are sufficiently different, then the quadratic rule is superior as it explicitly takes into account the different covariance matrices. Friedman (1989) points out that the sample covariance matrices S_ν are often biased – especially when the sample size is low: the large eigenvalues tend to be overestimated, whereas the small eigenvalues are underestimated. The latter can make the covariance matrices unstable or cause problems with matrix inversions. To overcome this instability, Friedman (1989) proposed to regularise the covariance matrices. I briefly describe his main ideas which apply directly to data.

Consider the classes $\mathcal{C}_1, \ldots, \mathcal{C}_\kappa$, and let S_ν be the sample covariance matrix pertaining to class $\mathcal{C}_\nu$. For $\alpha \in [0,1]$, the **regularised (sample) covariance matrix** $S_\nu(\alpha)$ of the νth class is

$$S_\nu(\alpha) = \alpha S_\nu + (1-\alpha) S_{pool},$$

where S_{pool} is the pooled sample covariance matrix calculated from all classes. The two special cases – $\alpha = 0$ and $\alpha = 1$ – correspond to linear discrimination and quadratic discrimination. The regularised covariance matrices $S_\nu(\alpha)$ replace the covariance matrices S_ν in the quadratic rule and thus allow a compromise between linear and quadratic discrimination. A second tuning parameter, $\gamma \in [0,1]$, controls the 'shrinkage' of the pooled covariance matrix S_{pool} towards a multiple of the identity matrix by putting

$$S_{pool}(\gamma) = \gamma S_{pool} + (1-\gamma) s^2 \mathbf{I},$$

where s^2 is a suitably chosen positive scalar. Substitution of S_{pool} by $S_{pool}(\gamma)$ leads to a flexible two-parameter family of covariance matrices $S_\nu(\alpha, \gamma)$.

For random vectors $\mathbf{X}$ from the classes $\mathcal{C}_\nu$, the **regularised discriminant rule** $\mathfrak{r}_{\alpha,\gamma}$ assigns $\mathbf{X}$ to class ℓ if

$$\left\| \mathbf{X}_{S_\ell(\alpha,\gamma)} \right\|^2 + \log[\det(S_\ell(\alpha,\gamma))] = \min_{1 \le \nu \le \kappa} \left\{ \left\| \mathbf{X}_{S_\nu(\alpha,\gamma)} \right\|^2 + \log[\det(S_\nu(\alpha,\gamma))] \right\},$$

where $\mathbf{X}_{S_\ell(\alpha,\gamma)}$ is the vector that is sphered with the matrix $S_\ell(\alpha,\gamma)$. A comparison with Theorem 4.16 motivates this choice of rule.

The choice of the tuning parameters is important as they provide a compromise between stability of the solution and adherence to the different class covariance matrices. Friedman (1989) suggested choosing that value for α which results in the 'best' performance. In the supervised context of his paper, his performance criterion was the cross-validation error $\mathcal{E}_{\text{cv}}$ (see (9.19) in Section 9.5), but other errors can be used instead. However, the reader should be aware that the value of α depends on the chosen error criterion.

Other regularisation approaches to discrimination are similar to those used in regression. The ridge estimator of (2.39) in Section 2.8.2 is one such example. Like the regularised covariance matrix of this section, the ridge estimator leads to a stable inverse of the sample covariance matrix.

4.7.4 Support Vector Machines

In 1974, Vapnik proposed the mathematical ideas of Support Vector Machines (SVMs) in an article in Russian which became available in 1979 in a German translation (see Vapnik and Chervonenkis 1979). I describe the basic idea of linear SVMs and indicate their relationship to Fisher's rule. For details and more general SVMs, see Cristianini and Shawe-Taylor (2000), Hastie, Tibshirani, and Friedman (2001) and Schölkopf and Smola (2002).

Consider data $\mathbb{X}$ from two classes with labels $\mathbf{Y}$. A label takes the values $Y_i = 1$ if $\mathbf{X}_i$ belongs to class 1 and $Y_i = -1$ otherwise, for $i \leq n$. Let $Y_{\text{diag}} = \text{diag}(Y_1, \ldots, Y_n)$ be the diagonal $n \times n$ matrix with diagonal entries Y_i. Put

$$\boldsymbol{\gamma} = Y_{\text{diag}} \mathbb{X}^{\mathsf{T}} \mathbf{w} + b \mathbf{Y}^{\mathsf{T}}, \tag{4.43}$$

and call $\boldsymbol{\gamma}$ the **margin** or the **residuals** with respect to $(\mathbf{w}, b)$, where $\mathbf{w}$ is a $d \times 1$-direction vector and b is a scalar. We want to find a vector $\widehat{\mathbf{w}}$ and scalar $\widehat{b}$ which make all entries of $\boldsymbol{\gamma}$ positive. If such $\widehat{\mathbf{w}}$ and $\widehat{b}$ exist, put $k(\mathbf{X}) = \mathbf{X}^{\mathsf{T}} \widehat{\mathbf{w}} + \widehat{b}$; and then

$$k(\mathbf{X}) \begin{cases} \geq 0 & \text{if } \mathbf{X} \text{ has label } Y = 1 \\ < 0 & \text{if } \mathbf{X} \text{ has label } Y = -1. \end{cases} \tag{4.44}$$

The function k in (4.44) is linear in $\mathbf{X}$. More general functions k, including polynomials, splines or the feature kernels of Section 12.2.1, are used in SVMs with the aim of separating the two classes maximally in the sense that the minimum residuals are maximised.

To achieve a maximal separation of the classes, for the linear k of (4.44), one introduces a new scalar Δ which is related to the margin $\boldsymbol{\gamma}$ and solves

$$\max \Delta \qquad \text{subject to} \qquad \boldsymbol{\gamma} \geq \Delta \mathbf{1}_n,$$

where $\mathbf{1}_n$ is a vector of 1s, and the vector inequality is interpreted as entrywise inequalities. In practice, an equivalent optimisation problem of the form

$$\min_{\mathbf{w}, b, \boldsymbol{\xi}} \frac{1}{2} \mathbf{w}^{\mathsf{T}} \mathbf{w} + C \mathbf{1}_n^{\mathsf{T}} \boldsymbol{\xi} \qquad \text{subject to} \qquad \boldsymbol{\gamma} + \boldsymbol{\xi} \geq \mathbf{1}_n \quad \text{and} \quad \boldsymbol{\xi} \geq \mathbf{0} \tag{4.45}$$

or its dual problem

$$\max_{\boldsymbol{\alpha}} -\frac{1}{2} \boldsymbol{\alpha}^{\mathsf{T}} Y_{\text{diag}} \mathbb{X}^{\mathsf{T}} \mathbb{X} Y_{\text{diag}} \boldsymbol{\alpha} + \mathbf{1}_n^{\mathsf{T}} \boldsymbol{\alpha} \quad \text{subject to} \quad \mathbf{Y}^{\mathsf{T}} \boldsymbol{\alpha} = 0 \quad \text{and} \quad \mathbf{0} \leq \boldsymbol{\alpha} \leq C \mathbf{1}_n \tag{4.46}$$

is solved. In (4.45) and (4.46), $C > 0$ is a penalty parameter, and $\boldsymbol{\xi}$ has non-negative entries. Powerful optimisation algorithms have been developed which rely on finding solutions to convex quadratic programming problems (see Vapnik 1995; Cristianini and Shawe-Taylor 2000; Hastie et al. 2001; and Schölkopf and Smola 2002), and under appropriate conditions, the two dual problems have the same optimal solutions.

Note. In SVMs we use the labels $Y = \pm 1$, unlike our customary labels, which take values $Y = \ell$ for the ℓth class.

The change in SVMs to labels $Y = \pm 1$ results in positive residuals $\boldsymbol{\gamma}$ in (4.43). Apart from this difference in the labels in (4.43), k in (4.44) is reminiscent of Fisher's decision function h of (4.17): the SVM direction $\mathbf{w}$ takes the role of the vector $\boldsymbol{\beta}$ in (4.17), and $\mathbf{w}$ is closely related to Fisher's direction vector $\boldsymbol{\eta}$. Ahn and Marron (2010) point out that the two methods are equivalent for classical data with $d \ll n$.

For HDLSS settings, both SVMs and Fisher's approach encounter problems, but the problems are of a different nature in the two approaches. I will come back to Fisher's linear discrimination for HDLSS settings in Section 13.3, where I discuss the problems, some solutions and related asymptotic properties.

As the number of variables increases a phenomenon called *data piling* occurs in a SVM framework. As a result, many very small residuals can make the SVM unstable. Marron,

Todd, and Ahn (2007) highlight these problems and illustrate data piling with examples. Their approach, Distance-Weighted Discrimination (DWD), provides a solution to the data piling for HDLSS data by including new criteria into their objective function. Instead of solving (4.45), DWD considers the individual entries γ_i^ξ, with $i \le n$, of the vector

$$\boldsymbol{\gamma}^\xi = Y_{\text{diag}} \mathbb{X}^\mathsf{T} \mathbf{w} + b\mathbf{Y}^\mathsf{T} + \boldsymbol{\xi}$$

and includes the reciprocals of γ_i^ξ in the optimisation problem to be solved:

$$\min_{\boldsymbol{\gamma}^\xi, \mathbf{w}, b, \boldsymbol{\xi}} \sum_{i=1}^{n} \frac{1}{\gamma_i^\xi} + C\mathbf{1}_n^\mathsf{T}\boldsymbol{\xi} \qquad \text{subject to} \qquad \|\mathbf{w}\| \le 1, \quad \boldsymbol{\gamma}^\xi \ge \mathbf{0} \quad \text{and} \quad \boldsymbol{\xi} \ge \mathbf{0}, \tag{4.47}$$

where $C > 0$ is a penalty parameter as in (4.45). This innocuous looking change in the objective function seems to allow a smooth transition to the classification of HDLSS data. For a fine-tuning of their method, including choices of the tuning parameter C and comparisons of DWD with SVMs, see Marron, Todd, and Ahn (2007).

4.8 Principal Component Analysis, Discrimination and Regression

4.8.1 Discriminant Analysis and Linear Regression

There is a natural link between Discriminant Analysis and Linear Regression: both methods have predictor variables, the data $\mathbb{X}$, and outcomes, the labels or responses $\mathbf{Y}$. An important difference between the two methods is the type of variable the responses $\mathbf{Y}$ represent; in a discrimination setting, the responses are discrete or categorical variables, whereas the regression responses are continuous variables. This difference is more important than one might naively assume.

Although Discriminant Analysis may appear to be an easier problem – because we only distinguish between κ different outcomes – it is often harder than solving linear regression problems. In Linear Regression, the estimated responses are the desired solution. In contrast, the probabilities obtained in Logistic Regression are not the final answer to the classification problem because one still has to make a decision about the correspondence between probabilities and class membership. Table 4.8 highlights the similarities and differences between the two methods. The table uses Fisher's direction vector $\widehat{\boldsymbol{\eta}}$ in order to work with a concrete example, but other rules could be used instead.

The predictors or the data $\mathbb{X}$ are the same in both approaches and are used to estimate a vector, namely, $\widehat{\boldsymbol{\eta}}$ or $\widehat{\boldsymbol{\beta}}$, in the table. These vectors contain the weights for each variable and thus contain information about the relative importance of each variable in the estimation and subsequent prediction of the outcomes. The vector $\widehat{\boldsymbol{\beta}}$ in Linear Regression is estimated directly from the data. For the direction vectors in Discriminant Analysis, we use

- the first eigenvector of $W^{-1}B$ when we consider Fisher's discrimination;
- the vectors $\Sigma^{-1}\boldsymbol{\mu}_\nu$ in normal linear discrimination; or
- the coefficient vector $\boldsymbol{\beta}$ in logistic regression, and so on.

As the number of variables of $\mathbb{X}$ increases, the problems become harder, and variable selection becomes increasingly more important. In regression, variable selection techniques are well established (see Miller 2002). I introduce approaches to variable selection and variable ranking in the next sections and elaborate on this topic in Section 13.3.

Table 4.8 *Similarities and Differences in Discrimination and Regression*

	Discriminant Analysis	Linear Regression
Predictors	$\mathbb{X} = \begin{bmatrix}\mathbf{X}_1 \cdots \mathbf{X}_n\end{bmatrix}$	$\mathbb{X} = \begin{bmatrix}\mathbf{X}_1 \cdots \mathbf{X}_n\end{bmatrix}$
Outcomes	Labels with κ values	Continuous responses
	$\mathbf{Y} = \begin{bmatrix}Y_1 \cdots Y_n\end{bmatrix}$	$\mathbf{Y} = \begin{bmatrix}Y_1 \cdots Y_n\end{bmatrix}$
Relationship	$\mathbf{X}_i \longmapsto Y_i$ (non)linear	$Y_i = \boldsymbol{\beta}^{\mathsf{T}}\mathbf{X}_i + \epsilon_i$
Projection	Discriminant direction $\widehat{\boldsymbol{\eta}}$	Coefficient vector $\widehat{\boldsymbol{\beta}}$
Estimation	First eigenvector of $W^{-1}B$;	$\widehat{\boldsymbol{\beta}}^{\mathsf{T}} = \mathbf{Y}\mathbb{X}^{\mathsf{T}}(\mathbb{X}\mathbb{X}^{\mathsf{T}})^{-1}$;
	$\widehat{\boldsymbol{\eta}}^{\mathsf{T}}\mathbf{X}_i \longmapsto \widehat{Y}_i$	$\widehat{Y}_i = \widehat{\boldsymbol{\beta}}^{\mathsf{T}}\mathbf{X}_i$

4.8.2 Principal Component Discriminant Analysis

The discriminant rules considered so far in this chapter use all variables in the design of the discriminant rule. For high-dimensional data, many of the variables are *noise* variables, and these can adversely affect the performance of a discriminant rule. It is therefore advisable to eliminate such variables prior to the actual discriminant analysis. We consider two ways of reducing the number of variables:

1. Transformation of the original high-dimensional data, with the aim of keeping the first few transformed variables.
2. Variable selection, with the aim of keeping the most important or relevant original variables.

A natural candidate for the first approach is the transformation of Principal Component Analysis, which, combined with Discriminant Analysis, leads to **Principal Component Discriminant Analysis**. I begin with this approach and then explain the second approach in Section 4.8.3.

Consider the d-dimensional random vector $\mathbf{X} \sim (\boldsymbol{\mu}, \Sigma)$ with spectral decomposition $\Sigma = \Gamma\Lambda\Gamma^{\mathsf{T}}$ and k-dimensional principal component vector $\mathbf{W}^{(k)} = \Gamma_k^{\mathsf{T}}(\mathbf{X} - \boldsymbol{\mu})$ as in (2.2) in Section 2.2, where $k \leq d$. Similarly, for data $\mathbb{X}$ of size $d \times n$, let $\mathbb{W}^{(k)}$ be the principal component data of size $k \times n$ as defined in (2.6) in Section 2.3.

In Section 2.8.2, the principal component data $\mathbb{W}^{(k)}$ are the predictors in Linear Regression. In Discriminant Analysis, they play a similar role.

For $k \leq d$ and a rule $\mathfrak{r}$ for $\mathbb{X}$, the **derived (discriminant) rule** $\mathfrak{r}_k$ for the lower-dimensional PC data $\mathbb{W}^{(k)}$ is constructed in the same way as $\mathfrak{r}$ using the obvious substitutions. If $\mathfrak{r}$ is Fisher's discriminant rule, then $\mathfrak{r}_k$ uses the first eigenvector $\widehat{\boldsymbol{\eta}}_k$ of $\widehat{W}_k^{-1}\widehat{B}_k$, where $\widehat{W}_k$ and $\widehat{B}_k$ are the within-sample covariance matrix and the between-sample covariance matrix of $\mathbb{W}^{(k)}$, namely, for the ith sample $\mathbf{W}_i^{(k)}$ from $\mathbb{W}^{(k)}$,

$$\mathfrak{r}_k(\mathbf{W}_i^{(k)}) = \ell \qquad \text{if } \left|\widehat{\boldsymbol{\eta}}_k^{\mathsf{T}}\mathbf{W}_i^{(k)} - \widehat{\boldsymbol{\eta}}^{\mathsf{T}}\overline{\mathbf{W}}_\ell^{(k)}\right| = \min_{1 \leq \nu \leq \kappa} \left|\widehat{\boldsymbol{\eta}}^{\mathsf{T}}\mathbf{W}_i^{(k)} - \widehat{\boldsymbol{\eta}}^{\mathsf{T}}\overline{\mathbf{W}}_\nu^{(k)}\right|.$$

For other discriminant rules $\mathfrak{r}$ for $\mathbb{X}$, the derived rule $\mathfrak{r}_k$ is constructed analogously.

A natural question is how to choose k. Folklore-based as well as more theoretically founded choices exist for the selection of k in Principal Component Analysis, but they may not be so suitable in the discrimination context. A popular choice, though maybe not the best, is $\mathbb{W}^{(1)}$, which I used in Principal Component Regression in Section 2.8.2. Comparisons with Canonical Correlation methods in Section 3.7 and in particular the results shown in Table 3.11 in Section 3.7.3 indicate that the choice $k = 1$ may result in a relatively poor performance.

The discriminant direction $\widehat{\eta}$ of a linear or quadratic rule incorporates the class membership of the data. If we reduce the data to one dimension, as in $\mathbb{W}^{(1)}$, then we are no longer able to choose the best discriminant direction. Algorithm 4.2 overcomes this inadequacy by finding a data-driven choice for k.

Algorithm 4.2 ***Principal Component Discriminant Analysis***
Let $\mathbb{X}$ be data which belong to κ classes. Let r be the rank of the sample covariance matrix of $\mathbb{X}$. Let $\mathbb{W}^{(r)}$ be the r-dimensional PC data. Let $\mathfrak{r}$ be a discriminant rule, and let $\mathcal{E}$ be a classification error.

Step 1. For $p \leq r$, consider $\mathbb{W}^{(p)}$.
Step 2. Construct the rule $\mathfrak{r}_p$, derived from $\mathfrak{r}$, for $\mathbb{W}^{(p)}$, classify the observations in $\mathbb{W}^{(p)}$ and calculate the derived error $\mathcal{E}_p$ for $\mathbb{W}^{(p)}$, which uses the same criteria as $\mathcal{E}$.
Step 3. Put $p = p + 1$, and repeat steps 1 and 2.
Step 4. Find the dimension p^* which minimises the classification error:

$$p^* = \underset{1 \leq p \leq r}{\operatorname{argmin}} \mathcal{E}_p.$$

Then p^* is the PC dimension that results in the best classification with respect to the performance measure $\mathcal{E}$. ■

In the PC-based approach, we cannot use more than r components. Typically, this is not a serious drawback because an aim in the analysis of high-dimensional data is to reduce the dimension in order to simplify and improve discrimination. We will return to the PC approach to discrimination in Section 13.2.1 and show how Principal Component Analysis may be combined with other feature-selection methods.

At the end of the next section I apply the approach of this and the next section to the *breast cancer* data and compare the performances of the two methods.

4.8.3 Variable Ranking for Discriminant Analysis

For high-dimensional data, and in particular when $d > n$, Principal Component Analysis automatically reduces the dimension to at most n. If the main aim is to simplify the data, then dimension reduction with Principal Component Analysis is an obvious and reasonable step to take. In classification and regression, other considerations come into play:

- Variables which contribute strongly to variance may not be important for classification or regression.
- Variables with small variance contribution may be essential for good prediction.

By its nature, Principal Component Discriminant Analysis selects the predictor variables entirely by their variance contributions and hence may not find the 'best' variables for predicting responses accurately. In this section we look at ways of choosing lower-dimensional subsets which may aid in the prediction of the response variables.

When data are collected, the order of the variables is *arbitrary* in terms of their contribution to prediction. If we select variables according to their *ability* to predict the response variables accurately, then we hope to obtain better prediction than that achieved with all variables. In gene expression data, t-tests are commonly used to find suitable predictors from among a very large number of genes (see Lemieux et al. 2008). The t-tests require normal data, but in practice, not much attention is paid to this requirement. Example 3.11 in Section 3.7.3 successfully employs pairwise t-tests between the response variable and each predictor in order to determine the most significant variables for regression, and Table 3.10 in Example 3.11 lists the selected variables in decreasing order of their significance.

Definition 4.23 Let $\mathbf{X} = \begin{bmatrix} X_1 \cdots X_d \end{bmatrix}^{\mathsf{T}}$ be a d-dimensional random vector. A **variable ranking scheme** for $\mathbf{X}$ is a permutation of the variables of $\mathbf{X}$ according to some rule $\boldsymbol{\nu}$. Write

$$X_{\nu_1}, X_{\nu_2}, \ldots, X_{\nu_p}, \ldots, X_{\nu_d} \tag{4.48}$$

for the ranked variables, where X_{ν_1} is best, and call

$$\mathbf{X}_{\boldsymbol{\nu}} = \begin{bmatrix} X_{\nu_1} \\ X_{\nu_2} \\ \vdots \\ X_{\nu_d} \end{bmatrix} \quad \text{and} \quad \mathbf{X}_{\boldsymbol{\nu},p} = \begin{bmatrix} X_{\nu_1} \\ X_{\nu_2} \\ \vdots \\ X_{\nu_p} \end{bmatrix} \quad \text{for } p \leq d$$

the **ranked vector** and the ***p*-ranked vector**, respectively.

For data $\mathbb{X} = \begin{bmatrix} \mathbf{X}_1 \cdots \mathbf{X}_n \end{bmatrix}$ with d-dimensional random vectors $\mathbf{X}_i$ and some rule $\boldsymbol{\nu}$, put

$$\mathbb{X}_{\boldsymbol{\nu}} = \begin{bmatrix} \mathbf{X}_{\bullet,\nu_1} \\ \mathbf{X}_{\bullet,\nu_2} \\ \vdots \\ \mathbf{X}_{\bullet,\nu_d} \end{bmatrix} \quad \text{and} \quad \mathbb{X}_{\boldsymbol{\nu},p} = \begin{bmatrix} \mathbf{X}_{\bullet,\nu_1} \\ \mathbf{X}_{\bullet,\nu_2} \\ \vdots \\ \mathbf{X}_{\bullet,\nu_p} \end{bmatrix} \quad \text{for } p \leq d, \tag{4.49}$$

and call $\mathbb{X}_{\boldsymbol{\nu}}$ the **ranked data** and $\mathbb{X}_{\boldsymbol{\nu},p}$ the ***p*-ranked data**.

Let $\boldsymbol{\rho} = \begin{bmatrix} \rho_1 \cdots \rho_d \end{bmatrix} \in \mathbb{R}^d$, and consider the order statistic $|\rho_{(1)}| \geq |\rho_{(2)}| \geq \cdots \geq |\rho_{(d)}|$. Define a ranking rule $\boldsymbol{\nu} \in \mathbb{R}^d$ by $\nu_j = \rho_{(j)}$ for $j \leq d$. Then $\boldsymbol{\rho}$ induces a variable ranking scheme for $\mathbf{X}$ or $\mathbb{X}$, and $\boldsymbol{\rho}$ is called the **ranking vector**. □

Note that the p-ranked vector $\mathbf{X}_{\boldsymbol{\nu},p}$ consists of the first p ranked variables of $\mathbf{X}_{\boldsymbol{\nu}}$ and is therefore a p-dimensional random vector. Similarly, the column vectors of $\mathbb{X}_{\boldsymbol{\nu},p}$ are p-dimensional, and the $\mathbf{X}_{\bullet,\nu_j}$ are $1 \times n$ row vectors containing the ν_jth ranked variable for each of the n observations.

Variable ranking refers to the permutation of the variables. The next step, the decision on how many of the ranked variables to use, should be called **variable selection**, but the two terms are often used interchangeable. Another common synonym is **feature selection**, and the permuted variables are sometimes called **features**. Variable ranking is typically applied if we are interested in a few variables and want to separate those variables from the rest. It then makes sense if we can pick the 'few best' variables from the top, so the first-ranked variable is the 'best' with respect to the ranking.

For two-class problems with sample class mean $\overline{\mathbf{X}}_\nu$ for the νth class and common covariance matrix Σ, let S be the (pooled) sample covariance matrix, and let $D = S_{\text{diag}}$ be the matrix consisting of the diagonal entries of S. Put

$$\mathbf{d} = D^{-1/2}(\overline{\mathbf{X}}_1 - \overline{\mathbf{X}}_2) \qquad \text{and} \qquad \mathbf{b} = S^{-1/2}(\overline{\mathbf{X}}_1 - \overline{\mathbf{X}}_2). \tag{4.50}$$

The vectors $\mathbf{d}$ and $\mathbf{b}$ induce ranking schemes for $\mathbb{X}$ obtained from the ordered (absolute) entries $|d_{(1)}| \geq |d_{(2)}| \geq \cdots \geq |d_{(d)}|$ of $\mathbf{d}$ and similarly for $\mathbf{b}$. The two ranking schemes have a natural interpretation for two-class problems which are distinguished by their class means: they correspond to scaled and sphered vectors, respectively. The definitions imply that $\mathbf{b}$ takes into account the joint covariance structure of all variables, whereas $\mathbf{d}$ is based on the marginal variances only.

For normal data, $\mathbf{d}$ is the vector which leads to t-tests. The ranking induced by $\mathbf{d}$ is commonly used for HDLSS two-class problems and forms the basis of the theoretical developments of Fan and Fan (2008), which I describe in Section 13.3.3. In Example 4.12, I will use both ranking schemes of (4.50).

Another natural ranking scheme is that of Bair et al. (2006), which the authors proposed for a latent variable model in regression. They motivated their approach by arguing that in high-dimensional data, and in particular, in gene expression data, only a proper subset of the variables contributes to the regression relationship. Although the ranking scheme of Bair et al. (2006) is designed for regression responses, I will introduce it here and then return to it in the regression context of Sections 13.3.1 and 13.3.2.

Let $\begin{bmatrix}\mathbb{X}\\ \mathbf{Y}\end{bmatrix}$ be $d+1$-dimensional regression data with univariate responses $\mathbf{Y}$. In the latent variable model of Bair et al. (2006), there is a subset $\Theta = \{\nu_1, \ldots, \nu_t\} \subset \{1, \ldots, d\}$ of size $t = |\Theta|$ and a $k \times t$ matrix A with $k \leq t$ such that

$$\mathbf{Y} = \boldsymbol{\beta}^{\mathsf{T}} A \mathbb{X}_\Theta + \epsilon, \tag{4.51}$$

and $\mathbf{Y}$ does not depend on the variables X_j with $j \in \{1, \ldots, d\} \setminus \Theta$, where $\boldsymbol{\beta}$ is the k-dimensional vector of regression coefficients, and ϵ are error terms. The variable ranking of Bair et al. (2006) is based on a ranking vector $\mathbf{s} = [s_1 \cdots s_d]$, where the s_j are the marginal coefficients

$$s_j = \frac{\left|\mathbf{X}_{\bullet j}\mathbf{Y}^{\mathsf{T}}\right|}{\left\|\mathbf{X}_{\bullet j}\right\|} \qquad \text{for } j = 1, \ldots, d. \tag{4.52}$$

The ranking vector $\mathbf{s}$ yields the ranked row vectors $\mathbf{X}_{\bullet(j)}$ such that $\mathbf{X}_{\bullet(1)}$ has the strongest normalised correlation with $\mathbf{Y}$. Because their model is that of linear regression, this ranking clearly reflects their intention of finding the variables that contribute most to the regression relationship.

Bair et al. (2006) determined the subset Θ by selecting vectors $\mathbf{X}_{\bullet\nu_j}$ for which s_{ν_j} is sufficiently large and taking $A = \widehat{\Gamma}_k^{\mathsf{T}}$ in (4.51), so consider the first k eigenvectors of the sample covariance matrix S_Θ of $\mathbb{X}_\Theta$. Typically, in their approach, $k = 1$. For discrimination, the response vector $\mathbf{Y}$ is the $1 \times n$ vector of labels, and we calculate the ranking vector $\mathbf{s}$ as in (4.52) with the regression responses replaced by the vector of labels. I return to their regression framework in Section 13.2.1.

The following algorithm integrates variable ranking into discrimination and determines the number of relevant variables for discrimination. It applies to arbitrary ranking schemes and in particular to those induced by (4.50) and (4.52).

Algorithm 4.3 ***Discriminant Analysis with Variable Ranking***

Let $\mathbb{X}$ be data which belong to κ classes. Let r be the rank of the sample covariance matrix S of $\mathbb{X}$. Let $\mathfrak{r}$ be a discriminant rule, and let $\mathcal{E}$ be a classification error.

Step 1. Consider a variable-ranking scheme for $\mathbb{X}$ defined by a rule $\boldsymbol{\nu}$, and let $\mathbb{X}_{\boldsymbol{\nu}}$ be the ranked data.

Step 2. For $2 \le p \le r$, consider the p-ranked data

$$\mathbb{X}_{\boldsymbol{\nu},p} = \begin{bmatrix} \mathbf{X}_{\bullet,\nu_1} \\ \vdots \\ \mathbf{X}_{\bullet,\nu_p} \end{bmatrix},$$

which consist of the first p rows of $\mathbb{X}_{\boldsymbol{\nu}}$.

Step 3. Derive the rule $\mathfrak{r}_p$ from $\mathfrak{r}$ for the data $\mathbb{X}_{\boldsymbol{\nu},p}$, classify these p-dimensional data and calculate the error $\mathcal{E}_p$ based on $\mathbb{X}_{\boldsymbol{\nu},p}$.

Step 4. Put $p = p + 1$, and repeat steps 2 and 3.

Step 5. Find the value p^* which minimises the classification error:

$$p^* = \underset{2 \le p \le d}{\operatorname{argmin}} \, \mathcal{E}_p.$$

Then p^* is the number of variables that results in the best classification with respect to the error $\mathcal{E}$.

An important difference between Algorithms 4.2 and 4.3 is that the former uses the p-dimensional PC data, whereas the latter is based on the first p ranked variables.

Example 4.12 We continue with a classification of the **breast cancer** data. I apply Algorithm 4.3 with four variable-ranking schemes, including the original order of the variables, which I call the *identity ranking*, and compare their performance to that of Algorithm 4.2. The following list explains the notation of the five approaches and the colour choices in Figure 4.10. Items 2 to 5 of the list refer to Algorithm 4.3. In each case, p is the number of variables, and $p \le d$.

1. The PCs $\mathbb{W}^{(p)}$ of Algorithm 4.2. The results are shown as a blue line with small dots in the top row of Figure 4.10.
2. The identity ranking, which leaves the variables in the original order, leads to subsets $\mathbb{X}_{I,p}$. The results are shown as a maroon line in the top row of Figure 4.10.
3. The ranking induced by the scaled difference of the class means, $\mathbf{d}$ of (4.50). This leads to $\mathbb{X}_{\mathbf{d},p}$. The results are shown as a black line in all four panels of Figure 4.10.
4. The ranking induced by the sphered difference of the class means, $\mathbf{b}$ of (4.50). This leads to $\mathbb{X}_{\mathbf{b},p}$. The results are shown as a red line in the bottom row of Figure 4.10.
5. The ranking induced by $\mathbf{s}$ with coefficients (4.52) and $\mathbf{Y}$ the vector of labels. This leads to $\mathbb{X}_{\mathbf{s},p}$. The results are shown as a blue line in the bottom row of Figure 4.10.

Table 4.9 *Number of Misclassified Observations, # Misclass, Best Dimension, p^*, and Colour in Figure 4.10 for the Breast Cancer Data of Example 4.12.*

Approach		Raw		Scaled		Colour
#	Data	# Misclass	p^*	# Misclass	p^*	in Fig 4.10
1	$\mathbb{W}^{(p)}$	12	16	12	25	Blue and dots
2	$\mathbb{X}_{I,p}$	12	25	12	25	Maroon
3	$\mathbb{X}_{\mathbf{d},p}$	11	21	11	21	Black
4	$\mathbb{X}_{\mathbf{b},p}$	12	11	12	16	Red
5	$\mathbb{X}_{\mathbf{s},p}$	15	19	11	21	Solid blue

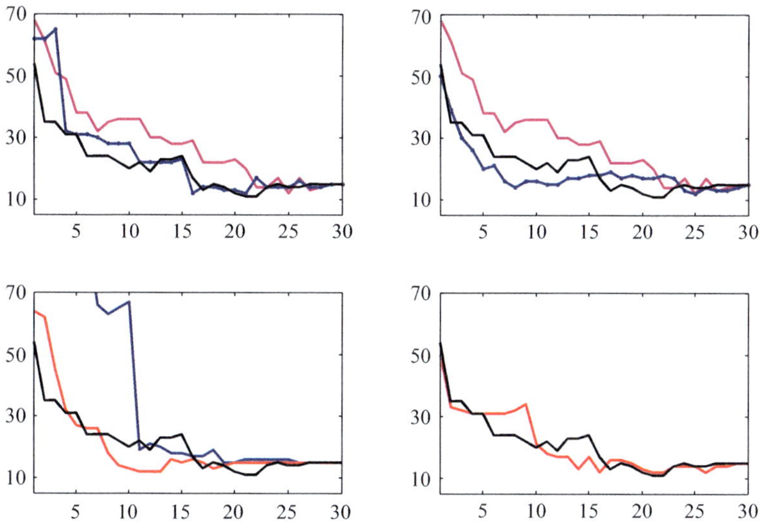

Figure 4.10 Number of misclassified observations versus dimension p (on the x-axis) for the breast cancer data of Example 4.12: raw data (*left*) and scaled data (*right*); $\mathbb{X}_{\mathbf{d},p}$ in black in all panels, $\mathbb{W}^{(p)}$ (*blue with dots*) and $\mathbb{X}_{I,p}$ (*maroon*) in the top two panels; $\mathbb{X}_{\mathbf{b},p}$ (*red*) and $\mathbb{X}_{\mathbf{s},p}$ (*solid blue*) in the bottom two panels. In the bottom-right panel, the blue and black graphs agree.

In Example 4.6 we saw that Fisher's rule misclassified fifteen observations, and the normal rule misclassified eighteen observations. We therefore use Fisher's rule in this analysis and work with differently selected subsets of the variables.

Table 4.9 reports the results of classification with the five approaches. Separately for the raw and scaled data, it lists p^*, the optimal number of observations for each method, and the number of observations that are misclassified with the best p^*. The last column 'Colour' refers to the colours used in Figure 4.10 for the different approaches.

Figure 4.10 complements the information provided in Table 4.9 and displays the number of misclassified observations for the five approaches as a function of the dimension p shown on the x-axis. The left subplots show the number of misclassified observations for the raw data, and the right subplots give the same information for the scaled data. The colours in the plots are those given in the preceding list and in the table. There is considerable discrepancy for the raw data plots but much closer agreement for the scaled data, with identical ranking vectors for the scaled data of approaches 3 and 5.

The black line is the same for the raw and scaled data because the ranking uses standardised variables. For this reason, it provides a benchmark for the other schemes. The performance of the identity ranking (in maroon) is initially worse than the others and needs a larger number p^* than the other approaches in order to reach the small error of twelve misclassified observations.

An inspection of the bottom panels shows that the blue line on the left does not perform so well. Here we have to take into account that the ranking of Bair et al. (2006) is designed for regression rather than discrimination. For the scaled data, the blue line coincides with the black line and is therefore not visible. The red line is comparable with the black line.

The table shows that the raw and scaled data lead to similar performances, but more variables are required for the scaled data before the minimum error is achieved. The ranking of $\mathbb{X}_{\mathbf{d},p}$ (in black) does marginally better than the others and coincides with $\mathbb{X}_{\mathbf{s},p}$ for the scaled data. If a small number of variables is required, then $\mathbb{X}_{\mathbf{b},p}$ (in red) does best. The compromise between smallest error and smallest dimension is interesting and leaves the user options for deciding which method is preferable. Approaches 3 and 4 perform better than 1 and 2: The error is smallest for approach 3, and approach 4 results in the most parsimonious model.

In summary, the analysis shows that

1. variable ranking prior to classification
   (a) reduced misclassification, and
   (b) leads to a more parsimonious model and that
2. the PC data with sixteen PCs reduces the classification error from fifteen to twelve. ■

The analysis of the *breast cancer* data shows that no single approach yields best results, but variable ranking and a suitably chosen number of variables can improve classification. There is a trade-off between the smallest number of misclassified observations and the most parsimonious model which gives the user extra flexibility. If time and resources allow, I recommend applying more than one method and, if appropriate, applying different approaches to the raw and scaled data.

In Section 13.3 we continue with variable ranking and variable selection and extend the approaches of this chapter to HDLSS data.

Problems for Part I

Principal Component Analysis

1. Show part 2 of Proposition 2.1.
2. For $n = 100, 200, 500$ and $1,000$, simulate data from the true distribution referred to in Example 2.3. Separately for each n carry out parts (a)–(c).
   (a) Calculate the sample eigenvalues and eigenvectors.
   (b) Calculate the two-dimensional PC data, and display them.
   (c) Compare your results with those of Example 2.3 and comment.
   (d) For $n = 100$, repeat the simulation 100 times, and save the eigenvalues you obtained from each simulation. Show a histogram of the eigenvalues (separately for each eigenvalue), and calculate the sample means of the eigenvalues. Compare the means to the eigenvalues of the true covariance matrix and comment.
3. Let $\mathbf{X}$ satisfy the assumptions of Theorem 2.5. Let A be a non-singular matrix of size $d \times d$, and let $\mathbf{a}$ be a fixed d-dimensional vector. Put $\mathbf{T} = A\mathbf{X} + \mathbf{a}$, and derive expressions for
   (a) the mean and covariance matrix of T, and
   (b) the mean and covariance matrix of the PC vectors of T.
4. For each of the fourteen subjects of the *HIV flow cytometry* data described in Example 2.4,
   (a) calculate the first and second principal component scores for all observations,
   (b) display scatterplots of the first and second principal component scores and
   (c) show parallel coordinate views of the first and second PCs together with their respective density estimates.

   Based on your results of parts (a)–(c), determine whether there are differences between the HIV$^+$ and HIV$^-$ subjects.
5. Give an explicit proof of part 2 of Theorem 2.12.
6. The MATLAB commands *svd, princomp* and *pcacov* can be used to find eigenvectors and singular values or eigenvalues of the data $\mathbb{X}$ or its covariance matrix S. Examine and explain the differences between these MATLAB functions, and illustrate with data that are at least of size 3×5.
7. Let $\mathbf{X}$ and $\mathbf{p}_j$ satisfy the assumptions of Theorem 2.12, and let H_j be the matrix of (2.9) for $j \le d$. Let λ_k be the eigenvalues of Σ.
   (a) Show that H_j satisfies

$$\Sigma H_j = \lambda_j H_j, \qquad H_j \Sigma H_j = \lambda_j H_j \qquad \text{and} \qquad \operatorname{tr}(H_j H_j) = 1.$$

(b) Assume that Σ has rank d. Put $H = \sum_{k=1}^{d} \lambda_k H_k$. Show that $H = \Gamma\Lambda\Gamma^{\mathsf{T}}$, the spectral decomposition of Σ as defined in (1.19) of Section 1.5.2.

8. Prove Theorem 2.17.
9. Let $\mathbb{X}$ be $d \times n$ data with $d > n$. Let S be the sample covariance matrix of $\mathbb{X}$. Put $Q = (n-1)^{-1}\mathbb{X}^{\mathsf{T}}\mathbb{X}$.

   (a) Determine the relationship between the spectral decompositions of S and Q.
   (b) Relate the spectral decomposition of Q to the singular value decomposition of $\mathbb{X}$.
   (c) Explain how Q instead of S can be used in a principal component analysis of HDLSS data.

10. Let $\mathbb{X}$ be $d \times n$ data. Explain the difference between scaling and sphering of the data. Compare the PCs of raw, scaled and sphered data theoretically, and illustrate with an example of data with $d \geq 3$ and $n \geq 5$.
11. Consider the *breast cancer* data and the *Dow Jones returns*.

    (a) Calculate the first eight principal components for the Dow Jones returns and show PC score plots of these components.
    (b) Repeat part (a) for the raw and scaled breast cancer data.
    (c) Comment on the figures obtained in parts (a) and (b).

12. Under the assumptions of Theorem 2.20, give an explicit expression for a test statistic which tests whether the jth eigenvalue λ_j of a covariance matrix Σ with rank r is zero. Apply this test statistic to the *abalone* data to determine whether the smallest eigenvalue could be zero. Discuss your inference results.
13. Give a proof of Corollary 2.27.
14. For the *abalone* data, derive the classical Linear Regression solution, and compare it with the solution given in Example 2.19:

    (a) Determine the best linear least squares model, and calculate the least squares residuals.
    (b) Determine the best univariate predictor, and calculate its least squares residuals.
    (c) Consider and calculate a ridge regression model as in (2.39).
    (d) Use the covariance matrix $\widehat{C}_p$ of (2.34) and the PC-based estimator of (2.43) to determine the least-squares residuals of this model.
    (e) Compare the different approaches, and comment on the results.

15. Give an explicit derivation of the regression estimator $\widehat{\boldsymbol{\beta}}_P$ shown in the diagram in Section 2.8.2, and determine its relationship with $\widehat{\boldsymbol{\beta}}_W$.
16. For the *abalone* data of Example 2.19, calculate the least-squares estimate $\widehat{\boldsymbol{\beta}}_{\mathrm{LS}}$, and carry out tests of significance for all explanatory variables. Compare your results with those obtained from Principal Component Regression and comment.
17. Consider the model given in (2.45) and (2.46). Assume that the matrix of latent variables $\mathbb{F}$ is of size $p \times n$ and p is smaller than the rank of $\mathbb{X}$. Show that $\mathbb{F}$ is the matrix which consists of the first p sphered principal component vectors of $\mathbb{X}$.
18. For the *income* data described in Example 3.7, use the nine variables described there with the income variable as response. Using the first 1,000 records, calculate and compare the regression prediction obtained with least squares and principal components. For the latter case, examine $\kappa = 1,\ldots,8$, and discuss how many predictors are required for good prediction.

Canonical Correlation Analysis

19. Prove part 3 of Proposition 3.1.
20. For the *Swiss bank notes* data of Example 2.5, use the length and height variables to define $\mathbb{X}^{[1]}$ and the distances of frame and the length of the diagonal to define $\mathbb{X}^{[2]}$.

    (a) For $\mathbb{X}^{[1]}$ and $\mathbb{X}^{[2]}$, calculate the between covariance matrix S_{12} and the matrix of canonical correlations C.
    (b) Calculate the left and right eigenvectors $\mathbf{p}$ and $\mathbf{q}$ of C, and comment on the weights for each variable.
    (c) Calculate the three canonical transforms. Compare the weights of these canonical projections with the weights obtained in part (b).
    (d) Calculate and display the canonical correlation scores, and comment on the strength of the correlation.

21. Give a proof of Proposition 3.8.
22. Consider (3.19). Find the norms of the canonical transforms $\boldsymbol{\varphi}_k$ and $\boldsymbol{\psi}_k$, and derive the two equalities. Determine the singular values of Σ_{12}, and state the relationship between the singular values of Σ_{12} and C.
23. Consider the *abalone* data of Examples 2.13 and 2.17. Use the three length measurements to define $\mathbb{X}^{[1]}$ and the weight measurements to define $\mathbb{X}^{[2]}$. The abalone data have 4,177 records. Divide the data into four subsets which consist of the observations 1–1,000, 1,001–2,000, 2,001–3,000 and 3,001–4177. For each of the data, subsets and for the complete data

    (a) calculate the eigenvectors of the matrix of canonical correlations,
    (b) calculate and display the canonical correlation scores, and
    (c) compare the results from the subsets with those of the complete data and comment.

24. Prove part 3 of Theorem 3.10.
25. For the *Boston housing* data of Example 3.5, calculate and compare the weights of the eigenvectors of C and those of the canonical transforms for all four CCs and comment.
26. Let $\mathbf{T}^{[1]}$ and $\mathbf{T}^{[2]}$ be as in Theorem 3.11.

    (a) Show that the between covariance matrix of $\mathbf{T}^{[1]}$ and $\mathbf{T}^{[2]}$ is $A_1 \Sigma_{12} A_2^{\top}$.
    (b) Prove part 3 of Theorem 3.11.

27. Consider all eight variables of the *abalone* data of Example 2.19.

    (a) Determine the correlation coefficients between the *number of rings* and each of the other seven variables. Which variable is most strongly correlated with the number of rings?
    (b) Consider the set-up of $\mathbb{X}^{[1]}$ and $\mathbb{X}^{[2]}$ as in Problem 23. Let $\mathbb{X}^{[1a]}$ consist of $\mathbb{X}^{[1]}$ and the number of rings. Carry out a canonical correlation analysis for $\mathbb{X}^{[1a]}$ and $\mathbb{X}^{[2]}$. Next let $\mathbb{X}^{[2a]}$ consist of $\mathbb{X}^{[2]}$ and the number of rings. Carry out a canonical correlation analysis for $\mathbb{X}^{[1]}$ and $\mathbb{X}^{[2a]}$.
    (c) Let $\mathbb{X}^{[1]}$ be the number of rings, and let $\mathbb{X}^{[2]}$ be all other variables. Carry out a canonical correlation analysis for this set-up.
    (d) Compare the results of parts (a)–(c).
    (e) Compare the results of part (c) with those obtained in Example 2.19 and comment.

28. A canonical correlation analysis for scaled data is described in Section 3.5.3. Give an explicit expression for the canonical correlation matrix of the scaled data. How can we interpret this matrix?
29. For $\rho = 1, 2$, consider random vectors $\mathbf{X}^{[\rho]} \sim \left(\boldsymbol{\mu}_\rho, \Sigma_\rho\right)$. Assume that Σ_1 has full rank d_1 and spectral decomposition $\Sigma_1 = \Gamma_1 \Lambda_1 \Gamma_1^{\mathsf{T}}$. Let $\mathbf{W}^{[1]}$ be the d_1-dimensional PC vector of $\mathbf{X}^{[1]}$. Let C be the matrix of canonical correlations, and assume that $\operatorname{rank}(C) = d_1$. Further, let $\mathbf{U}$ be the d_1-variate CC vector derived from $\mathbf{X}^{[1]}$. Show that there is an orthogonal matrix E such that

$$\mathbf{U} = E\mathbf{W}_{\Lambda}^{[1]}, \tag{4.53}$$

where $\mathbf{W}_{\Lambda}^{[1]}$ is the vector of sphered principal components. Give an explicit expression for E. Is E unique?
30. For $\rho = 1, 2$, consider $\mathbf{X}^{[\rho]} \sim \left(\boldsymbol{\mu}_\rho, \Sigma_\rho\right)$. Fix $k \le d_1$ and $\ell \le d_2$. Let $\mathbf{W}^{(k,\ell)}$ be the $(k+\ell)$-variate vector whose first k entries are those of the k-dimensional PC vector $\mathbf{W}^{[1]}$ of $\mathbf{X}^{[1]}$ and whose remaining ℓ entries are those of the ℓ-dimensional PC vector $\mathbf{W}^{[2]}$ of $\mathbf{X}^{[2]}$.
    (a) Give an explicit expression for $\mathbf{W}^{(k,\ell)}$, and determine its covariance matrix in terms of the covariance matrices of the $\mathbf{X}^{[\rho]}$.
    (b) Derive explicit expressions for the canonical correlation matrix $C_{k,\ell}$ of $\mathbf{W}^{[1]}$ and $\mathbf{W}^{[2]}$ and the corresponding scores $\mathbf{U}_k$ and $\mathbf{V}_\ell$ in terms of the corresponding properties of the $\mathbf{X}^{[\rho]}$.
31. Let $\mathbf{X} = \left[\mathbf{X}^{[1]}, \mathbf{X}^{[2]}\right]^{\mathsf{T}} \sim \mathcal{N}(\boldsymbol{\mu}, \Sigma)$ with Σ as in (3.1), and assume that Σ_2^{-1} exists. Consider the conditional random vector $T = (\mathbf{X}^{[1]} \mid \mathbf{X}^{[2]})$. Show that
    (a) $\mathbb{E}T = \boldsymbol{\mu}_1 + \Sigma_{12}\Sigma_2^{-1}(\mathbf{X}^{[2]} - \boldsymbol{\mu}_2)$, and
    (b) $\operatorname{var}(T) = \Sigma_1 - \Sigma_{12}\Sigma_2^{-1}\Sigma_{12}^{\mathsf{T}}$.

    Hint: Consider the joint distribution of $A(\mathbf{X} - \boldsymbol{\mu})$, where

$$A = \begin{pmatrix} \mathbf{I}_{d_1 \times d_1} & -\Sigma_{12}\Sigma_2^{-1} \\ \mathbf{0}_{(d_2 \times d_1)} & \mathbf{I}_{d_2 \times d_2} \end{pmatrix}.$$

32. Consider the hypothesis tests for Principal Component Analysis of Section 2.7.1 and for Canonical Correlation Analysis of Section 3.6.
    (a) List and explain the similarities and differences between them, including how the eigenvalues or singular values are interpreted in each case.
    (b) Explain how each hypothesis test can be used.
33. (a) Split the *abalone* data into two groups as in Problem 23. Carry out appropriate tests of independence at the $\alpha = 2$ per cent significance level – first for all records and then for the first 100 records only. In each case, state the degrees of freedom of the test statistic. State the conclusions of the tests.
    (b) Suppose that you have carried out a test of independence for a pair of canonical correlation scores with $n = 100$, and suppose that the null hypothesis of this test was accepted at the $\alpha = 2$ per cent significance level. Would the conclusion of the test remain the same if the significance level changed to $\alpha = 5$ per cent? Would you expect the same conclusion as in the first case for $n = 500$ and $\alpha = 2$ per cent? Justify your answer.

34. Let $\mathbb{X}$ and $\mathbb{Y}$ be data matrices in d and q dimensions, respectively, and assume that $\mathbb{X}\mathbb{X}^{\mathsf{T}}$ is invertible. Let $\widehat{C}$ be the sample matrix of canonical correlations for $\mathbb{X}$ and $\mathbb{Y}$. Show that (3.37) holds.
35. For the *income* data, compare the strength of the correlation resulting from PCR and CCR by carrying out the following steps. First, find the variable in each group which is best predicted by the other group, where 'best' refers to the absolute value of the correlation coefficient. Then carry out analyses analogous to those described in Example 3.8. Finally, interpret your results.
36. For $\kappa \leq r$, prove the equality of the population expressions corresponding to (3.38) and (3.42). *Hint:* Consider $\kappa = 1$, and prove the identity.

Discriminant Analysis

37. For the *Swiss bank notes* data, define class $\mathcal{C}_1$ to be the genuine notes and class $\mathcal{C}_2$ to be the counterfeits. Use Fisher's rule to classify these data. How many observations are misclassified? How many genuine notes are classified as counterfeits, and how many counterfeits are regarded as genuine?
38. Explain and highlight the similarities and differences between the best directions chosen in Principal Component Analysis, Canonical Correlation Analysis and Discriminant Analysis. What does each of the directions capture? Demonstrate with an example.
39. Consider the *wine recognition* data of Example 4.5. Use all observations and two cultivars at a time for parts (a) and (b).

    (a) Apply Fisher's rule to these data. Compare the performance on these data with the classification results reported in Example 4.5.
    (b) Determine the leave-one-out performance based on Fisher's rule. Compare the results with those obtained in Example 4.5 and in part (a).

40. Theorem 4.6 is a special case of the generalised eigenvalue problem referred to in Section 3.7.4. Show how Principal Component Analysis and Canonical Correlation Analysis fit into this framework, and give an explicit form for the matrices A and B referred to in Section 3.7.4. *Hint:* You may find the paper by Borga, Knutsson, and Landelius (1997) useful.
41. Generate 500 random samples from the first distribution given in Example 4.3. Use the normal linear rule to calculate decision boundaries for these data, and display the boundaries together with the data in a suitable plot.
42. (a) Consider two classes $\mathcal{C}_\ell = \mathcal{N}(\mu_\ell, \sigma^2)$ with $\ell = 1, 2$ and $\mu_1 < \mu_2$. Determine a likelihood-based discriminant rule $\mathfrak{r}_{\mathrm{norm}}$ for a univariate X as in (4.14) which assigns X to class $\mathcal{C}_1$.
    (b) Give a proof of Theorem 4.10 for the multivariate case. *Hint:* Make use of the relationship

$$2\mathbf{X}^{\mathsf{T}}\left[\Sigma^{-1}(\boldsymbol{\mu}_1 - \boldsymbol{\mu}_2)\right] > \boldsymbol{\mu}_1^{\mathsf{T}}\Sigma^{-1}\boldsymbol{\mu}_1 - \boldsymbol{\mu}_2^{\mathsf{T}}\Sigma^{-1}\boldsymbol{\mu}_2. \tag{4.54}$$

43. Consider the *breast cancer* data. Calculate and compare the leave-one-out error based on both the normal rule and Fisher's rule. Are the conclusions the same as for the classification error?

44. Consider classes $\mathcal{C}_\nu = \mathcal{N}(\boldsymbol{\mu}_\nu, \Sigma)$ with $1 \le \nu \le \kappa$ and $\kappa \ge 2$, which differ in their means but have the same covariance matrix Σ.
    (a) Show that the likelihood-based rule (4.26) is equivalent to the rule defined by
    $$\left[\mathbf{X} - \frac{1}{2}\boldsymbol{\mu}_\ell\right]^{\mathsf{T}} \Sigma^{-1}\boldsymbol{\mu}_\ell = \max_{1\le\nu\le\kappa} \left[\mathbf{X} - \frac{1}{2}\boldsymbol{\mu}_\nu\right]^{\mathsf{T}} \Sigma^{-1}\boldsymbol{\mu}_\nu.$$
    (b) Deduce that the two rules $\mathfrak{r}_{\text{norm}}$ of (4.25) and $\mathfrak{r}_{\text{norm1}}$ of (4.26) are equivalent if the random vectors have a common covariance matrix Σ.
    (c) Discuss how the two rules apply to data, and illustrate with an example.
45. Consider the *glass* identification data set from the Machine Learning Repository (http://archive.ics.uci.edu/ml/datasets/Glass+Identification) The glass data consist of 214 observations in nine variables. These observations belong to seven classes.
    (a) Use Fisher's rule to classify the data.
    (b) Combine classes in the following way. Regard *building windows* as one class, *vehicle windows* as one class, and the remaining three as a third class. Perform classification with Fisher's rule on these data, and compare your results with those of part (a).
46. Consider two classes $\mathcal{C}_1$ and $\mathcal{C}_2$ which have different means but the same covariance matrix Σ. Let $\mathfrak{r}_{\text{F}}$ be Fisher's discriminant rule for these classes.
    (a) Show that B and W equal the expressions given in (4.16).
    (b) Show that $\boldsymbol{\eta}$ given by (4.16) is the maximiser of $W^{-1}B$.
    (c) Determine an expression for the decision function h_F which is given in terms of $\boldsymbol{\eta}$ as in (4.16), and show that it differs from the normal linear decision function by a constant. What is the value of this constant?
47. Starting from the likelihood function of the random vector, give an explicit proof of Corollary 4.17.
48. Use the first two dimensions of Fisher's *iris* data and the normal linear rule to determine decision boundaries, and display the boundaries together with the data. *Hint:* For calculation of the rule, use the sample means, and the pooled covariance matrix.
49. Prove part 2 of Theorem 4.19.
50. Consider two classes given by Poisson distributions with different values of the parameter λ. Find the likelihood-based discriminant rule which assigns a random variable the value 1 if $L_1(X) > L_2(X)$.
51. Prove part 2 of Theorem 4.21.
52. Consider the *abalone* data. The first variable, *sex*, which we have not considered previously, has three groups, M, F and I (for infant). It is important to distinguish between the mature abalone and the infant abalone, and it is therefore natural to divide the data into two classes: observations M and F belong to one class, and observations with label I belong to the second class.
    (a) Apply the normal linear and the quadratic rules to the abalone data.
    (b) Apply the rule based on the regularised covariance matrix $S_\nu(\alpha)$ of Friedman (1989) (see Section 4.7.3) to the abalone data, and find the optimal α.
    (c) Compare the results of parts (a) and (b) and comment.
53. Consider the *Swiss bank notes* data, with the two classes as in Problem 37.

(a) Classify these data with the nearest-neighbour rule for a range of values k. Which k results in the smallest classification error?
(b) Classify these data with the logistic regression approach. Discuss which values of the probability are associated with each class.
(c) Compare the results in parts (a) and (b), and also compare these with the results obtained in Problem 37.

54. Consider the three classes of the *wine recognition* data.
(a) Apply Principal Component Discriminant Analysis to these data, and determine the optimal number p^* of PCs for Fisher's rule.
(b) Repeat part (a), but use the normal linear rule.
(c) Compare the results of parts (a) and (b), and include in the comparison the results obtained in Example 4.7.

Part II

Factors and Groupings

5

Norms, Proximities, Features and Dualities

Get your facts first, and then you can distort them as much as you please (Mark Twain, 1835–1910).

5.1 Introduction

The first part of this book dealt with the three classical problems: finding structure within data, determining relationships between different subsets of variables and dividing data into classes. In Part II, we focus on the first of the problems, finding structure – in particular, groups or factors – in data. The three methods we explore, Cluster Analysis, Factor Analysis and Multidimensional Scaling, are classical in their origin and were developed initially in the behavioural sciences. They have since become indispensable tools in diverse areas including psychology, psychiatry, biology, medicine and marketing, as well as having become mainstream statistical techniques. We will see that Principal Component Analysis plays an important role in these methods as a preliminary step in the analysis or as a special case within a broader framework.

Cluster Analysis is similar to Discriminant Analysis in that one attempts to partition the data into groups. In biology, one might want to determine specific cell subpopulations. In archeology, researchers have attempted to establish taxonomies of stone tools or funeral objects by applying cluster analytic techniques. Unlike Discriminant Analysis, however, we do not know the class membership of any of the observations. The emphasis in Factor Analysis and Multidimensional Scaling is on the interpretability of the data in terms of a small number of meaningful descriptors or dimensions. Based, for example, on the results of intelligence tests in mathematics, language and so on, we might want to dig a little deeper and find *hidden* kinds of intelligence. In Factor Analysis, we would look at scores from test results, whereas Multidimensional Scaling starts with the proximity between the observed objects, and the structure in the data is derived from this knowledge.

Historically, Cluster Analysis, Factor Analysis and Multidimensional Scaling dealt with data of moderate size, a handful of variables and maybe some hundreds of samples. Renewed demand for these methods has arisen in the analysis of very large and complex data such as gene expression data, with thousands of dimensions and a much smaller number of samples. The resulting challenges have led to new and exciting research within these areas and have stimulated the development of new methods such as Structural Equation Modelling, Projection Pursuit and Independent Component Analysis.

As in Chapter 1, I give definitions and state results without proof in this preliminary chapter of Part II. We start with norms of a random vector and a matrix and then consider measures of closeness, or proximity, which I define for pairs of random vectors. The final section summarises notation and relates the matrices $\mathbb{X}\mathbb{X}^\top$ and $\mathbb{X}^\top\mathbb{X}$.

Unless otherwise specified, we use the convention

$$\mathbf{X}_i = \begin{bmatrix} X_{i1} \cdots X_{id} \end{bmatrix}^\top \quad \text{for a } d\text{-dimensional random vector with } i = 1, 2, \ldots, n, \text{ and}$$

$$A = \begin{bmatrix} \mathbf{a}_1 \cdots \mathbf{a}_n \end{bmatrix} = \begin{bmatrix} \mathbf{a}_{\bullet 1} \\ \vdots \\ \mathbf{a}_{\bullet d} \end{bmatrix}$$

for a $d \times n$ matrix with entries a_{ij}, column vectors $\mathbf{a}_i$ and row vectors $\mathbf{a}_{\bullet j}$.

5.2 Vector and Matrix Norms

Definition 5.1 Let $\mathbf{X}$ be a d-dimensional random vector.

1. For $p = 1, 2, \ldots$, the ℓ_p **norm** of $\mathbf{X}$ is $\|\mathbf{X}\|_p = \left(\sum_{k=1}^d |X_k|^p\right)^{1/p}$. Special cases are
   (a) the **Euclidean norm** or ℓ_2 **norm**: $\|\mathbf{X}\|_2 = \left(\mathbf{X}^\top\mathbf{X}\right)^{1/2}$, and
   (b) the ℓ_1 **norm**: $\|\mathbf{X}\|_1 = \sum_{k=1}^d |X_k|$.
2. The ℓ_∞ **norm** or the **sup norm** of $\mathbf{X}$ is $\|\mathbf{X}\|_\infty = \max_{1\le k\le d} |X_k|$. □

We write $\|\mathbf{X}\|$ instead of $\|\mathbf{X}\|_2$ for the Euclidean norm of $\mathbf{X}$ and use the subscript notation only to avoid ambiguity.

For vectors $\mathbf{a} \in \mathbb{R}^d$, Definition 5.1 reduces to the usual ℓ_p-norms.

Definition 5.2 Let A be a $d \times n$ matrix.

1. The **sup norm** of A is $\|A\|_{\text{sup}} = |\lambda_1|$, the largest singular value, in modulus, of A.
2. The **Frobenius norm** of A is $\|A\|_{\text{Frob}} = [\operatorname{tr}(AA^\top)]^{1/2}$. □

Observe that

$$\|A\|_{\text{sup}} = \sup_{\{\mathbf{e}\in R^n:\, \|\mathbf{e}\|=1\}} \|A\mathbf{e}\|_2,$$

$$\|A\|_{\text{Frob}} = \left[\operatorname{tr}(A^\top A)\right]^{1/2},$$

$$\|A\|_{\text{Frob}}^2 = \sum_{i=1}^n \|\mathbf{a}_i\|_2^2 = \sum_{j=1}^d \left\|\mathbf{a}_{\bullet j}\right\|_2^2. \tag{5.1}$$

The square of the Frobenius norm is the sum of all squared entries of the matrix A. This follows from (5.1).

5.3 Measures of Proximity

5.3.1 Distances

In Section 4.7.1 we considered k-nearest neighbour methods based on the Euclidean distance. We begin with the definition of a distance and then consider common distances.

Definition 5.3 For $i = 1, 2, \ldots$, let $\mathbf{X}_i$ be d-dimensional random vectors. A **distance** Δ is a map defined on pairs of random vectors $\mathbf{X}_i, \mathbf{X}_j$ such that $\Delta(\mathbf{X}_i, \mathbf{X}_j)$ is a positive random variable which satisfies

1. $\Delta(\mathbf{X}_i, \mathbf{X}_j) = \Delta(\mathbf{X}_j, \mathbf{X}_i) \geq 0$ for all i, j,
2. $\Delta(\mathbf{X}_i, \mathbf{X}_j) = 0$ if $i = j$, and
3. $\Delta(\mathbf{X}_i, \mathbf{X}_j) \leq \Delta(\mathbf{X}_i, \mathbf{X}_k) + \Delta(\mathbf{X}_k, \mathbf{X}_j)$ for all i, j and k.

Let $\mathbb{X} = \begin{bmatrix}\mathbf{X}_1 \dots \mathbf{X}_n\end{bmatrix}$ be d-dimensional data. We call Δ a **distance** for $\mathbb{X}$ if Δ is defined for all pairs of random vectors belonging to $\mathbb{X}$.

A distance Δ is called a **metric** if 2 is replaced by $\Delta(\mathbf{X}_i, \mathbf{X}_j) = 0$ if and only if $i = j$. □

For vectors $\mathbf{a} \in \mathbb{R}^d$, Definition 5.3 reduces to the usual real-valued distance function.

In the remainder of Section 5.3, I will refer to pairs of random vectors as $\mathbf{X}_1$ and $\mathbf{X}_2$ or sometimes as $\mathbf{X}_i$ and $\mathbf{X}_j$ without explicitly defining these pairs each time I use them.

Probably the two most common distances in statistics are

- the **Euclidean distance** Δ_E, which is defined by

$$\Delta_E(\mathbf{X}_1, \mathbf{X}_2) = \left[(\mathbf{X}_1 - \mathbf{X}_2)^{\mathsf{T}}(\mathbf{X}_1 - \mathbf{X}_2)\right]^{1/2}, \tag{5.2}$$

- the **Mahalanobis distance** Δ_M, which requires an invertible common covariance matrix Σ, and is defined by

$$\Delta_M(\mathbf{X}_1, \mathbf{X}_2) = \left[(\mathbf{X}_1 - \mathbf{X}_2)^{\mathsf{T}}\Sigma^{-1}(\mathbf{X}_1 - \mathbf{X}_2)\right]^{1/2}. \tag{5.3}$$

From Definition 5.1, it follows that

$$\Delta_E(\mathbf{X}_1, \mathbf{X}_2) = \|\mathbf{X}_1 - \mathbf{X}_2\|_2.$$

The Euclidean distance is a special case of the Mahalanobis distance. The two distances coincide for random vectors with identity covariance matrix, and hence they agree for sphered vectors. When the true covariance matrix is not known, we use the sample covariance matrix but keep the name Mahalanobis distance.

The following are common distances.

- For $p = 1, 2, \dots$ and positive w_j, the **weighted p-distance** or **Minkowski distance** Δ_p is

$$\Delta_p(\mathbf{X}_1, \mathbf{X}_2) = \left[\sum_{j=1}^{d} w_j |X_{1j} - X_{2j}|^p\right]^{1/p}. \tag{5.4}$$

- The **max distance** or **Chebychev distance** $\Delta_{\max}$ is

$$\Delta_{\max}(\mathbf{X}_1, \mathbf{X}_2) = \max_{j=1,\dots,d} \left|X_{1j} - X_{2j}\right|. \tag{5.5}$$

- The **Canberra distance** Δ_{Canb} is defined for random vectors with positive entries by

$$\Delta_{\text{Canb}}(\mathbf{X}_1, \mathbf{X}_2) = \sum_{j=1}^{d} \frac{|X_{1j} - X_{2j}|}{X_{1j} + X_{2j}}.$$

- The **Bhattacharyya distance** Δ_{Bhat} is

$$\Delta_{\text{Bhat}}(\mathbf{X}_1, \mathbf{X}_2) = \sum_{j=1}^{d} (X_{1j}^{1/2} - X_{2j}^{1/2})^2.$$

For $p = 1$, (5.4) is called the **Manhattan metric**. If, in addition, all weights are one, then (5.4) is called the **city block distance**. If the weights $w_j = \sigma_j^{-2}$, where the σ_j^2, are the diagonal entries of Σ and $p = 2$, then (5.4) is a special case of the Mahalanobis distance (5.3) and is called the **Pearson distance**.

The next group of distances is of special interest as the dimension increases.

- The **cosine distance** $\Delta_{\cos}$ is

$$\Delta_{\cos}(\mathbf{X}_1, \mathbf{X}_2) = 1 - \cos(\mathbf{X}_1, \mathbf{X}_2), \tag{5.6}$$

  where $\cos(\mathbf{X}_1, \mathbf{X}_2)$ is the cosine of the included angle of the two random vectors.
- The **correlation distance** Δ_{cor} is

$$\Delta_{\mathrm{cor}}(\mathbf{X}_1, \mathbf{X}_2) = 1 - \rho(\mathbf{X}_1, \mathbf{X}_2), \tag{5.7}$$

  where $\rho(\mathbf{X}_1, \mathbf{X}_2)$ is the correlation coefficient of the random vectors.
- The **Spearman distance** Δ_{Spear} is

$$\Delta_{\mathrm{Spear}}(\mathbf{X}_1, \mathbf{X}_2) = 1 - \rho_S(\mathbf{X}_1, \mathbf{X}_2), \tag{5.8}$$

  where $\rho_S(\mathbf{X}_1, \mathbf{X}_2)$ is Spearman's ranked correlation of the two random vectors.
- For binary random vectors with entries 0 and 1, the **Hamming distance** Δ_{Hamm} is

$$\Delta_{\mathrm{Hamm}}(\mathbf{X}_1, \mathbf{X}_2) = \frac{\#\{X_{1j} \neq X_{2j} : j \leq d\}}{d}. \tag{5.9}$$

The list of distances is by no means exhaustive but will suffice for our purposes.

5.3.2 Dissimilarities

Distances are symmetric in the two random vectors. If, however, $\Delta(\mathbf{X}_1, \mathbf{X}_2) \neq \Delta(\mathbf{X}_2, \mathbf{X}_1)$, then a new notion is required.

Definition 5.4 For $i = 1, 2, \ldots$, let $\mathbf{X}_i$ be d-dimensional random vectors. A **dissimilarity** or **dissimilarity measure** ϱ is a map which is defined for pairs of random vectors $\mathbf{X}_i, \mathbf{X}_j$ such that $\varrho(\mathbf{X}_i, \mathbf{X}_j) = 0$ if $i = j$.

For d-dimensional data $\mathbb{X}$, ϱ is a **dissimilarity** for $\mathbb{X}$ if ϱ is defined for all pairs of random vectors belonging to $\mathbb{X}$. □

Distances are dissimilarities, but the converse does not hold, because dissimilarities can be negative. Hartigan (1967) and Cormack (1971) list frameworks which give rise to very general dissimilarities, which I shall not pursue.

Dissimilarities ϱ which are not distances:

- For random vectors $\mathbf{X}_1$ and $\mathbf{X}_2$ with $\|\mathbf{X}_1\|_2 \neq 0$, put

$$\varrho(\mathbf{X}_1, \mathbf{X}_2) = \frac{\|\mathbf{X}_1 - \mathbf{X}_2\|_2}{\|\mathbf{X}_1\|_2}. \tag{5.10}$$

- For d-variate vectors $\mathbf{X}_1$ and $\mathbf{X}_2$ with strictly positive entries, put

$$\varrho(\mathbf{X}_1, \mathbf{X}_2) = \sum_{j=1}^{d} X_{1j} \log\left(\frac{X_{1j}}{X_{2j}}\right). \tag{5.11}$$

For random vectors $\mathbf{X}_1, \ldots, \mathbf{X}_n$ belonging to any of a number of distinct groups,

$$\varrho(\mathbf{X}_i, \mathbf{X}_j) = \begin{cases} 0 & \text{if } \mathbf{X}_i \text{ and } \mathbf{X}_j \text{ belong to the same group,} \\ 1 & \text{otherwise} \end{cases}$$

defines a dissimilarity, which is a distance but not a metric. If the groups are ordered and $\mathbf{X}_i$ belongs to group ℓ, then

$$\varrho(\mathbf{X}_i, \mathbf{X}_j) = \begin{cases} k & \text{if } \mathbf{X}_j \text{ belongs to group } \ell + k, \\ -k & \text{if } \mathbf{X}_j \text{ belongs to group } \ell - k. \end{cases} \tag{5.12}$$

It is easy to verify that (5.12) is a dissimilarity but not a distance.

5.3.3 Similarities

Distances and dissimilarities are measures which tell us how 'distant' or far apart two random vectors are. Two of the distances, the cosine distance $\Delta_{\cos}$ of (5.6) and the correlation distance Δ_{cor} of (5.7), measure distance by the deviation of the cosine and the correlation coefficient from the maximum value of 1. Thus, instead of considering the distance $1 - \cos(\mathbf{X}_1, \mathbf{X}_2)$, we could directly consider $\cos(\mathbf{X}_1, \mathbf{X}_2)$.

Definition 5.5 For $i = 1, 2, \ldots$, let $\mathbf{X}_i$ be d-dimensional random vectors. A **similarity** or **similarity measure** ς is defined for pairs of random vectors $\mathbf{X}_i, \mathbf{X}_j$. The random variable $\varsigma(\mathbf{X}_i, \mathbf{X}_j)$ satisfies

1. $\varsigma(\mathbf{X}_i, \mathbf{X}_i) \geq 0$, and
2. $[\varsigma(\mathbf{X}_i, \mathbf{X}_j)]^2 \leq \varsigma(\mathbf{X}_i, \mathbf{X}_i)\varsigma(\mathbf{X}_j, \mathbf{X}_j)$. □

Often the similarity ς is symmetric in the two arguments.
Examples of similarities are

- The **cosine similarity**

$$\varsigma(\mathbf{X}_1, \mathbf{X}_2) = \cos(\mathbf{X}_1, \mathbf{X}_2). \tag{5.13}$$

- The **correlation similarity**

$$\varsigma(\mathbf{X}_1, \mathbf{X}_2) = \rho(\mathbf{X}_1, \mathbf{X}_2). \tag{5.14}$$

- The **polynomial similarity** which is defined for $k = 1, 2, \ldots$ by

$$\varsigma(\mathbf{X}_1, \mathbf{X}_2) = \langle \mathbf{X}_1, \mathbf{X}_2 \rangle^k = (\mathbf{X}_1^{\mathsf{T}} \mathbf{X}_2)^k. \tag{5.15}$$

- The **Gaussian similarity** which is defined for $a > 0$ by

$$\varsigma(\mathbf{X}_1, \mathbf{X}_2) = \exp(-\|\mathbf{X}_1 - \mathbf{X}_2\|^2 / a). \tag{5.16}$$

For $k = 1$, (5.15) reduces to $\varsigma_k(\mathbf{X}_1, \mathbf{X}_2) = \mathbf{X}_1^{\mathsf{T}} \mathbf{X}_2$, the scalar product or inner product of the random vectors. If we desire that the similarity has a maximum of 1, then we put

$$\varsigma(\mathbf{X}_1, \mathbf{X}_2) = \frac{\langle \mathbf{X}_1, \mathbf{X}_2 \rangle^k}{\langle \mathbf{X}_1, \mathbf{X}_1 \rangle^{k/2} \langle \mathbf{X}_2, \mathbf{X}_2 \rangle^{k/2}}.$$

As these examples show, similarities award *sameness* with large values. Similarities and dissimilarities are closely related. The relationship is explicit when we restrict both to positive random variables with values in the interval $[0,1]$. In this case, we find that

$$\varsigma(\mathbf{X}_1,\mathbf{X}_2) = 1 - \varrho(\mathbf{X}_1,\mathbf{X}_2).$$

Similarities and dissimilarities are also called **proximities**, that is, measures that describe the nearness of pairs of objects.

I have defined proximities for pairs of random vectors, but the definition naturally extends to pairs of functions and then provides a measure of the nearness of two functions. The Kullback-Leibler divergence (9.14) of Section 9.4 is such an example and looks similar to the dissimilarity (5.11). The scalar product of two functions defines a similarity, but not every similarity is a scalar product.

5.4 Features and Feature Maps

The concept of a feature has almost become folklore in pattern recognition and machine learning. Yet, much earlier, feature-like quantities were described in Multidimensional Scaling, where they are defined via embeddings.

Definition 5.6 Let $\mathbb{X}$ be d-dimensional data. Let $\mathcal{F}$ be a function space, and let $\mathfrak{f}$ be a map from $\mathbb{X}$ into $\mathcal{F}$. For $\mathbf{X}$ from $\mathbb{X}$, we call

- $\mathfrak{f}(\mathbf{X})$ a **feature** or **feature vector** of $\mathbf{X}$,
- the collection $\mathfrak{f}(\mathbb{X})$ the **feature data**, and
- $\mathfrak{f}$ a **feature map**.

If $\mathfrak{f}(\mathbf{X})$ is a vector, then its components are also called **features**; an injective feature map is also called an **embedding**. □

Examples of features, with real-valued feature maps, are individual variables of $\mathbf{X}$ or combinations of variables. For $j,k \le d$, the principal component score $\boldsymbol{\eta}_j^{\mathsf{T}}(\mathbf{X}-\boldsymbol{\mu})$ is a feature, and the k-dimensional principal component vector $\mathbf{W}^{(k)}$ is a feature vector. The corresponding feature map is

$$\mathfrak{f}\colon \mathbf{X} \longmapsto \mathfrak{f}(\mathbf{X}) = \Gamma_k^{\mathsf{T}}(\mathbf{X}-\boldsymbol{\mu}).$$

Other features we have met include $\boldsymbol{\eta}^{\mathsf{T}}\mathbf{X}$, the projection of $\mathbf{X}$ onto Fisher's discriminant direction $\boldsymbol{\eta}$. In Section 8.2 we will meet feature maps of objects describing the data $\mathbb{X}$, and the corresponding feature data are the configurations of Multidimensional Scaling. In Section 12.2.1 we meet feature maps which are functions into L_2 spaces, and we also meet features arising from non-linear maps.

The purpose of features, feature vectors or feature data is to represent the original vector or data in a form that is more amenable to the intended analysis: the d-dimensional vector $\mathbf{X}$ is summarised by the one-dimensional score $\Gamma_1^{\mathsf{T}}(\mathbf{X}-\boldsymbol{\mu})$. Associated with features are the notions of feature extraction and feature selection. *Feature extraction* often refers to the process of constructing features for $\mathbf{X}$ in fewer dimensions than $\mathbf{X}$.

Feature selection is the process which determines a particular feature map $\mathfrak{f}$ for $\mathbf{X}$ or $\mathbb{X}$ and implicitly the feature. The choice of feature map is determined by the type of analysis one wants to carry out: we might want to find the largest variability in the data, or we might want to determine the cluster structure in data. In Section 10.8 we combine principal component and independent component results in the selection of feature vectors. The distinction between feature extraction and feature selection is not always clear. In later chapters I will refer to feature selection only.

5.5 Dualities for $\mathbb{X}$ and $\mathbb{X}^\top$

Many of the ideas in this section apply to arbitrary matrices, here we consider data matrices only. For notational convenience, I assume that the $d \times n$ data $\mathbb{X} = \begin{bmatrix}\mathbf{X}_1\ \mathbf{X}_2 \cdots \mathbf{X}_n\end{bmatrix}$ are *centred*. Let r be the rank of $\mathbb{X}$ or, equivalently, of the sample covariance matrix S of $\mathbb{X}$, and put

$$Q^{\langle d \rangle} = \mathbb{X}\mathbb{X}^\top \qquad \text{and} \qquad Q_{\langle n \rangle} = \mathbb{X}^\top\mathbb{X}. \tag{5.17}$$

We regard $Q_{\langle n \rangle}$ as a **dual** of $Q^{\langle d \rangle}$, and vice versa, and refer to these matrices as the $Q^{\langle d \rangle}$-matrix and the $Q_{\langle n \rangle}$-matrix of $\mathbb{X}$. The $d \times d$ matrix $Q^{\langle d \rangle}$ is related to the sample covariance matrix by $Q^{\langle d \rangle} = (n-1)S$. The dual matrix $Q_{\langle n \rangle}$ is of size $n \times n$ and is to $\mathbb{X}^\top$ what $Q^{\langle d \rangle}$ is to $\mathbb{X}$. Both matrices are positive semidefinite. By Proposition 3.1 of Section 3.2, the spectral decompositions of $Q^{\langle d \rangle}$ and $Q_{\langle n \rangle}$ are

$$Q^{\langle d \rangle} = U D^2 U^\top \qquad \text{and} \qquad Q_{\langle n \rangle} = V D^2 V^\top,$$

where U and V are r-orthogonal matrices consisting of the eigenvectors $\mathbf{u}_j$ and $\mathbf{v}_j$ of $Q^{\langle d \rangle}$ and $Q_{\langle n \rangle}$, respectively, and D^2 is the common diagonal matrix with r non-zero entries d_j^2 in decreasing order. The link between the two representations is the singular value decomposition of $\mathbb{X}$, which is given in Result 1.13 in Section 1.5.3, namely,

$$\mathbb{X} = U D V^\top, \tag{5.18}$$

with U the matrix of left eigenvectors, V the matrix of right eigenvectors, and D the diagonal matrix of singular values. Because the rank of $\mathbb{X}$ is r, U is a $d \times r$ matrix, D is a diagonal $r \times r$ matrix and V is a $n \times r$ matrix. As in (3.5) in Section 3.2, the left and right eigenvectors of $\mathbb{X}$ satisfy

$$\mathbb{X}\mathbf{v}_j = d_j \mathbf{u}_j \qquad \text{and} \qquad \mathbb{X}^\top \mathbf{u}_j = d_j \mathbf{v}_j \qquad \text{for } j = 1, \ldots, r. \tag{5.19}$$

The relationship between $Q^{\langle d \rangle}$ and $Q_{\langle n \rangle}$ allows us to work with the matrix that is of smaller size: in a classical setting, we work with $Q^{\langle d \rangle}$, whereas $Q_{\langle n \rangle}$ is preferable in a HDLSS setting.

In the following, I use the subscript convention of (1.21) in Section 1.5.2, so D_k is a $k \times k$ (diagonal) matrix and V_k is the $n \times k$ matrix.

Result 5.7 *Consider the $d \times n$ centred data $\mathbb{X} = \begin{bmatrix}\mathbf{X}_1 \cdots \mathbf{X}_n\end{bmatrix}$ and the matrix $Q_{\langle n \rangle} = \mathbb{X}^\top\mathbb{X}$. Let r be the rank of $Q_{\langle n \rangle}$, and write $Q_{\langle n \rangle} = V D^2 V^\top$. For $k \le r$, put $\mathbb{W}_Q^{(k)} = D_k V_k^\top$. Then $\mathbb{W}_Q^{(k)}$ is a $k \times n$ matrix.*

1. *For $i \le n$, the columns $\mathbf{W}_i$ of $\mathbb{W}_Q^{(k)}$ are*

$$\mathbf{W}_i = \begin{bmatrix} d_1 \upsilon_{1i}, \ldots, d_k \upsilon_{ki} \end{bmatrix}^\top,$$

where υ_{ji} is the ith entry of jth column vector $\mathbf{v}_j$ of V.

2. *The sample covariance matrix of* $\mathbb{W}_Q^{(k)}$ *is*

$$\mathrm{var}(\mathbb{W}_Q^{(k)}) = \frac{1}{n-1} D_k^2.$$

3. *Let* $q_{i\ell}$ *be the entries of* $Q_{\langle n \rangle}$*, and put* $\mathbb{W}_Q = \mathbb{W}_Q^{(r)}$*, the following hold:*
   (a) *for any two column vectors* $\mathbf{W}_i$ *and* $\mathbf{W}_\ell$ *of* $\mathbb{W}_Q$*,*

$$q_{i\ell} = \mathbf{W}_i^{\mathsf{T}} \mathbf{W}_\ell = \mathbf{X}_i^{\mathsf{T}} \mathbf{X}_\ell,$$

   (b)

$$\mathbb{W}_Q^{\mathsf{T}} \mathbb{W}_Q = Q_{\langle n \rangle}, \qquad \textit{and}$$

   (c) *for any orthogonal* $r \times r$ *matrix* E*,*

$$\left(E\mathbb{W}_Q\right)^{\mathsf{T}} \left(E\mathbb{W}_Q\right) = Q_{\langle n \rangle}.$$

6

Cluster Analysis

There is no sense in being precise when you don't even know what you're talking about (John von Neumann, 1903–1957).

6.1 Introduction

Cluster Analysis is an exploratory technique which partitions observations into different clusters or groupings. In medicine, biology, psychology, marketing or finance, multivariate measurements of objects or individuals are the data of interest. In biology, human blood cells of one or more individuals – such as the *HIV flow cytometry* data – might be the objects one wants to analyse. Cells with similar multivariate responses are grouped together, and cells whose responses differ considerably from each other are partitioned into different clusters. The analysis of cells from a number of individuals such as $\text{HIV}^+$and HIV^-individuals may result in different cluster patterns. These differences are informative for the biologist and might allow him or her to draw conclusions about the onset or progression of a disease or a patient's response to treatment.

Clustering techniques are applicable whenever a *mountain* of data needs to be grouped into manageable and meaningful *piles*. In some applications we know that the data naturally fall into two groups, such as HIV^+or HIV^-, but in many cases the number of clusters is not known. The goal of Cluster Analysis is to determine

- the cluster allocation for each observation, and
- the number of clusters.

For some clustering methods – such as k-means – the user has to specify the number of clusters prior to applying the method. This is not always easy, and unless additional information exists about the number of clusters, one typically explores different values and looks at potential interpretions of the clustering results.

Central to any clustering approach is the notion of *similarity* of two random vectors. We measure the degree of similarity of two multivariate observations by a distance measure. Intuitively, one might think of the Euclidean distance between two vectors, and this is typically the first and also the most common distance one applies in Cluster Analysis. In this chapter we consider a number of distance measures, and we will explore their effect on the resulting cluster structure. For high-dimensional data in particular, the cosine distance can be more meaningful and can yield more interpretable results than the Euclidean distance.

We think of multivariate measurements as continuous random variables, but attributes of objects such as colour, shape or species are relevant and should be integrated into an analysis as much as possible. For some data, an additional variable which assigns colour or species type a numerical value might be appropriate. In medical applications, the pathologist might know that there are four different cancer types. If such extra knowledge is available, it should inform our analysis and could guide the choice of the number of clusters.

The strength of Cluster Analysis is its exploratory nature. As one varies the number of clusters and the distance measure, different cluster patterns appear. These patterns might provide new insight into the structure of the data. For many data sets, the statistician works closely with a collector or owner of the data, who typically knows a great deal about the data. Different cluster patterns can indicate the existence of unexpected substructures, which, in turn, can lead to further or more in-depth investigations of the data. For this reason, where possible, the interpretation of a cluster analysis should involve a subject expert.

In the machine learning world, Cluster Analysis is referred to as *Unsupervised Learning*, and like Discriminant Analysis, it forms part of *Statistical Learning*. There are many parallels between Discriminant Analysis and Cluster Analysis, but there are also important differences. In Discriminant Analysis, the random vectors have labels, which we use explicitly to construct discriminant rules. In addition, we know the number of classes, and we have a clear notion of *success*, which can be expressed by a classification error. In Cluster Analysis, we meet vectors or data without labels. There is no explicit assessment criterion because *truth* is not known; instead, the closeness of observations drives the process.

There are scenarios between Discriminant Analysis and Cluster Analysis which arise from partially labelled data or data with unreliable labels. In this chapter we are interested primarily in data which require partitioning in the absence of labels. Although the goal, namely, allocation of observations to groupings, is the same in Discriminant Analysis and Cluster Analysis, the word *classify* is reserved for labelled data, whereas the term *cluster* refers to unlabelled data.

The classes in Discriminant Analysis are sometimes modelled by distributions which differ in their means. Similar approaches exist in Cluster Analysis: The random vectors are modelled by Gaussian mixtures or by mixtures of t-distributions – usually with an unknown number of terms – and ideas from Markov chain Monte Carlo (MCMC) are used in the estimation of model parameters. The underlying assumption of these models, namely, that the data in the different parts are Gaussian, is not easy to verify and may not hold. For this reason, I prefer distribution-free approaches and will not pursue mixture-model approaches here but refer the interested reader to McLachlan and Basford (1988) and McLachlan and Peel (2000) as starting points for this approach to Cluster Analysis.

A large – and growing – number of clustering approaches exist, too many to even start summarising them in a single chapter. For a survey of algorithms which include fuzzy clustering and approaches based on neural networks, see Xu and Wunsch II (2005). In Discriminant Analysis, we consider the fundamental linear methods, in appreciation of the comments of Hand (2006) that 'simple methods typically yield performances almost as good as more sophisticated methods.' Although I am not aware of similar guidelines for clustering methods, I focus on fundamental methods and consider statistical issues such as the effect of dimension reduction and the choice of the number of clusters.

The developments in Cluster Analysis are driven by data, and a population set-up is therefore not so useful. For this reason, I will only consider samples of random vectors, the data, in this chapter. We begin with the two fundamental methods which are at the heart of Cluster Analysis: hierarchical agglomerative clustering in Section 6.2, and k-means clustering in Section 6.3. Hierarchical agglomerative clustering starts with singleton clusters and merges clusters, whereas the k-means method divides the data into k groups. In both approaches we explore and examine the effect of employing different measures of distance. The centres of clusters are similar to modes, and Cluster Analysis has clear links to density estimation. In Section 6.4 we explore polynomial histogram estimators, and in particular second-order polynomial histogram estimators (SOPHE), as a tool for clustering large data sets. For data in many dimensions, we look at Principal Component Clustering in Section 6.5. In Section 6.6 we consider methods for finding the number of clusters which include simple and heuristic ideas as well as the more sophisticated approaches of Tibshirani, Walther, and Hastie (2001) and Tibshirani and Walther (2005). The last approach mimics some of the ideas of Discriminant Analysis by comparing the cluster allocations of different parts of the same data and uses these allocations to 'classify' new observations into existing clusters. Problems for this chapter are listed at the end of Part II.

Terminology. Unlike classes, clusters are not well-defined or unique. I begin with a heuristic definition of clusters, which I refine in Section 6.3. We say that the data $\mathbb{X} = \left[\mathbf{X}_1 \cdots \mathbf{X}_n\right]$ split into disjoint clusters $\mathcal{C}_1, \mathcal{C}_2, \ldots, \mathcal{C}_m$ if the observations $\mathbf{X}_i$, which make up cluster $\mathcal{C}_k$, satisfy one or more of the following:

1. The observations $\mathbf{X}_i$ in $\mathcal{C}_k$ are close with respect to some distance.
2. The observations $\mathbf{X}_i$ in $\mathcal{C}_k$ are close to the **centroid** of $\mathcal{C}_k$, the average of the observations in the cluster.
3. The observations $\mathbf{X}_i$ in $\mathcal{C}_k$ are closer to the centroid of $\mathcal{C}_k$ than to any other cluster centroid.

In Cluster Analysis, and in particular, in agglomerative clustering, the starting points are the singleton clusters (the individual observations) which we merge into larger groupings. We use the notation $\mathcal{C}_k$ for clusters, which we used in Chapter 4 for classes, because this notation is natural. If an ambiguity arises, I will specify whether we deal with classes or clusters.

6.2 Hierarchical Agglomerative Clustering

To allocate the random vectors $\mathbf{X}_i$ from the data $\mathbb{X} = \left[\mathbf{X}_1 \cdots \mathbf{X}_n\right]$ to clusters, there are two complementary approaches.

1. In **agglomerative clustering**, one starts with n singleton clusters and merges clusters into larger groupings.
2. In **divisive clustering**, one starts with a single cluster and divides it into a number of smaller clusters.

A clustering approach is **hierarchical** if one observation at a time is merged or separated. Both divisive and agglomerative clustering can be carried out hierarchically, but we will only consider hierarchical agglomerative clustering. Divisive clustering is used more commonly in the form of k-means clustering, which we consider in Section 6.3. For either of the clustering approaches – dividing or merging – we require a notion of distance to tell us which observations are close. We use the distances defined in Section 5.3.1 and recall that a distance Δ satisfies $\Delta(\mathbf{X}_i, \mathbf{X}_j) = \Delta(\mathbf{X}_j, \mathbf{X}_i)$.

Hierarchical agglomerative clustering is natural and conceptually simple: at each step we consider the current collection of clusters – which may contain singleton clusters – and we merge the two clusters that are closest to each other. The distances of Section 5.3.1 relate to pairs of vectors; here we also require a measure of closeness for clusters. Once the two measures of closeness – for pairs of vectors and for pairs of clusters – have been chosen, hierarchical agglomerative clustering is well-defined and transparent and thus a natural method to begin with.

Definition 6.1 Let $\mathbb{X} = \begin{bmatrix}\mathbf{X}_1 \cdots \mathbf{X}_n\end{bmatrix}$ be d-dimensional data. Let Δ be a distance for $\mathbb{X}$, and assume that $\mathbb{X}$ partitions into κ clusters $\mathcal{C}_1, \ldots, \mathcal{C}_\kappa$, for $\kappa < n$. For any two of these clusters, say $\mathcal{C}_\alpha$ and $\mathcal{C}_\beta$, put

$$\mathcal{D}_{\alpha,\beta} = \{\Delta_{\alpha_i,\beta_j} = \Delta(\mathbf{X}_{\alpha_i}, \mathbf{X}_{\beta_j}) \colon \mathbf{X}_{\alpha_i} \in \mathcal{C}_\alpha \text{ and } \mathbf{X}_{\beta_j} \in \mathcal{C}_\beta\}.$$

A **linkage** $\mathcal{L}$ is a measure of closeness of – or separation between – pairs of sets and depends on the distance Δ between the vectors of $\mathbb{X}$. We write $\mathcal{L}_{\alpha,\beta}$ for the linkage between $\mathcal{C}_\alpha$ and $\mathcal{C}_\beta$ and consider the following linkages:

1. The **single linkage**

$$\mathcal{L}^{\min}_{\alpha,\beta} = \min_{\alpha_i,\beta_j}\{\Delta_{\alpha_i,\beta_j} \in \mathcal{D}_{\alpha,\beta}\}.$$

2. The **complete linkage**

$$\mathcal{L}^{\max}_{\alpha,\beta} = \max_{\alpha_i,\beta_j}\{\Delta_{\alpha_i,\beta_j} \in \mathcal{D}_{\alpha,\beta}\}.$$

3. The **average linkage**

$$\mathcal{L}^{\text{mean}}_{\alpha,\beta} = \frac{1}{n_\alpha n_\beta} \sum_{\alpha_i} \sum_{\beta_j} \Delta_{\alpha_i \beta_j} \qquad (\Delta_{\alpha_i,\beta_j} \in \mathcal{D}_{\alpha,\beta}),$$

   where n_α and n_β are the numbers of observations in clusters $\mathcal{C}_\alpha$ and $\mathcal{C}_\beta$, respectively.

4. The **centroid linkage**

$$\mathcal{L}^{\text{cent}}_{\alpha,\beta} = ||\overline{\mathbf{X}}_\alpha - \overline{\mathbf{X}}_\beta||,$$

   where $\overline{\mathbf{X}}_\alpha$ and $\overline{\mathbf{X}}_\beta$ are the centroids of $\mathcal{C}_\alpha$ and $\mathcal{C}_\beta$, that is, the averages of the observations in $\mathcal{C}_\alpha$ and $\mathcal{C}_\beta$, respectively. □

If the clusters are singletons, the four linkages reduce to the distance Δ between pairs of vectors. A linkage may not extend to a distance; it is defined for pairs of sets, so the triangle inequality may not apply. The preceding list of linkages is not exhaustive but suffices for our purpose. Definition 6.1 tells us that the linkage depends on the distance between observations, and a particular linkage therefore can lead to different cluster arrangements, as we shall see in Example 6.1.

Algorithm 6.1 *Hierarchical Agglomerative Clustering*
Consider the data $\mathbb{X} = \left[\mathbf{X}_1 \cdots \mathbf{X}_n\right]$. Fix a distance Δ and a linkage $\mathcal{L}$. Suppose that the data are allocated to κ clusters for some $\kappa \leq n$. Fix $K < \kappa$. To assign the data to K clusters, put $\nu = \kappa$, $m = \nu(\nu-1)/2$, and let $\mathcal{G}_\nu$ be the collection of clusters $\mathcal{C}_1, \ldots, \mathcal{C}_\kappa$.

Step 1. For $k, \ell = 1, \ldots, \nu$ and $k < \ell$, calculate pairwise linkages $\mathcal{L}_{k,\ell}$ between the clusters of $\mathcal{G}_\nu$.
Step 2. Sort the linkages, and rename the ordered sequence of linkages $\mathcal{L}_{(1)} \leq \mathcal{L}_{(2)} \leq \cdots \leq \mathcal{L}_{(m)}$.
Step 3. Let α and β be the indices of the two clusters $\mathcal{C}_\alpha$ and $\mathcal{C}_\beta$ such that $\mathcal{L}_{\alpha,\beta} = \mathcal{L}_{(1)}$. Merge the clusters $\mathcal{C}_\alpha$ and $\mathcal{C}_\beta$ into a new cluster $\mathcal{C}^*$.
Step 4. Let $\mathcal{G}_{\nu-1}$ be the collection of clusters derived from $\mathcal{G}_\nu$ by replacing the clusters $\mathcal{C}_\alpha$ and $\mathcal{C}_\beta$ by $\mathcal{C}^*$ and by keeping all other clusters the same. Put $\nu = \nu - 1$.
Step 5. If $\nu > K$, repeat steps 1 to 4. ■

These five steps form the essence of the agglomerative nature of this hierarchical approach. Typically, one starts with the n singleton clusters and reduces the number until the target number is reached. We only require m linkage calculations in step 1 because distances and linkages are symmetric.

The final partition of the data depends on the chosen distance and linkage, and different partitions may result if we vary either of these measures, as we shall see in Example 6.1. In each cycle of steps 2 to 4, two clusters are replaced by one new one. Ties in the ordering of the distances do not affect the clustering, and the order in which the two merges are carried out does not matter. The process is repeated at most $n-1$ times, by which time there would be only a single cluster. Generally, one stops earlier, but there is not a unique stopping criterion. One could stop when the data are partitioned into K clusters for some predetermined target K. Or one could impose a bound on the volume or size of a cluster.

The merges resulting from the agglomerative clustering are typically displayed in a binary cluster tree called a **dendrogram**. The x-axis of a dendrogram orders the original data in such a way that merging observations are adjacent, starting with the smallest distance. Merging observations are connected by an upside-down U-shape, and the height of the 'arms' of each U is proportional to the distance between the two merging objects. Dendrograms may be displayed in a different orientation from the one described here, and the trees are sometimes displayed as growing downwards or sideways. The information is the same, and the interpretation is similar whichever orientation is used.

Example 6.1 We return to the **iris** data. The data are labelled, but in this example we cluster subsets without using the labels, and we compare the resulting cluster membership with the class labels.

The iris data consist of three different species which are displayed in different colours in Figure 1.3 of Section 1.2. The red class is completely separate from the other two classes, and we therefore restrict attention to the overlapping green and black classes. We explore the hierarchical agglomerative clustering and the resulting dendrogram for the first ten and twenty observations from each of these two species, so we work with twenty and forty observations, respectively. The first three variables $\left[\mathbf{X}_{\bullet,1}, \mathbf{X}_{\bullet,2}, \mathbf{X}_{\bullet,3}\right]^{\mathsf{T}}$ of the twenty and forty observations used in this analysis are shown in Figure 6.1. The maroon points replace the green points of Figure 1.3. In the calculations, however, I use all four variables.

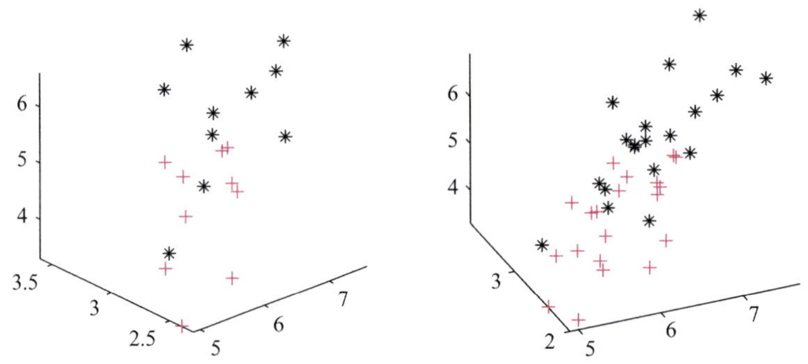

Figure 6.1 Variables 1, 2 and 3 of two species of iris data from Example 6.1; twenty observations (*left*) and forty observations (*right*), with different colours for each species.

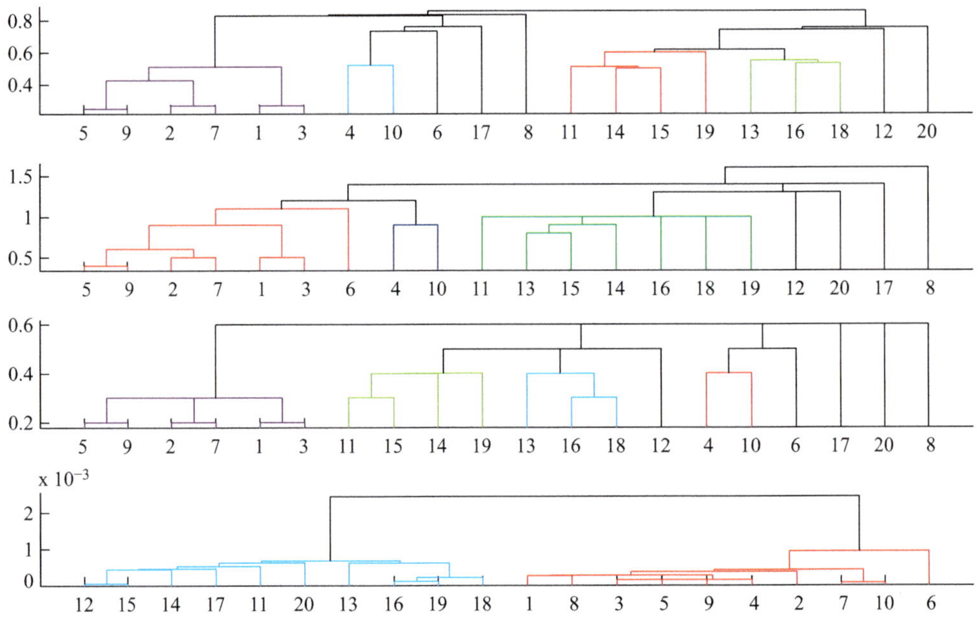

Figure 6.2 Dendrogram of twenty samples from two species of Example 6.1; single linkage with Euclidean distance (*top plot*), ℓ_1 distance (second plot), ℓ_∞ distance (*third plot*), and cosine distance (*bottom*).

Figure 6.2 shows dendrograms for the twenty observations displayed in the left subplot of Figure 6.1. I use the single linkage for all plots and different distances, which are defined in Section 5.3.1. The top subplot uses the Euclidean distance. This is followed by the ℓ_1 distance (5.4) in the second subplot, the ℓ_∞ distance (5.5) in the third, and the cosine distance (5.6) in the last subplot. The observations with numbers $1, \ldots, 10$ on the x-axis are the first ten observations of the 'maroon' species, and the observations with numbers $11, \ldots, 20$ refer to the first ten from the 'black' species. In the top subplot, observations 5 and 9 are closest, followed by the pairs (1, 3) and (2, 7). The colours of the arms other than black show observations that form a cluster during the first few steps. Links drawn in black correspond to much larger distances and show observations which have no close neighbours: for the Euclidean distance, these are the five observations 6, 17, 8, 12 and 20.

A comparison of the top three dendrograms shows that their first three merges are the same, but the cluster formations change thereafter. For the ℓ_1 distance, observation 6 merges with the red cluster to its left. Observations 17, 8, 12 and 20 do not have close neighbours for any of the three distances.

The cosine distance in the bottom subplot results in a different cluster configuration from those obtained with the other distances. It is interesting to observe that its final two clusters (shown in red and blue) are the actual iris classes, and these two clusters are well separated, as the long black lines show.

In Figure 6.3, I use the Euclidean distance as the common distance but vary the linkage. This figure is based on the forty observations shown in the right subplot of Figure 6.1. The top dendrogram employs the single linkage, followed by the complete linkage, the average, and finally the centroid linkage. The linkage type clearly changes the cluster patterns. The single linkage, which I use in all subplots of Figure 6.2, yields very different results from the other three linkages: The numbering of close observations on the x-axis differs considerably, and the number and formation of tight clusters – which are shown by colours other than black – vary with the linkage. For the single linkage, there are two large clusters: the green cluster contains only points from the 'maroon' species, and the red cluster contains mainly points from the 'black' species. The other three linkage types do not admit such clear interpretations.

Figure 6.4 shows the last thirty steps of the algorithm for all 100 observations of the 'green' and 'black' species. For these dendrograms, the numbers on the x-axis refer to the

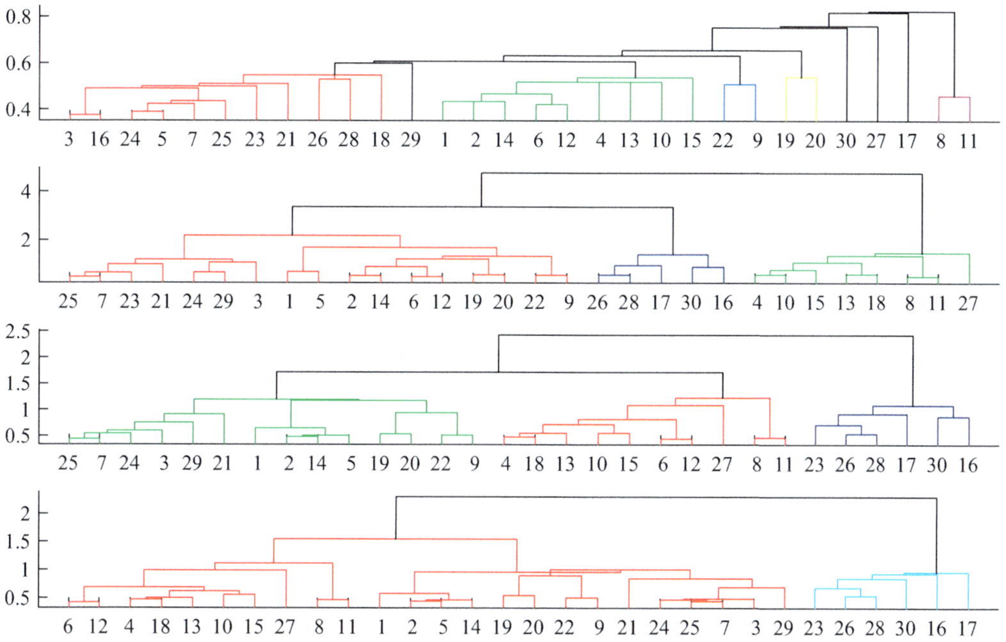

Figure 6.3 Dendrogram of Example 6.1; first twenty observations of each species based on the Euclidean distance and different linkages: single (*top*), complete (*second*), average (*third*) and centroid (*bottom*).

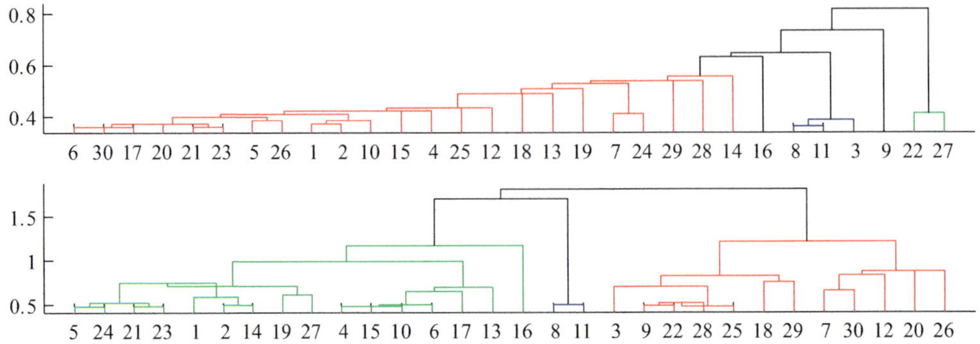

Figure 6.4 Dendrogram of Example 6.1; 100 observations from two species with Euclidean distance and single linkage (*top*) and centroid linkage (*bottom*).

thirty clusters that exist at that stage of the algorithm rather than the individual observations that we met in the dendrograms of Figures 6.2 and 6.3. As in the preceding figure, all calculations are based on the Euclidean distance. In the top dendrogram, the single linkage is used, and this is compared with the centroid linkage in the bottom subplot. Unlike the preceding figure, for this bigger data set, the centroid linkage performs better than the single linkage when compared with the class labels. The centroid linkage results in two large clusters and the two isolated points 8 and 11.

A closer inspection of Figure 6.4 allows us to determine the samples for which the species and the cluster membership differ. Figure 1.3 in Section 1.2 shows that the two species overlap. In the purely distance-based cluster analysis, each observation is pulled to the cluster it is nearest to, as we no longer experience the effect of the class label which partially counteracts the effect of the pairwise distances. ■

To understand the effect different distances and linkages can have on cluster formation, I have chosen a small number of observations in Example 6.1. The Euclidean, ℓ_1 and ℓ_∞ distances result in similar cluster arrangements. This is a consequence of the fact that they are based on equivalent norms; that is, any pair of distances Δ_ℓ and Δ_k of these three satisfy

$$c_1 \Delta_k(\mathbf{X}_i, \mathbf{X}_j) \le \Delta_\ell(\mathbf{X}_i, \mathbf{X}_j) \le c_2 \Delta_k(\mathbf{X}_i, \mathbf{X}_j),$$

for constants $0 < c_1 \le c_2$. The cosine and correlation distances are not equivalent to the three preceding ones, a consequence of their different topologies and geometries. As we have seen, the cosine distance results in a very different cluster formation.

For the *iris* data, we do know the labels. Generally, we do not have such information. For this reason, I recommend the use of more than one linkage and distance. In addition, other information about the data, such as scatterplots of subsets of the variables, should be incorporated in any inferences about the cluster structure.

The appeal of the hierarchical agglomerative approach is that it is intuitive and easy to construct. Because the process is entirely based on the nearest distance at every step, observations which are not close to any cluster are disregarded until the last few steps of the algorithm. This fact obscures the need to search for the right number of clusters and may be regarded as a limitation of the hierarchical approach. The k-means approach, which we consider in the next section, overcomes this particular problem.

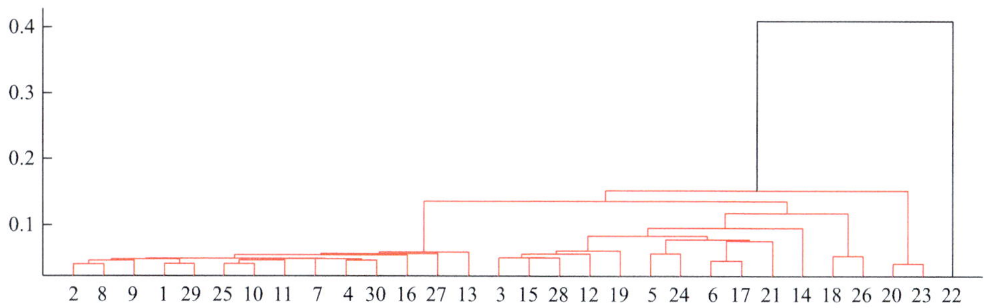

Figure 6.5 Dendrogram of Example 6.2 for the first 200 observations.

If we want to find more than two clusters of comparable size with the hierarchical agglomerative approach, decisions need to be made about the *left-over* observations or outliers. There is no general best rule for dealing with the left-over points, although applications may dictate whether the remaining points are negligible or relevant. I illustrate these points in the next example and then address the issue further in the Problems at the end of Part II.

Example 6.2 We return to the **abalone** data, which contain 4,177 samples in eight variables. For these data, one wants to distinguish between young and mature abalone. Clustering the data is therefore a natural choice. We will not consider all data but for illustration purposes restrict attention to the first 200 observations and the first four variables.

Figure 6.5 shows the last thirty steps of the hierarchical approach based on the single linkage and the Euclidean distance. To gain some insight into the cluster tree, we need to investigate the branches of the tree in a little more detail. From the figure, we conclude that two main clusters appear to emerge at height $y = 0.1$, which have merged into one cluster at $y = 0.15$.

An inspection of the clusters with x-labels 20, 23 and 22 in the figure reveals that they are all singletons. In contrast, the clusters with x-labels 1, 2, 25 and 8 contain about 80 per cent of the 200 observations which have been assigned to these clusters in earlier steps. The information regarding the size of the clusters and the individual association of observations to clusters is not available from Figure 6.5 but is easily obtained from the output of the analysis. ■

The dendrogram in Example 6.2 shows the cluster allocation of the *abalone* data. Although the final link in the dendrogram combines all clusters into one, we can use this initial investigation to examine the cluster structure and to make a decision about outliers which could be left out in a subsequent analysis. I will not do so here but instead introduce the k-means approach, which handles outlying points in a different manner.

6.3 k-Means Clustering

Unlike hierarchical agglomerative clustering, k-means clustering partitions all observations into a specified number of clusters. There are many ways of dividing data into k clusters, and it is not immediately clear what constitutes the 'best' or even 'good' clustering. We could require that clusters do not exceed a certain volume so that observations that are too far away from a cluster centre will not belong to that cluster. As in Discriminant Analysis, we let the variability within clusters drive the partitioning, but this time without knowledge of the class membership. The key idea is to partition the data into k clusters in such a way that the variability within the clusters is minimised.

At the end of Section 6.1, I gave an intuitive definition of clusters. The next definition refines this concept and provides criteria for 'good' clusters.

Definition 6.2 Let $\mathbb{X} = \begin{bmatrix}\mathbf{X}_1 \cdots \mathbf{X}_n\end{bmatrix}$ be d-dimensional data. Let Δ be a distance for $\mathbb{X}$. Fix $k < n$. Assume that $\mathbb{X}$ have been partitioned into k clusters $\mathcal{C}_1, \ldots, \mathcal{C}_k$. For $\nu \leq k$, let n_ν be the number of $\mathbf{X}_i$ which belong to cluster $\mathcal{C}_\nu$, and let $\overline{\mathbf{X}}_\nu$ be the cluster centroid of $\mathcal{C}_\nu$. Put

$$\mathcal{P} = \mathcal{P}(\mathbb{X}, k) = \{\, \mathcal{C}_\nu : \nu = 1, \ldots, k\},$$

and call $\mathcal{P}$ a ***k*-cluster arrangement** for $\mathbb{X}$. The **within-cluster variability** $W_\mathcal{P}$ of $\mathcal{P}$ is

$$W_\mathcal{P} = \sum_{\nu=1}^{k} \sum_{\{\mathbf{X}_i \in \mathcal{C}_\nu\}} \Delta(\mathbf{X}_i, \overline{\mathbf{X}}_\nu)^2. \tag{6.1}$$

A k-cluster arrangement $\mathcal{P}$ is **optimal** if $W_\mathcal{P} \leq W_{\mathcal{P}'}$ for every k-cluster arrangement $\mathcal{P}'$.

For fixed k and distance Δ, ***k*-means clustering** is a partitioning of $\mathbb{X}$ into an optimal k-clusters arrangement. □

If the number of clusters is fixed, we refer to a k-cluster arrangement as a *cluster arrangement*. The within-cluster variability depends on the distance, and the optimal cluster arrangement will vary with the distance.

The scalar quantity $W_\mathcal{P}$ and the sample covariance matrix $\widehat{W}$ of Corollary 4.9 in Section 4.3.2 are related: they compare an observation with the mean of its cluster and, respectively, of its class. But they differ in that $\widehat{W}$ is a matrix based on the class covariance matrices, whereas $W_\mathcal{P}$ is a scalar. When all clusters or classes have the same size n_c, with $n_c = n/k$, and $W_\mathcal{P}$ is calculated from the Euclidean distance, the two quantities satisfy

$$W_\mathcal{P} = (n_c - 1)\, \operatorname{tr}(\widehat{W}). \tag{6.2}$$

There are many different ways of partitioning data into k clusters. The k-means clustering of Definition 6.2 is optimal, but there is no closed-form solution for finding an optimal cluster arrangement. Many k-means algorithms start with k user-supplied centroids or with random disjoint sets and their centroids and iterate to improve the location of the centroids. User-supplied centroids lead to a unique arrangement, but this cluster arrangement may not be optimal. Unless otherwise specified, in the examples of this chapter I use random starts in the k-means algorithm. This process can result in non-optimal solutions; a local optimum instead of a global optimum is found. To avoid such locally optimal solutions, I recommend running the k-means algorithm a number of times for the chosen k and then picking the arrangement with the smallest within-cluster variability.

Optimal cluster arrangements for different values of k typically result in clusters that are not contained in each other because we do not just divide one cluster when going from k to $k+1$ clusters.

How should we choose the number of clusters? A natural way of determining the number of clusters is to look for big improvements, that is, big reductions in within-cluster variability, and to increase the number of clusters until the reduction in the within-cluster variability becomes small. This criterion is easy to apply by inspecting graphs of within-cluster variability against the number of clusters and shows some similarity to the scree plots in Principal

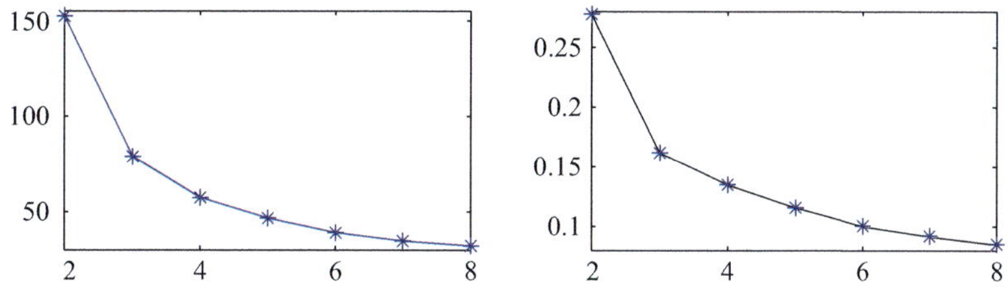

Figure 6.6 Within-cluster variability for Example 6.3 against number of clusters: with Euclidean distance (*left*) and with cosine distance (*right*).

Component Analysis. Like the scree plots, it is a subjective way of making a decision, and big improvements may not exist. Section 6.6 looks at criteria and techniques for determining the number of clusters.

Example 6.3 We apply k-means clustering to the three species of the **iris** data without making use of the class membership of the species. I determine the cluster arrangements separately for the Euclidean distance and the cosine distance.

For each distance, and for $k \leq 8$, I calculate multiple k-cluster arrangements with a k-means algorithm. I found that ten repetitions of the k-means algorithm suffice for obtaining good cluster arrangements and therefore have calculated ten runs for each k. We select the k-cluster arrangement with the smallest within-cluster variability over the ten runs and refer to this repeated application of the k-means algorithm as 'finding the best in ten runs'.

Figure 6.6 shows the within-cluster variability on the y-axis as a function of the number of clusters on the x-axis, starting with two clusters. The left graph shows the results for the Euclidean distance, and the right graph shows the results for the cosine distance. The actual within-cluster variabilities of the two distances are not comparable; for the cosine distance, the data are normalised because one is only interested in the angles subtended at the origin. As a consequence, the resulting variability will be much smaller. Important is the shape of the two graphs rather than the individual numbers.

How should we choose the number of clusters on the basis of these graphs? For both distances, I stop at eight clusters because the number of samples in the clusters becomes too small to obtain meaningful results. As the two graphs show, there is a big improvement for both distances when going from two to three clusters – the within-cluster variability is halved, with much smaller percentage improvements for higher numbers of clusters. Thus the naive choice is three clusters.

For the Euclidean distance, Table 6.1 gives details for the optimal cluster arrangements with $k = 1, \ldots, 5$ and the 'best in ten runs' for each k. The table shows the within-cluster variabilities, the coordinates of the cluster centroids and the number of observations in each cluster. For the case $k = 1$, the centroid is the average of all observations. I include these numbers for comparison only.

For $k = 2$, three observations from the second and third species appear in the first cluster, which is consistent with Figure 1.3 in Section 1.2. For the three-cluster arrangement, the second and third classes differ from the second and third clusters. Because these two classes overlap, this result is not surprising. For three and more clusters, the cluster I list first in each case remains centered at (5.01, 3.42, 1.46, 0.24) and consists of the fifty observations of the first (red) species in Figures 1.3 and 1.4. A fourth cluster appears by splitting the largest

Table 6.1 *Within-Cluster Variability for Optimal k-Cluster Arrangements from Example 6.3*

k	$W(k)$	Centroids	No. in cluster
1	680.82	(5.84, 3.05, 3.76, 1.20)	150
2	152.37	(5.01, 3.36, 1.56, 0.29)	53
		(6.30, 2.89, 4.96, 1.70)	97
3	78.94	(5.01, 3.42, 1.46, 0.24)	50
		(5.90, 2.75, 4.39, 1.43)	62
		(6.85, 3.07, 5.74, 2.07)	38
4	57.32	(5.01, 3.42, 1.46, 0.24)	50
		(5.53, 2.64, 3.96, 1.23)	28
		(6.25, 2.86, 4.81, 1.63)	40
		(6.91, 3.10, 5.85, 2.13)	32
5	46.54	(5.01, 3.42, 1.46, 0.24)	50
		(5.51, 2.60, 3.91, 1.20)	25
		(6.21, 2.85, 4.75, 1.56)	39
		(6.53, 3.06, 5.51, 2.16)	24
		(7.48, 3.13, 6.30, 2.05)	12
True	89.39	(5.01, 3.42, 1.46, 0.24)	50
3		(5.94, 2.77, 4.26, 1.33)	50
		(6.59, 2.97, 5.55, 2.03)	50

cluster into two parts and absorbing a few extra points. The fifth cluster essentially splits the cluster centred at (6.85, 3.07, 5.74, 2.07) into two separate clusters. As k increases, the clusters become less well separated with little improvement in within-cluster variability.

The last three rows in Table 6.1 show the corresponding values for the three species and the sample means corresponding to the three classes. Note that the centroid of the first cluster in the three-cluster arrangement is the same as the sample mean of the first species. It is interesting to observe that the within-class variability of 89.38 is higher than the optimal within-cluster variability for $k = 3$. The reason for this discrepancy is that the optimal cluster arrangement minimises the variability and does not take anything else into account, whereas the class membership uses additional information. In this process, two observations from class 2 and fourteen from class 3 are 'misclassified' in the clustering with the Euclidean distance.

For the cosine distance, a different picture emerges: in the three-cluster scenario, five observations from class 2 are assigned to the third cluster, and all class 3 observations are 'correctly' clustered. A reason for the difference in the two cluster arrangements is the difference in the underlying geometries of the distances. Thus, as in Example 6.3, the performance with the cosine distance is closer to the classification results than that of the Euclidean distance. ■

This example highlights some important differences between classification and clustering: class membership may be based on physical or biological properties, and these properties cannot always be captured adequately by a distance. We are able to compare the cluster results with the class membership for the *iris* data, and we can therefore decide

Table 6.2 *Cluster Centres and Number of Points in Each Cluster from Example 6.4*

Cluster	Centres	No. in cluster
1	(5.00, 3.42, 1.46, 0.24, 3.20)	250
2	(5.00, 3.20, 1.75, 0.80, 2.10)	200
3	(6.52, 3.05, 5.50, 2.15, 2.30)	200
4	(6.22, 2.85, 4.75, 1.57, 2.50)	150
5	(7.48, 3.12, 6.30, 2.05, 2.90)	100
6	(5.52, 2.62, 3.94, 1.25, 3.00)	100

which distance is preferable. When we do not know the class membership, we consider a number of distances and make use of the comparison in the interpretation of the cluster arrangements.

The next example looks at *simulated* data and illustrates the exploratory rather than rule-based nature of clustering as the dimension of the data changes.

Example 6.4 We consider **simulated data** arising from six clusters in five, ten and twenty dimensions. The cluster centres of the five-dimensional data are given in Table 6.2. For the ten- and twenty-dimensional data, the cluster centres of the first five variables are those given in the table, and the cluster centres of the remaining five and fifteen variables, respectively, are 0. The clusters of the ten- and twenty-dimensional data are therefore determined by the first five variables.

Separately for five, ten and twenty dimensions, I generate 1,000 independent random variables from the normal distribution with means the cluster centres of Table 6.2 or 0 for variables 6 to 20. We look at an easier case, for which the marginal standard deviation is $\sigma = 0.25$ for all variables, and at a harder case, with $\sigma = 0.5$. The numbers of observations in each cluster are given in the last column of Table 6.2 and are the same for the five-, ten- and twenty-dimensional data.

For each of the three dimensions and the two standard deviations, I use the Euclidean distance and the cosine distance to find the smallest within-cluster variability over 100 runs of a k-means algorithm for $k = 2, \ldots, 20$. Figure 6.7 shows within-cluster variability graphs as a function of the number of clusters. I mark the curves corresponding to five-, ten- and twenty-dimensional data by different plotting symbols: blue asterisks for the five-dimensional data, red plus signs for the ten-dimensional data, and maroon circles for the twenty-dimensional data. The graphs on the left show the results for $\sigma = 0.5$, and those on the right show analogous results for $\sigma = 0.25$. The graphs in the top row are obtained with the Euclidean distance and those in the bottom row with the cosine distance.

We note that the within-cluster variability increases with the dimension and the standard deviation of the data. As mentioned in Example 6.3, the within-cluster variability based on the Euclidean distance is much larger than that based on the cosine distance. From this fact alone, we cannot conclude that one distance is better than the other. The overall shapes of all twelve variability curves are similar: The variability decreases with cluster number, with no kinks in any of the plots.

In the six plots on the left, the within-cluster variability is gently decreasing without any marked improvements for a particular number of clusters. For the larger σ used in these data,

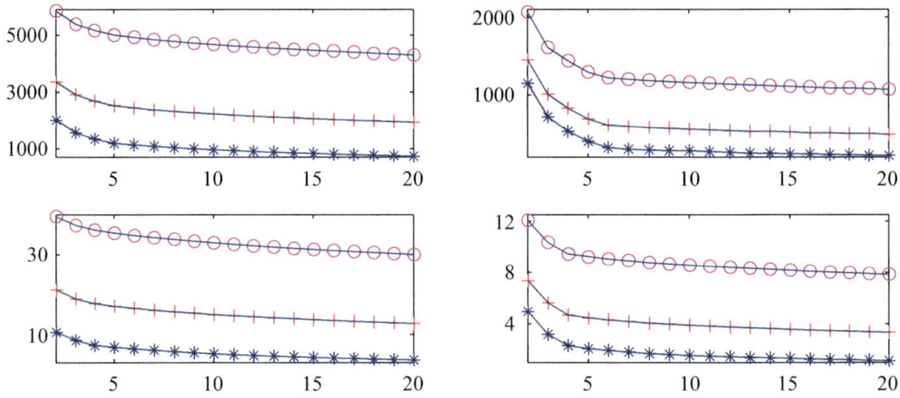

Figure 6.7 Within-cluster variability for Example 6.4 against number of clusters: with Euclidean distance (*top*) and cosine distance (*bottom*); in five (blue asterisks), ten (red plus signs) and twenty (maroon circles) dimensions and standard deviations: $\sigma = 0.5$ (*left*) and $\sigma = 0.25$ (*right*).

we cannot draw any conclusion about a suitable number of clusters from these graphs. The plots on the right are based on the smaller standard deviation $\sigma = 0.25$. There is possibly a hint of a kink at $k = 6$ in the top graphs and at $k = 4$ in the bottom graphs. The evidence for six or four clusters is weak, and because the two distances result in different values for k, we cannot convincingly accept either number as the number of clusters.

These simulations illustrate the dependence of the within-cluster variability on the distance measure, the dimension and the variability of the data. They also show that the naive approach of picking the number of clusters as the kink in the variability plot is not reliable because often there is no kink, or the kink is not the same for different distances. ■

In the *iris* data example, the number of clusters is easy to determine, but for the simulated data of Example 6.4, a naive visual inspection which looks for a kink does not yield any useful information about the number of clusters. In some applications, an approximate cluster arrangement could suffice, and the exact number of clusters may not matter. In other applications, the number of clusters is important, and more sophisticated methods will need to be employed.

In the next example we know that the data belong to two clusters, but we assume that we do not know which observation belongs to which group. After a cluster analysis into two groups, we again compare the cluster allocation with the class labels.

Example 6.5 We continue with the thirty-dimensional **breast cancer** data, which consist of 357 benign and 212 malignant observations, and allocate the raw observations to clusters without using the labels. Within-cluster variabilities based on the Euclidean distance are shown in Figure 6.8. As we go from one to two clusters, the within-cluster variability reduces to about one-third, whereas the change from two to three clusters is very much smaller. The shape of the within-cluster variability graph based on the cosine distance looks very similar, but the values on the y-axis differ considerably.

For the Euclidean and cosine distances, we explore the optimal two- and three-cluster arrangements in more detail, where optimal refers to 'best in ten runs'. Table 6.3 summarises

Table 6.3 *Two- and Three-Cluster Arrangements for Example 6.5*

		Two clusters		Three clusters		
		Cluster 1	Cluster 2	Cluster 1	Cluster 2	Cluster 3
Euclidean	Benign	356	1	317	40	0
	Malignant	82	130	23	105	84
	Total	438	131	340	145	84
Cosine	Benign	351	6	331	26	0
	Malignant	58	154	31	134	47
	Total	409	160	362	160	47

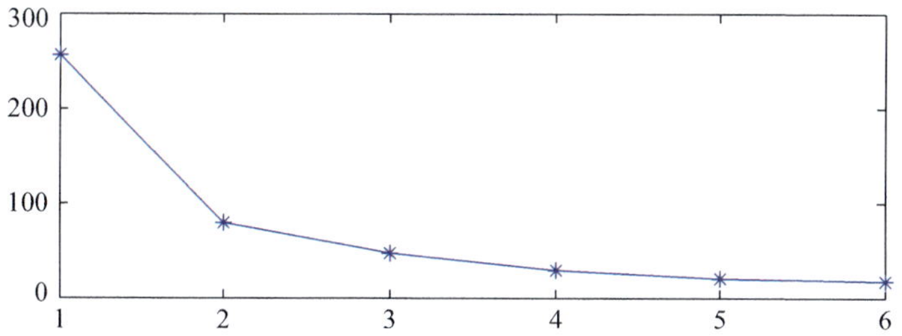

Figure 6.8 Within-cluster variability on the y-axis in units of 10^6 against the number of clusters for Example 6.5.

the results and compares the cluster allocations with the two classes of benign and malignant observations. For the two-cluster arrangement, the cosine distance agrees better with the classes; for sixty-four observations or 11.25 per cent of the data, the clusters and classes differ, whereas the Euclidean distance results in eighty-three observations or 14.6 per cent of the data with different class and cluster allocations.

How can we interpret the three-cluster arrangements? For both distances, the pattern is similar. A new cluster has appeared which draws from both previous clusters, and this new cluster has a majority of malignant observations. The proportion of malignant observations in the benign cluster (cluster 1) has decreased, and the smallest cluster consists entirely of malignant observations.

The three-cluster arrangement with two malignant clusters may capture the classes better than the two-cluster arrangement. The new cluster – cluster 2 in the three-cluster arrangement of the table – is of interest and could express a grey area. An explanation and interpretation of this new cluster are beyond the scope of this book as they require a subject expert. ■

One reason the cluster allocations of the *breast cancer* data do not agree well with the classes is the relatively large number of variables – thirty in this case. Some of the variables are *nuisance* variables which do not contribute to effective clustering. If the effect of these variables is reduced or eliminated, cluster allocations will be more efficient and may improve the agreement of cluster allocation and class membership.

For the *iris* data in Example 6.3 and the *breast cancer* data, the cosine distance results in cluster allocations that are closer to the classes than the allocations produced by the Euclidean distance. In both cases, the data clouds overlap, and the Euclidean distance appears to be worse at separating the overlapping points correctly. The Euclidean distance is often the default in k-means algorithms, and I therefore recommend going beyond the default and exploring other distances.

In Example 6.4, the data are simulated from the multivariate Gaussian distribution. For Gaussian mixtures models with different means, Roweis and Ghahramani (1999) present a generic population model that allows them to derive Cluster Analysis based on entropy maximisation, Principal Component Analysis, Factor Analysis and some other methods as special cases. The assumption of a Gaussian model lends itself to more theoretical developments, including probability calculations, which exploit the Gaussian likelihood. Further, if we know that the data are mixtures of Gaussians, then their ideas can lead to simpler algorithms and potentially more appropriate cluster identifications. Thus, again, the more knowledge we have about the data, the better are the analyses if we exploit this knowledge judiciously. For many data sets, such knowledge is not available, or we may know from other sources that the data deviate from Gaussian mixtures. Hierarchical or k-means clustering apply to such data as both are based on distances between random vectors without reference to underlying distributions.

Instead of minimising the within-cluster variability, Witten and Tibshirani (2010) suggested minimising a cluster analogue of the between-class variability in the form of a weighted between-cluster sum of squares. Their inclusion of weights with an ℓ_1 constraint leads to a sparse selection of variables to be included into the clustering. However, care is required in the choice of the weights and the selection of their tuning parameter, the bound for the ℓ_1 norm of the weight vector. As in the k-means clustering of this section, Witten and Tibshirani assume that the number of clusters is fixed. In Section 6.6.1, I define the between-cluster variability and then explain how Calinski and Harabasz (1974) combined this variability with the within-cluster variability in their selection of the number of clusters. The idea of sparse weights is picked up in Section 13.4 again in connection with choosing sparse weights in Principal Component Analysis.

I conclude this section with Table 6.4, which summarises the hierarchical agglomerative (HA) and the k-means clustering approaches and shows some of the features that are treated differently. In particular:

- Outliers are easily picked with HA but disappear in the k-means approach. As a consequence, in HA clustering, one can eliminate these points and re-run the algorithm.
- Because HA is based on distance, it produces the same answer in each run, whereas the cluster allocations from k-means may vary with the initially chosen values for the cluster centres.
- The within-cluster variabilities provide a quantitative assessment and allow a visual interpretation of the improvement of adding an extra cluster.
- Dendrograms show which observations belong to which cluster, but they become less useful as the number of observations increases.

Neither of the two approaches is clearly superior, and a combination of both may lead to more meaningful and interpretable results.

Table 6.4 *Hierarchical Agglomerative (HA) and k-Means Clustering*

	HA	k-Means
Hierarchical	Yes	No
Outliers	Easy to detect	Allocated to clusters
No. of clusters	Not determined	User provides
Cluster allocation	Same each run	Depends on initial cluster centres
Visual tools	Dendrograms	Within-cluster variability plots

6.4 Second-Order Polynomial Histogram Estimators

Cluster Analysis and Density Estimation are closely related; they share the common goals of determining the shape of the data and finding the cluster centres or modes. Kernel density estimators have good theoretical foundations, especially for univariate and low-dimensional data (see Silverman 1986; Scott 1992; Wand and Jones 1995). With increasing dimension, the emphasis shifts from estimating the density everywhere to estimating modes and regions of high density, and as in the low-dimensional case, we still have to find the number of clusters and modes. For one-dimensional data, Silverman (1981), Mammen, Marron, and Fisher (1991) and Minotte (1997) proposed hypothesis tests for the number of modes. Chaudhuri and Marron (1999) complemented these tests with their 'feature-significance'. The feature-significance ideas have been extended to two dimensions in Chaudhuri and Marron (2000) and to three and more dimensions in Duong et al. (2008). A combination of feature significance and hypothesis tests performs well, and these methods can, in theory, be applied to an arbitrary number of dimensions. The computational complexity and cost of estimating densities with smooth, non-parametric kernel estimators, however, make these approaches less feasible in practice as the dimension increases.

Unlike smooth kernel estimators, multivariate histogram estimators are computationally efficient, but their performance as density estimators is inferior to that of smooth kernel estimators. Motivated by the computational simplicity of histogram estimators, Sagae, Scott, and Kusano (2006) suggested estimating the density of binned univariate and bivariate data by polynomials, separately in each bin. Jing, Koch, and Naito (2012) extended these ideas to general d-dimensional polynomial histogram estimators. They derived explicit expressions for first- and second-order polynomial histogram estimators and determined their asymptotic properties. For moderate to large sample sizes and up to about twenty variables, their method provides a simple and efficient way of calculating the number and location of modes, high-density regions and clusters in practice.

The second-order polynomial histogram estimator (SOPHE) of Jing, Koch, and Naito (2012) has many desirable properties, both theoretical and computational, and provides a good compromise between computational efficiency and accurate shape estimation. Because of its superiority over their first-order estimator, I focus on the SOPHE. I describe the approach of Jing, Koch, and Naito (2012), state some theoretical properties, and then illustrate how this estimator clusters data.

Let $\mathbb{X} = \left[\mathbf{X}_1 \cdots \mathbf{X}_n\right]$ be d-dimensional data from a probability density function f, and assume that the data are centred. We divide the range of the data into L bins B_ℓ ($\ell \leq L$), d-dimensional cubes of size h^d for some binwidth $h > 0$, and we let $\mathbf{t}_\ell$ be the bin centres. For $\mathbf{X}$ from f, the **second-order polynomial histogram estimator** (SOPHE) g of f is of the form

$$g(\mathbf{X}) = a_0 + \mathbf{a}^{\mathsf{T}}\mathbf{X} + \mathbf{X}^{\mathsf{T}} A\mathbf{X}, \tag{6.3}$$

where, for bin B_ℓ,

$$\int_{B_\ell} g(\mathbf{x})d\mathbf{x} = \frac{n_\ell}{n}, \qquad \int_{B_\ell} \mathbf{x}g(\mathbf{x})d\mathbf{x} = \frac{n_\ell \overline{\mathbf{X}}_\ell}{n} \qquad \text{and} \qquad \int_{B_\ell} \mathbf{x}\mathbf{x}^{\mathsf{T}} g(\mathbf{x})d\mathbf{x} = \frac{n_\ell M_\ell}{n},$$

with n_ℓ the number of observations, $\overline{\mathbf{X}}_\ell$ the sample mean, and M_ℓ the second moment for bin B_ℓ. Jing, Koch, and Naito (2012) put

$$\widehat{a}_{0,\ell} = \widehat{b}_{0,\ell} - \widehat{\mathbf{b}}_\ell^{\mathsf{T}}\mathbf{t}_\ell + \mathbf{t}_\ell^{\mathsf{T}}\widehat{A}_\ell\mathbf{t}_\ell, \qquad \widehat{\mathbf{a}}_\ell = \widehat{\mathbf{b}}_\ell - 2\widehat{A}_\ell^{\mathsf{T}}\mathbf{t}_\ell \qquad \text{and}$$

$$\widehat{A}_\ell = \frac{1}{h^{d+4}}\frac{n_\ell}{n}\left[72 S_\ell + 108\ \mathrm{diag}(S_\ell) - 15h^2\mathbf{I}\right],$$

where

$$S_\ell = \frac{1}{n_\ell}\sum_{\mathbf{X}_k \in B_\ell} (\mathbf{X}_k - \mathbf{t}_\ell)(\mathbf{X}_k - \mathbf{t}_\ell)^{\mathsf{T}},$$

$$\widehat{\mathbf{b}}_\ell = \frac{12}{h^{d+2}}\frac{n_\ell}{n}(\overline{\mathbf{X}}_\ell - \mathbf{t}_\ell),$$

$$\widehat{b}_{0,\ell} = \frac{1}{h^{d+2}}\frac{n_\ell}{n}\left[\frac{4+5d}{4}h^2 - 15\,\mathrm{tr}(S_\ell)\right].$$

For $\mathbf{X} \in B_\ell$, they define the SOPHE $\widehat{f}$ of f by $\widehat{f}(\mathbf{X}) = \widehat{a}_{0,\ell} + \widehat{\mathbf{a}}_\ell^{\mathsf{T}}\mathbf{X} + \mathbf{X}^{\mathsf{T}}\widehat{A}_\ell\mathbf{X}$, so

$$\widehat{f}(\mathbf{X}) = \frac{1}{h^{d+4}}\frac{n_\ell}{n}\left\{\frac{(4+5d)}{4}h^4 - 15h^2\,\mathrm{tr}(S_\ell) + 12h^2(\mathbf{X} - \mathbf{t}_\ell)^{\mathsf{T}}(\overline{\mathbf{X}}_\ell - \mathbf{t}_\ell)\right.$$
$$\left. + (\mathbf{X} - \mathbf{t}_\ell)^{\mathsf{T}}\left[72S_\ell + 108\ \mathrm{diag}(S_\ell) - 15h^2\mathbf{I}\right](\mathbf{X} - \mathbf{t}_\ell)\right\}. \tag{6.4}$$

In their proposition 4, Jing, Koch, and Naito (2012), give explicit expressions for the bias and variance of $\widehat{f}$, which I leave out because our focus is Cluster Analysis. Their asymptotic properties of $\widehat{f}$ include the optimal binwidth choice and fast rate of convergence of $\widehat{f}$ to f. Before quoting these results, we require some notation for kth-order derivatives (see chapter 8 of Schott 1996 and chapter 1 of Serfling 1980). For vectors $\mathbf{x}, \mathbf{y}, \mathbf{t} \in \mathbb{R}^d$ and a function $F\colon \mathbb{R}^d \to \mathbb{R}$, the **$k$th-order derivative** at $\mathbf{t}$ is

$$\mathbf{y}^k D^k F(\mathbf{t}) = \sum_{C(d,k)} y_1^{i_1}\cdots y_d^{i_d} \left.\frac{\partial^k F(\mathbf{x})}{\partial x_1^{i_1}\cdots\partial x_d^{i_d}}\right|_{\mathbf{x}=\mathbf{t}} \tag{6.5}$$

with $C(d,k) = \left\{(i_1,\ldots,i_d)\colon i_j \geq 0\ \&\sum_{j=1}^d i_j = k\right\}$. For $k = 1,2$, the notation reduces to the gradient vector $\nabla F(\mathbf{t})$ and the Hessian matrix evaluated at $\mathbf{t}$. For $k = 3$, the case of interest in Theorem 6.3, put $F_{ijk} = (\partial^3 F)/(\partial x_i \partial x_j \partial x_k)$, for $1 \leq i, j, k \leq d$, and write

$$(F_{ujk})_{(u=1,\ldots,d)}(\mathbf{t}) = [F_{1jk}(\mathbf{t})\cdots F_{djk}(\mathbf{t})]^{\mathsf{T}} \tag{6.6}$$

for the vector of partial derivatives, where one partial derivative varies over all dimensions.

Theorem 6.3 [Jing, Koch, and Naito (2012)] *Assume that the probability density function f has six continuous derivatives and that the second- and third-order partial derivatives are* *square integrable. Let $\widehat{f}$ be the second-order polynomial histogram estimator (6.4) for f. The asymptotic integrated squared bias and variance of $\widehat{f}$ are*

$$\mathit{AISB}(\widehat{f}) = h^6 C_S(f) \qquad \text{and} \qquad \mathit{AIV}(\widehat{f}) = \frac{1}{nh^d}\left[\frac{(d+1)(d+2)}{2}\right],$$

where $C_S(f) = \int G(\mathbf{X})\, d\mathbf{X}$, and in the notation of (6.5) and (6.6), G is given on bin B_ℓ by

$$G(\mathbf{X}) = \left\{\frac{1}{h^3}\left[\frac{h^2}{12}(\mathbf{X}-\mathbf{t}_\ell)^{\mathsf{T}}\left(\frac{\sum_i f_{uii}}{2} - \frac{f_{uuu}}{5}\right)_{(u=1,\ldots,d)}(\mathbf{t}_\ell) - \frac{1}{6}(\mathbf{X}-\mathbf{t}_\ell)^3 D^3 f(\mathbf{t}_\ell)\right]\right\}^2.$$

Further,

$$\mathit{AMISE}(\widehat{f}) = h^6 C_S(f) + \frac{1}{nh^d}\left[\frac{(d+1)(d+2)}{2}\right],$$

and the optimal binwidth h_{opt} and the rate of convergence are

$$h_{opt} = \left[\frac{(d+1)(d+2)}{2C_S(f)}\right]^{1/(d+6)} n^{-1/(d+6)} \qquad \text{and} \qquad \mathit{AMISE} = O\left(n^{-6/(d+6)}\right).$$

A proof of Theorem 6.3 is given in Jing, Koch, and Naito (2012). The variance of the SOPHE $\widehat{f}$ has asymptotically the same rate $(nh^d)^{-1}$ as many kernel estimators. The bias term is derived from a third derivative, which results in an $O(h^6)$-term for squared bias, the large binwidth and the fast rate of convergence. Table 1 in Jing, Koch, and Naito (2012) summarises asymptotic properties of histogram estimators and smooth kernel estimators.

Theorem 6.3 concerns the d-dimensional density estimate $\widehat{f}$. The capability of SOPHE for clustering relies on the fact that it estimates the density separately for each d-dimensional bin. As d grows, an increasing proportions of bins will be empty. Empty bins are ignored by the SOPHE, whereas non-empty neighbouring bins are collected into clusters. In this sense, SOPHE is a cluster technique.

As d increases, the relatively large binwidth of the SOPHE is desirable as it directly affects the computational effort. Indeed, the SOPHE provides a good compromise: it combines smoothness and accuracy with an efficient estimation of modes and clusters.

Algorithm 6.2 is adapted from Jing, Koch, and Naito (2012) and shows how to apply the SOPHE to data.

Algorithm 6.2 ***Mode and Cluster Tracking with the SOPHE***

Let $\mathbb{X}$ be d-dimensional data, and assume that the variables have been scaled to an interval $[0, R]$, for some $R > 0$. Fix a threshold θ_0 and an integer $\nu > 1$.

Step 1. Define the binwidth h_ν to be $h_\nu = R/\nu$. Let $L_\nu = \nu^d$ be the total number of bins, and let B_ℓ (with $\ell = 1, \ldots, L_\nu$) be d-dimensional cubes with length h_ν.

Step 2. Find bins with high density. Find the number of observations n_ℓ in each bin, and put $\mathcal{B}_{\theta_0} = \{B_\ell \colon n_\ell > \theta_0\}$.

Step 3. Track modes. Calculate $\widehat{f}$ for each bin $B_\ell \in \mathcal{B}_{\theta_0}$. If $\widehat{f}$ has a local maximum for some B_k, then B_k contains a mode. Let $\mathcal{B}_{\text{modes}}$ be the set of modal bins.

Step 4. Determine clusters: For each modal bin $B_k \in \mathcal{B}_{\text{modes}}$, find the neighbouring non-empty bins, starting with the largest modal bin. If a modal bin has no non-empty neighbours, it does not give rise to a cluster.

Step 5. Increase ν to $\nu+1$, and repeat steps 1 to 4 for the new number of bins while the number of modes and clusters increases. ■

Experience with real and simulated data has shown that for increasing values of ν, the number of modes initially increases and then decreases, so starting with a small value of ν is recommended. A natural choice of the final binwidth is that which results in the maximum number of clusters. If this maximum is obtained for more than one binwidth, then the largest binwidth (and smallest ν) is recommended. This choice appears to work well for simulated data. Further, comparisons of the cluster results for *flow cytometry* data show good agreement with the analyses of the biologists (see Zaunders et al., 2012).

The choice of the threshold parameter θ_0 may affect the number of modes. Zaunders et al. (2012) described the analysis of real flow cytometry data sets with the SOPHE algorithm for data with five to thirteen variables and about 20,000 to 700,000 observations. Based on their analyses, they recommended starting with a value of $\theta_0 \in [20, 100]$, but larger values may be more appropriate, depending on the number of variables and the sample size.

The SOPHE algorithm works well for data with plenty of observations, as the next example shows.

Example 6.6 We consider **PBMC flow cytometry** data measured on peripheral blood mononuclear cells purified from blood, which are described and analysed in Zaunders et al. (2012). The acquisition of these flow cytometry data involves centrifuging blood to yield a concentrate of 30×10^6 cells in 1 ML of blood. As common in flow cytometry, *forward scatter (FS)* and *side scatter (SS)* are the first two variables of the data. The PBMC data consist of a total of ten variables and $n = 709{,}086$ cells. The cells are stained with the eight antibodies *CD8-APC (R670), CD4-AF-700 (R730), CD20-APC-CY7 (R780), CD45-FITC (B525), CD3-PERCP-CY5.5 (B695), CD16-PAC BLUE (V450), CCR3-PE (G585)* and *CD33-PE-CY7 (G780)*, and these antibodies correspond to the eight remaining variables.

Flow cytometry technology is improving rapidly through the introduction of new markers and antibodies, and each additional antibody or marker could result in a split of a subpopulation into smaller parts. The discovery of new subpopulations or clusters may lead to a link between markers and diseases.

After a log10 transformation of all variables (except *FS* and *SS*), I partition the ten-dimensional data into clusters using Algorithm 6.2 with $\nu = 4, \ldots, 8$ bins for each variable and the threshold $\theta_0 = 100$. The number of clusters initially increases and then decreases, as Table 6.5 shows. The maximum number of clusters, here fourteen, occurs for seven bins in each variable, so the data are divided into a total of $L_7 = 7^{10}$ bins. A large proportion of these bins are empty.

A summary of the clusters in terms of the centres of the modal bins offers the biologists or medical experts an intuitive interpretation of the cluster arrangement. Each bin is characterised by a sequence of ten integers in the range $1, \ldots, 7$, which show the bin number of each coordinate. The rows of Table 6.6 show the locations of the modes in their bin coordinates. We start with mode *m1* in the first row, which corresponds to the largest cluster. Here mode *m1* falls in the bin with bin coordinates [2 1 1 4 1 5 4 1 1 1], so the first variable, *FS*,

Table 6.5 *Number of Bins in Each Dimension and Number of Clusters from Example 6.6*

No. of bins	4	5	6	7	8
No. of clusters	8	11	11	14	12

Table 6.6 *Location of Modes m1–m14 in Bin Coordinates from Example 6.6*

	Variables										No. of cells in		% in
Mode	1	2	3	4	5	6	7	8	9	10	Mode	Cluster	cluster
m1	2	1	1	4	1	5	4	1	1	1	9,105	176,126	24.84
m2	3	2	6	1	2	5	4	1	1	1	7,455	131,863	18.60
m3	4	4	1	3	2	5	1	1	2	5	673	126,670	17.86
m4	3	2	6	3	2	5	4	1	1	1	4,761	64,944	9.16
m5	2	1	1	1	4	4	1	1	1	1	10,620	58,124	8.20
m6	3	2	1	1	1	5	5	3	1	1	4,057	33,190	4.68
m7	3	2	1	1	1	5	5	1	1	1	1,913	26,475	3.73
m8	3	2	4	1	1	5	1	4	1	1	1,380	25,793	3.64
m9	3	2	1	1	1	5	1	4	1	1	1,446	18,279	2.58
m10	3	2	4	1	1	5	5	3	1	1	619	12,515	1.76
m11	4	3	1	3	1	5	1	5	2	3	133	11,366	1.60
m12	3	2	1	1	1	3	1	1	6	3	182	8,670	1.22
m13	2	2	1	1	1	4	1	1	1	1	213	8,085	1.14
m14	3	2	4	5	1	5	4	1	1	1	177	6,986	0.99

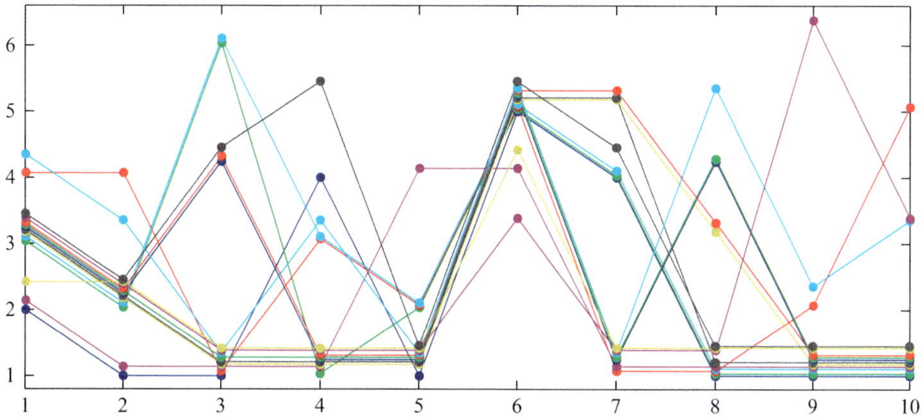

Figure 6.9 Parallel coordinate view of locations of modes from Table 6.6, Example 6.6, with variables on the x-axis, and bin numbers on the y-axis.

of mode *m1* is in bin 2 along the *FS* direction, the second variable, *SS*, is in bin 1 along the *SS* direction, and so on. For these data, no mode fell into bin 7 in any of the ten variables, so only bin numbers 1 to 6 occur in the table. The table further lists the number of cells in each modal bin and the number of cells in the cluster to which the modal bin belongs. The last column shows the percentage of cells in each cluster. The clusters are listed in decreasing order, with the first cluster containing 24.84 per cent of all cells. The smallest cluster contains a mere 0.99 per cent of all cells.

Figure 6.9 shows parallel coordinate views of the data in Table 6.6: the ten variables are shown on the x-axis, the y-axis shows the bin number, and each line connects the bin numbers of one mode. Unlike the usual parallel coordinate plots, I have increased the y-value for the second and later modes by a small amount as this makes it easier to see which modes share a bin and for which variables: Vertical dots in an interval $[\ell, \ \ell + 0.455]$ belong to the same bin number ℓ. With this interpretation, Table 6.6 and Figure 6.9 show that modes *m2, m4, m6–m10, m12* and *m14* fall into bin 3 for the first variable, *FS*, and into bin 2 for the second variable, *SS*. These modes therefore collapse into one when we consider the two variables *FS* and *SS* only. Indeed, all variables are required to divide the data into fourteen clusters.

From the summaries, the biologists were able to correctly identify the main clusters. These agree well with the major clusters found by the custom-designed flow cytometry software FlowJo, which requires an experienced biologist as it allows views of two variables at a time only. For details, see Zaunders et al. (2012). ■

Our second example deals with time series data from gene expression experiments. Unlike the HDLSS *breast tumour gene expression* data of Example 2.15 in Section 2.6.2, in our next example the genes are the observations, and the time points are the variables.

Example 6.7 The **South Australian gravevine** data of Davies, Corena, and Thomas (2012)[1] consist of observations taken from two vineyards in South Australia in different years: the Willunga region in 2004 and the Clare Valley in 2005. Assuming that the plants of one vineyard are biologically indistinguishable during a development cycle, the biologists record the expression levels of predetermined genes from about twenty vines – at weekly intervals during the development cycle of the grape berries – and then take the average of the twenty or so measurements at each time point. There are 2,062 genes from the vines of each vineyard – the observations – and the weekly time points are the variables.

The development stages of the grapes determine the start and end points of the recordings. Because of different environmental and treatment conditions in the two regions, there are nineteen weekly measurements for the Willunga vineyard and seventeen for the Clare Valley vineyard. Of interest to the biologist is the behaviour of the genes over time and the effect of different conditions on the development of the grapes. The biologists are interested in analysing and interpreting groupings of genes into clusters, and in particular, they want to find out whether the same genes from the two vineyards belong to the same cluster.

As a first step in an analysis of the grapevine data, I cluster the data with the SOPHE algorithm. I analyse the data from each vineyard separately and also cluster the combined data. As the Willunga data have more variables, we need to drop two variables before we can combine the two data sets. The approach I take for selecting the variables of the Willunga data is to leave out two variables at a time and to find combinations which are consistent with the cluster structures of the two separate data sets. It turns out that there are several

[1] For these data, contact Chris Davies, CSIRO Plant Industry, Glen Osmond, SA 5064, Australia.

Table 6.7 *Percentage of Genes in Each Cluster for the Combined Data, the Willunga and Clare Valley Data from Example 6.7, Shown Separately for the Two Combinations of Variables of the Willunga Data*

	Without time points 5 and 7			Without time points 5 and 16	
Cluster	Combined	Willunga	Clare Valley	Combined	Willunga
C1	43.23	47.14	44.42	43.89	47.24
C2	36.06	38.60	33.22	32.47	38.80
C3	11.11	14.26	14.11	14.82	13.97
C4	5.36		5.67	5.24	
C5	4.24		2.57	3.59	

such combinations; I show two which highlight differences that occur: the data without variables 5 and 7 and without variables 5 and 16. For each vineyard, the data are now of size $17 \times 2,062$, and the combined data contain all 4,124 observations.

In the SOPHE algorithm, I use the threshold $\theta_0 = 10$ because the sparsity of the seventeen-dimensional data results in 0.0001 to 0.001 per cent of non-empty bins, and of these, less than 1.5 per cent contain at least ten genes. For all three data sets, I use two to four bins, and in each case, three bins per dimension produce the best result. Table 6.7 shows the results for three bins in the form of the percentage of genes in each cluster. The table shows that most of the data belong to one of two clusters. The size of these large clusters varies slightly within the combined and Willunga data, a consequence of the fact that different subsets of variables of the Willunga data are used in the analyses.

Figure 6.10 shows parallel coordinate views of mode locations for all three data sets and the two subsets of variables used in the Willunga data. Variables 1 to 17 are shown on the x-axis, and the y-axis shows the bin number which takes values 1 to 3. As in Figure 6.9, I have shifted the bin numbers of the second and later modes in the vertical direction as this allows a clearer view of the locations of each mode. In the left panels of Figure 6.10, time points 5 and 7 are left out from the Willunga data, and in the right panel, time points 5 and 16 are left out. The top panels show the results for the combined data, the middle panels refer to the Clare Valley data, and the bottom panels refer to the Willunga results. The combined data and the Clare Valley data divide into five clusters, whereas the Willunga data separate into three clusters only.

In each panel, the mode of the largest cluster is shown in solid black and the second largest in solid blue. The third largest has red dots, the fourth largest is shown as a dash with black dots and the smallest can be seen as a dash with blue dots. The Clare Valley clusters are the same in both panels as they are calculated from the same data. For the Willunga data, there is a time shift between the left and right panels which reflects that time point 7 is dropped in the left panel and time point 16 in the right panel. For the combined data in the top panels, this time shift is noticeable in the largest black cluster but is not apparent in the second- and third-largest clusters. However, the second-smallest cluster on the left with 5.36 per cent of the data – see Table 6.7 – has become the smallest-cluster in the right panel with 3.59 per cent of the data belonging to it.

A comparison of the combined data and the Clare Valley data shows the close agreement between the five modes in each. A likely interpretation is that the clusters of the Clare Valley

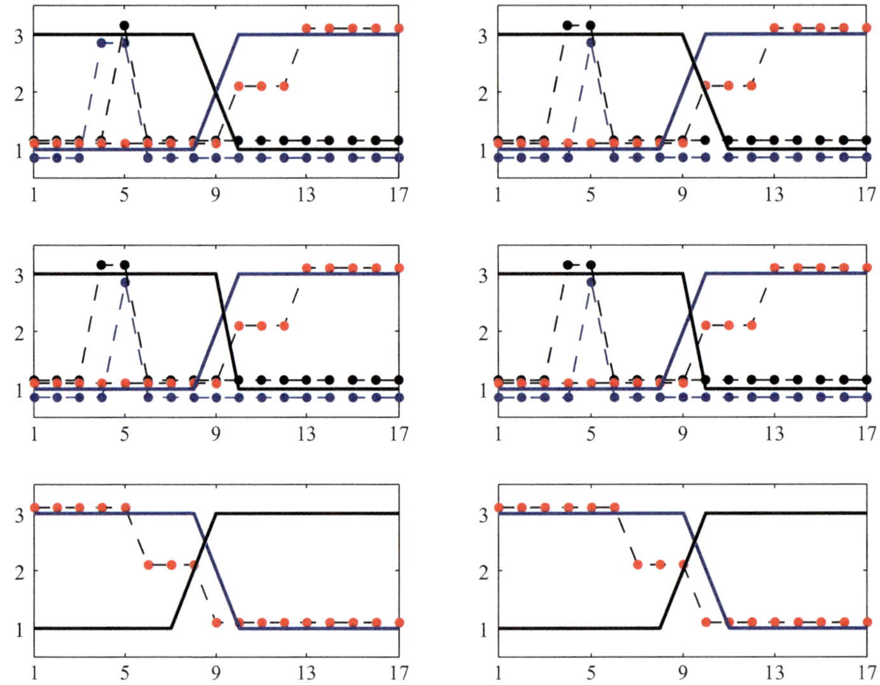

Figure 6.10 Parallel coordinate views of locations of modes from Example 6.7 with variables on the x-axis. Panels on the left and right use different subsets of variables of the Willunga data. Top panels show the five modes of the combined data, the middle panels show those of the Clare Valley data and the lower panels show the three modes of the Willunga data.

data are sufficiently distinct from each other that they do not merge when the Willunga data are added.

The two large clusters in each data set correspond to modes whose locations are very similar: the mode of the largest cluster for the combined data lies in bin 3 until variable 8 or 9 and then drops to the lowest level, bin 1, whereas the mode of the second-largest cluster is low until variable 8 and then stays in bin 3 and so remains high. A closer inspection reveals that the largest cluster of the combined data and Clare Valley data is that of the 'high-then-low' mode, whereas the mode of the largest Willunga cluster, containing about 47 per cent of genes, is of the form 'low-then-high'. The reason for this discrepancy is visible in Figure 6.10. The third mode of the Willunga data is not very different from the 'high-then-low' blue mode; in the combined data, this third cluster seems to have been absorbed into the cluster belonging to the black mode and is no longer discernible as a separate cluster.

The analyses highlight similarities and differences in the cluster structure of the combined data and the separate data of the two vineyards. All data sets divide into two large clusters which account for about 80 per cent of all genes, and some smaller clusters. The two large clusters are similar in size and location for all data sets, but there are differences in the number and location of the smaller clusters. The two large clusters are complementary: one cluster has high initial expression levels (black in the combined data) and around time 9 drops to the lowest level, whereas the other large cluster starts with low levels and then rises to high levels later. The smaller clusters have a less clear interpretation but might be of interest to the biologist. ■

In the two examples of this section, I use different thresholds. If the threshold is too high, then we will not be able to find all modes, especially for larger dimensions, because the number of bins increases as ν^d with the dimension – a nice illustration of the 'curse of dimensionality'.

Although originally intended for density estimation, the SOPHE performs well in finding modes and their clusters. The SOPHE can be used to find regions of high density by simply adding a second threshold parameter, $\theta_1 > \theta_0$, which allows only those bins to be joined to modes and modal regions that have more than θ_1 observations. Unlike k-means clustering, the SOPHE does not require the user to specify the number of clusters but relies instead on a suitable choice of the bin threshold θ_0.

6.5 Principal Components and Cluster Analysis

As in Principal Component Discriminant Analysis in Section 4.8.2, we want to include a dimension-reduction step into Cluster Analysis. We may use Principal Component Analysis in different ways, namely,

1. to reduce the dimension of the original data to the first few principal component data, or
2. to use the shape of the principal components as an exploratory tool for choosing the number of clusters.

In Principal Component Discriminant Analysis, the rule is based on the principal component data. For the thirty-dimensional *breast cancer* data, Example 4.12 in Section 4.8.3 shows that Fisher's rule performs better for lower-dimensional PC data than for the original data. Is this also the case in Cluster Analysis? Because Cluster Analysis does not have a natural error criterion which assesses the performance of a method, this question does not have a conclusive answer. However, dimension reduction prior to clustering will make clustering more efficient – an aspect which becomes increasingly important for high-dimensional data.

The main idea of **Principal Component Cluster Analysis** is to replace the original data by lower-dimensional PC data $\mathbb{W}^{(q)}$ and to partition the derived data $\mathbb{W}^{(q)}$ into clusters. Any of the basic clustering methods – hierarchical agglomerative clustering, k-means clustering or SOPHE – can be used in Principal Component Cluster Analysis. We consider a k-means approach for PC data, and I refer the reader to the Problems at the end of Part II for an analogous approach based on hierarchical agglomerative clustering or SOPHE.

6.5.1 *k-Means Clustering for Principal Component Data*

Let $\mathbb{X} = \begin{bmatrix}\mathbf{X}_1\ \mathbf{X}_2 \cdots \mathbf{X}_n\end{bmatrix}$ be d-dimensional data. Fix $k > 1$, and let

$$\mathcal{P} = \mathcal{P}(\mathbb{X}, k) = \{\, \mathcal{C}_\nu = \mathcal{C}(\overline{\mathbf{X}}_\nu, n_\nu) \colon \nu = 1, \ldots, k\} \tag{6.7}$$

be an optimal k-cluster arrangement for $\mathbb{X}$, where n_ν is the number of observations in cluster $\mathcal{C}_\nu$ and $n = \sum n_\nu$. We write $\mathcal{C}(\overline{\mathbf{X}}_\nu, n_\nu)$ for the cluster $\mathcal{C}_\nu$ when we want to emphasise the cluster centroid $\overline{\mathbf{X}}_\nu$ and the number of observations n_ν in the cluster.

Fix $q < d$, and let $\mathbb{W} = \mathbb{W}^{(q)}$ be the q-dimensional principal component data. Assume that $\mathbb{W}$ have been partitioned into k clusters $\widetilde{\mathcal{C}}_\nu$ with cluster centroids $\overline{\mathbf{W}}_\nu$ and m_ν vectors belonging to $\widetilde{\mathcal{C}}_\nu$, for $\nu \le k$. Then

$$\mathcal{P}_{\mathbb{W}} = \mathcal{P}(\mathbb{W}, k) = \{\, \widetilde{\mathcal{C}}_\nu = \widetilde{\mathcal{C}}(\overline{\mathbf{W}}_\nu, m_\nu) \colon \nu = 1, \ldots, k\} \tag{6.8}$$

defines a k-**cluster arrangement** for $\mathbb{W}$.

We define an optimal k-cluster arrangement for $\mathbb{W}$ as in Definition 6.2, and calculate within-cluster variabilities similar to those of (6.1) by replacing the original random vectors by the PC vectors. As before, k-**means clustering** for $\mathbb{W}$ refers to an optimal partitioning of $\mathbb{W}$ into k clusters.

For a fixed k, optimal cluster arrangements for $\mathbb{X}$ and $\mathbb{W}$ may not be the same; the cluster centres of the PC clusters $\overline{\mathbf{W}}_\nu$ can differ from the PC projections $\widehat{\Gamma}_q^{\mathsf{T}}(\overline{\mathbf{X}}_\nu - \overline{\mathbf{X}})$, and the numbers in each clusters may change. The next example illustrates what can happen.

Example 6.8 We continue with the six clusters **simulated data** but only use the ten- and twenty-dimensional data. Table 6.2 gives details for the cluster means of the first five variables which are responsible for the clusters. As in Example 6.4, the remaining five and fifteen variables, respectively, are 'noise' variables which are generated as $\mathcal{N}(\mathbf{0}, D)$ samples, where D is a diagonal matrix with constant entries σ^2, and $\sigma = 0.25$ and 0.5.

A principal component analysis prior to clustering can reduce the noise and result in a computationally more efficient and possibly more correct cluster allocation. As expected, the first eigenvector of the sample covariance matrix S of $\mathbb{X}$ puts most weight on the first five variables; these are the variables with distinct cluster centroids and thus have more variability.

To gain some insight into the PC-based approach, I apply the k-means algorithm with $k = 6$ to the PC data $\mathbb{W}^{(1)}$, $\mathbb{W}^{(3)}$ and $\mathbb{W}^{(5)}$ and compare the resulting 6-means cluster arrangements with that of the raw data. In the calculations, I use the Euclidean distance and the best in twenty runs. Table 6.8 shows the number of observations in each cluster for $\sigma = 0.5$ and in parentheses for $\sigma = 0.25$. The clusters are numbered in the same way as in Table 6.2. The column 'True' lists the number of observations for each simulated cluster.

The results show that the cluster allocation improves with more principal components and is better for the smaller standard deviation – compared with the actual numbers in each cluster. The five-dimensional PC data $\mathbb{W}^{(5)}$ with $\sigma = 0.25$, which are obtained from the twenty-dimensional data, show good agreement with the actual numbers; k-means clustering does better for these data than the original twenty-dimensional data. The results show that eliminating noise variables improves the cluster allocation. The one-dimensional PC data $\mathbb{W}^{(1)}$ had the poorest performance with respect to the true clusters, which sugests that the first PC does not adequately summarise the information and structure in the data.

The calculations I have presented deal with a single data set for each of the four cases – two different variances and the ten- and twenty-dimensional data. It is of interest to see how PC-based clustering performs on average. We turn to such calculations in the Problems at the end of Part II. ■

The example illustrates that dimension reduction with principal components can improve a cluster allocation. In practice, I recommend calculating optimal cluster arrangements for PC data over a range of dimensions. As the dimension of the PC data increases, the numbers

Table 6.8 *Number of Observations in Each Cluster for PC Data* $\mathbb{W}^{(\ell)}$*, with* $\ell = 1,3,5$*, from Example 6.8 and* $\sigma = 0.5$

Cluster	True	$\mathbb{W}^{(1)}$	$\mathbb{W}^{(3)}$	$\mathbb{W}^{(5)}$	All data
Number in clusters for ten-dimensional data					
1	250	192 (240)	224 (252)	223 (251)	227 (251)
2	200	249 (225)	225 (198)	225 (199)	223 (199)
3	200	178 (189)	160 (194)	176 (195)	148 (197)
4	150	175 (182)	182 (152)	154 (150)	174 (149)
5	100	119 (105)	121 (100)	120 (101)	117 (100)
6	100	87 (88)	88 (104)	102 (104)	111 (104)
Number in clusters for twenty-dimensional data					
1	250	196 (227)	235 (250)	231 (250)	233 (251)
2	200	253 (223)	211 (200)	214 (200)	214 (199)
3	200	190 (196)	196 (201)	191 (200)	187 (200)
4	150	139 (143)	161 (148)	166 (150)	156 (149)
5	100	144 (108)	108 (99)	112 (99)	115 (100)
6	100	78 (103)	89 (102)	86 (101)	95 (101)

Note: Results in parentheses are for the data with $\sigma = 0.25$.

in the clusters will become more stable. In this case I recommend using the smallest number of PCs for which the allocation becomes stable. In the preceding example, this choice results in $\mathbb{W}^{(5)}$.

We return to the thirty-dimensional *breast cancer* data and investigate k-means clustering for the PC data. Although we know the classes for these data, I will calculate optimal k-cluster arrangements for $2 \le k \le 4$. The purpose of this example is to highlight differences between Discriminant Analysis and Cluster Analysis.

Example 6.9 We continue with the **breast cancer** data and explore clustering of the principal component data. For the PC data $\mathbb{W}^{(1)}, \ldots, \mathbb{W}^{(10)}$ obtained from the raw data, I calculate k-means clusters for $2 \le k \le 4$ over ten runs based on the Euclidean distance. The within-cluster variability and the computational effort increase as the dimension of the principal component data increases. The efficiency of clustering becomes more relevant for large data, and thus improvements in efficiency at no cost in accuracy are important.

For $k = 2$, the optimal cluster arrangements are the same for all PC data $\mathbb{W}^{(1)}, \ldots, \mathbb{W}^{(10)}$ and agree with the 2-means clustering of $\mathbb{X}$ (see Example 6.5). For $k = 3$, the optimal cluster arrangements for $\mathbb{X}$ and $\mathbb{W}^{(q)}$ agree for $q = 1,2,3,9,10$ but differ by as many as twenty-two observations for the other values of q. For $k = 4$, the results are very similar and differ by at most two observations.

When we divide the data into four clusters, essentially each of the clusters of the 2-cluster arrangement breaks into two clusters. This is not the case when we consider three clusters, as the calculations in Example 6.5 show.

For $k = 2$, the calculations show that low-dimensional PC data suffice: the same cluster arrangements are obtained for the one-dimensional PC data as for the thirty-dimensional raw data. One reason why dimension reduction prior to clustering works so well for these data is the fact that PC_1 carries most of the variability. ■

The analysis of the *breast cancer* data shows that dimension reduction prior to a cluster analysis can reduce the computational effort without affecting optimal cluster arrangements.

6.5.2 Binary Clustering of Principal Component Scores and Variables

In Discriminant Analysis, two-class problems are of particular interest, and the same applies to Cluster Analysis. In this section we borrow the idea of a rule and adapt it to separate data into two clusters.

Definition 6.4 Let $\mathbb{X} = \begin{bmatrix}\mathbf{X}_1 \cdots \mathbf{X}_n\end{bmatrix}$ be data from two subpopulations. Let $\overline{\mathbf{X}}$ be the sample mean and S the sample covariance matrix of $\mathbb{X}$. Write $S = \widehat{\Gamma}\widehat{\Lambda}\widehat{\Gamma}^{\mathsf{T}}$ for the spectral decomposition of S. For $\ell \leq d$, let $\mathbf{W}_{\bullet\ell} = \widehat{\eta}_\ell^{\mathsf{T}}(\mathbb{X} - \overline{\mathbf{X}})$ be the ℓth principal component score of $\mathbb{X}$. Define the **PC$_\ell$ sign cluster rule** $\mathfrak{r}_{\mathrm{PC}_\ell}$ by

$$\mathfrak{r}_{\mathrm{PC}_\ell}(\mathbf{X}_i) = \begin{cases} 1 & \text{if } W_{i\ell} > 0, \\ 2 & \text{if } W_{i\ell} < 0, \end{cases} \qquad (6.9)$$

which partitions the data into the two clusters

$$\mathcal{C}_1 = \{\mathbf{X}_i : \mathfrak{r}_{\mathrm{PC}_\ell}(\mathbf{X}_i) = 1\} \qquad \text{and} \qquad \mathcal{C}_2 = \{\mathbf{X}_i : \mathfrak{r}_{\mathrm{PC}_\ell}(\mathbf{X}_i) = 2\}.$$

□

The name *rule* invites a comparison with the rule and its decision function in Discriminant Analysis as it plays a similar role. Typically, we think of the PC$_1$ or PC$_2$ sign cluster rule because the first few PCs contain more of the variability and structure of the data. Instead of the eigenvector of S and the resulting PC$_\ell$ data, one can use other direction vectors, such as those in Section 13.2.2, as the starting point to a cluster rule.

I illustrate cluster rule (6.9) with the *illicit drug market* data. The two clusters we obtain will improve the insight into the forces in the Australian illicit drug market.

Example 6.10 In the previous analysis of the **illicit drug market** data, I used a split of the seventeen series into two groups, the 'direct' and 'indirect' measures of the market. The seventeen series or observations are measured over sixty-six consecutive months. In the current example, we do not use this split. Instead, I apply cluster rule (6.9) to the data – scaled as described in Example 2.14 of Section 2.6.2.

The first eigenvalue contributes more than 45 per cent to the total variance and so is relatively large. From (2.7) in Section 2.3, the first principal component projections

$$\mathbb{P}_{\bullet 1} = \widehat{\Gamma}_1\widehat{\Gamma}_1^{\mathsf{T}}(\mathbb{X} - \overline{\mathbf{X}}) = \widehat{\Gamma}_1\mathbf{W}_{\bullet 1}$$

represent the contribution of each observation in the direction of the first eigenvector. A parallel coordinate view of these projections is shown in the left subplot of Figure 6.11. Following the differently coloured traces, we see that the data polarise into two distinct groups. Thus, the PC$_1$ data are a candidate for cluster rule (6.9). The entries of the first principal component score, $\mathbf{W}_{\bullet 1}$, are shown in the right subplot of Figure 6.11, with the x-axis representing the observation number. This plot clearly shows the effect of the cluster rule and the two parts the data divide into. Cluster 1 corresponds to the series with positive $\mathbf{W}_{\bullet 1}$ values in Figure 6.11. Table 6.9 shows which cluster each series belongs to. It is interesting to note that the same cluster arrangement is obtained with the 2-means clustering of the scaled data.

Table 6.9 *Two Clusters Obtained with Cluster Rule (6.9) for Example 6.10*

	Cluster 1		Cluster 2
1	Heroin possession offences	2	Amphetamine possession offences
5	Heroin overdoses (ambulance)	3	Cocaine possession offences
6	ADIS heroin	4	Prostitution offences
8	PSB new registrations	7	PSB reregistrations
9	Heroin deaths	10	ADIS cocaine
14	Drug psychoses	11	ADIS amphetamines
16	Break and enter dwelling	12	Robbery 1
		13	Amphetamine overdoses
		15	Robbery 2
		17	Steal from motor vehicles

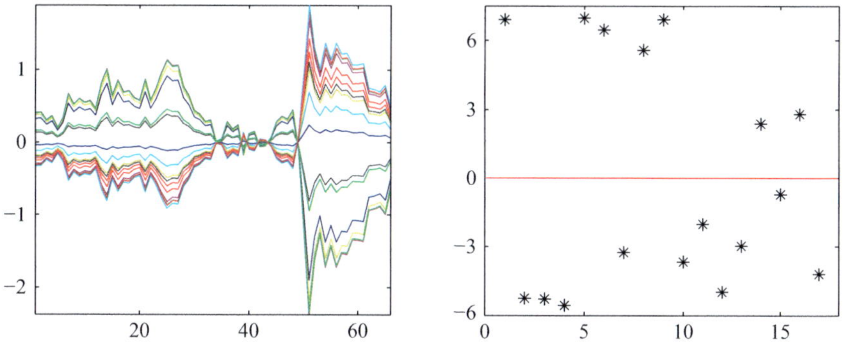

Figure 6.11 First principal component projections $\mathbb{P}_{\bullet 1}$ for Example 6.10 and 2-cluster arrangement with cluster rule (6.9).

The series numbers in Table 6.9 are the same as those of Table 3.2 in Section 3.3. The partition we obtain with rule (6.9) separates the data into heroin-related series and all others. For example, *PSB new registrations* are primarily related to heroin offences, whereas the *PSB reregistrations* apply more generally. This partition differs from that of the direct and indirect measures of the market (see Example 3.3), and the two ways of partiting the data focus on different aspects. ■

It is important to realise that there is not a 'right' and 'wrong' way to partition data, instead, the different partitions provide deeper insight into the structure of the data and provide a more complete picture of the drug market.

Our second example is a variant of cluster rule (6.9); it deals with clustering variables rather than observations. Clustering variables is of interest when subsets of variables show different behaviours.

Example 6.11 The variables of the **Dow Jones returns** are thirty stocks observed daily over a ten year period starting in January 1991. Example 2.9 of Section 2.4.3 shows the first two projection plots of these data. An interesting question is whether we can divide the stocks into different groups based on the daily observations.

The PC sign cluster rule applies to the observations. Here we require a rule that applies to the variables $\mathbf{X}_{\bullet 1} \cdots \mathbf{X}_{\bullet d}$. Let S be the sample covariance matrix of $\mathbb{X}$. For $j \leq d$, let $\widehat{\boldsymbol{\eta}}_j = \left[\widehat{\eta}_{j1}, \ldots, \widehat{\eta}_{jd}\right]^{\mathsf{T}}$ be the jth eigenvector of S. Put

$$\mathfrak{r}_j(\mathbf{X}_{\bullet k}) = \begin{cases} 1 & \text{if } \widehat{\eta}_{jk} > 0 \\ 2 & \text{if } \widehat{\eta}_{jk} < 0. \end{cases} \tag{6.10}$$

Then $\mathfrak{r}_j$ assigns variables into two groups which are characterised by the positive and negative entries of the jth eigenvector.

All entries of the first eigenvector of S have the same sign; this eigenvector is therefore not appropriate. The second eigenvector has positive and negative entries. A visual verification of the sign change of the entries is apparent in the cross-over of the lines in the bottom-left projection plot of Figure 2.8 in Example 2.9.

Using the second eigenvector for rule (6.10), the variables 3, 13, 16, 17 and 23 have negative weights, whereas the other twenty-five variables have positive eigenvector weights. The five variables with negative weights correspond to the information technology (IT) companies in the list of stocks, namely, *AT&T, Hewlett-Packard, Intel Corporation, IBM and Microsoft*. Thus the second eigenvector is able to separate the IT stocks from all other stocks. ■

The two examples of this section show how we can adapt ideas from Discriminant Analysis to Cluster Analysis. The second example illustrates two points:

- If the first eigenvector does not lead to interesting information, we need to look further, either at the second eigenvector or at other directions.
- Instead of grouping the observations, the variables may divide into different groups. It is worth exploring such splits as they can provide insight into the data.

6.5.3 Clustering High-Dimensional Binary Data

As the dimension increases, it may not be possible to find tight clusters, but if natural groupings exist in the data, it should be possible to discover them. In (2.25) of Section 2.7.2 we consider the angle between two (direction) vectors as a measure of closeness for high-dimensional data. The cosine distance is closely related to the angle and is often more appropriate than the Euclidean distance for such data.

The ideas I describe in this section are based on the research of Marron, Koch and Gustafsson and are reported as part of his analysis of proteomics data of mass spectrometry profiles in Gustafsson (2011).

Dimension reduction is of particular interest for high-dimensional data, and Principal Component Analysis is the first tool we tend to consider. For twenty-dimensional data, Example 6.8 illustrates that care needs to be taken when choosing the number of principal components. The first principal component scores may not capture enough information about the data, especially as the dimension grows, and typically it is not clear how to choose an appropriate number of principal components. In this section we explore a different way of decreasing the complexity of high-dimensional data.

Let $\mathbb{X} = \left[\mathbf{X}_1 \cdots \mathbf{X}_n\right]$ be d-dimensional data. Let $\tau \geq 0$ be a fixed threshold. Define the **(derived) binary data**

$$\mathbb{X}^{\{0,1\}} = \left[\mathbf{X}_1^{\{0,1\}}, \ldots, \mathbf{X}_n^{\{0,1\}}\right],$$

where the jth entry $X_{ij}^{\{0,1\}}$ of $\mathbf{X}_i^{\{0,1\}}$ is

$$X_{ij}^{\{0,1\}} = \begin{cases} 1 & \text{if } X_{ij} > \tau \\ 0 & \text{if } X_{ij} \leq \tau. \end{cases}$$

The binary data $\mathbb{X}^{\{0,1\}}$ have the same dimension as $\mathbb{X}$ but are simpler, and one can group the data by exploiting where the data are non-zero.

A little reflection tells us that the cosine distance Δ_{cos} of (5.6) in Section 5.3.1 separates binary data: if $\mathbf{e}_k$ and $\mathbf{e}_\ell$ are unit vectors in the direction of the kth and ℓth variables, then $\Delta_{\text{cos}}(\mathbf{e}_k, \mathbf{e}_\ell) = 1$ if $k = \ell$, and $\Delta_{\text{cos}}(\mathbf{e}_k, \mathbf{e}_\ell) = 0$ otherwise.

The Hamming distance Δ_{Ham} of (5.9) in Section 5.3.1 is defined for binary data: it counts the number of entries for which $\mathbf{X}_k^{\{0,1\}}$ and $\mathbf{X}_\ell^{\{0,1\}}$ differ. Binary random vectors whose zeroes and ones agree mostly will have a small Hamming distance, and random vectors with complementary zeroes and ones will have the largest distance. Note that the Euclidean distance applied to binary data is equivalent to the Hamming distance.

I use the cosine and the Hamming distances in a cluster analysis of mass spectrometry curves.

Example 6.12 The observations of the **ovarian cancer proteomics** data of Example 2.16 in Section 2.6.2 are mass spectrometry curves or profiles. Each of the 14,053 profiles corresponds to (x, y) coordinates of an ovarian cancer tissue sample, and each profile is measured at 1,331 mass-to-charge m/z values, our variables. The top-left panel of Figure 6.12 is the same as Figure 2.17 in Example 2.16 and shows a stained tissue sample with distinct tissue types. To capture these different regions by different clusters, it is natural to try to divide the data into three or more clusters.

For each m/z value, the counts of peptide ion species are recorded. Figure 2.18 in Example 2.16 shows a subset of the profiles. A closer inspection of the data reveals that many entries are zero; that is, no peptide ions are detected. It is natural to summarise the data into 'ions detected' and 'no ions detected' for each m/z value. This process results in binary data $\mathbb{X}^{\{0,1\}}$ with threshold $\tau = 0$, and we now focus on these data rather than the original data.

We do not use Principal Component Analysis for dimension reduction of the binary data. Instead we observe that the PC_1 and PC_2 data yield interesting results in connection with the PC sign cluster rule of Section 6.5.2. Each profile arises from a grid point and corresponding (x, y) coordinates on the tissue sample. It is convenient to display the cluster membership of each profile as a coloured dot at its (x, y) coordinates; this results in a **cluster map** or **cluster image** of the data. Figure 6.12 shows the cluster maps obtained with the PC_1 sign cluster rule in the top-middle panel and those obtained with PC_2 in the top-right panel. Coordinates with green dots – 6,041 for PC_1 and 5,894 for PC_2 – belong to the first cluster, and those with blue dots belong to the second.

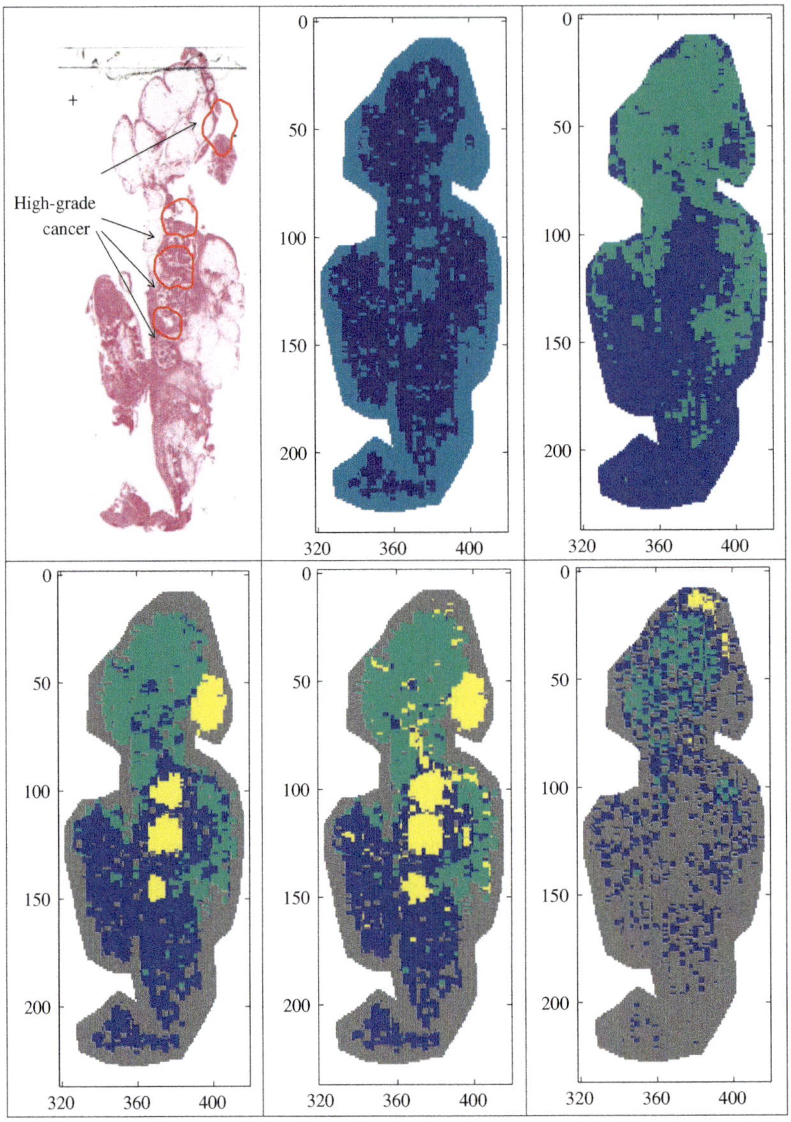

Figure 6.12 Stained tissue sample and cluster maps from Example 6.12. (*Top*): Maps from the PC_1 (*middle*) and PC_2 (*right*) sign cluster rule; (*bottom*): 4-means clustering with the Hamming distance (*left*), cosine distance (*middle*) and Euclidean distance applied to the raw data (*right*).

In the cluster map of the PC_1 data, the cancer regions merge with the background of the tissue matrix, and these regions are distinguished from the non-cancerous regions. In contrast, PC_2 appears to divide the data into the two distinct non-cancerous tissue types, and the cancer tissue has been merged with the peritoneal stroma, shown in bright red in the top-left panel. These first groupings of the data are promising and are based merely on the PC_1 and PC_2 data.

In a deeper analysis of the binary data, I apply k-means clustering with the Hamming and cosine distances and $k = 3, 4$ and 5. Figure 6.12 shows the resulting cluster images for $k = 4$; the bottom-left panel displays the results based on the Hamming distance, and

Table 6.10 *Numbers of Profiles in Four Clusters for Example 6.12*

Data	Distance	Grey	Blue	Green	Yellow
Binary	Hamming	4,330	5,173	3,657	893
Binary	Cosine	4,423	4,301	3,735	1,594
Raw	Euclidean	9,281	3,510	1,117	145

Note: Cluster colours as in Figure 6.12.

the bottom-middle panel shows those obtained with the cosine distance. For comparison, the bottom-right panel shows the results of 4-means clustering of the raw data with the Euclidean distance. The cluster map of the original (raw) data does not lead to interpretable results.

The cluster images arising from the Hamming and cosine distances are similar, and the blue, green and yellow regions agree well with the three tissue types shown in the top-left panel of the stained tissue sample. The high-grade cancer area – shown in yellow – corresponds to the circled regions of the stained tissue sample. Green corresponds to the adipose tissue, which is shown in the light colour in the top-left panel, and blue corresponds to peritoneal stroma shown in bright red. The grey regions show the background, which arises from the MALDI-IMS. This background region is not present in the stained tissue sample. For details, see Gustafsson (2011).

The partitioning obtained with the Hamming distance appears to agree a little better with the tissue types in the stained tissue sample than that obtained with the cosine distance. The bottom-right panel, whose partitioning is obtained directly from the raw data, does not appear to have any interpretable pattern.

Table 6.10 shows the number of profiles that belong to each of the four clusters. The clusters are described by the same colour in the table and the lower panels of Figure 6.12. The numbers in Table 6.10 confirm the agreement of the two cluster distances used on the binary data. The main difference is the much larger number of profiles in the yellow cosine cluster. A visual comparison alone will not allow us to decide whether these additional regions are also cancerous; expert knowledge is required to examine these regions in more detail.

The figure and table show that 4-means clustering of the binary data leads to a partitioning into regions of different tissue types. In particular, regions containing the high-grade cancer are well separated from the other tissue types. A comparison of the partitioning of the original and binary data demonstrates that the simpler binary data contain the relevant information. ■

This example illustrates that transforming high-dimensional data to binary data can lead to interpretable partitionings of the data. Alternatively, dimension reduction or other variable selection prior to clustering such as the sparse k-means ideas of Witten and Tibshirani (2010) could lead to interpretable cluster allocations.

If reliable information about tissue types is available, then classification rather than a cluster analysis is expected to lead to a better partitioning of the data. Classification, based on the binary data, could make use of the more exploratory results obtained in a preliminary cluster analysis such as that presented in Example 6.12.

6.6 Number of Clusters

The analyses of the *breast cancer* data in Examples 6.5 and 6.9 suggest that care needs to be taken when we choose the number of clusters. It may be naive to think that the correct partitioning of the data are two clusters just because the data have two different class labels. In the introduction to this chapter I refer to the label as an 'extra variable'. This extra variable is essential in Discriminant Analysis. In Cluster Analysis, we make choices without this extra variable; as a consequence, the class allocation which integrates the label and the cluster allocation which does not may differ.

For many data sets, we do not know the number of components that make up the data. For k-means clustering, we explore different values of k and consider the within-cluster variability for each k. Making inferences based on within-cluster variability plots alone may not be adequate.

In this section we look at methods for determining the number of components. There are many different approaches in addition to those we look at in the remainder of this chapter. For example, the problem of determining the number of clusters can be cast as a model-selection problem: the number of clusters corresponds to the order of the model, and the parameters are the means, covariance matrices and the number of observations in each component. Finite mixture models and *expectation-maximisation* (EM) algorithms represent a standard approach to this model-selection problem; see Roweis and Ghahramani (1999), McLachlan and Peel (2000), Fraley and Raftery (2002) and Figueiredo and Jain (2002) and references therein for good accounts of model-based approaches and limitations of these methods. The EM methods are generally likelihood-based – which normally means the Gaussian likelihood – are greedy and may be slow to converge. Despite these potential disadvantages, EM methods enjoy great popularity as they relate to classical Gaussian mixture models. Figueiredo and Jain (2002) included a selection of the number of clusters in their EM-based approach. Other approaches use ideas from information theory such as entropy (see, e.g., Gokcay and Principe 2002).

6.6.1 Quotients of Variability Measures

The within-cluster variability W of (6.1) decreases to zero as the number of clusters increases. A naive minimisation of the within-cluster variability is therefore not appropriate as a means to finding the number of clusters. In Discriminant Analysis, Fisher (1936) proposed to partition the data in such a way that the variability in each class is minimised and the variability between classes is maximised. Similar ideas apply to Cluster Analysis.

In this section it will be convenient to regard the within-cluster variability W of (6.1) as a function of k and to use the notation $W(k)$.

Definition 6.5 Let $\mathbb{X}$ be d-dimensional data. Let Δ be the Euclidean distance. Fix $k > 0$, and let $\mathcal{P}$ be a partition of $\mathbb{X}$ into k clusters with cluster centroids $\overline{\mathbf{X}}_\nu$, and $\nu \leq k$. The **between-cluster variability**

$$B(k) = \sum_{\nu=1}^{k} \Delta(\overline{\mathbf{X}}_\nu, \overline{\mathbf{X}})^2.$$

□

The between-cluster variability B is closely related to the trace of the between-sample covariance matrix $\widehat{B}$ of Corollary 4.9 in Section 4.3.2. The two concepts, however, differ in that B is based on the sample mean $\overline{\mathbf{X}}$, whereas $\widehat{B}$ compares the class means to their average.

Milligan and Cooper (1985) reviewed thirty procedures for determining the number of clusters. Their comparisons are based on two to five distinct non-overlapping clusters consisting of fifty points in four, six and eight dimensions. Most data sets we consider are much more complex, and therefore, many of the procedures they review have lost their appeal. However, the best performer according to Milligan and Cooper (1985), the method of Calinski and Harabasz (1974), is still of interest.

In chronological order, I list criteria for choosing the number of clusters. For notational convenience, I mostly label the criteria by the initials of the authors who proposed them.

1. Calinski and Harabasz (1974)

$$\mathrm{CH}(k) = \frac{B(k)/(k-1)}{W(k)/(n-k)} \qquad \text{with } \widehat{k}_{\mathrm{CH}} = \operatorname{argmax} \mathrm{CH}(k). \tag{6.11}$$

2. Hartigan (1975)

$$\mathrm{H}(k) = \left\{ \frac{W(k)}{W(k+1)} - 1 \right\} (n-k-1) \qquad \text{with } \widehat{k}_{\mathrm{H}} = \min\{k\colon H(k) < 10\}, \tag{6.12}$$

which is based on an approximate F distribution cutoff.
3. Krzanowski and Lai (1988):

$$\mathrm{KL}(k) = \left| \frac{\mathrm{Diff}(k)}{\mathrm{Diff}(k+1)} \right| \qquad \text{with } \widehat{k}_{\mathrm{KL}} = \operatorname{argmax} \mathrm{KL}(k)\,, \tag{6.13}$$

where $\mathrm{Diff}(k) = (k-1)^{2/d} W(k-1) - k^{2/d} W(k)$.
4. The within-cluster variability quotient

$$\mathrm{WV}(k) = \frac{W(k)}{W(k+1)} \qquad \text{with } \widehat{k}_{\mathrm{WV}} = \max\{k\colon \mathrm{WV}(k) > \tau\}, \tag{6.14}$$

where $\tau > 0$ is a suitably chosen threshold, typically $1.2 \le \tau \le 1.5$.

Common to these four criteria is the within-cluster variability – usually in the form of a difference or quotient. The expression of Calinski and Harabasz (1974) mimics Fisher's quotient. Hartigan (1975) is the only expression that explicitly refers to the number of observations. The condition $H(k) < 10$, however, poses a severe limitation on the applicability of Hartigan's criterion to large data sets. I have included what I call the WV quotient as this quotient often works well in practice. It is based on Hartigan's suggestion but does not depend on n. In general, I recommend considering more than one of the $\widehat{k}$ statistics.

Before I apply these four statistics to data, I describe the methods of Tibshirani, Walther, and Hastie (2001) and Tibshirani and Walther (2005) in Sections 6.6.2 and 6.6.3, respectively.

6.6.2 The Gap Statistic

The SOPHE of Section 6.4 exploits the close connection between density estimation and Cluster Analysis, and as the dimension grows, this relationship deepens because we focus more on the modes or centres of clusters and high-density regions rather than on estimating the density everywhere. The question of the correct number of modes remains paramount. The idea of testing for the number of clusters is appealing and has motivated the approach of Tibshirani, Walther, and Hastie (2001), although their method does not actually perform statistical tests.

The idea behind the *gap statistic* of Tibshirani, Walther, and Hastie (2001) is to find a benchmark or a *null distribution* and to compare the observed value of W with the expected value under the null distribution: large deviations from the mean are evidence against the null hypothesis.

Instead of working directly with the within-cluster variability W, Tibshirani, Walther, and Hastie (2001) consider the deviation of $\log W$ from the expectation under an appropriate null reference distribution.

Definition 6.6 Let $\mathbb{X}$ be d-dimensional data, and let W be the within-cluster variability. For $k \geq 1$, put

$$\mathrm{Gap}(k) = \mathbb{E}\{\log[W(k)]\} - \log[W(k)],$$

and define the **gap statistic** $\widehat{k}_{\mathrm{G}}$ by

$$\widehat{k}_{\mathrm{G}} = \underset{k}{\operatorname{argmax}}\, \mathrm{Gap}(k). \tag{6.15}$$

□

It remains to determine a *suitable* reference distribution for this statistic, and this turns out to be the difficult part. For univariate data, the uniform distribution is the preferred single-component distribution, as theorem 1 of Tibshirani, Walther, and Hastie (2001) shows, but a similar result for multivariate data does not exist. Their theoretical results lead to two choices for the reference distribution:

1. The uniform distribution defined on the product of the ranges of the variables,
2. The uniform distribution defined on the range obtained from the principal component data.

In the first case, the data are generated directly from product distributions with uniform marginals. In the second case, the marginal uniform distributions are found from the (uncentred) principal components $\widehat{\Gamma}_r^{\mathsf{T}}\mathbb{X}$, where r is the rank of the sample covariance matrix. Random samples $\mathbf{V}_i^*$ of the same size as the data are generated from this product distribution, and finally, the samples are backtransformed to produce the reference data $\mathbf{V}_i = \widehat{\Gamma}_r \mathbf{V}_i^*$.

The first method is simpler, but the second is better at integrating the shape of the distribution of the data. In either case, the expected value $E\{\log[W(k)]\}$ is approximated by an average of b copies of the random samples.

In Algorithm 6.3 I use the term 'cluster strategy', which means a clustering approach. Typically, Tibshirani, Walther, and Hastie (2001) refer to hierarchical (agglomerative) clustering or k-means clustering with the Euclidean distance, but other approaches could also be used.

Algorithm 6.3 ***The Gap Statistic***

Let $\mathbb{X}$ be d-dimensional data. Fix a cluster strategy for $\mathbb{X}$. Let $\kappa > 0$.

Step 1. For $k = 1, \ldots, \kappa$, partition $\mathbb{X}$ into k clusters, and calculate the within-cluster variabilities $W(k)$.

Step 2. Generate b data sets $\mathbb{V}_j$ $(j \leq b)$ of the same size as $\mathbb{X}$ from the reference distribution. Partition $\mathbb{V}_j$ into k clusters, and calculate their within-cluster variabilities $W_j^*(k)$. Put

$$\omega_k = \frac{1}{b} \sum_{j=1}^{b} \log[W_j^*(k)],$$

and compute the estimated gap statistic

$$\mathrm{Gap}(k) = \frac{1}{b}\sum_{j=1}^{b} \log[W_j^*(k)] - \log[W(k)] = \omega_k - \log[W(k)].$$

Step 3. Compute the standard deviation $\mathrm{sd}_k = [(1/b)\sum_j\{\log[W_j^*(k)] - \omega_k\}^2]^{1/2}$ of ω_k, and set $s_k = \mathrm{sd}_k(1+1/b)^{1/2}$. The estimate $\widehat{k}_G$ of the number of clusters is

$$\widehat{k}_G = \min\{k \le \kappa\colon \mathrm{Gap}(k) \ge \mathrm{Gap}(k+1) - s_{k+1}\}.$$

■

Tibshirani, Walther, and Hastie (2001) showed in simulations and for real data that the gap statistic performs well. For simulations, they compared the results of the gap statistic with the statistics $\widehat{k}_G$ of (6.11)–(6.13).

6.6.3 The Prediction Strength Approach

This method of Tibshirani and Walther (2005) for estimating the number of clusters adapts ideas from Discriminant Analysis. Their approach has an intuitive interpretation, provides information about cluster membership of pairs of observations, does not depend on distributional assumptions of the data and has a rigorous foundation. I present their method within our framework. As a consequence, the notation will differ from theirs.

Definition 6.7 Let $\mathbb{X} = [\mathbf{X}_1\ \mathbf{X}_2 \cdots \mathbf{X}_n]$ be d-dimensional data. Fix $k > 0$, and let $\mathcal{P}(\mathbb{X},k)$ be a k-cluster arrangement of $\mathbb{X}$. The **co-membership matrix** D corresponding to $\mathbb{X}$ has entries

$$D_{ij} = \begin{cases} 1 & \text{if } \mathbf{X}_i \text{ and } \mathbf{X}_j \text{ belong to the same cluster,} \\ 0 & \text{otherwise.} \end{cases}$$

Let $\mathbb{X}'$ be d-dimensional data with k-cluster arrangement $\mathcal{P}(\mathbb{X}',k)$. The **co-membership matrix** $D[\mathcal{P}(\mathbb{X},k),\mathbb{X}']$ of $\mathbb{X}'$ relative to the cluster arrangement $\mathcal{P}(\mathbb{X},k)$ of $\mathbb{X}$ is

$$D[\mathcal{P}(\mathbb{X},k),\mathbb{X}']_{ij} = \begin{cases} 1 & \text{if } \mathbf{X}'_i \text{ and } \mathbf{X}'_j \text{ belong to the same cluster in } \mathcal{P}(\mathbb{X},k), \\ 0 & \text{otherwise.} \end{cases}$$

□

The $n \times n$ matrix D contains information about the membership of pairs of observations from $\mathbb{X}$. For a small number of clusters, many of the entries of D will be 1s, and as the number of clusters increases, entries of D will change from 1 to 0.

The matrix $D[\mathcal{P}(\mathbb{X},k),\mathbb{X}']$ summarises information about pairs of observations across different cluster arrangements. The extension of the co-membership matrix to two data sets of the same number and type of variables and the same k records what would happen to data from $\mathbb{X}'$ under the arrangement $\mathcal{P}(\mathbb{X},k)$. If k is small and the two data sets have similar shapes and structures, then the co-membership matrix $D[\mathcal{P}(\mathbb{X},k),\mathbb{X}']$ consists of many entries 1, but if the shape and the number of components of the two data sets differ, then this co-membership matrix contains more 0 entries.

The notion of the co-membership matrix is particularly appealing when we want to compare two cluster arrangements because the proportion of 1s is a measure of the 'sameness' of clusters in the two arrangements. This is the key idea which leads to the **prediction strength**

$$\mathrm{PS}(k) = \min_{\{1\le \ell \le k\}} \left\{ \frac{1}{n'_\ell(n'_\ell - 1)} \sum_{\mathbf{X}'_i \neq \mathbf{X}'_j \in \mathcal{C}'_\ell} D[\mathcal{P}(\mathbb{X},k),\mathbb{X}']_{ij} \right\}, \tag{6.16}$$

where n'_ℓ is the number of elements in the cluster $\mathcal{C}'_\ell$. As the number of observations across clusters varies, the prediction strength provides a robust measure because it errs on the side of low values by looking at the worst case.

In practice, $\mathbb{X}$ and $\mathbb{X}'$ are two randomly chosen parts of the same data, so the structure in both parts is similar. The idea of prediction strength relies on the intuition that for the true number of clusters k_0, the k_0-cluster arrangements of the two parts of the data will be similar, and $\mathrm{PS}(k_0)$ will therefore be high. As the number of clusters increases, observations are more likely to belong to different clusters in the cluster arrangements of $\mathbb{X}$ and $\mathbb{X}'$, and $\mathrm{PS}(k)$ becomes smaller. This reasoning leads to the statistic

$$\widehat{k}_{\mathrm{PS}} = \operatorname{argmax} \mathrm{PS}(k)$$

as the estimator for the number of clusters.

If a true cluster membership of the data exists, say, the k-cluster arrangement $\mathcal{P}^*(\mathbb{X})$, then the **prediction error loss** of the k-cluster arrangement $\mathcal{P}(\mathbb{X},k)$ is

$$\mathcal{E}_{\mathcal{P}}(k) = \frac{1}{n^2} \sum_{i,j=1}^{n} \left| D[\mathcal{P}^*(\mathbb{X}),\mathbb{X}]_{ij} - D[\mathcal{P}(\mathbb{X},k),\mathbb{X}]_{ij} \right|,$$

so the error counts the number of pairs for which the two co-membership matrices disagree. The error can be decomposed into two parts: the proportion of pairs that $\mathcal{P}(\mathbb{X},k)$ erroneously assigns to the same cluster and the proportion of pairs that $\mathcal{P}(\mathbb{X},k)$ erroneously assigns to different clusters. These two terms are similar to the conventional squared bias and variance decomposition of a prediction error.

Tibshirani and Walther (2005) examine the $\widehat{k}_{\mathrm{PS}}$ statistic on the simulations they describe in Tibshirani, Walther, and Hastie (2001), and they reach the conclusion that the performance of the $\widehat{k}_{\mathrm{PS}}$ statistic is very similar to that of the gap statistic $\widehat{k}_{\mathrm{G}}$ of Algorithm 6.3.

6.6.4 Comparison of $\widehat{k}$-Statistics

In this final section I illustrate the performance of the different statistics on data. There is no uniformly *best* statistic for choosing the number of clusters. The aim is to try out different statistics and, in the process, gain intuition into the applicability, advantages and disadvantages of the different approaches. A comprehensive comparison of the six estimators $\widehat{k}$ of Sections 6.6.1–6.6.3 requires a range of data sets and simulations, which is more than I want to do here. In the Problems at the end of Part II, we apply these statistics to simulations with different combinations of sample size, dimension and number of clusters.

Example 6.13 We return to the **HIV flow cytometry** data sets I introduced in Chapter 1 and analysed throughout Chapter 2. We consider the first HIV^+ and the first HIV^- data sets from the collection of fourteen such sets of flow cytometry measurements. Each data set consists of 10,000 observations and five different variables; plots of subsets of the five-dimensional data are displayed in Figures 1.1, 1.2 and 2.3. The plots suggest that there are differences in the cluster configurations of the HIV$^+$ and HIV$^-$ data sets. We aim to quantify these visually perceived differences.

The following analysis is based on k-means clustering with the Euclidean distance. The cosine distance leads to similar cluster arrangements, I will therefore only report the results for the Euclidean distance.

For each of the two data sets we consider the first 100, the first 1,000 and all 10,000 observations, so a total of six combinations. For $k \leq 10$, I calculate the optimal k-cluster arrangement over 100 runs and then evaluate H(k), KL(k) and WV(k) of (6.12)–(6.14) from this optimal arrangement. For the WV quotient I use the threshold $\tau = 1.2$. For CH(k) of (6.11) I use the parameters of the optimal $W(k)$ for the calculation of the between-cluster variability $B(k)$. Finally, I determine the statistics $\widehat{k}$ for each of these four criteria as described in (6.11)–(6.14). The results are shown in Table 6.11. Note that H(k) < 10 holds for the two small data sets only, which consist of 100 observations each. For $n = 1,000$ and $n = 10,000$ I could not obtain a value for $\widehat{k}$ while assuming that $k \leq 10$. I have indicated the missing value by an en dash in the table.

For calculation of the gap statistic, I restrict attention to the uniform marginal distribution of the original data. I use $b = 100$ simulations and then calculate the optimal $W(k)$ over ten runs. Table 6.11 shows the values for the statistic $\widehat{k}_G$ of (6.15).

For the prediction-strength statistic I split the data into two parts of the same size. For the two small data sets I frequently encountered some empty clusters, and I therefore considered the first 200 samples of the HIV$^+$and HIV$^-$as the smallest data sets. After some experimenting with dividing the data into two parts in different ways, I settled on two designs:

1. taking the first half of the data as $\mathbb{X}$ and the second half as $\mathbb{X}'$; and
2. taking the second half of the data as $\mathbb{X}$ and the first half as $\mathbb{X}'$.

The optimal k for the first design is shown first, and the optimal k for the second design is given in parentheses. I fixed these two designs because randomly chosen parts gave values fluctuating between 2 and 6.

The last row of the table shows the results I obtain with the SOPHE, Algorithm 6.2. The two smallest data sets are two small for the SOPHE to find clusters. For the data sets of size 1,000 observations, I use $\theta_0 = 10$, and for the data consisting of 10,000 observations, I use $\theta_0 = 20$.

An inspection of Table 6.11 shows that statistics, $\widehat{k}_{KL}$, $\widehat{k}_{WV}$ and $\widehat{k}_G$ agree on the HIV$^+$data. For the HIV$^-$data, the results are not so consistent: $\widehat{k}_{KL}$, $\widehat{k}_{WV}$ and $\widehat{k}_G$ find different values for k as the sample size varies, and they differ from each other. The statistic $\widehat{k}_{CH}$ results in larger values than any of the other statistics. The PS-statistic produces rather varied results for the two data sets, so it appears to work less well for these data. The SOPHE finds four clusters for 1,000 observations but five clusters for the 10,000 observations, in agreement with the $\widehat{k}_{KL}$, $\widehat{k}_{WV}$ and $\widehat{k}_G$ approaches.

The agreement in the number of clusters of different approaches is strong evidence that this common number of clusters is appropriate for the data. For the **HIV** data, the analyses show that five clusters represent the data well.

Table 6.11 *Number of Clusters of* HIV^+ *and* HIV^- *Data from Example 6.13 for 100 [200 for PS], 1,000 and 10,000 Observations*

	HIV^+			HIV^-		
	100	1,000	10,000	100	1,000	10,000
CH	10	10	10	8	9	10
H	6	–	–	8	–	–
KL	5	5	5	4	8	5
WV	5	5	5	7	5	5
Gap	5	5	5	7	5	7
PS	2 (5)	5 (6)	5 (2)	3 (5)	5 (6)	2 (3)
SOPHE	–	4	5	–	4	5

Having detected the clusters for each of these data sets opens the way for the next step of the analysis; the size and location of the clusters. This second step should clarify that the patterns of the HIV^+ and HIV^- data differ: the variable *CD8* increases and the variable *CD4* decreases with the onset of HIV^+, as I mentioned in Example 2.4 in Section 2.3. The output of the SOPHE analysis shows that this change occurs: a comparison of the two largest clusters arising from all observations reveals that the mode location of the largest HIV^- cluster is identical to the mode location of the second largest HIV^+ cluster, and the two mode locations differ in precisely the *CD4* and *CD8* variables. ■

The comparisons in the example illustrate that there may not be a unique or right way for choosing the number of clusters. Some criteria, such as CH and H, are less appropriate for larger sample sizes, whereas the SOPHE does not work for small sample sizes but works well for large samples. The results shown in the example are consistent with the simulation results presented in Tibshirani and Walther (2005) in that both the gap statistic and the PS statistic obtain values for $\widehat{k}$ that are lower or similar to those of CH, H and KL.

So how should we choose the number of clusters? To answer this question, it is important to determine how the number of clusters is to be used. Is the cluster arrangement the end of the analysis, or is it an intermediate step? Are we interested in finding large clusters only, or do we want to find clusters that represent small subpopulations containing 1 to 5 per cent of the data? And how crucial is it to have the *best* answer? I can only offer some general guidelines and recommendations.

- Apply more than one criterion to obtain a comparison, and use the agreed number of clusters if an agreement occurs.
- Consider combining the results of the comparisons, depending on whether one is prepared to err at the low or high end of the number of clusters.
- Calculate multiple cluster allocations which differ in the number of clusters for use in subsequent analyses.

The third option may have merit, in particular, if the partitioning is an intermediate step in the analysis. Cluster allocations which differ in their number of clusters for the same data might lead to new insight in further analysis and potentially could open a new way of determining the number of clusters.

7

Factor Analysis

I am not bound to please thee with my answer (William Shakespeare, *The Merchant of Venice*, 1596–1598).

7.1 Introduction

It is not always possible to measure the quantities of interest directly. In psychology, intelligence is a prime example; scores in mathematics, language and literature, or comprehensive tests are used to describe a person's intelligence. From these measurements, a psychologist may want to derive a person's intelligence. Behavioural scientist Charles Spearman is credited with being the originator and pioneer of the classical theory of mental tests, the theory of intelligence and what is now called *Factor Analysis*. In 1904, Spearman proposed a two-factor theory of intelligence which he extended over a number of decades (see Williams et al., 2003). Since its early days, Factor Analysis has enjoyed great popularity and has become a valuable tool in the analysis of complex data in areas as diverse as behavioural sciences, health sciences and marketing. The appeal of Factor Analysis lies in the ease of use and the recognition that there is an association between the hidden quantities and the measured quantities.

The aim of Factor Analysis is

- to exhibit the relationship between the measured and the underlying variables, and
- to estimate the underlying variables, called the *hidden* or *latent variables*.

Although many of the key developments have arisen in the behavioural sciences, Factor Analysis has an important place in statistics. Its model-based nature has invited, and resulted in, many theoretical and statistical advances. The underlying model is the Gaussian model, and the use of likelihood methods in particular has proved valuable in allowing an elegant description of the underlying structure, the hidden variables and their relationship with the measured quantities. In the last few decades, important new methods such as Structural Equation Modelling and Independent Component Analysis have evolved which have their origins in Factor Analysis.

Factor Analysis applies to data with fewer hidden variables than measured quantities, but we do not need to know explicitly the number of hidden variables. It is not possible to determine the hidden variables uniquely, and typically, we focus on expressing the observed variables in terms of a smaller number of uncorrelated factors. There is a strong connection with Principal Component Analysis, and we shall see that a principal component decomposition leads to one possible Factor Analysis solution. There are important distinctions between

Principal Component Analysis and Factor Analysis, however, which I will elucidate later in this chapter.

We begin with the factor model for the population in Section 7.2 and a model for the sample in Section 7.3. There are a number of methods which link the measured quantities and the hidden variables. We restrict attention to two main groups of methods in Section 7.4: a non-parametric approach, based on ideas from Principal Component Analysis, and methods based on maximum likelihood. In Section 7.5 we look at asymptotic results, which naturally lead to hypothesis tests for the number of latent variables. Section 7.6 deals with different approaches for estimating the latent variables and explains similarities and differences between the different approaches. Section 7.7 compares Principal Component Analysis with Factor Analysis and briefly outlines how Structural Equation Modelling extends Factor Analysis. Problems for this chapter can be found at the end of Part II.

7.2 Population k-Factor Model

In Factor Analysis, the observed vector has more variables than the vector of hidden variables, and the two random vectors are related by a linear model. We assume that the dimension k of the hidden vector is known.

Definition 7.1 Let $\mathbf{X} \sim (\boldsymbol{\mu}, \Sigma)$. Let r be the rank of Σ. Fix $k \leq r$. A **k-factor model** of $\mathbf{X}$ is

$$\mathbf{X} = A\mathbf{F} + \boldsymbol{\mu} + \boldsymbol{\epsilon}, \tag{7.1}$$

where $\mathbf{F}$ is a k-dimensional random vector, called the **common factor**, A is a $d \times k$ linear transformation, called the **factor loadings**, and the **specific factor** $\boldsymbol{\epsilon}$ is a d-dimensional random vector.

The common and specific factors satisfy

1. $\mathbf{F} \sim (\mathbf{0}, \mathbf{I}_{k\times k})$,
2. $\boldsymbol{\epsilon} \sim (\mathbf{0}, \Psi)$, with Ψ a diagonal matrix, and
3. $\operatorname{cov}(\mathbf{F}, \boldsymbol{\epsilon}) = \mathbf{0}_{k\times d}$. □

We also call the common factor the **factor scores** or the **vector of latent** or **hidden variables**. The various terminologies reflect the different origins of Factor Analysis, such as the behavioural sciences with interest in the factor loadings and the methodological statistical developments which focus on the latent (so not measured) nature of the variables. Some authors use the model $\operatorname{cov}(\mathbf{F}) = \Phi$ for some diagonal $k \times k$ matrix Φ. In Definition 7.1, the common factor is sphered and so has the identity covariance matrix.

As in Principal Component Analysis, the driving force is the covariance matrix Σ and its decomposition. Unless otherwise stated, in this chapter r is the rank of Σ. I will not always explicitly refer to this rank. From model (7.1),

$$\Sigma = \operatorname{var}(\mathbf{X}) = \operatorname{var}(A\mathbf{F}) + \operatorname{var}(\boldsymbol{\epsilon}) = AA^{\mathsf{T}} + \Psi$$

because $\mathbf{F}$ and $\boldsymbol{\epsilon}$ are uncorrelated. If $\operatorname{cov}(\mathbf{F}) = \Phi$, then $\Sigma = A\Phi A^{\mathsf{T}} + \Psi$.

Of primary interest is the unknown transformation A which we want to recover. In addition, we want to estimate the common factor.

To begin with, we examine the relationship between the entries of Σ and those of AA^{T}. Let $A = (a_{jm})$, $j \le d$, $m \le k$, and put

$$T = AA^{\mathsf{T}} = (\tau_{jm}) \qquad \text{with } \tau_{jm} = \textstyle\sum_{\ell=1}^{k} a_{j\ell} a_{m\ell}.$$

The diagonal entry $\tau_{jj} = \sum_{\ell=1}^{k} a_{j\ell} a_{j\ell}$ is called the ***j*th communality**. The communalities satisfy

$$\sigma_j^2 = \sigma_{jj} = \tau_{jj} + \psi_j \qquad \text{for } j \le d, \tag{7.2}$$

where the ψ_j are the diagonal elements of Ψ. The off-diagonal elements of Σ are related only to the elements of AA^{T}:

$$\sigma_{jm} = \tau_{jm} \quad \text{for } j, m \le d. \tag{7.3}$$

Further relationships between $\mathbf{X}$ and the common factor are listed in the next proposition.

Proposition 7.2 *Let* $\mathbf{X} \sim (\boldsymbol{\mu}, \Sigma)$ *satisfy the k-factor model* $\mathbf{X} = A\mathbf{F} + \boldsymbol{\mu} + \boldsymbol{\epsilon}$, *for* $k \le r$, *with* r *being the rank of* Σ. *The following hold:*

1. *The covariance matrix of* $\mathbf{X}$ *and* $\mathbf{F}$ *satisfies*

$$\operatorname{cov}(\mathbf{X}, \mathbf{F}) = A.$$

2. *Let* E *be an orthogonal* $k \times k$ *matrix, and put*

$$\widetilde{A} = AE \qquad \text{and} \qquad \widetilde{\mathbf{F}} = E^{\mathsf{T}}\mathbf{F}.$$

Then

$$\mathbf{X} = \widetilde{A}\widetilde{\mathbf{F}} + \boldsymbol{\mu} + \boldsymbol{\epsilon}$$

is a k-factor model of $\mathbf{X}$.

The proof follows immediately from the definitions.

The second part of the proposition highlights the non-uniqueness of the factor model. At best, the factor loadings and the common factor are unique up to an orthogonal transformation because an orthogonal transformation of the common factor is a common factor, and similar relationships hold for the factor loadings. But matters are worse; we cannot uniquely recover the factor loadings or the common factor from knowledge of the covariance matrix Σ, as the next example illustrates.

Example 7.1 We consider a **one-factor model**. The factor might be physical fitness, which we could obtain from the performance of two different sports, or the talent for language, which might arise from oral and written communication skills. We assume that the two-dimensional (2D) random vector $\mathbf{X}$ has mean zero and covariance matrix

$$\Sigma = \begin{pmatrix} 1.25 & 0.5 \\ 0.5 & 0.5 \end{pmatrix}.$$

The one-factor model is

$$\mathbf{X} = \begin{bmatrix} X_1 \\ X_2 \end{bmatrix} = \begin{bmatrix} a_1 \\ a_2 \end{bmatrix} F + \begin{bmatrix} \epsilon_1 \\ \epsilon_2 \end{bmatrix},$$

so F is a scalar. From (7.2) and (7.3) we obtain conditions for A:

$$a_1 a_2 = \sigma_{12} = 0.5, \qquad a_1^2 < \sigma_{11} = 1.25, \qquad \text{and} \qquad a_2^2 < \sigma_{22} = 0.5.$$

The inequalities are required because the covariance matrix of the specific factor is positive. A solution for A is

$$a_1 = 1, \qquad a_2 = 0.5.$$

A second solution for A is given by

$$a_1^* = \frac{3}{4}, \qquad a_2^* = \frac{2}{3}.$$

Both solutions result in positive covariance matrices Ψ for the specific factor $\boldsymbol{\epsilon}$. The two solutions are not related by an orthogonal transformation.

Unless other information is available, it is not clear which solution one should pick. We might prefer the solution which has the smallest covariance matrix Ψ, as measured by the trace of Ψ. Write Ψ and Ψ^* for the covariance matrices of $\boldsymbol{\epsilon}$ pertaining to the first and second solutions, then

$$tr(\Psi) = 0.25 + 0.25 = 0.5 \qquad \text{and} \qquad tr(\Psi^*) = \frac{11}{16} - \frac{1}{18} = 0.7431.$$

In this case, the solution with $a_1 = 1$ and $a_2 = 0.5$ would be preferable. ■

Kaiser (1958) introduced a criterion for distinguishing between factor loadings which is easy to calculate and interpret.

Definition 7.3 Let $A = (a_{j\ell})$ be a $d \times k$ matrix with $k \leq d$. The **(raw) varimax criterion** (VC) of A is

$$VC(A) = \sum_{\ell=1}^{k} \left[\frac{1}{d} \sum_{j=1}^{d} a_{j\ell}^4 - \left(\frac{1}{d} \sum_{j=1}^{d} a_{j\ell}^2 \right)^2 \right]. \tag{7.4}$$

□

The varimax criterion reminds us of a sample variance – with the difference that it is applied to the squared entries $a_{j\ell}^2$ of A. Starting with factor loadings A, Kaiser considers rotated factor loadings AE, where E is an orthogonal $k \times k$ matrix, and then chooses that orthogonal transformation $\widetilde{E}$ which maximises the varimax criterion:

$$\widetilde{E} = \underset{E}{\operatorname{argmax}}\, VC(AE).$$

As we shall see in Figure 7.1 in Example 7.4 and in later examples, varimax optimal rotations lead to visualisations of the factor loading that admit an easier interpretation than unrotated factor loadings.

In addition to finding the optimal orthogonal transformation, the VC can be used to compare two factor loadings.

Example 7.2 For the **one-factor model** of Example 7.1, an orthogonal transformation is of size 1×1, so trivial. An explicit VC calculation for A and A^* gives

$$VC(A) = 0.1406 \qquad \text{and} \qquad VC(A^*) = 0.0035.$$

If we want the factor loadings with the larger VC, then A is preferable.

The two ways of choosing the factor loadings, namely, finding the matrix A with the smaller trace of Ψ or finding that with the larger VC, are not equivalent. In this example, however, both ways resulted in the same answer. ■

For three-dimensional random vectors and a one-factor model, it is possible to derive analytical values for a_1, a_2, and a_3. We calculate these quantities in the Problems at the end of Part II.

If the marginal variances σ_j^2 differ greatly along the entries j, we consider the scaled variables. The scaled random vector $\Sigma_{diag}^{-1/2}(\mathbf{X} - \boldsymbol{\mu})$ of (2.17) in Section 2.6.1 has covariance matrix R, the matrix of correlations coefficients (see Theorem 2.17 of Section 2.6.1). The k-factor model for the scaled random vector $\mathbf{T} = \Sigma_{diag}^{-1/2}(\mathbf{X} - \boldsymbol{\mu})$ with covariance matrix R is

$$\mathbf{T} = \Sigma_{diag}^{-1/2} A\mathbf{F} + \Sigma_{diag}^{-1/2}\boldsymbol{\epsilon} \qquad \text{and} \qquad R = \Sigma_{diag}^{-1/2} A A^{\mathsf{T}} \Sigma_{diag}^{-1/2} + \Sigma_{diag}^{-1/2} \Psi \Sigma_{diag}^{-1/2}. \tag{7.5}$$

The last term, $\Sigma_{diag}^{-1/2} \Psi \Sigma_{diag}^{-1/2}$, is a diagonal matrix with entries $\psi_j' = \psi_j / \sigma_j^2$. Working with R instead of Σ is advantageous because the diagonal entries of R are 1, and the entries of R are correlation coefficients and so are easy to interpret.

7.3 Sample k-Factor Model

In practice, one estimates the factor loadings from information based on data rather than the true covariance matrix, if the latter is not known or available.

Definition 7.4 Let $\mathbb{X} = \begin{bmatrix}\mathbf{X}_1 \, \mathbf{X}_2 \cdots \mathbf{X}_n\end{bmatrix}$ be a random sample with sample mean $\overline{\mathbf{X}}$ and sample covariance matrix S. Let r be the rank of S. For $k \le r$, a **(sample) k-factor model** of $\mathbb{X}$ is

$$\mathbb{X} = A\mathbb{F} + \overline{\mathbf{X}} + \mathfrak{N}, \tag{7.6}$$

where $\mathbb{F} = \begin{bmatrix}\mathbf{F}_1 \, \mathbf{F}_2 \cdots \mathbf{F}_n\end{bmatrix}$ are the **common factors**, and each $\mathbf{F}_i$ is a k-dimensional random vector, A is the $d \times k$ matrix of **sample factor loadings** and the **specific factor** is the $d \times n$ matrix $\mathfrak{N}$.

The common and specific factors satisfy

1. $\mathbb{F} \sim (\mathbf{0}, \mathbf{I}_{k\times k})$,
2. $\mathfrak{N} \sim (\mathbf{0}, \Psi)$, with diagonal sample covariance matrix Ψ, and
3. $\operatorname{cov}(\mathbb{F}, \mathfrak{N}) = \mathbf{0}_{k\times d}$. □

In (7.6) and throughout this chapter I use $\overline{\mathbf{X}}$ instead of the matrix $\overline{\mathbf{X}}\mathbf{1}_n^{\mathsf{T}}$ for the matrix of sample means – in analogy with the notational convention established in (1.9) in Section 1.3.2.

The definition implies that for each observation $\mathbf{X}_i$ there is a common factor $\mathbf{F}_i$. As in the population case, we write the sample covariance matrix in terms of the common and specific factors:

$$S = \operatorname{var}(\mathbb{X}) = \operatorname{var}(A\mathbb{F}) + \operatorname{var}(\mathfrak{N}) = AA^{\mathsf{T}} + \Psi.$$

For the scaled data $S_{diag}^{-1/2}(\mathbb{X}-\overline{\mathbf{X}})$ and the corresponding matrix of sample correlation coefficients R_S, we have

$$R_S = S_{diag}^{-1/2} S S_{diag}^{-1/2} = S_{diag}^{-1/2} A A^{\mathsf{T}} S_{diag}^{-1/2} + S_{diag}^{-1/2} \Psi S_{diag}^{-1/2}.$$

The choice between using the raw data and the scaled data is similar in Principal Component Analysis and Factor Analysis. Making this choice is often the first step in the analysis.

Example 7.3 We consider the five-dimensional **car** data with three physical variables, *displacement*, *horsepower*, and *weight*, and two performance-related variables, *acceleration* and *miles per gallon* (*mpg*). Before we consider a k-factor model of these data, we need to decide between using the raw or scaled data. The sample mean

$$\overline{\mathbf{X}} = (194.4\ 104.5\ 2977.6\ 15.5\ 23.4)^{\mathsf{T}}$$

has entries of very different sizes. The sample covariance matrix and the matrix of sample correlation coefficients are

$$S = 10^3 \times \begin{pmatrix} 10.95 & 3.61 & 82.93 & -0.16 & -0.66 \\ 3.61 & 1.48 & 28.27 & -0.07 & -0.23 \\ 82.93 & 28.27 & 721.48 & -0.98 & -5.52 \\ -0.16 & -0.07 & -0.98 & 0.01 & 0.01 \\ -0.66 & -0.23 & -5.52 & 0.01 & 0.06 \end{pmatrix},$$

and

$$R_S = \begin{pmatrix} 1.000 & 0.897 & 0.933 & -0.544 & -0.805 \\ 0.897 & 1.000 & 0.865 & -0.689 & -0.778 \\ 0.933 & 0.865 & 1.000 & -0.417 & -0.832 \\ -0.544 & -0.689 & -0.417 & 1.000 & 0.423 \\ -0.805 & -0.778 & -0.832 & 0.423 & 1.000 \end{pmatrix}.$$

The third raw variable of the data has a much larger mean and variance than the other variables. It is therefore preferable to work with R_S. In Section 7.4.2 we continue with these data and examine different methods for finding the common factors for the scaled data. ∎

7.4 Factor Loadings

The two main approaches to estimating the factor loadings divide into non-parametric methods and methods which rely on the normality of the data. For normal data, we expect the latter methods to be better; in practice, methods based on assumptions of normality still work well if the distribution of the data does not deviate too much from the normal distribution. It is not possible to quantify precisely what 'too much' means; the simulations in Example 4.8 of Section 4.5.1 give some idea how well a normal model can work for non-normal data.

7.4.1 Principal Components and Factor Analysis

A vehicle for finding the factors is the covariance matrix. In a non-parametric framework, there are two main methods for determining the factors:

- Principal Component Factor Analysis, and

- Principal Factor Analysis.

In Principal Component Factor Analysis, the underlying covariance matrix is Σ or S. In contrast, Principal Factor Analysis is based on the scaled covariance matrix of the common factors. I begin with a description of the two methods for the population.

Method 7.1 ***Principal Component Factor Analysis***
Let $\mathbf{X} \sim (\boldsymbol{\mu}, \Sigma)$. Fix $k \leq r$, with r the rank of Σ, and let

$$\mathbf{X} = A\mathbf{F} + \boldsymbol{\mu} + \boldsymbol{\epsilon}$$

be a k-factor model of $\mathbf{X}$. Write $\Sigma = \Gamma\Lambda\Gamma^{\mathsf{T}}$ for its spectral decomposition, and put

$$\widehat{A} = \Gamma_k \Lambda_k^{1/2}, \tag{7.7}$$

where Γ_k is the $d \times k$ submatrix of Γ, and Λ_k is the $k \times k$ diagonal matrix whose entries are the first k eigenvalues of Σ.

For the factor loadings $\widehat{A}$ of (7.7), the common factor

$$\widehat{\mathbf{F}} = \Lambda_k^{-1/2}\Gamma_k^{\mathsf{T}}(\mathbf{X} - \boldsymbol{\mu}), \tag{7.8}$$

and the covariance matrix $\widehat{\Psi}$ of the specific factor $\boldsymbol{\epsilon}$ is

$$\widehat{\Psi} = \Sigma_{diag} - (\Gamma_k \Lambda_k \Gamma_k^{\mathsf{T}})_{diag} \quad \text{with diagonal entries } \widehat{\psi}_j = \sum_{\ell=k+1}^{d} \lambda_\ell \eta_{\ell j}^2 \quad j \leq d.$$

□

We check that properties 1 to 3 of Definition 7.1 are satisfied. The k entries of the factor $\widehat{\mathbf{F}}$ are

$$\widehat{F}_j = \lambda_j^{-1/2}\boldsymbol{\eta}_j^{\mathsf{T}}(\mathbf{X} - \boldsymbol{\mu}) \qquad \text{for } j \leq k.$$

These entries are uncorrelated by Theorem 2.5 of Section 2.5. Calculations show that the covariance matrix of $\widehat{\mathbf{F}}$ is the identity. Further,

$$\widehat{A}\widehat{A}^{\mathsf{T}} = \Gamma_k \Lambda_k \Gamma_k^{\mathsf{T}}$$

follows from (7.7). From (7.2), we obtain

$$\widehat{\psi}_j = \sigma_j^2 - \widehat{\tau}_{jj},$$

and in the Problems at the end of Part II we show that $\widehat{\psi}_j = \sum_{m>k} \lambda_m \eta_{mj}^2$.

Method 7.2 ***Principal Factor Analysis or Principal Axis Factoring***
Let $\mathbf{X} \sim (\boldsymbol{\mu}, \Sigma)$. Fix $k \leq r$, with r the rank of Σ, and let

$$\mathbf{X} = A\mathbf{F} + \boldsymbol{\mu} + \boldsymbol{\epsilon}$$

be a k-factor model of $\mathbf{X}$. Let R be the matrix of correlation coefficients of $\mathbf{X}$, and let Ψ be the covariance matrix of $\boldsymbol{\epsilon}$. Put

$$R_A = R - \Sigma_{diag}^{-1/2}\Psi\Sigma_{diag}^{-1/2}, \tag{7.9}$$

and write $R_A = \Gamma_{R_A} \Lambda_{R_A} \Gamma_{R_A}^{\mathsf{T}}$ for its spectral decomposition. The factor loadings are

$$\widehat{A} = \Gamma_{R_A,k} \Lambda_{R_A,k}^{1/2},$$

where $\Gamma_{R_A,k}$ is the $d \times k$ matrix which consists of the first k eigenvectors of Γ_{R_A}, and similarly, $\Lambda_{R_A,k}$ is the $k \times k$ diagonal matrix which consists of the first k eigenvalues of Λ_{R_A}. □

Because R is the covariance matrix of the scaled vector $\Sigma_{diag}^{-1/2}(\mathbf{X} - \boldsymbol{\mu})$, R_A is the covariance matrix of the scaled version of $A\mathbf{F}$, and the diagonal elements of R_A are $1 - \psi_j$.

Methods 7.1 and 7.2 differ in two aspects:

1. the choice of the covariance matrix to be analysed, and
2. the actual decomposition of the chosen covariance matrix.

The first of these relates to the covariance matrix of the raw data versus that of the scaled data. As we have seen in Principal Component Analysis, if the observed quantities are of different orders of magnitude, scaling prior to Principal Component Analysis is advisable.

The second difference is more important. Method 7.1 makes use of the available information – the covariance matrix Σ or its sample version S – whereas Method 7.2 relies on available knowledge of the covariance matrix Ψ of the specific factor. In special cases, good estimates might exist for Ψ, but this is not the rule, and the first method tends to be more useful in practice.

The expression for the factor loadings in Method 7.1 naturally leads to expressions for the factors; in the second method, the factor loadings are defined at the level of the scaled data, and therefore, no natural expression for the common factors is directly available. To overcome this disadvantage, Method 7.1 is often applied to the scaled data.

I include relevant sample expressions for Method 7.1. Similar expressions can be derived for Method 7.2. Let $\mathbb{X} = \begin{bmatrix}\mathbf{X}_1 \ \mathbf{X}_2 \cdots \mathbf{X}_n\end{bmatrix}$ be data with sample mean $\overline{\mathbf{X}}$ and sample covariance matrix $S = \widehat{\Gamma}\widehat{\Lambda}\widehat{\Gamma}^{\mathsf{T}}$. For $k \leq r$, and r the rank of S, let

$$\mathbb{X} = A\mathbb{F} + \overline{\mathbf{X}} + \mathfrak{N}$$

be a k-factor model of $\mathbb{X}$. The factor loadings and the matrix of common factors are

$$\widehat{A} = \widehat{\Gamma}_k \widehat{\Lambda}_k^{1/2} \quad \text{and} \quad \widehat{\mathbb{F}} = \widehat{\Lambda}_k^{-1/2} \widehat{\Gamma}_k^{\mathsf{T}} (\mathbb{X} - \overline{\mathbf{X}}). \tag{7.10}$$

The next example shows how Methods 7.1 and 7.2 work in practice.

Example 7.4 We continue with the **illicit drug market** data which are shown in Figure 1.5 in Section 1.2.2. The seventeen direct and indirect measures, listed in Table 3.2 in Section 3.3, are the variables in this analysis. Based on these observed quantities, we want to find factor loadings, and hidden variables of the drug market.

As in previous analyses, we work with the scaled data because the two variables *break and enter dwelling* and *steal from motor vehicles* are on a much larger scale than all others. We consider a three-factor model and the matrix of sample correlation coefficients R_S and compare the factor loadings of Methods 7.1 and 7.2 visually.

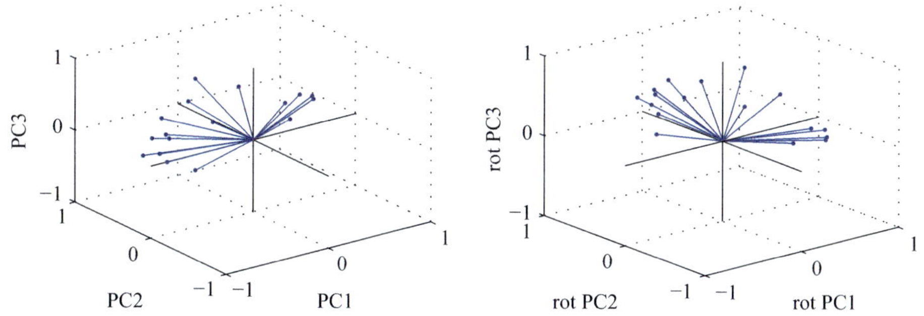

Figure 7.1 Biplots of factor loadings of a three-factor model from Example 7.4. PC factor loadings in the left plot and their varimax optimal loadings in the right plot.

Figure 7.1 shows the row vectors of the $d \times 3$ factor loadings $\widehat{A}$. Each row of $\widehat{A}$ is represented as a point in $\mathbb{R}^3$: the x-coordinates correspond to the loadings of the first factor, the y-coordinates are those of the second factor and the z-coordinates are the loadings of the third factor. The matrix A has d rows corresponding to the d variables of $\mathbb{X}$, and the axes are the directions of PC_1, PC_2 and PC_3, respectively. Plots of factor loadings of this type are called **biplots**.

Biplots exhibit pattern in the variables: here five vectors are close, as expressed by the narrow solid angle that includes them, whereas all other vectors show a greater spread.

The left subplot of Figure 7.1 shows the factor loadings calculated from (7.7), and their VC optimal versions are shown in the right subplot. The overall impression of the two plots is similar, but the orientations of the vectors differ. The five vectors that are close in the left panel are still close in the right panel, and they remain well separated from the rest. In the varimax optimal rotation on the right, these five vectors lie predominantly in the span of the first two eigenvectors of S, whereas the remaining twelve vectors are almost symmetrically spread out in a cone about the third eigenvector.

The five vectors within a narrow solid angle correspond to the first five variables of cluster 1 in Table 6.9 in Section 6.5.2. A comparison with Figure 6.11 in Section 6.5.2 shows that these five variables are the five points in the top-left corner of the right panel. In the cluster analysis of these data in Example 6.10, a further two variables, *drug psychoses* and *break and enter dwelling*, are assigned to cluster 1, but Figure 6.11 shows that the magnitude of these last two variables is much smaller than that of the first five. Thus, there could be an argument in favour of assigning these last two variables to a different cluster in line with the spatial distribution in the biplots.

For the factor loadings based on Method 7.2 and R_A as in (7.9), I take $\Psi = \sigma_\epsilon^2 \mathbf{I}_{d\times d}$, so Ψ is constant along the diagonal, and $\sigma_\epsilon^2 = 0.3$. This value of σ_ϵ^2 is the *left-over* variance after removing the first three factors in Method 7.1. Calculations of the factor loadings with values of $\sigma_\epsilon^2 \in [0.3, 1]$ show that the loadings are essentially unchanged; thus, the precise value of σ_ϵ^2 is not important in this example.

The factor loadings obtained with Method 7.2 are almost the same as those obtained with Principal Component Factor Analysis and are not shown here. For both methods, I start with the scaled data. After scaling, the difference between the two methods of calculating factor loadings is negligible. Method 7.1 avoids having to guess Ψ and is therefore preferable for these data. ■

In the example, I introduced biplots for three-factor models, but they are also a popular visualisation tool for two-factor models. The plots show how the data are 'loaded' onto the first few components and which variables are grouped together.

Example 7.5 The **Dow Jones returns** are made up of thirty stocks observed daily over a ten-year period. In Example 2.9 of Section 2.4.3 we have seen that five of these stocks, all IT companies in the list of stocks, are distinguished from the other stocks by negative PC_2 entries. A biplot is a natural tool for visualising the two separate groups of stocks.

For the raw data and Method 7.1, I calculate the loadings. The second PC is of special interest, and I therefore calculate a two-factor model of the Dow Jones returns. Figure 7.2 shows the PC factor loadings in the left panel and the varimax optimal loadings in the right panel. The first entries of the factor loadings are shown on the x-axis and the second on the y-axis.

It may seem surprising at first that all loading vectors have positive x-values. There is a good reason: the first eigenvector has positive entries only (see Example 2.9 and Figure 2.8).

The entries of the second eigenvector of the five IT companies are negative, but in the biplots the longest vector is shown with a positive direction. The vectors, which correspond to the five IT companies, are therefore shown in the positive quadrant of the left panel. Four of the vectors in the positive quadrant are close, whereas the fifth one – here indicated by an arrow – seems to be closer to the tight group of vectors with negative y-values. Indeed, in the panel on the right, this vector – also indicated by an arrow – belongs even more to the non-IT group.

The biplots produce a different insight into the grouping of the stocks from that obtained in the principal component analysis. What has happened? Stock 3 (*AT&T*) has a very small negative PC_2 weight, whereas the other four IT stocks have much larger negative PC_2 weights. This small negative entry is closer to the entries of the stocks with positive entries, and in the biplot it is therefore grouped with the non-IT companies. ■

For the *illicit drug market* data and the *Dow Jones returns*, we note that the grouping of the vectors is unaffected by rotation into varimax optimal loadings. However, the vectors

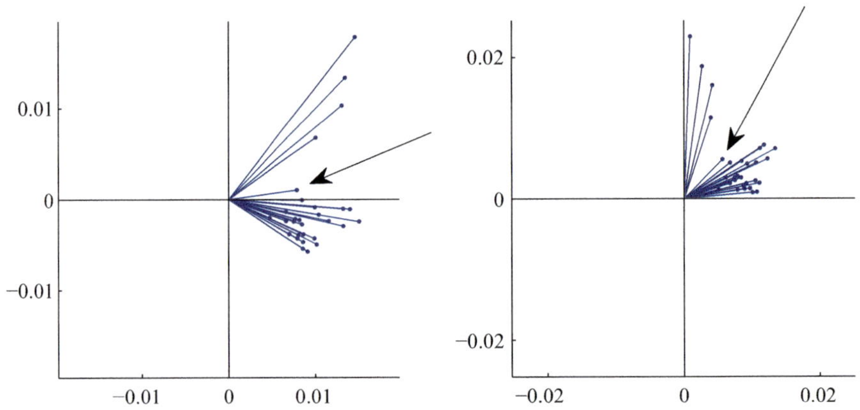

Figure 7.2 Biplots of factor loadings of a two-factor model from Example 7.5. PC factor loadings in the left plot and their varimax optimal loadings in the right plot.

align more closely with the axes as a result of such rotations. This view might be preferable visually.

7.4.2 Maximum Likelihood and Gaussian Factors

Principal Component Analysis is a non-parametric approach and can therefore be applied to data without requiring knowledge of the underlying distribution of the data. If we know that the data are Gaussian or not very different from Gaussian, exploiting this extra knowledge may lead to better estimators for the factor loadings. We consider Gaussian k-factor models separately for the population and the sample.

Definition 7.5 Let $\mathbf{X} \sim \mathcal{N}(\boldsymbol{\mu}, \Sigma)$. Let r be the rank of Σ. For $k \leq r$, a **normal** or **Gaussian** k**-factor model** of $\mathbf{X}$ is

$$\mathbf{X} = A\mathbf{F} + \boldsymbol{\mu} + \boldsymbol{\epsilon},$$

where the common factor $\mathbf{F}$ and specific factor $\boldsymbol{\epsilon}$ are normally distributed and satisfy 1 to 3 of Definition 7.1.

Let $\mathbb{X} = \begin{bmatrix}\mathbf{X}_1 \ \mathbf{X}_2 \cdots \mathbf{X}_n\end{bmatrix}$ be a sample of random vectors $\mathbf{X}_i \sim \mathcal{N}(\boldsymbol{\mu}, \Sigma)$. Let $\overline{\mathbf{X}}$ be the sample mean. For $k \leq r$, and r the rank of S, a **normal** or **Gaussian (sample)** k**-factor model** of $\mathbb{X}$, is

$$\mathbb{X} = A\mathbb{F} + \overline{\mathbf{X}} + \mathfrak{N},$$

where the columns of the common factor $\mathbb{F}$ and the columns of the specific factor $\mathfrak{N}$ are normally distributed, and $\mathbb{F}$ and $\mathfrak{N}$ satisfy 1 to 3 of Definition 7.4. □

Consider $\mathbb{X} = \begin{bmatrix}\mathbf{X}_1 \ \mathbf{X}_2 \cdots \mathbf{X}_n\end{bmatrix}$, with $\mathbf{X}_i \sim \mathcal{N}(\boldsymbol{\mu}, \Sigma)$, for $i = 1, \ldots, n$. For the $\mathbf{X}_i$ and their likelihood L given in (1.16) of Section 1.4, the parameters of interest are $\theta_1 = \boldsymbol{\mu}$ and $\theta_2 = \Sigma$. Consider the maximum likelihood estimators (MLEs)

$$\widehat{\theta}_1 = \widehat{\boldsymbol{\mu}} \qquad \text{and} \qquad \widehat{\theta}_2 = \widehat{\Sigma} \qquad \text{with} \qquad \widehat{\boldsymbol{\mu}} = \overline{\mathbf{X}} \quad \text{and} \quad \widehat{\Sigma} = \frac{n-1}{n} S.$$

Then

$$L(\widehat{\boldsymbol{\mu}}) = (2\pi)^{-nd/2} \det(\Sigma)^{-n/2} \exp\left[-\frac{n-1}{2} tr(\Sigma^{-1} S)\right], \tag{7.11}$$

$$L(\widehat{\boldsymbol{\mu}}, \widehat{\Sigma}) = (2\pi)^{-nd/2} \left(\frac{n-1}{n}\right)^{-nd/2} \det(S)^{-n/2} \exp\left(-\frac{nd}{2}\right). \tag{7.12}$$

We derive these identities in the Problems at the end of Part II. Note that (7.12) depends on the data only through the determinant of S.

In the next theorem we start with (7.11) and estimate Σ by making use of the k-factor model.

Theorem 7.6 *Assume that* $\mathbb{X} \sim \mathcal{N}(\boldsymbol{\mu}, \Sigma)$ *has a normal* k*-factor model with* $k \leq r$*, where* r *is the rank of* Σ*. Write* $\Sigma = AA^{\mathsf{T}} + \Psi$*, and let* $\theta = (\boldsymbol{\mu}, A, \Psi)$*.*

1. *The Gaussian likelihood* $L(\theta|\mathbb{X})$ *is maximised at the maximum likelihood estimator (MLE)* $\widehat{\theta} = (\widehat{\mu}, \widehat{A}, \widehat{\Psi})$ *subject to the constraint that* $\widehat{A}^{\mathsf{T}}\widehat{\Psi}^{-1}\widehat{A}$ *is a diagonal* $k \times k$ *matrix.*

2. *Let* $\widehat{\theta} = (\widehat{\mu}, \widehat{A}, \widehat{\Psi})$ *be the MLE for* $L(\theta|\mathbb{X})$. *If* E *is an orthogonal matrix such that* $\widetilde{A} = \widehat{A}E$ *maximises the varimax criterion (7.4), then* $\widetilde{\theta} = (\widehat{\mu}, \widetilde{A}, \widehat{\Psi})$ *also maximises* $L(\theta|\mathbb{X})$.
3. *Let* S *be the sample covariance matrix of* $\mathbb{X}$, *and write* $S = \widehat{\Gamma}\widehat{\Lambda}\widehat{\Gamma}^{\mathsf{T}}$ *for its spectral decomposition. If* $\Psi = \sigma^2 \mathbf{I}_{d\times d}$, *for some unknown* σ, *then* $\widehat{\theta}$, *the MLE of part 1, reduces to*

$$\widehat{\theta} = (\widehat{\mu}, \widehat{A}, \widehat{\sigma}^2) \quad \text{with } \widehat{A} = \widehat{\Gamma}_k(\widehat{\Lambda}_k - \widehat{\sigma}^2\mathbf{I}_{k\times k})^{1/2} \quad \text{and} \quad \widehat{\sigma}^2 = \frac{1}{d-k}\sum_{j>k}\widehat{\lambda}_j,$$

where the $\widehat{\lambda}_j$ *are the diagonal elements of* $\widehat{\Lambda}$.

Proof Parts 1 and 2 of the theorem follow from the invariance property of maximum likelihood estimators; for details, see chapter 7 of Casella and Berger (2001). The third part of the theorem is a result of Tipping and Bishop (1999), which I stated as Theorem 2.26 in Section 2.8. ∎

Because the covariance matrix Ψ is not known, it is standard to regard it as a multiple of the identity matrix. We have done so in Example 7.4.

Parts 1 and 2 of Theorem 7.6 show ways of obtaining solutions for A and Ψ. Part 1 lists a sufficient technical condition, and part 2 shows that for any solution $\widehat{A}$, a varimax optimal $\widetilde{A}$ is also a solution. The solutions of parts 1 and 2 differ, and a consequence of the theorem is that solutions based on the likelihood of the data are not unique. The more information we have, the better our solution will be: in part 3 we assume that Ψ is a multiple of the identity matrix, and then the MLE has a simple and explicit form.

For fixed k, let $\widehat{A}_{\mathrm{PC}}$ be the estimator of A obtained with Method 7.1, and let $\widehat{A}_{\mathrm{TB}}$ be the estimator obtained with part 3 of Theorem 7.6. A comparison of the two estimators shows that $\widehat{A}_{\mathrm{TB}}$ can be regarded as a regularised version of $\widehat{A}_{\mathrm{PC}}$ with tuning parameter $\widehat{\sigma}^2$, and the respective communalities are linked by

$$\widehat{A}_{\mathrm{PC}}\widehat{A}_{\mathrm{PC}}^{\mathsf{T}} - \widehat{A}_{\mathrm{TB}}\widehat{A}_{\mathrm{TB}}^{\mathsf{T}} = \widehat{\sigma}^2\widehat{\Gamma}_k\widehat{\Gamma}_k^{\mathsf{T}}.$$

A Word of Caution. In Factor Analysis, the concept of **factor rotation** is commonly used. Unlike a rotation matrix, which is an orthogonal transformation, a factor rotation is a linear transformation which need not be orthogonal but which satisfies certain criteria, such as goodness-of-fit. Factor rotations are also called *oblique rotations*. In the remainder of this chapter we use the Factor Analysis convention: a rotation means a factor rotation, and an orthogonal rotation is a factor rotation which is an orthogonal transformation.

Example 7.6 We calculate two-factor models for the five-dimensional **car** data and estimate the factor loadings with Method 7.1, and the MLE-based loadings. We start from the matrix of sample correlation coefficients R_S which is given in Example 7.3.

For the PC and ML methods, we calculate factor loadings and rotated factor loadings with rotations that maximise the VC in (7.4) over

1. orthogonal transformations E, and
2. oblique or factor rotations G.

The oblique rotations G violate the condition that $GG^{\mathsf{T}} = \mathbf{I}_{k\times k}$ but aim to load the factors onto single eigenvector directions, so they aim to align them with the axes. Such a view can enhance the interpretation of the variables. The factor loadings are given in Table 7.1, and components 1 and 2 list the entries of the vectors shown in the biplots in Figure 7.3.

Table 7.1 *Factor Loadings for Example 7.6 with VC Values in the Last Row*

PC factor loadings and rotations			ML factor loadings and rotations		
None	Orthogonal	Oblique	None	Orthogonal	Oblique
Component 1					
−0.9576	−0.8967	−0.9165	0.9550	0.8773	0.8879
−0.9600	−0.7950	−0.7327	0.9113	0.7618	0.6526
−0.9338	−0.9525	−1.0335	0.9865	0.9692	1.0800
0.6643	0.2335	−0.0599	−0.5020	−0.2432	0.0964
0.8804	0.8965	0.9717	−0.8450	−0.7978	−0.8413
Component 2					
−0.1137	0.3547	0.0813	−0.0865	0.3871	−0.1057
0.1049	0.5482	0.3409	−0.3185	0.5930	−0.4076
−0.2754	0.2012	−0.1157	0.1079	0.2129	0.1472
−0.7393	−0.9661	−1.0262	0.7277	−0.8500	0.9426
0.2564	−0.1925	0.1053	0.0091	−0.2786	0.0059
VC values					
0.0739	0.2070	0.3117	0.1075	0.1503	0.2652

Columns 1 and 4, labelled 'None' in the table, refer to no rotation. For the ML results, the 'None' solutions are obtained from part 1 of Theorem 7.6. Columns 2, 3, 5 and 6 refer to VC-optimal orthogonal and oblique rotations, respectively. The last row of the table shows the VC value for each of the six factor loadings. The PC loadings have the smallest VC values, but rotations of the PC loadings exceed those of ML. The two oblique rotations have higher VC values than the orthogonal rotations, showing that VC increases for non-orthogonal transformations.

Figure 7.3 shows biplots of PC factor loadings in the top row and of ML factor loadings in the bottom row. The left panels show the 'no rotation' results, the middle panels refer to the VC-optimal loadings, and the results on the right show the biplots of the best oblique rotations. For component 1, the PC and ML loadings 'None' are very similar, and the same holds for the orthogonal and oblique rotations apart from having opposite signs.

The biplots show that variables 1 to 3, the physical properties *displacement*, *horsepower* and *weight*, are close in all six plots in Figure 7.3, whereas the performance-based properties 4 and 5, *acceleration* and *miles per gallon*, are spread out. For the oblique rotations, the last two variables agree closely with the new axes: *acceleration* (variable 4) is very close to the positive second eigenvector direction, and *miles per gallon* aligns closely with the negative first direction; the actual numbers are given in Table 7.1.

Because factor loadings corresponding to part 3 of Theorem 7.6 are very similar to the PC loadings, I have omitted them. ■

From a mathematical point of view, oblique rotations are not as elegant as orthogonal rotations, but many practitioners prefer oblique rotations because of the closer alignment of the loadings with the new axes.

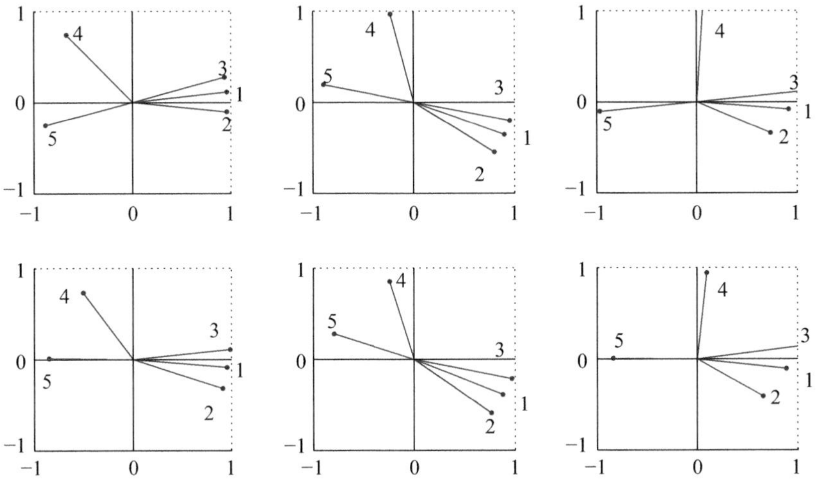

Figure 7.3 Biplots of (rotated) factor loadings of Example 7.6. PC factor loadings (*top row*), ML loadings (*bottom row*); the left column has no rotation, and rotations are orthogonal in the middle column and oblique in the right column.

7.5 Asymptotic Results and the Number of Factors

So far we explored methods for calculating factor loadings for a given number of factors. In some applications, there may be reasons for choosing a particular k, for example, if the visual aspect of the biplots is crucial to the analysis, in this case we take $k = 2$ or 3. Apart from special cases, the question arises: How many factors should we choose?

In this section we consider normal models and formulate hypothesis tests for the number of factors. In Section 2.7.1 we considered hypothesis tests based on the eigenvalues of the covariance matrix. Very small eigenvalues are evidence that the corresponding PC score is negligible. In Factor Analysis, the likelihood of the data drives the testing. Using Theorem 7.6 part 1, a natural hypothesis test for a k-factor model against all alternatives is

$$H_0 \colon \Sigma = AA^\top + \Psi, \tag{7.13}$$

where A is a $d \times k$ matrix and $A^\top \Psi^{-1} A$ is diagonal, versus

$$H_1 \colon \Sigma \neq AA^\top + \Psi.$$

For data $\mathbb{X}$, assume that the $\mathbf{X}_i \sim \mathcal{N}(\boldsymbol{\mu}, \Sigma)$. We consider the likelihood L of $\mathbb{X}$ as a function of Σ. The likelihood-ratio test statistic Λ_{H_0} for testing H_0 against H_1 is

$$\Lambda_{\mathrm{H}_0}(\mathbb{X}) = \frac{\sup_{H_0} L(\Sigma|\mathbb{X})}{\sup L(\Sigma|\mathbb{X})}, \tag{7.14}$$

where $\sup_{H_0}$ refers to the supremum under the null hypothesis, and $\sup L$ refers to the unrestricted supremum. Thus, we compare the likelihood at the MLE under H_0 and the likelihood at the unrestricted MLE in the denominator. The MLE for the mean $\boldsymbol{\mu}$ is independent of this maximisation, and it thus suffices to consider the likelihood at the MLE $\widehat{\boldsymbol{\mu}} = \overline{\mathbf{X}}$ as in (7.11). For notational convenience, I omit the dependence of L on $\widehat{\boldsymbol{\mu}}$ in the following discussion.

Let $\widehat{A}\widehat{A}^{\top} + \widehat{\Psi}$ be the maximiser of the likelihood under H_0, where $\widehat{A}$ and $\widehat{\Psi}$ satisfy part 1 of Theorem 7.6, and let $\widehat{\Sigma}$ be the unrestricted MLE for Σ. From (7.12), it follows that

$$\Lambda_{\mathrm{H}_0}(\mathbb{X}) = \frac{L(\widehat{A}\widehat{A}^{\top} + \widehat{\Psi}|\mathbb{X})}{L(\widehat{\Sigma}|\mathbb{X})}$$

$$= \left(\frac{\det(\widehat{A}\widehat{A}^{\top} + \widehat{\Psi})}{\det(\widehat{\Sigma})}\right)^{-n/2} \exp\left\{\frac{nd}{2} - \frac{n-1}{2} tr\left[(\widehat{A}\widehat{A}^{\top} + \widehat{\Psi})^{-1} S\right]\right\}. \qquad (7.15)$$

A check of the number of parameters in the likelihood-ratio test statistic shows that the unrestricted covariance matrix $\widehat{\Sigma}$ is based on $d(d+1)/2$ parameters because it is positive definite. For the restricted covariance matrix $\widehat{A}\widehat{A}^{\top} + \widehat{\Psi}$, we count dk parameters for $\widehat{A}$, d for $\widehat{\Psi}$, but we also have $k(k-1)/2$ constraints on $\widehat{A}$ and $\widehat{\Psi}$ arising from (7.13). A final count of the number of parameters – those of the unrestricted case minus those of the restricted case – yields the degrees of freedom:

$$\nu = \frac{1}{2}d(d+1) - \left[dk + d - \frac{1}{2}k(k-1)\right] = \frac{1}{2}\left[(d-k)^2 - (d+k)\right]. \qquad (7.16)$$

Theorem 7.7 *Let $\mathbb{X} = \begin{bmatrix}\mathbf{X}_1 & \cdots & \mathbf{X}_n\end{bmatrix}$ be d-dimensional data with $\mathbf{X}_i \sim \mathcal{N}(\boldsymbol{\mu}, \Sigma)$ and sample covariance matrix S with rank d. Fix $k < d$. Let Λ_{H_0} as in (7.14) be the likelihood-ratio test statistic for testing the k-factor model (7.13) against all alternatives. The asymptotic distribution of Λ_{H_0} is*

$$-2\log \Lambda_{H_0}(\mathbb{X}) \to \chi^2_{\nu} \qquad \text{as } n \to \infty, \qquad \text{where } \nu = \tfrac{1}{2}\left[(d-k)^2 - (d+k)\right].$$

Details on the likelihood-ratio test statistic Λ_{H_0} and the approximate distribution of $-2\log\Lambda_{\mathrm{H}_0}$ are provided in chapter 8 of Casella and Berger (2001). For a proof of the theorem and the asymptotic distribution of Factor Analysis models, see Amemiya and Anderson (1990).

A number of names are associated with early inference results for Factor Analysis, including Rao (1955), Lawley (1940, 1953), Anderson and Rubin (1956), Lawley and Maxwell (1971) and references therein. Theorem 7.7 – given here without a proof – applies to the traditional set-up of part 1 of Theorem 7.6.

Theorem 7.7 is stated for normal data, but extensions to larger classes of common and specific factors exist. Anderson and Amemiya (1988) showed that under some regularity conditions on the common and specific factors and by requiring that the factors are independent, the asymptotic distribution and the asymptotic covariance matrix of the parameters of the model are the same as in the normal case. For common factors with a general covariance matrix which may differ from the identity, let $\boldsymbol{\theta}$ be the model parameter which depends on the factor loadings A. If $\widehat{\boldsymbol{\theta}}_n$ is an estimator of $\boldsymbol{\theta}$, then, by Anderson and Amemiya (1988),

$$\sqrt{n}\left[\widehat{\boldsymbol{\theta}}_n - \boldsymbol{\theta}\right] \xrightarrow{D} \mathcal{N}(\mathbf{0}, V_0) \qquad \text{in distribution, for some matrix } V_0.$$

Theorem 7.7 is stated for a fixed k, but in applications we consider a sequence of null hypotheses $H_{0,k}$, one for each k, and we start with the smallest value for k. If the null hypothesis is rejected, we increase k until $H_{0,k}$ is accepted or until k_{max} is reached, and note

Table 7.2 *Correlation Coefficients of the Five Variables from Example 7.7*

ρ	2	3	4	5
1	0.5543	0.4104	0.3921	0.5475
2		0.4825	0.4312	0.6109
3			0.6069	0.7073
4				0.6597

Table 7.3 *MLE-Based Hypothesis Test for Example 7.7*

k	ν	χ^2_ν	p-Value	Decision
1	5	12.1211	0.0332	Reject $H_{0,1}$
2	1	0.1422	0.7061	Accept $H_{0,2}$

Table 7.4 *Factor Loadings with One and Two Factors for Example 7.7*

Variable	A_1	$A_{2,ML}$		$A_{2,VC}$	
1	0.6021	0.6289	0.3485	0.2782	0.6631
2	0.6686	0.6992	0.3287	0.3456	0.6910
3	0.7704	0.7785	−0.2069	0.7395	0.3194
4	0.7204	0.7246	−0.2070	0.6972	0.2860
5	0.9153	0.8963	−0.0473	0.7332	0.5177

that k_{max} satisfies

$$(d-k)^2 - (d+k) > 0. \qquad (7.17)$$

The next example shows how the tests work in practice.

Example 7.7 The five-dimensional **exam grades** data consist of 120 observations. The five variables are two scores for mathematics, two scores for literature and one comprehensive test score. The correlation coefficients of the five variables are displayed in Table 7.2. The correlation is positive in all instances and ranges from 0.3921 (between the first mathematics and the first literature scores) to 0.7073 (between the second literature score and the comprehensive test score).

For $k < d$, we test the adequacy of the k-factor model $\mathbb{X} = A\mathbb{F} + \overline{\mathbf{X}} + \mathfrak{N}$ with the tests

$$H_{0,k}\colon \Sigma = AA^{\mathsf{T}} + \Psi \qquad \text{versus} \qquad H_{1,k}\colon \Sigma \neq AA^{\mathsf{T}} + \Psi,$$

starting with $k = 1$. The likelihood-ratio test statistic is approximately χ^2_ν by Theorem 7.7, and (7.17) restricts k to $k = 1, 2$. Table 7.3 shows that at the 5 per cent significance level, $H_{0,1}$ is rejected, but $H_{0,2}$ is not.

For $k = 1$, there is a unique vector of factor loadings. Table 7.4 shows the entries of A_1 corresponding to $k = 1$. The one-factor model has roughly equal weights on the mathematics and literature scores but ranks the comprehensive score much higher.

For $k = 2$, we compare the unrotated ML loadings $A_{2,\mathrm{ML}}$ with the varimax optimal loadings $A_{2,\mathrm{VC}}$ based on orthogonal rotations. The weights of the first component of $A_{2,\mathrm{ML}}$ are similar to A_1 but differ from those of $A_{2,\mathrm{VC}}$. The second components of the two-factor loadings differ considerably, including a sign change in the literature loadings. Despite these differences in the factor loadings, the results of the hypothesis tests are independent of a rotation of the factors.

The conclusion from the hypothesis tests is a two-factor model. This model is appropriate because it allows a separation of the mathematics scores from the literature and comprehensive test scores. With a single factor, this is not possible. ■

A Word of Caution. The sequence of hypothesis tests may not always result in conclusive answers, as the following examples illustrate.

1. For the seven-dimensional *abalone* data, $k \le 3$, by (7.17). The null hypotheses $H_{0,k}$ with $k \le 3$ are rejected for the raw and scaled data. This result is not surprising in light of the consensus for five principal components which we obtained in Section 2.8.
2. For the thirty-dimensional *breast cancer* data, $k \le 22$, by (7.17). As for the *abalone* data, the null hypotheses are rejected for the raw and scaled data, and all values of $k \le 22$.

Further, one needs to keep in mind that the hypothesis tests assume a normal model. If these model assumptions are violated, the results may not apply or may only apply approximately.

In conclusion:

- The asymptotic results of Theorem 7.7 are useful, but their applicability is limited because of the restricted number of factors that can be tested.
- Determining the number of principal components or the number of factors is a difficult problem, and no single method will produce the right answer for all data. More methods are required which address these problems.

7.6 Factor Scores and Regression

Knowledge of the factor loadings and the number of factors may be all we require in some analyses – in this case, the hard work is done. If a factor analysis is the first of a number of steps, then we may require the common factor as well as the loadings. In this section I describe different approaches for obtaining common factors which include candidates offered by Principal Component Analysis and those suitable for a maximum likelihood framework. In analogy with the notion of scores in Principal Component Analysis, I will also refer to the common factors as **factor scores**.

7.6.1 Principal Component Factor Scores

In Principal Component Analysis, the kth principal component vector $\mathbf{W}^{(k)} = \Gamma_k^{\mathsf{T}}(\mathbf{X} - \boldsymbol{\mu})$ with jth score $W_j = \boldsymbol{\eta}_j^{\mathsf{T}}(\mathbf{X} - \boldsymbol{\mu})$ and $j \le k$ are the quantities of interest. By Theorem 2.5 of Section 2.5, the covariance matrix of $\mathbf{W}^{(k)}$ is the diagonal matrix Λ_k. Factor scores have an identity covariance matrix. For the population, a natural candidate for the factor scores of a k-factor model is thus (7.8). To distinguish these factor scores from later ones, I use the notation

$$\widehat{\mathbf{F}}^{(\mathrm{PC})} = \Lambda_k^{-1/2}\Gamma_k^{\mathsf{T}}(\mathbf{X}-\boldsymbol{\mu}) \tag{7.18}$$

and refer to $\widehat{\mathbf{F}}^{(\mathrm{PC})}$ as the **principal component factor scores** or the **PC factor scores**. For the PC factor loadings of Method 7.1 with $\widehat{A}$ the $d \times k$ matrix of (7.7), we obtain

$$\widehat{A}\widehat{\mathbf{F}}^{(\mathrm{PC})} = \Gamma_k\Lambda_k^{1/2}\Lambda_k^{-1/2}\Gamma_k^{\mathsf{T}}(\mathbf{X}-\boldsymbol{\mu}) = \Gamma_k\Gamma_k^{\mathsf{T}}(\mathbf{X}-\boldsymbol{\mu}).$$

The term $\widehat{A}\widehat{\mathbf{F}}^{(\mathrm{PC})}$ approximates $\mathbf{X}-\boldsymbol{\mu}$ and equals $\mathbf{X}-\boldsymbol{\mu}$ when $k=d$. If the factor loadings include an orthogonal matrix E, then appropriate PC factor scores for the population random vector $\mathbf{X}$ are

$$\widehat{\mathbf{F}}^{(\mathrm{E})} = E^{\mathsf{T}}\Lambda_k^{-1/2}\Gamma_k^{\mathsf{T}}(\mathbf{X}-\boldsymbol{\mu}),$$

and $\widehat{A}E\widehat{\mathbf{F}}^{(\mathrm{E})} = \widehat{A}\widehat{\mathbf{F}}^{(\mathrm{PC})}$ follows.

For data $\mathbb{X} = \begin{bmatrix}\mathbf{X}_1 \cdots \mathbf{X}_n\end{bmatrix}$, we refer to the sample quantities, and in a k-factor model, we define the **(sample) PC factor scores** by

$$\widehat{\mathbf{F}}_i^{(\mathrm{PC})} = \widehat{\Lambda}_k^{-1/2}\widehat{\Gamma}_k^{\mathsf{T}}(\mathbf{X}_i-\overline{\mathbf{X}}), \quad i \le n. \tag{7.19}$$

A simple calculation shows that the sample covariance matrix of the $\widehat{\mathbf{F}}_i^{(\mathrm{PC})}$ is the identity matrix $\mathbf{I}_{k\times k}$. By Proposition 7.2, rotated sample factor scores give rise to a k-factor model if the orthogonal transform is appropriately applied to the factor loadings. Oblique rotations, as in Example 7.6, can be included in PC factor scores, but the pair (loadings, scores) would not result in $\widehat{A}\widehat{\mathbf{F}}$.

If Method 7.2 is used to derive factor loadings, then the primary focus is the decomposition of the matrix of correlation coefficients. Should factor scores be required, then (7.19) can be modified to apply to scaled data.

The ML factor loadings constructed in part 3 of Theorem 7.6 are closely related to those obtained with Method 7.1. In analogy with PC factor scores, we define **(sample) ML factor scores** by

$$\widehat{\mathbf{F}}_i^{(\mathrm{ML})} = (\widehat{\Lambda}_k - \widehat{\sigma}^2\mathbf{I}_{k\times k})^{-1/2}\widehat{\Gamma}_k^{\mathsf{T}}(\mathbf{X}_i-\overline{\mathbf{X}}). \tag{7.20}$$

The sample covariance matrix of the $\widehat{\mathbf{F}}_i^{(\mathrm{ML})}$ is not the identity matrix. However, for the factor loadings $\widehat{A}_{\mathrm{ML}} = \widehat{\Gamma}_k(\widehat{\Lambda}_k - \widehat{\sigma}^2\mathbf{I}_{k\times k})^{1/2}$ of part 3 of Theorem 7.6,

$$\widehat{A}_{\mathrm{ML}}\widehat{\mathbf{F}}_i^{(\mathrm{ML})} = \widehat{\Gamma}_k\widehat{\Gamma}_k^{\mathsf{T}}(\mathbf{X}_i-\overline{\mathbf{X}}) = \widehat{A}_{\mathrm{PC}}\widehat{\mathbf{F}}_i^{(\mathrm{PC})},$$

with $\widehat{\mathbf{F}}_i^{(\mathrm{PC})}$ and $\widehat{A}_{\mathrm{PC}} = \widehat{A}$ as in (7.19) and (7.7), respectively. Although the ML factor scores differ from the PC factor scores, in practice, there is little difference between them, especially if $\widehat{\sigma}^2$ is small compared with the trace of $\widehat{\Lambda}_k$. We compare the different factor scores at the end of Section 7.6, including those described in the following sections.

7.6.2 Bartlett and Thompson Factor Scores

If factor loadings are given explicitly, as in part 3 of Theorem 7.6, then factor scores can be defined similarly to the PC or ML factor scores. In the absence of explicit expressions for factor loadings – as in parts 1 and 2 of Theorem 7.6 – there is no natural way to define factor scores. In this situation we return to the basic model (7.1). We regard the specific factors as error terms, with the aim of minimising the error. For a k-factor population model, consider

$$(\mathbf{X} - \boldsymbol{\mu} - A\mathbf{F})^{\mathsf{T}}\Psi^{-1}(\mathbf{X} - \boldsymbol{\mu} - A\mathbf{F}), \tag{7.21}$$

and regard the covariance matrix Ψ of the $\boldsymbol{\epsilon}$ as a scaling matrix of the error terms. Minimising (7.21) over common factors $\mathbf{F}$ yields

$$A^{\mathsf{T}}\Psi^{-1}(\mathbf{X} - \boldsymbol{\mu}) = A^{\mathsf{T}}\Psi^{-1}A\mathbf{F}.$$

If the $k \times k$ matrix $A^{\mathsf{T}}\Psi^{-1}A$ is invertible, then we define the **Bartlett factor scores**

$$\widehat{\mathbf{F}}^{(\mathrm{B})} = \left(A^{\mathsf{T}}\Psi^{-1}A\right)^{-1} A^{\mathsf{T}}\Psi^{-1}(\mathbf{X} - \boldsymbol{\mu}). \tag{7.22}$$

If $A^{\mathsf{T}}\Psi^{-1}A$ is not invertible, then we define the **Thompson factor scores**

$$\widehat{\mathbf{F}}^{(\mathrm{T})} = \left(A^{\mathsf{T}}\Psi^{-1}A + \zeta\mathbf{I}_{k\times k}\right)^{-1} A^{\mathsf{T}}\Psi^{-1}(\mathbf{X} - \boldsymbol{\mu}), \tag{7.23}$$

for some $\zeta \geq 0$. This last definition is analogous to the ridge estimator (2.39) in Section 2.8.2. Some authors reserve the name *Thompson factor scores* for the special case $\zeta = 1$, but I do not make this distinction. We explore theoretical properties of the two estimators in the Problems at the end of Part II.

To define Bartlett and Thompson factor scores for data $\mathbb{X}$, we replace population quantities by the appropriate sample quantities. For the ith observation $\mathbf{X}_i$, put

$$\widehat{\mathbf{F}}_{\zeta,i} = \left(\widehat{A}^{\mathsf{T}}\widehat{\Psi}^{-1}\widehat{A} + \zeta\mathbf{I}_{k\times k}\right)^{-1} \widehat{A}^{\mathsf{T}}\widehat{\Psi}^{-1}(\mathbf{X}_i - \overline{\mathbf{X}}) \qquad \text{for } \zeta \geq 0 \text{ and } i \leq n. \tag{7.24}$$

If $\zeta = 0$ and $\widehat{A}^{\mathsf{T}}\widehat{\Psi}^{-1}\widehat{A}$ is invertible, then put $\widehat{\mathbf{F}}_i^{(\mathrm{B})} = \widehat{\mathbf{F}}_{\zeta,i}$, and call the $\widehat{\mathbf{F}}_i^{(\mathrm{B})}$ the **(sample) Bartlett factor scores**. If $\zeta > 0$, put $\widehat{\mathbf{F}}_i^{(\mathrm{T})} = \widehat{\mathbf{F}}_{\zeta,i}$, and call the $\widehat{\mathbf{F}}_i^{(\mathrm{T})}$ the **(sample) Thompson factor scores**. These sample scores do not have an identity covariance matrix.

The expression (7.24) requires estimates of A and Ψ, and we therefore need to decide how to estimate A and Ψ. Because the factor scores are derived via least squares, any of the estimators of this and the preceding section can be used; in practice, ML estimators are common as they provide a framework for estimating both A and Ψ.

In applications, the choice between the Bartlett and Thompson factor scores depends on the rank of $A_\Psi = A^{\mathsf{T}}\Psi^{-1}A$ and the size of its smallest eigenvalue. If A_Ψ or its estimator in part 1 of Theorem 7.6 is diagonal and the eigenvalues are not too small, then the Bartlett factor scores are appropriate. If A_Ψ is not invertible, then the Thompson factor scores will need to be used. A common choice is $\zeta = 1$. The Thompson scores allow the integration of extra information in the tuning parameter ζ, including Bayesian prior information.

7.6.3 Canonical Correlations and Factor Scores

In this and the next section we take a different point of view and mimic canonical correlation and regression settings for the derivation of factor scores. In each case we begin with the population.

Consider the k-factor population model $\mathbf{X} - \boldsymbol{\mu} = A\mathbf{F} + \boldsymbol{\epsilon}$. Let

$$\mathbf{X}^* = \begin{bmatrix} \mathbf{X} - \boldsymbol{\mu} \\ \mathbf{F} \end{bmatrix} \qquad \text{with } \boldsymbol{\mu}^* = \mathbf{0}_{d+k} \text{ and } \Sigma^* = \begin{pmatrix} \Sigma & A \\ A^{\mathsf{T}} & \mathbf{I}_{k\times k} \end{pmatrix} \tag{7.25}$$

the mean and covariance matrix of the $(d+k)$-dimensional vector $\mathbf{X}^*$.

Canonical Correlation Analysis provides a framework for determining the relationship between the two parts of the vector $\mathbf{X}^*$, and (3.35) of Section 3.7 establishes a link between the parts of $\mathbf{X}^*$ and a regression setting: the matrix of canonical correlations C. Write Σ_{F} for the covariance matrix of $\mathbf{F}$ and Σ_{XF} for the between covariance matrix $\operatorname{cov}(\mathbf{X}, \mathbf{F})$ of $\mathbf{X}$ and $\mathbf{F}$. Because $\Sigma_{\mathrm{XF}} = A$ by (7.25) and $\Sigma_{\mathrm{F}} = \mathbf{I}_{k\times k}$, the matrix of canonical correlations of $\mathbf{X}$ and $\mathbf{F}$ is

$$C = C(\mathbf{X}, \mathbf{F}) = \Sigma^{-1/2}\Sigma_{\mathrm{XF}}\Sigma_{\mathrm{F}}^{-1/2} = \Sigma^{-1/2}A. \tag{7.26}$$

Another application of (3.35) in Section 3.7, together with (7.26), leads to the definition of the **canonical correlation factor scores** or the **CC factor scores**

$$\widehat{\mathbf{F}}^{(\mathrm{CC})} = A^{\mathsf{T}}\Sigma^{-1}(\mathbf{X} - \boldsymbol{\mu}), \tag{7.27}$$

where we used that $\Sigma_{\mathrm{F}} = \mathbf{I}_{k\times k}$. If $A = \Gamma_k\Lambda_k^{1/2}$, as is natural in the principal component framework, then

$$\widehat{\mathbf{F}}^{(\mathrm{CC})} = \Lambda_k^{1/2}\Gamma_k^{\mathsf{T}}\Sigma^{-1}(\mathbf{X} - \boldsymbol{\mu}) = \mathbf{I}_{k\times d}\Lambda^{-1/2}\Gamma^{\mathsf{T}}(\mathbf{X} - \boldsymbol{\mu}). \tag{7.28}$$

For data similar arguments lead to the **(sample) CC factor scores**. For the ith observation, we have

$$\widehat{\mathbf{F}}_i^{(\mathrm{CC})} = \mathbf{I}_{k\times d}\widehat{\Lambda}^{-1/2}\widehat{\Gamma}^{\mathsf{T}}(\mathbf{X}_i - \overline{\mathbf{X}}) \qquad \text{for } i \le n. \tag{7.29}$$

A comparison of (7.28) and (7.18) or (7.19) and (7.29) shows that the PC factor scores and the CC factor scores agree. This agreement is further indication that (7.18) is a natural definition of the factor scores.

7.6.4 Regression-Based Factor Scores

For the $(d+k)$-dimensional vector $\mathbf{X}^*$ of (7.25), we consider the conditional vector $[\mathbf{F}|(\mathbf{X} - \boldsymbol{\mu})]$. From Result 1.5 of Section 1.4.2, it follows that the conditional mean of $[\mathbf{F}|(\mathbf{X} - \boldsymbol{\mu})]$ is

$$\mathbb{E}\left[\mathbf{F}|(\mathbf{X} - \boldsymbol{\mu})\right] = A^{\mathsf{T}}\Sigma^{-1}(\mathbf{X} - \boldsymbol{\mu}).$$

The left-hand side $\mathbb{E}\left[\mathbf{F}|(\mathbf{X} - \boldsymbol{\mu})\right]$ can be interpreted as a regression mean, and it is therefore natural to define the regression-like estimator

$$\widehat{\mathbf{F}}^{(\mathrm{Reg})} = A^{\mathsf{T}}\Sigma^{-1}(\mathbf{X} - \boldsymbol{\mu}). \tag{7.30}$$

The underlying Factor Analysis model is $\Sigma = AA^{\mathsf{T}} + \Psi$, and we now express $\widehat{\mathbf{F}}^{(\mathrm{Reg})}$ in terms of A and Ψ.

Proposition 7.8 *Let* $\mathbf{X} - \boldsymbol{\mu} = A\mathbf{F} + \boldsymbol{\epsilon}$ *be a k-factor model with* $\Sigma = AA^{\top} + \Psi$. *Assume that* Σ, Ψ *and* $\left(A^{\top}\Psi^{-1}A + \mathbf{I}_{k\times k}\right)$ *are invertible matrices. Then*

$$A^{\top}\left(AA^{\top} + \Psi\right)^{-1} = \left(A^{\top}\Psi^{-1}A + \mathbf{I}_{k\times k}\right)^{-1} A^{\top}\Psi^{-1}, \tag{7.31}$$

and the $\widehat{\mathbf{F}}^{(Reg)}$ *of (7.30) are*

$$\widehat{\mathbf{F}}^{(\mathrm{Reg})} = \left(A^{\top}\Psi^{-1}A + \mathbf{I}_{k\times k}\right)^{-1} A^{\top}\Psi^{-1}(\mathbf{X} - \boldsymbol{\mu}). \tag{7.32}$$

We call $\widehat{\mathbf{F}}^{(\mathrm{Reg})}$ the **regression factor scores**.

Proof We first prove (7.31). Put $A_{\Psi} = A^{\top}\Psi^{-1}A$, and write $\mathbf{I}$ for the $k \times k$ identity matrix. By assumption,

$$\Sigma^{-1} = \left(AA^{\top} + \Psi\right)^{-1} \qquad \text{and} \qquad (A_{\Psi} + \mathbf{I})^{-1}$$

exist, and hence

$$(A_{\Psi} + \mathbf{I})(A_{\Psi} + \mathbf{I})^{-1} A_{\Psi} - (A_{\Psi} + \mathbf{I})\left[\mathbf{I} - (A_{\Psi} + \mathbf{I})^{-1}\right] = A_{\Psi} - [(A_{\Psi} + \mathbf{I}) - \mathbf{I}] = 0,$$

which implies that

$$(A_{\Psi} + \mathbf{I})^{-1} A_{\Psi} = \left(\mathbf{I} - (A_{\Psi} + \mathbf{I})^{-1}\right). \tag{7.33}$$

Similarly, it follows that

$$(AA^{\top} + \Psi)^{-1}(AA^{\top} + \Psi) - \left[\Psi^{-1} - \Psi^{-1}A(A_{\Psi} + \mathbf{I})^{-1}A^{\top}\Psi^{-1}\right](AA^{\top} + \Psi) = 0,$$

and because $(AA^{\top} + \Psi)$ is invertible, we conclude that

$$(AA^{\top} + \Psi)^{-1} = \Psi^{-1} - \Psi^{-1}A(A_{\Psi} + \mathbf{I})^{-1}A^{\top}\Psi^{-1}. \tag{7.34}$$

From (7.33) and (7.34), it follows that

$$(AA^{\top} + \Psi)^{-1}A = \Psi^{-1}A(A_{\Psi} + \mathbf{I})^{-1}.$$

The desired expression for $\widehat{\mathbf{F}}^{(\mathrm{Reg})}$ follows from (7.30) and (7.31). ∎

Formally, $\widehat{\mathbf{F}}^{(\mathrm{Reg})}$ of (7.30) is the same as $\widehat{\mathbf{F}}^{(\mathrm{CC})}$ of (7.27), but (7.32) exploits the underlying model for Σ. Under the assumptions of part 1 of Theorem 7.6, $A^{\top}\Psi^{-1}A$ is a diagonal matrix, making (7.32) computationally simpler than (7.30). To calculate the regression factor scores, we require expressions $\widehat{A}$ and $\widehat{\Psi}$.

For data $\mathbb{X}$, the analogous **(sample) regression factor scores** are

$$\widehat{\mathbf{F}}_i^{(\mathrm{Reg})} = \left(\widehat{A}^{\top}\widehat{\Psi}^{-1}\widehat{A} + \mathbf{I}_{k\times k}\right)^{-1} \widehat{A}^{\top}\widehat{\Psi}^{-1}(\mathbf{X}_i - \overline{\mathbf{X}}) \qquad i \le n.$$

A comparison with the Thompson factor scores shows that the regression factor scores $\widehat{\mathbf{F}}^{(\mathrm{Reg})}$ correspond to the special case $\zeta = 1$. For data, the regression factor scores are therefore a special case of the scores $\widehat{\mathbf{F}}_{\zeta,i}$ of (7.24).

7.6.5 Factor Scores in Practice

In Sections 7.6.1 to 7.6.4 we explored and compared different factor scores. The PC factor scores of Section 7.6.1 agree with the CC factor scores of Section 7.6.3 and formally also with the regression scores of Section 7.6.4. The main difference is that the sample PC and CC factor scores are calculated directly from the sample covariance matrix, whereas the regression factor scores make use of the ML estimators for A and Ψ.

In an ML setting with explicit expressions for the ML estimators as in part 3 of Theorem 7.6, the scores are similar to the PC factor scores. If the ML estimators are not given explicitly, the Bartlett and Thompson factor scores of Section 7.6.2 and the regression scores of Section 7.6.4 yield factor scores. Furthermore, the Bartlett scores and the regression scores can be regarded as special cases of the Thompson scores, namely,

$$\widehat{\mathbf{F}}^{(\mathrm{T})} = \begin{cases} \widehat{\mathbf{F}}^{(\mathrm{Reg})} & \text{if } \zeta = 1, \\ \widehat{\mathbf{F}}^{(\mathrm{B})} & \text{if } \zeta = 0. \end{cases}$$

I will not delve into theoretical properties of the variously defined factor scores but examine similarities and differences of scores in an example with a small number of variables as this makes it easier to see what is happening.

Example 7.8 We consider the five-dimensional **exam grades** data. The hypothesis tests of Table 7.3 in Example 7.7 reject a one-factor model but accept a two-factor model. We consider the latter.

By part 2 of Theorem 7.6, transformations of ML factor loadings by orthogonal matrices result in ML estimates. It is interesting to study the effect of such transformations on the factor scores. In addition, I include oblique rotations, and I calculate the Bartlett and Thompson factor scores (with $\zeta = 1$) for each of the three scenarios: unrotated factor loadings, varimax optimal factor loadings and varimax optimal loadings obtained from oblique rotations.

Figure 7.4 shows scatterplots of the factor scores. In all plots, the x-axis refers to component 1 and the y-axis refers to component 2 of the two-factor model. The top row shows the results for the Barlett factor scores, and the bottom row – with the exception of the leftmost panel – shows the corresponding results for the Thompson scores. The top left panel displays the results pertaining to the unrotated ML factor loadings, the panels in the middle display those of the varimax optimal ML factor loadings and the right panels refer to oblique rotations. For the unrotated factor loadings, the Barlett and Thompson factor scores look identical, apart from a scale factor. For this reason, I substitute the Thompson scores with the PC factor scores of the unrotated PC factor loadings – shown in red in the figure.

Principal component factor scores and ML factor scores without rotation produce uncorrelated scores. The sample correlation coefficients are zero; this is consistent with the random pattern of the scatterplots. The factor scores of the varimax optimal factor loadings are correlated: -0.346 for the Bartlett scores and 0.346 for the Thompson scores. This correlation may be surprising at first, but a closer inspection of (7.22) and (7.23) shows that inclusion of a rotation in the factor loadings correlates the scores. As we move to the right panels in the figure, the correlation between the two components of the factor scores increases to 0.437 for the Bartlett scores and to a high 0.84 for the Thompson scores.

The scatterplots in Figure 7.4 show that there is a great variety of factor scores to choose from. There is no clear 'best' set of factor scores, and the user needs to think carefully about

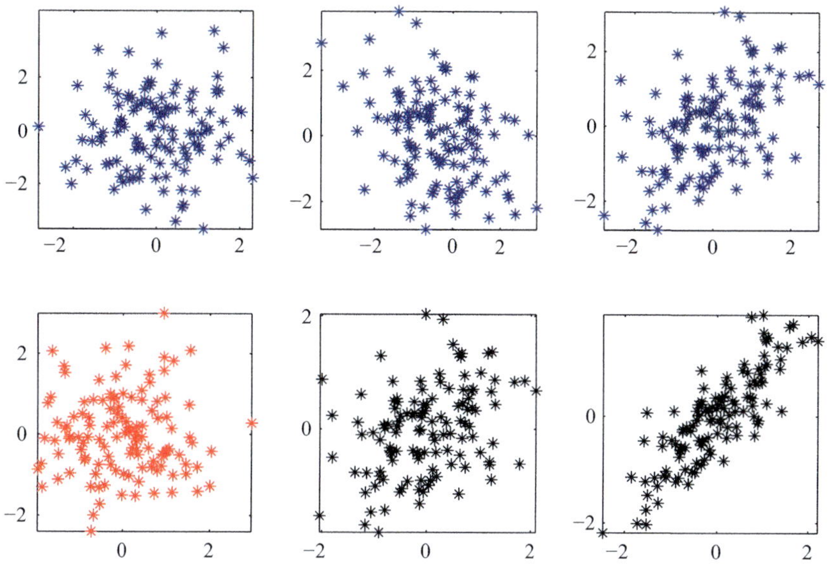

Figure 7.4 Factor scores for Example 7.8. Barlett scores (*top*), Thompson scores (*bottom*). Raw factors (*left*) with ML (*top*), and PC (*bottom, in red*). (*Middle*): ML with varimax optimal rotations; (*right*) ML with oblique optimal rotations.

the purpose and aim of the analysis before making a choice of the types of factor scores that best suit the aims of the analysis. ■

This example highlights the fact that factor scores are far from uniquely defined and vary greatly. This difference is mainly due to the variously defined factor loadings rather than the different definitions of factor scores. Indeed, there is a trade-off between optimising the factor loadings in terms of their VC value and the degree of uncorrelatedness of the scores: the higher the VC, the more correlated are the resulting scores. If ease of interpretability of the factor loadings is the main concern, then rotated factor loadings perform better, but this is at a cost, as the scores become correlated. If one requires uncorrelated scores, then the principal component factor scores or scores obtained from raw ML factor loadings without rotation are preferable.

7.7 Principal Components, Factor Analysis and Beyond

There are many parallels between Principal Component Analysis and Factor Analysis, but there are also important differences. The common goal is to represent the data in a smaller number of simpler components, but the two methods are based on different strategies and models to achieve these goals. Table 7.5 lists some of the differences between the two approaches.

Despite apparent differences, Principal Component Analysis is one of the main methods for finding factor loadings, but at the same time, it is *only one* of a number of methods. Example 7.8 shows differences that occur in practice. On a more theoretical level, Schneeweiss and Mathes (1995) and Ogasawara (2000) compared Principal Component Analysis and Factor Analysis; they stated conditions under which the principal component

Table 7.5 *Comparison of Principal Component Analysis and Factor Analysis*

	PCA	**FA**
Aim	Fewer and simpler components	Few factor loadings
Model	Not explicit	Strict k-factor model $\mathbf{X}-\boldsymbol{\mu}=A\mathbf{F}+\boldsymbol{\epsilon}$
Distribution	Non-parametric	Normal model common
Covariance matrix	Spectral decomposition $\Sigma=\Gamma\Lambda\Gamma^{\mathsf{T}}$	Factor decomposition $\Sigma=AA^{\mathsf{T}}+\Psi$
Solution method	Spectral decomposition of Σ; components ranked by variance; orthogonal projections only	PC-based loadings or ML-based loadings with orthogonal or oblique rotations
Scores	Projection onto eigenvectors ranked by variance; uniquely defined uncorrelated component scores	PC-based scores or ML regression scores; not unique, include rotations; rotated scores correlated
Data description	Approximate $\mathbb{X}\approx\sum_{j=1}^{k}\widehat{\boldsymbol{\eta}_j\boldsymbol{\eta}_j}^{\mathsf{T}}\mathbb{X}$	Complete, includes specific factor $\mathbb{X}-\overline{\mathbf{X}}=\widehat{A}_{d\times k}\mathbb{F}+\mathfrak{N}$

solutions provide 'adequate' factor loadings and scores. I do not want to quantify what 'adequate' means but mention these papers to indicate that there is some fine-tuning and control in ML solutions of the factor loadings that are not available in the PC solutions.

An approach which embraces principal components, factor loadings and factor scores is that of Tipping and Bishop (1999): for normal data, their ML solution is essentially the principal component solution with a tuning parameter $\widehat{\sigma}$. The explicit expressions for the factor loadings and the covariance matrix further lead to a method for determining the number of factors. Their strategy is particularly valuable when classical hypothesis testing, as described in Theorem 7.7, becomes too restrictive. If the normality assumptions of Tipping and Bishop's model apply approximately, then their loadings and scores have an easy interpretation. Sadly, the applicability of their method is restricted to data that are not too different from the Gaussian. For the *breast cancer* data of Example 2.9 in Section 2.8, we have seen that their method does not produce an appropriate choice for the number of factors.

Factor Analysis continues to enjoy great popularity in many areas, including psychology, the social sciences and marketing. As a result, new methodologies which extend Factor Analysis have been developed. Two such developments are

- Structural Equation Modelling and
- Independent Component Analysis.

Loosely speaking, Structural Equation Modelling combines Factor Analysis and multivariate regression. It takes into account interactions and non-linearities in the modelling and focuses on the confirmatory or model testing aspects of Factor Analysis. As done in Factor Analysis, in Structural Equation Modelling one commonly assumes a normal model and makes use of ML solutions. In addition, Structural Equation Modelling extends Factor Analysis to a regression-like model $(\mathbf{X},\mathbf{Y})$ with different latent variables for the predictors and responses

$$\mathbf{X} = A_X\mathbf{F} + \boldsymbol{\epsilon}, \qquad \mathbf{Y} = A_Y\mathbf{G} + \boldsymbol{\varepsilon} \qquad \text{and} \qquad \mathbf{T} = B_G\mathbf{G} + B_F\mathbf{F} + \boldsymbol{\delta}. \tag{7.35}$$

The error terms $\boldsymbol{\epsilon}, \boldsymbol{\varepsilon}$ and $\boldsymbol{\delta}$ satisfy conditions similar to those which $\boldsymbol{\epsilon}$ satisfies in a k-factor model, and the models for **X** and **Y** are k-factor and m-factor models for some values of k and m. The last equation links the latent variables **F** and **G**, called *cause* and *effect*, respectively.

Structural Equation Models have been used by sociologists since at least the 1960s in an empirical way. Jöreskog (1973) laid the foundations for the formalism (7.35) and prepared a framework for statistical inference. LISREL, one of Jöreskog's achievements which stands for 'Linear Structural Relationships', started as a model for these processes but has long since become synonymous with possibly the main software used for analysing structural equation models. A possible starting point for the interested reader is the review of Anderson and Gerbing (1988) and references therein, but there are also many books on almost any aspect of the subject.

The other extension, Independent Component Analysis, has moved into a very different direction. Its starting point is the model

$$\mathbf{X} = A\mathbf{S},$$

which is similar to the factor model in that both the transformation matrix A and the latent variables **S** are unknown. Unlike the Factor Analysis models, the underlying structure in Independent Component Analysis is non-Gaussian, and theoretical and practical solutions for A and **S** exist under the assumption of non-Gaussian vectors **X**. Independent Component Analysis is the topic of Chapter 10, and I postpone a description and discussion until then.

8

Multidimensional Scaling

In mathematics you don't understand things. You just get used to them (John von Neumann, 1903–1957; in Gary Zukav (1979), *The Dancing Wu Li Masters*).

8.1 Introduction

Suppose that we have n objects and that for each pair of objects a numeric quantity or a ranking describes the relationship between objects. The objects could be geographic locations, with a distance describing the relationship between locations. Other examples are different types of food or drink, with judges comparing items pairwise and providing a score for each pair. Multidimensional Scaling combines such pairwise information into a whole picture of the data and leads to a visual representation of the relationships.

Visual and geometric aspects have been essential parts of Multidimensional Scaling. For geographic locations, they lead to a map (see Figure 8.1). From comparisons and rankings of foods, drinks, perfumes or laptops, one typically reconstructs low-dimensional representations of the data and displays these representations graphically in order to gain insight into the relationships between the different objects of interest. In addition to these graphical representations, in a ranking of wines, for example, we might want to know which features result in wines that will sell well; the type of grape, the alcohol content and the region might be of interest. Based on information about pairs of objects, the aims of Multidimensional Scaling are

1. to construct vectors which represent the objects such that
2. the relationship between pairs of the original objects is preserved as much as possible in the new pairs of vectors. In particular, if two objects are close, then their corresponding new vectors should also be close.

The origins of Multidimensional Scaling go back to Young and Householder (1938) and Richardson (1938) and their interest in psychology and the behavioural sciences. The method received little attention until the seminal paper of Torgerson (1952), which was followed by a book (Torgerson, 1958). In the early 1960s, Multidimensional Scaling became a fast-growing research area with a series of major contributions from Shepard and Kruskal. These include Shepard (1962a, 1962b), Kruskal (1964a, 1964b, 1969, 1972), and Kruskal and Wish (1978). Gower (1966) was the first to formalise a concrete framework for this exciting discipline, and since then, other approaches have been developed. Apart from the classical book by Kruskal and Wish (1978), the books by Cox and Cox (2001) and Borg and Groenen (2005) deal with many different issues and topics in Multidimensional Scaling.

Many of the original definitions and concepts have undergone revisions or refinements over the decades. The notation has stabilised, and a consistent framework has emerged, which I use in preference to the historical definitions. The distinction between the three different approaches, referred to as *classical*, *metric* and *non-metric scaling*, is useful, and I will therefore describe each approach separately.

Multidimensional Scaling is a dimension-reduction method – in the same way that Principal Component Analysis and Factor Analysis are: We assume that the original objects consist of d variables, we attempt to represent the objects by a smaller number of meaningful variables, and we ask the question: How many dimensions do we require? If we want to represent the new configuration as a map or shape, then two or three dimensions are required, but for more general configurations, the answer is not so clear. In this chapter we focus on the following:

- We explore the main ideas of Multidimensional Scaling in their own right.
- We relate Multidimensional Scaling to other dimension-reduction methods and in particular to Principal Component Analysis.

Section 8.2 sets the framework and introduces classical scaling. We find out about principal coordinates and the loss criteria 'stress' and 'strain'. Section 8.3 looks at metric scaling, a generalisation of classical scaling, which admits a range of proximity measures and additional stress measures. Section 8.4 deals with non-metric scaling, where rank order replaces quantitative measures of distance. This section includes an extension of the strain criterion to the non-metric environment. Section 8.5 considers data and their configurations from different perspectives: I highlight advantages of the duality between $\mathbb{X}$ and $\mathbb{X}^\top$ for high-dimensional data and show how to construct a configuration for multiple data sets. We look at results from Procrustes Analysis which explain the relationship between multiple configurations. Section 8.6 starts with a relative of Multidimensional Scaling for count or quantitative data, Correspondence Analysis. It looks at Multidimensional Scaling for data that are known to belong to different classes, and we conclude with developments of embeddings that integrate local information such as cluster centres and landmarks. Problems pertaining to the material of this chapter are listed at the end of Part II.

8.2 Classical Scaling

Multidimensional Scaling is driven by data. It is possible to formulate Multidimensional Scaling for the population and pairs of random vectors, but I will focus on n random vectors and data.

The ideas of scaling have existed since the 1930s, but Gower (1966) first formalised them and derived concrete solutions in a transparent manner. Gower's framework – known as **Classical Scaling** – is based entirely on distance measures.

We begin with a general framework: the observed objects, dissimilarities and criteria for measuring closeness. The dissimilarities of the non-classical approaches are more general than the Euclidean distances which Gower used in classical scaling. Measures of proximity, including dissimilarities and distances, are defined in Section 5.3.2. Unless otherwise specified, Δ_{E} is the Euclidean distance or norm in this chapter.

Table 8.1 *Ten Cities from Example 8.1*

1	Tokyo	6	Jakarta
2	Sydney	7	Hong Kong
3	Singapore	8	Hiroshima
4	Seoul	9	Darwin
5	Kuala Lumpur	10	Auckland

Definition 8.1 Let $\mathbb{X} = \begin{bmatrix}\mathbf{X}_1 \cdots \mathbf{X}_n\end{bmatrix}$ be data. Let

$$\mathcal{O} = \{O_1, \ldots, O_n\}$$

be a set of **objects** such that the object O_i is derived from or related to $\mathbf{X}_i$. Let ϱ be a dissimilarity for pairs of objects from $\mathcal{O}$. We call

$$\varrho_{ik} = \varrho(O_i, O_k)$$

the **dissimilarity** of O_i and O_k and $\{\mathcal{O}, \varrho\}$ the **observed data** or the **(pairwise) observations**. □

The dissimilarities between objects are the quantities we observe. The objects are binary or categorical variables, random vectors or just names, as in Example 8.1. The underlying d-dimensional random vectors $\mathbf{X}_i$ are generally not observable or not available.

Example 8.1 We consider the **ten cities** listed in Table 8.1. The cities, given by their names, are the objects. The dissimilarities ϱ_{ik} of the objects are given in (8.1) as distances in kilometres between pairs of cities. In the matrix $D = \{\varrho_{ik}\}$ of (8.1), the entry $\varrho_{2,5} = 6{,}623$ refers to the distance between the second city, Sydney, and the fifth city, Kuala Lumpur. Because $\varrho_{ik} = \varrho_{ki}$, I show the ϱ_{ik} for $i \le k$ in (8.1) only:

$$D = \begin{pmatrix} 0 & 7825 & 5328 & 1158 & 5329 & 5795 & 2893 & 682 & 5442 & 8849 \\ & 0 & 6306 & 8338 & 6623 & 5509 & 7381 & 7847 & 3153 & 2157 \\ & & 0 & 4681 & 317 & 892 & 2588 & 4732 & 3351 & 8418 \\ & & & 0 & 4616 & 5299 & 2100 & 604 & 5583 & 9631 \\ & & & & 0 & 1183 & 2516 & 4711 & 3661 & 8736 \\ & & & & & 0 & 3266 & 5258 & 2729 & 7649 \\ & & & & & & 0 & 2236 & 4274 & 9148 \\ & & & & & & & 0 & 5219 & 9062 \\ & & & & & & & & 0 & 5142 \\ & & & & & & & & & 0 \end{pmatrix} \quad (8.1)$$

The aim is to construct a map of these cities from the distance information alone. We construct this map in Example 8.2. ■

Definition 8.2 Let $\{\mathcal{O}, \varrho\}$ be the observed data, with $\mathcal{O} = \{O_1, \ldots, O_n\}$ a set of n objects. Fix $p > 0$. An **embedding of objects** from $\mathcal{O}$ into $\mathbb{R}^p$ is a one-to-one map $\mathfrak{f}\colon \mathcal{O} \to \mathbb{R}^p$. Let Δ be a distance, and put $\Delta_{ik} = \Delta[\mathfrak{f}(O_i), \mathfrak{f}(O_k)]$. For functions g and h and positive weights

$w = w_{ik}$, with $i,k \le n$, the **(raw) stress**, regarded as a function of ϱ and $\mathfrak{f}$, is

$$\mathcal{S}\text{tress}(\varrho, \mathfrak{f}) = \left\{ \sum_{i<k}^{n} w_{ik} \left[g(\varrho_{ik}) - h(\Delta_{ik}) \right]^2 \right\}^{1/2}. \tag{8.2}$$

If $\mathfrak{f}^*$ minimises $\mathcal{S}$tress over embeddings into $\mathbb{R}^p$, then

$$\mathbb{W} = \left[\mathbf{W}_1 \cdots \mathbf{W}_n \right] \qquad \text{with } \mathbf{W}_i = \mathfrak{f}^*(O_i)$$

is a ***p*-dimensional (stress) configuration** for $\{\mathcal{O}, \varrho\}$, and the $\Delta(\mathbf{W}_i, \mathbf{W}_k)$ are the **(pairwise) configuration distances**. □

The stress extends the idea of the squared error, and the (stress) configuration, the minimiser of the stress, corresponds to the least-squares solution. In Section 5.4 we looked at feature maps and embeddings. A glance back at Definition 5.6 tells us that $\mathbb{R}^p$ corresponds to the space $\mathcal{F}$, $\mathfrak{f}$ and $\mathfrak{f}^*$ are feature maps and the configuration $\mathfrak{f}^*(O_i)$ with $i \le n$ represents the feature data.

A configuration depends on the dissimilarities ϱ, the functions g and h and the weights w_{ik}. In **classical scaling**, we have

$$\varrho = \Delta_{\text{E}} \text{ the Euclidean distance}, \qquad g, h \text{ identity functions} \quad \text{and} \quad w_{ik} = 1.$$

Indeed, $\varrho = \Delta_{\text{E}}$ characterises classical scaling. We take g to be the identity function, as done in the original classical scaling. This simplification is not crucial; see Meulman (1992, 1993), who allows more general functions g in an otherwise classical scaling framework.

8.2.1 Classical Scaling and Principal Coordinates

Let $\mathbb{X}$ be data of size $d \times n$. Let $\{\mathcal{O}, \varrho, \mathfrak{f}\}$ be the observed data together with an embedding $\mathfrak{f}$ into $\mathbb{R}^p$ for some $p \le d$. Put

$$\varrho_{ik} = \Delta_E(\mathbf{X}_i, \mathbf{X}_k) \qquad \text{and} \qquad \Delta_{ik} = \Delta_E\left[\mathfrak{f}(O_i), \mathfrak{f}(O_k)\right] \qquad \text{for } i,k \le n; \tag{8.3}$$

then the stress (8.2) leads to the notions

raw classical stress:

$$\mathcal{S}\text{tress}_{\text{clas}}(\varrho, \mathfrak{f}) = \left\{ \sum_{i<k}^{n} \left[\varrho_{ik} - \Delta_{ik} \right]^2 \right\}^{1/2} \qquad \text{and}$$

classical stress:

$$\mathcal{S}\text{tress}^*_{\text{clas}}(\varrho, \mathfrak{f}) = \frac{\mathcal{S}\text{tress}_{\text{clas}}(\varrho, \mathfrak{f})}{\left\{ \sum_{i<k}^{n} \varrho_{ik}^2 \right\}^{1/2}} = \left\{ \frac{\sum_{i<k}^{n} \left[\varrho_{ik} - \Delta_{ik} \right]^2}{\sum_{i<k}^{n} \varrho_{ik}^2} \right\}^{1/2}, \tag{8.4}$$

The classical stress is a standardised version of the raw stress. This standardisation is not part of the generic definition (8.2) but is commonly used in classical scaling. If there is no ambiguity, I will drop the subscript *clas*.

For the classical setting, Gower (1966) proposed a simple construction of a p-dimensional configuration $\mathbb{W}$.

Theorem 8.3 *Let $\mathbb{X} = [\mathbf{X}_1 \cdots \mathbf{X}_n]$ be centred data, and let $\{\mathcal{O}, \varrho\}$ be the observed data with $\varrho_{ik}^2 = (\mathbf{X}_i - \mathbf{X}_k)^\top(\mathbf{X}_i - \mathbf{X}_k)$. Put $Q_{\langle n\rangle} = \mathbb{X}^\top\mathbb{X}$, and let r be the rank of $Q_{\langle n\rangle}$.*

1. *The $n \times n$ matrix $Q_{\langle n\rangle}$ has entries*

$$q_{ik} = -\frac{1}{2}\left(\varrho_{ik}^2 - \frac{2}{n}\sum_{k=1}^n \varrho_{ik}^2 + \frac{1}{n^2}\sum_{i,k=1}^n \varrho_{ik}^2\right).$$

2. *If $Q_{\langle n\rangle} = VD^2V^\top$ is the spectral decomposition of $Q_{\langle n\rangle}$, then*

$$\mathbb{W} = DV^\top \tag{8.5}$$

defines a configuration in r variables which minimises $\mathcal{S}tress_{clas}$. The configuration (8.5) is unique up to multiplication on the left by an orthogonal $r \times r$ matrix.

Gower (1966) coined the phrases **principal coordinates** for the configuration vectors $\mathbf{W}_i$ and **Principal Coordinate Analysis** for his method of constructing the configuration. The term *Principal Coordinate Analysis* has become synonymous with classical scaling.

In Theorem 8.3, the dissimilarities ϱ_{ik} are assumed to be Euclidean distances of pairs of random vectors $\mathbf{X}_i$ and $\mathbf{X}_k$. In practice, the ϱ_{ik} are observed quantities, and we have no way of checking whether they **are** pairwise Euclidean distances. This lack of information does not detract from the usefulness of the theorem, which tells us how to exploit the ϱ_{ik} in the construction of the principal coordinates. Further, if we know $\mathbb{X}$, then the principal coordinates are the principal component data.

Proof From the definition of the ϱ_{ik}^2 it follows that $\varrho_{ik}^2 = \mathbf{X}_i^\top\mathbf{X}_i + \mathbf{X}_k^\top\mathbf{X}_k - 2\mathbf{X}_i^\top\mathbf{X}_k = \varrho_{ki}^2$. The last equality follows by symmetry. The $\mathbf{X}_i$ are centred, so taking sums over indices $i, k \le n$ leads to

$$\frac{1}{n}\sum_k^n \varrho_{ik}^2 = \mathbf{X}_i^\top\mathbf{X}_i + \frac{1}{n}\sum_k^n \mathbf{X}_k^\top\mathbf{X}_k \quad \text{and} \quad \frac{1}{n^2}\sum_{i,k}^n \varrho_{ik}^2 = \frac{2}{n}\sum_{i=1}^n \mathbf{X}_i^\top\mathbf{X}_i. \tag{8.6}$$

Combine (8.6) with $q_{ik} = \mathbf{X}_i^\top\mathbf{X}_k$, and then it follows that

$$\begin{aligned} q_{ik} &= -\frac{1}{2}\left(\varrho_{ik}^2 - \mathbf{X}_i^\top\mathbf{X}_i - \mathbf{X}_k^\top\mathbf{X}_k\right) \\ &= -\frac{1}{2}\left(\varrho_{ik}^2 - \frac{1}{n}\sum_{k=1}^n \varrho_{ik}^2 + \frac{1}{n}\sum_{k=1}^n \mathbf{X}_k^\top\mathbf{X}_k - \frac{1}{n}\sum_{i=1}^n \varrho_{ik}^2 + \frac{1}{n}\sum_{i=1}^n \mathbf{X}_i^\top\mathbf{X}_i\right) \\ &= -\frac{1}{2}\left(\varrho_{ik}^2 - \frac{2}{n}\sum_{k=1}^n \varrho_{ik}^2 + \frac{1}{n}\sum_{i,k=1}^n \varrho_{ik}^2\right). \end{aligned}$$

To show part 2, we observe that the spectral decomposition $Q_{\langle n\rangle} = VD^2V^\top$ consists of a diagonal matrix D^2 with $n - r$ zero eigenvalues because $Q_{\langle n\rangle}$ has rank r. It follows that $Q_{\langle n\rangle} = V_r D_r^2 V_r^\top$, where V_r consists of the first r eigenvectors of $Q_{\langle n\rangle}$, and D_r^2 is the diagonal $r \times r$ matrix. The matrix $\mathbb{W}$ of (8.5) is a configuration in r variables because $\mathbb{W} = \mathfrak{f}(\mathcal{O})$, and $\mathfrak{f}$ minimises the classical stress $\mathcal{S}tress_{\text{clas}}$ by Result 5.7 of Section 5.5. The uniqueness of $\mathbb{W}$ – up to pre-multiplication by an orthogonal matrix – also follows from Result 5.7. ∎

In Theorem 8.3, the rank r of $Q_{\langle n\rangle}$ is the dimension of the configuration. In practice, a smaller dimension p may suffice or even be preferable. The following algorithm tells us how to find p-dimensional coordinates.

Algorithm 8.1 ***Principal Coordinate Configurations in*** p ***Dimensions***
Let $\{\mathcal{O},\varrho\}$ with $\varrho=\Delta_{\mathrm{E}}$ be the observed data derived from $\mathbb{X}=\left[\mathbf{X}_1\cdots\mathbf{X}_n\right]$.

Step 1. Construct the matrix $Q_{\langle n\rangle}$ from ϱ as in part 1 of Theorem 8.3.

Step 2. Determine the rank r of $Q_{\langle n\rangle}$ and its spectral decomposition $Q_{\langle n\rangle}=VD^2V^{\top}$.

Step 3. For $p\le r$, put

$$\mathbb{W}_p=D_pV_p^{\top}, \tag{8.7}$$

where D_p is the $p\times p$ diagonal matrix consisting of the first p diagonal elements of D, and V_p is the $n\times p$ matrix consisting of the first p eigenvectors of $Q_{\langle n\rangle}$. ■

If the $\mathbb{X}$ were known and centred, then the p-dimensional principal coordinates $\mathbb{W}_p$ would equal the principal component data $\mathbb{W}^{(p)}$ of (2.6) in Section 2.3. To see this, we use the relationship between the singular value decomposition of $\mathbb{X}$ and the spectral decompositions of the $Q^{\langle d\rangle}$- and $Q_{\langle n\rangle}$-matrix of $\mathbb{X}$. Starting with the principal components $U^{\top}\mathbb{X}$ as in Section 5.5, for $p\le d$, we obtain

$$U_p^{\top}\mathbb{X}=U_p^{\top}UDV^{\top}=\mathbf{I}_{p\times d}DV^{\top}=D_pV_p^{\top}=\mathbb{W}_p.$$

The calculation shows that we can construct the principal components of $\mathbb{X}$, but we cannot – even approximately – reconstruct $\mathbb{X}$ unless we know the orthogonal matrix U.

We construct principal coordinates for the ten cities.

Example 8.2 We continue with the **ten cities** and construct the two-dimensional configuration (8.7) as in Algorithm 8.1 from the distances (8.1). The five non-zero eigenvalues of $Q_{\langle n\rangle}$ are

$$\lambda_1=10{,}082\times10^4 \qquad \lambda_2=3{,}611\times10^4 \qquad \lambda_3=48\times10^4$$
$$\lambda_4=6\times10^4 \qquad \lambda_5=0.02\times10^4.$$

The eigenvalues decrease rapidly, and $\mathcal{S}\mathrm{tress}_{\mathrm{clas}}=419.8$ for $r=5$. Compared with the size of the eigenvalues, this stress is small and shows good agreement between the original distances and the distances calculated from the configuration.

Figure 8.1 shows a configuration for $p=2$. The left subplot shows the first coordinates of $\mathbb{W}_2$ on the x-axis and the second on the y-axis. The cities Tokyo, Hiroshima and Seoul form the group of points on the top left, and the cities Kuala Lumpur, Singapore and Jakarta form another group on the bottom left.

For a more natural view of the map, we consider the subplot on the right, which is obtained by a 90-degree rotation. Although the geographic north is not completely aligned with the vertical direction, the map on the right gives a good impression of the location of the cities. For $p=2$, the raw classical stress is 368.12, and the classical stress is 0.01. This stress shows that the reconstruction is excellent, and we have achieved an almost perfect map from the configuration $\mathbb{W}_2$. ■

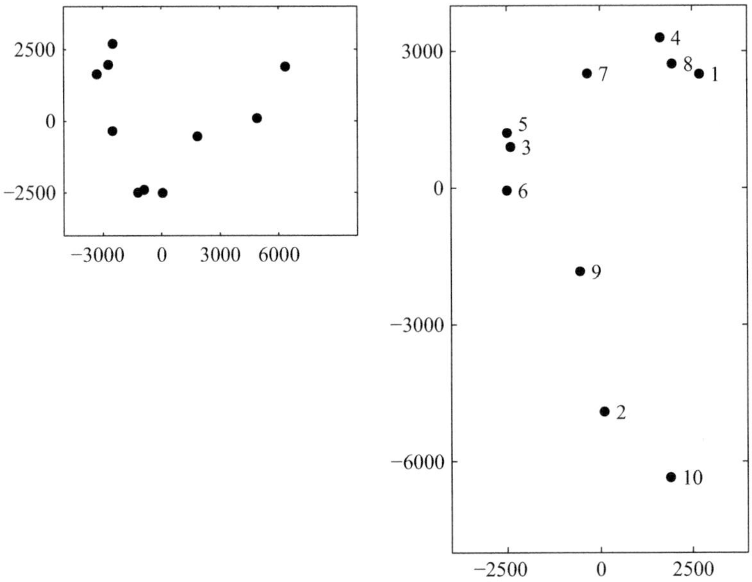

Figure 8.1 Map of locations of ten cities from Example 8.2. Principal coordinates (*left*) and rotated principal coordinates (*right*).

8.2.2 Classical Scaling with Strain

The $\mathcal{S}$tress compares dissimilarities of pairs (O_i, O_k) with distances of the pairs $[\mathfrak{f}(O_i), \mathfrak{f}(O_k)]$ and does not require knowledge of the $(\mathbf{X}_i, \mathbf{X}_k)$. In a contrasting approach, Torgerson (1952) starts with the observations $\mathbf{X}_1, \ldots, \mathbf{X}_n$ and defines a loss criterion called *strain*. The notion of strain was revitalised and extended in the 1990s in a number of papers, including Meulman (1992, 1993) and Trosset (1998). The following definition of strain is informed by these developments rather than the original work of Torgerson.

Definition 8.4 Let $\mathbb{X} = \begin{bmatrix}\mathbf{X}_1 \cdots \mathbf{X}_n\end{bmatrix}$ be centred data, and let r be the rank of $\mathbb{X}$. For $\kappa \leq r$, let A be a $\kappa \times d$ matrix, and write $A\mathbb{X}$ for the κ-dimensional transformed data. Fix $p \leq r$. Let $\mathfrak{f}$ be an embedding of $\mathbb{X}$ into $\mathbb{R}^p$. The **(classical) strain** between $A\mathbb{X}$ and $\mathfrak{f}(\mathbb{X})$ is

$$\mathcal{S}\text{train}\left[A\mathbb{X}, \mathfrak{f}(\mathbb{X})\right] = \left\| (A\mathbb{X})^{\mathsf{T}} A\mathbb{X} - \left[\mathfrak{f}(\mathbb{X})\right]^{\mathsf{T}} \mathfrak{f}(\mathbb{X}) \right\|_{\text{Frob}}^2, \tag{8.8}$$

where $\| \, . \, \|_{\text{Frob}}$ is the Frobenius norm of Definition 5.2 in Section 5.3.

If $\mathfrak{f}^*$ minimises the $\mathcal{S}$train over all embeddings into $\mathbb{R}^p$, then $\mathbb{W} = \mathfrak{f}^*(\mathbb{X})$ is the **p-dimensional (strain) configuration** for the transformed data $A\mathbb{X}$. □

The strain criterion starts with the transformed data $A\mathbb{X}$ and searches for the best p-dimensional embedding. Because $A\mathbb{X}$ and $\mathfrak{f}(\mathbb{X})$ have different dimensions, they are not directly comparable. However, their respective $Q_{(n)}$-matrices $(A\mathbb{X})^{\mathsf{T}} A\mathbb{X}$ and $\left[\mathfrak{f}(\mathbb{X})\right]^{\mathsf{T}} \mathfrak{f}(\mathbb{X})$ (see (5.17) in Section 5.5), are both of size $n \times n$ and can be compared. To arrive at a configuration, we fix the transformation A and then find the embedding $\mathfrak{f}$ which minimises the loss function.

I do not exclude the case $A\mathbb{X} = \mathbb{X}$ in the definition of strain. In this case, one wants to find the best p-dimensional approximation to the data, where 'best' refers to the loss defined by the $\mathcal{S}$train. Let $\mathbb{X}$ be centred, and let UDV^{T} be the singular value decomposition of $\mathbb{X}$. For

$\mathbb{W} = D_p V_p^\top$, as in (8.7), it follows that

$$\mathcal{S}\text{train}(\mathbb{X}, \mathbb{W}) = \sum_{j=p+1}^{r} d_j^4, \tag{8.9}$$

where the d_j values are the singular values of $\mathbb{X}$. Further, if $\mathbb{W}_2 = E D_p V_p^\top$ and E is an orthogonal $p \times p$ matrix, then $\mathcal{S}\text{train}(\mathbb{X}, \mathbb{W}_2) = \mathcal{S}\text{train}(\mathbb{X}, \mathbb{W})$. I defer the proof of this result to the Problems at the end of Part II.

If the rank of $\mathbb{X}$ is d and the variables of $\mathbb{X}$ have different ranges, then we could take A to be the scaling matrix $S_{\text{diag}}^{-1/2}$ of Section 2.6.1.

Proposition 8.5 *Let $\mathbb{X}$ be centred data with singular value decomposition $\mathbb{X} = U D V^\top$. Let r be the rank of the sample covariance matrix S of $\mathbb{X}$. Put $A = S_{diag}^{-1/2}$, where S_{diag} is the diagonal matrix (2.18) of Section 2.6.1. For $p < r$, put*

$$\mathbb{W} = \mathfrak{f}(\mathbb{X}) = \left(S_{diag}^{-1/2} D\right)_p V_p^\top.$$

Then the strain between $A\mathbb{X}$ and $\mathbb{W}$ is

$$\mathcal{S}train(A\mathbb{X}, \mathbb{W}) = (r - p)^2.$$

Proof The result follows from the fact that the trace of the covariance matrix of the scaled data equals the rank of $\mathbb{X}$. See Theorem 2.17 and Corollary 2.19 of Section 2.6.1. ■

Similar to the $\mathcal{S}$tress function, $\mathcal{S}$train is a function of two variables, and each loss function finds the embedding or feature map which minimises the respective loss. Apart from the fact that the strain is defined as a squared error, there are other differences between the two criteria. The stress compares real numbers: for pairs of objects, we calculate the difference between their dissimilarities and the distance of their feature vectors. This requires the choice of a dissimilarity. The strain compares $n \times n$ matrices: the transformed data matrix and the feature data, both at the level of the $Q_{\langle n \rangle}$-matrices of (5.17) in Section 5.5. The strain requires a transformation A and does not use dissimilarities between objects. However, if we think of the $A\mathbb{X}$ as the observed objects and use the Euclidean norm as the dissimilarity, then the two loss criteria are essentially the same.

The next example explores the connection between observed data $\{\mathcal{O}, \varrho\}$ and $A\mathbb{X}$ and examines the configurations which result for different $\{\mathcal{O}, \varrho\}$.

Example 8.3 The **athletes** data were collected at the Australian Institute of Sport by Richard Telford and Ross Cunningham (see Cook and Weisberg 1999). For the 102 male and 100 female athletes, the twelve variables are listed in Table 8.2, including the variable number and an abbreviation for each variable.

We take $\mathbb{X}$ to be the 11×202 data consisting of all variables except variable 9, *sex*. A principal component analysis of the scaled data (not shown here) reveals that the first two principal components separate the male and female data almost completely. A similar split does not occur for the raw data, and I therefore work with the scaled data, which I refer to as $\mathbb{X}$ in this example.

The first principal component eigenvector of the scaled data has highest absolute weight for *LBM* (variable 7), second highest for *Hg* (variable 5), and third highest for *Hc* (variable

Table 8.2 *Athletes Data from Example 8.3*

1	Bfat	Body fat	7	LBM	Lean body mass
2	BMI	Body mass index	8	RCC	Red cell count
3	Ferr	Plasma ferritin concentration	9	Sex	0 for male 1 for female
4	Hc	Haematocrit	10	SSF	Sum of skin folds
5	Hg	Haemoglobin	11	WCC	White cell count
6	Ht	Height	12	Wt	Weight

Table 8.3 *Combinations of* $\{\mathcal{O}, \varrho\}$, κ *and* p *from Example 8.3*

	$\mathcal{O}$	Variables	κ	p	Fig 8.2, position
1	$\mathcal{O}_1$	5 and 7	2	2	Top left
2	$\mathcal{O}_2$	10 and 11	2	2	Bottom left
3	$\mathcal{O}_3$	4, and 5 and 7	3	2	Top middle
4	$\mathcal{O}_3$	4, and 5 and 7	3	3	Bottom middle
5	$\mathcal{O}_4$	Except 9	11	2	Top right
6	$\mathcal{O}_4$	Except 9	11	3	Bottom right

4); *SSF* (variable 10) and *WCC* (variable 11) have the two lowest PC_1 weights in absolute value. We want to examine the relationship between PC weights and the effectiveness of the corresponding configurations in splitting the male and female athletes. For this purpose, I choose four sets of objects $\mathcal{O}_1, \ldots, \mathcal{O}_4$ which are constructed from subsets of variables of the scaled data, as shown in Table 8.3. For example, the set $\mathcal{O}_1$ is made up of the two 'best' PC_1 variables, so $\mathcal{O}_1 = \{O_i = (X_{i,5}, X_{i,7}) \textit{ and } i \leq n\}$.

We construct matrices A_ℓ such that $A_\ell \mathbb{X} = \mathcal{O}_\ell$ for $\ell = 1, \ldots, 4$ and $\mathbb{X}$ the scaled data. Thus, A_ℓ projects the scaled data $\mathbb{X}$ onto the scaled variables that define the set of objects $\mathcal{O}_\ell$. To obtain the set $\mathcal{O}_1$ of Table 8.3, we take $A_1 = (a_{ij})$ to be the 2×11 matrix with entries $a_{1,5} = a_{2,7} = 1$, and all other entries equal zero. In the table, κ is the dimension of $A\mathbb{X}$, as in Definition 8.4.

As standard in classical scaling, I choose the Euclidean distances as the dissimilarities, and for the four $\{\mathcal{O}, \varrho\}$ combinations, I calculate the classical stress (8.4) and the strain (8.8) and find the configurations for $p = 2$ or 3, as shown in the Table 8.3.

The stress and strain configurations look very similar, so I only show the strain configurations. Figure 8.2 shows the 2D and 3D configurations with the placement of the configurations as listed in Table 8.3. The first coordinate of each configuration is shown on the x-axis, the second on the y-axis and the third – when applicable – on the z-axis. The red points show the female observations, and the blue points show the male observations.

A comparison of the three configurations in the top row of the figure shows that the male and female points become more separate as we progress from $\mathcal{O}_1$ on the left to $\mathcal{O}_4$ on right, with hardly any overlap in the last case. Similarly, the right-most configuration in the bottom row is better than the middle one at separating males and females. The bottom-left plot shows the configuration obtained from $\mathcal{O}_2$. These variables, which have the lowest PC_1 weights, do not separate the male and female data.

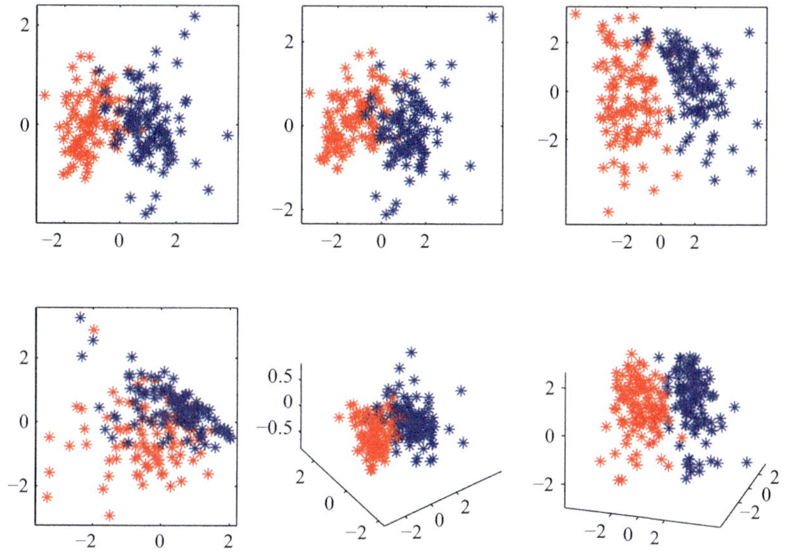

Figure 8.2 Two- and three-dimensional configurations for different $\{\mathcal{O}, \varrho\}$ from Table 8.3 and Example 8.3; female data in red, male in blue.

The classical stress and strain are small in the configurations shown in the left and middle columns. For the configurations in the right column, the strain decreases from 9 to 8, and the normalised classical stress decreases from 0.0655 to 0.0272 when going from the 2D to the 3D configuration.

The calculations show the dependence of the configuration on the objects. A 'careless' choice of objects, as in $\mathcal{O}_2$, leads to configurations which may hide the structure of the data. ■

For the *athletes* data, the four sets of objects $\mathcal{O}_1, \ldots, \mathcal{O}_4$ result in different configurations. A judicious choice of $\mathcal{O}$ can reveal the structure of the data, whereas the hidden structure may remain obscure for other choices of $\mathcal{O}$. In the Problems at the end of Part II, we consider other choices $\mathcal{O}$ in order to gain a better understanding of the change in configurations with the number of variables. Typically, the objects are given, and we have no control how they are obtained or how well the represent the structure in the data. As a consequence, we may not be able to find structure that is present in the data.

In Example 8.3, I have fixed A, but one could fix the row size κ of A only and then optimise the strain for both A and embeddings $\mathfrak{f}$. Meulman (1992) considered this latter case and optimised the strain in a two-step process: find A and update, find $\mathfrak{f}$ and the configuration $\mathbb{W}$ and update and then repeat the two steps. The sparse principal component criterion of Zou, Hastie, and Tibshirani (2006), which I describe in Definition 13.10 of Section 13.4.2, relies on the same two-step updating of their norms, which are closely related to the Frobenius norm in the strain criterion.

8.3 Metric Scaling

In classical scaling, the dissimilarities are Euclidean distances, and principal coordinate configurations provide lower-dimensional representations of the data. These configurations are

Table 8.4 *Common Forms of Stress*

Name	g	h	Squared stress
Classical stress	$g(t)=t$	$h(t)=t$	$\sum_{i<k}^{n}(\varrho_{ik}-\Delta_{ik})^2$
Linear stress	$g(t)=a+bt$	$h(t)=t$	$\sum_{i<k}^{n}[(a+b\varrho_{ik})-\Delta_{ik}]^2$
Metric stress	$g(t)=t$	$h(t)=t$	$\left[\sum_{i<k}^{n}\varrho_{ik}^2\right]^{-1}\sum_{i<k}^{n}(\varrho_{ik}-\Delta_{ik})^2$
Sstress	$g(t)=t^2$	$h(t)=t^2$	$\left[\sum_{i<k}^{n}\varrho_{ik}^4\right]^{-1}\sum_{i<k}^{n}\left(\varrho_{ik}^2-\Delta_{ik}^2\right)^2$
Sammon stress	$g(t)=t$	$h(t)=t$	$\left[\sum_{i<k}^{n}\varrho_{ik}\right]^{-1}\sum_{i<k}^{n}\frac{(\varrho_{ik}-\Delta_{ik})^2}{\varrho_{ik}}$

linear in the data. Metric and non-metric scaling include non-linear solutions, and this fact distinguishes Multidimensional Scaling from Principal Component Analysis. The availability of non-linear optimisation routines, in particular, has led to renewed interest in Multidimensional Scaling.

The distinction between metric and non-metric scaling is not always made, and this may not matter to the practitioner who wants to obtain a configuration for his or her data. I follow Cox and Cox (2001), who use the term **Metric Scaling** for quantitative dissimilarities and **Non-metric Scaling** for rank-order dissimilarities. We begin with metric scaling and consider non-metric scaling in Section 8.4.

8.3.1 Metric Dissimilarities and Metric Stresses

We begin with the observed data $\{\mathcal{O},\varrho\}$ consisting of n objects O_i and dissimilarities ϱ_{ik} between pairs of objects. For embeddings $\mathfrak{f}$ from $\mathcal{O}$ into $\mathbb{R}^p$ for some $p \geq 1$ and distances $\Delta_{ik} = \Delta[\mathfrak{f}(O_i),\mathfrak{f}(O_k)]$, we measure the disparity between the ϱ_{ik} and the Δ_{ik} with the stress (8.2).

In classical scaling, essentially a single stress function, the raw classical stress or its standardised version (8.4) are used, whereas metric scaling incorporates a number of stresses which differ in their functions g, h and in their normalising factors. Table 8.4 lists stress criteria that are common in metric scaling. For notational convenience, I give expressions for the *squared* stress in Table 8.4; this avoids having to include square roots in each expression. For completeness, I have included the classical stress in this list.

The classical stress and metric stress differ by the normalising constant $\sum\varrho_{ik}^2$. The linear stress represents an attempt at comparing non-Euclidean dissimilarities and configuration distances while preserving linearity. Improved computing facilities have made this stress less interesting in practice. The sstress is also called *squared stress*, but I will not use the latter term because the sstress is not the square of the stress. The non-linear Sammon stress of Sammon (1969) incorporates the dissimilarities in a non-standard form and, as a consequence, can lead to interesting results.

Unlike the classical stress, the stresses in metric scaling admit general dissimilarities. We explore different dissimilarities and stresses in examples. As we shall see, some data give rise to very different configurations when we vary the dissimilarities or stresses, whereas for other data different stresses hardly affect the configurations.

Example 8.4 We return to the **illicit drug market** data which have seventeen different series measured over sixty-six months. We consider the months as the variables, so we have a high-dimension low sample size (HDLSS) problem. Multidimensional Scaling is particularly suitable for HDLSS problems because it replaces the $d \times d$ covariance-like matrix $Q^{(d)}$ of (5.17) in Section 5.5 with the much smaller $n \times n$ dual matrix $Q_{(n)}$. I will return to this duality in Section 8.5.1.

For the scaled data, I calculate 1D and 2D sstress configurations $\mathbb{W}_1$ and $\mathbb{W}_2$ using the Euclidean distance for the dissimilarities. Figure 8.3 shows the resulting configurations: the graph in the top left shows $\mathbb{W}_1$ and that in the top right the first dimension of $\mathbb{W}_2$, both against the series number on the x-axis. At first glance, these plots look very similar. I show them both because a closer inspection reveals that they are not the same. Series 15 is negative in the left plot but has become positive in the plot on the right. A change such as this would not occur for the principal component solution of classical scaling. Here it is a consequence of the non-linearity of the sstress. Further, the sstress of the 2D configuration is smaller than that of the 1D configuration: it decreases from 0.1974 for $\mathbb{W}_1$ to 0.1245 for $\mathbb{W}_2$.

Calculations with different dissimilarities and stresses reveal that the resulting configurations are similar for these data. For this reason, I do not show the other configurations.

The two panels in the lower part of Figure 8.3 show scatterplots of the dissimilarities on the x-axis versus the configuration distances on the y-axis, so a point for each pair of

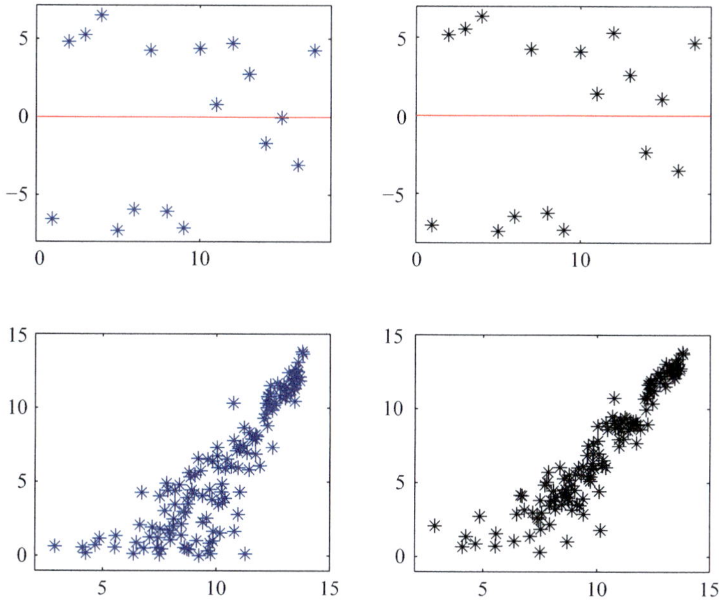

Figure 8.3 1D configuration $\mathbb{W}_1$ and first coordinate of $\mathbb{W}_2$ in the top row. Corresponding plots of configuration distances versus dissimilarities for Example 8.4 below.

observations. The left panel refers to $\mathbb{W}_1$, and the right panel refers to $\mathbb{W}_2$. The black scatterplot on the right is tighter, and as a result, the sstress is smaller. In both plots there is a larger spread for small distances because they contribute less to the overall loss and are therefore not as important.

The plot in the top-right panel of Figure 8.3 is the same as the right panel of Figure 6.11 in Section 6.5.2, apart from a sign change. These two plots divide the seventeen observations into the same two groups, although different techniques are used in the two analyses.

The recurrence of the same split of the data by two different methods shows the robustness of this split, a further indication that the two groups are really present in the data. ■

The next example shows the diversity of 2D configurations from different dissimilarities and stresses.

Example 8.5 We continue with the **athletes** data, which consist of 100 female and 102 male athletes. In Example 8.3 we compared the classical stress and the strain configurations, and we observed that the 2D configurations are able to separate the male and female athletes. In these calculations I used Euclidean distances as dissimilarities. Now we explore different dissimilarities and stresses and examine whether the new combinations are able to separate the male and female athletes.

As in the preceding example, I work with the scaled data and use all variables except the variable *sex*. Figure 8.4 displays six different configurations. The top row is based on the cosine distances (5.6) of Section 5.3.1 as the dissimilarities, and different stresses: the (metric) stress in the left plot, the sstress in the middle plot and the Sammon stress in the right plot. Females are shown in red, males in blue. In the bottom row I keep the stress fixed

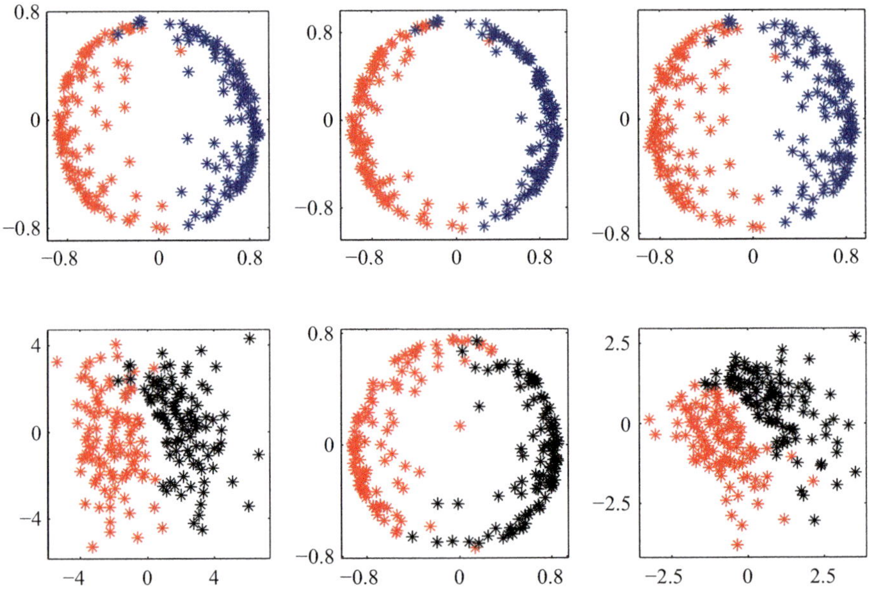

Figure 8.4 Configurations for Example 8.5, females in red, males in blue or black. (*Top row*): Cosine distances; from left to right: stress, sstress and Sammon stress. (*Bottom row*): Stress; from left to right: Euclidean, correlation and ℓ_∞ distance.

and vary the dissimilarities: the Euclidean distance in the left plot, the correlation distance (5.7) in the middle plot and the ℓ_∞ distance (5.5), both from Section 5.3.1, on the right. The female athletes are shown in red and the male athletes in black. The figure shows that the change in dissimilarity affects the pattern of the configurations more than a change in the type of stress. The cosine distances separate the males and females more clearly than the norm-based distances. The sstress (middle top) results in the tightest configuration but takes six times longer to calculate than the stress. The Sammon stress takes about 2.5 times as long as the stress. For these relatively small data sets, the computation time may not matter, but it may become important for large data sets.

There is no 'right' or 'wrong' answer; the different dissimilarities and stresses result in configurations which expose different information inherent in the data. It is a good idea to calculate more than one configuration and to look for interpretations these configurations allow. ■

When we vary the dissimilarities or the loss, we each time solve a different problem. There is no single dissimilarity or stress that produce the 'best' result, and I therefore recommend the use of different dissimilarities and stresses. The cosine dissimilarity often leads to insightful interpretations of the data. We noticed a similar phenomenon in Cluster Analysis (see Figure 6.2 in Example 6.1 of Section 6.2). As the dimension of the data increases, the angle between two vectors provides a measure of closeness, and for HDLSS problems in particular, the angle between vectors has become a standard tool for assessing the convergence of observations to a given vector (see Johnstone and Lu 2009 and Jung and Marron 2009).

8.3.2 Metric Strain

So far we have focused on different types of stress for measuring the loss between dissimilarities and configuration distances. Interestingly, the stress and sstress criteria were proposed initially for non-metric scaling and were later adapted to metric scaling. In contrast, the strain criterion (8.8) is mostly associated with classical and metric scaling. Indeed, the strain is a natural measure of discrepancy for metric scaling and relies on results from distance geometry. I present the version given in Trosset (1998), but it is worth noting that the original ideas go back to Schoenberg (1935) and Young and Householder (1938).

Theorem 8.7 requires additional notation, which we establish first.

Definition 8.6 Let $A = (a_{ik})$ and $B = (b_{ik})$ be $m \times m$ matrices. The **Hadamard product** or **Schur product** of A and B is the matrix $A \circ B$ whose elements are defined by the elementwise product

$$(A \circ B)_{ik} = (a_{ik} b_{ik}). \tag{8.10}$$

Let $\mathbf{1}_{m \times 1}$ be the column vector of 1s. The $m \times m$ **centring matrix** Θ is defined by

$$\Theta = \Theta_{m \times m} = \left[\mathbf{I}_{m \times m} - \frac{1}{m} \mathbf{1}_{m \times 1} (\mathbf{1}_{m \times 1})^{\mathsf{T}} \right] = \frac{1}{m} \begin{pmatrix} m-1 & -1 & \cdots & -1 \\ -1 & m-1 & \cdots & -1 \\ \vdots & \vdots & \ddots & \vdots \\ -1 & -1 & \cdots & m-1 \end{pmatrix}. \tag{8.11}$$

The **centring transformation** τ maps A to a matrix $\tau(A)$ defined by

$$A \longmapsto \tau(A) = \Theta A \Theta.$$

□

Using the newly defined terms, the matrix $Q_{\langle n\rangle}$ of part 1 in Theorem 8.3 becomes

$$Q_{\langle n\rangle} = -\frac{1}{2}\tau(P \circ P) = -\frac{1}{2}\Theta(P \circ P)\Theta \qquad \text{where } P = (\varrho_{ik}). \tag{8.12}$$

Theorem 8.7 [Trosset (1998)] *Let $\mathbb{X} = \begin{bmatrix}\mathbf{X}_1 \cdots \mathbf{X}_n\end{bmatrix}$ be data of rank r. Let ϱ be dissimilarities defined on $\mathbb{X}$, and let $P = (\varrho_{ik})$ be the $n \times n$ matrix of dissimilarities.*

1. *There is a configuration $\mathbb{W}$ whose distance matrix Δ equals P if and only if $\tau(P \circ P)$ is a symmetric positive semidefinite matrix of rank at most r.*
2. *If the $r \times n$ configuration $\mathbb{W}$ satisfies $\tau(P \circ P) = \mathbb{W}^{\mathsf{T}}\mathbb{W}$, then $\Delta = P$, where Δ is the distance matrix of $\mathbb{W}$.*

The theorem provides conditions for the existence of a configuration with the required properties. If these conditions do not hold, then we can still use the *ideas* of the theorem and find configurations which minimise the difference between $\tau(P \circ P)$ and $\tau(\Delta \circ \Delta)$.

Definition 8.8 Let $\mathbb{X} = \begin{bmatrix}\mathbf{X}_1 \cdots \mathbf{X}_n\end{bmatrix}$ be data with a dissimilarity matrix P. Let r be the rank of $\mathbb{X}$. Let Δ be an $n \times n$ matrix whose elements can be realised as pairwise distances of points in $\mathbb{R}^r$. The **metric strain** $\mathcal{S}\text{train}_{\text{met}}$, regarded as a function of P and Δ, is

$$\mathcal{S}\text{train}_{met}(P, \Delta) = \|\tau(P \circ P) - \tau(\Delta \circ \Delta)\|^2_{\text{Frob}}. \tag{8.13}$$

□

Unlike the classical strain (8.8), which is a function of the transformed data $A\mathbb{X}$ and the feature data $\mathfrak{f}(\mathbb{X})$, it is more natural to base the metric strain on the dissimilarity matrix P. The matrix Δ corresponds to the matrix of distances of $\mathfrak{f}(\mathbb{X}) \subset \mathbb{R}^r$. With this identification, we want to find an embedding $\mathfrak{f}^*$ with $W = \mathfrak{f}^*(\mathbb{X})$ and a distance matrix Δ^* which minimises (8.13).

I have defined the metric strain as a function of Δ. The definition shows that $\mathcal{S}\text{train}_{\text{met}}$ depends on $\tau(\Delta \circ \Delta)$. Putting $B = \tau(\Delta \circ \Delta)$, it is convenient to write $\mathcal{S}\text{train}_{\text{met}}$ as a function of B, namely,

$$\mathcal{S}\text{train}_{met}(P, B) = \|\tau(P \circ P) - B\|^2_{\text{Frob}}. \tag{8.14}$$

Theorem 8.9 [Trosset (1998)] *Let $\mathbb{X} = \begin{bmatrix}\mathbf{X}_1 \cdots \mathbf{X}_n\end{bmatrix}$ be data of rank r. Let ϱ be dissimilarities defined on $\mathbb{X}$, $P = (\varrho_{ik})$, and let $P \circ P$ be the Hadamard product. Assume that $\tau(P \circ P)$ has spectral decomposition $\tau(P \circ P) = V^* D^{*2} V^{*\mathsf{T}}$, with eigenvectors $\mathbf{v}_i^*$ and $i \le n$. Write $D^*_{(r)}$ for the $n \times n$ diagonal matrix whose first r diagonal elements $d_1^*, \ldots, d_r^*$ agree with those of D^* and whose remaining $n - r$ diagonal elements are zero. Then*

$$B^* = V^* D^{*\,2}_{(r)} V^{*\mathsf{T}}$$

is the minimiser of (8.14), and

$$\mathbb{W} = \begin{bmatrix} d_1^* \mathbf{v}_1^* \\ \vdots \\ d_r^* \mathbf{v}_r^* \end{bmatrix}$$

is an $r \times n$ configuration whose matrix of pairwise distances minimises the strain (8.13).

Theorem 8.9 combines theorem 2 and corollary 1 of Trosset (1998). For a proof of the theorem, see Trosset (1997, 1998), and chapter 14 of Mardia, Kent, and Bibby (1992).

The configurations obtained in Theorems 8.3 and 8.9 look similar. There are, however, differences. Theorem 8.3 refers to the classical set-up; the matrix $Q = VD^2V^\top$ is constructed from Euclidean distances and minimises the classical stress (8.4). The matrix B^* of Theorem 8.9 admits more general dissimilarities and minimises (8.14). The resulting configurations therefore will differ because they solve different problems. Meulman (1992) noted that strain optimal configurations underestimate configuration distances, whereas this is not the case for stress optimal configurations. In practice, these differences may not be apparent visually and can be negligible, as we have seen in Example 8.5.

8.4 Non-Metric Scaling

8.4.1 Non-Metric Stress and the Shepard Diagram

Shepard (1962a, 1962b) first formulated the ideas of non-metric scaling in an attempt to capture processes that cannot be described by distances or dissimilarities. The starting point is the observed data $\{\mathcal{O}, \varrho\}$ – as in metric scaling – but the dissimilarities are replaced by rankings, also called **rank orders**. The pairwise rankings of objects are the available observations. We might like to think of these rank orders as preferences in wine, breakfast foods, perfumes or other merchandise. Because the dissimilarities are replaced by rankings, we use the same notation ϱ, but refer to them as **rankings** or **ranked dissimilarities**, and write them in increasing order:

$$\varrho_1 = \min_{i,k;\, i\neq k} \varrho_{ik} \leq \cdots \max_{i,k;\, i\neq k} \varrho_{ik} = \varrho_N \qquad \text{with } N = n(n-1)/2. \tag{8.15}$$

Shepard's aim was to construct distances with the rank order for each pair of observations informed by that of the ϱ_{ik}. To achieve this goal, he placed the N points ϱ_ℓ at the vertices of a regular simplex in $\mathbb{R}^{N-1}$. He calculated Euclidean distances Δ_{ik} between all vertices, ranked the distances and compared the ranking of the dissimilarities with that of the distances Δ_{ik}. Points that are in the *wrong* rank order are moved in or out, and a new ranking of the distances is determined. The process is iterated until no further improvements in ranking are achieved. *Cosmetics* such as a rotation of the coordinates are applied so that the points agree with the principal axes in $\mathbb{R}^{N-1}$. For a p-dimensional configuration, Shepard proposed to take the first p principal coordinates as the desired configuration.

Shepard achieved the monotonicity of the ranked dissimilarities and the corresponding distances essentially by trial and error. It turns out that the monotonicity is the key to making non-metric scaling work. A few years later, Kruskal (1964a, 1964b) placed these intuitive ideas on a more rigorous basis by defining a measure of loss, the non-metric stress.

Definition 8.10 Let $\{\mathcal{O}, \varrho\}$ be the observed data, with $\mathcal{O} = \{O_1, \ldots, O_n\}$ and ranked dissimilarities ϱ_{ik}. For $p > 0$, let $\mathfrak{f}$ be an embedding from $\mathcal{O}$ into $\mathbb{R}^p$. Let Δ be a distance, and put $\Delta_{ik} = \Delta(\mathfrak{f}(O_i), \mathfrak{f}(O_k))$. **Disparities** are real-valued functions, defined for pairs of objects O_i and O_k and denoted by $\widehat{d}_{ik} = \widehat{d}(O_i, O_k)$, which satisfy the following:

- There is a monotonic function f such that

$$\widehat{d}_{ik} = f(\Delta_{ik})$$

for every pair (i,k) with $i,k \leq n$.
- For pairs of coefficients (i,k) and (i',k'),

$$\widehat{d}_{ik} \leq \widehat{d}_{i'k'} \tag{8.16}$$

whenever $\varrho_{ik} < \varrho_{i'k'}$.

The **non-metric stress** $\mathcal{S}\text{tress}_{\text{nonmet}}$, regarded as a function of $\widehat{d}$ and $\mathfrak{f}$, is

$$\mathcal{S}\text{tress}_{\text{nonmet}}(\widehat{d},\mathfrak{f}) = \left\{ \frac{\sum_{i<k}^{n} (\Delta_{ik} - \widehat{d}_{ik})^2}{\sum_{i<k}^{n} \Delta_{ik}^2} \right\}^{1/2}, \tag{8.17}$$

and the **non-metric sstress** $\mathcal{SS}\text{tress}_{\text{nonmet}}$ of $\widehat{d}$ and $\mathfrak{f}$ is

$$\mathcal{SS}\text{tress}_{\text{nonmet}}(\widehat{d},\mathfrak{f}) = \left\{ \frac{\sum_{i<k}^{n} (\Delta_{ik}^2 - \widehat{d}_{ik}^2)^2}{\sum_{i<k}^{n} \Delta_{ik}^4} \right\}^{1/2}. \tag{8.18}$$

The matrix $\mathbb{W} = \left[\mathfrak{f}(O_1)\cdots\mathfrak{f}(O_n)\right]$ which minimises $\mathcal{S}\text{tress}_{\text{nonmet}}$ (or $\mathcal{SS}\text{tress}_{\text{nonmet}}$) is called the ***p*-dimensional configuration** for the non-metric stress (or non-metric sstress, respectively). □

I will omit the subscript nonmet if there is no ambiguity. For classical and metric scaling, the stress is a function of the dissimilarities, and for given dissimilarities, we want to find the embedding and the distances Δ which minimise the stress. In non-metric scaling, the stress is defined from the disparities and uses the dissimilarities only via the monotonicity relationship (8.16). A consequence of this definition is that we are searching for distances and disparities which jointly minimise the stress or sstress.

The specific form of the monotone functions f, which was of paramount importance in the early days, is no longer the main object of interest because there are good algorithms which minimise the non-metric stress and sstress and calculate the p-dimensional configurations, the distances and the disparities. The following examples show how these ideas work and perform in practice.

Example 8.6 The **cereal** data consist of seventy-seven brands of breakfast cereals, for which eleven quantities are recorded. I use the first ten variables, which are listed in Table 8.5, and exclude the binary variable *type*.

The top-left panel of Figure 8.5 shows a parallel coordinate plot of the data with the variable numbers on the x-axis. I take the Euclidean distances of pairs of observations as the ranked dissimilarities and calculate the 2D configuration of these data with the non-metric sstress (8.18). The red dots in the top-right panel of the figure show this configuration, with the first component on the x-axis and the second on the y-axis. Superimposed in blue is the 2D PC data, with the first PC on the x-axis. The two plots look similar but are not identical.

The bottom row of Figure 8.5 depicts **Shepard diagrams**, plots which show the configuration distances or disparities on the y-axis against the ranked dissimilarities on the x-axis.

Table 8.5 *Cereal Data from Example 8.6*

1	# of calories	6	Carbohydrates (g)
2	Protein (g)	7	Sugars (g)
3	Fat (g)	8	Display shelf (1,2 etc)
4	Sodium (mg)	9	Potassium (mg)
5	Fibre (g)	10	Vitamins (% enriched)

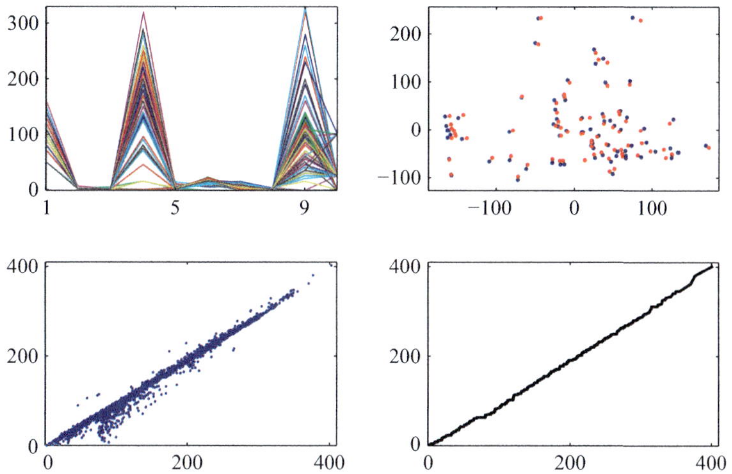

Figure 8.5 Cereal data (*top left*) from Example 8.6. (*Top right*): Non-metric configuration in red and first two PCs in blue. (*Bottom*): Configuration distances (*blue*) and disparities (*black*) versus ranked dissimilarities on the x-axis.

Shepard diagrams give an insight into the degree of monotonicity between distances or disparities and rankings. The bottom-left panel displays the configuration distances against the ranked dissimilarities, and the bottom-right panel displays the disparities against the ranked dissimilarities. There is a wider spread in the blue plot on the left than in the black plot, indicating that the smaller distances are not estimated as well as the larger distances. Overall both plots show clear monotonic behaviour.

Following this initial analysis of all data, I separately analyse two subgroups of the seventy-seven brands: the twenty-three brands of Kellogg's cereals and the twenty-one brands from General Mills. An analysis of the Kellogg's sample is also given in Cox and Cox (2001). The purpose of the separate analyses is to determine differences between the Kellogg's and General Mills brands of cereals which may not otherwise be apparent.

The stress and sstress configurations are very similar for both subsets, so I only show the sstress-related results in Figures 8.6 and 8.7. The top panels of Figure 8.6 refer to Kellogg's, the bottom panels to General Mills. Parallel coordinate plots of the data – displayed in the left panels – show that the Kellogg's cereals contain more *sodium*, variable 4, and *potassium*, variable 9, than General Mills' cereals. The plots on the right in Figure 8.6 display Shepard diagrams: the blue scatter plot shows the configuration distances, and the red line the disparities, both against the ranked dissimilarities, which are shown on the x-axis. We note that the range of distances and dissimilarities is larger for the Kellogg's sample than for the General Mills' sample, but both show good monotone behaviour.

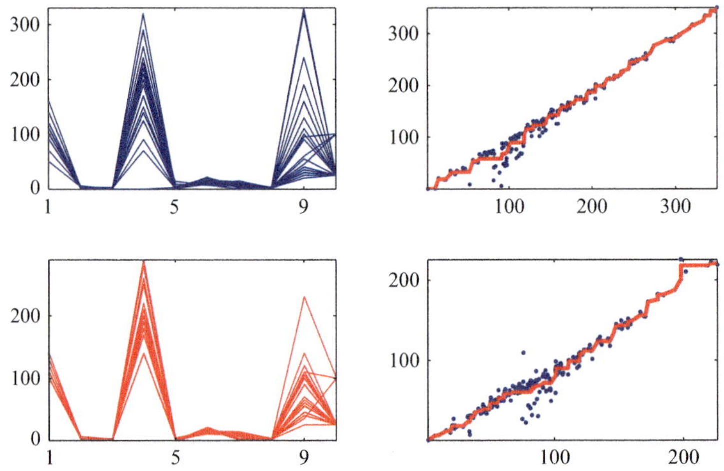

Figure 8.6 Kellogg's (*blue*) and General Mills' (*red*) brands of cereal data from Example 8.6 and their Shepard diagrams on the right.

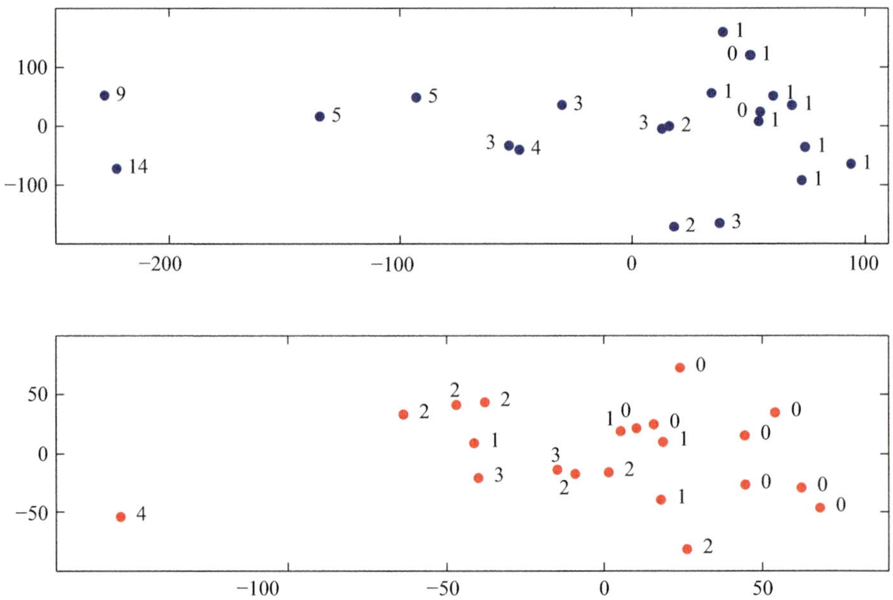

Figure 8.7 Configurations of Kellogg's (*blue*) and General Mills' (*red*) samples from Example 8.6 with fibre content given by the numbers next to the dots.

Figure 8.7 shows the 2D sstress configurations for Kellogg's in the top panel in blue and General Mills in the lower panel in red. The numbers next to each red or blue dot show the fibre content of the particular brand: 0 means no fibre, and the higher the number, the higher is the fibre content of the brand. The Kellogg's cereals have higher fibre content than the General Mills' cereals, but there is a 'low-fibre cluster' in both sets of configurations. The clustering behaviour of these low-fibre cereals indicates that they have other properties in common; in particular, they have low potassium content. These properties of the cereals brands are clearly of interest to buyers and are accessible in these configurations. ■

The example shows that Multidimensional Scaling can discover information that is not exhibited in a principal component analysis, in this case the cluster of brands with low fibre content. Variable 5, *fibre*, has a much smaller range than variables 4 and 9, *sodium* and *potassium*. The latter two contribute most to the first two principal components, unlike *fibre*, which has much smaller weights in the first and second PCs and is therefore not noticeable in the first two PCs.

Next we look at an example where ranking of the dissimilarities and subsequent non-metric scaling lead to poor results.

Example 8.7 Instead of using the Euclidean distances for the **ten cities** of Example 8.1, I now use rank orders obtained from the distances (8.1). Thus, $\varrho_1 = \varrho_{(3,5)}$, the rank obtained from the smallest distance between the third city, Singapore, and the fifth city, Kuala Lumpur, in Table 8.1 and (8.1). Similarly, for $N = 45$, the largest rank $\varrho_{45} = \varrho_{(4,10)}$ is obtained for the fourth and tenth cities, Seoul and Auckland.

The top-left panel of Figure 8.8 shows the 2D configuration calculated with the non-metric stress (8.17). The right panel shows the rotated configuration with a 90-degree rotation as in Figure 8.1, and the numbers next to the dots refer to the cities in Table 8.1. The bottom-left panel shows a Shepard diagram: the blue points show the configuration distances, and the black points show disparities, plotted against the ranked dissimilarities. The disparities are monotonic with the dissimilarities; the configuration distances less so.

A comparison with Figure 8.1 shows that the map of the cities does not bear any relationship to the real geographic layout of these ten cities, and rank order based non-metric scaling produces less satisfactory results for these data than classical scaling. ■

This example shows the superiority of the classical dissimilarities over the rank orders and confirms that we get better results when we use all available information. If the rank orders are all the available information, then configurations using non-metric scaling may be the best we can do.

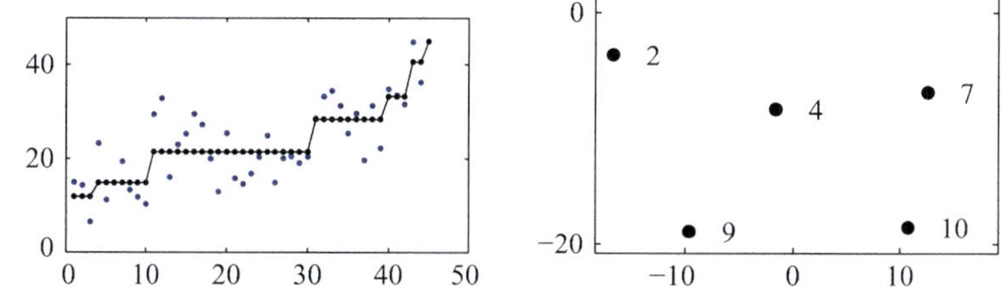

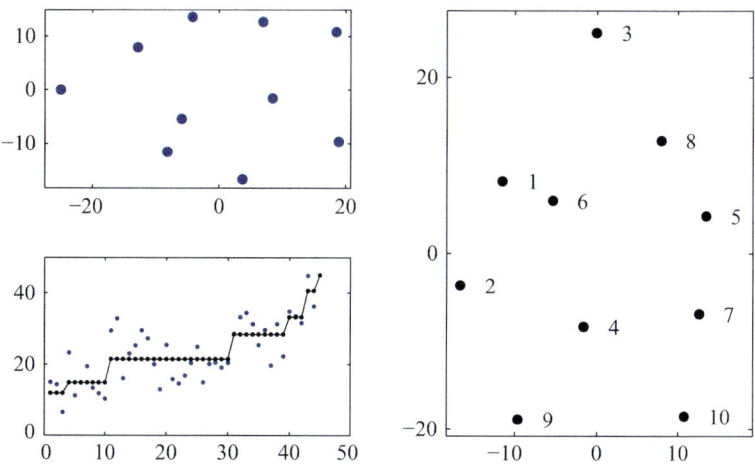

Figure 8.8 Non-metric configuration from rank orders for Example 8.7 and Shepard diagram.

8.4.2 Non-Metric Strain

In metric scaling we considered two types of loss: stress and strain. Historically, Kruskal's stress and sstress were the non-metric loss criteria. Meulman (1992) popularised the notion of strain in metric scaling, and Trosset (1998) proposed a non-metric version of strain which I outline now.

Let $\mathbb{X} = \left[\mathbf{X}_1 \cdots \mathbf{X}_n\right]$ be data with dissimilarities ϱ^0. Put $P^0 = (\varrho^0_{ik})$, and let

$$\varrho^0_1 \leq \cdots \leq \varrho^0_N \quad \text{with } N = n(n-1)/2$$

be the ordered ranks as in (8.15). Instead of constructing disparities from the configuration distances, we start with P^0 and consider dissimilarity matrices P for $\mathbb{X}$ whose entries ϱ_{ik} satisfy for pairs i,k and i',k':

$$\varrho_{ik} \leq \varrho_{i'k'} \qquad \text{whenever} \qquad \varrho^0_{ik} \leq \varrho^0_{i'k'}. \tag{8.19}$$

Put

$$M(P^0) = \{P\colon P \text{ is a dissimilarity matrix for } \mathbb{X} \text{ whose entries satisfy (8.19)}\}.$$

Then $M(P^0)$ furnishes a rich supply of dissimilarity matrices compatible with P^0.

Definition 8.11 Let $\mathbb{X} = \left[\mathbf{X}_1 \cdots \mathbf{X}_n\right]$ be data of rank r with an $n \times n$ dissimilarity matrix $P^0 = (\varrho^0_{ik})$. Put $r_0^2 = \left\|P^0\right\|^2_{\text{Frob}}$. For $p \leq r$, let $\mathbf{\Delta}_n(p)$ be the set of $n \times n$ matrices whose elements can be realised as pairwise distances of points in $\mathbb{R}^p$. For $P \in M(P^0)$ such that $\|P\|^2_{\text{Frob}} \geq r_0^2$ and for $\Delta \in \mathbf{\Delta}_n(p)$, the **non-metric strain** $\mathcal{S}\text{train}_{\text{nonmet}}$ is

$$\mathcal{S}\text{train}_{\text{nonmet}}(P, \Delta) = \|\tau(P \circ P) - \tau(\Delta \circ \Delta)\|^2_{\text{Frob}}. \tag{8.20}$$

□

Putting $B = \tau(\Delta \circ \Delta)$ and using Theorem 8.7, the non-metric strain between symmetric positive $n \times n$ matrices B of rank at most r and $P \in M(P^0)$ which satisfies $\|P\|^2_{\text{Frob}} \geq r_0^2$ becomes

$$\mathcal{S}\text{train}_{\text{nonmet}}(P, B) = \|\tau(P \circ P) - B\|^2_{\text{Frob}}. \tag{8.21}$$

The minimiser (Δ^*, P^*) of (8.20) is the desired distance and dissimilarity matrix, respectively, for $\mathbb{X}$. The definition of the non-metric strain $\mathcal{S}\text{train}_{\text{nonmet}}$ is very similar to that of the metric strain $\mathcal{S}\text{train}_{\text{met}}$ of (8.13), apart from the extra condition $\|P\|^2_{\text{Frob}} \geq r_0^2$ which is required to avoid degenerate solutions. Trosset (1998) discussed the non-linear optimisation problem (8.21) in more detail and showed how to obtain a minimiser.

The distance matrices which minimise the non-metric strain are different from the minimisers of the metric strain, the stress and sstress. In applications, the different solutions may be similar, and often it is a question of personal preference on the user's side which criterion is fitted. However, if computing facilities permit, it is advisable to experiment with more than one loss criterion as they may lead to different insights into a problem.

8.5 Data and Their Configurations

Multidimensional Scaling started as an exploratory technique that focused on reconstructing data from partial information and on representing data visually. Since its beginnings in the late 1930s, attempts have been made to formulate models suitable for statistical inferences. Ramsay (1982) and the discussion of his paper by some fourteen experts gave an understanding of the issues involved. A number of these discussants share Silverman's reservations, of feeling a 'little uneasy about the use of Multidimensional Scaling as a model-based inferential technique, rather than just an exploratory or presentational method' (Ramsay 1982, p. 307). These reservations need not detract from the merits of the method, and indeed, many of the newer non-linear approaches in Multidimensional Scaling successfully extend or complement linear dimension-reduction methods such as Principal Component Analysis.

In this section we consider two specific topics: scaling for high-dimensional data and relationships between different configurations of the same data.

8.5.1 HDLSS Data and the $\mathbb{X}$ and $\mathbb{X}^\top$ Duality

In classical scaling, Theorem 8.3 tells us how to obtain the $Q_{\langle n\rangle}$ matrix from the dissimilarities ϱ without requiring $\mathbb{X}$. Traditionally, $n > d$, and if $\mathbb{X}$ are available, it is clearly preferable to work directly with $\mathbb{X}$.

For HDLSS data $\mathbb{X}$, we exploit the duality between $Q_{\langle n\rangle} = \mathbb{X}^\top\mathbb{X}$ and $Q^{\langle d\rangle} = \mathbb{X}\mathbb{X}^\top$, see Section 5.5 for details. The $Q^{\langle d\rangle}$ matrix is closely related to the sample covariance matrix of $\mathbb{X}$ and thus can be used instead of S to derive the principal component data. We now explore the relationship between $Q_{\langle n\rangle}$ and the principal components of $\mathbb{X}$.

Proposition 8.12 *Let* $\mathbb{X} = \begin{bmatrix}\mathbf{X}_1 \cdots \mathbf{X}_n\end{bmatrix}$ *be* d*-dimensional centred data, with* $d > n$ *and rank* r. *Put* $Q_{\langle n\rangle} = \mathbb{X}^\top\mathbb{X}$, *and write*

$$\mathbb{X} = UDV^\top \qquad \textit{and} \qquad Q_{\langle n\rangle} = VD^2V^\top$$

for the singular value decomposition of $\mathbb{X}$ *and the spectral decomposition of* $Q_{\langle n\rangle}$, *respectively. For* $k \leq r$, *put*

$$\mathbb{W}^{(k)} = U_k^\top\mathbb{X} \qquad \textit{and} \qquad \mathbb{V}^{(k)} = D_k V_k^\top. \tag{8.22}$$

Then $\mathbb{W}^{(k)} = \mathbb{V}^{(k)}$.

This proposition tells us that the principal component data $\mathbb{W}^{(k)}$ can be obtained from the left eigenvectors of $\mathbb{X}$, the columns of U, or equivalently from the right eigenvectors of $\mathbb{X}$, the columns of V. Classically, for $n > d$, one calculates the left eigenvectors, which coincide with the eigenvectors of the sample covariance matrix S of $\mathbb{X}$. However, if $d > n$, finding the eigenvectors of the smaller $n \times n$ matrix $Q_{\langle n\rangle}$ is computationally much faster.

Proof Fix $k \leq r$. Using the singular value decomposition of $\mathbb{X}$, we have

$$\mathbb{W}^{(k)} = U_k^\top UDV^\top,$$

where U is of size $d \times r$, D is the diagonal $r \times r$ matrix of singular values and V is of size $n \times r$. Because U is r-orthogonal, $U_k^\top U = \mathbf{I}_{k\times r}$, and from this equality it follows that

$$\mathbb{W}^{(k)} = \mathbf{I}_{k\times r} DV^\top = D_k V_r^\top = D_k V_k^\top = \mathbb{V}^{(k)}.$$

The equality $D_k V_r^\top = D_k V_k^\top$ holds because the last $r - k$ columns of V_r are zero. ∎

The *illicit drug market* data of Example 8.4 fit into the HDLSS framework, but in the calculations of the stress and sstress configurations, I did not mention the duality between $\mathbb{X}$ and $\mathbb{X}^{\mathsf{T}}$. Our next example makes explicit use of Proposition 8.12.

Example 8.8 We continue with the **breast tumour gene expression** data of Example 2.15 of Section 2.6.2, which consist of 4,751 genes, the variables, and seventy-eight patients. These data clearly fit into the HDLSS framework. The rank of the data is at most seventy-eight, and it therefore makes sense to calculate the principal component data from the eigenvectors of the matrix $Q_{(n)}$.

In a standard principal component analysis, the eigenvectors of S contain the weights for each variable. I calculate the eigenvectors of $Q_{(n)}$, and we therefore obtain weights for the individual observations. The analysis shows that observation 54 has much larger absolute weights for the first three eigenvectors of $Q_{(n)}$ than any of the other observations. Observation 54 is shown in blue in the bottom-right corner of the top-left plot of Figure 8.9. The top row shows the 2D and 3D PC data, with $\mathbb{V}^{(k)}$ as in (8.22). The 2D scatterplot in the top left, in particular, marks observation 54 as an outlier.

For each observation, the data contain a survival time in months and a binary outcome which is 1 for patients surviving five years and 0 otherwise. There are forty-four patients who survived five years – shown in black in the figure – and thirty-four who did not – shown in blue. As we can see, in a principal component analysis the blue and black points are not separated; instead, outliers are exposed. These contribute strongly to the total variance.

The scatterplots in the bottom row of Figure 8.9 result from a calculation of $\mathbb{V}^{(k)}$ in (8.22) after observation 54 has been removed. A comparison between the 2D scatterplots shows that after removal of observation 54, the PC directions are aligned with the x- and y-axis. This is not the case in the top-left plot. Because the sample size is small, I am not advocating

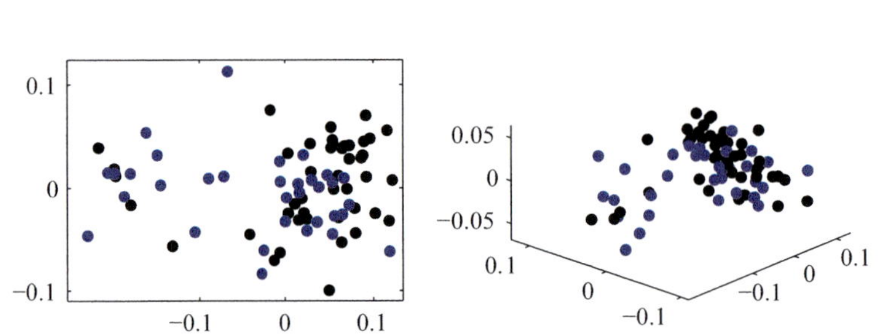

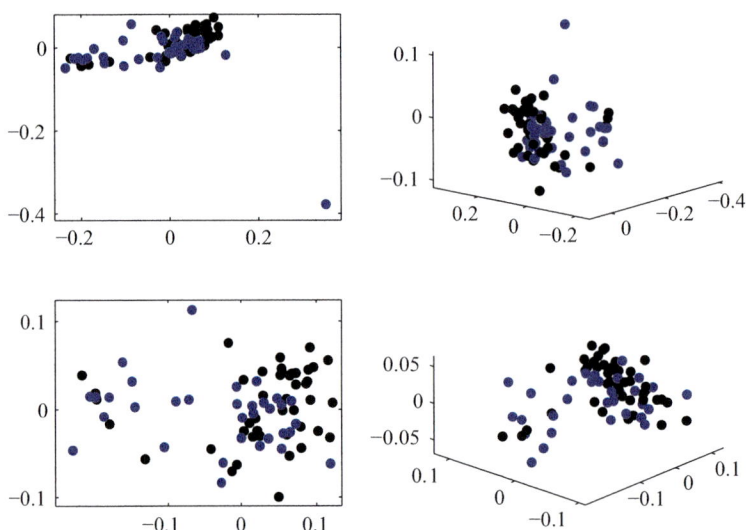

Figure 8.9 PC configurations in two and three dimensions from Example 8.8. (*Top*): all seventy-eight samples. (*Bottom*): Without the blue observation 54. (*Black dots*): Survived more than five years. (*Blue dots*): survived less than five years.

leaving out observation 54 but suggest instead carrying out any subsequent analysis with and without this particular observation and then comparing the results.

As we have seen, the eigenvectors of $Q_{\langle n \rangle}$ provide useful information about the observations. A real benefit of using $Q_{\langle n \rangle}$ – instead of $Q^{\langle d \rangle}$ or S – is the computational efficiency. Calculation of the 78×78 matrix $Q_{\langle n \rangle}$ takes about 3 per cent of the time it takes to calculate $Q^{\langle d \rangle}$. The computational advantage increases further when we try to calculate the eigenvalues and eigenvectors of $Q_{\langle n \rangle}$ and $Q^{\langle d \rangle}$, respectively. ∎

The duality of $\mathbb{X}$ and $\mathbb{X}^{\mathsf{T}}$ goes back at least as far as Gower (1966) and classical scaling. For HDLSS data, it results in computational efficiencies. In the theoretical development of Principal Component Analysis – see Theorem 2.25 of Section 2.7.2 – Jung and Marron (2009) exploited this duality to prove the convergence of the sample eigenvalues to the population counterparts. In addition, the duality of the Principal Component Analysis for $\mathbb{X}$ and $\mathbb{X}^{\mathsf{T}}$ can be regarded as a forerunner or a special case of Kernel Principal Component Analysis, which we consider in Section 12.2.2.

8.5.2 Procrustes Rotations

In the earlier parts of this chapter we constructed p-dimensional configurations based on different dissimilarities and loss criteria. As the dissimilarities and loss criteria vary, so do the resulting configurations. In this section we quantify the difference between two configurations.

Let $\mathbb{X}$ be d-dimensional data. If $\mathbb{W}$ is a p-dimensional configuration for $\mathbb{X}$ with $p \le d$ and E is an orthogonal $p \times p$ matrix, then $E\mathbb{W}$ is also a configuration for $\mathbb{X}$ which has the same distance matrix Δ as $\mathbb{W}$. Theorem 8.3 states this fact for classical scaling with stress, but it holds more generally. As a consequence, configurations are unique up to orthogonal transformations. Following Gower (1971), we consider the difference between two configurations $\mathbb{W}$ and $E\mathbb{V}$ and find the orthogonal matrix E which minimises this difference. As it turns out, E has a simple form.

Theorem 8.13 *Let $\mathbb{W}$ and $\mathbb{V}$ be matrices of size $p \times n$ whose rows are centred. Let E be an orthogonal matrix of size $p \times p$, and put*

$$\chi(E) = \|\mathbb{W} - E\mathbb{V}\|^2_{Frob}.$$

Put $T = \mathbb{V}\mathbb{W}^{\mathsf{T}}$, and write $T = UDV^{\mathsf{T}}$ for its singular value decomposition. If $E^* = VU^{\mathsf{T}}$, then

$$E^* = \operatorname{argmin} \chi(E) \qquad \text{and} \qquad \chi(E^*) = \|\mathbb{W}\|^2_{Frob} + \|\mathbb{V}\|^2_{Frob} - 2\,\mathrm{tr}(D).$$

The orthogonal matrix E^* is the **Procrustes rotation** of $\mathbb{V}$ relative to $\mathbb{W}$.

Proof For any orthogonal E,

$$\mathrm{tr}(E\mathbb{V}\mathbb{V}^{\mathsf{T}}E^{\mathsf{T}}) = \mathrm{tr}(\mathbb{V}\mathbb{V}^{\mathsf{T}}) = \|\mathbb{V}\|^2_{\mathrm{Frob}},$$

and hence we have the identity

$$\chi(E) = \|\mathbb{W}\|^2_{\mathrm{Frob}} + \|\mathbb{V}\|^2_{\mathrm{Frob}} - 2\,\mathrm{tr}(E\mathbb{V}\mathbb{W}^{\mathsf{T}}).$$

Minimising $\chi(E)$ is therefore equivalent to maximising $\operatorname{tr}(E\mathbb{V}\mathbb{W}^\top)$. We determine the maximiser by introducing a symmetric $p \times p$ matrix L of Lagrange multipliers and then find the maximiser of

$$\tilde{\chi}(E) = \operatorname{tr}\left[ET - \frac{1}{2}L\left(EE^\top - \mathbf{I}_{p\times p}\right)\right] \qquad \text{with } T = \mathbb{V}\mathbb{W}^\top.$$

Differentiating $\tilde{\chi}$ with respect to E and setting the derivative equal to the zero matrix lead to

$$T = LE^\top.$$

Using the singular value decomposition of T and the symmetry of L, we have

$$L^2 = (TE)(TE)^\top = TT^\top = UDV^\top VDU^\top = UD^2U^\top.$$

By the uniqueness of the square root, we obtain $L = UDU^\top$. From $T = LE^\top$ it follows that

$$UDV^\top = UDU^\top E^\top,$$

and hence $V = EU$.

The desired expression for the maximiser follows. Further, $\operatorname{tr}(E^*\mathbb{V}\mathbb{W}^\top) = \operatorname{tr}(D)$ by the trace property, and hence the expression for $\chi(E^*)$ holds. ■

Theorem 8.13 relates matrices of the same size, and applying it to two configuration matrices of $\mathbb{X}$ tells us how different these configurations are. For configurations with a different number of dimensions p_1 and p_2, we 'pad' the smaller configuration matrix with zeroes and then apply Theorem 8.13 to the two matrices of the same size. The padding can also be applied when we want to compare data with a lower-dimensional configuration. I summarise the results in the following corollary.

Corollary 8.14 *Let $\mathbb{X}$ be a centred $d \times n$ data of rank r, and write $\mathbb{X} = UDV^\top$ for the singular value decomposition. For $p \le r$, let $\mathbb{W}_p = D_pV_p^\top$ be the $p \times n$ configuration obtained in step 3 of Algorithm 8.1. Let $U_{p,0}$ be the $d \times d$ matrix whose first p columns coincide with U_p and whose remaining $d - p$ columns are zero vectors. The following hold.*

1. *The matrix $\mathbb{W}_{p,0} = U_{p,0}^\top\mathbb{X}$ agrees with $\mathbb{W}_p$ in the first p rows, and its remaining $d - p$ rows are zeroes.*
2. *The Procrustes rotation E^* of $\mathbb{W}_{p,0}$ relative to $\mathbb{X}$ is $E^* = U_{p,0}$, and*

$$E^*\mathbb{W}_{p,0} = U_{p,0}U_{p,0}^\top\mathbb{X} = U_pU_p^\top\mathbb{X}.$$

Proof From Proposition 8.12, $\mathbb{W}_p = U_p^\top\mathbb{X}$, and thus $\mathbb{W}_{p,0} = U_{p,0}^\top\mathbb{X}$ is a centred $d \times n$ matrix satisfying part 1. For $T = \mathbb{W}_{p,0}\mathbb{X}^\top = U_{p,0}^\top\mathbb{X}\mathbb{X}^\top$, we obtain the decomposition

$$T = U_{p,0}^\top UD^2U^\top.$$

The Procrustes rotation E^* of Theorem 8.13 is $E^* = U\left(U_{p,0}^\top U\right)^\top = U_{p,0}$. ■

This corollary asserts that $E^* = U_{p,0}$ is the minimiser of χ in Theorem 8.13. Part 2 of the theorem thus tells us that $U_pU_p^\top\mathbb{X}$ is the best approximation to $\mathbb{X}$ with respect χ. A comparison with Corollary 2.14 in Section 2.5.2 shows that the two corollaries arrive at the same conclusion but by different routes.

The Procrustes rotation of Theorem 8.13 is a special case of the more general transformations

$$\mathbb{V} \longmapsto P(\mathbb{V}) = cE\mathbb{V} + \mathbf{b},$$

where c is a scale factor, E is an orthogonal matrix and $\mathbf{b}$ is a vector. Extending χ of Theorem 8.13, to transformations of this form, we may want to compare two configurations $\mathbb{W}$ and $\mathbb{V}$ and then find the optimal transformation parameters of $\mathbb{V}$ with respect to $\mathbb{W}$. These parameters minimise

$$\chi(E, c, \mathbf{b}) = \|\mathbb{W} - P(\mathbb{V})\|^2_{\text{Frob}}.$$

Problems of this type and their solutions are the topic of **Procrustes Analysis**, which derives its name from Procrustes, the 'stretcher' of Greek mythology who viciously stretched or scaled each passer-by to the size of his iron bed. An introduction to Procrustes Analysis is given in Cox and Cox (2001). Procrustes Analysis is not restricted to configurations; it is an important method for matching or registering matrices, images or general shapes in high-dimensional space in shape analysis. For details, see Dryden and Mardia (1998).

8.5.3 Individual Differences Scaling

Multidimensional Scaling traditionally deals with one data matrix $\mathbb{X}$ and the observed data $\{\mathcal{O}, \varrho\}$ and constructs configurations for $\mathbb{X}$ from $\{\mathcal{O}, \varrho\}$. If we have repetitions of an experiment, or if we have a number of 'judges', each of whom creates a dissimilarity matrix for the given data, then extensions of this traditional set-up are required, such as

1. treating multiple sets $\mathbb{X}_\ell$ and $\{\mathcal{O}_\ell, \varrho_\ell\}$ simultaneously, and
2. considering multiple sets $\{\mathcal{O}, \varrho_\ell\}$ for a single $\mathbb{X}$, where $\ell = 1, \ldots, M$.

Analyses for these extensions of Multidimensional Scaling are sometimes also called **Three-Way Scaling**.

For the two scenarios – repetitions of $\mathbb{X}$ and $\{\mathcal{O}, \varrho\}$ and multiple 'judges' $\{\mathcal{O}, \varrho_\ell\}$ for a single $\mathbb{X}$ – one wants to construct configurations. The aims and methods of solution differ depending on whether one wants to construct a single configuration from the M sets $\{\mathcal{O}_\ell, \varrho_\ell\}$ or M separate configurations. I will not describe solutions but mention some approaches that address these problems.

Tucker and Messick (1963) worked directly with the dissimilarities and defined an *augmented* matrix of dissimilarities based on the dissimilarities $\varrho_{ik,\ell}$, where $\ell \leq M$ refers to the judges. The augmented matrix P_{TM} of dissimilarities is

$$P_{\text{TM}} = \begin{bmatrix} \varrho_{12,1} & \cdots & \varrho_{12,M} \\ \vdots & \vdots & \vdots \\ \varrho_{(n-1)n,1} & \cdots & \varrho_{(n-1)n,M} \end{bmatrix} \qquad (8.23)$$

so P_{TM} has $n(n-1)/2$ rows and M columns. The ℓth column contains the dissimilarities of judge ℓ, and the kth row contains the dissimilarities for a specific pair of observations across all judges. Using a stress loss, individual configurations or an average configuration are constructed.

Carroll and Chang (1970) used the augmented matrix (8.23) and assigned to judge ℓ a weight vector $\boldsymbol{\omega}_\ell \in \mathbb{R}^d$ such that the entries $\omega_{\ell j}$ of $\boldsymbol{\omega}_\ell$ corresponded to the d variables of $\mathbb{X}$. Using the weights, the configuration distances for the observations $\mathbf{X}_i$ and $\mathbf{X}_k$ and judge ℓ are

$$\Delta_{ik,\ell} = \left[\sum_{j=1}^{d} \omega_{\ell j}(X_{ij} - X_{kj})^2\right]^{1/2}.$$

The configuration distances give rise to configurations $\mathbb{W}_{p,\ell}$ as in Algorithm 8.1. The optimal configuration minimises a stress loss based on the average over the M judges. Many algorithms have been developed which build on the work of Carroll and Chang (1970); see, for example, Davies and Coxon (1982).

Meulman (1992) extended the strain criterion of Definition 8.4 to data sets $\mathbb{X}_1, \ldots, \mathbb{X}_M$. The desired configuration $\mathbb{W}^*$ minimises the average strain over $A\mathbb{X}_\ell$, where A is a $\kappa \times d$ matrix and

$$\mathcal{S}\mathrm{train}(A\mathbb{X}_1, \ldots, A\mathbb{X}_M, \mathbb{W}) = \frac{1}{m}\sum_{j=1}^{M} \mathcal{S}\mathrm{train}(A\mathbb{X}_j, \mathbb{W}).$$

Of special interest is the case $M = 2$, which has connections to Canonical Correlation Analysis. In a subsequent paper, Meulman (1996) examined this connection between the two methods. This research area is known as as **Homogeneity Analysis**.

8.6 Scaling for Grouped and Count Data

In this last section we look at a number of methods that have evolved from Multidimensional Scaling in response to the needs of different types of data, such as categorical and count data or data that are characterised locally, for example, through their class membership.

8.6.1 Correspondence Analysis

Multidimensional Scaling was proposed for continuous d-dimensional data. The mathematical equations and developments governing Multidimensional Scaling, however, also apply to count or categorical data. We now look at data that are available in the form of contingency tables, and we consider a scaling-like development for such data, known as **Correspondence Analysis**. I give a brief introduction to Correspondence Analysis in the language and notation we have developed in this chapter. There are many books on Correspondence Analysis, and Greenacre (1984) and the more recent Greenacre (2007) are some that provide a generic view and detail on this topic.

Definition 8.15 Let $\mathbb{X}$ be matrix of size $r \times c$ whose entries $X_{\ell k}$ are the counts for the cell corresponding to column ℓ and row k. We present data of this form in a table, called a **contingency table**, which is shown in Table 8.6. The centre part of the table contains the counts $X_{\ell k}$ in the $r \times c$ cells. The entries of the first column are row indices, written $\mathbf{r}_j$ for the jth row, and the entries of the first row are column indices, written $\mathbf{c}_i$ for the ith column. Further the last row and column contain the column and row totals, respectively. □

Table 8.6 *Contingency Table of Size $r \times c$*

	$\mathbf{c}_1$	$\mathbf{c}_2$	$\cdots$	$\mathbf{c}_c$	Row totals
$\mathbf{r}_1$	X_{11}	X_{21}	$\cdots$	X_{c1}	$\sum_k X_{k1}$
$\mathbf{r}_2$	X_{12}	X_{22}	$\cdots$	X_{c2}	$\sum_k X_{k2}$
$\vdots$	$\vdots$	$\vdots$	$\ddots$	$\vdots$	$\vdots$
$\mathbf{r}_r$	X_{1r}	X_{2r}	$\cdots$	X_{cr}	$\sum_k X_{kr}$
Column totals	$\sum_\ell X_{1\ell}$	$\sum_\ell X_{2\ell}$	$\cdots$	$\sum_\ell X_{c\ell}$	$N = \sum_{k\ell} X_{k\ell}$

The rows of a contingency table are observations in c variables, and the columns are observations in r variables. It is common practice to give the row and column totals in a contingency table and to let N be the sum total of all cell counts.

There is an abundance of examples which can be represented in contingency tables, including political preferences in different locations, treatment and outcomes and ranked responses in marketing or psychology. Typically, one examines and tests – by means of suitably chosen χ^2 statistics – whether rows are independent or have the same proportions. Voter preference in different states could be examined in this way. The role of rows and columns can be interchanged, so one might want to test for independence of columns instead of rows.

In a correspondence analysis of contingency table data, the aims include the construction of low-dimensional configurations but are not restricted to these. Although Correspondence Analysis (CA) makes use of ideas from Multidimensional Scaling (MDS), there are differences between the two methods that go beyond those relating to the type of data, as the following summary shows:

(MDS) For $d \times n$ data $\mathbb{X}$ in Multidimensional Scaling, the columns $\mathbf{X}_i$ of $\mathbb{X}$ are the observations or random vectors, and the row vector $\mathbf{X}_{\bullet j}$ is the jth variable or dimension across all n observations.

(CA.a) For $r \times c$ count data $\mathbb{X}$, we may regard the rows as the observations (in c variables) or the columns as the observations (in r variables). As a consequence, a scaling-like analysis of $\mathbb{X}$ or $\mathbb{X}^\top$ or both is carried out in Correspondence Analysis.

(CA.b) For the $r \times c$ matrix of count data $\mathbb{X}$, we want to examine the sameness of rows of $\mathbb{X}$.

We first look at (CA.a), and then turn to (CA.b).

Let r and c be positive integers, and let $\mathbb{X}$ be given by the $r \times c$ matrix of Table 8.6. Let r_0 be the rank of $\mathbb{X}$, and write $\mathbb{X} = UDV^\top$ for its singular value decomposition. We treat the rows and columns of $\mathbb{X}$ as separate vector spaces. The $r \times r_0$ matrix U of left eigenvectors of $\mathbb{X}$ forms part of a basis for the r variables of the columns. Similarly, the $c \times r_0$ matrix V of right eigenvectors of $\mathbb{X}$ forms part of a basis for the c variables of the rows of $\mathbb{X}$. For $p \le r_0$, we construct two separate configurations based on the ideas of classical scaling:

$$\begin{aligned} &D_p U_p^\top \qquad \text{a configuration for the } r\text{-dimensional columns, and} \\ &D_p V_p^\top \qquad \text{a configuration for the } c\text{-dimensional rows.} \end{aligned} \tag{8.24}$$

Because there is no preference between the rows and the columns, one calculates both configurations, and typically displays the first two or three dimensions of each configuration in one figure.

Example 8.9 We consider the **assessment marks** of a class of twenty-three students in a second-year statistics course I taught. The total assessment consisted of five assignments and a final examination, and we therefore have $r = 6$ and $c = 23$. For each student, the mark he or she achieved in each assessment task is an integer, and in this example I regard these integer values as counts.

The raw data are shown in the two parallel coordinate plots in the top panels of Figure 8.10: the twenty-three curves in the left plot show the students' marks with the assessment tasks $1, \ldots, 6$ on the x-axis. The six curves in the right plot show the marks of the six assessment tasks, with the student number on the x-axis. In the left plot I highlight the performance of three students (with indices 1, 5 and 16) in red and of another (with index 9) in black. The three red curves are very similar, whereas the black curve is clearly different from all other curves. In the plot on the right, the assignment marks are shown in blue and the examination marks in black. The 'black' student from the left panel is the student with x-value 9 in the right panel.

The bottom-left panel of Figure 8.10 shows the 2D configurations $D_2 U_2^{\mathrm{T}}$ and $D_2 V_2^{\mathrm{T}}$ of (8.24) in the same figure and so differs from the configurations we are used to seeing in Multidimensional Scaling. Here U_2 is a 6×2 matrix, and V_2 is a 23×2 matrix. The dots represent $D_2 V_2^{\mathrm{T}}$, and the red and black dots show the same student indices as earlier. The diamond symbols represent $D_2 U_2^{\mathrm{T}}$, with the black diamond on the far right the examination mark. The configurations show which rows or columns are alike. The three red dots are very close together, whereas the black dot is clearly isolated from the rest of the data. Two diamonds are close to each other – with coordinates close to (20,20), so two assessment tasks had similar marks. The rest of the diamonds are not clustered, which shows that the distribution of assignment marks and examination marks may not be the same.

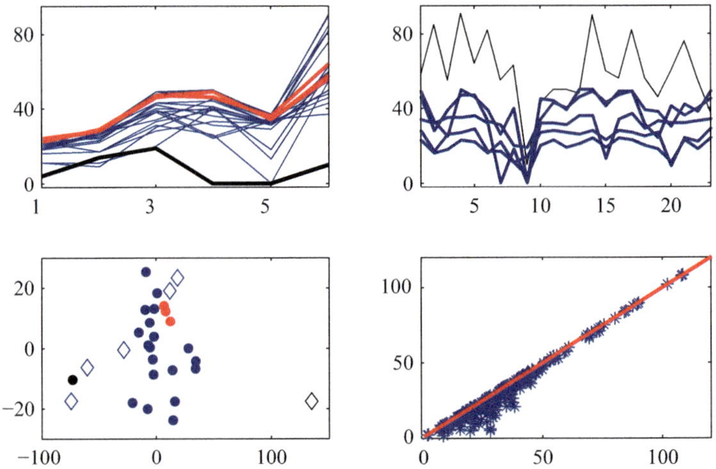

Figure 8.10 Parallel coordinate plots of Example 8.9 and corresponding 2D configurations in the lower panels.

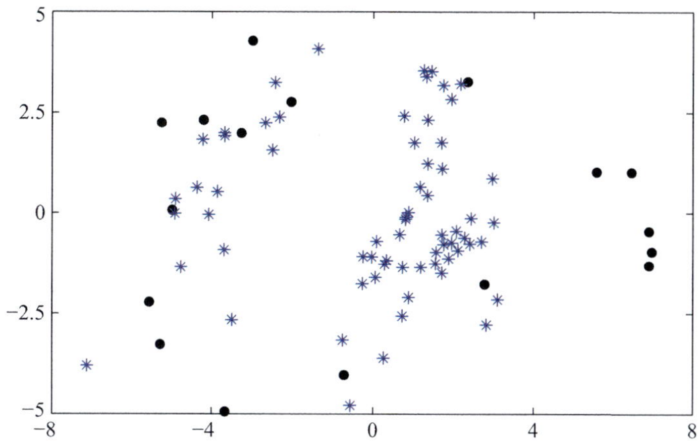

Figure 8.11 2D configurations of Example 8.10.

The scatterplot in the bottom right compares dissimilarities with configuration distances for the configuration $D_2 V_2^{\mathsf{T}}$; here I use the Euclidean distance in both cases. The dissimilarities of the observations are shown on the x-axis against the configuration distances on the y-axis, and the red line is the diagonal $y = x$. Note that most distances are small, and more points are below the red line than above. The latter shows that the pairwise distances of the columns are larger than those of the 2D configurations. Overall, the points are close to the line. ■

The *illicit drug market* data of Example 8.4 also fit into the realm of Correspondence Analysis: The entries of $\mathbb{X}$ are the seventeen series of count data observed over sixty-six months.

Example 8.10 In the analysis of the **illicit drug market** data in Example 8.4, I construct 1D and 2D configurations for the seventeen series and compare the first components of the two configurations. In this example I combine the 2D configuration of Example 8.4 with the 2D configuration (8.24) for the sixty-six months. The two configurations are shown in the same plot in Figure 8.11: black dots refer to the seventeen series and blue asterisks to the sixty-six months.

The figure shows a clear gap which splits the data into two parts. The months split into the first forty-nine on the right-hand side and the remaining later months on the left-side of the plot. This split agrees with the two parts in Figure 2.14 in Example 2.14 in Section 2.6.2. In Figure 2.14, all PC_1 weights before month 49 are positive, and all PC_1 weights after month 49 are negative. Gilmour et al. (2006) interpreted this change in the sign of the weights as a consequence of the heroin shortage in early 2001.

The configuration for the series emphasises the split into two parts further. The black dots in the right part of the figure are those listed as cluster 1 in Table 6.9 in Example 6.10 in Section 6.5.2 but also include series 15, *robbery 2*, whereas the black dots on the left are essentially the series belonging to cluster 2.

The visual effect of the combined configuration plot is a clearer view of the partitions that exist in these data than is apparent in separate configurations. ■

These examples illustrate that cluster structure may become more obvious or pronounced when we combine the configurations for rows and columns. This is not typically done in Multidimensional Scaling but is standard in Correspondence Analysis.

We now turn to the second analysis (CA.b), and consider distances between observations which are given as counts. Although the second analysis can be done for the columns or rows, I will mostly focus on columns and, without loss of generality, regard each column as an observation.

Definition 8.16 Let $\mathbb{X}$ be a matrix of size $r \times c$ whose entries are counts. The **(column) profile** $\mathbf{X}_i$ of $\mathbb{X}$ consists of the r entries $\{X_{i1}, \ldots, X_{ir}\}$ of $\mathbf{X}_i$. Two profiles $\mathbf{X}_i$ and $\mathbf{X}_k$ are **equivalent** if $\mathbf{X}_i = \lambda \mathbf{X}_k$ for some constant λ.

For $i \leq c$ and $C_i = \sum_\ell X_{i\ell}$, the column total of profile $\mathbf{X}_i$, put $\widetilde{X}_{ij} = X_{ij}/C_i$ for $j \leq r$, and call $\widetilde{\mathbf{X}}_i = [\widetilde{X}_{i1} \cdots \widetilde{X}_{ir}]^\mathsf{T}$ the ith **normalised profile**.

The χ^2 **distance** or **profile distance** of two profiles $\mathbf{X}_k$ and $\mathbf{X}_\ell$ is the distance $\varrho_{k\ell}$ of the normalised profiles $\widetilde{\mathbf{X}}_k$ and $\widetilde{\mathbf{X}}_\ell$, defined by

$$\varrho_{k\ell} = \left[\sum_{j=1}^{r} \frac{1}{\sum_{i=1}^{c} \widetilde{X}_{ij}} \left(\widetilde{X}_{kj} - \widetilde{X}_{\ell j} \right)^2 \right]^{1/2}. \tag{8.25}$$

□

When the data are counts, it is often more natural to compare proportions. The definitions imply that equivalent profiles result in normalised profiles that agree.

The name χ^2 *distance* is used because of the connection of the $\varrho_{k\ell}^2$ with the χ^2 statistic in the analysis of contingency tables. In Correspondence Analysis we do not make distributional assumptions about the data – such as the multinomial assumption inherent in the analysis of contingency tables – thus we cannot take recourse to the inferential interpretation of the χ^2 distribution. Instead, the graphical displays of Correspondence Analysis provide a visual tool which shows which profiles are alike. The connection between the profile distances and the configurations is summarised in the following proposition.

Proposition 8.17 *Let $\mathbb{X}$ be a matrix of size $r \times c$ whose entries are counts and whose columns are profiles. Let $\widetilde{\mathbb{X}}$ be the matrix of normalised profiles. Assume that $r \leq c$ and that $\widetilde{\mathbb{X}}$ has rank r. Write $\widetilde{\mathbb{X}} = \widetilde{U}\widetilde{D}\widetilde{V}^\mathsf{T}$ for the singular value decomposition of $\widetilde{\mathbb{X}}$. Let $\varrho_{k\ell}$ be the profile distance (8.25) of profiles $\mathbf{X}_k$ and $\mathbf{X}_\ell$. Then*

$$\varrho_{k\ell}^2 = (\mathbf{e}_k - \mathbf{e}_\ell)^\mathsf{T} \widetilde{U} \widetilde{D}^2 \widetilde{U}^\mathsf{T} (\mathbf{e}_k - \mathbf{e}_\ell), \tag{8.26}$$

where the $\mathbf{e}_k$ are unit vectors with a 1 in the kth entry and zero otherwise.

A proof of this proposition is deferred to the Problems at the end of Part II. The proposition elucidates how Correspondence Analysis combines ideas of Multidimensional Scaling and the analysis of contingency tables and shows the mathematical links between them.

To illustrate the second type of Correspondence Analysis (CA.b), I calculate profile distances for the proteomics profiles and propose an interpretation of these distances.

Example 8.11 The curves of the **ovarian cancer proteomics** data are measurements taken at points of a tissue sample. In Example 6.12 of Section 6.5.3 we explored clustering of the binary curves using the Hamming and the cosine distances. Figure 6.12 and Table 6.10

in Example 6.12 show that the data naturally fall into four different tissue types, but they also alert us to the difference in the grouping by the two approaches. From the figures and numbers in the table alone, it is not clear how different the methods really are and how big the differences are between various tissue types.

To obtain a more quantitative assessment of the sameness and the differences between the curves in the different cluster arrangements, I calculate the distances (8.25). Starting with the binary data, as in Example 6.12, we first need to define the count data. Each binary curve has a 0 or 1 at each of the 1,331 m/z values. It is natural to consider the total counts within a cluster for each m/z value. Let $\mathcal{C}_1 = ham1, \ldots, \mathcal{C}_4 = ham4$ be the four clusters obtained with the Hamming distance, and let $\mathcal{C}_5 = cos1, \ldots, \mathcal{C}_8 = cos4$ be the four clusters obtained with the cosine distance. The clusters lead to $c = 8$ profiles, which I refer to as the **counts profiles** or **cluster profiles**. The cluster profiles are given by

$$\mathbf{X}^{(k)} = \sum_{\mathbf{X}_\iota \in \mathcal{C}_k} \mathbf{X}_\iota \qquad \text{for } k \leq 8, \tag{8.27}$$

and each cluster profile has $r = 1,331$ entries, the counts at the 1,331 m/z values.

For the counts data $\left[\mathbf{X}^{(1)} \cdots \mathbf{X}^{(8)}\right]$ and their normalised profiles $\left[\widetilde{\mathbf{X}}^{(1)} \cdots \widetilde{\mathbf{X}}^{(8)}\right]$ I calculate the pairwise profile distances $\varrho_{k\ell}$ of (8.25). The $\varrho_{k\ell}$ values are shown in Table 8.7.

Table 8.7 compares eight counts profiles; *ham1* refers to the cancer clusters obtained with the Hamming distance and shown in yellow in Figure 6.12, *ham2* refers to the adipose tissue which is shown in green in the figure, *ham3* refers to peritoneal stroma shown in blue and *ham4* refers to the grey background cluster. The cosine-based clusters are listed in the same order. The information presented in the table is shown by differently coloured squares in Figure 8.12, with *ham1*,..., *cos4* squares arranged from left to right and from top to bottom in the same order as in the table. Thus, the second square from the left in the top row of Figure 8.12 represents $\varrho_{(1,2)}$, the distance between the cancer cluster *ham1* and the adipose cluster *ham2*. The darker the colour of the square, the smaller is the distance $\varrho_{k\ell}$.

Table 8.7 and Figure 8.12 show that there is good agreement between the Hamming and cosine counts profiles for each tissue type. The match agrees least for the cancer counts, with a ϱ-value of 0.17. The three non-cancerous tissue types are closer to each other than to the cancer tissue type, and the cancer tissue and background tissue differ most.

The relative lack of agreement between the two cancer-counts profiles *ham1* and *cos1* may be indicative of the fact that the groups are obtained from a cluster analysis rather than a discriminant analysis and point to the need for good classification rules for these data. ∎

8.6.2 Analysis of Distance

Gower and Krzanowski (1999) combined ideas from the Analysis of Variance (ANOVA) and Multidimensional Scaling and proposed a technique called **Analysis of Distance** which applies to a larger class of data than the Analysis of Variance. A fundamental assumption in their approach is that the data partition into κ distinct groups. I describe the main ideas of Gower and Krzanowski, adjusted to our framework. As we shall see, the Analysis of Distance has strong connections with Discriminant Analysis and Multidimensional Scaling.

Table 8.7 *Profile Distances $\varrho_{k\ell}$ for the Hamming and Cosine Counts Data from Example 8.11*

ham1	*ham2*	*ham3*	*ham4*	*cos1*	*cos2*	*cos3*	*cos4*	
0	0.55	0.54	0.62	0.17	0.55	0.56	0.62	*ham1*
	0	0.40	0.49	0.44	0.01	0.42	0.49	*ham2*
		0	0.53	0.41	0.40	0.04	0.53	*ham3*
			0	0.53	0.49	0.55	0.03	*ham4*
				0	0.44	0.44	0.53	*cos1*
					0	0.42	0.49	*cos2*
						0	0.55	*cos3*
							0	*cos4*

Figure 8.12 Profile distances $\varrho_{k\ell}$ based on the cluster profiles of the *ham1*, ..., *cos4* clusters of Table 8.7. The smaller the distance, the darker is the colour.

Let $\mathbb{X} = \left[\mathbf{X}_1 \cdots \mathbf{X}_n\right]$ be d-dimensional centred data. Let ϱ_{ik} be classical Euclidean dissimilarities for pairs of columns of $\mathbb{X}$. Let K be the $n \times n$ matrix with elements $-\varrho_{ik}^2/2$. In the notation of (8.10) to (8.12), $K = -P \circ P/2$, and

$$Q_{\langle n\rangle} = \mathbb{X}^{\mathsf{T}}\mathbb{X} = \Theta K \Theta,$$

where Θ is the centring matrix of (8.11).

Gower and Krzanowski collapsed the $d \times n$ data matrix into a $d \times \kappa$ matrix based on the κ groups $\mathbb{X}$ partitions into and then applied Multidimensional Scaling to this much smaller matrix.

Definition 8.18 Let $\mathbb{X} = \left[\mathbf{X}_1 \cdots \mathbf{X}_n\right]$ be d-dimensional centred data. Let ϱ_{ik} be classical Euclidean dissimilarities, and let K be the matrix with entries $-\varrho_{ik}^2/2$. Assume that the data belong to κ distinct groups, with $\kappa < d$; the kth group has n_k members from $\mathbb{X}$, and $\sum n_k = n$, for $k \le \kappa$. Let G be the $\kappa \times n$ matrix with elements

$$g_{ik} = \begin{cases} 1 & \text{if } \mathbf{X}_i \text{ belongs to the } k\text{th group} \\ 0 & \text{otherwise,} \end{cases}$$

and let N be the $\kappa \times \kappa$ diagonal matrix with diagonal elements n_k. The **matrix of group means** $\mathbb{X}_G$ and the matrix K_G derived from the group dissimilarities are

$$\mathbb{X}_G = \mathbb{X} G^{\mathsf{T}} N^{-1} \qquad \text{and} \qquad K_G = N^{-1} G K G^{\mathsf{T}} N^{-1}. \tag{8.28}$$

□

A little reflection shows that $\mathbb{X}_G$ is of size $d \times \kappa$ and K_G is of size $\kappa \times \kappa$, and thus, the transformation $\mathbb{X} \to \mathbb{X}_G$ reduces the $d \times n$ raw data to the $d \times \kappa$ matrix of group means. Put

$$Q_{<G,\kappa>} = \mathbb{X}_G^{\mathsf{T}} \mathbb{X}_G.$$

Using (8.12) and Theorem 8.3, it follows that

$$Q_{<G,\kappa>} = \Theta K_G \Theta. \tag{8.29}$$

If $V_G D_G^2 V_G^{\mathsf{T}}$ is the spectral decomposition of $Q_{<G,\kappa>}$, then $\mathbb{W}_G = D_G V_G^{\mathsf{T}}$ is an r-dimensional configuration for $\mathbb{X}_G$, where r is the rank of $\mathbb{X}_G$.

So far we have applied ideas from classical scaling to the group data. To establish a connection with the Analysis of Variance, Gower and Krzanowski (1999) started with the $n \times n$ matrix K of Definition 8.18 and considered submatrices K_k of size $n_k \times n_k$ which contain all entries of K relating to the kth group. Let $\overline{\varrho}_{ik}$ be the Euclidean distance between the ith and kth group means, and let $\overline{K}$ be the matrix of size $\kappa \times \kappa$ with entries $(-\overline{\varrho}_{ik}^2/2)$. For

$$\begin{aligned} T &= -\frac{1}{n}(\mathbf{1}_{n\times 1})^{\mathsf{T}} K \mathbf{1}_{n\times 1} \\ W &= -\sum_{k=1}^{\kappa} \frac{1}{n_k}(\mathbf{1}_{n_k\times 1})^{\mathsf{T}} K_k \mathbf{1}_{n_k\times 1} \\ B &= -\frac{1}{n}[n_1 \ldots n_\kappa]\overline{K}[n_1 \ldots n_\kappa]^{\mathsf{T}}, \end{aligned}$$

Gower and Krzanowski showed the fundamental identity of the Analysis of Distance, namely,

$$T = W + B, \tag{8.30}$$

which reminds us of the decomposition of the total variance into between-class and within-class variances.

Example 8.12 The **ovarian cancer proteomics** data from Example 8.11 are the type of data suitable for an Analysis of Distance; the groups correspond to clusters or classes. A little thought tells us that the matrix of group means (8.28) consists of the normalised profiles calculated from the counts profiles.

Instead of the profile distances which I calculated in the preceding analysis of these data, in an Analysis of Distance, one is interested in lower-dimensional profiles and the identity

(8.30), which highlights the connection with Discriminant Analysis. We pursue these calculations in the Problems at the end of Part II and compare the results with those obtained in the correspondence analysis of these data in the preceding example. The two approaches could be combined to form the basis of a classification strategy for such count data. ■

In the Analysis of Variance, F-tests allow us to draw inferences. Becaues the Analysis of Distance is defined non-parametrically, F-tests are not appropriate; however, one can still calculate T, W and B. The quantities B and W are closely related to similar quantities in Discriminant Analysis: the between-class matrix $\widehat{B}$ and the within-class matrix $\widehat{W}$, which are defined in Corollary 4.9 in Section 4.3.2. In Discriminant Analysis we typically use the projection onto the first eigenvector of $\widehat{W}^{-1}\widehat{B}$. In contrast, in an Analysis of Distance, we construct low-dimensional configurations of the matrix of group means, similar to the low-dimensional configurations in Multidimensional Scaling.

The Analysis of Distance reduces the computational burden of Multidimensional Scaling by replacing $\mathbb{X}$ and $Q_{\langle n\rangle}$ with $\mathbb{X}_G$, the matrix of sample means of the groups, and the matrix $Q_{<G,\kappa>}$ of (8.29). Both $\mathbb{X}_G$ and $Q_{<G,\kappa>}$ are typically of much smaller size than $\mathbb{X}$ and $Q_{\langle n\rangle}$. Based on $Q_{<G,\kappa>}$, one calculates configurations for the grouped data. To extend the configurations to $\mathbb{X}$, one makes use of the group structure of the data and adds points as in Gower (1968) which are consistent with the existing configuration distances of the group means.

8.6.3 Low-Dimensional Embeddings

Classical scaling with Euclidean dissimilarities has become the starting point for a number of new research directions. In Section 8.2 we considered objects and dissimilarities and constructed configurations without knowledge of the underlying data $\mathbb{X}$. The more recent developments have a different focus: they start with d-dimensional data $\mathbb{X}$, where both d and n may be large, and the goal is that of embedding the data into a low-dimensional space which contains the important structure of the data. The embeddings rely on Euclidean dissimilarities of pairs of observations, and they are often non-linear and map into manifolds. *Local Multidimensional Scaling*, *Non-Linear Dimension Reduction*, and *Manifold Learning* are some of the names associated with these post-classical scaling methods.

To give a flavour of the research in this area, I focus on common themes and outline three approaches:

- the Distributional Scaling of Quist and Yona (2004),
- the Landmark Multidimensional Scaling of de Silva and Tenenbaum (2004), and
- the Local Multidimensional Scaling of Chen and Buja (2009).

We begin with common notation for the three approaches. Let $\mathbb{X} = \left[\mathbf{X}_1 \cdots \mathbf{X}_n\right]$ be d-dimensional data. For $p < r$, and r the rank of $\mathbb{X}$, let $\mathfrak{f}$ be an embedding of $\mathbb{X}$ or subsets of $\mathbb{X}$ into $\mathbb{R}^p$. The dissimilarities ϱ on $\mathbb{X}$ and distances Δ on $\mathfrak{f}(\mathbb{X})$ are defined for pairs of observations $\mathbf{X}_i$ and $\mathbf{X}_k$ by

$$\varrho_{ik} = \|\mathbf{X}_i - \mathbf{X}_k\| \qquad \text{and} \qquad \Delta_{ik} = \|\mathfrak{f}(\mathbf{X}_i) - \mathfrak{f}(\mathbf{X}_k)\| .$$

Section 8.5.1 highlights advantages of the duality between $\mathbb{X}$ and $\mathbb{X}^{\mathsf{T}}$ for HDLSS data. For arbitrary data with large sample sizes, the dissimilarity matrix becomes very large, and the

$Q_{\langle n\rangle}$-matrix approach becomes less desirable for constructing low-dimensional configurations. For data that partition into κ groups, Section 8.6.2 describes an effective reduction in sample size by replacing the raw data with their group means. To be able to do this, it is necessary that the data belong to κ groups or classes and that the group or class membership is known and distance-based. If the data belong to different clusters, with an unknown number of clusters, then the approach of Gower and Krzanowski (1999) does not work.

Group membership is a local property of the observations. We now consider other local properties that can be exploited and integrated into scaling. Suppose that the data $\mathbb{X}$ have some local structure $\mathcal{L} = \{\ell_1, \ldots, \ell_\kappa\}$ consistent with pairwise distances ϱ_{ik}. The task of finding a good global embedding $\mathfrak{f}$ which preserves the local structure splits into two parts:

1. finding a good embedding for each of the local units ℓ_k, and
2. extending the embedding to all observations so that the structure $\mathcal{L}$ is preserved.

Distributional Scaling. Quist and Yona (2004) used clusters as the local structures and assumed that the cluster assignment is known or can be estimated. The first step of the global embedding is similar to that of Gower and Krzanowski (1999) in Section 8.6.2. For the second step, Quist and Yona proposed using a penalised stress loss. For pairs of clusters $\mathcal{C}_k$ and $\mathcal{C}_m$, they considered a function $\rho_{km}\colon \mathbb{R} \to \mathbb{R}$ which reflects the distribution of the distances ϱ_{ij} between points of the two clusters:

$$\rho_{km}(x) = \frac{\sum_{\mathbf{X}_i \in \mathcal{C}_k} \sum_{\mathbf{X}_j \in \mathcal{C}_m} w_{ij}\, \delta(x - \varrho_{ij})}{\sum_{\mathbf{X}_i \in \mathcal{C}_k} \sum_{\mathbf{X}_j \in \mathcal{C}_m} w_{ij}},$$

where the w_{ij} are weights, and δ is the Kronecker delta function. Similarly, they defined a function $\widetilde{\rho}_{km}$ for the embedded clusters $\mathfrak{f}(\mathcal{C}_k)$ and $\mathfrak{f}(\mathcal{C}_m)$. For a tuning parameter $0 \le \alpha \le 1$, the penalised stress $\mathcal{S}\text{tress}_{QY}$ is regarded as a function of the distances ϱ and the embedding $\mathfrak{f}$:

$$\mathcal{S}\text{tress}_{QY}(\varrho, \mathfrak{f}) = (1-\alpha)\mathcal{S}\text{tress}(\varrho, \mathfrak{f}) + \alpha \sum_{k \le m} W_{km}\, D(\rho_{km}, \widetilde{\rho}_{km}).$$

The last sum is taken over pairs of clusters, the W_{km} are weights for pairs of clusters $\mathcal{C}_k$ and $\mathcal{C}_m$, and $D(\rho_{km}, \widetilde{\rho}_{km})$ is a measure of the dissimilarity of ρ_{km} and its embedded version $\widetilde{\rho}_{km}$. Examples and parameter choices are given in their paper. Instead of the stress $\mathcal{S}\text{tress}(\varrho, \mathfrak{f})$, they also used the Sammon stress (see Table 8.4).

The success of the method depends on how much is known about the cluster structure or how well the cluster structure – the number of clusters and the cluster membership – can be estimated. The final embedding $\mathfrak{f}$ is obtained iteratively, and the authors suggested starting with an embedding that results in low stress. Their method depends on a large number of parameters. It is not easy to see how the choices of the tuning parameter α, the weights w_{ij} between points and the weights W_{km} between clusters affect the performance of the method. Although the number of parameters may seem overwhelming, a judicious choice can lead to a low-dimensional configuration which contains the important structure of the data.

Landmark Multidimensional Scaling. de Silva and Tenenbaum (2004) used κ landmark points as the local structure and calculated distances from each of the n observations to each of the landmark points. As a consequence, the dissimilarity matrix P is reduced to a dissimilarity matrix P_κ of size $\kappa \times n$. Classical scaling is carried out for the landmark points.

The landmark points are usually chosen randomly from the n observations, but other choices are possible. If there are real landmarks in the data, such as cluster centres, they can be used. This approach results in the $p \times \kappa$ configuration

$$\mathbb{W}_{(\kappa)} = D_p V_p^{\mathsf{T}}, \tag{8.31}$$

which is based on the $Q_{\langle n \rangle}$ matrix derived from the weights of P_κ as in Algorithm 8.1, and p is the desired dimension of the configuration. The subscript κ indicates the number of landmark points.

In a second step – the extension of the embedding to all observations – de Silva and Tenenbaum (2004) defined a distance-based procedure which determines where the remaining $n-\kappa$ observations should be placed. For notational convenience, one assumes that the first κ column vectors of P_κ are the distances between the landmark points. Let $P_\kappa \circ P_\kappa$ be the Hadamard product, and let $\mathbf{p}_k$ be the kth column vector of $P_\kappa \circ P_\kappa$. The remaining $n-\kappa$ observations are represented by the column vectors $\mathbf{p}_m$ of $P_\kappa \circ P_\kappa$, where $\kappa < m \le n$. Put

$$\overline{\varrho} = \frac{1}{\kappa} \sum_{k=1}^{\kappa} \mathbf{p}_k,$$

$$\mathbb{W}^{\#}_{(\kappa)} = D_p^{-1} V_p^{\mathsf{T}},$$

and define the embedding $\mathfrak{f}$ for non-landmark observations $\mathbf{X}_i$ and columns $\mathbf{p}_i$ by

$$\mathbf{X}_i \longmapsto \mathfrak{f}(\mathbf{X}_i) = -\frac{1}{2} \mathbb{W}^{\#}_{(\kappa)} (\mathbf{p}_i - \overline{\varrho}).$$

In a final step, de Silva and Tenenbaum applied a principal component analysis to the configuration axes in order to align the axes with those of the data.

As in Quist and Yona (2004), the dissimilarities of the data are taken into account in the second step of the approach of de Silva and Tenenbaum (2004). Because the landmark points may be randomly chosen, de Silva and Tenenbaum suggested running the method multiple times and discarding bad choices of landmarks. According to de Silva and Tenenbaum, poorly chosen landmarks have low correlation with good landmark choices.

Local Multidimensional Scaling. Chen and Buja (2009) worked with the stress loss, which they split into a 'local' part and a 'non-local' part, where 'local' refers to nearest neighbours. A naive restriction of stress to pairs of observations which are close does not lead to meaningful configurations (see Graef and Spence 1979). For this reason, care needs to be taken when dealing with non-local pairs of observations. Chen and Buja (2009) considered symmetrised k-nearest neighbourhood (kNN) graphs based on sets

$$N_{\text{sym}} = \{(i,m)\colon \mathbf{X}_i \in N(\mathbf{X}_m, k) \ \& \ \mathbf{X}_m \in N(\mathbf{X}_i, k)\},$$

where i and m are the indices of the observations $\mathbf{X}_i$ and $\mathbf{X}_m$, and the sets $N(\mathbf{X}_m, k)$ are those defined in (4.34) of Section 4.7.1. The key idea of Chen and Buja (2009) was to modify the stress loss for pairs of observations which are not in S_{sym} and to consider, for a fixed c,

$$\mathcal{S}\text{tress}_{CB}(\varrho, \mathfrak{f}) = \sum_{(i,m)\in N_{\text{sym}}} (\Delta_{im} - \varrho_{im})^2 - c \sum_{(i,m)\notin N_{\text{sym}}} \Delta_{im}.$$

Chen and Buja (2009) call the first sum in $\mathcal{S}\text{tress}_{CB}$ the 'local stress' and the second the 'repulsion'. Whenenver $(i,m) \in N_{\text{sym}}$, the local stress forces the configuration distances to be small, and thus preserves the local structure of neighbourhoods. Chen and Buja (2009) have developed a criterion, called *local continuity*, for choosing the parameter c in $\mathcal{S}\text{tress}_{CB}$. With regards to the choice of k, their computations show that a number of values should be tried in applications.

Chen and Buja (2009) reviewed a number of methods that have evolved from classical scaling. These methods include the Isomap of Tenenbaum, de Silva, and Langford (2000), the Local Linear Embeddings of Roweis and Saul (2000) and Kernel Principal Component Analysis of Schölkopf, Smola, and Müller (1998). We consider Kernel Principal Component Analysis in Section 12.2.2, a non-linear dimension-reduction method which extends Principal Component Analysis and Multidimensional Scaling. For more general approaches to non-linear dimension reduction, see Lee and Verleysen (2007) or chapter 16 in Izenman (2008).

In this chapter we only briefly touched on how to choose the dimension of a configuration. If visualisation is the aim, then two or three dimensions is the obvious choice. Multidimensional Scaling is a dimension-reduction method, in particular, for high-dimensional data, whose more recent extensions focus on handling and reducing the dimension and taking into account local or localised structure. Finding dimension-selection criteria for non-linear dimension reduction may prove to be a bigger challenge than that posed by Principal Component Analysis, a challenge which will require mathematics we may not understand but have to get used to.

Problems for Part II

Cluster Analysis

1. Consider the *Swiss bank notes* data of Example 2.5 in Section 2.4.1.
   (a) Use hierarchical agglomerative clustering with the Euclidean distance, and determine the membership of the two clusters just before the final merge.
   (b) Repeat part (a) with the cosine distance. Are the results the same?
   (c) Use a different linkage, and repeat parts (a) and (b).
   (d) Find optimal 2-cluster arrangements based on the k-means approach using both the Euclidean and the cosine distances.
   (e) Compare the results in parts (a)–(d) and comment on your findings.
2. Hierarchical agglomerative clustering assigns observations to clusters based on the nearest distance at every step in the algorithm and results in a tree with a final single cluster.
   (a) Describe a way of modifying hierarchical agglomerative clustering such that it finds k clusters with cluster sizes at least ν, where ν is given and not too small compared with the total number n of $\mathbb{X}$.
   (b) Illustrate your ideas with the *Swiss bank notes* data from the preceding problem using $k = 3$ and $\nu > 5$.
3. Consider the *glass* identification data set from the Machine Learning Repository (http://archive.ics.uci.edu/ml/datasets/Glass+Identification) without the class label. Base all calculations on the single linkage.
   (a) Use the Euclidean distance, the ℓ_1 distance, the ℓ_∞ distance, and the cosine distance to calculate dendrograms. Display the four dendrograms, and compare the results.
   (b) Use the Euclidean distance with the four linkages single, complete, average, and centroid of Definition 6.1. Calculate and display the four dendrograms, and comment on the results.
4. Show that the within-cluster variability cannot increase when an extra cluster is added optimally.
5. Consider the *income* data of Example 3.7 of Section 3.5.2. Use all observations and all variables except the variable *income* as the data $\mathbb{X}$.
   (a) Use the k-means algorithm with the Euclidean distance, and determine the best cluster arrangements for $k = 2, \ldots, 10$. Comment on your findings, and discuss how many clusters lead to a suitable representation of the data.
   (b) Repeat the analysis of part (a) for the cosine distance. Compare the results obtained using the two distance measures.

6. Prove the relationship (6.2) between the within-cluster variability $W_{\mathcal{P}}$ and the sample covariance matrix $\widehat{W}$ of Corollary 4.9 (for $n_c = n/k$), and illustrate with an example.
7. Consider the collection of *HIV flow cytometry* data sets from Example 6.13 in Section 6.6.4. Previously we analysed the first HIV$^+$ and the first HIV$^-$subject in the collection. Now we consider all fourteen subjects and for each subject the variables *FS*, *SS*, *CD4*, *CD8* and *CD3*.

   (a) Apply the SOPHE Algorithm 6.2 to partition the data sets of the fourteen subjects using seven bins for each variable and threshold $\theta_0 = 20$.
   (b) Show parallel coordinate views of the mode locations for the cluster results.
   (c) Compare the results for the fourteen data sets. Can we tell from these results whether a data set is HIV$^+$and HIV$^-$? If so, how do they differ?

8. Let $\mathbb{X}$ be d-dimensional data. Consider the Euclidean distance. Fix the number of clusters $k > 1$. Let $\mathcal{P}$ be the optimal k-cluster arrangement of $\mathbb{X}$, and let $\mathcal{P}_W$ be the optimal k-cluster arrangement of the PC$_1$ data $\mathbb{W}^{(1)}$.

   (a) Show that the minimum within-cluster variability of the PC$_1$ data is not larger than that of the original data. *Hint:* Consider first the cluster arrangement the PC$_1$ data inherit from the optimal cluster arrangement of the original data.
   (b) For the athletes data from the *Australian Institute of Sports*, determine the optimal 2-cluster arrangement of the original data. Derive the cluster arrangement that the PC$_1$ data inherit from the original data.
   (c) Calculate the optimal cluster arrangement $\mathcal{P}_W$ for the PC$_1$ data, and compare it with cluster arrangement obtained in part (b).

9. Describe principal component–based hierarchical agglomerative clustering, and explain how the dendrogram is constructed in this case. Illustrate with an example.
10. In Example 6.13 in Section 6.6.4 we analysed the first subject of the HIV$^+$and the HIV$^-$data sets, respectively. For each of the fourteen subjects in the collection, there are four data sets which consist of different CD variables. Previously, we looked at the second subset with CD variables, *CD4*, *CD8* and *CD3*. Now we also consider the other three subsets for each subject is the collection.

    (a) Use the k-means algorithm to find the number of clusters that best describes these four data sets for each subjets.
    (b) Repeat part (a), but this time use the four variability measures described in Section 6.6.1, and report your findings.
    (c) Compare the results of parts (a) and (b), and determine the number of clusters for each of the four subsets of the HIV$^+$data.

11. We consider the PC sign cluster rule (6.9), Fisher's rule (4.13) in Section 4.3.2, the within-cluster variability (6.1) and the criterion (6.11).

    (a) Compare rules (6.9) and (4.13) in terms of their aims and the criteria they maximise.
    (b) Discuss the relationship between rule (6.9) and the within-cluster variability (6.1). What are the aims of each criterion?
    (c) Relate Fisher's rule (4.13) and expression (6.11). *Hint:* Consider CH(k) for fixed values of k.
    (d) Consider the *Swiss bank notes* data of Example 2.5 in Section 2.4.1. Apply and calculate the four rules and criteria listed at the beginning of the problem, using the

genuine and counterfeit samples as the two classes for Fisher's rule. Comment on the results obtained with the four criteria.

12. Consider the mixture of *compensation beads* stained with monoclonal antibodies in flow cytometry experiments and described in Zaunders et al. (2012). There are a total of 6,5031 cells and eleven variables.
    (a) Use k-means clustering to partition these data, and determine the optimal number of clusters using the four criteria (6.11) to (6.14).
    (b) Use the gap statistic of Section 6.6.2, and determine the optimal number of clusters.
    (c) Use the SOPHE algoritm with six bins and $\theta_0 = 50$, and partition the data.
    (d) Determine the optimal number of clusters k_0^* based on your results in parts (a) to (c). Compare the clusters obtained with k_0^*-means clustering with those obtained with the SOPHE, and comment on the agreement and differences between these cluster allocations.

Factor Analysis

13. Prove Proposition 7.2.
14. Let $\mathbb{X}$ be d-dimensional data with sample covariance matrix S.
    (a) For the sample case, give explicit expressions for the terms R_A and $\widehat{A}$ in Method 7.2.
    (b) For the data in Example 7.3, calculate the one-, two-, and three-factor models using Method 7.2. Which factor model best describes the data? Give reasons for your choice.
15. Let $\mathbf{X}$ be a three-dimensional random vector with correlation matrix R, and assume that $\mathbf{X}$ is given by a one-factor model.
    (a) Determine relationships for the correlation coefficients ρ_{jm} in terms of the scaled factors, and compare with (7.2) and (7.3).
    (b) Find explicit expressions for the three entries of $\Sigma_{diag}^{-1/2}$ in terms of the correlation coefficients by using the equalities derived in part (a).
    (c) Determine the matrices Ψ and $\Sigma_{diag}^{-1/2}\Psi\Sigma_{diag}^{-1/2}$ based on the results in parts (a) and (b).
16. Consider the HIV$^+$and HIV$^-$ *flow cytometry* data sets of Example 6.13 in Section 6.6.4. Separately consider the first 1,000 samples and the total data for each of the two data sets.
    (a) Determine the two- and three-dimensional PC factors, and show your results in biplots.
    (b) Determine two- and three-dimensional factors based on ML, and display the results in biplots.
    (c) For the methods in (a) and (b), calculate the optimal orthogonal and oblique rotations.
    (d) Compare the results of the previous parts, and comment.
17. Based on the factor loadings and common factor of Method 7.1, show that
    (a) $\widehat{\tau}_{jj} = \sum_{m=1}^{k} \lambda_m \eta_{mj}^2$ and
    (b) hence derive that $\widehat{\psi}_j = \sum_{m>k} \lambda_m \eta_{mj}^2$.
18. For the normal k-factor model, prove (7.11) and (7.12). *Hint:* Calculate $tr(\widehat{\Sigma}^{-1}S)$.

19. Assume that the normal k-factor model satisfies the assumptions of part 3 of Theorem 7.6. Show that A and Ψ satisfy the constraints of part 1 of the theorem.
20. Consider the *Dow Jones returns* data of Example 7.5.
    (a) Calculate ML-based factor loadings for k=2,3 and display them in biplots.
    (b) Compare your results with the PC results and biplots of Example 7.5, and comment.
21. Consider the *Swiss bank notes* data of Example 2.2 in Section 2.4.1.
    (a) Carry out appropriate hypothesis tests for the number of factors using the methods of Section 7.5. Interpret the results of the test.
    (b) Determine the number of factors using the method of Tipping and Bishop (1999) described in Theorem 2.26 of Section 2.8.1.
    (c) Discuss the similarities and differences of the two approaches of parts (a) and (b).
    (d) Based on Theorem 2.20 of Section 2.7.1, determine a suitable test statistics for testing $\lambda_j = 0$ for some $j \leq d$. Discuss how these tests differ from the tests in part (a), and comment on their suitability for determining the number of factors.
22. Explain the most important differences between Principal Component Analysis and Factor Analysis, and illustrate these differences with the *ionosphere* data and the *Dow Jones returns*. *Hint:* In your comparison, calculate factor loadings and rotations for more than three factors, and comment on how many factors are necessary to describe the data appropriately. Give reasons for your choice.
23. Derive expressions for factor scores $\mathbf{F}_{\mathrm{R}}$ based on Method 7.2 such that the scaled vectors satisfy a k-factor model. Determine the relationship between the factor scores $\mathbf{F}_{\mathrm{R}}$ and those of (7.8).
24. Derive expression (7.29) using $A = \Gamma_k \Lambda_k^{1/2}$.
25. Assume that the data $\mathbb{X}$ are normal. Consider the factor scores $\widehat{\mathbf{F}}^{(\mathrm{ML})}$ of (7.20). Calculate an expression for the sample covariance matrix of $\widehat{\mathbf{F}}^{(\mathrm{ML})}$. Find conditions under which the sample covariance matrix converges to the identity matrix as the sample size increases.
26. Assume that $\mathbf{X} \sim \mathcal{N}(\boldsymbol{\mu}, \Sigma)$ is given by a k-factor model. Put $\Delta = A^{\top}\Psi^{-1}A$, and assume that Δ is a diagonal matrix with diagonal entries δ_j for $j \leq k$.
    (a) Let $\mathbf{F}^{(\mathrm{B})}$ be the Bartlett factor scores (7.22) of $\mathbf{X}$. Show that the covariance matrix of $\mathbf{F}^{(\mathrm{B})}$ is $\mathbf{I}_{k\times k} + \Delta^{-1}$ and hence that the jth diagonal entry is $(1+\delta_j^{-1})$.
    (b) Let $\mathbf{F}^{(\mathrm{T})}$ be the Thompson factor scores (7.23) of $\mathbf{X}$. Put $\Delta_\zeta = \Delta + \zeta \mathbf{I}_{k\times k}$. Show that the covariance matrix of $\mathbf{F}^{(\mathrm{T})}$ is $\Delta_\zeta^{-1}\Delta\Delta_1\Delta_\zeta^{-1}$. Further, show that for $\zeta = 1$, the diagonal entries of the covariance matrix are $\delta_j/(1+\delta_j)$ for $j = 1,\ldots,k$.
    (c) Compare the covariance matrices of parts (a) and (b). *Hint:* Consider the trace or matrix norms in your comparisons.

Multidimensional Scaling

27. Let $Q_{\langle n\rangle}$ be the $n \times n$ matrix defined from Euclidean dissimilarities as in Theorem 8.3. Let r be the rank of $Q_{\langle n\rangle}$. For $p < r$, let $\mathbb{W}_p$ of (8.7) be the classical principal coordinates.
    (a) Show that $\mathbb{W}_p$ minimises the classical error $\mathcal{S}\mathrm{tress}_{\mathrm{clas}}$ over all p-dimensional configurations.

(b) Prove the following. If U_p is an orthogonal $p \times p$ matrix, then $U_p \mathbb{W}_p$ is also a minimiser for $\mathcal{S}\text{tress}_{\text{clas}}$.

28. The centre of the earth is about 6,350 km from each point on the surface. Add this centre to the matrix D of distances in (8.1). Construct a map in two and three dimensions for these eleven points, using principal coordinates. Compare your two- and three-dimensional maps with Figure 8.1, and comment.
29. Let $\mathbb{X}$ be d-dimensional centred data of rank r. Write $\mathbb{X} = U\Lambda^{1/2}V^{\mathsf{T}}$ for its singular value decomposition. For $p \leq r$ let $\mathbb{W} = \Lambda_p^{1/2} V_p^{\mathsf{T}}$ be a p-dimensional configuration for $\mathbb{X}$. Put $D_{\mathrm{X,W}} = \mathbb{X}^{\mathsf{T}}\mathbb{X} - \mathbb{W}^{\mathsf{T}}\mathbb{W}$.

(a) Show that

$$D_{\mathrm{X,W}} = V\Lambda \begin{bmatrix}\mathbf{0}_{n\times p} & \mathbf{v}_{p+1} & \cdots & \mathbf{v}_r\end{bmatrix}^{\mathsf{T}},$$

where $\mathbf{0}_{n\times p}$ is the $n \times p$ zero matrix and $\mathbf{v}_j$ is the jth column vector of V.

(b) Put $\widetilde{V}_p = \begin{bmatrix}\mathbf{0}_{n\times p} & \mathbf{v}_{p+1} & \cdots & \mathbf{v}_r\end{bmatrix}^{\mathsf{T}}$, and show that

$$D_{\mathrm{X,W}}^{\mathsf{T}} D_{\mathrm{X,W}} = \widetilde{V}_p \Lambda^2 \widetilde{V}_p^{\mathsf{T}}.$$

30. For the *athletes* data of Example 8.3, consider objects $\mathcal{O}_k$ for $k \geq 5$, where $\mathcal{O}_5$ contains the variables of $\mathcal{O}_3$ and the variable with the fourth largest PC_1 weight (in absolute value), $\mathcal{O}_6$ contains the variables of $\mathcal{O}_5$ and the variable with the fifth largest PC_1 weight (in absolute value) and so on.

(a) Give an explicit expression for the matrices A_5 and A_8 which correspond to $\mathcal{O}_5$ and $\mathcal{O}_8$.

(b) Calculate the two-dimensional strain configurations for $\mathcal{O}_k$ with $k = 5,\ldots,8$, and give the numerical value for each strain. Display the configurations in the colours used in Figure 8.2.

(c) Comment on the configurations, and determine whether a subset of variables suffices to obtain a configuration which separates the male and female data well.

31. Let the $d \times n$ data $\mathbb{X}$ have rank r. For $p \leq r$, define $\mathbb{W} = \mathbb{W}_p$ as in (8.7). Let E be an orthogonal $p \times p$ matrix, and put $\mathbb{W}_2 = E\mathbb{W}$.

(a) Show (8.9).

(b) Show that $\mathcal{S}\text{train}(\mathbb{X}, \mathbb{W}_2) = \mathcal{S}\text{train}(\mathbb{X}, \mathbb{W})$.

32. Consider the *wine recognition* data of Example 4.5 in Section 4.4.2. Calculate and display two- and three-dimensional configurations for these data using the cosine distance and the Euclidean norm as the dissimilarities and the metric stress, sstress and Sammon stress as loss measures. For each configuration, show Shepard diagrams. Compare and interpret the results.
33. Verify that (8.14) reduces to the strain (8.8).
34. Consider the *athletes* data of Example 8.5 and the four stress-optimal configurations of Figure 8.4 shown in the top-left panel and the three bottom panels. Let $\mathbb{W}_{(1)}$ to $\mathbb{W}_{(4)}$ be the optimal configurations derived from the Euclidean, cosine, correlation and max distance.

(a) For pairs of optimal configurations $\mathbb{W}_{(i)}$ and $\mathbb{W}_{(k)}$, with $i,k = 1,\cdots,4$ calculate the squared Frobenius norm of $\mathbb{W}_{(i)} - \mathbb{W}_{(k)}$, and find $\chi(E^*)$ of Theorem 8.13.

(b) Repeat the calculations of part (a) for the three-dimensional configurations (using the same four dissimilarities as in part (a)).
(c) Compare and interpret the results of parts (a) and (b).

35. Give a proof of Proposition 8.17.
36. Consider the *exam grades* data of Example 7.7 in Section 7.5. Let $\mathbb{X}^{(1)}$ be the subset of these data, which consists of the first thirty observations.
    (a) Calculate the two-dimensional configurations $D_2 U_2^{\mathsf{T}}$ and $D_2 V_2^{\mathsf{T}}$ of (8.24) for $\mathbb{X}^{(1)}$, and display.
    (b) Calculate the five row profiles $\mathbf{X}_{\bullet j}^{(1)}$ and their normalised versions $\widetilde{\mathbf{X}}_{\bullet j}^{(1)}$, and display appropriate plots of these data.
    (c) Define profile distances for row profiles, and derive and prove Proposition 8.17 for row profiles.
    (d) Interpret and comment on the results.
37. To explore an Analysis of Distance on Fisher's *iris* data, carry out the following steps, which are based on the notation of Section 8.6.2.
    (a) Give an explicit formula of and a numerical expression for $\mathbb{X}_G$, K_G, $Q_{<G,\kappa>}$ and $\mathbb{W}_G$.
    (b) Using $\mathbb{W}_G$, calculate the two- and three-dimensional configurations.
    (c) Calculate T, W and B, and show that (8.30) holds.
38. For an Analysis of Distance, make use of Gower (1968), and describe how the configuration of the grouped data is extended to all data. Illustrate on Fisher's iris data.
39. For the *ovarian cancer proteomics* data of Examples 8.11 and 8.12, use the groups defined by the Hamming distance.
    (a) Give explicit expression – including dimensions – for $\mathbb{X}_G$, K_G, $Q_{<G,\kappa>}$ and $\mathbb{W}_G$.
    (b) Determine the singular values of $\mathbb{X}_G$, and compare the singular values of $N\mathbb{X}_G$ with the first κ singular values of $\mathbb{X}$.
    (c) Explain the relationship between the matrix of left eigenvectors U_G of $\mathbb{X}_G$ and the matrix of left eigenvectors $\widetilde{U}$ of Proposition 8.17.

Part III

Non-Gaussian Analysis

9

Towards Non-Gaussianity

Man denkt an das, was man verließ; was man gewohnt war, bleibt ein Paradies (Johann Wolfgang von Goethe, *Faust II*, 1749–1832). *We think of what we left behind; what we are familiar with remains a paradise.*

9.1 Introduction

Gaussian random vectors are special: uncorrelated Gaussian vectors are independent. The difference between independence and uncorrelatedness is subtle and is related to the deviation of the distribution of the random vectors from the Gaussian distribution.

In Principal Component Analysis and Factor Analysis, the variability in the data drives the search for low-dimensional projections. In the next three chapters the search for direction vectors focuses on independence and deviations from Gaussianity of the low-dimensional projections:

- Independent Component Analysis in Chapter 10 explores the close relationship between independence and non-Gaussianity and finds directions which are as independent and as non-Gaussian as possible;
- Projection Pursuit in Chapter 11 ignores independence and focuses more specifically on directions that deviate most from the Gaussian distribution.
- the methods of Chapter 12 attempt to find characterisations of independence and integrate these properties in the low-dimensional direction vectors.

As in Parts I and II, the introductory chapter in this final part collects and summarises ideas and results that we require in the following chapters.

We begin with a visual comparison of Gaussian and non-Gaussian data. For a non-zero random vector $\mathbf{X}$, the **direction of X**

$$\operatorname{dir}(\mathbf{X}) = \frac{\mathbf{X}}{\|\mathbf{X}\|} \tag{9.1}$$

is a unit vector. Similarly, the **direction of the centred vector** $\mathbf{X} - \boldsymbol{\mu}$, or $\mathbf{X} - \overline{\mathbf{X}}$ in the sample case, is

$$\widetilde{\mathbf{X}} = \operatorname{dir}(\mathbf{X} - \boldsymbol{\mu}) = \frac{\mathbf{X} - \boldsymbol{\mu}}{\|\mathbf{X} - \boldsymbol{\mu}\|} \quad \text{or} \quad \widetilde{\mathbf{X}} = \operatorname{dir}(\mathbf{X} - \overline{\mathbf{X}}) = \frac{\mathbf{X} - \overline{\mathbf{X}}}{\|\mathbf{X} - \overline{\mathbf{X}}\|}, \tag{9.2}$$

provided that the centred vector is non-zero. The directions of vectors produce patterns on the surface of the unit sphere. Figure 9.1 shows the directions of 2,000 samples of three-dimensional centred data. Starting from the left, we have mixtures of one, two and three

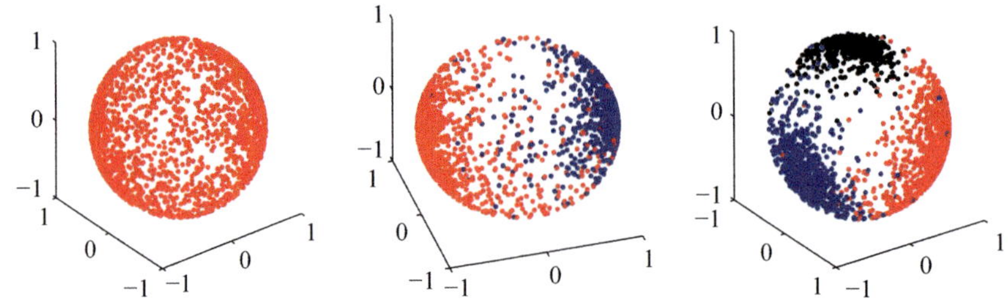

Figure 9.1 Mixtures of one (*left*), two (*middle*) and three Gaussians (*right*); each Gaussian in the mixtures is shown in a different colour; the data are directions of the centred vectors.

Gaussians, respectively. For the bimodal data, 60 per cent are from the 'red' population with mean $(0,0,0)$, and the remaining 40 per cent are from the 'blue' population with mean $(5,0,0)$. The covariance matrices are given in Example 11.3 of Section 11.4.3: Σ_1 describes the variation in the 'red' population and Σ_2 that of the 'blue' and 'black' populations. It is clear that there is almost a void between these two regions. In the right subplot, the red data make up 40 per cent, the blue 35 per cent and the black the remaining 25 per cent, with red centred at the $(0,0,0)$, blue at $(-5,0,0)$ and black at $(0,-4,3)$. These *continents of colour* are well separated.

A little reflection tells us that the directions of centred Gaussian vectors are uniformly distributed on the surface of the unit sphere if the Gaussian distribution is spherical, whereas sphered Gaussian vectors lie on concentric spheres.

If we wanted to choose one or more directions or projections for these three data sets, an obvious choice would be vectors which point to the centres of the clusters in the mixture cases. But no such option exists for the Gaussian data on the left. Diaconis and Freedman (1984) stated that 'heuristically, a projection will be uninteresting if it is random or unstructured', and quoting Huber (1985): 'interesting projections are far from the Gaussian.'

Since the 1980s, the pursuit of non-Gaussian projections, directions and subspaces has blossomed: Projection Pursuit and Projection Pursuit Regression had their beginnings in the statistical sciences, and Independent Component Analysis and Blind Source Separation originated in engineering and the computer sciences. The exploratory aspects of these methods, the search for interesting directions and low-dimensional structure in data, have been complemented by theoretical developments which describe properties of interesting projections.

9.2 Gaussianity and Independence

Gaussian models play an important role in the development of statistical methodology and theory. Under a Gaussian model, the probability density function or the likelihood allows us to make inferences, and we can derive asymptotic properties of sample-based estimators. Theoretical developments in Multivariate Analysis in particular rely heavily on Gaussian assumptions. Unfortunately, the Gaussian model may not be realistic when we consider high-dimensional data.

It is important to understand what we lose when the assumption of Gaussianity is dropped. An obvious loss is the equivalence of independent and uncorrelated variables. The statements in the next result are specific to Gaussian vectors and are discussed in Comon (1994).

Result 9.1 [Comon (1994)] *Let* $\mathbf{S} \sim \left(\mathbf{0}, \mathbf{I}_{p\times p}\right)$ *be a p-variate Gaussian random vector. For* $d \geq p$, *let* A *be a* $d \times p$ *matrix. Put* $\mathbf{X} = A\mathbf{S}$, *and assume that the variables of* $\mathbf{X}$ *are independent.*

1. *Then* $A = \Lambda E \Delta$, *where* Λ *and* Δ *are diagonal* $d \times d$ *and* $p \times p$ *matrices, respectively, and the* $d \times p$ *matrix* E *is p-orthogonal and so satisfies* $E^{\mathsf{T}} E = \mathbf{I}_{p\times p}$.
2. *If* $d = p$, *then* Λ *is invertible and* E *is orthogonal.*
3. *If* $d = p$ *and* $\mathbf{X}$ *has the identity covariance matrix, then* A *is orthogonal.*

Part 3 of this result tells us that independent Gaussian vectors are related by an orthogonal matrix. Comon (1994) noted that this result is specific to Gaussian random vectors. In Section 10.2 we consider a non-Gaussian counterpart of this result.

9.3 Skewness, Kurtosis and Cumulants

In addition to the mean and variance, we consider the skewness and kurtosis, the third- and fourth-order central moments of random vectors. For Gaussian vectors, the skewness is zero, and the kurtosis is adjusted to be zero, so generally these two quantities do not play an important role in Gaussian theory. However, skewness and kurtosis both have been used as measures of the deviation from normality. A number of definitions for multivariate skewness and kurtosis are in use, but a commonly accepted 'winner' has not yet emerged. I follow Malkovich and Afifi (1973) as they propose a natural generalisation of the one-dimensional definitions. We start with the population case.

Definition 9.2 Let $\mathbf{X} \sim (\boldsymbol{\mu}, \Sigma)$ be a d-variate random vector. Let $\boldsymbol{\alpha}$ be a unit vector in $\mathbb{R}^d$. The **third central moment** B_3 **of X in the direction** $\boldsymbol{\alpha}$ is

$$B_3(\mathbf{X}, \boldsymbol{\alpha}) = \mathbb{E}\left\{\left[\frac{\boldsymbol{\alpha}^{\mathsf{T}}(\mathbf{X} - \boldsymbol{\mu})}{\sqrt{\boldsymbol{\alpha}^{\mathsf{T}}\Sigma\boldsymbol{\alpha}}}\right]^3\right\}, \tag{9.3}$$

and the **skewness** $\boldsymbol{\beta}_3$ **of X** is

$$\beta_3(\mathbf{X}) = \max_{\{\boldsymbol{\alpha}:\, \|\boldsymbol{\alpha}\|=1\}} |B_3(\mathbf{X}, \boldsymbol{\alpha})|. \tag{9.4}$$

where the maximum is taken over all unit vectors $\boldsymbol{\alpha} \in \mathbb{R}^d$.

The **fourth central moment** B_4 **of X in the direction** $\boldsymbol{\alpha}$ is

$$B_4(\mathbf{X}, \boldsymbol{\alpha}) = \mathbb{E}\left\{\left[\frac{\boldsymbol{\alpha}^{\mathsf{T}}(\mathbf{X} - \boldsymbol{\mu})}{\sqrt{\boldsymbol{\alpha}^{\mathsf{T}}\Sigma\boldsymbol{\alpha}}}\right]^4 - 3\right\}, \tag{9.5}$$

and the **kurtosis** $\boldsymbol{\beta}_4$ **of X** is

$$\beta_4(\mathbf{X}) = \max_{\{\boldsymbol{\alpha}:\, \|\boldsymbol{\alpha}\|=1\}} |B_4(\mathbf{X}, \boldsymbol{\alpha})|, \tag{9.6}$$

where the maximum is taken over all unit vectors $\boldsymbol{\alpha} \in \mathbb{R}^d$. □

Originally, the multivariate skewness was defined as the square of β_3, but many authors now use it in the form given here. In the definition of the kurtosis I have included the term -3 as this results in the multivariate normal distribution having zero kurtosis.

In the univariate case, skewness and kurtosis are the standardised central third and fourth moments, respectively; and standardisation refers to dividing by the appropriate power of the variance σ^2 as in (9.3) and (9.5).

Definition 9.3 Let $\mathbb{X} = \begin{bmatrix}\mathbf{X}_1 \cdots \mathbf{X}_n\end{bmatrix}$ be a sample of d-dimensional random vectors with sample mean $\overline{\mathbf{X}}$ and sample covariance matrix S. Let $\boldsymbol{\alpha}$ be a unit vector in $\mathbb{R}^d$.

The **sample third central moment $\widehat{\mathbf{B}}_3$ of $\mathbb{X}$ in the direction $\boldsymbol{\alpha}$** is

$$\widehat{B}_3(\mathbb{X},\boldsymbol{\alpha}) = \widehat{B}_3([\mathbf{X}_1,\ldots,\mathbf{X}_n],\boldsymbol{\alpha}) = \frac{1}{n}\sum_{i=1}^{n}\left[\frac{\boldsymbol{\alpha}^{\mathsf{T}}(\mathbf{X}_i - \overline{\mathbf{X}})}{\sqrt{\boldsymbol{\alpha}^{\mathsf{T}}S\boldsymbol{\alpha}}}\right]^3, \tag{9.7}$$

and the **sample skewness b_3** of $\mathbb{X}$ is

$$b_3(\mathbb{X}) = b_3(\mathbf{X}_1,\ldots,\mathbf{X}_n) = \max_{\{\boldsymbol{\alpha}:\,\|\boldsymbol{\alpha}\|=1\}}\left|\widehat{B}_3([\mathbf{X}_1,\ldots,\mathbf{X}_n],\boldsymbol{\alpha})\right|. \tag{9.8}$$

The **sample fourth central moment $\widehat{\mathbf{B}}_4$ of $\mathbb{X}$ in the direction $\boldsymbol{\alpha}$** is

$$\widehat{B}_4(\mathbb{X},\boldsymbol{\alpha}) = \widehat{B}_4([\mathbf{X}_1,\ldots,\mathbf{X}_n],\boldsymbol{\alpha}) = \frac{1}{n}\sum_{i=1}^{n}\left[\frac{\boldsymbol{\alpha}^{\mathsf{T}}(\mathbf{X}_i - \overline{\mathbf{X}})}{\sqrt{\boldsymbol{\alpha}^{\mathsf{T}}S\boldsymbol{\alpha}}}\right]^4 - 3, \tag{9.9}$$

and the **sample kurtosis $\mathbf{b}_4$** of $\mathbb{X}$ is

$$b_4(\mathbb{X}) = b_4(\mathbf{X}_1,\ldots,\mathbf{X}_n) = \max_{\{\boldsymbol{\alpha}:\,\|\boldsymbol{\alpha}\|=1\}}\left|\widehat{B}_4([\mathbf{X}_1,\ldots,\mathbf{X}_n],\boldsymbol{\alpha})\right|. \tag{9.10}$$

□

In analogy with the notation for the sample mean and the sample covariance matrix, we use b_3 and b_4 for the sample skewness and kurtosis – rather than $\widehat{\beta}_3$ and $\widehat{\beta}_4$, which suggest estimators of the population parameters. Both skewness and kurtosis are scalar quantities; these are easier to work with than the 'raw' skewness or kurtosis, which are defined without using the direction vectors and turn out to be tensors.

The scalar quantities β_k are closely related to third- and fourth- order cumulants. We restrict attention to centred random vectors in the following definitions. For non-centred random vectors, similar moments exist, but we do not require these.

Definition 9.4 Let $\mathbf{X} = [X_1 \cdots X_d]^{\mathsf{T}}$ be a mean zero random vector, and assume that the following moment quantities are finite:

1. For entries X_i, X_j of $\mathbf{X}$, put

$$C_{ij}(\mathbf{X}) = \mathbb{E}(X_i X_j).$$

   The C_{ij} are the **second-order cumulants** of the entries of $\mathbf{X}$.
2. For entries X_i, X_j and X_k of $\mathbf{X}$, put

$$C_{ijk}(\mathbf{X}) = \mathbb{E}(X_i X_j X_k).$$

   The C_{ijk} are the **third-order cumulants** of the entries of $\mathbf{X}$.

3. For any four entries X_i, X_j, X_k and X_l of $\mathbf{X}$, put

$$C_{ijkl}(\mathbf{X}) = \mathbb{E}(X_i X_j X_k X_l) - \mathbb{E}(X_i X_j)\mathbb{E}(X_k X_l) - \mathbb{E}(X_i X_k)\mathbb{E}(X_j X_l) - \mathbb{E}(X_i X_l)\mathbb{E}(X_j X_k).$$

The C_{ijkl} are the **fourth-order cumulants** of the entries of $\mathbf{X}$. □

The next result relates moments and cumulants. For details, see McCullagh and Kolassa (2009).

Result 9.5 *Let* $\mathbf{X} = [X_1 \cdots X_d]^{\mathsf{T}}$ *be a mean zero random vector with marginal variances* σ_j^2. *For* $k = 3,4$ *and* $j \leq d$, *let* $\beta_{k,j}(\mathbf{X})$ *be the skewness, if* $k = 3$ *or the kurtosis of* X_j *when* $k = 4$. *The following hold:*

1. $$\beta_{3,j}(\mathbf{X}) = \frac{C_{jjj}(\mathbf{X})}{[\mathrm{var}(X_j)]^{3/2}} = \frac{\mathbb{E}(X_j^3)}{\sigma_j^3}.$$

2. $$\beta_{4,j}(\mathbf{X}) = \frac{C_{jjjj}(\mathbf{X})}{[\mathrm{var}(X_j)]^2} = \frac{\mathbb{E}(X_j^4) - 3\left[\mathbb{E}(X_j^2)\right]^2}{\sigma_j^4}.$$

3. *If* $\mathbf{X}$ *has independent entries, then*

$$C_{ijkl}(\mathbf{X}) = \beta_{4,i}(\mathbf{X})\sigma_j^4 \delta_{ijkl},$$

where δ *is the Kronecker delta function in four indices which satisfies* $\delta_{ijkl} = 1$ *if all four indices are the same and* $\delta_{ijkl} = 0$ *otherwise.*

9.4 Entropy and Mutual Information

The concepts I introduce in this section make explicit use of the probability density function f of a random vector. Details and proofs of the results quoted in this section can be found in chapters 2 and 8 of Cover and Thomas (2006).

Definition 9.6 Let $\mathbf{X} \sim (\boldsymbol{\mu}, \Sigma)$ be a d-dimensional random vector with probability density function f. For $j \leq d$, let f_j be marginals, or marginal probability density functions, of f. Let $\mathbf{X}_{\mathrm{Gauss}} \sim \mathcal{N}(\boldsymbol{\mu}, \Sigma)$ be a d-dimensional Gaussian random vector with the same mean and covariance matrix as $\mathbf{X}$, and let f_G be the probability density function of $\mathbf{X}_{\mathrm{Gauss}}$.

1. The **entropy** $\mathcal{H}$ of f is

$$\mathcal{H}(f) = -\int f \log f. \tag{9.11}$$

2. The **negentropy** $\mathcal{J}$ of f is

$$\mathcal{J}(f) = \mathcal{H}(f_G) - \mathcal{H}(f). \tag{9.12}$$

3. The **mutual information** $\mathcal{I}$ of f is

$$\mathcal{I}(f) = \sum_{j=1}^{d} \mathcal{H}(f_j) - \mathcal{H}(f). \tag{9.13}$$

The **entropy** of $\mathbf{X}$ is the entropy of its probability density function f, and similarly, the **negentropy** and **mutual information** of $\mathbf{X}$ are those of f. □

The entropy (9.11) is defined for continuous random vectors and is given by an integral. For discrete random vectors, the integrals are replaced by sums. To distinguish the entropy for continuous and discrete random vectors, the former is sometimes called the **differential entropy**. For discrete random vectors, the negentropy and the mutual information are also sums. The following results are not affected by the transition from continuous to discrete random vectors.

Some authors define entropy as a function of the random vector $\mathbf{X}$ rather than its probability density function f, as I have done. The definitions are equivalent but present different ways of looking at the entropy, and both approaches have merit. Viewing entropy as a function of the random vectors naturally leads to $\mathcal{H}(\mathbf{X}) = \mathbb{E}[\log f(\mathbf{X})]$, the expectation of the random vector $\log f(\mathbf{X})$.

Result 9.7 *Let $\mathbf{X} \sim (\boldsymbol{\mu}, \Sigma)$ be a d-dimensional random vector with probability density function f and marginals f_j.*

1. *Let $\mathcal{F}_{\mathbf{I}}$ be the set of probability density functions of d-dimensional random vectors with the identity covariance matrix $\mathbf{I}_{d\times d}$. The maximiser f of the entropy $\mathcal{H}$ over the set $\mathcal{F}_{\mathbf{I}}$ is Gaussian.*
2. *Let A be an invertible $d \times d$ matrix. Define a transformation $\tau : \mathbf{X} \mapsto A\mathbf{X}$. If f_τ is the probability density function of $A\mathbf{X}$, then*
$$\mathcal{H}(f_\tau) = \mathcal{H}(f) + \log[\det(A)].$$
3. *Assume that Σ is invertible. Let f_Σ be the probability density function of the sphered random vector $\mathbf{X}_\Sigma = \Sigma^{-1/2}(\mathbf{X} - \boldsymbol{\mu})$. The entropies of f and f_Σ are related by*
$$\mathcal{H}(f_\Sigma) = \mathcal{H}(f) - \frac{1}{2}\log[\det(\Sigma)].$$
4. *The negentropy $\mathcal{J} \geq 0$; and $\mathcal{J} = 0$ if and only if f is Gaussian.*
5. *The mutual information $\mathcal{I} \geq 0$; and $\mathcal{I} = 0$ if and only if $f = \prod f_j$.*

This result shows that entropy is the building block. The negentropy compares the entropy of f with that of its Gaussian counterpart and so provides a measure of the deviation from the Gaussian. The mutual information compares the entropy of f with that of the marginals of f and so measures the deviation from independence. Part 3 of Result 9.7 is a special case of part 2. Because the result is commonly used in the form given in part 3, I have explicitly quoted this version.

The functions $\mathcal{H}, \mathcal{J}$ and $\mathcal{I}$ relate f to its marginals or to its Gaussian counterpart. Our next concept compares two arbitrary functions in d variables.

Definition 9.8 Let f and g be probability density functions of d-dimensional random vectors. The **Kullback-Leibler divergence** $\mathcal{K}$ of f from g is
$$\mathcal{K}(f,g) = \int f \log\left(\frac{f}{g}\right). \tag{9.14}$$
□

The Kullback-Leibler divergence is also called the **Kullback-Leibler distance** or the **relative entropy**. I prefer to use the term *divergence*, because $\mathcal{K}$ is not symmetric in the two arguments and therefore not a distance. The Kullback-Leibler divergence is used as a measure of the difference of two functions and applies to a wide class of measurable functions, a fact we do not require. We are more interested in the relationship between $\mathcal{K}$ and the mutual information $\mathcal{I}$.

Result 9.9 *Let f and g be probability density functions of d-dimensional random vectors. Let f_j be the marginals of f. The following hold.*

1. *The Kullback-Leibler divergence $\mathcal{K} \geq 0$, and $\mathcal{K}(f,g) = 0$ if and only if $f = g$.*
2. *The mutual information and the Kullback-Leibler divergence satisfy $\mathcal{I}(f) = \mathcal{K}(f, \prod f_j)$.*
3. *If g^* is the minimiser of $\mathcal{K}(f,g)$ over product densities $g = \prod g_j$, then $g^* = \prod f_j$.*
4. *Let $\mathbf{X}$ and $\mathbf{Y}$ be d-dimensional random vectors with probability density functions f and g, respectively. Consider the transformation $\tau : \mathbf{Z} \mapsto A\mathbf{Z} + \mathbf{a}$, where $\mathbf{Z}$ is $\mathbf{X}$ or $\mathbf{Y}$ as appropriate, A is a non-singular $d \times d$ matrix and $\mathbf{a} \in \mathbb{R}^d$. Write f_τ and g_τ for the probability density functions corresponding to the transformed random vectors. Then*

$$\mathcal{K}(f_\tau, g_\tau) = \mathcal{K}(f,g).$$

Part 2 of this result shows the relationship between $\mathcal{I}$ and $\mathcal{K}$, and part 4 shows the invariance of the Kullback-Leibler divergence under invertible affine transformations. Part 3 exhibits the best approximation to f by product densities with respect to the Kullback-Leibler divergence. Of interest is also the best Gaussian approximation to f. I give this result in part 1 of Theorem 10.9 in Section 10.4.1, which explains the relationship between independence and non-Gaussianity further.

9.5 Training, Testing and Cross-Validation

The notion of *training and testing* is probably as old as classification and regression and occurs naturally in prediction. In classification or regression prediction, we consider individual new observations; in contrast, a large proportion of the data may need to be tested in Supervised Learning.

Classically, we begin with data $\mathbb{X}$ that belong to distinct classes or have continuous responses. Call the class labels or the responses $\mathbb{Y}$. In the **learning phase**, we use $\mathbb{X}$ and $\mathbb{Y}$ to derive a rule such as Fisher's discriminant rule or the coefficients in linear regression. In the **prediction phase**, we apply the previously established classification or regression rule to new data $\mathbb{X}_{\text{new}}$ in order to predict responses $\mathbb{Y}_{\text{new}}$.

The classification error and the sum of squared errors are typical criteria for assessing the error during the learning phase. In Statistical Learning, we divide $\mathbb{X}$ into two disjoint subsets called the **training set** $\mathbb{X}_0$ and the **testing set** $\mathbb{X}_{\text{tt}}$, with subscript *tt* for testing. We calculate the classification errors or sum of squared errors as appropriate for the training set; the split of the data now allows us to calculate an error for the testing set, and we can thus assess how well the rule performs 'out of sample'. The split of the data into training and testing subsets can be done in many different ways. I will not discuss strategies for these splits but assume that the data have been divided into two parts.

9.5.1 Rules and Prediction

Let $\mathbb{X} = \begin{bmatrix}\mathbf{X}_1 \cdots \mathbf{X}_n\end{bmatrix}$ be $d \times n$ data, and let $\mathbb{Y} = \begin{bmatrix}\mathbf{Y}_1 \cdots \mathbf{Y}_n\end{bmatrix}$ be $q \times n$ responses. For regression, we allow vector-valued responses, so $q \geq 1$. For classification, $q = 1$, and the labels have values $1, \ldots, \kappa$, where κ is the number of classes. I refer to both the classification labels and the regression responses as *responses* and use Fisher's discriminant rule and linear regression to explain the notion of rules and prediction. For details on Fisher's discriminant rule, see Section 4.3.2. Working with specific rules allows us to see how the classification and regression rules are constructed and how they apply to new data. The two rules can be substituted by other discriminant and regression rules. The pattern will be similar, but the details will vary.

Let $\mathbb{X}_0$ be the training data, which consist of n_0 samples from $\mathbb{X}$, and let $\mathbb{X}_{\text{tt}}$, the testing set, consist of the remaining $n_p = n - n_0$ samples. Let $\mathbb{Y}_0$ be the responses belonging to $\mathbb{X}_0$, and let $\mathbb{Y}_{\text{tt}}$ be the corresponding responses for $\mathbb{X}_{\text{tt}}$.

Classification. Assume that $\mathbb{X} = \begin{bmatrix}\mathbf{X}_1 \cdots \mathbf{X}_n\end{bmatrix}$ belong to κ classes. For $\nu \leq \kappa$, let $\overline{\mathbf{X}}_{0,\nu}$ be the sample mean of the random vectors from $\mathbb{X}_0$ which belong to the νth class. For $\mathbf{X}_{0,i}$ from the training set $\mathbb{X}_0$, Fisher's discriminant rule $\mathfrak{r}_F$ satisfies

$$\mathfrak{r}_F(\mathbf{X}_{0,i}) = \ell \qquad \text{if } \left|\widehat{\eta}_0^{\mathsf{T}}\mathbf{X}_{0,i} - \widehat{\eta}_0^{\mathsf{T}}\overline{\mathbf{X}}_{0,\ell}\right| = \min_{\nu \leq \kappa} \left|\widehat{\eta}_0^{\mathsf{T}}\mathbf{X}_{0,i} - \widehat{\eta}_0^{\mathsf{T}}\overline{\mathbf{X}}_{0,\nu}\right|,$$

where $\widehat{\eta}_0$ is the eigenvector of $\widehat{W}_0^{-1}\widehat{B}_0$ which corresponds to the largest eigenvalue of $\widehat{W}_0^{-1}\widehat{B}_0$. The matrices $\widehat{B}_0$ and $\widehat{W}_0$ are defined as in Corollary 4.9 in Section 4.3.2, and the subscript 0 indicates that the matrices are calculated from $\mathbb{X}_0$ only.

For an observation $\mathbf{X}_{\text{new}} = \mathbf{X}_{\text{tt},i}$ from $\mathbb{X}_{\text{tt}}$, Fisher's rule leads to the prediction

$$\mathfrak{r}_F(\mathbf{X}_{\text{new}}) = m \qquad \text{if } \left|\widehat{\eta}_0^{\mathsf{T}}\mathbf{X}_{\text{new}} - \widehat{\eta}_0^{\mathsf{T}}\overline{\mathbf{X}}_{0,m}\right| = \min_{\nu \leq \kappa} \left|\widehat{\eta}_0^{\mathsf{T}}\mathbf{X}_{\text{new}} - \widehat{\eta}_0^{\mathsf{T}}\overline{\mathbf{X}}_{0,\nu}\right|. \tag{9.15}$$

Linear Regression. Assume that the $\mathbb{X}$ are centred. Let χ be a transformation defined on $\mathbb{X}$, and put $\widetilde{\mathbb{X}} = \chi(\mathbb{X})$. The transformation χ could be the identity map, or it could be the projection of $\mathbb{X}$ onto canonical correlation projections, as in (3.41) of Section 3.7.2.

Assume that $\widetilde{\mathbb{X}}$ and $\mathbb{Y}$ are linearly related so that $\mathbb{Y} = B^{\mathsf{T}}\widetilde{\mathbb{X}}$. Put $\widetilde{\mathbb{X}}_0 = \chi(\mathbb{X}_0)$ for the training data $\mathbb{X}_0$. For $\widetilde{\mathbf{X}}_{0,i}$ from $\widetilde{\mathbb{X}}_0$, the least squares regression rule yields

$$\mathfrak{r}_{LS}(\widetilde{\mathbf{X}}_{0,i}) = \widehat{\mathbf{Y}}_i,$$

where

$$\widehat{\mathbf{Y}}_i = \widehat{B}_0^{\mathsf{T}}\widetilde{\mathbf{X}}_{0,i} \qquad \text{and} \qquad \widehat{B}_0 = (\widetilde{\mathbb{X}}_0\widetilde{\mathbb{X}}_0^{\mathsf{T}})^{-1}\widetilde{\mathbb{X}}_0\mathbb{Y}_0^{\mathsf{T}}.$$

For an observation $\mathbf{X}_{\text{new}} = \mathbf{X}_{\text{tt},i}$ belonging to the testing set $\mathbb{X}_{\text{tt}}$, put $\widetilde{\mathbf{X}}_{\text{new}} = \chi(\mathbf{X}_{\text{new}})$. The predicted value is

$$\mathfrak{r}_{LS}(\mathbf{X}_{\text{new}}) = \widehat{B}_0^{\mathsf{T}}\widetilde{\mathbf{X}}_{\text{new}} = \left[\mathbb{Y}_0\widetilde{\mathbb{X}}_0^{\mathsf{T}}(\widetilde{\mathbb{X}}_0\widetilde{\mathbb{X}}_0^{\mathsf{T}})^{-1}\right]\widetilde{\mathbf{X}}_{\text{new}}. \tag{9.16}$$

9.5.2 Evaluating Rules with the Cross-Validation Error

In classification and regression, (9.15) and (9.16) show explicitly how the training data are used for prediction. To find out how good this prediction is, we explore criteria for assessing the performance of a rule on the testing data.

For data $\mathbb{X} = [\mathbf{X}_1 \ldots \mathbf{X}_n]$, responses $\mathbb{Y}$, and the rule $\mathfrak{r} = \mathfrak{r}_F$ or $\mathfrak{r} = \mathfrak{r}_{LS}$ as in the preceding section, the error e_i of the ith observation $\mathbf{X}_i$ is

$$e_i = \begin{cases} c_i \left| \operatorname{sgn}[\mathbf{Y}_i - \mathfrak{r}(\mathbf{X}_i)] \right| & \text{for class labels } \mathbf{Y}_i, \\ w_i \|\mathbf{Y}_i - \mathfrak{r}(\mathbf{X}_i)\|_p^p & \text{for regression responses } \mathbf{Y}_i, \end{cases} \tag{9.17}$$

where $\mathbf{c} = [c_1, \ldots, c_n]$ is the cost factor of Definition 4.12 in Section 4.4.2, $\operatorname{sgn}(\mathbf{0}) = 0$ for the zero vector and the w_i are positive weights. The regression error is given by an ℓ_p-norm (see Definition 5.1 in Section 5.2), and typically $p = 2$, or $p = 1$.

Definition 9.10 Let $\mathbb{X} = [\mathbf{X}_1 \cdots \mathbf{X}_n]$ be data with class labels or regression responses. Let $\mathfrak{r}$ be a discriminant or regression rule for $\mathbb{X}$, as appropriate. Let k and m be integers such that $km = n$. For $j \le m$, define the ***j*th training set** $\mathbb{X}_{0,j}$ and the ***j*th testing set** $\mathbb{X}_{\text{tt},j}$ by

$$\begin{aligned} \mathbb{X}_{0,j} &= \left[\mathbf{X}_1 \cdots \mathbf{X}_{(j-1)k} \quad \mathbf{X}_{jk+1} \cdots \mathbf{X}_n\right] \\ \mathbb{X}_{\text{tt},j} &= \left[\mathbf{X}_{(j-1)k+1} \cdots \mathbf{X}_{jk}\right]. \end{aligned}$$

Let $\mathfrak{r}_j$ be defined as $\mathfrak{r}$ but derived on the training data $\mathbb{X}_{0,j}$ only, whereas $\mathfrak{r}$ is obtained from $\mathbb{X}$. Use the rule $\mathfrak{r}_j$, instead of $\mathfrak{r}$, to calculate the classification or regression error e_i as in (9.17) for each $\mathbf{X}_i$ from the testing set $\mathbb{X}_{\text{tt},j}$. The error $\mathcal{E}_j$ on $\mathbb{X}_{\text{tt},j}$ is

$$\mathcal{E}_j = \sum_{i=(j-1)k+1}^{jk} e_i. \tag{9.18}$$

The ***m*-fold cross-validation error** $\mathcal{E}_{\text{cv}}$ is

$$\mathcal{E}_{\text{cv}} = \frac{1}{n} \sum_{j=1}^{m} \mathcal{E}_j, \tag{9.19}$$

and ***m*-fold cross-validation** is the m-fold partitioning of the data $\mathbb{X}$ into training and testing pairs $\left\{\mathbb{X}_{0,j}, \mathbb{X}_{\text{tt},j}\right\}$ together with their rules $\mathfrak{r}_j$ and errors $\mathcal{E}_j$ and $\mathcal{E}_{cv}$. □

In m-fold cross-validation, the data are systematically divided into training and testing sets, and in each partitioning of $\mathbb{X}$, the testing set consists of k random vectors. The testing sets for different values of $j \le m$ are disjoint. As a consequence, each observation $\mathbf{X}_i$ from the original data $\mathbb{X}$ contributes exactly once to the m-fold cross-validation error, and the error over the jth testing set depends on the rule $\mathfrak{r}_j$ only.

A little reflection shows that the leave-one-out approach of Definition 4.13 in Section 4.4.2 is the special case of m-fold cross-validation with $k = 1$ and $m = n$. For this reason, the leave-one-out approach is also called *n-fold cross-validation*.

The number m determines the number of partitions and rules one has to calculate; large values m result in more computational effort. On the other hand, if k becomes too large, then the rules are derived on small training sets, which affects the accuracy of the performance. A reasonable compromise is ten- to twenty-fold cross-validation so that 10 to 20 per cent of the data are left for testing in each partitioning.

Cross-validation can be highly computer-intensive. For this reason, users sometimes select a single partitioning of $\mathbb{X}$ into training and testing sets. Such an approach is not so appropriate if one wants to make comparisons between different algorithms or learning methods. The relative size of the training and testing sets depends on a number of factors, including the total number of observations. In classification, the number of classes and the proportion of samples in each class also play a role. If we consider data from two classes, where one class is very much larger than the other, then a judicious choice of the training and testing sets is important. Unequal proportions in the two classes arise, for example, in bank loan data, where customers are classified according to whether they repay on time or default.

Chapter 4 focused mostly on basic discrimination rules which can be used to construct more advanced rules. Discriminant Analysis and, more recently, Statistical Learning are fast-growing areas, and many excellent approaches have become available. At the time of this writing, WEKA data-mining software (see Witten and Frank 2005), and CRAN the R software libraries for statistical computing (see R Development Core Team 2005), contain freely available Statistical Learning software.

Below is a list of some commonly used approaches which the interested reader might find helpful in a search for other methods. Decision trees have an important place in classification and regression. Breiman et al. (1998) first published their approach, Classification and Regression Trees (CART), in 1984. For regression models and continuous responses, Friedman (1991) proposed Multivariate Additive Regression Splines (MARS). Bagging by Breiman (1996) is another classifier method and a powerful extension of CART, as are the Random Forests of Breiman (2001). The Random Forests have quickly gained great popularity because of their accuracy and theoretical foundation, but they are computationally expensive. Terms that have evolved for these methods are *learners* for a discriminant or regression rule and *ensemble learning* for the broader topic.

10

Independent Component Analysis

The truth is rarely pure and never simple (Oscar Wilde, *The Importance of Being Ernest*, 1854–1900).

10.1 Introduction

In the Factor Analysis model $\mathbf{X} = A\mathbf{F} + \boldsymbol{\mu} + \boldsymbol{\epsilon}$, an essential aim is to find an expression for the unknown $d \times k$ matrix of factor loadings A. Of secondary interest is the estimation of $\mathbf{F}$. If $\mathbf{X}$ comes from a Gaussian distribution, then the principal component (PC) solution for A and $\mathbf{F}$ results in independent scores, but this luxury is lost in the PC solution of non-Gaussian random vectors and data. Surprisingly, it is not the search for a generalisation of Factor Analysis, but the departure from Gaussianity that has paved the way for new developments.

In psychology, for example, scores in mathematics, language and literature or comprehensive tests are used to describe a person's intelligence. A Factor Analysis approach aims to find the underlying or hidden kinds of intelligence from the test scores, typically under the assumption that the data come from the Gaussian distribution. Independent Component Analysis, too, strives to find these hidden quantities, but under the assumption that the data are non-Gaussian. This assumption precludes the use of the Gaussian likelihood, and the independent component (IC) solution will differ from the maximum-likelihood (ML) Factor Analysis solution, which may not be appropriate for non-Gaussian data.

To get some insight into the type of solution one hopes to obtain with Independent Component Analysis, consider, for example, the superposition of sound tracks. Example 10.1 gives details of the two-dimensional data which arise from the sound tracks of a trumpet and an organ. The IC solution should be the original two tracks. For this example, we can compare the IC solution with the trumpet and organ tracks; in general, we do not have the 'originals', nor may such an explicitly interpretable decomposition exist. Instead, Independent Component Analysis may lead us to find interesting structure in the data that may not be explicit in a Principal Component or Factor Analysis solution. Example 13.3 in Section 13.2.2, for example, shows how we can exploit the IC solution to split the data into meaningful groups.

How do we arrive at IC solutions? For non-Gaussian random vectors $\mathbf{X}$, Independent Component Analysis aims to find non-Gaussian random vectors $\mathbf{S}$ with independent entries such that $\mathbf{X} = A\mathbf{S}$ for some unknown matrix A. Because its components are independent, $\mathbf{S}$ is called the *source* in the signal-processing literature, and $\mathbf{X}$ is regarded as the *signal*. In their ground-breaking papers, Hérault and Ans (1984) and Hérault, Jutten, and Ans (1985) proposed a powerful method, *Blind Source Separation*, for finding random vectors $\mathbf{S}$ with non-Gaussian independent components. Their approach stems from the signal-processing and neural-network environments, and aims to decode complex signals in an unsupervised way. Around 1986, Hérault and Jutten coined the phrase *Independent Component Analysis*, and both terms are common and are now used interchangeably. I prefer the name Independent Component Analysis, which emphasises the concepts 'independence' and 'component analysis'. As we shall see, much of our focus is directed towards an understanding of

- the relationship between non-Gaussianity and independence, and
- the connection between Principal Component Analysis and the new method.

Since the late 1980s, Independent Component Analysis has become a well-established method. Engineers, computer scientists and statisticians have contributed to the wealth of knowledge and algorithmic developments. The article 'Independent component analysis, A new concept?' by Comon (1994) was closely followed by Bell and Sejnowski (1995) and many papers separately or jointly by Amari and Cardoso, starting with Amari and Cardoso (1997). Among the other important early contributions are Lee (1998), Donoho (2000), Lee et al. (2000) and Hyvärinen, Karhunen, and Oja (2001). Parallel to the methodological developments, many algorithms have emerged; some are of a very general nature, whereas others are more application-specific. ICA Central (1999) started as an agency for promoting research on Independent Component Analysis, for collecting data and algorithms, and for making these available. The review article by Choi et al. (2005) includes a list of contributors and applications, and Klemm, Haueisen, and Ivanova (2009) compare the performance of twenty-two IC-based algorithms for electrical brain activity.

Independent of the developments in the signal-processing community, a search for interesting features and projections in data took place in the statistics community and lead to *Projection Pursuit*. There is a substantial overlap between Independent Component Analysis and Projection Pursuit, but the two methods are not the same. I treat them separately: Independent Component Analysis in this chapter and Projection Pursuit in Chapter 11.

Section 10.2 starts with a signal-source model for Independent Component Analysis and a motivating example. Section 10.3 looks at sources and the identification of sources. Section 10.4 focuses on mutual information and the closely related concepts of negentropy and Kullback-Leibler divergence as they hold the key to a differentiation between independence and uncorrelatedness and between Gaussian and non-Gaussian distributions. We consider theoretical developments and explore approximations to the mutual information which can be calculated from data. Section 10.5 introduces a semi-parametric framework for estimation of the mixing matrix and contains theoretical developments. Section 10.6 looks at real and simulated data for moderate dimensions. In Section 10.7 we turn to high-dimensional data and dimension reduction as a preliminary step in the analysis. The section also contains properties of low-dimensional projections as the dimension of the data grows. In our final Section 10.8, I explain how we can use ideas from Independent Component Analysis in dimension selection and dimension reduction. As done in previous chapters, Problems are listed at the end of Part III.

Many properties and theorems we learn about in this and later chapters are quite recent, and their proofs are too detailed to give in this exposition. For this reason, I will usually only name the author(s) and leave it to the reader to explore the proofs in the original papers.

10.2 Sources and Signals

10.2.1 Population Independent Components

Independent Component Analysis originated in the signal-processing community, and the terminology reflects these origins. Like Factor Analysis, Independent Component Analysis is model-based. We begin with the model for the population.

Definition 10.1 Let $\mathbf{X} \sim (\mathbf{0}, \Sigma)$ be a d-dimensional random vector. For $p \leq d$, a **p-dimensional independent component model** for $\mathbf{X}$ is

$$\mathbf{X} = A\mathbf{S}, \tag{10.1}$$

where $\mathbf{S} \sim \left(\mathbf{0}, \mathbf{I}_{p\times p}\right)$ is a p-dimensional random vector with independent entries, and A is a $d \times p$ matrix. We call $\mathbf{S}$ the **source (vector)**, A the **mixing matrix** and $\mathbf{X}$ the **(mixed) signal**. □

The source is pure in the sense that all its entries or variables are independent and have unit variance. Let us take a look at these two properties: a vector of independent entries has a diagonal covariance matrix which is not in general a multiple of the identity. If the covariance matrix of a random vector is the identity, then the variables are uncorrelated but typically not independent; an exception are vectors from the Gaussian distribution. Result 9.1 of Section 9.2 explores Gaussian sources together with independent signals. The independence requirement of the source is a departure from Factor Analysis, where the common factors have uncorrelated variables.

Originally, the number of signal and source variables was the same. If we want the mixing matrix A to be invertible – which simplifies the mathematical arguments – then we require $p = d$. We restrict attention to $p = d$ for now, and consider $p < d$ in later sections.

The goal is to recover the source $\mathbf{S}$ without knowledge of the transformation A, hence the name *Blind Source Separation*. We construct estimators $\widehat{B}$ of $B = A^{-1}$ and $\widehat{\mathbf{S}}$ of $\mathbf{S}$ such that

$$\widehat{B} \approx A^{-1} \quad \text{and} \quad \widehat{\mathbf{S}} = \widehat{B}\mathbf{X} \quad \text{has independent components.} \tag{10.2}$$

We call B the **unmixing** or **separating** matrix.

Before we look at details of Independent Component Analysis, a comparison of the basic assumptions in Principal Component Analysis, Factor Analysis and Independent Component Analysis will highlight similarities and differences between these methods. For convenience, we assume that $\mathbf{X}$ is centred. In Table 10.1, $\mathfrak{f}(\mathbf{X})$ represents the principal component vector, the common factor or the source, as appropriate for each method.

The Gaussian assumption, marked with an asterisk in Table 10.1, is not part of the definition of a factor model, but it is often tacitly made. I include this assumption here in order to emphasise the difference between the methods. If the Gaussian assumption is dropped in the factor model, then the components of the feature vector will not be independent but only uncorrelated. In Independent Component Analysis, we deviate from the Gaussian assumption but still desire independent sources.

For independent Gaussian sources, Result 9.1 in Section 9.2 tells us that the mixing matrix is of the form $A = \Lambda E \Delta$, where Λ and Δ are diagonal and E is orthogonal if $\mathbf{X}$ has independent components. For almost non-Gaussian sources, a tighter result holds.

Theorem 10.2 [Comon (1994)] *Let* $\mathbf{S}$ *be a* d*-dimensional source. Assume that at most one of the components of* $\mathbf{S}$ *is Gaussian and that the probability density functions of the*

Table 10.1 *Comparison of PCA, FA and ICA*

	PCA	FA	ICA
Model	Not explicit	$\mathbf{X} = A\mathbf{F} + \epsilon$	$\mathbf{X} = A\mathbf{S}$
Distribution of $\mathbf{X}$	No assumption	Gaussian*	Non-Gaussian
Vector $\mathfrak{f}(\mathbf{X})$	PC vector $\mathbf{W}^{(p)}$	Common factor $\mathbf{F}$	source $\mathbf{S}$
Dimension p of $\mathfrak{f}(\mathbf{X})$	$p \le d$	$p < d$	$p = d$
Components of $\mathfrak{f}(\mathbf{X})$	Uncorrelated	Independent	Independent

components are not point mass functions. Let E be an orthogonal $d \times d$ matrix. Put $\mathbf{T} = E\mathbf{S}$. Then statements (a)–(c) are equivalent.

(a) $\mathbf{T}$ has independent components.
(b) The components of $\mathbf{T}$ are pairwise independent.
(c) $E = \Delta P$, where Δ is a diagonal matrix, and P is a permutation matrix.

Comon's theorem is a uniqueness result which asserts that the only orthogonal transformation $E: \mathbf{S} \mapsto \mathbf{T}$, which leaves the non-Gaussian independent vector independent, is a permutation of the variables. Recall that a permutation matrix is derived from the identity matrix by permuting its columns or, equivalently, its rows.

A comparison of Result 9.1 in Section 9.2 and Theorem 10.2 reveals that any orthogonal transformation of a Gaussian independent vector results in another independent vector. For the Factor Analysis setting, Proposition 7.2 in Section 7.2 tells us that $\mathbf{F}$ and $E^{\top}\mathbf{F}$ are factor loadings provided that E is an orthogonal matrix. In light of Theorem 10.2, it follows that Proposition 7.2 does not extend to non-Gaussian sources and independent $\mathbf{T}$.

To prove Theorem 10.2, Comon (1994) used a subtle argument involving the relationship between Gaussian components and the form of the matrix E. He showed that if (b) holds and the matrix E is of more general form than $E = \Delta P$ in (c), then $\mathbf{S}$ has more than one Gaussian component. We explore Comon's theorem further in the Problems at the end of Part III.

10.2.2 Sample Independent Components

For the sample, we start with analogues of the sources and signals.

Definition 10.3 Let $\mathbb{X} = \left[\mathbf{X}_1\ \mathbf{X}_2 \cdots \mathbf{X}_n\right]$ be $d \times n$ data, with $\mathbf{X}_i \sim (\mathbf{0}, \Sigma)$ and $i \le n$. For $p \le d$, a ***p*-dimensional independent component model** for $\mathbb{X}$ is

$$\mathbb{X} = A\mathbb{S}, \tag{10.3}$$

where $\mathbb{S} = \left[\mathbf{S}_1\ \mathbf{S}_2 \cdots \mathbf{S}_n\right]$, each $\mathbf{S}_i \sim \left(\mathbf{0}, \mathbf{I}_{p\times p}\right)$ is a p-dimensional random vector with independent entries and A is a $d \times p$ matrix. We call $\mathbb{S}$ the **sources** or the **source data**, A the **mixing matrix** and $\mathbb{X}$ the **(matrix of mixed) signals**. □

The signals $\mathbb{X}$ of (10.3) correspond to the multivariate or high-dimensional data, and in the language of Factor Analysis, the sources or source data are the unknown common factors. I illustrate sources and signals with a simple example and will return to a description of the IC estimates of $\mathbb{S}$ in Section 10.6.

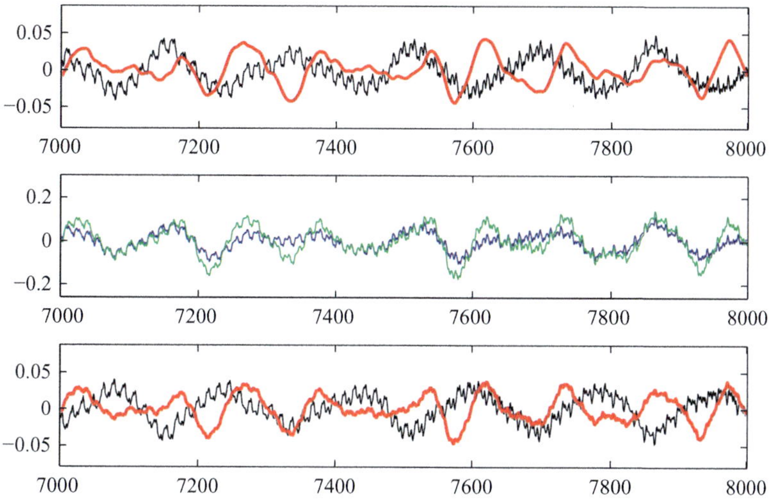

Figure 10.1 Sources (*top*), signals (*middle*) and IC estimates (*bottom*) for Example 10.1. Trumpet in black, and organ in red.

Example 10.1 We consider two **sound tracks** from Strauss' 'Also sprach Zarathustra', which was used as the film music of *2001: A Space Odyssey*. For the calculations, I use 24,000 samples recorded in time, which correspond to about 0.5 second of sound. The sound tracks are those of the trumpet and organ. The top panel of Figure 10.1 shows a subsample of 1,000 observations of both tracks, starting at time point 7,001. The trumpet is shown in black, and the organ in red.

I combine the sources with a mixing matrix

$$A_{(1)} = \begin{bmatrix} 1 & 2 \\ -1 & 1 \end{bmatrix}.$$

The signals are shown in the middle panel of Figure 10.1. The bottom panel of the figure shows the estimated $\widehat{\mathbb{S}}$ in the same colours as the sources in the top panel. The organ estimate looks similar to the source data, whereas the trumpet estimate differs by a sign from its source data. It is interesting to see how close the source estimates are to the actual sources. Example 10.3 describes how the estimates are calculated.

To illustrate the uniqueness result stated in Theorem 10.2, we consider a second mixing matrix

$$A_{(2)} = \begin{bmatrix} 2 & 1 \\ 2 & 3 \end{bmatrix},$$

and calculate the corresponding signals and IC source estimates. The signals (not shown) differ from those in the middle panel of Figure 10.1, but the source estimates they yield are almost identical to those estimated from the signals with the mixing matrix $A_{(1)}$. The top row of Figure 10.2 shows the source and the two IC estimates for the trumpet in the left panel and the organ in the right panel. The estimates are unique up to scale and permutation only, and for easier visual comparison, I have normalised the estimates to the scale of the sources. In both panels the source is shown in black, the estimate resulting from the mixing matrix $A_{(1)}$ is shown in red, and that from $A_{(2)}$ in blue. This time I have only displayed 500 observations (corresponding to the observations 7,001 to 7,500) for an easier visual comparison.

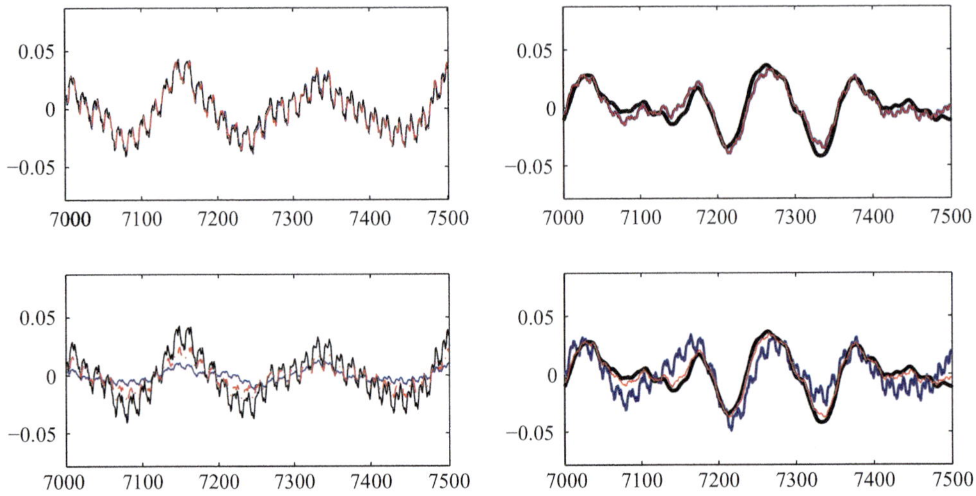

Figure 10.2 Source (*black*) and two estimates (*red* from $A_{(1)}$ and *blue* from $A_{(2)}$). IC estimates are shown in the top row and PC estimates in the bottom row, with the trumpet on the left and the organ on the right, from the sound tracks of Example 10.1.

The blue estimates are drawn before the red ones, and curves or lines disappear when later-drawn estimates are almost the same. In the top panels of Figure 10.2, the blue estimates are hardly visible because they are practically identical to the red ones, which suggests that there is essentially only one solution.

The bottom two panels in Figure 10.2 show the two principal component scores for comparison. The red scores are obtained from the signals generated with $A_{(1)}$, and the blue scores are those of the signals generated with $A_{(2)}$. As in the top row, I have normalised the scores. Because the sample covariance matrices of the two signals differ, the two sets of PC scores are distinct, and they deviate from the IC estimates. ■

The *sound tracks* example exhibits interesting properties of non-Gaussian sources:

1. There is essentially only one IC source estimate, and this estimate does not seem to depend on the mixing matrix.
2. PC estimates of the same independent sources differ from each other when the signals are obtained from different mixing matrices.
3. The IC estimates differ from the uncorrelated (but not independent) PC estimates.

A Word of Caution. Theorem 10.2 states the uniqueness of independent sources for the population. For data, these *ideal* conditions are not completely met. For example, the IC estimates displayed in Figure 10.2 are close to the sources, but they are not the same as the sources. This non-uniqueness is largely a consequence of estimating A^{-1} rather than using the correct quantity A^{-1}.

10.3 Identification of the Sources

In the *sound tracks* example, the sources are *estimated* from the signals. In Principal Component Analysis, algebraic solutions exist, and one uses the mean and covariance matrix of

$\mathbf{X}$ to find the solutions. In Independent Component Analysis algebraic solutions do not exist; instead, one employs properties of the signal beyond the mean and covariance matrix.

There are two main paths for finding independent component solutions which focus on the different unknowns:

- the identification of the source $\mathbf{S}$; and
- the estimation of the mixing matrix A.

The two paths are related because the estimation of one quantity allows the calculation of the other, however, the emphasis of the two paths differs. The second path is akin to the estimation problem in Factor Analysis, where we find the factor loadings A, but the approaches for estimating A differ greatly in Factor Analysis and Independent Component Analysis. We consider approaches to estimating A in Section 10.5.

In Sections 10.3 and 10.4, I focus on the identification of $\mathbf{S}$. The ideas and most of the results of these two sections are most strongly connected with the work of Comon (1994) and Cardoso (1998, 1999, 2003). I acknowledge their contributions by giving explicit references for each proposition or theorem I state.

To estimate the source from the signal, we make explicit use of two properties of the source:

1. the covariance matrix of the source is the identity matrix, and
2. the components of the source are independent.

We first concentrate on 1, the identity covariance matrix of the source, and consider both the population and sample case. As we have seen in previous chapters, if $\mathbf{X} \sim (\boldsymbol{\mu}, \Sigma)$, then $\mathbf{X}_\Sigma = \Sigma^{-1/2}(\mathbf{X} - \boldsymbol{\mu})$ has the identity covariance matrix. But this is not the only way one can construct random vectors with an identity covariance matrix.

Definition 10.4 Let $\mathbf{X} \sim (\mathbf{0}, \Sigma)$ be a d-dimensional random vector. If Σ is the identity matrix, then $\mathbf{X}$ is called **(spatially) white**.

Let $\mathbb{X}$ be centred data of size $d \times n$, and let S be the sample covariance matrix of $\mathbb{X}$. We call $\mathbb{X}$ **(spatially) white**, if S is the identity.

A $d \times d$ matrix Ξ is called a **whitening matrix** for $\mathbf{X}$ or $\mathbb{X}$ if the covariance matrix of $\Xi\mathbf{X}$ or the sample covariance matrix of $\Xi\mathbb{X}$, respectively, is the identity matrix. We put $\mathbf{X}^\diamond = \Xi\mathbf{X}$ and $\mathbb{X}^\diamond = \Xi\mathbb{X}$ for the **whitened signal** and **whitened data**, respectively. □

The term *white* comes from the signal-processing community and refers to uncorrelated random vectors or observations.

If $\mathbf{X} \sim (\mathbf{0}, \Sigma)$ with $\Sigma = \Gamma\Lambda\Gamma^{\mathsf{T}}$, then $\Xi_1 = \Sigma^{-1/2}$ and $\Xi_2 = \Lambda^{-1/2}\Gamma^{\mathsf{T}}$ are whitening matrices for $\mathbf{X}$, but many other whitening matrices exist. If Ξ is a whitening matrix for $\mathbf{X}$ and U is an orthogonal matrix of the same size as Ξ, then $U\Xi$ is a whitening matrix for $\mathbf{X}$. A simple calculation shows that the orthogonal matrix Γ of eigenvectors of Σ links the whitening matrices $\Xi_1 = \Sigma^{-1/2}$ and $\Xi_2 = \Lambda^{-1/2}\Gamma^{\mathsf{T}}$. Similar arguments apply to whitening matrices for $\mathbb{X}$.

Whitening a signal $\mathbf{X}$ prior to decomposing it into independent components is sensible in view of the next result.

Proposition 10.5 [Cardoso (1998)] *Let* $\mathbf{X} \sim (\mathbf{0}, \Sigma)$ *be a d-dimensional signal satisfying (10.1) with* $p = d$ *and some mixing matrix A. If* Ξ *is a whitening matrix for* $\mathbf{X}$, *then*

$$U = \Xi A$$

is an orthogonal matrix.

Applying Cardoso's result to $\mathbf{X}^{\diamond} = \Xi \mathbf{X}$ implies that

$$\mathbf{X}^{\diamond} = U\mathbf{S}, \tag{10.4}$$

with $U = \Xi A$ orthogonal, and hence the unmixing matrix simply becomes $U^{\top}$. We call (10.4) the **white** or **whitened independent component model**. The corresponding white or whitened model for data $\mathbb{X}$, a whitening matrix Ξ, and white data $\mathbb{X}^{\diamond} = \Xi \mathbb{X}$ is

$$\mathbb{X}^{\diamond} = U\mathbb{S}. \tag{10.5}$$

Cardoso (1999) proposed the term **orthogonal approach** to Independent Component Analysis for the model (10.4) and its solution. The original development of Independent Component Analysis dealt with raw and white signals, but the analysis of white signals has clear advantages, and we thus restrict attention to the latter. In the complete analysis, of course, the whitening step is an important first step in the analysis, and working with the white vector or data thus just shifts one arduous task away from the Independent Component Analysis of the random vector or data.

The first mentioned property of a source – the identity covariance matrix – can be accomplished by whitening the signal. The second property of a source – the independence of its entries – is harder to satisfy. A natural candidate for assessing independence is the mutual information. By Result 9.7 of Section 9.4, the mutual information $\mathcal{I} = 0$ if and only if the random vector has independent components.

A combination of Result 9.7 in Section 9.4, and Proposition 10.5 leads to an identification of the independent source.

Definition 10.6 Let $\mathbf{X}$ be a d-dimensional white signal satisfying (10.4). An **independent component solution** for $\mathbf{X}$ is a pair $(U, \mathbf{S}_0)$ consisting of an orthogonal matrix U and a white d-dimensional random vector $\mathbf{S}_0$ from a probability density function π_0 such that

$$\pi_0 = \underset{\pi \in \mathcal{F}_{\mathbf{I}}}{\operatorname{argmin}}\, \mathcal{I}(\pi) \qquad \text{subject to} \qquad \mathbf{X} = U\mathbf{S}_0,$$

where $\mathcal{I}$ is the mutual information of (9.13), and $\mathcal{F}_{\mathbf{I}}$ is the set of probability density functions, defined in part 1 of Result 9.7 in Section 9.4.

Let $\mathbb{X}$ be d-dimensional white data satisfying (10.5). An **independent component solution** for $\mathbb{X}$ is a pair $(U, \mathbb{S}_0)$ consisting of an orthogonal matrix U and white $d \times n$ data $\mathbb{S}_0$ from a probability density function π_0 such that

$$\pi_0 = \underset{\pi \in \mathcal{F}_{\mathbf{I}}}{\operatorname{argmin}}\, \mathcal{I}(\pi) \qquad \text{subject to} \qquad \mathbb{X} = U\mathbb{S}_0. \tag{10.6}$$

□

The mutual information is a natural tool for assessing independence. However, it requires knowledge of the underlying distribution of the source, which is generally not available; and neither are canonical data-based estimators for $\mathcal{I}$. We get around this difficulty by

(i) finding a suitable approximation $\mathfrak{I}$ to $\mathcal{I}$ such that
(ii) $\mathfrak{I}$ admits a natural sample-based estimator $\widehat{\mathfrak{I}}$, and
(iii) replacing $\mathcal{I}$ by $\widehat{\mathfrak{I}}$ in (10.6).

It is convenient to refer to $\widehat{\mathfrak{I}}$ as an *estimator* for $\mathcal{I}$. Candidates for $\mathfrak{I}$ are moments and cumulants of order 3 and higher (see Definitions 9.2 and 9.4 in Section 9.3). Similar to the covariance matrix, the moments and cumulants have natural data-based estimators and so satisfy (ii). In Section 10.4.2 we encounter explicit expressions for $\widehat{\mathfrak{I}}$ which are based on combinations of higher-order moments. For such $\widehat{\mathfrak{I}}$, the minimiser $\widehat{\mathbb{S}}_0 = \operatorname{argmin} \widehat{\mathfrak{I}}(\mathbb{S})$ will not have independent components. These reflections lead to the following definition.

Definition 10.7 Let $\mathbb{X}$ be $d \times n$ white data which satisfy (10.5). Let $\widehat{\mathfrak{I}}$ be an estimator for $\mathcal{I}$. An **almost independent component solution** for $\mathbb{X}$ is a pair $(U, \mathbb{S}_0)$ consisting of an orthogonal matrix U and white $d \times n$ data $\mathbb{S}_0$ such that

$$\mathbb{S}_0 = \underset{\mathbb{S}\text{ white}}{\operatorname{argmin}} \widehat{\mathfrak{I}}(\mathbb{S}) \qquad \text{subject to} \qquad \mathbb{X} = U\mathbb{S}_0. \tag{10.7}$$

The solution $\mathbb{S}_0$ is **almost independent** or **as independent as possible** given $\widehat{\mathfrak{I}}$. □

Definitions 10.6 and 10.7 differ from those of Comon (1994) and Cardoso (1998) in that these authors optimised a criterion called the *contrast* or *contrast function*. Various definitions exist for a contrast, with the mutual information the motivating example of a contrast function. The end result is the same in their approaches and the one I have chosen here because the $\widehat{\mathfrak{I}}$ turn out to be contrast functions. We arrive at $\widehat{\mathfrak{I}}$ in two steps. The first step results in $\mathfrak{I}$, a function which approximates $\mathcal{I}$ for the population, and the second step shows how we estimate the approximation $\mathfrak{I}$ from data and hence find a solution in practice. This two-step path avoids having to define the somewhat vague concept of a contrast function.

For an independent component solution $\mathbf{S}_0$, we cannot directly compare the solution with the source $\mathbf{S}$; instead, we use the Kullback-Leibler divergence $\mathcal{K}$ of (9.14) in Section 9.4, which provides a measure of the closeness of the underlying distributions.

Proposition 10.8 [Cardoso (2003)] *Let $\pi = \prod \pi_j$ and f be d-dimensional probability density functions. For $j \le d$, let f_j be the marginals of f. The Kullback-Leibler divergence between f and π is*

$$\mathcal{K}(f, \pi) = \mathcal{I}(f) + \sum_{j=1}^{d} \mathcal{K}(f_j, \pi_j). \tag{10.8}$$

The result is based on (6) to (8) in Cardoso (2003) and also follows from part 2 of Result 9.9 in Section 9.4.

If f is the probability density function of $\mathbb{S}_0$ and $\pi = \prod \pi_j$ is that of the true source data $\mathbb{S}$, then, by Proposition 10.8, the error or mismatch between f and π splits into two parts:

- $\mathcal{I}(f)$ measures the remaining dependence in $\mathbb{S}_0$, and
- $\mathcal{K}(f_j, \pi_j)$ measures the mismatch between the marginals f_j and π_j.

In the next section we explore the relationship (10.8) further.

10.4 Mutual Information and Gaussianity

In Principal Component Analysis, the variance drives the search for the optimal directions; non-Gaussianity plays a similar role in Independent Component Analysis. Result 9.1 in Section 9.2 and Theorem 10.2 exhibit differences between Gaussian and non-Gaussian random vectors, and Theorem 10.2 further establishes the uniqueness of independent non-Gaussian sources. As we shall see in Section 10.4.1, the mutual information holds the key to the difference between Gaussian and non-Gaussian random vectors.

10.4.1 Independence, Uncorrelatedness and Non-Gaussianity

For Gaussian random vectors, independence is equivalent to uncorrelatedness. In this section we disentangle the two concepts for non-Gaussian vectors. Instead of regarding dependence as a quantity with a binary outcome dependent or independent, mutual information measures the extent of dependence. There are many connections between entropy, negentropy, mutual information and the Kullback-Leibler divergence which bring out the link between independence and Gaussianity. I summarise the connections in two theorems: The first states results for arbitrary random vectors and their distributions, and the second focuses on white random vectors.

Theorem 10.9 [Comon (1994), Cardoso (2003)] *Let f be a d-dimensional probability density function with marginals f_j, and let Σ be the covariance matrix of f. Let f_G be the d-dimensional Gaussian probability density function with the same mean and covariance matrix as f, with marginal variances σ_j^2 and let $f_{G,j}$ be the marginals of f_G.*

1. *The Gaussian probability density function that is closest to f with respect to the Kullback-Leibler divergence is f_G, and*

$$\mathcal{K}(f, f_G) = \min \mathcal{K}(f, \phi), \tag{10.9}$$

   where the minimum is taken over all Gaussian d-dimensional probability density functions ϕ.
2. *If Σ is invertible, then*

$$\mathcal{I}(f) = \mathcal{J}(f) - \sum_{j=1}^{d} \mathcal{J}(f_j) + \frac{1}{2} \log \left[\frac{\prod_{j=1}^{d} \sigma_j^2}{\det(\Sigma)} \right], \tag{10.10}$$

   where $\mathcal{J}$ is the negentropy of (9.12) in Section 9.4.
3. *The mutual information of f and f_G are related by*

$$\mathcal{I}(f) = \mathcal{I}(f_G) + \mathcal{K}(f, f_G) - \sum_{j=1}^{d} \mathcal{K}(f_j, f_{G,j}). \tag{10.11}$$

Part 2 of the theorem is taken from Comon (1994), the other parts are from Cardoso (2003). The different parts in the theorem embrace the three ideas: dependence, correlatedness and deviation from Gaussianity.

Deviation from Gaussianity. This idea is captured by two related concepts: the negentropy $\mathcal{J}(f)$ and the Kullback-Leibler divergence $\mathcal{K}(f, f_G)$. Rewriting both expressions as expectations of appropriate random variables, we note that

$$\mathcal{J}(f) = \mathbb{E}_f(\log f) - \mathbb{E}_{f_G}(\log f_G) \quad \text{and} \quad \mathcal{K}(f, f_G) = \mathbb{E}_f(\log f) - \mathbb{E}_f(\log f_G), \tag{10.12}$$

where $\mathbb{E}_g$ is the expectation with respect to the probability density function g. The expressions $\mathcal{J}(f)$ and $\mathcal{K}(f, f_G)$ measure the deviation from Gaussianity in a similar way; the main difference is that the negentropy has an expectation with respect to f_G for the $\log f_G$ term, whereas the Kullback-Leibler divergence uses $\mathbb{E}_f$ for both $\log f$ and $\log f_G$. As f becomes more non-Gaussian, the difference between the two expressions increases.

Dependence and Correlatedness. It is well known that the components of a random vector can be uncorrelated but not independent. Part 3 of Theorem 10.9 quantifies the disparity between dependence and correlatedness:

- $\mathcal{I}(f)$ measures the *dependence* of f on its marginals.
- $\mathcal{I}(f_G)$ measures the *correlatedness* between f_G and its marginals. The correlatedness exploits the equivalence of independence and zero correlation for Gaussian random vectors.
- $\mathcal{K}(f, f_G)$ and $\mathcal{K}(f_j, f_{G,j})$ measure the *non-Gaussianity* in f and its marginals f_j.
- For a non-Gaussian f, the difference $\mathcal{I}(f) - \mathcal{I}(f_G)$ between dependence and correlatedness is the difference between the non-Gaussianity $\mathcal{K}(f, f_G)$ of f and the marginal non-Gaussianities $\mathcal{K}(f_j, f_{G,j})$ of the f_js. This equality shows the interplay between the deviation of f from independence and the deviations of f and its marginals f_j from the associated Gaussian distributions.

I give an outline of the proof of parts 2 and 3 of Theorem 10.9 because these proofs show the interplay and relationship between the subtle concepts dependence, uncorrelatedness and deviation from the Gaussian.

Proof of Theorem 10.9 We consider a proof of part 1 in the Problems at the end of Part III. To show part 2, observe that

$$\mathcal{I}(f) = \mathcal{J}(f) - \sum_{j=1}^{d} \mathcal{J}(f_j) + \sum_{j=1}^{d} \mathcal{H}(f_{G,j}) - \mathcal{H}(f_G).$$

By theorem 8.14 of Cover and Thomas (2006), a bound for the entropy is given by

$$\mathcal{H}(f) \le \frac{1}{2}\{d + d\log(2\pi) + \log[\det(\Sigma)]\},$$

and equality is attained when $f = f_G$. This last equality leads to the desired result.

Part 3 is based on sections 2.2 and 2.3 of Cardoso (2003). We first calculate the Kullback-Leibler divergence between f and the product of the Gaussian marginals $f_{G,j}$. There are two natural paths we can follow to obtain this divergence: these are shown in diagram (10.13).

$$\begin{array}{ccc} f & \longrightarrow & f_G \\ \downarrow & & \downarrow \\ \prod_{j=1}^{d} f_j & \longrightarrow & \prod_{j=1}^{d} f_{G,j} \end{array} \tag{10.13}$$

The resulting expressions for $\mathcal{K}\left(f,\prod_{j=1}^{d} f_{G,j}\right)$ are

$$\mathcal{K}\left(f,\prod_{j=1}^{d} f_{G,j}\right)=\mathcal{K}(f,f_G)+\mathcal{K}\left(f_G,\prod_{j=1}^{d} f_{G,j}\right)$$

and

$$\mathcal{K}\left(f,\prod_{j=1}^{d} f_{G,j}\right)=\mathcal{K}\left(f,\prod_{j=1}^{d} f_j\right)+\mathcal{K}\left(\prod_{j=1}^{d} f_j,\prod_{j=1}^{d} f_{G,j}\right),$$

and from these equalities, (10.11) follows because $\mathcal{I}(f)=\mathcal{K}(f,\prod_{j=1}^{d} f_j)$. ■

If we replace f in Theorem 10.9 by the probability density function of a white random vector, then $\mathcal{I}(f_G)=0$, and consequently, some of the statements in Theorem 10.9 simplify. The white independent component model (10.4) is of practical interest, and I therefore state expressions for the mutual information of white random vectors.

Corollary 10.10 *Let* $\mathbf{X}\sim(\mathbf{0},\Sigma)$ *be a* d*-dimensional random vector with probability density function* f. *Let* Ξ *be an whitening matrix for* $\mathbf{X}$, *and assume that* Ξ *is invertible. Let* $f_\diamond$ *be the probability density function of* $\mathbf{X}^\diamond=\Xi\mathbf{X}$, *and let* $f_{G^\diamond}$ *be the Gaussian probability density function with mean zero and identity covariance matrix. Let* $f_{\diamond,j}$ *and* $f_{G^\diamond,j}$ *be the marginals of* $f_\diamond$ *and* $f_{G^\diamond}$, *respectively. The following hold:*

$$\mathcal{I}(f_\diamond)=\mathcal{K}(f_\diamond,f_{G^\diamond})-\sum_{j=1}^{d}\mathcal{K}(f_{\diamond,j},f_{G^\diamond,j}) \tag{10.14}$$

and

$$\mathcal{I}(f_\diamond)=\sum_{j=1}^{d}\mathcal{H}(f_{\diamond,j})-\log\left[\det(\Xi)\right]-\mathcal{H}(f). \tag{10.15}$$

The proof of the corollary follows from Result 9.7 in Section 9.4, Proposition 10.8 and Theorem 10.9.

Remark 1. (10.14) is a special case of (10.11) because $\mathcal{I}(f_{G^\diamond})=0$. Next, consider $\mathcal{K}(f_\diamond,f_{G^\diamond})$. Because Ξ is invertible, part 4 of Result 9.9 in Section 9.4 leads to

$$\mathcal{K}(f_\diamond,f_{G^\diamond})=\mathcal{K}(f,f_G),$$

where f_G is the minimiser of (10.9). Because $\mathbf{X}$ is given, and therefore implicitly also f, we treat $\mathcal{K}(f,f_G)$ as a constant. The expression (10.14) reduces to

$$\mathcal{I}(f_\diamond)=-\sum_{j=1}^{d}\mathcal{K}(f_{\diamond,j},f_{G^\diamond,j})+c_1 \qquad \text{and} \qquad c_1=\mathcal{K}(f,f_G).$$

It follows that white random vectors, which are as independent as possible, are as non-Gaussian as possible. Cardoso (2003) made similar observations for signals with arbitrary covariance matrix Σ, but for our purpose, white random vectors suffice.

Remark 2. The signal $\mathbf{X}$ is given, and we may thus treat $\mathcal{H}(f)$ as a constant. Consequently, (10.15) expresses the degree of dependence of $f_\diamond$ in terms of the entropies of its marginals

$f_{\diamond,j}$ and the log of Ξ. If we write $\Xi = E\Sigma^{-1/2}$, where E is an orthogonal matrix, then (10.15) becomes

$$\begin{aligned}\mathcal{I}(f_\diamond) &= \sum_{j=1}^{d} \mathcal{H}(f_{\diamond,j}) + \frac{1}{2}\log[\det(\Sigma)] - \log[\det(E)] - \mathcal{H}(f) \\ &= \sum_{j=1}^{d} \mathcal{H}(f_{\diamond,j}) - \log[\det(E)] + c_2 \quad (\text{for some } c_2 \geq 0),\end{aligned}$$

because Σ does not change. The last equality implies that the minimisation of $\mathcal{I}$ over white $f_\diamond$ reduces to a minimisation over orthogonal matrices E rather than whitening matrices Ξ, and from this minimisation, we obtain a solution $\mathbf{S}_0$. Cao and Liu (1996) examined the matrix which relates the source $\mathbf{S}$ to $\mathbf{S}_0$, and they derived distributional properties of $\mathbf{S}$ and $\mathbf{S}_0$.

This section shows how the mutual information integrates deviations from independence and deviations from Gaussianity. It remains to determine suitable approximations $\mathfrak{I}$ to $\mathcal{I}$ which are also adequate measures of the deviation from the Gaussian distribution.

10.4.2 Approximations to the Mutual Information

The three functions which link independence and non-Gaussianity are the mutual information, the negentropy and the Kullback-Leibler divergence. At the beginning of Section 10.3, I mentioned that independent component solutions require higher-order moments than the mean and covariance matrix. As we shall see in this section, third- and fourth-order moments and combinations of these yield good approximations to the negentropy and the mutual information.

Comon (1994) detailed approximations to the negentropy using cumulants. Some of his results were derived via a different path by Cardoso (1998), who approximated the Kullback-Leibler divergence, and by Lee et al. (1999, 2000) who presented estimates of the negentropy. The close relationship of the negentropy and Kullback-Leibler divergence is apparent in (10.12), and approximations to either of these quantities therefore will be similar.

The third- and fourth-order cumulants are related to the skewness and kurtosis (see Section 9.3). The skewness measures asymmetry in the distribution, and the kurtosis measures deviations from the normal distribution, such as bimodality, flatness or sharpness of the peak. For the Gaussian distribution, skewness and kurtosis are zero, and non-zero skewness and kurtosis are therefore indicators of non-Gaussianity.

Note. The skewness and kurtosis of Section 9.3 are defined as functions of the random vectors, as is customary for moments. In contrast, in Section 9.4, I define the entropy, the negentropy, the Kullback-Leibler divergence and the mutual information as functions of probability density functions. This apparent difference does not cause any problems; a definition of the entropy based on random vectors is equivalent to the one we use. To avoid confusion, I will state the correspondence between the random vector and its probability density function where appropriate.

We begin with an approximation to the negentropy given in Comon (1994). The relationship between the negentropy and the mutual information, which is stated in part 2 of Theorem 10.9, allows us to derive an approximation to the mutual information.

Theorem 10.11 [Comon (1994)] *Let* $\mathbf{X} = [X_1 \cdots X_d]^\top$ *be a mean zero white random vector with probability density function* f *and marginals* f_j *corresponding to the* X_j. *For* $j \leq d$, *assume that* X_j *is the sum of* m *independent random variables with finite cumulants for some positive* m.

1. *Let* $\beta_{3,j}(\mathbf{X})$ *and* $\beta_{4,j}(\mathbf{X})$ *be the skewness and kurtosis of* X_j *as in Result 9.5 of Section 9.3. The negentropy of* f_j *is*

$$\mathcal{J}(f_j) = \frac{1}{48}\left\{4\,[\beta_{3,j}(\mathbf{X})]^2 + [\beta_{4,j}(\mathbf{X})]^2 + 7\,[\beta_{3,j}(\mathbf{X})]^4 - 6\,[\beta_{3,j}(\mathbf{X})]^2\,\beta_{4,j}(\mathbf{X})\right\} + O(m^{-2}).$$

2. *Let* $\mathbf{S}$ *be a mean zero white random vector with probability density function* g, *and let* A *be a non-singular matrix such that* $\mathbf{X} = A\mathbf{S}$. *Let* $\beta_{3,j}(\mathbf{S}) = C_{jjj}(\mathbf{S})$ *and* $\beta_{4,j}(\mathbf{S}) = C_{jjjj}(\mathbf{S})$ *be the cumulants of the* j*th entry of* $\mathbf{S}$ *as in Definition 9.4 of Section 9.3. Then*

$$\mathcal{I}(g) = \mathcal{J}(f) - \frac{1}{48}\sum_{j=1}^{d}\left\{4\,[\beta_{3,j}(\mathbf{S})]^2 + [\beta_{4,j}(\mathbf{S})]^2 + 7\,[\beta_{3,j}(\mathbf{S})]^4 - 6\,[\beta_{3,j}(\mathbf{S})]^2\,\beta_{4,j}(\mathbf{S})\right\} + O(m^{-2}). \qquad (10.16)$$

The theorem provides a step towards identifying suitable approximations $\mathfrak{I}$ to $\mathcal{I}$.

The approximation in part 1 of the theorem is based on an Edgeworth expansion, that is, an expansion of the distribution function about the standard normal distribution function (see Abramowitz and Stegun 1965 or Kendall, Stuart, and Ord 1983). The notation $\beta_{\rho,j}(\mathbf{S})$ in part 2 reflects that the moments equal the cumulants for the white $\mathbf{S}$. Part 2 also uses the invariance of the negentropy under non-singular transformations, and thus, $\mathcal{J}(g) = \mathcal{J}(f)$. We return to this property in the Problems at the end of Part III.

Comon (1994) and Cardoso (1998) realised that the approximation (10.16) is still too complicated for efficient computations and looked at further simplifications. Their next steps differ. We look at both their approximations and start with that of Comon.

For $\rho = 3, 4$ and $\mathbf{S}$ a d-variate mean zero white random vector, let $\beta_{\rho,j}(\mathbf{S})$ be the cumulants of part 2 in Theorem 10.11. Put

$$\mathcal{G}_\rho(\mathbf{S}) = \sum_{j=1}^{d}\left[\beta_{\rho,j}(\mathbf{S})\right]^2. \qquad (10.17)$$

Comon (1994) considered the skewness criterion $\mathcal{G}_3(\mathbf{S})$ and the kurtosis criterion $\mathcal{G}_4(\mathbf{S})$ separately instead of the sum of the four terms in (10.16). Each $\mathcal{G}_\rho(\mathbf{S})$ results in a less close approximation to the mutual information than the four summands of (10.16), but the simpler expressions $\mathcal{G}_\rho(\mathbf{S})$ provide a compromise between computational feasibility and accuracy. Comon's approach was adopted in Hyvärinen (1999), and we return to it in Section 10.6.

Cardoso (1998) started with the Kullback-Leibler divergence and considered approximations by even-order cumulants. In Theorem 10.12, we may think of $\mathbf{X}_0$ as a candidate $\widehat{\mathbf{S}}$ for the source $\mathbf{S}$; the theorem establishes estimates of the difference between the $\mathbf{X}_0$ and $\mathbf{S}$ and tells us how close $\mathbf{X}_0$ is to the $\mathbf{S}$.

Theorem 10.12 [Cardoso (1998)] *Let* $\mathbf{S}$ *be a* d*-variate source with probability density function* $\pi = \prod \pi_j$. *Let* $\mathbf{X}_0 = [X_{0,1} \cdots X_{0,d}]^\top$ *be a* d*-variate random vector with probability density function* f *and marginals* f_j. *For* $\mathbf{T} = \mathbf{X}_0$ *or* $\mathbf{T} = \mathbf{S}$, *let* $C_{ij}(\mathbf{T})$ *and* $C_{ijkl}(\mathbf{T})$ *be the* *second- and fourth-order cumulants as in Definition 9.4 of Section 9.3. Put*

$$\mathcal{D}_2(\mathbf{X}_0,\mathbf{S}) = \sum_{i,j} [C_{ij}(\mathbf{X}_0) - C_{ij}(\mathbf{S})]^2 \qquad \textit{and}$$

$$\mathcal{D}_4(\mathbf{X}_0,\mathbf{S}) = \sum_{i,j,k,l} [C_{ijkl}(\mathbf{X}_0) - C_{ijkl}(\mathbf{S})]^2.$$

1. *The Kullback-Leibler divergence between f and π is given approximately by*

$$\mathcal{K}(f,\pi) \approx \frac{1}{48}[12\mathcal{D}_2(\mathbf{X}_0,\mathbf{S}) + \mathcal{D}_4(\mathbf{X}_0,\mathbf{S})]. \tag{10.18}$$

2. *If $\mathbf{X}_0$ is white, then $\mathcal{D}_2(\mathbf{X}_0,\mathbf{S}) = 0$, and*

$$\mathcal{D}_4(\mathbf{X}_0,\mathbf{S}) = \mathcal{D}_4^{\mathbf{1}}(\mathbf{X}_0,\mathbf{S}) + c_3,$$

where

$$\mathcal{D}_4^{\mathbf{1}}(\mathbf{X}_0,\mathbf{S}) = -2\sum_{j=1}^{d} \beta_{4,j}(\mathbf{S}) C_{jjjj}(\mathbf{X}_0),$$

and c_3 does not depend on $\mathbf{X}_0$ or $\mathbf{S}$. Hence, the Kullback-Leibler divergence between f and π reduces to

$$\mathcal{K}(f,\pi) \approx \frac{1}{48}\mathcal{D}_4^{\mathbf{1}}(\mathbf{X}_0,\mathbf{S}) + c_3. \tag{10.19}$$

Cardoso (1998) expressed the probability density function f of $\mathbf{X}_0$ in terms of the leading terms of an Edgeworth expansion about the normal; see McCullagh (1987), which used zeroth- second- and fourth-order cumulants of $\mathbf{X}_0$. Cardoso did not provide an expression for the error term in (10.18).

Part 1 of the theorem approximates the Kullback-Leibler divergence by differences of cumulants in $\mathbf{X}_0$ and $\mathbf{S}$. The second-order terms represent correlatedness and disappear under spatial whiteness. The remaining fourth-order terms thus express the essence of independence. Because the kurtosis terms approximate $\mathcal{K}(f,\pi)$ but do not equal it, the resulting solution will be as independent as achievable with the fourth-order approximation.

In the next theorem and the rest of this section we require the following notation:

$$\begin{aligned}
\mathcal{I}_o &= \{(i,j,k,l) : 1 \le i,j,k,l \le d\},\\
\mathcal{I}_1 &= \{(i,j,k,l) \in \mathcal{I}_o : i = j = k = l\},\\
\mathcal{I}_2 &= \{(i,j,k,l) \in \mathcal{I}_o : \text{pairs of indices are the same}\},\\
\mathcal{I}_3 &= \{(i,j,k,l) \in \mathcal{I}_o : \text{exactly two indices are the same}\}.
\end{aligned}$$

In addition, we write $\mathcal{I}_o \setminus \mathcal{I}_1$ for indices in $\mathcal{I}_o$ but not in $\mathcal{I}_1$.

The next result focuses on properties of a candidate solution $\widehat{\mathbf{S}}$ for the source $\mathbf{S}$. Both Comon (1994) and Cardoso (1998) derived these results, but along a different paths. I present the version of Cardoso (1998) and use his approximations.

Theorem 10.13 [Cardoso (1998)] *Let $\mathbf{S}$ be a d-variate source with probability density functions $\pi = \prod \pi_j$. Let $\widehat{\mathbf{S}}$ be a d-variate random vector from the probability density function f*

with marginals f_j. If $\widehat{\mathbf{S}}$ is white, then

$$\sum_{\mathcal{I}_o} \left[C_{ijkl}(\widehat{\mathbf{S}})\right]^2 = c_2$$

for some c_2 which is constant in $\widehat{\mathbf{S}}$, and

$$\mathcal{I}(f) \approx \sum_{\mathcal{I}_o \setminus \mathcal{I}_1} \left[C_{ijkl}(\widehat{\mathbf{S}})\right]^2. \tag{10.20}$$

Cardoso (1998) did not provide an error bound for approximation (10.20). His interests were focused on practical issues: the tightness of an approximation is relaxed in favour of computational efficiency. I will not discuss computational advantages of various approximations as I am more interested in acquainting the reader with the two different expressions proposed in Cardoso (1998, 1999). Both expressions are based on cross-cumulants of subsets of the indices in (10.20):

$$\mathfrak{I}_{\text{JADE}}(\widehat{\mathbf{S}}) = \sum_{\mathcal{I}_o \setminus \mathcal{I}_3} \left[C_{ijkl}(\widehat{\mathbf{S}})\right]^2, \tag{10.21}$$

$$\mathfrak{I}_{\text{SHIBB}}(\widehat{\mathbf{S}}) = \sum_{\mathcal{I}_o \setminus \mathcal{I}_2} \left[C_{ijkl}(\widehat{\mathbf{S}})\right]^2.$$

The acronym JADE stands for the 'joint approximate diagonalisation of eigenmatrices', and SHIBB refers to the Bernstein-Bando-Shi (BBS) derivative estimates. Because the cross-cumulants carry the information about dependence, the subsets of indices which define JADE and SHIBBS are suitable for checking independence.

A comparison between (10.17) and (10.20) shows that the two approaches essentially differ in the sets of indices used in the calculations of their respective criteria. Other approximations to the negentropy, the Kullback-Leibler divergence and the mutual information exist and have been implemented in IC algorithms (see chapter 8 of Hyvärinen, Karhunen, and Oja 2001). For our purpose, it suffices to apply and compare the algorithms based on the expressions (10.17) and (10.21) of Hyvärinen (1999) and Cardoso (1999), respectively, and we will do so in Section 10.6.

10.5 Estimation of the Mixing Matrix

We now turn to the second solution path mentioned at the beginning of Section 10.3: estimation of the mixing matrix A. The ideas and results I present are mainly those of Amari and Cardoso (1997) and Amari (2002).

Throughout this section, we let $\mathbf{S}$ be a d-dimensional source and $\mathbf{X}$ a mean zero d-dimensional signal. We write π and f for the probability density function of $\mathbf{S}$ and $\mathbf{X}$, respectively, and π_j and f_j for their marginals, where $j \le d$. We assume that there exists an invertible matrix A_0 such that

$$\mathbf{X} = A_0\mathbf{S} \qquad \text{or equivalently} \qquad \mathbf{S} = B_0\mathbf{X}, \tag{10.22}$$

and $B_0 = A_0^{-1}$. In the following exposition, I reserve the symbols A_0 and B_0 for the true mixing and unmixing matrices; A and B are more general and may be candidates for A_0 and B_0.

10.5.1 An Estimating Function Approach

Sections 10.3 and 10.4 focus on identifying the source as the primary solution for the independent component models (10.1) and (10.3). Now we investigate the second solution path, which aims at estimating the mixing matrix. To do so, we recast the independent component model (10.1) in a semi-parametric framework.

It will be convenient to regard (B_0, π) as parameters of the model (10.22). Because I will freely move between A_0 and B_0, it makes sense to consider both as parameters of the model. We begin with the distribution of $\mathbf{X}$, and write

$$f(\mathbf{X}) = f(\mathbf{X}; A_0, \pi) = |\det(A_0)|^{-1}\pi(A_0^{-1}\mathbf{X})$$

to clarify that f is a function of the parameters A_0 and π. Both these quantities are unknown, but only the mixing matrix A_0 is of current interest. It is thus useful to regard π as a *nuisance parameter* and to attempt to separate the two parameters.

In Section 10.5, we consider general probability density functions f and their likelihoods, and therefore the likelihoods will also be general, rather than denoting the (normal) likelihood of the Gaussian distribution which we used throughout Chapter 7.

We treat the log-likelihood $\log L$ of f or $\mathbf{X}$ as a function of A_0 given $\mathbf{X}$, so

$$\log L(A_0|\mathbf{X}) = -\log[\det(A_0)] + \sum_{j=1}^{d} \log \pi_j(A_{0,j}^{-1}\mathbf{X}), \tag{10.23}$$

where $A_{0,j}^{-1}$ is the jth row of the inverse A_0^{-1} of A_0. The last expression in (10.23) follows because the source variables are independent. We cannot further separate A_0 and π in (10.23). An explicit algebraic expression for A_0 is therefore not available, and ways of estimating A_0 need to be considered.

In a log-likelihood framework, it is natural to consider the score function, the derivative of the log-likelihood. Here it will be convenient to regard the **score function** $\mathfrak{u}$ as a function of $\mathbf{X}$ and a matrix A, so $\mathfrak{u}$ is given by

$$\mathfrak{u}(\mathbf{X}, A) = \frac{\partial}{\partial A} \log L(A|\mathbf{X}). \tag{10.24}$$

The score function motivates the general implicit approach of Amari and Cardoso, which is based on estimating functions. As we shall see in Theorem 10.15, the score function gives rise to a commonly used type of estimating function. In addition, the score function holds the key to estimating the matrix A_0: Amari and Cardoso used $\mathfrak{u}$ to generate estimating functions and then showed how to construct estimating functions which yield the true B_0. For details on the score function and its use in statistical inference, see Cox and Hinkley (1974).

I start with a definition of estimating functions adapted to our framework and then introduce a particular class of estimating functions. In Section 10.5.2, we consider properties of the score function (10.24) and, in particular, relationships of the score function and the class of estimating functions described in (10.27).

Definition 10.14 Let $\mathbf{X}$ be a d-dimensional random vector with mean zero, and let B_0 be a $d \times d$ matrix. Assume that there exists a d-dimensional source $\mathbf{S}$ such that $\mathbf{S} = B_0\mathbf{X}$. An **estimating function** $\mathcal{U}$ for B_0 is a function $\mathcal{U} : \mathbb{R}^d \times \mathbb{R}^{d\times d} \to \mathbb{R}^{d\times d}$ which is defined for pairs of vectors and matrices $(\mathbf{X}, B)$ by

$$\mathcal{U}(\mathbf{X}, B) = B'$$

and satisfies

1. $\mathbb{E}[\mathcal{U}(\mathbf{X},B)] = \mathbf{0}_{d\times d}$ if and only if $B = B_0$, and
2. $\mathbb{E}\left[\frac{\partial}{\partial B}\mathcal{U}(\mathbf{X},\cdot)\right]$ is non-singular, (10.25)

where the expectation is taken with respect to the probability density function of $\mathbf{X}$. □

Typically, the matrix B which satisfies (10.25) cannot be derived explicitly but is obtained iteratively. A common updating algorithm, referred to as a *learning rule*, is

$$B_{i+1} = B_i - \delta_i \mathcal{U}(\mathbf{X}, B_i) \qquad \text{for } i = 1,2,\ldots, \tag{10.26}$$

for some scalar tuning parameter δ_i.

Amari and Cardoso (1997) considered estimating functions of the form

$$\mathcal{U}(\mathbf{X},B) = \left[\mathbf{I}_{d\times d} - \Theta(\mathbf{S}')\mathbf{S}'^{\mathsf{T}}\right](B^{\mathsf{T}})^{-1}, \tag{10.27}$$

where $\mathbf{S}' = B\mathbf{X}$ and $\Theta = [\theta_1 \cdots \theta_d]^{\mathsf{T}}$ is an d-valued function defined by

$$\Theta(\mathbf{S}') = [\theta_1(S_1')\cdots\theta_d(S_d')]^{\mathsf{T}},$$

with

$$\theta_j(S) = -\frac{d}{dS}\log \pi_j(S), \qquad \text{for } S = S_j' \text{ and } j \leq d, \tag{10.28}$$

and $\pi = [\pi_1,\ldots,\pi_d]$ is a parameter of the model.

Other non-linear functions θ_j and more general forms of estimating functions $\mathcal{U}$ and learning rules exist. A collection of nine rules is given in table 1 of Choi et al. (2005), which includes (10.26) as a special case. For our purpose, it suffices to consider estimating functions as in (10.27).

10.5.2 Properties of Estimating Functions

We use the definitions and the framework of the preceding section and begin with a result which establishes the close relationship between the score function and estimation functions as in (10.27).

Theorem 10.15 [Amari and Cardoso (1997)] *Let* $\mathbf{S}$ *and* $\mathbf{X}$ *be d-dimensional source and mean zero signal with probability density functions* π *and* f*, respectively. Let* A_0 *be an invertible matrix such that* $\mathbf{X} = A_0\mathbf{S}$*. Put* $B_0 = A_0^{-1}$*. Let L be the likelihood function of* $\mathbf{X}$ *as in (10.23), and let* $\mathfrak{u}$ *of (10.24) be the score function of* $\mathbf{X}$*. Let* $\Theta = [\theta_1 \cdots \theta_d]^{\mathsf{T}}$ *be the function defined in (10.28). Then the following hold:*

1. *The score function* $\mathfrak{u}$ *is an estimating function for* B_0*.*
2. *If B is an invertible* $d\times d$ *matrix and* $\mathbf{S}' = B\mathbf{X}$*, then*

$$\mathfrak{u}(\mathbf{X},B) = \left[\mathbf{I}_{d\times d} - \Theta(\mathbf{S}')\mathbf{S}'^{\mathsf{T}}\right](B^{\mathsf{T}})^{-1},$$

and hence

$$\mathfrak{u}(\mathbf{X},B) = \mathcal{U}(\mathbf{X},B),$$

where $\mathcal{U}$ *is the estimating function for* B_0 *defined in (10.27).*

3. *Let $\widetilde{\pi}$ be a d-dimensional probability density function with marginals $\widetilde{\pi}_j$ for $j \leq d$. Let $\widetilde{\Theta}$ be a function as in (10.28) but with the π_j replaced by the $\widetilde{\pi}_j$. For B and $\mathbf{S}'$ as in part 2, put*

$$\widetilde{\mathfrak{u}}(\mathbf{X}, B) = \left[\mathbf{I}_{d\times d} - \widetilde{\Theta}(\mathbf{S}')\mathbf{S}'^{\top}\right](B^{\top})^{-1}.$$

Then $\widetilde{\mathfrak{u}}$ is an estimating function for B_0.

This theorem tells us how to construct estimating functions. Because estimating function equations are solved iteratively, we want to know under what conditions a limit B exists and whether $B = B_0$.

Notation. In Theorems 10.16 and 10.17, the phrase '$\mathcal{U}(\mathbf{X}, B)$ *converges to* B^*' is shorthand for '*If the estimation function equation $\mathcal{U}(\mathbf{X}, B_i) = B_i'$ is applied iteratively using the learning rule (10.26), then the sequence B_i' converges to B^*.*'

Theorem 10.16 [Amari and Cardoso (1997)] *Let $\mathbf{S}$ and $\mathbf{X}$ be d-dimensional source and mean zero signal. Let A_0 be an invertible matrix such that $\mathbf{X} = A_0\mathbf{S}$. Put $B_0 = A_0^{-1}$. Let $\mathcal{U}$ be an estimating function for B_0 which satisfies (10.27). For a $d \times d$ invertible matrix B, put*

$$K(B) = \frac{\partial}{\partial B}\mathbb{E}[\mathcal{U}(\mathbf{X}, B)].$$

If $K(B)$ is invertible, put

$$\mathcal{U}^*(\mathbf{X}, B) = [K(B)]^{-1}\mathcal{U}(\mathbf{X}, B).$$

It follows that

1. *$\mathcal{U}^*$ is an estimating function for B_0, and*
2. *$\mathcal{U}^*(\mathbf{X}, B)$ converges to B_0.*

Theorem 10.16 is an adaptation of theorem 5 in Amari and Cardoso (1997). The theorem establishes sufficient conditions for B_0 to be the limit of $\mathcal{U}^*(\mathbf{X}, B)$. However, $\mathcal{U}^*$ depends on the unknown source $\mathbf{S}$ and its probability density function. Theorem 10.17 provides some guidance regarding a choice of estimating functions in practice.

Theorem 10.17 [Amari and Cardoso (1997)] *Let $\mathbf{S}$ and $\mathbf{X}$ be d-dimensional source and mean zero signal with probability density functions π and f, respectively. Let A_0 be an invertible matrix such that $\mathbf{X} = A_0\mathbf{S}$. Put $B_0 = A_0^{-1}$.*

1. *If $\mathfrak{u}$ is the score function of (10.24), then the sequence of solutions B_i' of $\mathfrak{u}$, defined as in (10.26), is asymptotically efficient.*
2. *Let $\widehat{\pi}$ be an estimator or π. If $\widehat{\mathfrak{u}}$ is defined as in (10.24) with $\widehat{\pi}$ instead of π, then $\widehat{\mathfrak{u}}(\mathbf{X}, B)$ converges to B_0.*

For the convergence statement in part 2, see the 'Notation' just before Theorem 10.16. Part 1 of the theorem tells us that for the true π, the sequence of solutions of $\mathfrak{u}$ is asymptotically efficient and so has minimum variance and attains the Cramer-Rao lower bound. If we start with an estimator $\widehat{\pi}$ of the true π in part 2, then the estimator $\widehat{\mathfrak{u}}(\mathbf{X}, \cdot)$ of B_0 is consistent. For details on consistency and asymptotic efficiency of an estimator, see chapter 7 of Casella and Berger (2001).

Amari and Cardoso (1997) stated that the estimator based on $\widehat{\pi}$ has good efficiency properties, provided that $\widehat{\pi}$ is close to the true π. This last statement highlights that a judicious choice of the source distribution is important in the definition of the score function. Amari (2002) recommended an adaptive choice for Θ of (10.28) and a parametric family for modelling the source distribution. Possible candidates for the source distribution are mixtures of Gaussian probability density functions and members of the exponential family. An expression for Θ, as in (10.28), can be derived from the chosen candidate for the source distribution.

A number of semi-parametric and likelihood-based approaches have been proposed in the literature. Vlassis and Motomura (2001) used kernel density estimation to maximise the likelihood, and Hastie and Tibshirani (2002) proposed a penalised likelihood approach for estimating B_0 and π. Chen and Bickel (2006) derived asymptotic properties of an estimator proposed in Eriksson and Koivunen (2003), which we will meet in Section 12.5.1. I will comment there on the developments and results in Chen and Bickel (2006).

The theorems of this section, the newer research mentioned in the preceding paragraph and the approaches to Independent Component Analysis which I discuss in Chapter 12 and, in particular, in Section 12.4 put Independent Component Analysis on a rigorous foundation. A burning question is therefore: How does Independent Component Analysis work in practice?

10.6 Non-Gaussianity and Independence in Practice

The motivating Example 10.1 in Section 10.2.2 has a large number of observations but only two variables. The examples in this section have a range of dimensions; when the signal dimension is small, we aim to find the same number of sources, but as d increases, dimension reduction as a preliminary step in the analysis is desirable or may even become essential. To see how Independent Component Analysis works, we explore and examine this method on real and simulated data.

10.6.1 Independent Component Scores and Solutions

We begin with notation pertaining to IC solutions. Sections 10.3 and 10.4 demonstrate the advantages of working with white random vectors, and I will therefore restrict the definition to such vectors and data. However, I will use notation which allows an easy transition to the original mixing and unmixing matrix should this be required.

The definitions for the population and sample are similar but have subtle differences; for completeness, I present both versions, starting with the population.

Definition 10.18 Let $\mathbf{X} \sim (\mathbf{0}, \Sigma)$ be a d-dimensional random vector, and assume that Σ is invertible. Let Ξ be a whitening matrix for $\mathbf{X}$, and put $\mathbf{X}^\diamond = \Xi\mathbf{X}$. Let U be the orthogonal mixing matrix of Proposition 10.5, and call the jth column $\boldsymbol{\upsilon}_j$ of U the jth **independent component (IC) direction**. Consider $k \le d$.

1. The kth **independent component score** is the scalar $S_k = \boldsymbol{\upsilon}_k^{\mathsf{T}}\mathbf{X}^\diamond$.
2. The k-dimensional **independent component vector** is $\mathbf{S}^{(k)} = \begin{bmatrix} S_1 \cdots S_k \end{bmatrix}^{\mathsf{T}}$.
3. The kth **independent component projection (vector)** is the d-dimensional vector

$$\mathbf{Q}_k = \boldsymbol{\upsilon}_k \boldsymbol{\upsilon}_k^{\mathsf{T}}\mathbf{X}^\diamond = S_k \boldsymbol{\upsilon}_k. \qquad \square$$

Table 10.2 *Comparison of Principal and Independent Components*

	PCA	ICA
Random vectors	$\mathbf{X} \sim (\boldsymbol{\mu}, \Sigma)$ $\Sigma = \Gamma\Lambda\Gamma^{\mathsf{T}}$	$\mathbf{X}^{\diamond} = U\mathbf{S} \sim (\mathbf{0}, \mathbf{I}_{d\times d})$ $\mathbf{X}^{\diamond} = \Lambda^{-1/2}\Gamma^{\mathsf{T}}\mathbf{X}$,
kth score	$W_k = \boldsymbol{\eta}_k^{\mathsf{T}}(\mathbf{X} - \boldsymbol{\mu})$ $\boldsymbol{\eta}_k$ column of Γ	$S_k = \boldsymbol{\upsilon}_k^{\mathsf{T}}\mathbf{X}^{\diamond}$ $\boldsymbol{\upsilon}_k$ column of U

Definition 10.19 Let $\mathbb{X} = \begin{bmatrix}\mathbf{X}_1 \cdots \mathbf{X}_n\end{bmatrix}$ be centred d-dimensional data, and assume that the covariance matrix S is invertible. Let Ξ be a whitening matrix for $\mathbb{X}$, and put $\mathbb{X}^{\diamond} = \Xi\mathbb{X}$. Let $\widehat{U}$ be an estimator for U of Definition 10.18, which has been obtained by one of the methods described in Sections 10.4 and 10.5. Let $\widehat{\boldsymbol{\upsilon}}_j$ be the jth column of $\widehat{U}$, and call it the jth **(sample) independent component (IC) direction**. Consider $k \le d$.

1. The kth **independent component score** of $\mathbb{X}$ is the $1 \times n$ row vector $\mathbf{S}_{\bullet k} = \widehat{\boldsymbol{\upsilon}}_k^{\mathsf{T}}\mathbb{X}^{\diamond}$.
2. The $k \times n$ **independent component data** $\mathbb{S}^{(k)}$ consist of the first k independent component scores $\mathbf{S}_{\bullet j}$, with $j \le k$:

$$\mathbb{S}^{(k)} = \begin{bmatrix}\widehat{\boldsymbol{\upsilon}}_1 \cdots \widehat{\boldsymbol{\upsilon}}_k\end{bmatrix}^{\mathsf{T}}\mathbb{X}^{\diamond} = \begin{bmatrix}\mathbf{S}_{\bullet 1}\\ \vdots \\ \mathbf{S}_{\bullet k}\end{bmatrix}. \tag{10.29}$$

3. The $d \times n$ matrix of the kth **independent component projections** or **projection vectors** are

$$\mathbb{Q}_{\bullet k} = \widehat{\boldsymbol{\upsilon}}_k\widehat{\boldsymbol{\upsilon}}_k^{\mathsf{T}}\mathbb{X}^{\diamond}. \tag{10.30}$$

□

A quick glance back to Sections 2.2 and 2.3 and the summary Table 2.2 reveals the similarity between the corresponding definitions for Principal Component Analysis and Independent Component Analysis. Table 10.2 captures the main features of the two approaches for random vectors and their scores. For notational convenience, I refer to the whitening matrix $\Xi = \Lambda^{-1/2}\Gamma^{\mathsf{T}}$.

In Principal Component Analysis, the spectral decomposition $S = \widehat{\Gamma}\widehat{\Lambda}\widehat{\Gamma}^{\mathsf{T}}$ of the sample covariance matrix S of $\mathbb{X}$ is unique – up to the sign of the eigenvectors – and this uniqueness leads to the scores being uniquely defined, too. In Independent Component Analysis, we do not have an explicit solution of an algebraic equation that results in the IC scores. Instead, we calculate approximate independent component solutions (10.7) which depend on an estimator $\widehat{\mathfrak{I}}$ of the mutual information $\mathcal{I}$. I restrict attention to the three approximations and estimators which I described in Section 10.4.2. In the calculations, I will state which estimator I use.

Practical experience has shown that for a given $\widehat{\mathfrak{I}}$, the IC solutions may not be unique, and in any particular run of the optimisation routine, the first independent component score, $\mathrm{IC}_1 = \mathbf{S}_{\bullet 1}$, may not be the most non-Gaussian. These features are consequences of the randomness of the starting point in the optimisation routines. To ensure that the IC_1 scores are derived from the most non-Gaussian direction and subsequent IC scores correspond to less non-Gaussian directions, we carry out the steps of Algorithm 10.1.

Algorithm 10.1 ***Practical Almost Independent Component Solutions***
Let $\mathbb{X}^{\diamond}$ be d-dimensional white data. Fix $K > 0$ for the number of iterations, and put $k = 1$.

Step 1. Fix an approximation $\mathfrak{I}$ from the three options
(a) FastICA of Hyvärinen (1999) with the skewness criterion $\mathcal{G}_3$ of (10.17),
(b) FastICA of Hyvärinen (1999) with the kurtosis criterion $\mathcal{G}_4$ of (10.17), and
(c) JADE of Cardoso (1999) with the criterion $\mathfrak{I}_{\text{JADE}}$ of (10.21).

Step 2. Calculate the orthogonal unmixing matrix $\widetilde{U}^{\mathsf{T}}$ which results from the chosen approximation in step 1. For $j \le d$, let $\widetilde{\boldsymbol{v}}_j$ be the jth column of $\widetilde{U}$. Calculate scores $\widetilde{\mathbf{S}}_{\bullet j} = \widetilde{\boldsymbol{v}}_j^{\mathsf{T}} \mathbb{X}^{\diamond}$.

Step 3. For $\widetilde{\mathbf{S}}_{\bullet j}$ and $j \le d$, calculate the statistic

$$\mathfrak{m}_j = \begin{cases} \text{absolute skewness of } \widetilde{\mathbf{S}}_{\bullet j} & \text{if step 1(a) is used,} \\ \text{absolute kurtosis of } \widetilde{\mathbf{S}}_{\bullet j} & \text{if step 1(b) or step 1(c) is used.} \end{cases}$$

Step 4. For $j \le d$, sort the $\mathfrak{m}_j$ in decreasing order $\mathfrak{m}_{(1)} \ge \mathfrak{m}_{(2)} \ge \cdots \mathfrak{m}_{(d)}$, where $\mathfrak{m}_{(1)}$ is the largest of the $\mathfrak{m}_j$. Use the order inherited from the $\mathfrak{m}_{(j)}$, and put

$$\widehat{\mathbb{S}}^{[k]} = \left[\mathbf{S}_{\bullet 1} \cdots \mathbf{S}_{\bullet d}\right]^{\mathsf{T}} \qquad \text{and} \qquad \widehat{U}^{[k]} = \left[\boldsymbol{v}_1 \ldots \boldsymbol{v}_d\right].$$

Step 5. If $k < K$, increase k by 1, and repeat steps 2–4 for this new value.

Step 6. Find the matrix $\mathbb{S}^* = \widehat{\mathbb{S}}^{[k]}$ such that $\mathbf{S}^*_{\bullet 1}$ maximises $\mathfrak{m}_{(1)}$ over all K runs. If there is a tie, sort among the ties by $\mathbf{S}^*_{\bullet 2}$ and so on until no further ties remain. Set

$$\widehat{\mathbb{S}} = \mathbb{S}^* \qquad \text{and} \qquad \widehat{U} = U^*.$$

■

Instead of the approximations in step 1, other approximations and estimators for $\mathcal{I}$ can be substituted. Provided that the new estimator depends on skewness or kurtosis, the most non-Gaussian directions are still obtained in a similar fashion. Typically, we start with the white data, but the algorithm applies to raw data by adding an initial whitening step.

The d scores of step 2 are found iteratively, starting with $\widetilde{\mathbf{S}}_{\bullet 1}$. Because the optimisation depends on a random start, the $(j+1)$st IC can have a larger $\mathfrak{m}$-value than the jth. This deficiency is remedied in step 4 for a single run of the optimisation and more globally over all K runs of the optimisation routine in step 6, where we choose the best solution: the most non-Gaussian.

10.6.2 Independent Component Solutions for Real Data

We explore Algorithm 10.1 on two real data sets: the classical *iris* data, which has four variables, and the seventeen-dimensional *illicit drug market* data. As we shall see, for the second example it is not always possible to find all seventeen sources, a common occurrence as the number of variables grows.

Table 10.3 *IC Solutions for Different Estimators* $\widehat{\mathfrak{I}}$

$\widehat{\mathfrak{I}}$	$\mathfrak{m}$	IC_1	IC_2	IC_3	IC_4
$\mathcal{G}_3$	Skewness	0.9977	0.6881	0.4206	0.0015
$\mathcal{G}_4$	Kurtosis	4.0977	3.7561	2.9428	1.5518

Example 10.2 Fisher's four-dimensional **iris** data have been analysed in many different ways. The purpose of this analysis is to illustrate how Independent Component Analysis works. In particular, we will see that the IC scores depend on the estimator $\widehat{\mathfrak{I}}$ and differ from the PC scores.

I use the skewness and kurtosis approximations $\mathcal{G}_3$ and $\mathcal{G}_4$ and Algorithm 10.1 to calculate the IC scores. For $K = 10$ runs, the algorithm results in the values shown in Table 10.3.

The skewness values are small, and IC_4 in particular is almost Gaussian. The ten runs result in the same solutions, and no sorting is necessary. This is not the case for the $\mathcal{G}_4$ solutions. The scores with absolute kurtosis 1.5518 are often found first, and a reordering of the scores $\widetilde{\mathbf{S}}_{\bullet j}$, as in step 4, is required. There is very little difference in the solutions for the K runs. We note that the kurtosis values differ more from the Gaussian than the skewness values, which suggests that even IC_4 contains non-Gaussian structure.

Figure 10.3 shows the scores against the observation number on the x-axis, starting with the most non-Gaussian, $IC_1 = \mathbf{S}_{\bullet 1}$, on the left. The scores obtained with skewness $\mathcal{G}_3$ are shown in the top row and those with kurtosis $\mathcal{G}_4$ in the second row. For comparison, I show the PC scores in black in the bottom row, again starting with PC_1 on the left. The skewness IC_1 scores show a greater range of values for the last species, which corresponds to the x-values 101–150. The kurtosis IC_1 scores exhibit spikes at observation numbers 42, 118 and 132. The spikes give rise to the relatively large kurtosis value.

A remarkable feature of these scores is that the IC_4 kurtosis, the right-most plot in the middle row, looks almost the same as PC_1, shown in the bottom-left panel. The main difference is a change in sign and a scale difference. The scale difference occurs because PC scores are not scaled to have an identity covariance matrix, whereas IC scores have the identity covariance matrix because they are derived from whitened data.

Figure 10.4 shows three-dimensional score plots of the first three IC and PC scores, respectively. In the 3D score plots, I use the same colours for the different iris species as in Figure 1.3 of Section 1.2: red for the first species, black for the second species, and green for the third species. The skewness IC scores are shown in the top-left panel, the kurtosis IC scores in the top-right panel, and the PC scores in the bottom-left panel. For comparison, I repeat the bottom-left panel of Figure 1.3 in the bottom-right panel. This 3D scatterplot shows variables 1, 3, and 4 of the raw data.

The skewness scores split into three groups that show excellent agreement with the three species. The kurtosis scores do not cluster the data; instead, the kurtosis plot emphasises outliers such as the 'red' observation at the top left and the two 'black' observations at the right end of the plot. A closer inspection of these three points reveals that they are the observations with numbers 42, 118 and 132, that is, the observations at which we observed the spikes in the kurtosis IC_1 plot in Figure 10.3. Because the plots are based on the ICs with highest kurtosis, IC_4 kurtosis, whose scores are almost the same as those of PC_1, is not shown.

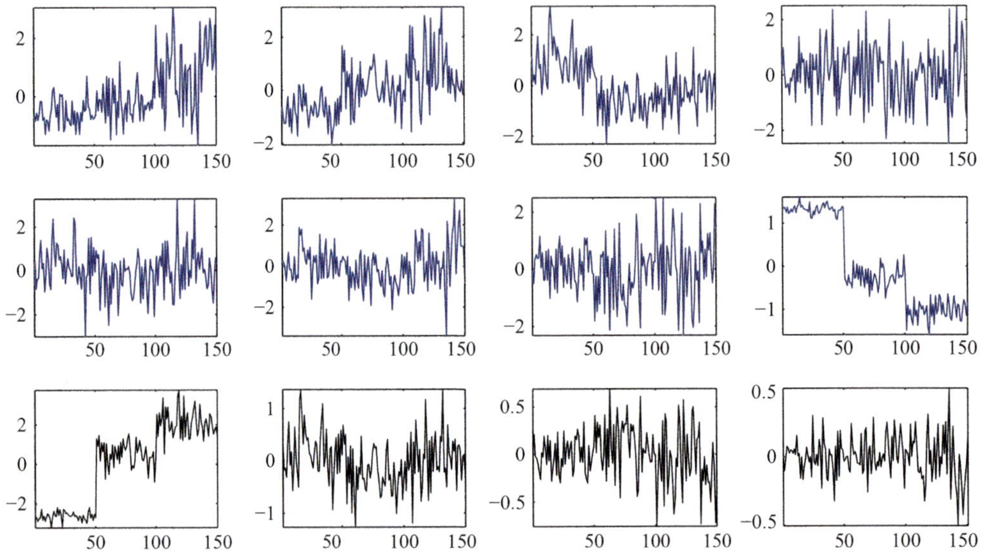

Figure 10.3 IC and PC scores of the iris data from Example 10.2. (*Top row*): IC_1 to IC_4 with skewness; (*middle row*): IC_1 to IC_4 with kurtosis; (*bottom row*) and PC_1 to PC_4.

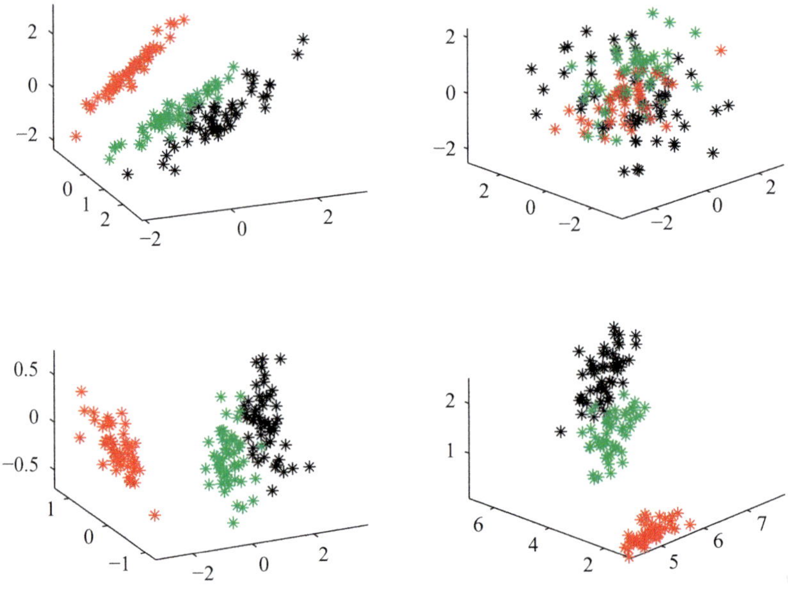

Figure 10.4 3D score plots for the iris data from Example 10.2. (*Top*): ICs with $\mathcal{G}_3$ (*left*), ICs with $\mathcal{G}_4$ (*right*); (*bottom*) PCs (*left*), variables 1, 3 and 4 of the raw data (*right*).

The graphs show that the independent component solutions obtained with the skewness and kurtosis estimators differ considerably and expose different structure in the data. ■

This example shows differences between the IC scores and the PC scores. The IC scores are not unique because different estimators give rise to different solutions. I deliberately have not posed or answered the question: Which analysis is best? Many factors influence what is a suitable analysis, and often more than one type of analysis is necessary to delve more deeply into the structure of the data.

Example 10.3 Figure 10.2 in Example 10.1 show the good agreement of the estimated and true sources for the **sound tracks** data. In the calculations I used the skewness approximation $\mathcal{G}_3$ in Algorithm 10.1. In this case, each run of the algorithm converged to the same estimates, a consequence of the low dimension – here two – and the large number of variables. ∎

Example 10.4 In this analysis of the **illicit drug market** data I use the seventeen series as variables; these are listed in Table 3.2 in Example 3.3 of Section 3.3. The data are shown in Figure 1.5 in Section 1.2.2. For Independent Component Analysis, scaling is irrelevant because the preliminary step whitens the data. However, the scaled data will allow more meaningful comparisons with the principal component scores, and I therefore work with the scaled data, which I refer to as $\mathbb{X}$ in the current analysis. Let $S = \widehat{\Gamma}\widehat{\Lambda}\widehat{\Gamma}^{\mathsf{T}}$ be the sample covariance matrix of $\mathbb{X}$ together with its spectral decomposition, and let $\mathbb{X}^{\diamond} = \widehat{\Lambda}^{-1/2}\widehat{\Gamma}^{\mathsf{T}}\mathbb{X}$ be the whitened data.

For the calculation of the IC scores, I use Algorithm 10.1 with $K = 10$ and all three approximations listed in step 1. The largest skewness is 3.4. The first and second IC scores obtained with the two kurtosis criteria are very similar and have kurtosis values 19.5 and 18.4 for $\mathcal{G}_4$ of (10.17) and 21 and 20 for $\mathfrak{I}_{\text{JADE}}$ of (10.21).

Figures 10.5 and 10.6 show projection plots, here as functions of the sixty-six months, which are shown on the x-axis. Row one in each figure shows normalised PC projections, rows two to four refer to IC projections, starting with skewness $\mathcal{G}_3$ in row two, kurtosis $\mathcal{G}_4$ in row three, and kurtosis $\mathfrak{I}_{\text{JADE}}$ in row four.

The main difference between the two sets of figures is that Figure 10.5 shows the first four PC and IC projections, starting with the first projections in the left panels, whereas Figure 10.6 shows the last four projection plots. The PC projections $\mathbb{P}_{\bullet k}$ are defined in (2.7) of Section 2.3. I have normalised the PC projections for easier comparison with the IC projections $\mathbb{Q}_{\bullet k}$ of (10.30). Both JADE's $\mathfrak{I}_{\text{JADE}}$ and the FastICA kurtosis approximation $\mathcal{G}_4$ yield all seventeen components, but FastICA skewness $\mathcal{G}_3$ only manages nine source components for these data, so in the second row we see skewness projections 6–9.

In Figure 10.5, the IC plots of columns one to three are very similar within a column but differ considerably from the corresponding PC projections. The first two graphs in rows two to four have sharp spikes at month 50 on the x-axis, the time when the heroin shortage occurred. The PC projections have no such spike; instead, PC_1 conveys a bimodality of the series. The spike in the IC projections is the most non-Gaussian feature that an independent component analysis finds.

The projections plots of Figure 10.6 show very little structure or pattern other than the first skewness projections in row two which still contain spikes. This first panel corresponds to IC_6, whereas the first panel in the other rows corresponds to component 14, by which stage most of the structure has been absorbed, as also shown in small kurtosis values – below 4 in the right-most panels. In the PC case, we know that the last few components contribute marginally to variance and so are often ignored. Similarly, the last few IC projections are negligible; they contain essentially Gaussian noise. ∎

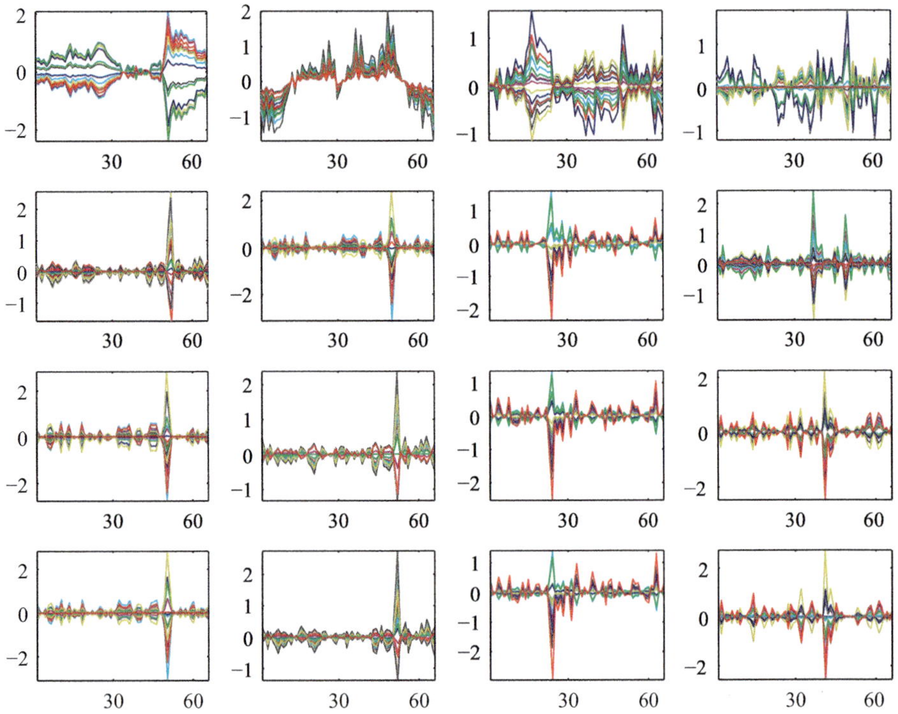

Figure 10.5 Illicit drug market data of Example 10.4. First four PC and IC projections: PC (*first row*), skewness $\mathcal{G}_3$ (*second row*), kurtosis $\mathcal{G}_4$ (*third row*), and $\mathfrak{I}_{\text{JADE}}$ (*fourth row*), with the observation number on the x-axis.

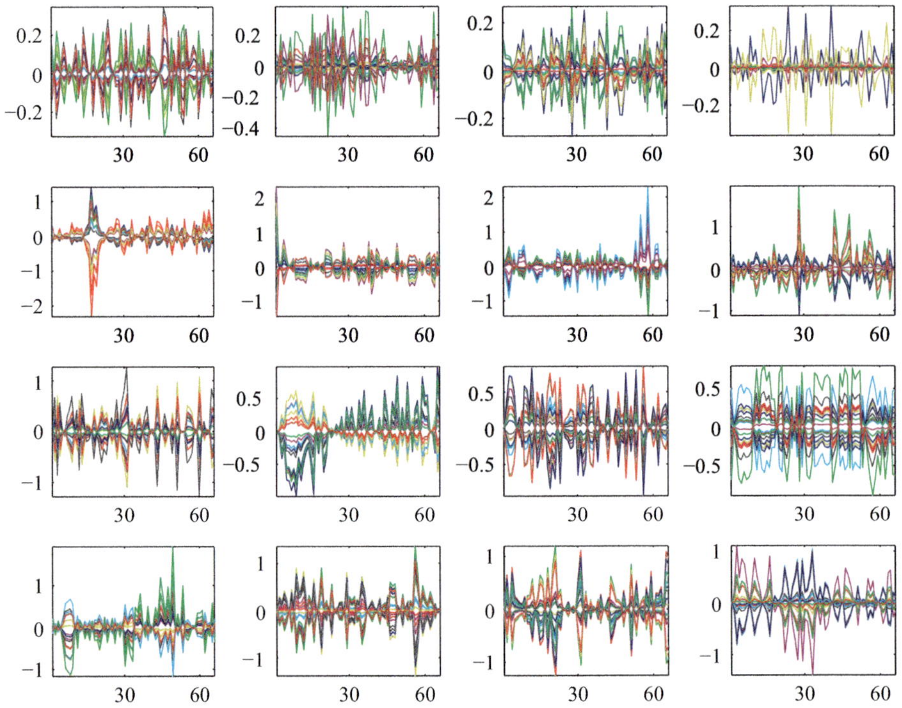

Figure 10.6 Illicit drug market data of Example 10.4. PC and IC projections as in Figure 10.5 but showing the last four projections in each case.

The analysis of the *illicit drug market* data illustrates three main points:

- The first few IC scores contain structural information about the data, whereas the last few scores do not contain much pattern and are associated with Gaussian random noise.
- The IC scores are not unique; they depend on the approximation $\mathfrak{I}$.
- The information that is exhibited by a principal component analysis differs from that found with an independent component analysis.

Because the different approaches furnish different solutions, I recommend trying a number of different IC criteria, as well as a principal component analysis, and combining the information available from the different analyses to reach a better understanding of the structure inherent in the data.

10.6.3 Performance of $\widehat{\mathfrak{I}}$ for Simulated Data

Initially, Independent Component Analysis was designed for a small number of dimensions and many samples, as in the **sound tracks** data. We now look at higher signal dimensions and sources which have specific non-Gaussian distributions. For these sources and their associated signals, we examine the potential of the three IC approximations for finding the non-Gaussian distributions.

In step 3 of Algorithm 10.1 we do not distinguished between positive and negative kurtosis but calculate absolute values only. If the shape of the source is known to be flat or multimodal, then the source has negative kurtosis and is called **sub-Gaussian**, whereas a source with sharper peaks and longer tails than the Gaussian is called **super-Gaussian** and has positive kurtosis. Example 10.5 covers both types of source models. For explicit expressions of sub- and super-Gaussian source models, see Rai and Singh (2004).

Example 10.5 For **simulated source data** $\mathbb{S}$ from different probability density functions and invertible matrices A, I calculate signals $\mathbb{X} = A\mathbb{S}$. I use the whitened signals $\mathbb{X}^{\diamond}$ and the relationship $\mathbb{X}^{\diamond} = \Xi A\mathbb{S} = U\mathbb{S}$ to determine estimates of $B = U^{\top}$.

Each source is a product of identical marginals which capture deviation from the Gaussian in a different way. The univariate distributions are

1. the uniform distribution on $[0,1]$,
2. the exponential distribution with mean 0.5,
3. the beta distribution $Beta(\alpha,\beta)$ with $\alpha = 5$ and $\beta = 1.2$, and
4. the bimodal, a mixture of two Gaussians, with 25 per cent from $\mathcal{N}(0,0.49)$ and 75 per cent from $\mathcal{N}(4,1.44)$.

The specific exponential, beta and bimodal probability density functions I use in the calculations are shown in Figure 10.7. The uniform is a special case of the beta distribution, with parameters $\alpha = \beta = 1$, and because its shape is well known, it is not shown in Figure 10.7. The uniform distribution is sub-Gaussian. The exponential is right-skewed and has a sharp peak, so it is super-Gaussian. Beta distributions have compact support $[0,1]$; the $Beta(5,1.2)$ is left-skewed. Our bimodal distribution has different proportions and variances defining each mode and belongs to the sub-Gaussians. The four source distributions differ

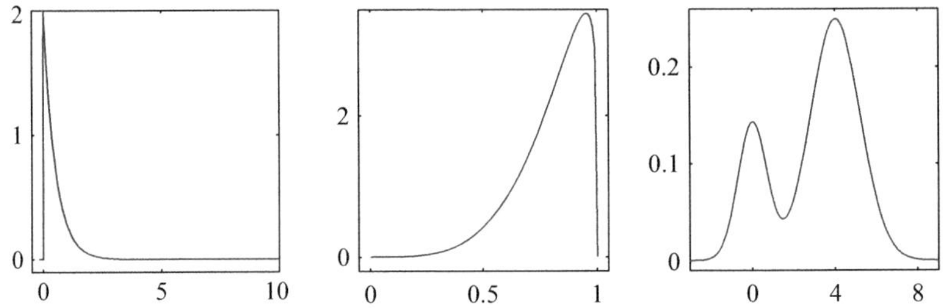

Figure 10.7 Marginal source distributions for Example 10.5; exponential (*left*), beta (*middle*) and bimodal (*right*).

from the Gaussian in that they have compact support, a sharp peak, asymmetry and bimodality, respectively. Throughout this example and its continuation, 'beta distribution' refers to $Beta(5, 1.2)$, and I use the term *uniform* for the $Beta(1, 1)$.

For each of the four marginal source models, I consider products of the same marginal with a different number of terms d, the dimension of the source. In the simulations I vary the sample size n for each source, and I consider three approximations $\mathfrak{I}$ to the mutual information. The following list summarises the different parameters used in the simulations:

- **Marginals** π_j: uniform, exponential, beta, bimodal
- **Dimension** d: 4, 8, 25, 50
- **Sample size** n: 100, 1,000, 10,000
- **Criterion** $\mathfrak{I}$: FastICA skewness $\mathcal{G}_3$, FastICA kurtosis $\mathcal{G}_4$, JADE kurtosis $\mathfrak{I}_{\text{JADE}}$

These combinations result in sixteen source distributions. The mixing matrix A is an invertible random matrix with normalised columns which belongs to the class of Higham test matrices (see Davies and Higham 2000). These matrices are generated with the MATLAB command gallery ('randcolu'). I generate four mixing matrices A, one for each dimension d.

For each source distribution and each sample size, I generate 100 repetitions of the source data $\mathbb{S}$. I calculate $\mathbb{X}$ and determine estimates $\widehat{B}$ of $B = A^{-1}$ with Algorithm 10.1 applied to $\mathbb{X}$ and the estimators for $\mathcal{G}_3, \mathcal{G}_4$ and $\mathfrak{I}_{\text{JADE}}$.

To assess the performance of the estimators for the different distributions, for each simulation I compare $\widehat{B}$ with B, using the sup norm – see Definition 5.2 of Section 5.2. An estimate $\widehat{B}$ of B will be close to B up to a permutation of the rows only, and thus we consider the error

$$\mathcal{E}(\widehat{B}) = \min \left\| B - \widehat{B} \right\|_{sup},$$

where the minimum is taken over permutations of the rows of $\widehat{B}$. The median error over all 100 repetitions is

$$q_{\text{med}} = \underset{\ell \le 100}{\text{median}}\, \mathcal{E}(\widehat{B}_\ell). \tag{10.31}$$

Figure 10.8 shows the performance of the estimators for four-dimensional signals in the top panel and for the eight-dimensional signals in the bottom panel. The values on the x-axis refer to the different distributions as follows:

- x-values 1–3 correspond to the uniform with sample sizes 100, 1,000, and 10,000, respectively.
- x-values 4–6 correspond to the exponential with sample sizes 100, 1,000, and 10,000, respectively.
- x-values 7–9 correspond to the beta with sample sizes 100, 1,000, and 10,000, respectively
- x-values 10–12 correspond to the bimodal with sample sizes 100, 1,000, and 10,000, respectively.

The y-axis shows q_{med} of (10.31): in red for the skewness $\widehat{\mathcal{G}}_3$, in black for the kurtosis $\widehat{\mathcal{G}}_4$ and in blue for the kurtosis $\widehat{\mathfrak{I}}_{\text{JADE}}$. The three q_{med} values for a single distribution are connected by a thick line which shows the dependence on the sample size: The thick blue line in the top plot shows that for each distribution the median performance of $\widehat{\mathfrak{I}}_{\text{JADE}}$ improves as the sample size increases. This observation agrees with the consistency results of Theorem 10.17.

Figure 10.8 shows that $\widehat{\mathfrak{I}}_{\text{JADE}}$ (in blue) performs generally better for $d = 4$ than the other two estimators but worse for $d = 8$. The skewness criterion (in red) performs very well for the exponential distribution (x-values 4–6), but not as well for the other three. The black $\widehat{\mathcal{G}}_4$ performs equally well for all four distributions and similarly to the red $\widehat{\mathcal{G}}_3$ for the distributions other than the exponential.

Implementations of the FastICA estimators $\widehat{\mathcal{G}}_3$ and $\widehat{\mathcal{G}}_4$ typically stop searching for components after a certain number of iterations, and the resulting $\widehat{B}$ therefore can end up with fewer than d rows. The user can increase the number of iterations, but I chose not to do so. The aim of JADE is to find a $d \times d$ matrix $\widehat{B}$, and this takes much longer for the higher dimensions. For the 50-dimensional data and $n = 100$, one run takes more than 30 minutes compared with 0.002 second for $d = 4$, 0.016 second for $d = 8$ and about 2 seconds for $d = 25$. Because of the large increase in computing time for the 50-dimensional data, I only calculated 10 JADE runs for $d = 50$.

For 25 and 50 dimensions, FastICA skewness typically results in a matrix $\widehat{B}$ with fewer than d rows. It is therefore not comparable with the other approaches, because q_{med} is

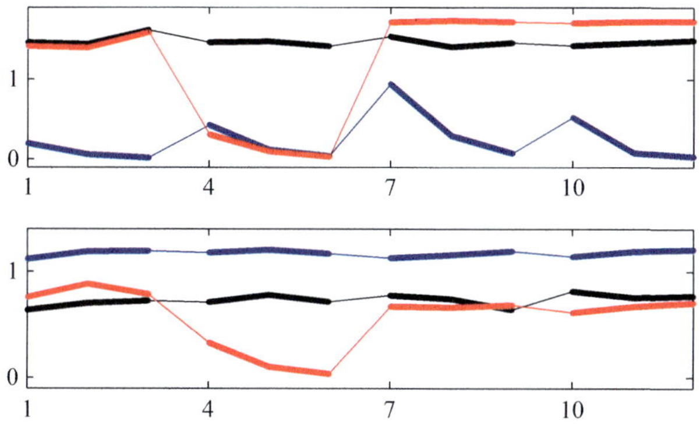

Figure 10.8 Median performance q_{med} for four-dimensional data (*top*) and eight-dimensional data (*bottom*) from Example 10.5. $\widehat{\mathcal{G}}_3$ (*red*), $\widehat{\mathcal{G}}_4$ (*black*) and $\widehat{\mathfrak{I}}_{JADE}$ (*blue*). Plots show results for $n = 100$, 1,000 and 10,000 at consecutive x-values: uniform (at x-values 1–3), exponential (at x-values 4–6), beta (at x-values 7–9) and bimodal distributions (at x-values 10–12).

Table 10.4 *Performance Ranking of Distributions for Each Estimator, Best=1*

d	Approximation	Uniform	Exponential	Beta	Bimodal
4	JADE $\widehat{\mathfrak{J}}_{\text{JADE}}$	1	2	4	3
	FastICA kurtosis $\widehat{\mathcal{G}}_4$	1	2	3/4	3/4
	FastICA skewness $\widehat{\mathcal{G}}_3$	4	1	2/3	2/3
8	JADE $\widehat{\mathfrak{J}}_{\text{JADE}}$	1	2	4	3
	FastICA kurtosis $\widehat{\mathcal{G}}_4$	1	2	4	3
	FastICA skewness $\widehat{\mathcal{G}}_3$	4	1	2/3	2/3
25	JADE $\widehat{\mathfrak{J}}_{\text{JADE}}$	3/4	1	3/4	2
	FastICA kurtosis $\widehat{\mathcal{G}}_4$	1/2	1/2	3/4	3/4
	FastICA skewness $\widehat{\mathcal{G}}_3$	4	1	2/3	2/3
50	JADE $\widehat{\mathfrak{J}}_{\text{JADE}}$	3	1	2	4
	FastICA kurtosis $\widehat{\mathcal{G}}_4$	2/3	1	4	2/3
	FastICA skewness $\widehat{\mathcal{G}}_3$	4	1/2	3	1/2

calculated for $d \times d$ matrices. Graphs for the two kurtosis approximations with $d = 25$ and $d = 50$ are similar to the lower graph with $d = 8$ in Figure 10.8 and therefore are not shown. Again FastICA kurtosis performs better than JADE, and the gap in the median performance increases as d increases.

Table 10.4 presents a different aspect of the performance of the three approximations and estimators. Separately for each dimension, the table shows which source distribution was easiest to distinguish from the Gaussian and which was the hardest. Here 'easiest' and 'hardest' refer to the median performance across the three sample sizes. A 1 means easiest to distinguish, and a 4 shows that the estimator performed worst for that distribution. Ties are shown by two numbers, such as 2/3. It is interesting to observe how the pattern in the table changes as the dimension increases. All three estimators easily distinguish between the exponential and the Gaussian. The skewness estimator is not good at detecting uniform sources, whereas the other two detect uniform sources easily. The kurtosis estimators typically find the $Beta(5, 1.2)$ and the bimodal harder to distinguish from the Gaussian than the uniform or the exponential.

In summary, we observe that the sharp peak of the exponential distribution is easy to distinguish from the Gaussian with all three estimators. Although the $Beta(5, 1.2)$ has support $[0, 1]$, its single mode makes it hard to distinguish from the Gaussian. The more surprising result is that the bimodal is relatively hard to distinguish from the Gaussian with the kurtosis estimators. It might be better to combine the criteria, but such an approach is computationally not as efficient. However, as a workable compromise, I recommend applying a skewness and a kurtosis estimator and comparing and combining the results. ■

Our examples focus on three approximations to the mutual information: the skewness and kurtosis approximations of Hyvärinen (1999) and JADE, the kurtosis approximation of Cardoso (1999). The estimators of all three approximations result in good general-purpose algorithms. There are many other algorithms, some generic, and some problem-specific (see the comments and references in the introduction to this chapter). My aim in this section has been to show how Independent Component Analysis works in practice.

There are many approaches to Independent Component Analysis other than those I presented. These include, but are not restricted to Attias (1999), who extended factor models to independent factor models using mixtures of Gaussians as the source models and an entropy maximisation (EM) algorithm to estimate the parameters, and the closely related Bayesian approach of Winther and Petersen (2007), which also relies on EM to find the sources. Many EM algorithms become computationally expensive as the dimension increases, and this may restrict their applicability in practice. Hyvärinen, Karhunen, and Oja (2001) considered noisy models, similar to the factor models, within an independent component framework. In addition, their book includes chapters which relate Independent Component Analysis to Bayesian methods and to time series.

In the early developments of Independent Component Analysis, the dimension of the source and the signal were the same. The examples of this section, however, show that we are not always able to detect all d source variables. There are possible explanations for this discrepancy: the source has a lower dimension than the signal, or some of the source variables are Gaussian noise variables and therefore cannot be detected with Independent Component Analysis. With an increasing dimension, one requires only the first few and most non-Gaussian source variables. In this case, Independent Component Analysis becomes a dimension-reduction tool – similar to the way Principal Component Analysis, Factor Analysis and Multidimensional Scaling are. In the remaining sections we explore this aspect of Independent Component Analysis.

10.7 Low-Dimensional Projections of High-Dimensional Data

When the dimension of source and signal are small or moderate, IC solutions are generally easy to find, especially for large sample sizes. The problems become harder when the dimension of the signals increases. For large d, the emphasis shifts from finding d sources to finding the most interesting projections. We begin with an illustrative example and then consider the theoretical developments of Hall and Li (1993).

10.7.1 Dimension Reduction and Independent Component Scores

In 'classical' Independent Component Analysis, the dimension of the signal and source are the same and are small to moderate. We tacitly assume that the sample covariance matrix of $\mathbb{X}$ is invertible so that we can calculate the whitened data.

As the dimension increases, we look at the four regimes $\mathbf{n} \succ \mathbf{d}$ to $\mathbf{n} \prec \mathbf{d}$ of Definition 2.22 in Section 2.7.2, which deal with different rates of growth of d and n. As long as $n > d$, which holds for $\mathbf{n} \succ \mathbf{d}$ and $\mathbf{n} \succeq \mathbf{d}$, the strategies of the preceding sections apply: we whiten the data and then determine d or fewer sources with Algorithm 10.1. Of course, we still need to decide how many sources we require.

For cases $\mathbf{n} \preceq \mathbf{d}$ and $\mathbf{n} \prec \mathbf{d}$, so $d > n$, the whitening step needs to be adjusted because the rank of the sample covariance matrix S is at most n. Whitening is closely linked to the covariance matrix Σ and its spectral decomposition $\Gamma\Lambda\Gamma^{\top}$, and it is therefore natural to whiten high-dimensional low sample size (HDLSS) random vectors with the help of Γ and Λ.

Definition 10.20 Let $\mathbf{X} \sim (\mathbf{0}, \Sigma)$. Let r be the rank of Σ, and write $\Sigma = \Gamma \Lambda \Gamma^{\mathsf{T}}$ for its spectral decomposition. For $p \leq r$, the ***p*-(dimensional) whitened** or the ***p*-white** random vector

$$\mathbf{X}(p)^{\diamond} = \Lambda_p^{-1/2} \Gamma_p^{\mathsf{T}} \mathbf{X}.$$

Let $\mathbb{X} \sim (\mathbf{0}, \Sigma)$ be data with sample covariance matrix S. Let r be the rank of S, and write $S = \widehat{\Gamma} \widehat{\Lambda} \widehat{\Gamma}^{\mathsf{T}}$ for its spectral decomposition. For $p \leq r$, the ***p*-(dimensional) whitened** or the ***p*-white** data

$$\mathbb{X}(p)^{\diamond} = \widehat{\Lambda}_p^{-1/2} \widehat{\Gamma}_p^{\mathsf{T}} \mathbb{X}. \tag{10.32}$$

□

An inspection of (10.32) shows that the p-white data are the sphered p-dimensional PC data $\mathbb{W}^{(p)}$. I could have defined the p-white data as $\widehat{\Gamma}_p \widehat{\Lambda}_p^{-1/2} \widehat{\Gamma}_p^{\mathsf{T}} \mathbb{X}$ in analogy with the sphered data $S^{-1/2}\mathbb{X}$. These two sets of p-whitened data are related by the orthogonal matrix $\widehat{\Gamma}_p$. From Theorem 10.2, we know that IC solutions of white data are unique up to a permutation of the rows, and the inclusion or exclusion of $\widehat{\Gamma}_p$ does not affect the IC solutions.

Whitening is often the first step in an Independent Component Analysis. If the rank r of Σ is smaller than the dimension d, then the map $\mathbf{X} \longmapsto \Gamma_r^{\mathsf{T}} \mathbf{X} = \mathbf{W}^{(r)}$ is a projection onto the r-dimensional PC vectors. For $p \leq r$, p-whitening of $\mathbf{X}$ can be regarded as a two-step dimension reduction:

$$\mathbf{X} \longmapsto \Gamma_r^{\mathsf{T}} \mathbf{X}$$

is followed by

$$\Gamma_r^{\mathsf{T}} \mathbf{X} \longmapsto \Lambda_p^{-1/2} \mathbf{I}_{p \times r} \Gamma_r^{\mathsf{T}} \mathbf{X} = \Lambda_p^{-1/2} \Gamma_p^{\mathsf{T}} \mathbf{X},$$

because $\mathbf{I}_{p \times r} \Gamma_r^{\mathsf{T}} = \Gamma_p^{\mathsf{T}}$. In the first step, the reduction to r dimensions, there is no choice. The second step requires the choice of a p. In the next section we explore IC solutions that result from different values of p, and then I describe a way of selecting a suitable p. Our next example, however, makes the choice for p easy because there is a natural candidate.

Example 10.6 The **data bank of kidneys** consists of thirty-six kidneys, and for each kidney, 264 measurements are available. Koch, Marron, and Chen (2005) investigate these healthy and normal-looking kidneys as a first step towards generating a synthetic model for healthy kidney populations. As an initial step in the investigation and modelling, they examined whether this sample could come from a normal distribution.

The bank of kidneys forms part of a wider investigation into the analysis of human organs (see Chen et al. 2002). Methods for segmentation of images of organs are laborious and expensive, and require medical experts. As a consequence, there has been a growing demand for good synthetic medical images and shapes for a characterisation of segmentation performance. Coronal views of four kidneys are shown in Figure 10.9. This and the next two figures are taken from Koch, Marron, and Chen (2005).

We explore two aspects of Independent Component Analysis for the kidney data:

1. an integration of problem-specific information into the choice of the whitening dimension p; and
2. an interpretation of different IC_1 scores.

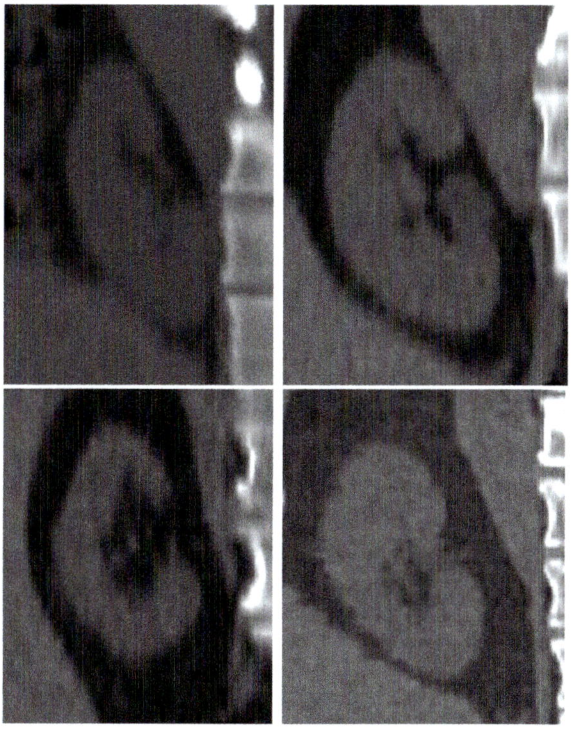

Figure 10.9 Coronal view of four kidneys from the study of thirty-six healthy-looking kidneys in Example 10.6.

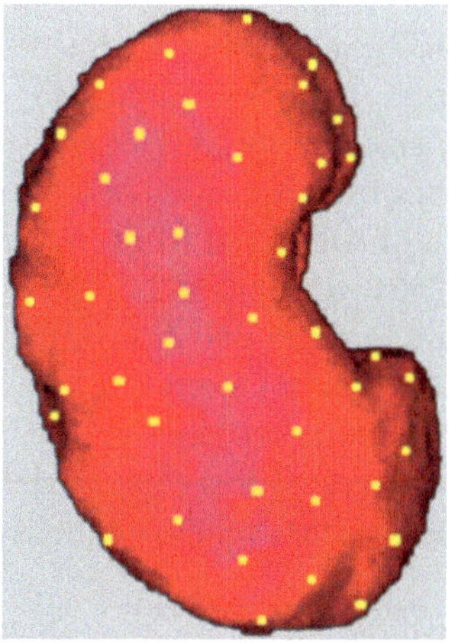

Figure 10.10 Single kidney with fiducial points from Example 10.6.

The shape of each kidney is characterised by eighty-eight 'fiducial' or landmark points on its surface, as shown schematically in Figure 10.10. These points are associated with salient geometric features on the surface of a kidney. The shape of a kidney can be reconstructed from its fiducial points, and shapes of different kidneys are compared via image registration using these points. The (x, y, z) coordinates of the eighty-eight fiducial points represent

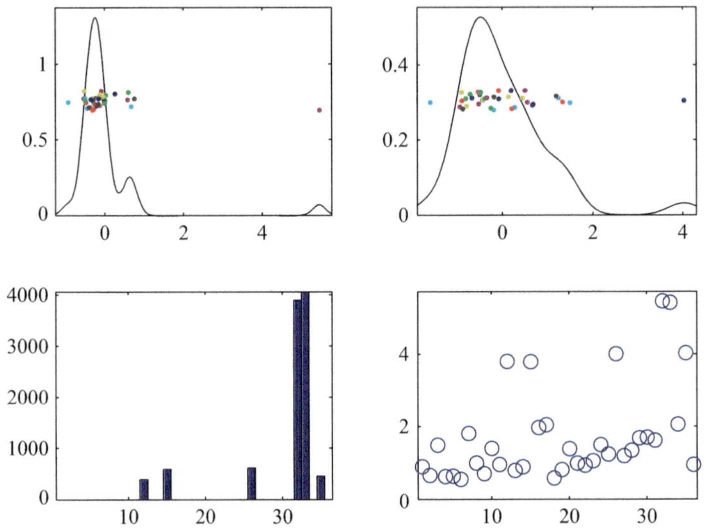

Figure 10.11 Two candidate IC_1 scores for Example 10.6. (*Bottom row*) Frequency of IC_1 scores in repeated runs (*left*), and skewness for each kidney sample (*right*), both versus observation number on the x-axis.

the 264 variables for each of the thirty-six kidneys. For these HDLSS data, the covariance matrix is highly singular. From the possible thirty-six variables available for principal components, Koch, Marron, and Chen (2005) restricted attention to the first seven PCs because the magnitude of the remaining components is smaller than the inter-user manual segmentation error, so for these data, the measurement error inherent in the design determines a suitable choice for p. From (10.32), we obtain the whitened data $\mathbb{X}(7)^\diamond = \widehat{\Lambda}_7^{-1/2}\widehat{\Gamma}_7^{\mathsf{T}}(\mathbb{X}-\overline{\mathbf{X}})$.

The first PC scores – shown as figure 5 in Koch, Marron, and Chen (2005) – could be normal, but because the sample size is small, further investigation is warranted. An obvious candidate for checking deviations from normality is IC_1. The FastICA skewness estimator $\widehat{\mathcal{G}}_3$ and the kurtosis estimator $\widehat{\mathcal{G}}_4$ of Algorithm 10.1 produce very similar results, so only the skewness results are reported. The two graphs in the top row of Figure 10.11 show smoothed histograms of IC_1 scores that are found in two different runs with $\widehat{\mathcal{G}}_3$. The coloured dots in each graph are the individual scores of the thirty-six kidneys plotted at random heights. In the top-left panel, the majority of the scores have a value, here shown as the x-value, close to zero, and there is one very large score with a value of about 5.5. The smoothed histograms show large deviations from the Gaussian which are the result of the outliers, the points at the right end of each panel.

Repeated runs with $\widehat{\mathcal{G}}_3$ lead to different IC_1 solutions. Step 3 of Algorithm 10.1 shows how to choose the 'best' scores. In this example, we look at the candidate IC_1 scores more closely. For $K = 10{,}000$ runs of FastICA $\widehat{\mathcal{G}}_3$, Koch, Marron, and Chen (2005) found six different IC_1 scores, and each is characterised by an outlier, here a large value for one particular kidney. The outliers of the six IC_1 solutions correspond to observations 32, 33, 12, 15, 26 and 35. The IC_1 scores corresponding to observation numbers 32 and 33 are found most commonly, and are shown with 32 in the left and 33 in the top-right panel of Figure 10.11. The bottom-left panel of the figure shows the frequency of these six IC_1 scores in 10,000 runs, with the observation number of the outlier causing the large score on the x-axis.

The picture in the bottom right shows the absolute skewness, on the y-axis, of all possible IC_1 scores, indexed by the observation number on the x-axis. The six IC_1 solutions arising from observations 32, 33, 12, 15, 26 and 35 are clearly distinct from the others by their much larger absolute skewness. These six IC_1 scores show a large deviation from normality, in strong contrast to PC_1.

The results illustrate that even for a severely reduced number of variables in the whitened data, one can find strong non-Gaussian structure, and furthermore, the IC scores differ substantially from the first PC scores. Koch, Marron, and Chen (2005) used these IC scores as the basis for a test of Gaussianity of the kidney data. We consider this aspect of their analysis in Section 13.2.3. ■

For the *bank of kidneys*, the segmentation error is used as a guide for choosing the whitening dimension p and leads to a dimension reduction from 264 variables to a mere seven PC variables, so much fewer than the rank of thirty-six of these data. If pertinent information of this kind is available, it should be used.

The IC_1 candidates for the kidney data are characterised by individual kidneys which appear as outliers for that particular IC. From $\mathrm{IC}_1 = \boldsymbol{v}_1^{\mathsf{T}}\mathbb{X}(p)^{\diamond}$, it follows that $\boldsymbol{v}_1$ has a large weight or a 'spike' for one particular observation, here kidney number 32 or 33, and small weights for all other kidneys. These spiked weights give rise to large absolute skewness and kurtosis. Figure 10.5 in Example 10.4 shows similar single spikes in the first few ICs of the *illicit drug market* data, and this pattern is typical for IC_1 solutions.

10.7.2 Properties of Low-Dimensional Projections

For a d-dimensional random vector $\mathbf{X}$ and a direction vector $\boldsymbol{\omega}$, we consider the projection $\boldsymbol{\omega}^{\mathsf{T}}\mathbf{X}$. Following Hall and Li (1993), we investigate the distribution of $\boldsymbol{\omega}^{\mathsf{T}}\mathbf{X}$ and examine whether univariate quantities of this form are linearly related for different directions $\boldsymbol{\omega}_1$ and $\boldsymbol{\omega}_2$. As we shall see, the deviation from linearity is closely related to the degree of non-Gaussianity of $\boldsymbol{\omega}^{\mathsf{T}}\mathbf{X}$, and the interplay between these two concepts leads to a type of Central Limit Theorem, but this time as $d \to \infty$.

The distribution of random variables $\boldsymbol{\omega}^{\mathsf{T}}\mathbf{X}$ is of great interest because candidates for $\boldsymbol{\omega}$ include the PC eigenvectors, Fisher's discriminant, the rows of the unmixing matrix B or U^T in Independent Component Analysis and projection pursuit directions, which we will meet in Chapter 11.

Definition 10.21 Let $\mathcal{S}^{d-1}$ be the unit sphere in $\mathbb{R}^d$. For $\mathbf{X} \sim (\mathbf{0}, \mathbf{I}_{d\times d})$, let f be its probability density function. Let $\boldsymbol{\omega}$ be a d-dimensional unit vector from the uniform distribution on $\mathcal{S}^{d-1}$, and let $f_{\boldsymbol{\omega}^{\mathsf{T}}\mathbf{X}}$ be the probability density function of $\boldsymbol{\omega}^{\mathsf{T}}\mathbf{X}$. Write ϕ_p for the p-variate standard normal probability density function, with $p \geq 1$. For $p = d$, define a random variable $\chi(\mathbf{X})$ by

$$\chi(\mathbf{X}) = \frac{f(\mathbf{X})}{\phi_d(\mathbf{X})}. \tag{10.33}$$

Consider a scalar t, and define the functions A_1 and A_2 by

$$A_1(t) = \mathbb{E}\left\{\left[\frac{f_{\boldsymbol{\omega}^{\mathsf{T}}\mathbf{X}}(t)}{\phi_1(t)} - 1\right]^2\right\} \tag{10.34}$$

and

$$A_2(t) = \mathbb{E}\left[\left\|\,\mathbb{E}\,(\mathbf{X}|\boldsymbol{\omega}\,:\boldsymbol{\omega}^{\mathsf{T}}\mathbf{X}=t)\right\|^2 - t^2\right]\left[\frac{f_{\boldsymbol{\omega}^{\mathsf{T}}\mathbf{X}}(t)}{\phi_1(t)}\right]^2, \tag{10.35}$$

where the expectation is taken with respect to the distribution of $\boldsymbol{\omega}$. □

The three terms, the random variable $\chi(\mathbf{X})$ and the functions A_1 and A_2, appear at first unrelated. As we shall see, $\chi(\mathbf{X})$ provides the link between the concepts the functions A_1 and A_2 capture. The function A_1 is a measure of the departure from normality of the one-dimensional projections. Linearity is captured by the convergence to zero of $\left\|\mathbb{E}(\mathbf{X}|\boldsymbol{\omega}:\boldsymbol{\omega}^{\mathsf{T}}\mathbf{X}=t)\right\|^2 - t^2$. Hall and Li (1993) could not derive an expression for this term and considered (10.35) instead.

The first step in the arguments of Hall and Li (1993) is recognition of the importance of the random variable $\chi(\mathbf{X})$ of (10.33). For a random vector $\mathbf{Z} \sim \phi_d$, Hall and Li considered $\chi(\mathbf{Z})$, and assuming that $\mathbf{Z}$ and $\boldsymbol{\omega}$ are independent, they showed that for $t \in \mathbb{R}$,

$$f_{\boldsymbol{\omega}^{\mathsf{T}}\mathbf{X}}(t) = \phi_1(t)\mathbb{E}\left[\chi(\mathbf{Z})|\boldsymbol{\omega}:\boldsymbol{\omega}^{\mathsf{T}}\mathbf{Z}=t\right]$$

and

$$\mathbb{E}(\mathbf{X}|\boldsymbol{\omega}\,:\boldsymbol{\omega}^{\mathsf{T}}\mathbf{X}=t) = \frac{\mathbb{E}\left[\mathbf{Z}\chi(\mathbf{Z})|\boldsymbol{\omega}:\boldsymbol{\omega}^{\mathsf{T}}\mathbf{Z}=t\right]}{\mathbb{E}\left[\chi(\mathbf{Z})|\boldsymbol{\omega}:\boldsymbol{\omega}^{\mathsf{T}}\mathbf{Z}=t\right]}.$$

The common term in the two equations is the conditional expectation $\mathbb{E}\left[\chi(\mathbf{Z})|\boldsymbol{\omega}:\boldsymbol{\omega}^{\mathsf{T}}\mathbf{Z}=t\right]$ which links the degree of non-Gaussianity in $f_{\boldsymbol{\omega}^{\mathsf{T}}\mathbf{X}}(t)$ and the deviation from linearity of the conditional expectation $\mathbb{E}\left[\mathbf{Z}\chi(\mathbf{Z})|\boldsymbol{\omega}:\boldsymbol{\omega}^{\mathsf{T}}\mathbf{Z}=t\right]$. Substitution of these expressions into (10.34) and (10.35) highlights the close relationship between these two ideas. We get

$$A_1(t) = \mathbb{E}\left(\left\{\mathbb{E}\left[\chi(\mathbf{Z})|\boldsymbol{\omega}:\boldsymbol{\omega}^{\mathsf{T}}\mathbf{Z}=t\right]\right\}^2\right) - 2\mathbb{E}\left\{\mathbb{E}\left[\chi(\mathbf{Z})|\boldsymbol{\omega}:\boldsymbol{\omega}^{\mathsf{T}}\mathbf{Z}=t\right]\right\} + 1$$

and

$$A_2(t) = \mathbb{E}\left\{\left\|\mathbb{E}\left[\mathbf{Z}\chi(\mathbf{Z})|\boldsymbol{\omega}:\boldsymbol{\omega}^{\mathsf{T}}\mathbf{Z}=t\right]\right\|^2\right\} - t^2\mathbb{E}\left(\left\{\mathbb{E}\left[\chi(\mathbf{Z})|\boldsymbol{\omega}:\boldsymbol{\omega}^{\mathsf{T}}\mathbf{Z}=t\right]\right\}^2\right).$$

Hall and Li converted the conditional expectations into unconditional ones by making use of an idea which they call the *rotational twin*. For $\boldsymbol{\omega}$ as in Definition 10.21 and independent d-dimensional standard normal random vectors $\boldsymbol{\nu}_\ell$, with $\ell = 1, 2$, such that

$$\boldsymbol{\nu}_\ell|\boldsymbol{\omega} \sim \mathcal{N}(\mathbf{0}, \mathbf{I} - \boldsymbol{\omega}\boldsymbol{\omega}^{\mathsf{T}}),$$

the **rotational twin** $(\mathbf{W}_1, \mathbf{W}_2)$ is defined for $t \in \mathbb{R}$ by

$$\mathbf{W}_1 = t\boldsymbol{\omega} + \boldsymbol{\nu}_1 \qquad \text{and} \qquad \mathbf{W}_2 = t\boldsymbol{\omega} + \boldsymbol{\nu}_2. \tag{10.36}$$

Hall and Li developed distributional theory for the pair $(\mathbf{W}_1, \mathbf{W}_2)$, which leads to distributional results for the one-dimensional projections $\boldsymbol{\omega}^{\mathsf{T}}\mathbf{X}$. I summarise their main results in the next two theorems.

Theorem 10.22 [Hall and Li (1993)] *Let* $\mathbf{X} \sim (\mathbf{0}, \mathbf{I}_{d\times d})$. *Let* θ *be the angle between the two vectors of the rotational twin* $(\mathbf{W}_1, \mathbf{W}_2)$ *of (10.36). If* $\mathbf{X}$ *and* θ *satisfy*

$$\mathbb{E}\left(\frac{d}{\|\mathbf{X}\|^2}\right) = O(1), \qquad \mathbb{E}\left(\frac{1}{\sin^2\theta}\right) = O(1)$$

and

$$\mathbb{P}\left\{\left|\frac{\|\mathbf{X}\|^2}{d} - 1\right| \geq c\right\} = o\left(d^{-1/2}\right) \qquad \textit{for any } c,$$

then, as $d \to \infty$,

$$\left|\frac{f_{\boldsymbol{\omega}^{\mathsf{T}}\mathbf{X}}(t)}{\phi_1(t)} - 1\right| \xrightarrow{p} 0$$

and

$$\int_{-M}^{M}\left[\frac{f_{\boldsymbol{\omega}^{\mathsf{T}}\mathbf{X}}(t)}{\phi_1(t)} - 1\right]^2 dt \to 0 \qquad \textit{for any } M > 0. \tag{10.37}$$

This theorem shows that for $\mathbf{X}$ from **any** distribution, the distribution of the scores $\boldsymbol{\omega}^{\mathsf{T}}\mathbf{X}$ becomes Gaussian as $d \to \infty$! We interpret this result as a Central Limit Theorem: the weighted sum $\boldsymbol{\omega}^{\mathsf{T}}\mathbf{X}$ becomes Gaussian as the number of terms, here the dimension d, grows.

Theorem 10.23 [Hall and Li (1993)] *Let* $\mathbf{X} \sim (\mathbf{0}, \mathbf{I}_{d\times d})$. *Let* θ *be the angle between the two vectors of the rotational twin* $(\mathbf{W}_1, \mathbf{W}_2)$ *of (10.36). For* $c > 0$, *put*

$$B_1(c) = \{\mathbf{X} : \frac{1}{d}\|\mathbf{X}\|^2 \leq 1 - c\} \qquad \textit{and} \qquad B_2(c) = \{\theta : |\sin\theta| \leq c\}.$$

Write I_B *for the indicator function of the set* B, *where* $B = B_1$ *or* $B = B_2$. *If* $\mathbf{X}$ *and* θ *satisfy*

$$\mathbb{E}\left(\frac{d}{\|\mathbf{X}\|^2} I_{B_1(c)}\right) = o\left(d^{-1}\right) \qquad \textit{for some } c > 0,$$

$$\mathbb{E}\left(|\arcsin\theta|\, I_{B_2(c)}\right) = o\left(d^{-1}\right) \qquad \textit{for some } c > 0, \textit{ and}$$

$$\mathbb{P}\left\{\left|\frac{\|\mathbf{X}\|^2}{d} - 1\right| \geq c\right\} = o\left(d^{-1}\right) \qquad \textit{for any } c,$$

then, as $d \to \infty$,

$$\left\|\,\mathbb{E}\left(\mathbf{X}|\boldsymbol{\omega} : \boldsymbol{\omega}^{\mathsf{T}}\mathbf{X} = t\right)\right\|^2 - t^2 \xrightarrow{p} 0$$

and

$$\int_{-M}^{M}\left[\left\|\,\mathbb{E}\left(\mathbf{X}|\boldsymbol{\omega} : \boldsymbol{\omega}^{\mathsf{T}}\mathbf{X} = t\right)\right\|^2 - t^2\right]^{1/2} f_{\boldsymbol{\omega}^{\mathsf{T}}\mathbf{X}}(t)\, dt \to 0 \qquad \textit{for any } M > 0.$$

Theorem 10.23 states that for $\mathbf{X}$ from any distribution, the scores $\boldsymbol{\omega}_\ell^{\mathsf{T}}\mathbf{X}$ based on different directions $\boldsymbol{\omega}_\ell$ become linearly related in the sense that for any two directions $\boldsymbol{\omega}_j$ and $\boldsymbol{\omega}_k$, the regression curve of $\boldsymbol{\omega}_j^{\mathsf{T}}\mathbf{X}$ on $\boldsymbol{\omega}_k^{\mathsf{T}}\mathbf{X}$ becomes nearly linear. As in Theorem 10.22, the results in

Theorem 10.23 hold no matter what the distribution $\mathbf{X}$ comes from. The deviation from linearity of one-dimensional projections also has been pursued in regression and visualisation (see Cook 1998; Cook and Yin 2001; and Li 1992).

Hall and Li (1993) generalised their results from one direction vector $\boldsymbol{\omega}$ to a matrix B consisting of k randomly selected orthonormal directions and showed that the conditional distribution of $B^{\mathsf{T}}\mathbf{X}|B$ is asymptotically normal as $d \to \infty$, where d is the dimension of $\mathbf{X}$. This generalisation is closely related to results of Diaconis and Freedman (1984), who showed that for high-dimensional data, almost all low-dimensional projections are nearly normal. The result of Hall and Li is stronger in the sense that it tells us how to select the directions. Instead of choosing them randomly, they should be selected to be as non-linear as possible. This choice further emphasises the close relationship between parts 1 and 2 of Theorem 10.22.

The next example illustrates the ideas of Theorem 10.22.

Example 10.7 We continue with the **simulated source data** and consider the first independent component scores $\boldsymbol{\upsilon}_1^{\mathsf{T}}\mathbf{X}^{\diamond}$ of the whitened data $\mathbb{X}^{\diamond}$. The purpose of this example is to observe the convergence of the scores $\boldsymbol{\upsilon}_1^{\mathsf{T}}\mathbf{X}^{\diamond}$ to the Gaussian as the dimension increases.

I use dimensions $d = 4$, 50 and 100 and the four source distributions listed in Example 10.5, products of uniform, exponential, beta and bimodal marginals, and generate $n = 100$ samples for each of the four source distributions and each of the three dimensions. Calculation of the signals and the IC_1 scores is the same as described in Example 10.5, but I only use the kurtosis approximation $\mathcal{G}_4$ of Algorithm 10.1 and find the solution in $K = 10$ runs.

Figure 10.12 shows the results in the form of 'qq-plots', plots of the empirical quantiles of the scores against the quantiles of the standard normal distribution. The top row shows results for the four-dimensional simulations, and the bottom row shows results for the 50-dimensional simulations. The graphs from the 100-dimensional simulations looked almost identical to those for 50 dimensions and therefore are not shown. The results from the uniform, exponential, beta and bimodal distributions are shown in columns 1 to 4, starting with the uniform on the left.

If the empirical quantiles agreed with those of the normal distribution, the blue points would closely agree with the red line. This is clearly not the case for the plots in the top row,

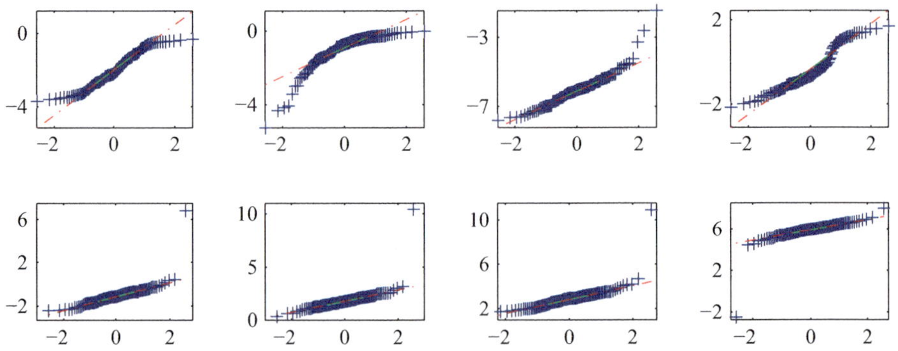

Figure 10.12 Quantile plots of IC_1 scores from four-dimensional data (*top*) and 50-dimensional data (*bottom*) from Example 10.7. The source distributions are from left to right: uniform, exponential, beta and bimodal.

which strongly deviate from the normal (red) line. Although we only use 100 samples on 50-dimensional data in the bottom row, the plots speak for themselves. We see that most of the scores follow the normal curve closely. However, in each panel there is one observation which gives rise to a large kurtosis and thus deviates from the normal (red) line.

I have chosen the worst-case scenario: the projections onto the most non-Gaussian directions. Even for such random vectors, we observe that the one-dimensional projections get closer to the Gaussian quite quickly as the dimension increases. ■

10.8 Dimension Selection with Independent Components

From the early days of Independent Component Analysis, researchers and practitioners recognised the advantages of working with white data, and this approach to Independent Component Analysis has become standard practice. As the dimension of the signal increases, so does the need to reduce the dimension of the white data prior to finding non-Gaussian directions. In Example 10.6, additional information about the data leads to a suitable choice for this reduced dimension p. In many data sets, additional information may not be available, yet we still need to make a decision regarding the number of components we use for the p-white data.

To appreciate that IC scores typically differ with p, consider $d \times n$ data $\mathbb{X} \sim (\mathbf{0}, \Sigma)$. Let S be the sample covariance matrix of $\mathbb{X}$, and write $S = \widehat{\Gamma}\widehat{\Lambda}\widehat{\Gamma}^{\mathsf{T}}$ for the spectral decomposition. For notational convenience, assume that $r = d$ is the rank of S. Suppose that we want the first κ ICs, for some $\kappa < d$. Consider p such that $\kappa \leq p < r$, and put

$$\mathbb{X}^{\diamond} = \widehat{\Lambda}^{-1/2}\widehat{\Gamma}^{\mathsf{T}}\mathbb{X} \qquad \text{and} \qquad \mathbb{X}(p)^{\diamond} = \widehat{\Lambda}_p^{-1/2}\widehat{\Gamma}_p^{\mathsf{T}}\mathbb{X}.$$

For the d- and p-dimensional whitened data, we desire the IC solutions

$$B\mathbb{X}^{\diamond} = \mathbb{S} \qquad \text{and} \qquad B_{(p)}\mathbb{X}(p)^{\diamond} = \mathbb{S}_{(p)},$$

where $B = U^{\mathsf{T}}$ is the $d \times d$ unmixing matrix for $\mathbb{X}^{\diamond}$, and $B_{(p)}$ is the corresponding $p \times p$ unmixing matrix for $\mathbb{X}(p)^{\diamond}$. The first κ ICs in each case are

$$\left[\boldsymbol{\upsilon}_1 \ldots \boldsymbol{\upsilon}_\kappa\right]^{\mathsf{T}} \widehat{\Lambda}^{-1/2}\widehat{\Gamma}^{\mathsf{T}}\mathbb{X} \qquad \text{and} \qquad \left[\boldsymbol{\upsilon}_{(p),1} \cdots \boldsymbol{\upsilon}_{(p),\kappa}\right]^{\mathsf{T}} \widehat{\Lambda}_p^{-1/2}\widehat{\Gamma}_p\mathbb{X},$$

where the $\boldsymbol{\upsilon}_j^{\mathsf{T}}$, the rows of B, are unit vectors in $\mathbb{R}^d$, and the rows of $B_{(p)}$, the $\boldsymbol{\upsilon}_{(p),j}^{\mathsf{T}}$, are p-dimensional unit vectors. These sets of vectors are found by minimising suitable d- and p-dimensional non-linear functions $\widehat{\mathfrak{I}}$, and as a consequence, the resulting IC scores will differ. Our next example illustrates this difference in the scores as p varies.

Example 10.8 We continue with the seventeen-dimensional **illicit drug market** data, and let $\mathbb{X}$ be the scaled data as in Example 10.4. For $p = 3, 7$ and 10, let $\mathbb{X}^{\diamond}(p)$ be the p-white data. For each p, I calculate IC_1 and IC_2 scores with Algorithm 10.1 and the skewness estimator $\widehat{\mathcal{G}}_3$. Figure 10.13 displays the ICs for $p = 3$ in the left plot, $p = 7$ in the middle and $p = 10$ on the right. The x-axis shows the observations, here the sixty-six months. The red lines show the IC_1 scores, and the black lines show the IC_2 scores. It is obvious that these estimated sources differ considerably. What is not so obvious is: Which p should we choose, and why? ■

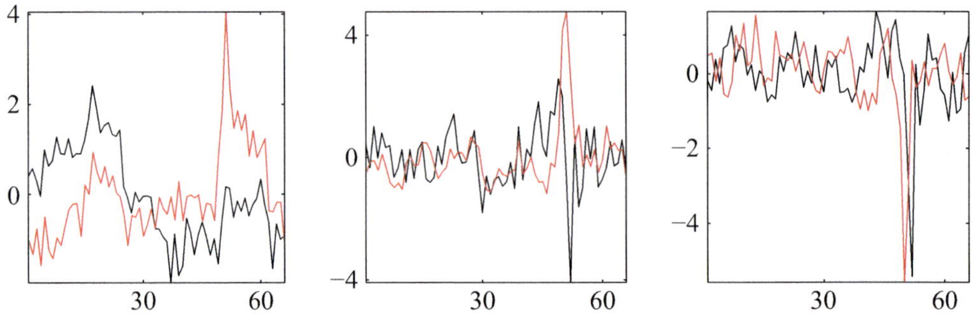

Figure 10.13 IC_1 and IC_2 scores for p-white data from Example 10.8 with $p = 3, 7$ and 10.

In the search for the 'best' p, we might want to choose the value p that results in the most non-Gaussian solution, as measured by absolute skewness or kurtosis. This choice is naive, because the absolute skewness and kurtosis increase with the number of components p of the data, and thus the 'best' choice would always be the original dimension d of the data.

Koch and Naito (2007) proposed to select the dimension p^* which makes the p^*-dimensional whitened data *relatively* most non-Gaussian. I outline this approach, which simultaneously treats the skewness and kurtosis criteria.

Let $\mathbb{X} \sim (\mathbf{0}, \Sigma)$ be data with sample covariance matrix $S = \widehat{\Gamma}\widehat{\Lambda}\widehat{\Gamma}^{\mathsf{T}}$ of rank r. For $p \leq r$, let

$$\mathbb{X}^{\diamond}(p) = \widehat{\Lambda}_p^{-1/2}\widehat{\Gamma}_p^{\mathsf{T}}\mathbb{X}$$

be the p-white data. Using the notation of Section 9.3, we write β_3 as in (9.4) for the skewness (so $\rho = 3$) and β_4 as in (9.6) for the kurtosis. Similarly, we let b_ρ as in (9.8) and (9.10) be the corresponding sample quantities. Consider the bias (p), the bias of $b_\rho[\mathbb{X}^{\diamond}(p)]$:

$$\mathrm{bias}(p) = \mathbb{E}\left\{b_\rho[\mathbb{X}^{\diamond}(p)]\right\} - \beta_\rho[\mathbb{X}^{\diamond}(p)], \tag{10.38}$$

where the expectation is taken with respect to $\mathbb{X}^{\diamond}(p)$. To assess deviations from the Gaussian distribution based on skewness or kurtosis, it is necessary to evaluate the performance of $b_\rho[\mathbb{X}^{\diamond}(p)]$ in a framework with null skewness or kurtosis. This suggests that the multivariate Gaussian distribution is the right null structure, and hence $\beta_\rho[\mathbb{X}^{\diamond}(p)] = 0$. As a consequence, (10.38) reduces to

$$\mathrm{bias}(p) = \mathbb{E}\left\{b_\rho[\mathbb{X}^{\diamond}(p)]\right\}.$$

Koch and Naito determined an expression for the bias under Gaussian assumptions and used this expression as a benchmark. For $p \leq r$, they calculated the sample bias and then determined the dimension p^* for which the sample bias deviated most from the bias of the Gaussian with the same dimension.

Let $\mathbb{X}_{\mathrm{G}} \sim \mathcal{N}(\mathbf{0}, \Sigma)$ be Gaussian data of size $d \times n$ and rank r. For $p \leq r$, let $\mathbb{X}_{\mathrm{G}}^{\diamond}(p)$ be the p-white data derived from $\mathbb{X}_{\mathrm{G}}$. For $\rho = 3, 4$, the sample skewness and kurtosis $b_\rho\left[\mathbb{X}_{\mathrm{G}}^{\diamond}(p)\right]$ are shown to satisfy

$$\sqrt{\frac{n}{\rho}}\, b_\rho\left[\mathbb{X}_{\mathrm{G}}^{\diamond}(p)\right] \xrightarrow{D} T_\rho(p) \quad \text{as } n \to \infty,$$

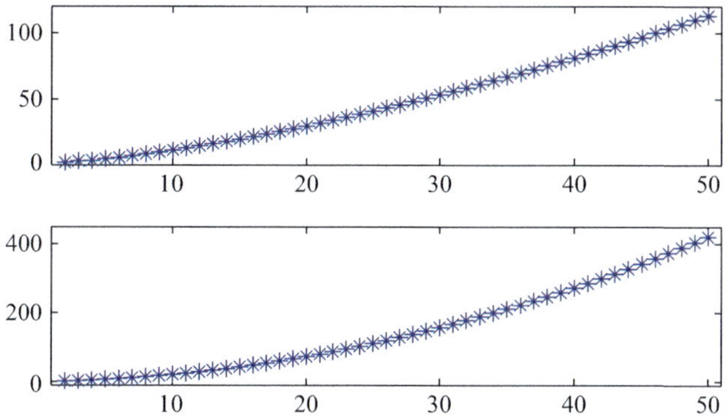

Figure 10.14 Skewness (*top*) and kurtosis (*bottom*) estimates $\widehat{\mathrm{UB}}_\rho(p)$ against dimension p.

where $T_\rho(p)$ is the maximum of a zero mean Gaussian random field on $\mathcal{S}^{p-1}$, the unit sphere in p dimensions. For large n, it follows that

$$\mathbb{E}\left\{\sqrt{\frac{n}{\rho}}\, b_\rho\left[\mathbb{X}^{\diamond}_{\mathrm{G}}(p)\right]\right\} \simeq \mathbb{E}\left[T_\rho(p)\right].$$

The expected values $\mathbb{E}\left[T_\rho(p)\right]$ for skewness and kurtosis are therefore the objects of interest. Koch and Naito were not able to give explicit expressions for these expectations; instead, they derived bounds for $\mathbb{E}\left[T_\rho(p)\right]$.

Theorem 10.24 [Koch and Naito (2007)] *Fix $p \geq 1$. For $\rho = 3, 4$, let $T_\rho(p)$ be the maximum of a zero mean Gaussian random field on the unit sphere in p dimensions. Then*

$$LB_\rho(p) \leq \mathbb{E}\left[T_\rho(p)\right] \leq UB_\rho(p),$$

where the upper bound

$$UB_\rho(p) = LB_\rho(p) + \zeta_\rho^{1/2}\,\mathbb{E}\,(\chi_\ell)[1 - \Psi(\theta_\rho, p)],$$

and

- $\zeta_\rho = \frac{2\rho-2}{3\rho-2}, \quad \theta_\rho = \arccos\left(\zeta_\rho^{1/2}\right)$;
- χ_ℓ *is a chi-distributed random variable with* $\ell = \binom{p+\rho-1}{\rho}$ *degrees of freedom*;
- $\Psi(\theta_\rho, p) = \sum_{e=0\ e:even}^{p-1} \omega_{p-e,\rho} \overline{\mathcal{B}}_{(p-e)/2,(\ell-p+e)/2}(\zeta_\rho)$;
- $\overline{\mathcal{B}}_{\alpha,\beta}(\cdot)$ *is the upper tail probability of the beta distribution* $Beta(\alpha,\beta)$; *and*
- $\omega_{p-e,\rho} = (-1)^{e/2}\,\rho^{(p-1)/2}\left(\frac{\rho-1}{\rho}\right)^{e/2} \frac{\Gamma[(p+1)/2]}{\Gamma[(p+1-e)/2](e/2)}$.

Koch and Naito gave an expression for the lower bound, which I have not included because the lower bound vanishes rapidly with increasing dimension and thus contains little information. The second term in the upper bound of $\mathrm{UB}_\rho(p)$ therefore quickly dominates the behaviour of $\mathrm{UB}_\rho(p)$. Figure 10.14 shows the estimated values $\widehat{\mathrm{UB}}_\rho(p)$ for skewness in the top panel and for kurtosis in the lower panel. The dimension p is displayed on the x-axis. These estimates are calculated with Mathematica, as described in Koch and Naito (2007), and are tabulated in their paper. The kurtosis values increase more rapidly than the skewness values, and both are non-linear in p.

We are now ready to define the 'best' p as in Koch and Naito (2007).

Definition 10.25 For $p \leq r$, let $\mathbb{X}^{\diamond}(p)$ be p-white data derived from data $\mathbb{X}$ with rank r. For $\rho = 3,4$, the **dimension selector** $\widehat{I}_\rho$ is the bias-adjusted version of the sample skewness or kurtosis

$$\widehat{I}_\rho(p) = \sqrt{\frac{n}{\rho}}\, b_\rho\left[\mathbb{X}^{\diamond}(p)\right] - \widehat{\mathrm{UB}}_\rho(p), \tag{10.39}$$

and $\widehat{I}_\rho$ results in the **most non-Gaussian dimension** p^*_ρ, which is defined by

$$p^*_\rho = \underset{2\leq p\leq r}{\operatorname{argmax}}\, \widehat{I}_\rho(p). \tag{10.40}$$

□

The dimension p^*_ρ is optimal in that it maximises $\widehat{I}_\rho$, the gap between a normalised version of the sample quantity $b_\rho\left[\mathbb{X}^{\diamond}(p)\right]$ and the bound for the Gaussian. The upper bounds for skewness and kurtosis grow with the dimension p of the whitened data, and the dimension selector $\widehat{I}_\rho$ chooses the dimension p^*_ρ which captures the non-Gaussian nature of the data best while reducing the dimensionality of the problem. In this sense, p^*_ρ is the most non-Gaussian dimension for $\mathbb{X}$.

Table 10.5 is extracted from table 1 of Koch and Naito (2007), but additionally includes the *Dow Jones returns* and the data *bank of kidneys* from Example 10.6, which are not covered in their table. For raw and scaled data, Table 10.5 lists values of p^* for skewness as p^*_3 and for kurtosis as p^*_4. The dimension selector $\widehat{I}_\rho$ of (10.39) depends on the sample size. To exhibit this dependence explicitly, Koch and Naito consider subsets of the *abalone* data and the *breast cancer* data and also calculate p^* for the subsets. The column 'per cent of variance' reports the contribution to total variance that is achieved with p^* PC dimensions. If the contribution to total variance differs for p^*_3 and p^*_4, then I list that arising from p^*_3 first.

Table 10.5 shows that the skewness dimension does not exceed the kurtosis dimension, and often the two values agree. Koch and Naito observed that kurtosis sometimes has a local maximum at p^*_3, whereas the argmax may be considerably larger. I have indicated the occurrence of such local maxima by the superscript dagger in the last column of the table.

We conclude this chapter with an example from Koch and Naito (2007) that explores the performance of the dimension selector for simulated data.

Example 10.9 In previous analyses of the **simulated source data**, we considered sources with identical marginals. In this example we include Gaussian marginals in the sources. For Independent Component Analysis, such marginals are uninteresting or 'noise'. The task will therefore be to find the number of non-Gaussian marginals.

Koch and Naito calculate the dimension p^*_ρ of (10.40) for five-dimensional data with three non-Gaussian variables and for ten-dimensional data with four non-Gaussian variables. For each source, the non-Gaussian variables have identical marginals from the uniform, exponential, beta or bimodal distribution, with the same parameters as in Example 10.5. As in that example, 'beta' refers to the $Beta(5, 1.2)$ distribution. They generated 1,000 repetitions for each source distribution and sample sizes $n = 50, 100, 500$ and $1,000$. The mixing matrices are the same as those in Example 10.5.

Table 10.5 *Dimension Selection with p^*_ρ of (10.40) for Real Data*

Data	$d \times n$	Data type	Percent of variance	p^*_3	p^*_4
Swiss bank notes	6×200	Raw	97.31	4	4
Abalone	8×4177	Raw	100	7	7
	8×1000	Raw, 1000 rand records	100	6	7
	8×4177	Scaled	100	6	7
	8×1000	Scaled, 1000 rand records	98.39; 99.26	4	5
Wine recognition	13×178	raw	100	8	10
	13×178	Scaled	89.34	7	7
Breast cancer	30×569	raw	100	15	$27^\dagger$
	30×100	Raw, first 100 records	99.9; 100	2	$24^\dagger$
	30×569	Scaled	100	25	26
	30×100	Scaled, first 100 records	96.62; 98.03	11	13
Dow Jones returns	30×2528	Raw	95.39	26	26
Illicit drug market	66×17	Raw	99.94	2	2
	66×17	Scaled	67.49	3	3
Bank of kidneys	264×36	Raw	85.49	5	5

Note: The symbol † indicates that a local maximum of the kurtosis criterion equals p^*_3.

Table 10.6 *Dimension p^*_ρ of (10.40) for Simulated Data*

	Kurtosis results			
Sample size	50	100	500	1,000
Uniform	2	2	2	2
Exponential	3	3	4	4
Beta	4	4	4	4
bimodal	2	2	3	3
	Skewness results			
Sample size	50	100	500	1,000
Uniform	2	2	2	2
Exponential	2	3	7	7
Beta	4	4	5	5
Bimodal	2	2	3	4

Note: Kurtosis results for three non-Gaussian dimensions, and skewness results for four non-Gaussian dimensions.

Koch and Naito (2007) used the skewness criterion $\mathcal{G}_3$ and kurtosis criterion $\mathcal{G}_4$ with $K = 10$ in Algorithm 10.1 to find the almost independent solutions. They calculated the skewness and kurtosis of the p-dimensional source estimates and found the optimal dimension p^*_ρ of (10.40). Table 10.6 reports the mode of the p^*_ρ values over the 1,000 repetitions separately for each non-Gaussian distribution and each sample size. The first part of the table shows the dimension selected with kurtosis for the five-dimensional data with three non-Gaussian dimensions, and the second part of the table shows the dimension selected with skewness for the ten-dimensional data with four non-Gaussian marginals.

The dimension selectors perform better for non-symmetric distributions; the symmetric uniform marginals in particular seem to be harder to distinguish from the Gaussian with the kurtosis and skewness criteria; hence, p^*_ρ is lower than the number of non-Gaussian dimensions. The best p^*_ρ is monotone with the sample size and may result in a number of values for a single distribution which typically contains the true value. The reason why p^*_ρ can be larger than the true number of non-Gaussian dimensions is the fact that FastICA with the skewness and kurtosis approximations $\mathcal{G}_3$ and $\mathcal{G}_4$ finds non-Gaussian directions in simulated Gaussian data.

Overall, the results obtained for the sources with contaminating Gaussian dimensions are consistent with those obtained for the completely non-Gaussian sources in Example 10.5. ■

Koch and Naito (2007) reported comparisons with the dimension selector of Tipping and Bishop (1999), which I describe in Theorem 2.26 in Section 2.8.1. It is worth recalling that the approach of Tipping and Bishop was intended for a Gaussian Factor Analysis framework – see Theorem 7.6 of Section 7.4.2. As the distributional assumptions of Koch and Naito differ considerably from those of Tipping and Bishop, it is not surprising that the resulting dimension selectors yield different 'best' dimensions. A comparison of the results in Examples 2.17 and 2.18 in Section 2.8.1 with those shown in Table 10.5 reveals that the p^*_ρ values are higher than those found with the selector of Tipping and Bishop. The p^* of Tipping and Bishop maximises the Gaussian likelihood over dimensions $p \leq d$, whereas the two p^*_ρ values of Koch and Naito maximise the deviation from the Gaussian distribution, and are appropriate for non-Gaussian data. The distributional assumptions underpinning each of the two methods are very different. As a consequence, interpretation of the selected dimension should take into account how well the data satisfy the assumptions of the chosen method.

The two methods are a beginning, and more methods are needed that make informed choices about the best dimension. In Corollary 13.6 in Section 13.3.4 we will meet the dimension-selection rule of Fan and Fan (2008), which makes use of the labels that are available in classification but does not apply to unlabelled data. Typically, for data we do not know the underlying truth and hence do not know which selector is better. It is, however, important to have different methods available, to apply more than one dimension selector and to compare and combine the results where possible or appropriate.

11

Projection Pursuit

'Which road do I take?' Alice asked. 'Where do you want to go?' responded the Cheshire Cat. 'I don't know,' Alice answered. 'Then,' said the Cat, 'it doesn't matter' (Lewis Carroll, *Alice's Adventures in Wonderland*, 1865).

11.1 Introduction

Its name, Projection Pursuit, highlights a key aspect of the method: the search for projections worth pursuing. Projection Pursuit can be regarded as embracing the classical multivariate methods while at the same time striving to find something 'interesting'. This invites the question of what we call interesting. For scores in mathematics, language and literature, and comprehensive tests that psychologists, for example, use to find a person's hidden indicators of intelligence, we could attempt to find as many indicators as possible, or one could try to find the most interesting or most informative indicator. In Independent Component Analysis, one attempts to find all indicators, whereas Projection Pursuit typically searches for the most interesting one.

In Principal Component Analysis, the directions or projections of interest are those which capture the variability in the data. The stress and strain criteria in Multidimensional Scaling variously broaden this set of directions. Of a different nature are the directions of interest in Canonical Correlation Analysis: they focus on the strength of the correlation between different parts of the data. Projection Pursuit covers a rich set of directions and includes those of the classical methods. The directions of interest in Principal Component Analysis, the eigenvectors of the covariance matrix, are obtained by solving linear algebraic equations. A deviation from the precise mathematical expression of the principal component (PC) directions is shared by Multidimensional Scaling and Projection Pursuit. This feature makes the latter methods richer but also less transparent.

The goals of Projection Pursuit are

- to pick 'interesting' directions in higher-dimensional data,
- to describe the low-dimensional projections by a statistic, called the *projection index*, and
- to find projections which maximise the index.

The higher the value of the projection index, the more interesting is the projection. The original projection index of Friedman and Tukey (1974) is the product of a robust measure of scale and a measure of interpoint distances, but this index has been superseded by proposals in Friedman (1987) and Jones and Sibson (1987).

Projection Pursuit is not a precisely defined technique; it allows and encourages the development of criteria which capture mathematically a notion of relevant or salient features and structure in data. Its aims include finding such features and mapping them onto low-dimensional manifolds which allow visual inspection.

Friedman and Tukey (1974) are credited with the name and the first successful implementation of the method. The ideas of Friedman and Tukey are based on those of Multidimensional Scaling and in particular on the work of Kruskal (1969, 1972). Early contributors to Projection Pursuit include Friedman and Stuetzle (1981), Friedman, Stuetzle, and Schroeder (1984), Diaconis and Freedman (1984), Huber (1985), Friedman (1987), Jones and Sibson (1987), and Hall (1988, 1989b). This list shows that, strictly speaking, Projection Pursuit precedes Independent Component Analysis. The two methods have much in common, but they differ in their underlying ideas and partly in their goals. Over the last few decades, the way we view Projection Pursuit has changed, and as a consequence, the computational and algorithmic aspects of Projection Pursuit require some comments. The approximations and their numerical implementations which formed an integral part of the original Projection Pursuit approaches have been superseded by computationally more feasible and more efficient algorithms, and this reality has informed our view of Projection Pursuit. We will focus on these developments and current practice in Section 11.4.

We begin with an exploration of 'interesting' directions in Section 11.2 and consider candidates for one-dimensional (1D) projection indices – separately for the population and the sample. Section 11.3 looks at extensions to two- and three-dimensional projections and their indices. Section 11.4 gives an overview of Projection Pursuit in practice. It starts with a comparison of Projection Pursuit and Independent Component Analysis and explains the developments that have influenced and informed the way we now calculate Projection Pursuit directions. Section 11.5 deals with theoretical and asymptotic properties of 1D indices, and the final Section 11.6 outlines the main ideas of Projection Pursuit in the context of density estimation and regression. Problems for this chapter can be found at the end of Part III.

11.2 One-Dimensional Projections and Their Indices

11.2.1 Population Projection Pursuit

The increasing complexity of data makes the task of summarising intrinsic features in a small number of projections ever more challenging. Unlike in Independent Component Analysis, which attempts to recover all d source variables from d signal components, in Projection Pursuit we typically focus on a small number of projections.

There is no general agreement about what features give rise to interesting projections.

If data contain a number of clusters, then projections which separate the data into clusters may be regarded as more interesting than those which do not. Huber (1985) states that 'a projection is less interesting the more normal it is' and supports this claim with a number of heuristic arguments:

1. A multivariate distribution is normal if and only if all its 1D projections are normal. So all of them are equally (un)interesting.
2. If the least normal projection is (almost) normal, we need not look at any other projection.

3. For high-dimensional point clouds, low-dimensional projections are approximately normal.

The last of these follows from Theorem 10.22 in Section 10.7.2 and is also stated in Diaconis and Freedman (1984).

Based on Huber's heuristic arguments, the pursuit of non-Gaussian structure in data can be likened to searching for interesting structure. If we accept Huber's claim, then we know what directions are *uninteresting*, but the negation of uninteresting is not so straightforward, as non-Gaussian structure is manifested in many different ways. The four distributions we consider in Example 10.5 of Section 10.6 variously differ from the Gaussian:

- distributions with compact support and no tails: the uniform and the *Beta*(5,1.2),
- asymmetric distributions: the *Beta*(5,1.2), the exponential and the bimodal, and
- distributions with more than one mode: the bimodal.

One could look for subtler differences such as those exhibited by t-distributions, which have fatter tails than the Gaussian but essentially agree with the Gaussian in the distinguishing properties just listed. Quoting Jones and Sibson (1987), 'it is easier to start from projections that are "not interesting" and then deviate from the non-interesting projections.'

Which non-Gaussian properties should we be targeting? Before answering this question, we look at hypothesis tests. The null hypothesis preserves the status quo. The alternative can be very specific – as in (3.30) in Section 3.6, where all correlation coefficients are zero under the null hypothesis, and the last and smallest one is non-zero under the alternative. A different alternative hypothesis is a general negation of the null (see (3.29) in Section 3.6). Friedman (1987) points out the parallels between hypothesis testing and deciding on non-Gaussian features in his statement that 'any test statistic for testing normality could serve as the basis for a projection index. Different test statistics have the property of being more (or less) sensitive to different alternative distributions.' We may want to negate one particular property captured by the Gaussian distribution, or we may want to be 'as far away as possible' from the normal. In view of all these comments, I will not give a definition of 'interestingness'. Following Jones and Sibson (1987), we start with the *uninteresting* Gaussian projections and attempt to maximise a deviation from uninteresting with respect to a suitably defined projection index.

Definition 11.1 Let $\mathbf{X}$ be a d-dimensional random vector, and let $\mathbf{a} \in \mathbb{R}^d$ be a direction vector. A **projection index** $\mathcal{Q}$ is a function which assigns a real number to pairs $(\mathbf{X}, \mathbf{a})$. □

Recall that a direction or a direction vector is a vector of norm one.

Example 11.1 In the framework of Principal Component Analysis and Discriminant Analysis, we consider suitable projection indices.

1. **PCA**. Let $\mathbf{X} \sim (\mu, \Sigma)$ be a d-dimensional random vector, and let $\mathbf{a} \in \mathbb{R}^d$ be a direction vector. A projection index for $\mathbf{X}$ and $\mathbf{a}$ is the map

$$\mathcal{Q}(\mathbf{X}, \mathbf{a}) = \operatorname{var}(\mathbf{a}^{\mathsf{T}}\mathbf{X}).$$

The maximiser of this projection index over direction vectors $\mathbf{a}$ is the eigenvector η_1 of Σ which corresponds to the largest eigenvalue λ_1 of Σ. Further,

$$\lambda_1 = \max_{\{\mathbf{a}:\|\mathbf{a}\|=1\}} \mathcal{Q}(\mathbf{X}, \mathbf{a}),$$

and the projection is $\eta_1^{\mathsf{T}}(\mathbf{X} - \boldsymbol{\mu}) = W_1$. This index is designed to find directions which contribute maximally to the variance. Instead of defining the projection index by means of the variance, we could have used the standard deviation. The maximum value would differ because it would result in $\sqrt{\lambda_1}$, but the maximiser would remain the same.

2. **DA**. Let $\mathbf{X}$ be a d-dimensional random vector which belongs to $\mathcal{C}_\nu = (\boldsymbol{\mu}_\nu, \Sigma_\nu)$, one of the κ classes with $\nu \le \kappa$. Let B and W be the matrices defined in (4.7) of Theorem 4.6 in Section 4.3. Let $\mathbf{a} \in \mathbb{R}^d$ be a direction. For $(\mathbf{X}, \mathbf{a})$, define the projection index

$$\mathcal{Q}(\mathbf{X}, \mathbf{a}) = \frac{\mathbf{a}^{\mathsf{T}} B \mathbf{a}}{\mathbf{a}^{\mathsf{T}} W \mathbf{a}}.$$

The maximum value of $\mathcal{Q}$ over directions $\mathbf{a}$ is achieved when $\mathbf{a} = \eta$, the eigenvector of $W^{-1}B$ which corresponds to the largest eigenvalue $\mathfrak{d}$ of Theorem 4.6. This maximiser η yields the projection $\eta^{\mathsf{T}}\mathbf{X}$, which is the essence of Fisher's discriminant rule (4.9) in Section 4.3.1. ■

In Principal Component Analysis we distinguish between the raw and the scaled data. For Projection Pursuit, Jones and Sibson (1987) proposed scaling *and* sphering the random vector or data and finding non-Gaussian directions for these whitened data. As we have seen throughout Chapter 10, working with white vectors simplifies the search for independent non-Gaussian directions. The same applies to projection indices, which measure the departure from Gaussianity.

Definition 11.2 Let $\mathbf{X} \sim (\mathbf{0}, \mathbf{I}_{d\times d})$. Let $\mathbf{a} \in \mathbb{R}^d$ be a direction, and put $X_{\mathbf{a}} = \mathbf{a}^{\mathsf{T}}\mathbf{X}$. A **projection index for measuring departure from Gaussianity** is a function $\mathcal{Q}$ which satisfies

$$\mathcal{Q} \ge 0 \qquad \text{and} \qquad \mathcal{Q}(\mathbf{X}, \mathbf{a}) = 0 \qquad \text{if } X_{\mathbf{a}} \text{ is Gaussian.}$$

□

The original projection index of Friedman and Tukey (1974) for measuring departure from Gaussianity is no longer used, and I will therefore not define it. Just a few years later, a number of different indices for measuring departure from Gaussianity were proposed and analysed in the literature. Some of the indices are originally defined for random vectors with arbitrary covariance matrix Σ. I have adjusted these definitions to random vectors with the identity covariance matrix for notational convenience and easier comparisons.

Candidates for projection indices which measure departure from Gaussianity. Consider $\mathbf{X} \sim (\mathbf{0}, \mathbf{I}_{d\times d})$. Let $\mathbf{a} \in \mathbb{R}^d$ be a direction vector, and let $f_{\mathbf{a}}$ be the probability density function of the projection $X_{\mathbf{a}} = \mathbf{a}^{\mathsf{T}}\mathbf{X}$. Let ϕ be the univariate standard normal probability density function, and let Φ be its distribution function. The following projection indices measure departure of $f_{\mathbf{a}}$ from the standard normal.

1. The **cumulant index** is

$$\mathcal{Q}_{\mathrm{C}}(\mathbf{X},\mathbf{a}) = \frac{1}{48}\left[4\beta_3^2(X_{\mathbf{a}}) + \beta_4^2(X_{\mathbf{a}})\right], \tag{11.1}$$

where β_3 and β_4 are the skewness and kurtosis of $X_{\mathbf{a}}$ (see Result 9.5 in Section 9.3).

2. The **entropy indices** are

$$\mathcal{Q}_{\mathrm{E}}(\mathbf{X},\mathbf{a}) = \int f_{\mathbf{a}} \log f_{\mathbf{a}} \qquad \text{and} \qquad \mathcal{Q}_{\mathrm{E}'}(\mathbf{X},\mathbf{a}) = \mathcal{Q}_{\mathrm{E}}(\mathbf{X},\mathbf{a}) + \log\left[(2\pi e)^{1/2}\right]. \tag{11.2}$$

3. Put

$$\Theta = 2\Phi(X_{\mathbf{a}}) - 1, \tag{11.3}$$

and let f_Θ be the probability density function of Θ. The **projection index** based on the **deviation from the uniform distribution** is

$$\mathcal{Q}_{\mathrm{U}}(\mathbf{X},\mathbf{a}) = \int_{-1}^{1}\left(f_\Theta - \frac{1}{2}\right)^2. \tag{11.4}$$

4. The **projection index** based on the **difference from the Gaussian** is

$$\mathcal{Q}_{\mathrm{D}}(\mathbf{X},\mathbf{a}) = \int (f_{\mathbf{a}} - \phi)^2. \tag{11.5}$$

5. Assume that the probability density function of **X** has compact support. Let ϕ_a be the marginal of the standard normal probability density function in the direction of **a**. The **projection index** based on the **ratio with the Gaussian** is

$$\mathcal{Q}_{\mathrm{R}}(\mathbf{X},\mathbf{a}) = E\left[\log\left(\frac{f_{\mathbf{a}}}{\phi_{\mathbf{a}}}\right)\right], \tag{11.6}$$

where the expectation is taken with respect to the probability density function of **X**.

6. The **Fisher information index** is

$$\mathcal{Q}_{\mathrm{F}}(\mathbf{X},\mathbf{a}) = \int \left(\frac{f'_{\mathbf{a}}}{f_{\mathbf{a}}}\right)^2 f_{\mathbf{a}} - 1, \tag{11.7}$$

where $f'_{\mathbf{a}}$ is the first derivative of $f_{\mathbf{a}}$.

The two versions of the entropy index differ by a constant. The first expression is commonly applied to data, and the second satisfies the properties of Definition 11.2. Huber (1985) and Jones and Sibson (1987) considered $\mathcal{Q}_{\mathrm{E}}$ and $\mathcal{Q}_{\mathrm{F}}$, Friedman (1987) proposed $\mathcal{Q}_{\mathrm{U}}$. The index $\mathcal{Q}_{\mathrm{U}}$ is described as a deviation from the uniform distribution, however, because of the transformation (11.3), it measures the deviation of $X_{\mathbf{a}}$ from the Gaussian. Hall (1988, 1989b) proposed $\mathcal{Q}_{\mathrm{R}}$ and $\mathcal{Q}_{\mathrm{D}}$, respectively, and derived theoretical properties of these indices. I explain the theoretical developments of Hall (1988, 1989b) in Section 11.5.

The cumulant index $\mathcal{Q}_{\mathrm{C}}$ differs from the others in that it is based on the skewness and kurtosis of $X_{\mathbf{a}} = \mathbf{a}^{\mathsf{T}}\mathbf{X}$ rather than the probability density function of $X_{\mathbf{a}}$. The population version $\mathcal{Q}_{\mathrm{C}}$, which I have given here, is not explicitly listed in any of the early contributions to Projection Pursuit, but Jones and Sibson (1987) considered a sample version of $\mathcal{Q}_{\mathrm{C}}$ which we will meet in the next section. The cumulant index $\mathcal{Q}_{\mathrm{C}}$ is based on the first two terms in the approximation to the negentropy (see Theorem 10.11 in Section 10.4.2) and is therefore a more accurate approximation to the negentropy than $\mathcal{G}_3$ or $\mathcal{G}_4$ of (10.17) in Section 10.4.2.

The six indices are functions of $\mathbf{X}$ and $\mathbf{a}$, but I have defined the indices by means of $f_{\mathbf{a}}$, the probability density function of $\mathbf{a}^{\mathsf{T}}\mathbf{X}$. Thus, the definition of the indices exploits the correspondence between a random vector or variable and its probability density function. In the definition of entropy (see Definition 9.6 in Section 9.4), we came across a similar correspondence between random vectors and their probability density functions.

To gain some insight into the non-Gaussian indices, we take a closer look at some of them and start with properties of $\mathcal{Q}_{\mathrm{E}}$ and $\mathcal{Q}_{\mathrm{F}}$.

Proposition 11.3 [Huber (1985)] *Let* $\mathbf{X} \sim (\mathbf{0}, \mathbf{I}_{d\times d})$. *Let* $\mathbf{a} \in \mathbb{R}^d$ *be a direction vector. Let* $f_{\mathbf{a}}$ *be the probability density function of the projection* $X_{\mathbf{a}} = \mathbf{a}^{\mathsf{T}}\mathbf{X}$, *and let* ϕ *be the univariate standard normal probability density function. Define* $\mathcal{Q}_{E'}$ *and* $\mathcal{Q}_F$ *as in (11.2) and (11.7). The following hold.*

$$\mathcal{Q}_{E'}(\mathbf{X},\mathbf{a}) = \int \log\left(\frac{f_{\mathbf{a}}}{\phi}\right) f_{\mathbf{a}},$$

$$\mathcal{Q}_F(\mathbf{X},\mathbf{a}) = \int \left(\frac{f'_{\mathbf{a}}}{f_{\mathbf{a}}} - \frac{\phi'}{\phi}\right)^2 f_{\mathbf{a}}. \tag{11.8}$$

Further,

$$\mathcal{Q}_{E'}(\mathbf{X},\mathbf{a}) = \mathcal{Q}_F(\mathbf{X},\mathbf{a}) = 0 \quad \Longleftrightarrow \quad \mathbf{a}^{\mathsf{T}}\mathbf{X} \textit{ is Gaussian.}$$

Note that $\mathcal{Q}_{\mathrm{E}'}(\mathbf{X},\mathbf{a}) - \log\left[(2\pi e)^{1/2}\right] = -\mathcal{H}(f_{\mathbf{a}})$ follows from (9.11) in Section 9.4. By part 1 of Result 9.7 in Section 9.4, the entropy has a maximum when $f_{\mathbf{a}}$ is Gaussian. A proof of Proposition 11.3 is the task of one of the Problems at the end of Part III.

The index $\mathcal{Q}_{\mathrm{R}}$ of (11.6) is defined for random vector $\mathbf{X} \sim (\boldsymbol{\mu}, \Sigma)$ in Hall (1988). To guarantee that the ratio in (11.6) makes sense, Hall (1988) required that $\mathbf{X}$ have compact support and that the expectation be defined on a suitably chosen subset of the support. The indices $\mathcal{Q}_{\mathrm{D}}$ and $\mathcal{Q}_{\mathrm{R}}$ are similar in that they consider the difference between $f_{\mathbf{a}}$ and a Gaussian, but they differ in the distance measure that is applied to the difference of the densities in each case.

Next, we turn to the index $\mathcal{Q}_{\mathrm{U}}$ proposed in Friedman (1987). If $X_{\mathbf{a}} \sim \mathcal{N}(0,1)$, then Θ in (11.3) is uniformly distributed on $[-1,1]$. The index $\mathcal{Q}_{\mathrm{U}}$ therefore portrays departure from the uniform distribution, a clever idea which is easy to assess visually. Because Θ is a transformation of $X_{\mathbf{a}}$, its probability density function is

$$f_\Theta(\Theta) = \frac{f_a(X_{\mathbf{a}})}{2\phi(X_{\mathbf{a}})}. \tag{11.9}$$

A simple calculation shows that

$$\mathcal{Q}_{\mathrm{U}}(\mathbf{X},\mathbf{a}) = \frac{1}{4}\int_{-1}^{1} \left(\frac{f_{\mathbf{a}}}{\phi} - 1\right)^2, \tag{11.10}$$

and thus the projection index $\mathcal{Q}_{\mathrm{U}}$ compares $f_{\mathbf{a}}$ with the normal probability density function. For details of the relationship between the probability density functions of X and a transformed variable $Y = g(X)$, see, for example, Casella and Berger (2001). Although $\mathcal{Q}_{\mathrm{U}}$ considers the difference between $f_{\mathbf{a}}$ and the normal, what we calculate is the difference of a transformed $f_{\mathbf{a}}$ from the uniform. The latter is much easier to assess visually, as Example 11.2 illustrates.

The expression (11.10) shows the similarity between $\mathcal{Q}_U$ and $\mathcal{Q}_D$ in (11.5) more clearly. The two indices have similar asymptotic properties, as we shall see in Section 11.5.2. Hall (1989b) pointed out that $\mathcal{Q}_D$ deals better with deviations in the tails and, in particular, with heavy-tailed distributions, but this difference may be more relevant in theoretical considerations.

Example 11.2 For univariate **non-Gaussian distributions**, I illustrate graphically aspects of the projection index $\mathcal{Q}_U$. I use the four 1D non-Gaussian distributions that serve as the marginal source distributions in Example 10.5 in Section 10.6.3, namely,

1. the uniform distribution on $[0,1]$,
2. the exponential distribution with mean 0.5,
3. the beta distribution $Beta(\alpha,\beta)$ with $\alpha = 5$ and $\beta = 1.2$, and
4. a mixture of two Gaussians, with 25 per cent from $\mathcal{N}(0, 0.49)$ and 75 per cent from $\mathcal{N}(4, 1.44)$.

The projection index $\mathcal{Q}_U$ is based on random variables with variance one, so I first transform the distributions. For $X \sim (\mu, \sigma^2)$, put $T = \sigma^{-1}(X - \mu)$. If X has the probability density function f_X, then the probability density function f_T of T satisfies $f_T(T) = \sigma f_X(\sigma T + \mu)$.

Figure 11.1 displays results pertaining to the four distributions listed at the beginning of this example. The ith row in the set of panels corresponds to the ith distribution in the list. The first column shows the probability density function f_T of the standardised T, and the

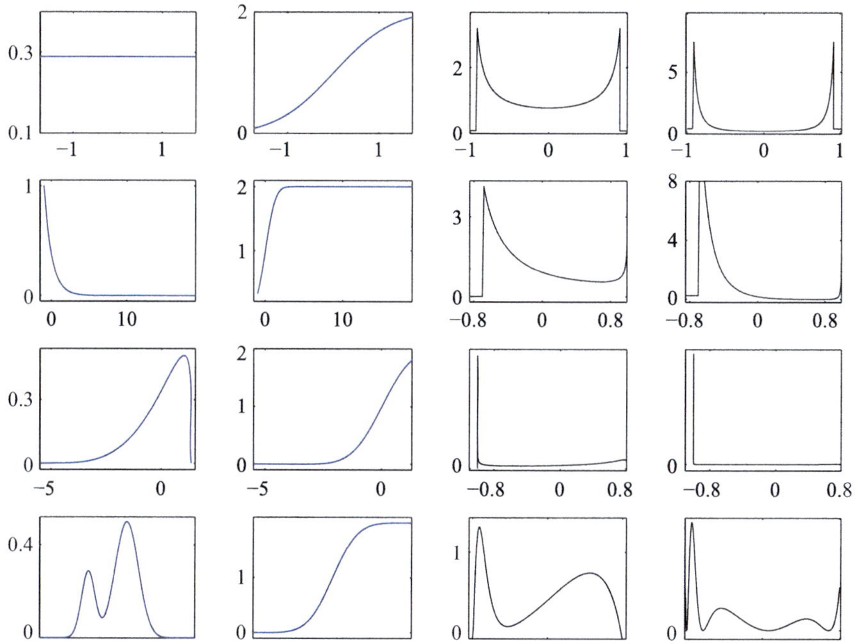

Figure 11.1 For the uniform (row 1), exponential (row 2), beta (row 3) and bimodal (row 4) distributions and T standardised, $f_T(T)$, $2\Phi(T)$, $f_\Theta(\Theta)$ and $(f_\Theta - 0.5)^2$ are shown in columns 1–4, respectively.

Table 11.1 *Projection Pursuit (PP) and Independent Component Analysis (ICA)*

	PP	ICA
Key idea	Non-Gaussianity	Independence of non-Gaussian sources
Statistic	Projection index	Approximation to mutual information
	$\mathcal{Q}$ of (11.1)–(11.7)	$\mathfrak{J}$ of step 1 in Algorithm 10.1

second column shows $2\Phi(T)$, both as functions of T. The third column shows f_Θ of (11.9), but now as a function of Θ. The last column shows the integrand $(f_\Theta - 0.5)^2$ of (11.4), again as a function of Θ.

If I had included results pertaining to T from the standard normal distribution, then its corresponding third column would be a horizontal line at $f_\Theta(\Theta) = 0.5$, and the display in the last column would be the constant zero. For the four non-Gaussian distributions, we see clear deviations from straight lines in the four figures in column three. For the beta distribution in row three, f_Θ has a sharp peak, and for the bimodal distribution in row four, f_Θ shows two bumps. In all four cases the index $\mathcal{Q}_\mathrm{U}$ deviates considerably from zero. We estimate $\mathcal{Q}_\mathrm{U}$ numerically for these four distributions in the Problems at the end of Part III. ∎

The index $\mathcal{Q}_\mathrm{U}$ takes different values for the four models in Example 11.2. The size of $\mathcal{Q}_\mathrm{U}$ for these models could be used to rank the deviation from Gaussianity of the distributions. We will not do so but rather regard $\mathcal{Q}_\mathrm{U}$ as a tool for finding deviations from the Gaussian.

A brief reflection on the connection between Projection Pursuit and Independent Component Analysis is appropriate before we look at projection indices for data. In Independent Component Analysis, $\mathfrak{J}$ of (i) in Section 10.3 and estimators $\widehat{\mathfrak{J}}$ are the key quantities. The projection indices $\mathcal{Q}$, (11.1)–(11.7), are the analogues of the statistics $\mathfrak{J}$ in Projection Pursuit. The functions $\mathfrak{J}$ are approximations to the mutual information $\mathcal{I}$ which measures the degree of independence; the analogue in Projection Pursuit is the concept of *non-Gaussianity*. We do not have a unique mathematical way of describing non-Gaussianity. There are advantages and disadvantages in this lack: it gives us greater flexibility to define projection indices, but at the same time, the choice of the *best* index is less focused. Table 11.1 shows this first comparison of the two methods.

11.2.2 Sample Projection Pursuit

For data, we consider the first five indices (11.1) to (11.6) in the list of candidates for measuring departure from Gaussianity. For notational convenience, I refer to these five indices as *projection indices* or merely as *indices* for the remainder of this chapter.

In Independent Component Analysis, we worked with approximations $\mathfrak{J}$ to the mutual information $\mathcal{I}$ and their estimators $\widehat{\mathfrak{J}}$. In this section we explore estimators $\widehat{\mathcal{Q}}$ for indices $\mathcal{Q}$ of (11.1) to (11.6). The estimators $\widehat{\mathfrak{J}}$ in Independent Component Analysis are designed with a focus on efficient computations. In contrast, in Projection Pursuit, more statistical issues are addressed, namely:

1. Does the estimator $\widehat{\mathcal{Q}}$ converge to $\mathcal{Q}$?
2. Is the maximiser of $\widehat{\mathcal{Q}}$ close to that of $\mathcal{Q}$?

We consider three approaches for defining the estimators $\widehat{\mathcal{Q}}$. Later, in Section 11.5, I describe properties of these estimators, including convergence of the estimators and their maximisers to the respective population quantities.

Consider data $\mathbb{X} = \begin{bmatrix}\mathbf{X}_1 \cdots \mathbf{X}_n\end{bmatrix}$ from a distribution F, where $\mathbf{X}_i \sim (\mathbf{0}, \mathbf{I}_{d\times d})$ for $i \leq n$. Let $\mathbf{a} \in \mathbb{R}^d$ be a direction vector, and let $f_{\mathbf{a}}$ be the probability density function of $\mathbf{X}_{\bullet\mathbf{a}} = \mathbf{a}^{\mathsf{T}}\mathbb{X}$. We estimate projection indices by approximations which are

- directly based on cumulants,
- based on orthogonal polynomials, and
- based on kernel density estimators of $f_{\mathbf{a}}$.

We look at each of these approaches separately and explore properties of the estimators.

The index $\widehat{\mathcal{Q}}_{\mathrm{C}}$ – based on cumulants. Jones and Sibson (1987) consider two different approximations to the entropy index (11.2). The following theorem contains the essence of their first approximation, which is based on moments and deals with probability density functions that do not deviate too much from the Gaussian.

Theorem 11.4 [Jones and Sibson (1987)] *Let X be a univariate random variable with probability density function f, and assume that f is of the form $f = (1+\epsilon)\phi$, where ϕ is the standard normal probability density function and ϵ is a function of X which satisfies*

$$\int \epsilon(x)x^m\phi(x)dx = 0 \qquad \textit{for } m = 0, 1, 2.$$

Let β_3 and β_4 be the skewness and kurtosis of X as in Result 9.5 in Section 9.3. If ϵ is small and decreases sufficiently fast at $\pm\infty$, then the entropy of f is

$$\mathcal{H}(f) \approx -\frac{1}{48}\left[4\beta_3^2(X) + \beta_4^2(X)\right]. \tag{11.11}$$

This theorem tells us that for probability density functions that are not too different from the Gaussian, the entropy is close to a weighted sum of skewness and kurtosis, and this approximation motivates their choice of the cumulant index $\mathcal{Q}_{\mathrm{C}}$. The estimators b_3 and b_4 of Definition 9.3 in Section 9.3 for skewness and kurtosis, respectively, naturally lead to

$$\widehat{\mathcal{Q}}_{\mathrm{C}}(\mathbb{X}, \mathbf{a}) = \frac{1}{48}\left[4b_3^2(\mathbf{X}_{\bullet\mathbf{a}}) + b_4^2(\mathbf{X}_{\bullet\mathbf{a}})\right] \quad \text{for} \quad \mathbf{X}_{\bullet\mathbf{a}} = \mathbf{a}^{\mathsf{T}}\mathbb{X}. \tag{11.12}$$

Equipped with Theorem 11.4, Jones and Sibson (1987) proposed (11.12) as an estimator for $\mathcal{Q}_{\mathrm{E}}$ of (11.2). The approximation (11.11) reminds us of the negentropy approximation by cumulants in Theorem 10.11 in Section 10.4.2. Projection Pursuit and Independent Component Analysis – though starting from different premises – use the same moments and arrive at similar results. The proof of Theorem 10.11 is based on Edgeworth expansions of f about the standard normal ϕ and so is the proof of Jones and Sibson (1987). Jones and Sibson further assumed that f is of the form $f = (1+\epsilon)\phi$.

The indices $\widehat{\mathcal{Q}}_{\mathrm{U}}$ and $\widehat{\mathcal{Q}}_{\mathrm{D}}$ – based on orthogonal polynomials. Friedman (1987) and Hall (1989b) approximated their estimators $\mathcal{Q}_{\mathrm{U}}$ of (11.4) and $\mathcal{Q}_{\mathrm{D}}$ of (11.5) by orthogonal functions. Friedman (1987) used orthonormal Legendre polynomials p_j and showed that for $m < \infty$, $\mathcal{Q}_{\mathrm{U}}$ can be approximated by

$$\widehat{\mathcal{Q}}_{\mathrm{U,m}}(\mathbb{X}, \mathbf{a}) = \sum_{j=1}^{m}\left\{\frac{1}{n}\sum_{i=1}^{n} p_j\left[2\Phi(\mathbf{a}^{\mathsf{T}}\mathbf{X}_i) - 1\right]\right\}^2, \tag{11.13}$$

where Φ is the standard normal distribution function. We return to this expression in Theorem 11.6 in Section 11.5. Friedman pointed out that even for small positive values of m, $\widehat{\mathcal{Q}}_{\rm U,m}$ measures departure from Gaussianity unless n is very small. The computational effort increases linearly with m, and Friedman recommended using a small value, such as $4 \le m \le 8$. A potential disadvantage of $\widehat{\mathcal{Q}}_{\rm U,m}$ is that distributions whose first m Legendre polynomial moments are zero are not distinguishable from the normal distribution with this index.

Hall (1989b) chose orthonormal Hermite polynomials h_j and showed that for $m < \infty$, $\mathcal{Q}_{\rm D}$ can be approximated by

$$\widehat{\mathcal{Q}}_{\rm D,m}(\mathbb{X},\mathbf{a}) = \sum_{j=0}^{m}\left[\frac{1}{n}\sum_{i=1}^{n} h_j(\mathbf{a}^{\mathsf T}\mathbf{X}_i)\right]^2 - \frac{\sqrt{2}}{\sqrt[4]{\pi}}\frac{1}{n}\sum_{i=1}^{n} h_0(\mathbf{a}^{\mathsf T}\mathbf{X}_i) + \frac{1}{2\sqrt{\pi}}. \tag{11.14}$$

Hall (1989b) was interested in the convergence of the estimators $\widehat{\mathcal{Q}}_{\rm D,m}$ and $\widehat{\mathcal{Q}}_{\rm U,m}$ to their population indices and gave conditions for the maximisers of the estimators to converge to those of the population indices. I report these results in Theorem 11.7.

The indices $\widehat{\mathcal{Q}}_{\rm E}$ and $\widehat{\mathcal{Q}}_{\rm R}$ – based on kernel density estimators. Kernel density estimation, especially for univariate densities, is a well-established technique (see, e.g., Wand and Jones 1995). A few years earlier, when Jones and Sibson (1987) and Hall (1988) proposed their kernel-based indices, these methods were only just being developed, with these authors being some of the main forces in these developments.

In addition to their cumulant-based index $\widehat{\mathcal{Q}}_{\rm C}$ of (11.12), which approximates the entropy, Jones and Sibson (1987) proposed an index $\widehat{\mathcal{Q}}_{\rm E}$ which estimates $f_{\mathbf{a}}$ by a kernel density estimator $\widehat{f}_{\mathbf{a}}$ and replaces $f_{\mathbf{a}}$ in (11.2) by $\widehat{f}_{\mathbf{a}}$. For a fixed kernel and bandwidth, they defined $\widehat{\mathcal{Q}}_{\rm E}$ by

$$\widehat{\mathcal{Q}}_{\rm E}(\mathbb{X},\mathbf{a}) = \frac{1}{n}\sum_{i=1}^{n} \widehat{f}_{\mathbf{a}}(\mathbf{a}^{\mathsf T}\mathbf{X}_i)\log\left[\widehat{f}_{\mathbf{a}}(\mathbf{a}^{\mathsf T}\mathbf{X}_i)\right]. \tag{11.15}$$

Depending on the choice of kernel and bandwidth, $\widehat{\mathcal{Q}}_{\rm E}$ may result in a better approximation to $\mathcal{Q}_{\rm E}$ than $\widehat{\mathcal{Q}}_{\rm C}$, but $\widehat{\mathcal{Q}}_{\rm C}$ is simpler to compute.

Hall (1988) proposed an empirical version of $\mathcal{Q}_{\rm R}$ in (11.6) which used kernel density estimators in the following way: for a fixed kernel K and bandwidth h, let $\widehat{g}$ be the kernel density estimator of a univariate probability density function g. Let $\left[X_1 \cdots X_n\right]$ be univariate random variables from g. For a random variable X, define the **leave-one-out** kernel density estimator

$$\widehat{g}_{-i}(X) = \frac{1}{n-1}\sum_{\ell \neq i} K_h\left(X - X_\ell\right) \quad\text{with}\quad K_h(X) = \frac{1}{h}K\left(\frac{X}{h}\right) \quad\text{and } i = 1,\ldots,n. \tag{11.16}$$

Hall (1988) preferred the leave-one-out estimator to the simpler (standard) density estimator $\widehat{g}$, which is defined from all X_ℓ but otherwise the same as $\widehat{g}_{-i}$, because the latter reduces bias. Hall defined the index

$$\widehat{\mathcal{Q}}_{\rm R}(\mathbb{X},\mathbf{a}) = \frac{1}{n}\sum_{i=1}^{n}\left\{\log\left[\frac{\widehat{f}_{\mathbf{a},-i}(\mathbf{a}^{\mathsf T}\mathbf{X}_i)}{\widehat{\phi}_{\mathbf{a},-i}(\mathbf{a}^{\mathsf T}\mathbf{X}_i)}\right]\right\}, \tag{11.17}$$

and provided a theoretical foundation for it, which we consider in Section 11.5.

Fisher's information index (11.7) for the sample requires estimation of the probability density function and its derivative. This estimation calls for two bandwidths, one for the function and a different one for its derivative. The extra complexity involved in the estimation of the two functions is the most likely reason why this index has not been explored much for the sample.

11.3 Projection Pursuit with Two- and Three-Dimensional Projections

To find interesting low-dimensional structure in data, it may be necessary to find more than one non-Gaussian direction. This section looks at ways of obtaining two or more non-Gaussian directions and projections.

There are two main approaches for obtaining 2D projections:

1. a natural generalisation of $\mathbf{a}$ to a $d \times 2$ matrix A, and optimisation of $\mathcal{Q}(\mathbf{X}, A)$, and
2. a removal of the structure of the first projection, which results in a transformed $\widetilde{\mathbf{X}}$, followed by a subsequent optimisation of $\mathcal{Q}$ applied to $\widetilde{\mathbf{X}}$ instead of $\mathbf{X}$.

In theory, a generalisation from $\mathbf{a}$ to A, and indeed to an arbitrary $d \times k$ matrix A with $k \leq d$, is straightforward; however practical and computational aspects become increasingly complex as k increases.

Of the five indices (11.1) to (11.6), three have been generalised to 2D indices. Jones and Sibson (1987) proposed an extension of their $\mathcal{Q}_{\mathrm{E}}$, which is in fact an extension of the cumulant index $\mathcal{Q}_{\mathrm{C}}$, and Friedman (1987) showed how to extend $\mathcal{Q}_{\mathrm{U}}$. In addition, Friedman (1987) proposed an approach based on removal of detected structure, which reduces the 2D problem to finding a 1D index for the remaining data.

11.3.1 Two-Dimensional Indices: $\mathcal{Q}_E$, $\mathcal{Q}_C$ and $\mathcal{Q}_U$

The extensions from 1D to 2D indices differ considerably. It is therefore more natural to look at the population and sample version of one index before moving on to the next rather than considering all population quantities first and then dealing with the sample quantities thereafter.

So far we have explored projection indices for a single direction $\mathbf{a}$. To extend this definition to two direction vectors $\mathbf{a}_1$ and $\mathbf{a}_2$, we need to impose relationships between the two vectors. In Principal Component Analysis, the scores are uncorrelated, and in Independent Component Analysis, the population scores satisfy the stronger independence constraint. In Projection Pursuit, we adopt the Principal Component approach and require that

- the directions $\mathbf{a}_1$ and $\mathbf{a}_2$ are orthonormal, and
- the projections $\mathbf{a}_1^{\mathsf{T}}\mathbf{X}$ and $\mathbf{a}_2^{\mathsf{T}}\mathbf{X}$ are uncorrelated,

and then we write $\mathcal{Q}(\mathbf{X}, \mathbf{a}_1, \mathbf{a}_2)$ for the **bivariate projection index**.

The bivariate entropy index. For two direction vectors $\mathbf{a}_1$ and $\mathbf{a}_2$, Jones and Sibson (1987) generalised the entropy index $\mathcal{Q}_{\mathrm{E}}$ to a bivariate index by replacing the univariate $f_{\mathbf{a}}$ by the joint probability density function $f_{(\mathbf{a}_1,\mathbf{a}_2)}$ of the random vector $\mathbf{S} = (\mathbf{a}_1^{\mathsf{T}}\mathbf{X}, \mathbf{a}_2^{\mathsf{T}}\mathbf{X})^{\mathsf{T}}$. This leads to the **bivariate entropy index**

$$\mathcal{Q}_{\mathrm{E}}(\mathbf{X}, \mathbf{a}_1, \mathbf{a}_2) = \int f_{(\mathbf{a}_1,\mathbf{a}_2)} \log f_{(\mathbf{a}_1,\mathbf{a}_2)}.$$

For data $\mathbb{X}$, the bivariate densities $f_{(\mathbf{a}_1,\mathbf{a}_2)}$ can be estimated non-parametrically using kernels, see Scott (1992) for multivariate density estimation and Duong and Hazelton (2005) for bandwidth selection in multivariate density estimation. For a given kernel and bandwidths, let $\widehat{f}_{(\mathbf{a}_1,\mathbf{a}_2)}$ be the kernel estimator of $f_{(\mathbf{a}_1,\mathbf{a}_2)}$. The bivariate entropy index for data is

$$\widehat{Q}_{\mathrm{E}}(\mathbb{X},\mathbf{a}_1,\mathbf{a}_2)=\frac{1}{n}\sum \widehat{f}_{(\mathbf{a}_1,\mathbf{a}_2)}\log \widehat{f}_{(\mathbf{a}_1,\mathbf{a}_2)},$$

where $\mathbb{X}$ are the $d\times n$ data, and the sum is taken over vectors $(\mathbf{a}_1^{\mathsf{T}}\mathbf{X}_i,\mathbf{a}_2^{\mathsf{T}}\mathbf{X}_i)^{\mathsf{T}}$. The computational complexity for general bandwidth choices becomes large very quickly. For this reason, Jones and Sibson (1987) pursued a generalisation of the simpler moment-based Q_{C}, which we look at now.

Consider the random vector $\mathbf{S}=(\mathbf{a}_1^{\mathsf{T}}\mathbf{X},\mathbf{a}_2^{\mathsf{T}}\mathbf{X})^{\mathsf{T}}$. Write $\boldsymbol{\beta}_{rs}=[\beta_r,\beta_s]^{\mathsf{T}}$ for the bivariate cumulant, where β_r is the rth central moment of $\mathbf{a}_1^{\mathsf{T}}\mathbf{X}$, and β_s is the sth central moment of $\mathbf{a}_2^{\mathsf{T}}\mathbf{X}$. The **bivariate cumulant index** is

$$\begin{aligned}Q_{\mathrm{C}}(\mathbf{X},\mathbf{a}_1,\mathbf{a}_2)=\frac{1}{48}\Big\{&4\left[\boldsymbol{\beta}_{30}^2(\mathbf{S})+3\boldsymbol{\beta}_{21}^2(\mathbf{S})+3\boldsymbol{\beta}_{12}^2(\mathbf{S})+\boldsymbol{\beta}_{03}^2(\mathbf{S})\right]\\&+\left[\boldsymbol{\beta}_{40}^2(\mathbf{S})+4\boldsymbol{\beta}_{31}^2(\mathbf{S})+6\boldsymbol{\beta}_{22}^2(\mathbf{S})+4\boldsymbol{\beta}_{13}^2(\mathbf{S})+\boldsymbol{\beta}_{04}^2(\mathbf{S})\right]\Big\}.\end{aligned}\qquad(11.18)$$

For data $\mathbb{X}$, the bivariate cumulant index is derived by replacing the population moments $\boldsymbol{\beta}_{rs}$ with the corresponding sample central moments $\mathbf{b}_{rs}=[b_r,b_s]^{\mathsf{T}}$, with b_r the rth sample central moment of $\mathbf{a}_1^{\mathsf{T}}\mathbb{X}$, and b_s the sth sample central moment of $\mathbf{a}_2^{\mathsf{T}}\mathbb{X}$. Using the notation $\mathbf{S}_i=(\mathbf{a}_1^{\mathsf{T}}\mathbf{X}_i,\mathbf{a}_2^{\mathsf{T}}\mathbf{X}_i)^{\mathsf{T}}$, the bivariate cumulant index for data becomes

$$\begin{aligned}\widehat{Q}_{\mathrm{C}}(\mathbf{X}_i,\mathbf{a}_1,\mathbf{a}_2)=\frac{1}{48}\Big\{&4\left[\mathbf{b}_{30}^2(\mathbf{S}_i)+3\mathbf{b}_{21}^2(\mathbf{S}_i)+3\mathbf{b}_{12}^2(\mathbf{S}_i)+\mathbf{b}_{03}^2(\mathbf{S}_i)\right]\\&+\left[\mathbf{b}_{40}^2(\mathbf{S}_i)+4\mathbf{b}_{31}^2(\mathbf{S}_i)+6\mathbf{b}_{22}^2(\mathbf{S}_i)+4\mathbf{b}_{13}^2(\mathbf{S}_i)+\mathbf{b}_{04}^2(\mathbf{S}_i)\right]\Big\}.\end{aligned}$$

As in the 1D case, the bivariate index $\widehat{Q}_{\mathrm{C}}$ is calculated directly from the data. For details of the derivation of $Q_{\mathrm{C}}(\mathbf{X},\mathbf{a}_1,\mathbf{a}_2)$ and computational issues, see Jones (1983). The complexity of the bivariate $\widehat{Q}_{\mathrm{C}}$ is much greater than that of the univariate version (11.12), yet it is computationally simpler than the sample index $\widehat{Q}_{\mathrm{E}}$, which is based on bivariate kernel density estimates. For computational reasons, Jones and Sibson (1987) preferred $\widehat{Q}_{\mathrm{C}}$ to $\widehat{Q}_{\mathrm{E}}$.

The direct bivariate extension of Q_{U}. Friedman (1987) proposed two different extensions of the index Q_{U} of (11.4) to two direction vectors. I explain his 'direct' extension first and describe his structure-removal approach in Section 11.3.2.

For $\ell=1,2$, directions $\mathbf{a}_\ell$, and random vector $\mathbf{X}$, put

$$X_\ell=\mathbf{a}_\ell^{\mathsf{T}}\mathbf{X}\qquad\text{and}\qquad\Theta_\ell=2\Phi(X_\ell)-1.$$

The random variables Θ_ℓ generalise the single transformed variable (11.3). Let $f_{(\Theta_1,\Theta_2)}$ be the joint probability density function of (Θ_1,Θ_2), and define the **bivariate index** based on **deviations from the uniform distribution** by

$$Q_{\mathrm{U}}(\mathbf{X},\mathbf{a}_1,\mathbf{a}_2)=\int_{-1}^{1}\int_{-1}^{1}\left[f_{(\Theta_1,\Theta_2)}(\Theta_1,\Theta_2)-\frac{1}{4}\right]^2,\qquad(11.19)$$

where the $\mathbf{a}_1$ and $\mathbf{a}_2$ are chosen such that X_1 and X_2 are uncorrelated.

For data $\mathbb{X} = [\mathbf{X}_1 \cdots \mathbf{X}_n]$ and $m < \infty$, Friedman approximated $\mathcal{Q}_{\mathrm{U}}$ by sums with at most m terms, where the terms are Legendre polynomials. For details of the derivation, see Friedman (1987). For fixed m, the bivariate sample index results in

$$\widehat{\mathcal{Q}}_{\mathrm{U,m}}(\mathbb{X}, \mathbf{a}_1, \mathbf{a}_2)$$
$$= \sum_{j=1}^{m} \sum_{k=1}^{m-j} \frac{(2j+1)(2k+1)}{4} \left\{ \frac{1}{n^2} \sum_{i,\ell=1}^{n} p_j \left[2\Phi(\mathbf{a}_1^{\mathsf{T}}\mathbf{X}_i) - 1\right] \left[2\Phi(\mathbf{a}_2^{\mathsf{T}}\mathbf{X}_\ell) - 1\right] \right\}^2$$
$$+ \sum_{j=1}^{m} \frac{2j+1}{4} \left(\left\{ \frac{1}{n} \sum_{i=1}^{n} p_j \left[2\Phi(\mathbf{a}_1^{\mathsf{T}}\mathbf{X}_i) - 1\right] \right\}^2 + \left\{ \frac{1}{n} \sum_{i=1}^{n} p_j \left[2\Phi(\mathbf{a}_2^{\mathsf{T}}\mathbf{X}_i) - 1\right] \right\}^2 \right). \tag{11.20}$$

A comparison of the univariate and bivariate sample indices $\widehat{\mathcal{Q}}_C$, $\widehat{\mathcal{Q}}_U$ and $\widehat{\mathcal{Q}}_E$ shows that the computational complexity is increased considerably in the bivariate case. In a search for the maximiser of an index, one has to satisfy the orthogonality constraint imposed on the direction vectors and the requirement that the projections be uncorrelated. These calculations are still feasible, as Jones and Sibson (1987) and Friedman (1987) illustrated with their software. The step from 2D to 3D projections becomes considerably more involved. In Projection Pursuit we do not solve algebraic equations as in Principal Component Analysis, and hence, any software needs to include good search algorithms in potentially high-dimensional spaces.

Other indices have been proposed in Cook, Buja, and Cabrera (1993) and Eslava and Marriott (1994). They are based on approximations to the mutual information similar to those which I described in Section 10.4.2. In particular, indices based on skewness or kurtosis alone can be defined, as has been done in Independent Component Analysis.

11.3.2 Bivariate Extension by Removal of Structure

The increased computational burden of the bivariate projection index $\mathcal{Q}_{\mathrm{U}}$ led Friedman (1987) to consider an iterative process which finds first the most non-Gaussian univariate projection. His key idea was to replace this most non-Gaussian projection essentially by a Gaussian projection and then to search for the second-most non-Gaussian projection.

I explain Friedman's proposal for the population, the setting he describes. I will comment on the steps that will need to be modified for the sample and briefly outline a structure-removal approach for data. Algorithm 11.1 in Section 11.4.3 tells us how we can calculate Friedman's bivariate directions based on structure removal in a computationally efficient way, and the examples in Section 11.4.3 illustrate these ideas.

Let $\mathbf{X} \sim (\mathbf{0}, \mathbf{I}_{d\times d})$, and let $\mathbf{a} \in \mathbb{R}^d$ be a direction vector. Consider $\mathcal{Q}_{\mathrm{U}}(\mathbf{X}, \mathbf{a})$ of (11.4), and let $\mathbf{a}_1$ be its maximiser. Put $X_1 = \mathbf{a}_1^{\mathsf{T}}\mathbf{X}$, and let F_1 be the distribution function of X_1. Use the univariate standard normal distribution function Φ, and define a function θ_1 by

$$\theta_1(X_1) = \Phi^{-1}\left[F_1(X_1)\right]. \tag{11.21}$$

It follows that $\theta_1(X_1) \sim \mathcal{N}(0,1)$. Let D_θ be the diagonal $d \times d$ matrix which is derived from the identity matrix $\mathbf{I}_{d\times d}$ by replacing the first entry 1 by $\theta_1(X_1)$. Let U be an orthogonal $d \times d$

matrix with first column $\mathbf{u}_1 = \mathbf{a}_1$, and choose the remaining columns such that $\mathbf{a}_1, \mathbf{u}_2, \ldots, \mathbf{u}_d$ form an orthonormal basis. Then

$$\mathbf{X} = \sum_{j=1}^{d} \mathbf{u}_j \mathbf{u}_j^{\top} \mathbf{X}. \tag{11.22}$$

The proof of this equality is similar to that of Theorem 2.12 in Section 2.5.2.

In his **structure removal**, Friedman (1987) replaced the original random vector $\mathbf{X}$ with a vector $\mathbf{X}^{\dagger}$ which is defined by

$$\mathbf{X}^{\dagger} = U D_{\theta} U^{\top} \mathbf{X}$$

and then uses $\mathbf{X}^{\dagger}$ instead of $\mathbf{X}$ in the search for the direction which results in the second-largest index $\mathcal{Q}_U$. Friedman showed that

$$\mathbf{X}^{\dagger} = \theta_1(X_1)\mathbf{a}_1 + \sum_{j=2}^{d} \mathbf{u}_j \mathbf{u}_j^{\top} \mathbf{X}. \tag{11.23}$$

How can we interpret $\mathbf{X}^{\dagger}$? The first term $\mathbf{a}_1 \mathbf{a}_1^{\top} \mathbf{X} = X_1 \mathbf{a}_1$ of (11.22) is calculated from the most non-Gaussian projection X_1. This term has been replaced by the Gaussian term $\theta_1(X_1)\mathbf{a}_1$, while the other terms in the expansion (11.22) remain the same. Because $\theta_1(X_1)\mathbf{a}_1 \sim \mathcal{N}(\mathbf{0}, \mathbf{I}_{d\times d})$, the term $\theta_1(X_1)\mathbf{a}_1$ is uninteresting in the search for non-Gaussian directions. This term could be interpreted as Gaussian noise and becomes negligible in the search for non-Gaussian structure.

The structure-removal process can be repeated until all interesting structure has been found or until sufficiently many interesting directions have been found.

To apply the structure-removal operations to data $\mathbb{X} = \begin{bmatrix}\mathbf{X}_1 \cdots \mathbf{X}_n\end{bmatrix}$, we replace the unknown distribution function F_1 of $\mathbf{a}_1^{\top}\mathbf{X}$ by the empirical distribution function $\widehat{F}_1$, which is defined for real X by

$$\widehat{F}_1(X) = \frac{1}{n} \#\{\, i \,:\, \mathbf{a}_1^{\top} \mathbf{X}_i \leq X \}. \tag{11.24}$$

By the Glivenko-Cantelli theorem (see theorem 11.4.2 of Dudley 2002),

$$\sup_{X = \mathbf{a}_1^{\top}\mathbf{X}} \left| \widehat{F}_1(X) - F_1(X) \right| \xrightarrow{a.s.} 0 \qquad \text{as } n \to \infty,$$

where a.s. refers to almost sure convergence, and thus $\widehat{F}_1$ is an appropriate estimator for F_1.

Apart from substitution of the distribution function with the empirical distribution function, the structure removal for data is analogous to that for the population. For each $\mathbf{X}_i$ of the data $\mathbb{X}$, one obtains a value $\theta_1(\mathbf{a}_1^{\top}\mathbf{X}_i)$ from (11.21) and a vector $\mathbf{X}_i^{\dagger}$ as in (11.23) which replaces $\mathbf{X}_i$. The new data $\mathbb{X}^{\dagger} = \begin{bmatrix}\mathbf{X}_1^{\dagger} \cdots \mathbf{X}_n^{\dagger}\end{bmatrix}$ now replace $\mathbb{X}$ in the search for the most non-Gaussian direction.

Friedman (1987) explained how to extend his structure removal to the removal of 2D structures. This is more difficult conceptually as well as computationally. For our purpose, the 1D structure removal suffices. The structure-removal idea is not restricted to the index $\mathcal{Q}_{\mathrm{U}}$ but can be applied to any of the indices of preceding sections. It offers an alternative to the 'direct' bivariate indices, but of course, the solutions will differ because the second-most non-Gaussian direction is found for the modified rather than for the original data. How much or whether this matters in practice will vary from one data set to another.

11.3.3 A Three-Dimensional Cumulant Index

Extending Projection Pursuit to three dimensions is conceptually attractive because we can visualise 3D projections. These may expose information in the data that is not available in 2D projections. Nason (1995) extended the cumulant index $\mathcal{Q}_{\rm C}$ of Jones and Sibson (1987) to three directions. I briefly outline the important features of Nason's approach.

In an extension of the 2D index to three directions $\mathbf{a}_k$, with $k = 1,2,3$, Nason (1995) required that the direction vectors be pairwise orthogonal. The orthogonality is not only advantageous computationally, it also allows easier interpretation of the 3D projections.

A 3D cumulant-based projection index makes use of trivariate skewness and kurtosis β_{rst}, where $r+s+t=m$, with $m=3$ for skewness and $m=4$ for kurtosis. The trivariate index has the form

$$\mathcal{Q}_{\rm C}(\mathbf{X},\mathbf{a}_1,\mathbf{a}_2,\mathbf{a}_3) = \sum_{\substack{r+s+t=3\\ r,s,t=0,\ldots,3}} 4c^{(3)}_{rst}\beta^2_{rst}(\mathbf{S}) + \sum_{\substack{r+s+t=4\\ r,s,t=0,\ldots,4}} c^{(4)}_{rst}\beta^2_{rst}(\mathbf{S}), \qquad (11.25)$$

where $\mathbf{S} = (\mathbf{a}_1^{\top}\mathbf{X},\mathbf{a}_2^{\top}\mathbf{X},\mathbf{a}_3^{\top}\mathbf{X})^{\top}$, and $c^{(3)}_{rst}$ and $c^{(4)}_{rst}$ are positive constants. Nason (1995) pointed out that the coefficients of the 2D index (11.18) are simply the coefficients of $x^r y^s$ in the expansion $(x+y)^m$, and similarly, the $c^{(m)}_{rst}$ are the coefficients of $x^r y^s z^t$ in the expansion of $(x+y+z)^m$.

A nice feature of the bivariate index $\mathcal{Q}_{\rm C}$ is its rotational invariance with respect to any choice of directions or axes. This property was established in Jones and Sibson (1987) and carries over to the trivariate $\mathcal{Q}_{\rm C}$ of (11.25). The original index of Friedman and Tukey (1974) does not possess this rotational invariance, nor does the index $\mathcal{Q}_{\rm U}$ of Friedman (1987).

The rotational invariance is not essential conceptually, but it assists in speeding up the optimisation of the index – something that becomes more relevant as the number of directions increases. Indeed, as in kernel density estimation, there is no intrinsic barrier to estimating projection indices in three or more dimensions other than the rapidly increasing computational effort. To decrease the computational burden, Nason (1995) made use of k-statistics in his estimation of the sample skewness and kurtosis. The k-statistics are unbiased estimators of the cumulants. As Nason showed, they can be expanded in power sums and give rise to efficient ways of calculating the sample index $\widehat{\mathcal{Q}}_{\rm C}$ by means of computer algebra. The interested reader can find a description of the package REDUCE, which accomplishes this expansion, in Nason (1995).

11.4 Projection Pursuit in Practice

The early contributors to Projection Pursuit, including Jones and Sibson (1987) and Friedman (1987), provided software – typically in Fortan – to calculate their indices. To appreciate the computational and practical developments in Projection Pursuit, it is important to understand the relationship between Projection Pursuit and Independent Component Analysis.

The next section gives an overview of the similarities and differences of Projection Pursuit and Independent Component Analysis which extends our first comparison in Table 11.1. Based on the understanding we glean from this deeper comparison, I appraise the way Projection Pursuit directions, scores and indices are computed at the time of writing this book.

11.4.1 Comparison of Projection Pursuit and Independent Component Analysis

The developments in the late 1980s in Projection Pursuit quickly attracted attention in the statistics community and beyond. Projection Pursuit originated at about the same time as Independent Component Analysis, and it is only natural that the similarities and common goals began to be appreciated by proponents of both camps.

Like Principal Component Analysis, Projection Pursuit and Independent Component Analysis aim to find structure in data, but whereas Principal Component Analysis uses a well-defined mathematical quantity – the covariance matrix – as the key to finding the structure, Projection Pursuit and Independent Component Analysis start with the sphered data and attempt to find deviations from the Gaussian.

Although Projection Pursuit and Independent Component Analysis originated and initially developed independently in the statistics and signal-processing communities, respectively, the two methods are remarkably similar and optimise similar criteria. In the following paragraphs I discuss a number of aspects of the two methods. We will see that there are differences at the conceptual level, but these differences begin to blur when it comes to practical data analysis. Table 11.2 summarises differences and similarities of the two methods.

Departure from Gaussianity. Conceptually, the two methods pursue different goals; the independence of the variables appears to drive Independent Component Analysis, whereas departure from Gaussianity is the key to Projection Pursuit. Projection Pursuit optimises the projection index $\mathcal{Q}$, whereas Independent Component Analysis optimises the mutual information $\mathcal{I}$. At first glance, the two criteria appear to be different. However, Theorem 10.9 in Section 10.4.1 highlights the close connection between the mutual information and the negentropy. The negentropy measures the departure from the Gaussian and can be approximated by polynomials in skewness and kurtosis. The cumulant index $\mathcal{Q}_{\mathrm{C}}$ is just one such approximation to the negentropy.

Independence versus uncorrelatedness. The independent components of the source are an integral part of the ICA model. In contrast, Projection Pursuit searches for the most non-Gaussian direction, so independence does not play a role. For two or more directions, the projections of the vector or data onto the distinct directions are uncorrelated. At the population level, Independent Component Analysis appears to be a more powerful method than Projection Pursuit because the scores are independent. But is this actually the case? Both approaches start with uncorrelated vectors or data. When we progress from the population to the sample, we do not know the probability density function of the sample. We approximate $\mathcal{I}$, typically by cumulants, and thus the scores are no longer independent, but only *as independent as possible*, given the particular approximation to $\mathcal{I}$ that is used.

All d sources versus the first and most important one(s). Projection Pursuit looks for low-dimensional structure, whereas Independent Component Analysis – in its original and pure form – seeks to find source vectors of the same dimension as the signal and thus differs from Projection Pursuit. However, for many data sets, and in particular for high-dimension low sample size data, we are only interested in the first few and most non-Gaussian directions which exhibit the structure in the data. Thus, the difference between the two approaches could be described more appropriately in the following way: Projection Pursuit stops at one or two directions, whereas Independent Component Analysis needs to be told how many directions to find.

Table 11.2 *Projection Pursuit (PP) and Independent Component Analysis (ICA)*

	PP	**ICA**
Aim	Most non-Gaussian direction(s) 1–2 projections	Independent components $\leq d$ source variables
Model	Not explicit	$\mathbf{X} = A\mathbf{S}$, A invertible
Data	Sphered data	Sphered data
Origin	Statistics	Signal processing
Statistic	Projection index	Mutual information
Approximations	Combined skewness and kurtosis	Skewness or kurtosis
Scores	Non-Gaussian and uncorrelated	As independent as possible and non-Gaussian

Table 11.2 and the preceding discussion highlight that the common theme, the departure from Gaussianity or, equivalently, the search for directions that are as non-Gaussian as possible, is the dominant feature in both methods. It is therefore natural to replace parts or aspects of one method with similar parts of the other method if this improves the method as a whole. This paradigm can be observed in the more recent developments of the computational aspects of Projection Pursuit.

11.4.2 From a Cumulant-Based Index to FastICA Scores

In the preceding section we observed that the similarities, the common concepts and aims of Projection Pursuit and Independent Component Analysis, vastly outweigh the differences between the two methods. The negentropy in particular plays a central part in both methods.

Direct computation of the negentropy depends on the unknown probability density function and becomes less feasible with increasing dimension. In the ICA world, we use an approximation to the negentropy, which can be estimated directly from the data and which exploits the relationship between the mutual information and the negentropy. The result is the function $\mathfrak{J}$ of Theorem 10.11 in Section 10.4.2, given by

$$\mathfrak{J} = \sum_{j=1}^{d} \left(4\beta_{3,j}^2 + \beta_{4,j}^2 + 7\beta_{3,j}^4 - 6\beta_{3,j}^2\beta_{4,j}\right), \tag{11.26}$$

where $\beta_{\rho,j}$ refers to the 1D cumulants skewness (for $\rho = 3$) and kurtosis (for $\rho = 4$). Instead of maximising $\mathfrak{J}$ over possible source vectors, Comon (1994) proposed maximising separately the skewness criterion $\mathcal{G}_3$ and the kurtosis criterion $\mathcal{G}_4$, which are given by

$$\mathcal{G}_\rho = \sum_{j=1}^{d} \beta_{\rho,j}^2 \qquad \text{with } \rho = 3, 4. \tag{11.27}$$

See (10.17) in Section 10.4.2 for details. The expressions $\mathcal{G}_\rho$ provide a compromise between computational feasibility and accuracy. The two separate criteria have been widely adopted in ICA calculations (see Hyvärinen 1999). Other criteria, including the JADE approximations of Cardoso (1999), and their software implementations are available for calculating IC directions and scores. Our present discussion, however, focuses on the FastICA approximations $\mathcal{G}_\rho$ of Hyvärinen (1999).

In Projection Pursuit, Jones and Sibson (1987) proposed the cumulant-based index (11.1), which is closely related to (11.26) and is estimated directly from data; see (11.12). It is given by

$$\mathcal{Q}_{\mathrm{C}} = \frac{1}{48}\left(4\beta_3^2 + \beta_4^2\right). \tag{11.28}$$

A comparison of (11.27) and (11.28) highlights the similarity of the approximations. The ICA approximations to $\mathfrak{I}$ are, by themselves, not as good as the PP approximation $\mathcal{Q}_{\mathrm{C}}$. However, the ICA approximations are implemented by more efficient algorithms. The computational advantage of the FastICA algorithm turned out to be more important than the more accurate approximation. And at the time of this writing, a version of FastICA had become the standard Projection Pursuit algorithm in R.

There are, however, differences between MATLAB's FastICA and the R version of FastICA. MATLAB's FastICA uses the kurtosis criterion $\mathcal{G}_4$ as default but supports $\mathcal{G}_3$ and the two criteria $\mathcal{G}_1(X) = -\exp(-X^2/2)$ and $\mathcal{G}_2(X) = \log[\cosh(cX)]/c$ which are defined for each entry of $\mathbf{X}$. Hyvärinen, Karhunen, and Oja (2001) considered $\mathcal{G}_1$ and $\mathcal{G}_2$ – the only two criteria used in the R version of FastICA – because of their robustness properties. I did not mention these two criteria in Chapter 10 because they do not measure deviations from the Gaussian as naturally as $\mathcal{G}_3$ and $\mathcal{G}_4$ and thus make an interpretation of the directions and scores less clear. Experience with the R algorithm shows that the IC directions differ from those obtained with the $\mathcal{G}_3$ and $\mathcal{G}_4$ criteria. This is not unexpected because different problems are solved. As we saw in Chapter 10, the directions obtained with $\mathcal{G}_3$ and $\mathcal{G}_4$ typically differ from each other, and they also differ from the JADE directions. Slightly different directions, such as those obtained with the criteria used in the R implementation of FastICA, will not improve our insight into Projection Pursuit, and for this reason, I will not apply these criteria in examples but leave it to the interested reader to compare the results for individual data sets.

As a consequence of the computational complexity of the indices and the relative efficiency in obtaining ICA directions, it is now generally accepted that the computations of PP directions and scores are replaced by the simpler computations of FastICA directions and scores. We will therefore regard Definitions 10.18 and 10.19 in Section 10.6.1 as the definitions of Projection Pursuit scores.

11.4.3 The Removal of Structure and FastICA

The computational burden posed by finding PP directions and the adoption of the FastICA software for calculating PP directions apply to univariate as well as jointly bi- or trivariate projection indices. There remains one development in Projection Pursuit that deserves more discussion: Friedman's construction of bivariate indices by removal of structure, which I describe in Section 11.3.2.

Friedman (1987) proposed a direct bivariate extension of the univariate projection index $\mathcal{Q}_{\mathrm{U}}$, as well as removal of the most non-Gaussian structure in the data, which is followed by a calculation of the most non-Gaussian direction for the modified data. Friedman implemented his ideas in Fortran. I will not use his Fortran code but will show how we can marry his structure-removal ideas with a calculation of 1D directions using FastICA.

In the FastICA algorithm, the directions are calculated iteratively starting from the most non-Gaussian direction. To find the second direction, FastICA optimises the relevant $\mathcal{G}_\rho$

criterion, subject to the new direction being orthogonal to the first direction. This iterative nature of finding consecutive non-Gaussian directions allows an integration of and adaptation to the structure removal ideas of Friedman. Algorithm 11.1 describes the necessary steps.

Algorithm 11.1 *Non-Gaussian Directions from Structure Removal and FastICA*
Let $\mathbb{X}^\diamond = \left[\mathbf{X}_1^\diamond \cdots \mathbf{X}_n^\diamond\right]$ be d-dimensional sphered data. Fix an approximation $\mathcal{G}_\rho$ for $\rho = 3, 4$.

Step 1. Calculate the first IC direction $\mathbf{a}_1 = \boldsymbol{\upsilon}_1$ and the most non-Gaussian scores as in steps 2 to 6 of Algorithm 10.1 in Section 10.6.1 (using $j = 1$).

Step 2. For $i \le n$, put $X_{i1} = \mathbf{a}_1^{\mathsf{T}} \mathbf{X}_i^\diamond$, and calculate

(a) the empirical distribution function $\widehat{F}_1$ of the X_{i1} as in (11.24), and
(b) the coefficients $\theta_{i1} = \theta_1(X_{i1})$ as in (11.21).

Step 3. Extend $\mathbf{a}_1$ to an orthonormal basis in $\mathbb{R}^d$, and call the new $d-1$ basis vectors $\boldsymbol{\upsilon}_2^0, \ldots, \boldsymbol{\upsilon}_n^0$. Put $U = \left[\mathbf{a}_1, \boldsymbol{\upsilon}_2^0, \ldots, \boldsymbol{\upsilon}_n^0\right]$. For $i \le n$, calculate

$$\mathbf{X}_i^\dagger = U D_{\theta,i} U^{\mathsf{T}} \mathbf{X}_i^\diamond,$$

where $D_{\theta,i}$ is the matrix obtained from the $d \times d$ identity matrix by replacing the first entry 1 with θ_{i1}. Put $\mathbb{X}^\dagger = \left[\mathbf{X}_1^\dagger \cdots \mathbf{X}_n^\dagger\right]$.

Step 4. Apply step 1 to $\mathbb{X}^\dagger$ with the same $\mathcal{G}_\rho$. Call the first IC direction of the modified data $\boldsymbol{\upsilon}_1^\dagger$.

Step 5. Orthogonalise $\boldsymbol{\upsilon}_1^\dagger$ with respect to $\boldsymbol{\upsilon}_1$, and call the resulting orthogonal direction $\mathbf{a}_2$. Calculate the second-most non-Gaussian scores by

$$X_{i2} = \mathbf{a}_2^{\mathsf{T}} \mathbf{X}_i^\diamond \qquad \text{for } i \le n,$$

and write

$$\mathbb{S}_{\text{SR}}^{(2)} = \begin{bmatrix} \mathbf{a}_1^{\mathsf{T}} \mathbb{X}^\diamond \\ \mathbf{a}_2^{\mathsf{T}} \mathbb{X}^\diamond \end{bmatrix}$$

for the 2D projection pursuit data. ■

I illustrate Algorithm 11.1 with two contrasting examples, *three-dimensional simulated* data and the thirteen-dimensional *athletes* data, and in each case I compare the projection pursuit data $\mathbb{S}_{\text{SR}}^{(2)}$ with the independent component data $\mathbb{S}^{(2)}$. It will be interesting to see how different the second rows of the two data sets are. The first rows will be the same, as they are calculated the same way.

Example 11.3 We consider the three-dimensional **simulated data** shown in the middle panel of Figure 9.1 in Section 9.1. A total of 2,000 vectors are generated from a mixture of two Gaussians; 60 per cent are from the population $\mathcal{N}(\boldsymbol{\mu}_1, \Sigma_1)$, and the remaining 40 per cent are from the population $\mathcal{N}(\boldsymbol{\mu}_2, \Sigma_2)$. The means and covariance matrices of the two populations are

$$\boldsymbol{\mu}_1 = \begin{bmatrix} 0 \\ 0 \\ 0 \end{bmatrix}, \quad \Sigma_1 = \begin{bmatrix} 2.4 & -0.5 & 0 \\ -0.5 & 1 & 0 \\ 0 & 0 & 1 \end{bmatrix}, \quad \boldsymbol{\mu}_2 = \begin{bmatrix} 5 \\ 0 \\ 0 \end{bmatrix} \quad \text{and} \quad \Sigma_1 = \begin{bmatrix} 2.4 & -0.5 & 0.3 \\ -0.5 & 1 & -0.4 \\ 0.3 & -0.4 & 1.5 \end{bmatrix}.$$

The middle panel of Figure 9.1 shows the directions of the centred data as in (9.1); the raw data are shown in the left panel of Figure 11.2. In both cases, vectors of the first population are shown in red and those of the second in blue.

I apply Algorithm 11.1 separately with the skewness and kurtosis criteria to these data. The modified data $\mathbb{X}^\dagger$ of step 3 are shown in the middle and right panels of Figure 11.2. The middle panel shows $\mathbb{X}^\dagger$ based on the skewness criterion, and the right panel shows $\mathbb{X}^\dagger$ for the kurtosis criterion. In both cases, the modified data are derived from the red IC directions shown in the left panels of Figure 11.3. The red and blue clusters are much less separated

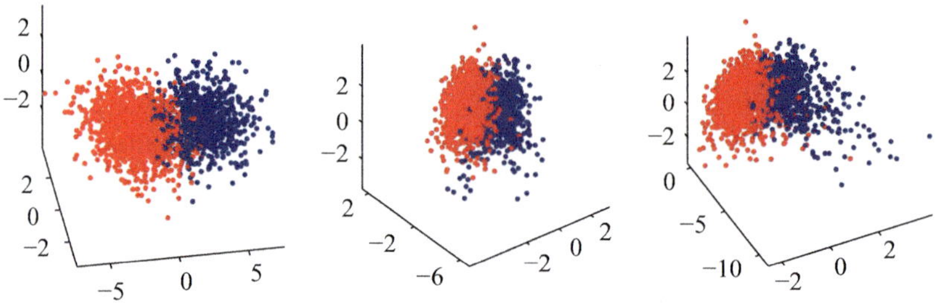

Figure 11.2 Original data (*left*) from Example 11.3. Modified data $\mathbb{X}^\dagger$ after removal of the most non-Gaussian structure: with skewness (*middle*) and with kurtosis (*right*).

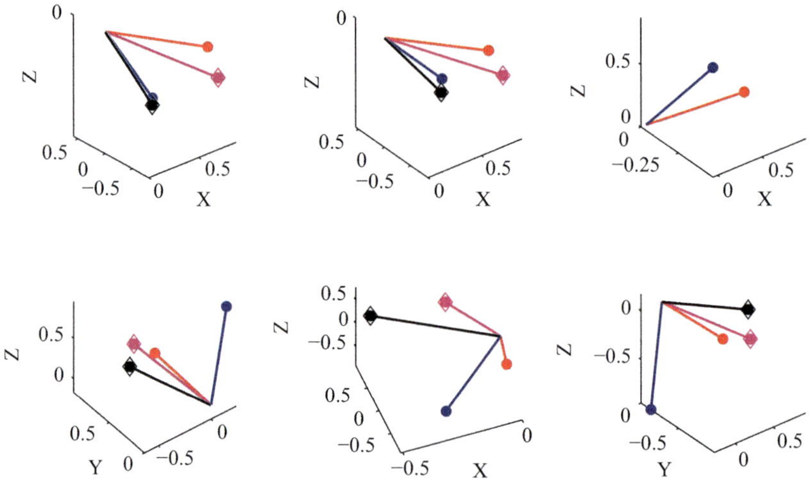

Figure 11.3 Most non-Gaussian directions from Example 11.3. With skewness criterion (*top left* and *middle*), with kurtosis criterion (*bottom*), and the first two PC directions (*top right*). First and second IC and PC directions are red and blue; first direction of the modified data is maroon and black after orthogonalisation.

in the middle and right panels than in the original data. In the right panel we note that some blue dots are more spread out than in the raw data.

Figure 11.3 shows the IC directions we obtain with the two algorithms. As is common in FastICA calculations, different runs of the algorithm can result in different solutions. For the skewness criterion, there are two distinct solutions, and for kurtosis, there are three. The left and middle panels in the top row show the two solutions obtained with the skewness criterion, and the three panels in the bottom row show the three solutions obtained with the kurtosis criterion. The top-right panel shows the first two PC directions, the eigenvectors of the covariance matrix of the data. In each panel the red vector refers to the IC_1 vector $\boldsymbol{\upsilon}_1 = \mathbf{a}_1$ (respectively, the first eigenvector $\widehat{\boldsymbol{\eta}}_1$). The blue vector refers to the IC_2 vector $\boldsymbol{\upsilon}_2$ obtained with Algorithm 10.1 and to the second eigenvector $\widehat{\boldsymbol{\eta}}_2$ for the top-right panel. The maroon direction shows $\boldsymbol{\upsilon}_1^{\dagger}$, and the black direction shows the direction $\mathbf{a}_2$ after orthogonalisation with $\mathbf{a}_1$.

For the skewness criterion, there is hardly any difference between the blue and black directions, so removal of structure results in very similar direction vectors. The three solutions found with the kurtosis criterion differ considerably, and there is little agreement between the second-most non-Gaussian directions obtained with and without removal of structure. Indeed, in the left and right panels in the bottom row, $\mathbf{a}_2$ is closer to $\mathbf{a}_1$ than to $\boldsymbol{\upsilon}_2$.

The first eigenvector $\widehat{\boldsymbol{\eta}}_1$ points in the direction of largest variance: here between the means of the two populations. The contribution to total variance in this direction is about 80 per cent. The direction of the second eigenvector does not have a clear interpretation, but it only contributes about 12 per cent to the total variance. These vectors differ considerably from the non-Gaussian directions.

How should we choose between the different IC approaches? Natural criteria for assessing the performance of the two algorithms are the skewness and kurtosis of the IC scores. Table 11.3 shows absolute skewness and kurtosis of the IC scores, as appropriate. The first column of numbers shows the angle between $\mathbf{a}_1$ and $\boldsymbol{\upsilon}_1^{\dagger}$ in the form of the inner product of the vectors; non-zero values indicate that the vectors are not orthogonal.

The table shows that the two IC skewness solutions are similar, and the absolute skewness of the scores obtained with the directions $\boldsymbol{\upsilon}_2$ and $\mathbf{a}_2$ are almost the same. The absolute kurtosis of the scores resulting from projections onto $\mathbf{a}_2$ are considerably smaller than those obtained from the corresponding projections onto $\boldsymbol{\upsilon}_2$.

Unlike the skewness results, the first kurtosis scores of the modified data $\mathbb{X}^{\dagger}$, so obtained with $\boldsymbol{\upsilon}_1^{\dagger}$, are very similar to the first IC scores. This could indicate that the most non-Gaussian direction of $\mathbb{X}^{\diamond}$ is still present in the modified data. The angles between $\mathbf{a}_1$ and $\boldsymbol{\upsilon}_1^{\dagger}$ confirm this hypothesis. After orthogonalisation, this effect disappears, and the scores obtained with $\mathbf{a}_2$ have considerably lower kurtosis than the scores obtained with $\boldsymbol{\upsilon}_2$.

The last two rows of the table list the absolute skewness and kurtosis of the PC scores. We note that the PC_1 scores have lower skewness and kurtosis than the PC_2 scores. This result is not surprising, because the PC directions pursue variability in the data rather than non-Gaussian structure.

The analysis shows that the two ways of calculating non-Gaussian directions and scores are very similar for the skewness criterion. The usual FastICA results with kurtosis yield more non-Gaussian scores than those obtained after structure removal for these simulated data. ■

Our next example deals with thirteen variables but only 202 observations.

Table 11.3 *Comparison of IC Directions for Example 11.3*

		Absolute skewness/kurtosis of scores from				
	$\mathbf{a}_1^{\mathsf{T}} \boldsymbol{\upsilon}_1^{\dagger}$	$\mathbf{a}_1$ (red)	$\boldsymbol{\upsilon}_2$ (blue)	$\boldsymbol{\upsilon}_1^{\dagger}$ (maroon)	$\mathbf{a}_2$ (black)	Figure panel
Skew. 1	0.8972	0.4004	0.3649	0.3071	0.3648	Top left
Skew. 2	0.8764	0.4010	0.3643	0.3042	0.3637	Top middle
Kurt. 1	0.9554	3.3925	3.1778	3.3170	2.5191	Bottom left
Kurt. 2	−0.9249	3.3450	3.2050	3.3450	2.6455	Bottom middle
Kurt. 3	0.9303	3.3928	3.1771	3.3237	1.9944	Bottom right
PCs (skew)		0.2257	0.1448			Top right
PCs (kurt)		2.1242	3.3517			

Example 11.4 We continue with the **athletes** data, which consist of measurements for 100 female and 102 male athletes. As in Example 11.3, I calculate the first two IC scores with Algorithms 10.1 and 11.1 and compare the performance of the two algorithms using absolute skewness and kurtosis of the scores.

In this application I work with the raw – rather than the scaled – data. I apply Algorithms 10.1 and 11.1 to the data separately for the skewness and kurtosis criteria, and use one iteration, so $K = 1$ in Algorithm 10.1, in each of twenty-five runs with each algorithm and criterion. I calculate the first two IC directions and scores with Algorithm 10.1, and order them as in step 4 of the algorithm. I use the first direction to accomplish the structure removal, which corresponds to steps 1 to 3 of Algorithm 11.1. Then I calculate the first two directions for $\mathbb{X}^{\dagger}$ and keep that which results in the more non-Gaussian scores. Step 6 of Algorithm 11.1 results in the 2D projection pursuit data.

Figure 11.4 shows the absolute skewness and kurtosis values of the scores on the y-axis, with the (ordered) run numbers on the x-axis. For easier visualisation, I have ordered the runs by the size of skewness (or kurtosis) of the first scores. The top panel depicts the twenty-five runs with the skewness criterion, and the bottom panel shows the kurtosis results. The absolute skewness and kurtosis of the first IC scores are shown in blue, those of the second IC scores in black, and the second scores – obtained after structure removal – are shown in red.

For the skewness results, the IC_1 scores are almost identical, whereas the IC_2 scores show considerable variation. In eight out of the twenty-five skewness runs, the red scores are higher than the black scores, and ten times they are about the same. The kurtosis results show more variability than the skewness results. In fourteen out of the twenty-five runs, the red scores are more non-Gaussian than their IC_2 counterparts. These results suggest that structure removal could be a viable alternative to calculating the second IC scores directly when searching for non-Gaussian structure. ∎

These examples illustrate that structure removal followed by a search for the most non-Gaussian direction leads to results comparable with those obtained with the direct search for two or more non-Gaussian directions. I have done the calculations with the FastICA software and the skewness and kurtosis criteria. Other software and criteria of non-Gaussianity might produce different results.

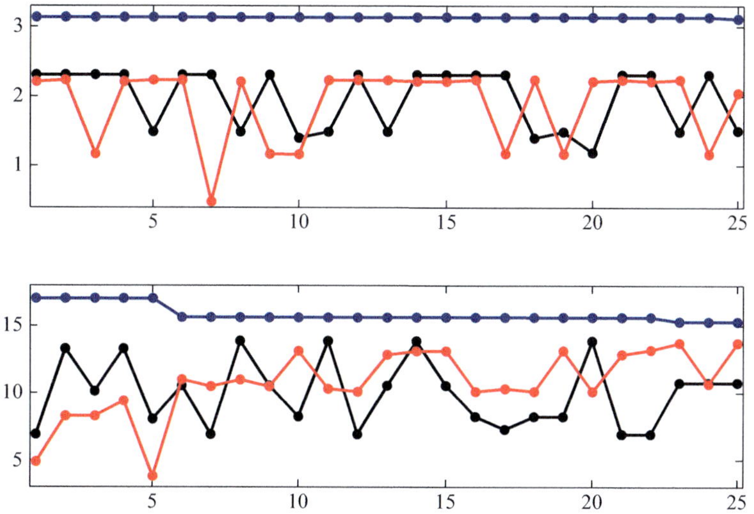

Figure 11.4 Absolute skewness (*top*) and kurtosis (*bottom*) of first and second IC scores for twenty-five runs of the data from Example 11.4, with the run number on the x-axis. Skewness (and kurtosis) of $\mathbb{S}^{(2)}$ and $\mathbb{S}^{(2)}_{SR}$; IC_1 blue, second components of $\mathbb{S}^{(2)}$ black and of $\mathbb{S}^{(2)}_{SR}$ red.

The inclusion of structure removal in the search for non-Gaussian directions is computationally more complex than the direct search for two or more directions, but the structure-removal idea and renewed search in modified data open new ways of looking at the pursuit of structure in data.

11.4.4 Projection Pursuit: A Continuing Pursuit

A possible interpretation of the discussion and the calculations of the last two sections is that Projection Pursuit does not work and that Projection Pursuit is obsolete. Is such a point of view justified, and is Projection Pursuit a past pursuit? My response is a definite 'No', which I attempt to justify in the next few paragraphs.

Section 11.4.2 explains that we calculate Projection Pursuit directions and scores with the FastICA algorithm. Instead of using the approximation $\mathcal{Q}_C$ and its estimator $\widehat{\mathcal{Q}}_C$, we calculate directions and scores based on a (slightly) different approximation to the negentropy which is computationally more efficient. One might argue that we do not need the machinery and the tools that are collectively called Projection Pursuit to be able to calculate the IC directions and scores. But this argument misses the real point of Projection Pursuit.

The main goal of Projection Pursuit is to find non-Gaussian structure. This goal requires an understanding of what we mean by non-Gaussian structure and answers to the following questions.

1. How is such structure manifested in a random vector or data?
2. Should we consider specific aspects of being non-Gaussian or a general deviation from the Gaussian?
3. How can we detect and measure non-Gaussianity or aspects of non-Gaussianity?
4. How good are the estimators of non-Gaussianity?

These questions are closely related, and this chapter provides partial answers to these questions. The fact that some of these questions remain unanswered, combined with the observation that our current answers lead to important new questions, tells us clearly that the subject is far from superseded.

Sections 11.2 and 11.3 provide answers to the second and third questions. The indices in Sections 11.2 and 11.3 variously capture aspects of non-Gaussianity. No single criterion or index can adequately express the notion of non-Gaussianity. As we discussed at the beginning of this chapter, it is easier to start with the Gaussian distribution and consider deviations from this distribution. One may want to think of the Gaussian as the assumption under the null hypothesis, and a general negation of the null hypothesis then corresponds to a deviation from the Gaussian. A deviation from the Gaussian distribution is a recurring theme in a number of the univariate indices, albeit in the form of differences or quotients of the normal and the data distribution.

Instead of negating the full Gaussian null hypothesis, indices could address specific aspects of being non-Gaussian, such as asymmetry or multimodality. Indices which embrace the idea of multimodality are of great interest. In non-parametric density estimation, hypothesis tests for the number of modes exist for low-dimensional and, in particular, univariate data (see the references at the beginning of Section 6.4). For data in up to fifty dimensions, Example 10.5 in Section 10.6.3 illustrates that the skewness or kurtosis criterion can capture bimodality, but there may be different measures which specifically address bi- or multimodality. Such measures could include appropriately chosen scatter matrices (see Section 12.4). The search for such measures continues.

Other and subtler aspects of non-normality such as different tail behaviour are also important. Such deviations may pose greater challenges and have yet to be addressed.

In this chapter we are concerned primarily with measures of non-Gaussianity that correspond to the general negation of the null hypothesis. Numerical algorithms for calculating the deviation from the Gaussian may be slow and expensive, especially if the unknown probability density function has to be estimated from the data. Whether this estimation is done non-parametrically or otherwise, the computational burden is far from negligible. In addition, it is not clear how accurately we are able to estimate the data distribution, especially as the dimension grows. Because of this computational burden, further approximations are made. As we have seen, these approximations include those inherent in the FastICA algorithm. Thus, the FastICA can be seen as a convenient vehicle for obtaining approximate directions in Projection Pursuit.

The last question in the preceding list deals with an assessment of the measures of non-Gaussianity and their estimators. Partial answers to this question are given in Section 11.5, which provides a theoretical foundation for a number of the indices described in Section 11.2.2. Although heuristic arguments are helpful, we do need to know under what conditions our estimators converge to the true index they estimate and what restrictions these assumptions place on the data. Checking the assumptions under which the theoretical results hold will help in assessing whether the methods are appropriate for our data and will lead to more meaningful interpretations of the results.

11.5 Theoretical Developments

Theorem 10.22 in Section 10.7 tells us that the distribution of the low-dimensional projections converges to the Gaussian distribution as the dimension of the random vector increases. This theorem applies equally to the projections in a Projection Pursuit setting. Furthermore, Hall (1988, 1989b) derived properties of the indices $\mathcal{Q}_{\mathrm{U}}$, $\mathcal{Q}_{\mathrm{D}}$ and $\mathcal{Q}_{\mathrm{R}}$ of (11.4) to (11.6) and answered the question: Is the maximiser of $\widehat{\mathcal{Q}}$ close to that of $\mathcal{Q}$?

The theoretical developments treat single directions and the original indices rather than the 2D and 3D projections and indices of Sections 11.3. We begin with the results of Hall (1988), which pertain to $\mathcal{Q}_{\mathrm{R}}$, and then consider the two indices $\mathcal{Q}_{\mathrm{U}}$ and $\mathcal{Q}_{\mathrm{D}}$, which are treated in Hall (1989b). Although both papers address the convergence of the maximiser of $\widehat{\mathcal{Q}}$ to that of $\mathcal{Q}$, the approaches and methods of proof differ; I therefore present the results separately.

11.5.1 Theory Relating to $\mathcal{Q}_R$

Hall (1988) employed kernel density estimation to derive properties of $\mathcal{Q}_{\mathrm{R}}$ and its estimator $\widehat{\mathcal{Q}}_{\mathrm{R}}$ of (11.17). For the same kernel K, Hall defined the two density estimators $\widehat{g}$ and $\widehat{g}_{-i}$ as in (11.16) but then used a smaller bandwidth h_0 for $\widehat{g}_{-i}$ than the optimal h for density estimation. The combination $\widehat{g}_{-i}$ and h_0 enabled him to find the optimal direction $\widehat{\mathbf{a}}_0$ of $\widehat{\mathcal{Q}}_{\mathrm{R}}$ and to derive properties of $\widehat{\mathbf{a}}_0$, including its convergence to $\mathbf{a}_0$, the optimal direction of the population index $\mathcal{Q}_{\mathrm{R}}$. In a second step, Hall constructed an estimator for the unknown probability density function f of $\mathbb{X}$. This step required a standard estimator $\widehat{g}$ and the 'correct' non-parametric bandwidth for a re-estimation of the univariate densities $f_{\mathbf{a}}$ and $\phi_{\mathbf{a}}$. From the properties of univariate density estimators, he derived the convergence of the density estimator of $\mathbb{X}$ to f.

Theorem 11.5 summarises results of Hall (1988). I have stated relevant results adjusted to our framework and given some results in a slightly weaker version than Hall did in his paper.

Theorem 11.5 [Hall (1988)] *Let $\mathbb{X} = \left[\mathbf{X}_1 \cdots \mathbf{X}_n\right]$ be data from a probability density function f. Let f have compact support and m derivatives, for some $m \geq 2$. Let $\mathbf{X}$ be a generic element of $\mathbb{X}$, and assume that $\mathbf{X} \sim (\mathbf{0}, \mathbf{I}_{d\times d})$. Let η be a function of n such that $\eta \to 0$ as $n \to \infty$. Let $\mathbf{a}$ be unit vector in $\mathbb{R}^d$. Let $\boldsymbol{\alpha}$ be a sequence of d-dimensional directions, indexed by n, such that $\|\mathbf{a} - \boldsymbol{\alpha}_n\| \leq \eta(n)$.*

1. *If $\mathcal{Q}_R$ is the projection index of (11.6), then*

$$\mathcal{Q}_R(\mathbf{X},\mathbf{a}) - \mathcal{Q}_R(\mathbf{X},\boldsymbol{\alpha}) = O(\eta) \qquad \textit{uniformly in } \boldsymbol{\alpha} \textit{ as } n \to \infty.$$

Let K be an mth-order univariate kernel, and assume that its fourth derivative $K^{(4)}$ is Hölder continuous. Assume that there is a sequence h_0, indexed by n, which satisfies

$$n^{-1/3} < h_0 < n^{-t} \quad \textit{for} \quad \frac{1}{2(m+\xi-1)} < t \quad \textit{and} \quad 0 \leq \xi \leq 1. \tag{11.29}$$

2. *If $\widehat{Q}_R$ is the sample index defined as in (11.16) and (11.17), and with bandwidth h_0 of (11.29), then*

$$\widehat{Q}_R(\mathbb{X}, \boldsymbol{\alpha}) \xrightarrow{a.s.} \widehat{Q}_R(\mathbb{X}, \mathbf{a}) \qquad \text{as } n \to \infty.$$

3. *If $\mathbf{a}_0$ is the maximiser of Q_R and $\widehat{\mathbf{a}}_0$ is the maximiser of $\widehat{Q}_R$, then*

$$\widehat{\mathbf{a}}_0 \xrightarrow{D} \mathbf{a}_0 \qquad \text{as } n \to \infty$$

and

$$\widehat{\mathbf{a}}_0 \text{ is an } \sqrt{n}\text{-consistent estimator of } \mathbf{a}_0.$$

4. *Assume that $\mathbf{a}_0$ and $\widehat{\mathbf{a}}_0$ of part 3 satisfy $\widehat{\mathbf{a}}_0 - \mathbf{a}_0 = o\left(n^{-m/(2m+1)}\right)$. Consider a sequence h_1, indexed by n, such that $h_1 \sim cn^{-1/(2m+1)}$ for some $c > 0$. If $\widehat{f}_{\mathbf{a}}$ is the kernel density estimator of $f_{\mathbf{a}}$ calculated from the kernel K and the bandwidth h_1, then*

$$\sup_{\mathbf{X} \in \mathbb{R}^d} \left| \widehat{f}_{\widehat{\mathbf{a}}_0}(\widehat{\mathbf{a}}_0^{\mathsf{T}}\mathbf{X}) - \widehat{f}_{\mathbf{a}_0}(\mathbf{a}_0^{\mathsf{T}}\mathbf{X}) \right| = o\left(n^{-m/(2m+1)}\right).$$

Parts 1 and 2 of the theorem are from theorem 3.1 of Hall (1988), part 3 corresponds to his corollary 3.1 and the last part is given in his theorem 4.1. The proofs of Theorem 11.5 can be adjusted from the corresponding proofs in Hall (1988).

Part 1 of the theorem shows that the population index is continuous in the direction vectors, and part 2 shows the analogue for the sample index. Because these results hold for any directions $\mathbf{a}$ and sequence $\boldsymbol{\alpha}_n$ satisfying $\|\mathbf{a} - \boldsymbol{\alpha}_n\| \le \eta(n)$, part 3 follows. Hall showed a result similar to part 4 for the standard normal density estimator $\widehat{\phi}_{\mathbf{a}}$ of $\phi_{\mathbf{a}}$ and combined the two marginal results with an estimator of f to show the convergence of the joint density estimator to f. The joint density estimator is constructed iteratively by an updating method similar to the marginal replacement of Kullback (1968). I briefly come back to this estimator in Section 11.6.1 in connection with Projection Pursuit density estimation.

In parts 2 and 3 of Theorem 11.5, a bandwidth h_0 that is smaller than the usual non-parametric bandwidth is used. This smaller bandwidth suffices to prove the convergence (in distribution) of $\widehat{\mathbf{a}}_0$ to $\mathbf{a}_0$ at the faster parametric rate of $O(n^{-1/2})$. The bandwidth h_0 is typically too small to produce a smooth non-parametric estimate of the univariate densities. For this reason, $f_{\mathbf{a}}$ in part 4 is re-estimated with the standard kernel and the 'correct' bandwidth h_1.

Of special interest is the case of a second-order kernel which is commonly used in density estimation, so $m = 2$. In this case the bounds for h_0 in part 1 reduce to $n^{-1/3} < h_0 < n^{-1/4}$. This bandwidth is smaller than the usual $h_1 = O(n^{-1/5})$ of part 4.

11.5.2 Theory Relating to Q_U and Q_D

The indices Q_{U} and Q_{D} appear to be different at first, but a closer inspection shows that they measure deviation from the Gaussian in a similar way. In addition, sample indices for these two indices are constructed in similar ways (see Section 11.2.2). In his proposal of Q_{U}, Friedman (1987) represented Q_{U} by Legendre polynomials p_j and focused on practical and computational aspects, so he was able to report that values of $m \le 8$ suffice in many applications. A little later, Hall (1989b) proposed the index Q_{D}, which he was able to represent by Hermite polynomials. Hall's interests complemented the practical considerations of

Friedman: he investigated convergence properties of the maximisers of $\widehat{Q}_{\mathrm{U}}$ and $\widehat{Q}_{\mathrm{D}}$, similar to the results for Q_{R} in Hall (1988). I start with Friedman's developments for Q_{U} and then describe Hall's.

Theorem 11.6 [Friedman (1987)] *Let* $\mathbf{X} \sim (\mathbf{0}, \mathbf{I}_{d\times d})$. *Let* $\mathbf{a} \in \mathbb{R}^d$ *be a unit vector, and put* $R = 2\Phi(\mathbf{a}^{\mathsf{T}}\mathbf{X}) - 1$ *as in (11.3). If* $Q_U = \int_{-1}^{1}\left(f_\Theta - \frac{1}{2}\right)^2$ *as in (11.4), then*

$$Q_U = \sum_{j=1}^{\infty}\left\{\mathbb{E}\left[p_j(R)\right]\right\}^2, \tag{11.30}$$

where the p_j *are orthonormal Legendre polynomials, and* $\mathbb{E}$ *is the expectation with respect to* R.

Let $\mathbb{X} = \left[\mathbf{X}_1 \cdots \mathbf{X}_n\right]$ *be data from the same distribution as* $\mathbf{X}$. *For* $m < \infty$, *the sample approximation to the infinite sum (11.30) is*

$$\widehat{Q}_{U,m}(\mathbb{X}, \mathbf{a}) = \sum_{j=1}^{m}\left\{\frac{1}{n}\sum_{i=1}^{n} p_j\left[2\Phi(\mathbf{a}^{\mathsf{T}}\mathbf{X}_i) - 1\right]\right\}^2, \tag{11.31}$$

and $\widehat{Q}_{U,m}$ *approximates* Q_U *as* m *increases.*

For the special case of the uniform distribution, $\mathbb{E}\left[p_j(R)\right] = 0$, for $j \geq 1$. Friedman (1987) extended (11.31) to the 2D index Q_{U} of (11.19) and showed (11.20). The proof is similar to his proof of Theorem 11.6 but more complex because the 2D analogue of Theorem 11.6 incurs a product of two infinite sums.

For the index Q_{D}, Hall (1989b) showed a similar result to that stated in Theorem 11.6, in which his approximation was based on Hermite polynomials. In addition, Hall (1989b) examined the behaviour of the maximisers of $\widehat{Q}_{\mathrm{U}}$ and $\widehat{Q}_D$ and stated conditions under which these maximisers converge to the maximisers of the population indices. Hall's results required a fine-tuning of the rate at which m is allowed to grow relative to the sample size n. As we shall see, this rate differs for the two indices.

Theorem 11.7 [Hall (1989b)] *Let* $\mathbb{X} = \left[\mathbf{X}_1 \cdots \mathbf{X}_n\right]$ *be a random sample from a probability density function* f. *Let* $\mathbf{X}$ *be a generic element of* $\mathbb{X}$, *and assume that* $\mathbf{X} \sim (\mathbf{0}, \mathbf{I}_{d\times d})$. *Assume that* f *satisfies the following:* f *has compact support; for some* $r \geq 2$, *the* r*th order derivatives of* f *are uniformly bounded; and all second-order derivatives are uniformly continuous. Let* $\mathbf{a} \in \mathbb{R}^d$ *be a unit vector, and let* $f_{\mathbf{a}}$ *be the probability density function of* $\mathbf{a}^{\mathsf{T}}\mathbf{X}$.

1. *Let* $\mathbf{a}_U$ *be the maximiser of the index* Q_U *of (11.4). Assume that all second-order derivatives of* Q_U *at* $\mathbf{a}_U$ *are negative. Let* $\widehat{Q}_{U,m}$ *be the sample index (11.31). Regard* m *as a function of* n. *If* $m \to \infty$ *in such a way that*

$$\frac{m^3}{n} \to 0 \quad \text{and} \quad \frac{m^{4(r-1)}}{n} \to \infty \qquad \text{as } n \to \infty,$$

then there exists a local maximiser $\widehat{\mathbf{a}}_U$ *of* $\widehat{Q}_{U,m}$ *such that*

$$\widehat{\mathbf{a}}_U - \mathbf{a}_U = O_p\left(n^{-1/2}\right) \qquad \text{as } n \to \infty. \tag{11.32}$$

2. *Assume that* $f_{\mathbf{a}}$ *and its derivatives satisfy suitable regularity conditions. Let* $\mathbf{a}_D$ *be the maximiser of the index* $\mathcal{Q}_D$ *of (11.5). Assume that the second-order derivatives of* $\mathcal{Q}_D$ *at* $\mathbf{a}_D$ *are negative. Let* $\widehat{\mathcal{Q}}_{D,m}$ *be the sample index (11.14). Regard* m *as a function of* n*. If* $m \to \infty$ *in such a way that*

$$\frac{m^{3/2}}{n} \to 0 \qquad \textit{and} \qquad \frac{m^{2(r-1)}}{n} \to \infty \qquad \textit{as } n \to \infty,$$

then there exists a local maximiser $\widehat{\mathbf{a}}_D$ *of* $\widehat{\mathcal{Q}}_{D,m}$ *such that*

$$\widehat{\mathbf{a}}_D - \mathbf{a}_D = O_p\left(n^{-1/2}\right) \qquad \textit{as } n \to \infty. \tag{11.33}$$

For a proof of this theorem, see Hall (1989b). Part 1 of Theorem 11.7 is Hall's theorem 3.1, and part 2 comes from his theorem 4.1, which lists regularity conditions that $f_{\mathbf{a}}$ needs to satisfy in part 2 of Theorem 11.7. Hall also derived the asymptotic distributions of the directions $\widehat{\mathbf{a}}_{\mathrm{U}}$ and $\widehat{\mathbf{a}}_{\mathrm{D}}$. A difference between the two parts of the theorem is the rate at which the optimal number of terms m in the orthogonal series expansion grows with the sample size. Here 'optimal' refers to the convergence of the sample direction to the population direction. We learn that the m for the index $\mathcal{Q}_{\mathrm{U}}$ grows at a rate $n^{1/3-\epsilon}$, whereas the corresponding m for $\mathcal{Q}_{\mathrm{D}}$ grows at the faster rate $n^{2/3-\delta}$ for $\epsilon, \delta > 0$.

The results of this section deal with projection indices that explicitly take into account the deviation of $f_{\mathbf{a}}$ from the normal. The indices are estimated by different methods – approximation by polynomials and kernel density estimation – however, under appropriate conditions, the estimators have in common that their 'best' directions are $\sqrt{n}$-consistent estimators of the populations parameters.

11.6 Projection Pursuit Density Estimation and Regression

The Projection Pursuit methodology readily extends to two important areas in statistics: density estimation and regression. I briefly outline these developments and show their relationship to the Projection Pursuit methodology as explored so far in this chapter. A red thread in the early Projection Pursuit proposals is Friedman; Friedman and Tukey (1974) first coined the term and formalised the ideas, Friedman and Stuetzle (1981) adapted the initial approach to a regression framework; Friedman, Stuetzle, and Schroeder (1984) applied Projection Pursuit to density estimation and Friedman (1987) revisited the original Projection Pursuit, proposed an intuitive index and showed how to extend the index to more than one dimension. Building on these early proposals, the method thrived, with contributions from many authors, as we have seen in this chapter.

11.6.1 Projection Pursuit Density Estimation

The aims of multivariate density estimation are to explore and exhibit the structure in multivariate data such as the spread of the data and the number and location of clusters or modes and to combine this information into an estimate of the density of the data. In non-parametric density estimation, the density is determined from the observations $\mathbf{X}_1, \ldots, \mathbf{X}_n$ without making assumptions regarding the true underlying density. To estimate the density using ideas from Projection Pursuit, Friedman, Stuetzle, and Schroeder (1984) proposed starting with a model of the d-dimensional probability density function and updating that model. The updating step incorporates the best direction obtained from the projection index.

Definition 11.8 Let $\mathbb{X} = [\mathbf{X}_1 \cdots \mathbf{X}_n]$ be d-dimensional data from a probability density function f. Let p_0 be a given d-dimensional probability density function. For $m = 1, \ldots, M$, the mth **projection pursuit density estimator** $\widehat{f}_m$ of f at $\mathbf{X}$ is

$$\widehat{f}_m(\mathbf{X}) = \widehat{f}_{m-1}(\mathbf{X})\widehat{g}_{\widehat{\mathbf{a}}_m}(\widehat{\mathbf{a}}_m^{\mathsf{T}}\mathbf{X}), \tag{11.34}$$

where $\widehat{f}_0 = p_0$, and for a unit vector $\mathbf{a} \in \mathbb{R}^d$, the **augmenting function**

$$\widehat{g}_{\mathbf{a}}(\mathbf{a}^{\mathsf{T}}\mathbf{X}) = \frac{\widehat{f}_{\mathbf{a}}(\mathbf{a}^{\mathsf{T}}\mathbf{X})}{\widehat{f}_{(m-1,\mathbf{a})}(\mathbf{a}^{\mathsf{T}}\mathbf{X})}.$$

Here $\widehat{f}_{\mathbf{a}}$ is a univariate density estimator calculated from $\mathbf{a}^{\mathsf{T}}\mathbb{X}$, and $\widehat{f}_{(m-1,\mathbf{a})}$ is the univariate marginal of $\widehat{f}_{m-1}$ in the direction of $\mathbf{a}$. Further, the unit vector $\widehat{\mathbf{a}}_m$ of (11.34) satisfies

$$\widehat{\mathbf{a}}_m = \underset{\mathbf{a}\in\mathbb{R}^d,\ \|\mathbf{a}\|=1}{\operatorname{argmax}}\ \widehat{q}_{\widehat{g}}(\mathbb{X},\mathbf{a}) \qquad \text{and} \qquad \widehat{q}_{\widehat{g}}(\mathbb{X},\mathbf{a}) = \left[\frac{1}{n}\sum_{i=1}^{n}\log\widehat{g}_{\mathbf{a}}(\mathbf{a}^{\mathsf{T}}\mathbf{X}_i)\right]. \tag{11.35}$$

□

The univariate augmenting function $\widehat{g}_{\widehat{\mathbf{a}}_m}$ is not a density estimator; it is the ratio of two marginal densities in the direction of $\widehat{\mathbf{a}}_m$. The direction $\widehat{\mathbf{a}}_m$ is chosen to be the minimiser of the sample entropy of $\log\widehat{g}_{\mathbf{a}}$. Criterion (11.35) is natural for finding non-Gaussian direction vectors because the Gaussian density maximises the entropy (see Result 9.7 in Section 9.4). Regarding a suitable choice for p_0, Friedman, Stuetzle, and Schroeder (1984) typically took $p_0 = \phi$, the standard Gaussian probability density function, and the aim was to move further away from the Gaussian in each updating step.

To determine whether M iterations suffice or further updating is required, Friedman, Stuetzle, and Schroeder (1984) proposed comparing the marginal of the current model with the actual marginal of f, both in the direction $\widehat{\mathbf{a}}_M$. In practice, f is unknown, and Friedman, Stuetzle, and Schroeder (1984) suggested an inspection of the graph of $\widehat{g}_{\widehat{\mathbf{a}}_M}$ versus $\widehat{\mathbf{a}}_M^{\mathsf{T}}\mathbb{X}$. If this graph shows a definite tendency or structure, a new direction should be included. If the graph shows noise without any systematic pattern, then further updating will not substantially improve the estimate.

For $m = 1$, the sample average $\widehat{q}_{\widehat{g}}$ of (11.35) is similar to $\widehat{Q}_R$ of (11.17), and they are essentially the same if p_0 is the normal probability density function. In this comparison, I have deliberately ignored the fact that Hall (1988) used a leave-one-out estimator because this is not important for the current discussion. For $m \geq 2$, the functions $\widehat{q}_{\widehat{g}}$ and $\widehat{Q}_R$ differ: $\widehat{q}_{\widehat{g}}$ compares the current marginal $\widehat{f}_{\mathbf{a}}$ with the previous marginal, whereas $\widehat{Q}_R$ measures the departure from Gaussianity and so compares the distribution of $\mathbf{a}^{\mathsf{T}}\mathbb{X}$ with that of the Gaussian.

In the comments following Theorem 11.5, I mentioned the estimator $\widehat{f}$ that Hall (1988) constructed from the marginal estimates. It is natural to compare Hall's estimator with that of Friedman, Stuetzle, and Schroeder (1984). In the notation of Theorem 11.5, the joint probability density f and the estimator $\widehat{f}$ of Hall (1988) are

$$\widehat{f}(\mathbf{X}) = \widehat{\phi}_d(\mathbf{X})\frac{\widehat{f}_{\widehat{\mathbf{a}}_0}(\widehat{\mathbf{a}}_0^{\mathsf{T}}\mathbf{X})}{\widehat{\phi}_{\widehat{\mathbf{a}}_0}(\widehat{\mathbf{a}}_0^{\mathsf{T}}\mathbf{X})} \qquad \text{and} \qquad f(\mathbf{X}) = \phi_d(\mathbf{X})\frac{f_{\mathbf{a}_0}(\mathbf{a}_0^{\mathsf{T}}\mathbf{X})}{\phi_{\widehat{\mathbf{a}}_0}(\widehat{\mathbf{a}}_0^{\mathsf{T}}\mathbf{X})},$$

where ϕ_d is the multivariate normal density. Although $\widehat{f}$ is based on $\widehat{Q}_R$, the usual kernel density estimates are employed in the definition of $\widehat{f}$ – as in part 3 of Theorem 11.5. The similarity between the two estimators is striking, and it is easy to see that Hall's estimator is essentially $\widehat{f}_1$ of Friedman, Stuetzle, and Schroeder (1984).

Unlike Friedman, Stuetzle, and Schroeder (1984), Hall's primary interest was the convergence at the respective maximisers; Hall (1988) showed that

$$\sup_{\mathbf{X} \in S^d} \left| \widehat{f}(\mathbf{X}) - f(\mathbf{X}) \right| = O_p\left(n^{-m/(2m+1)}\right),$$

where S^d is a suitably chosen subset of $\mathbb{R}^d$, and m is the same as in Theorem 11.5. Hall's estimator $\widehat{f}$ could be updated in the same way the projection pursuit density estimator $\widehat{f}_m$ of (11.34) is updated, and the projection pursuit directions or IC directions could be used.

11.6.2 Projection Pursuit Regression

So far we have defined projection indices for data $\mathbb{X}$. In a regression framework, we also have the responses $\mathbf{Y} = [Y_1 \cdots Y_n]$ and their relationships with $\mathbb{X}$. A projection index, based solely on $\mathbb{X}$, may not be suitable for determining the best regression relationship between the predictors and responses; it is therefore important to integrate the responses into the projection index. There are at least two ways we can take the responses into account. We can either consider the explicit relationship of the $\mathbf{X}_i$ and the Y_i or we can consider regression residuals. In Section 13.3 we consider explicit relationships. Friedman and Stuetzle (1981) make use of the residuals, and I outline their approach here.

Let $\mathbb{X} = \begin{bmatrix}\mathbf{X}_1 \cdots \mathbf{X}_n\end{bmatrix}$ be d-dimensional data, and let $\mathbf{Y} = \begin{bmatrix}Y_1 \cdots Y_n\end{bmatrix}$ be centred regression responses. Let g be a smooth function, and assume that the $(\mathbf{X}_i, Y_i)$ satisfy

$$Y_i = g(\mathbf{X}_i) + \epsilon_i \qquad \text{where} \qquad \mathbb{E}(\epsilon_i) = 0 \qquad \text{for } i \le n. \tag{11.36}$$

The aim is to estimate the smooth g from the pairs $(\mathbf{X}_i, Y_i)$ using ideas from Projection Pursuit.

Let $\mathbf{a}$ be a unit vector in $\mathbb{R}^d$. Let $S_\mathbf{a}$ be a smooth real-valued function defined on $\mathbb{R}$ which depends on $\mathbf{a}$, and consider the residuals $\rho_i = Y_i - S_\mathbf{a}(\mathbf{a}^\mathsf{T} X_i)$, for $i \le n$. If we can find a vector $\mathbf{a}^*$ and smooth function $S_{\mathbf{a}^*}$ such that $|\rho|$ goes to zero, then $S_{\mathbf{a}^*}(\mathbf{a}^{*\mathsf{T}}\mathbf{X})$ is a good estimator for $g(\mathbf{X})$, where $\mathbf{X}$ is a generic element of $\mathbb{X}$. The residuals ρ_i form the basis for a regression projection index which is constructed iteratively in Algorithm 11.2.

Algorithm 11.2 *An M-Step Regression Projection Index*
Let $\mathbb{X}$ be d-dimensional data. Let $\mathbf{Y} = \begin{bmatrix}Y_1 \cdots Y_n\end{bmatrix}$ be centred regression responses which satisfy $Y_i = g(\mathbf{X}_i) + \epsilon_i$ for some smooth function g and mean zero errors. Let $S_\mathbf{a}$ be a smooth function which depends on a unit vector $\mathbf{a} \in \mathbb{R}^d$. Fix a threshold $\tau > 0$.

Step 1. Define the initial vector of residuals $\boldsymbol{\rho}_0 = [\rho_{0,1}, \ldots, \rho_{0,n}]$ with $\rho_{0,i} = Y_i$. For $k = 1$, put

$$\widehat{\mathcal{R}}_k(\mathbb{X}, \mathbf{a}) = 1 - \frac{\sum_{i=1}^n \left[\rho_{k-1,i} - S_\mathbf{a}(\mathbf{a}^\mathsf{T}\mathbf{X}_i)\right]^2}{\boldsymbol{\rho}_{k-1}\boldsymbol{\rho}_{k-1}^\mathsf{T}}. \tag{11.37}$$

Step 2. For $k = 1, 2, \ldots$, find the direction vector $\widehat{\mathbf{a}}_k \in \mathbb{R}^d$ and the smooth function $S_{\widehat{\mathbf{a}}_k}$ which maximise $\widehat{\mathcal{R}}_k(\mathbb{X}, \mathbf{a})$ of step 1.

Step 3. If $\widehat{\mathcal{R}}_k(\mathbb{X}, \widehat{\mathbf{a}}_k) > \tau$, put $\boldsymbol{\rho}_k = [\rho_{k,1}, \ldots, \rho_{k,n}]$, where

$$\rho_{k,i} = \rho_{k-1,i} - S_{\widehat{\mathbf{a}}_k}(\widehat{\mathbf{a}}_k^{\mathsf{T}} \mathbf{X}_i) \qquad \text{for } i \leq n.$$

Step 4. Increase the value of k by 1, and repeat steps 2 to 4 until $\widehat{\mathcal{R}}_k(\mathbb{X}, \widehat{\mathbf{a}}_k) < \tau$. Put $M = m$.
Step 5. For $i \leq n$, put

$$\widehat{g}(\mathbf{X}_i) = \sum_{k=1}^{M} S_{\widehat{\mathbf{a}}_k}(\widehat{\mathbf{a}}_k^{\mathsf{T}} \mathbf{X}_i),$$

and define the ***M*-step regression projection index** by

$$\widehat{\mathcal{R}}_M(\mathbb{X}, \widehat{\mathbf{a}}_1, \ldots, \widehat{\mathbf{a}}_M) = 1 - \frac{\left[\sum_{i=1}^n Y_i - \widehat{g}(\mathbf{X}_i)\right]^2}{\boldsymbol{\rho}_{M-1}\boldsymbol{\rho}_{M-1}^{\mathsf{T}}}$$
$$= 1 - \frac{\sum_{i=1}^n \left[Y_i - \sum_{k=1}^M S_{\widehat{\mathbf{a}}_k}(\widehat{\mathbf{a}}_k^{\mathsf{T}} \mathbf{X}_i)\right]^2}{\sum_{i=1}^n \left[Y_i - \sum_{k=1}^{M-1} S_{\widehat{\mathbf{a}}_k}(\widehat{\mathbf{a}}_k^{\mathsf{T}} \mathbf{X}_i)\right]^2}. \qquad (11.38)$$

■

The expressions (11.37) and (11.38) measure the proportion of unexplained variance after one and M iterations, respectively. The regression index (11.38) depends on the smooth functions $S_{\widehat{\mathbf{a}}}$. The numerator of (11.38) takes into account the M smooth function $S_{\widehat{\mathbf{a}}_k}$, with $k \leq M$, whereas the denominator, which is based on the previous residuals, includes the first $M-1$ smooth functions only.

The smooth function g is found in an iterative procedure based on the index $\widehat{\mathcal{R}}$, but choices need to be made regarding the type of function that will be considered in step 2 of Algorithm 11.2. Candidates for smooth functions $S_{\mathbf{a}}$ range from local averages to sophisticated non-parametric kernel functions or splines. We will not explore these estimators of regression functions; instead, I refer the reader to Wand and Jones (1995) for details on kernel regression and to Venables and Ripley (2002) for an implementation in R.

The smooth function $\widehat{g}$ of step 5 of Algorithm 11.2 depends on the 1D projections $\widehat{\mathbf{a}}_k^{\mathsf{T}} \mathbf{X}_i$ and the smooth functions $S_{\widehat{\mathbf{a}}_k}$. Because the maximisation step involves the joint maximiser $(\widehat{\mathbf{a}}_m, S_{\widehat{\mathbf{a}}_m})$, it is not easy to separate the effect of the two quantities. The non-linear nature of the smooth function $\widehat{g}$ and the joint maximisation make it harder to interpret the results. The less interpretable result is the price we pay when we require a smooth function g rather than a simpler linear regression function, as is the case in Principal Component Regression or Canonical Correlation Regression.

In addition to his theoretical developments in Projection Pursuit, Hall (1989a) proved asymptotic convergence results for Projection Pursuit Regression. For the smooth regression function g of (11.36), a random vector $\mathbf{X}$, a unit vector $\mathbf{a}$ and scalar t, Hall (1989a) considered the quantities

$$\gamma_{\mathbf{a}}(t) = \mathbb{E}\left\{g(\mathbf{X}) \mid \mathbf{a}^{\mathsf{T}}\mathbf{X} = t\right\} \qquad \text{and} \qquad S(\mathbf{a}) = \mathbb{E}\left\{\left[g(\mathbf{X}) - \gamma_{\mathbf{a}}(\mathbf{a}^{\mathsf{T}}\mathbf{X})\right]^2\right\}$$

and defined the **first projective approximation** g_1 to g by

$$g_1(\mathbf{X}) = \gamma_{\mathbf{a}_0}(\mathbf{a}_0^{\mathsf{T}}\mathbf{X}) \qquad \text{where } \mathbf{a}_0 = \operatorname{argmax} S(\mathbf{a}).$$

Table 11.4 *Comparison of Projection Pursuit Procedures*

	PP	PPDE	PPR
Data	$\mathbf{X}_1 \cdots \mathbf{X}_n$ $\sim (\mathbf{0}, \mathbf{I})$	$\mathbf{X}_1 \cdots \mathbf{X}_n$ and initial density p_0	$(\mathbf{X}_1, Y_1) \cdots (\mathbf{X}_n, Y_n)$ and initial constant model
Direction **a**	Most non-Gaussian	Most different from p_0	Most different from model
Univariate estimate	density $f_{\mathbf{a}}$	Density $f_{\mathbf{a}}$ marginal $p_{0,\mathbf{a}}$	Regression function of residuals
Choice of index	Different $\mathcal{Q}$,	One only	One only
Sample index	e.g., entropy $\widehat{\mathcal{Q}}_E(\mathbb{X}, \mathbf{a})$	Relative entropy $\widehat{q}(\mathbb{X}, \mathbf{a})$	Unexplained variance $\widehat{\mathcal{R}}(\mathbb{X}, \mathbf{a})$

Hall (1989a) used non-parametric approximations to $\gamma_{\mathbf{a}}$ and S, similar to the kernel density estimates of (11.16), and proved that, asymptotically, the optimal direction vector of the non-parametric kernel estimate of $\gamma_{\mathbf{a}}$ converges to the maximiser of $\gamma_{\mathbf{a}}$. His regression results are similar to those of part 3 of Theorem 11.5. The proofs differ, however, partly because they relate to a non-parametric regression setting.

I conclude this chapter with a table which compares the generic Projection Pursuit with its extensions to density estimation, which I call PPDE in Table 11.4, and regression, which I call PPR.

12

Kernel and More Independent Component Methods

As we know, there are known knowns; there are things we know we know. We also know, there are known unknowns, that is to say, we know there are some things we do not know. But there are also unknown unknowns, the ones we do not know we do not know (Donald Rumsfeld, Department of Defense news briefing, 12 February 2002).

12.1 Introduction

The classical or pre-2000 developments in Independent Component Analysis focus on approximating the mutual information by cumulants or moments, and they pursue the relationship between independence and non-Gaussianity. The theoretical framework of these early independent component approaches is accompanied by efficient software, and the FastICA solutions, in particular, have resulted in these approaches being recognised as among the main tools for calculating independent and non-Gaussian directions. The computational ease of FastICA solutions, however, does not detract from the development of other methods that find non-Gaussian or independent components. Indeed, the search for new ways of determining independent components has remained an active area of research.

This chapter looks at a variety of approaches which address the independent component problem. It is impossible to do justice to this fast-growing body of research; I aim to give a flavour of the diversity of approaches by introducing the reader to a number of contrasting methods. The methods I describe are based on a theoretical framework, but this does not imply that heuristically based approaches are not worth considering. Unlike the early developments in Independent Component Analysis and Projection Pursuit, non-Gaussianity is no longer the driving force in these newer approaches; instead, they focus on different characterisations of independence. We will look at

- a non-linear approach which establishes a correlation criterion that is equivalent to independence,
- algebraic approaches which exploit pairs of covariance-like matrices in the construction of an orthogonal matrix U of the white independent component model (10.4), in Section 10.3, and
- approaches which estimate the probability density function or, equivalently, the characteristic function of the source non-parametrically.

Vapnik and Chervonenkis (1979) explored a special class of non-linear methods, later called *kernel (component) methods*, which do not deviate too far from linear methods and which allow the construction of low-dimensional projections of the data. In addition, kernel component methods exploit the duality of $\mathbb{X}$ and $\mathbb{X}^{\mathsf{T}}$, which makes these approaches suitable for high-dimension low sample size (HDLSS) data. Bach and Jordan (2002) recognised the rich structure inherent in the kernel component methods; they proposed a generalisation of the correlation coefficient which implies independence, and they applied this idea as a criterion for finding independent components.

Of a different nature is the second group of approaches, which solve the independent component model algebraically but require the (sample) covariance matrix to be invertible. The approaches of Oja, Sirkiä, and Eriksson (2006) and Tyler et al. (2009) differ in subtle ways, but they both focus on constructing a rotated coordinate system of the original vectors. These approaches do not look for low-dimensional projections and are not intended as dimension-reduction techniques. In this way, they are similar to the early developments in Independent Component Analysis, where the signal and source have the same dimension, and one wants to recover all source components.

The third group of approaches estimates the probability density function. Yeredor (2000) and Eriksson and Koivunen (2003) do not estimate the density directly but exploit the relationship between probability density functions and characteristic functions and then define a sample-based estimator for the latter. Learned-Miller and Fisher (2003) and Samarov and Tsybakov (2004) estimated the probability density function non-parametrically – using order statistics and kernel density estimation, respectively. The latter two approaches, in particular, may be of more theoretical interest and thus complement some of the other approaches.

Section 12.2 begins with embeddings of data into feature spaces – including infinite-dimensional spaces – and introduces feature kernels. Probably the most important representative among the kernel component methods is *Kernel Principal Component Analysis*, which exploits the duality between the covariance of the feature data and the kernel matrix. Section 12.2 also explains how Canonical Correlation Analysis fits into the kernel component framework. Section 12.3 describes the main ideas of Kernel Independent Component Analysis: we explore a new correlation criterion, the $\mathcal{F}$-correlation, and establish its equivalence to independence. The section considers estimates of the $\mathcal{F}$-correlation and illustrates how the method works in practice. Section 12.4 introduces scatter matrices and explains how to derive independent component solutions from pairs of scatter matrices. Section 12.5 outlines a further three solution methods to the independent component problem. Section 12.5.1 focuses on a characterisation of independence in terms of characteristic functions which can be estimated directly from data. Section 12.5.2 considers the mutual information via the entropy and constructs consistent non-parametric estimators for the entropy which are applied to the independent component model $\mathbf{X} = A\mathbf{S}$, and Section 12.5.3 uses ideas from kernel density estimation in the approximation of the unmixing matrix A^{-1}.

The methods in Sections 12.3 to 12.5 explore different aspects of independence and can be read in any order. Readers not familiar with kernel component methods should read Section 12.2 before embarking on Kernel Independent Component Analysis. The other independent component approaches, in Sections 12.4 and 12.5, do not require prior knowledge of Kernel Component Analysis.

12.2 Kernel Component Analysis

With the exception of Multidimensional Scaling, the dimension-reduction methods we considered so far are linear in the data. A departure from linearity is a major step; we leave behind the interpretability of linear methods and typically incur problems that are more complex, both mathematically and computationally. Such considerations affect the choice of potential non-linear methods. However, non-linear methods may lead to interesting directions in the data or provide further insight into the structure of the data.

In Principal Component Analysis, Canonical Correlation Analysis and Independent Component Analysis we construct direction vectors and projections which are linear in the data. In this section we broaden the search for structure in data to a special class of non-linear methods: non-linear embeddings into a feature space followed by linear projections of the feature data. At the level of the feature data, Vapnik and Chervonenkis (1979) constructed analogues of the dual matrices $Q^{\langle d\rangle}$ and $Q_{\langle n\rangle}$ – see (5.17) in Section 5.5 – which Gower (1966) employed in the construction of classical scaling configurations. Vapnik and Chervonenkis exploited the duality of the 'feature' analogues of $Q^{\langle d\rangle}$ and $Q_{\langle n\rangle}$ to obtain low-dimensional projections of the data. They called the analogue of the feature data matrix $Q_{\langle n\rangle}$ the *kernel matrix*, and their ideas and approach have become known as *Kernel Component Analysis*.

Bach and Jordan (2002) took advantage of the rich structure of non-linear embeddings into feature spaces and proposed a correlation criterion for the feature data which is equivalent to independence. This new criterion, together with the $Q^{\langle d\rangle}$ and $Q_{\langle n\rangle}$ duality for the feature data, enabled them to propose *Kernel Independent Component Analysis*, a novel approach for solving the independent component problem.

In this section I present kernels, feature spaces and their application in Principal Component and Canonical Correlation Analysis. Vapnik and Chervonenkis (1979) provided one of the first comprehensive accounts of these ideas. Schölkopf, Smola, and Müller (1998) developed the main ideas of Kernel Component Analysis and, in particular, Kernel Principal Component Analysis, a non-linear extension of Principal Component Analysis. Vapnik (1995, 1998), Cristianini and Shawe-Taylor (2000) and Schölkopf and Smola (2002) extended the kernel ideas to classification or supervised learning; this direction has since become known as *Support Vector Machines*. In Section 4.7.4 we touched very briefly on linear Support Vector Machines; I refer the interested reader to the books by Vapnik (1998), Cristianini and Shawe-Taylor (2000) and Schölkopf and Smola (2002) for details on Support Vector Machines.

Kernel component methods, like their precursor Multidimensional Scaling, are exploratory in nature. For this reason, I explain the new ideas for data and consider the population case only when necessary.

12.2.1 Feature Spaces and Kernels

We begin with an extension and examples of features and feature maps, which are defined in Section 5.4. For more details and more abstract definitions, see Schölkopf, Smola, and Müller (1998).

Definition 12.1 Let $\mathbb{X} = \begin{bmatrix}\mathbf{X}_1 \; \mathbf{X}_2 \cdots \mathbf{X}_n\end{bmatrix}$ be d-dimensional data, and let $\mathcal{X}$ be the span of the random vectors $\mathbf{X}_i$ in $\mathbb{X}$. Put

$$L_2(\mathcal{X}) = \left\{ g : \mathcal{X} \to \mathbb{R} \; : \; \int |g|^2 < \infty \right\}.$$

Let $\mathfrak{f}$ be a feature map from $\mathbb{X}$ into $L_2(\mathcal{X})$. Define an inner product $\langle\cdot,\cdot\rangle_{\mathfrak{f}}$ on elements of $L_2(\mathcal{X})$, such that $\langle\mathfrak{f}(\mathbf{X}_i),\mathfrak{f}(\mathbf{X}_i)\rangle_{\mathfrak{f}} < \infty$ for $\mathbf{X}_i$ from $\mathbb{X}$. Put

$$\mathcal{F} = \{g \in L_2(\mathcal{X}) \,:\, \langle g, g\rangle_{\mathfrak{f}} < \infty\},$$

and call the triple $\big(\mathcal{F},\mathfrak{f},\langle\cdot,\cdot\rangle_{\mathfrak{f}}\big)$ a **feature space** for $\mathbb{X}$.

Let k be a real-valued map defined on $\mathcal{X}\times\mathcal{X}$. Then k is a **feature kernel** or an **associated kernel** for $\mathfrak{f}$ if, for $\mathbf{X}_i,\mathbf{X}_j$ from $\mathbb{X}$ and $f \in \mathcal{F}$, k satisfies

$$\begin{aligned} k(\mathbf{X}_i,\cdot) &= k(\cdot,\mathbf{X}_i) = \mathfrak{f}(\mathbf{X}_i),\\ k(\mathbf{X}_i,\mathbf{X}_j) &= \big\langle \mathfrak{f}(\mathbf{X}_i),\mathfrak{f}(\mathbf{X}_j)\big\rangle_{\mathfrak{f}}, \end{aligned} \tag{12.1}$$

and the **reproducing property**

$$\langle k(\mathbf{X}_i,\cdot), f\rangle_{\mathfrak{f}} = f(\mathbf{X}_i). \tag{12.2}$$

□

We often write $\mathcal{F}$ for the feature space instead of $\big(\mathcal{F},\mathfrak{f},\langle\cdot,\cdot\rangle_{\mathfrak{f}}\big)$. Many feature maps are embeddings, but they do not have to be injective maps. If $q < d$ and $\mathfrak{f}:\mathbb{R}^d \to \mathbb{R}^q$ is the projection onto the first q variables, then $\mathfrak{f}$ is a feature map for d-dimensional data $\mathbb{X} = \big[\mathbf{X}_1\ \mathbf{X}_2\cdots\mathbf{X}_n\big]$, and the associated kernel k is defined by $k(\mathbf{X}_i,\mathbf{X}_j) = \mathfrak{f}(\mathbf{X}_i)^{\mathsf{T}}\mathfrak{f}(\mathbf{X}_j)$. Commonly used feature spaces $\mathcal{F}$ augment $\mathbb{X}$ by including non-linear combinations of variables of $\mathbb{X}$ as additional variables. The inner product on $\mathcal{F}$ then takes these extra variables into account.

There are many candidates for $\mathfrak{f}$ and its feature space. Surprisingly, however, the choice of the actual feature space does not matter; mostly, the feature kernel is the quantity of interest. Theorem 12.3 shows why it suffices to know and work with the kernel.

The next two examples and Table 12.1 give some insight into the variety of feature maps, feature spaces and kernels one can choose from.

Example 12.1 We consider the feature space of **rapidly decreasing functions**, which is a subspace of $L_2(\mathcal{X})$. Let $\mathbb{X} = \big[\mathbf{X}_1\ \mathbf{X}_2\cdots\mathbf{X}_n\big]$ be d-dimensional data. Define $\mathcal{X}$ and $L_2(\mathcal{X})$ as in Definition 12.1.

Let $\langle f,g\rangle = \int |fg|^2$ be the usual inner product on $L_2(\mathcal{X})$, where $f,g \in L_2(\mathcal{X})$. To obtain a feature space which is a subspace of $L_2(\mathcal{X})$, let $\{\varphi_j,\ j=1,\ldots\}$ be an orthonormal basis for $L_2(\mathcal{X})$, so the φ_j satisfy $\int \varphi_j\varphi_k = \delta_{jk}$, where δ_{jk} is the Kronecker delta function. Consider the sequence of positive numbers $\{\lambda_j : \lambda_1 \ge \lambda_2 \ge \ldots > 0\}$ which have the same cardinality as the basis and which satisfy $\sum_\ell \lambda_\ell < \infty$. For $\mathbf{X}_i$ from $\mathbb{X}$, define a feature map

$$\mathfrak{f}(\mathbf{X}_i) = \sum_\ell \gamma_\ell\varphi_\ell \quad \text{with} \quad \gamma_\ell = \lambda_\ell\varphi_\ell(\mathbf{X}_i). \tag{12.3}$$

Put

$$f = \sum_\ell \alpha_\ell\varphi_\ell \qquad \text{and} \qquad g = \sum_\ell \beta_\ell\varphi_\ell, \tag{12.4}$$

where $\{\alpha_\ell,\beta_\ell,\ \ell = 1,2,\ldots\}$ are real coefficients which satisfy $\sum_\ell |\alpha_\ell|^2 < \infty$ and similarly $\sum_\ell |\beta_\ell|^2 < \infty$. We define an $\mathfrak{f}$-inner product by

$$\langle f,g\rangle_{\mathfrak{f}} = \sum_{\ell,m} \frac{\alpha_\ell\beta_m}{\lambda_\ell}\langle\varphi_m,\varphi_\ell\rangle = \sum_\ell \frac{\alpha_\ell\beta_\ell}{\lambda_\ell}$$

and then define a feature space

$$\mathcal{F} = \left\{ f \in L_2(\mathcal{X}) \,:\, \langle f, f \rangle_{\mathfrak{f}} < \infty \right\}.$$

For $\mathbf{X}_i$, $\mathbf{X}_j$ from $\mathbb{X}$ and $f \in \mathcal{F}$ as in (12.4), put $k(\mathbf{X}_i, \cdot) = \mathfrak{f}(\mathbf{X}_i)$; then k is the feature kernel for $\mathfrak{f}$ which satisfies (12.1) and (12.2), namely,

$$k(\mathbf{X}_i, \mathbf{X}_j) = \left\langle k(\mathbf{X}_i, \cdot), k(\mathbf{X}_j, \cdot) \right\rangle_{\mathfrak{f}} = \sum_{\ell} \lambda_\ell \varphi_\ell(\mathbf{X}_i) \varphi_\ell(\mathbf{X}_j)$$

and

$$\langle k(\mathbf{X}_i, \cdot), f \rangle_{\mathfrak{f}} = \sum_{\ell} \alpha_\ell \varphi_\ell(\mathbf{X}_i) = f(\mathbf{X}_i). \tag{12.5}$$

■

Our next example is similar to the first example in Schölkopf, Smola, and Müller (1998).

Example 12.2 Let $\mathbb{X} = \begin{bmatrix} \mathbf{X}_1 \ \mathbf{X}_2 \cdots \mathbf{X}_n \end{bmatrix}$ be d-dimensional data and $\mathcal{X}$ as in Definition 12.1. Consider the set

$$\mathcal{F}_0(\mathcal{X}) = \{ f : \mathcal{X} \to \mathbb{R} \},$$

and assume that $\mathcal{F}_0(\mathcal{X})$ has an inner product $\langle \cdot, \cdot \rangle$. Let $\mathfrak{f}$ be an embedding of $\mathbb{X}$ into $\mathcal{F}_0(\mathcal{X})$. We use the $\mathfrak{f}(\mathbf{X}_i)$ to generate the feature space

$$\mathcal{F} = \left\{ f \in \mathcal{F}_0(\mathcal{X}) \,:\, f = \sum_{i=1}^{n} \alpha_i \mathfrak{f}(\mathbf{X}_i) \ \text{ for } \ \alpha_i \in \mathbb{R} \right\}.$$

Define a map $\mathbf{X}_j \mapsto [\mathfrak{f}(\mathbf{X}_i)](\mathbf{X}_j)$ by

$$[\mathfrak{f}(\mathbf{X}_i)](\mathbf{X}_j) = \langle \mathfrak{f}(\mathbf{X}_i), \mathfrak{f}(\mathbf{X}_j) \rangle \equiv \langle \mathfrak{f}(\mathbf{X}_i), \mathfrak{f}(\mathbf{X}_j) \rangle_{\mathfrak{f}}.$$

Then this map is symmetric in $\mathbf{X}_i$ and $\mathbf{X}_j$, and an associated kernel k is given by

$$k(\mathbf{X}_i, \mathbf{X}_j) = \langle \mathfrak{f}(\mathbf{X}_i), \mathfrak{f}(\mathbf{X}_j) \rangle_{\mathfrak{f}}.$$

Finally, for $f = \sum_{i=1}^{n} \alpha_i \mathfrak{f}(\mathbf{X}_i)$, it follows that $\langle k(\mathbf{X}_i, \cdot), f \rangle_{\mathfrak{f}} = f(\mathbf{X}_i)$. ■

Table 12.1 lists some commonly used kernels. In the table, $\mathbf{X}_i$ and $\mathbf{X}_j$ are d-dimensional random vectors.

12.2.2 Kernel Principal Component Analysis

Kernel Principal Component Analysis has been an active area of research, especially in the statistical learning and machine learning communities. I describe the ideas of Schölkopf, Smola, and Müller (1998) in a slightly less general framework than they did. In particular, I will assume that the feature map leads to a feature covariance operator with an essentially discrete spectrum. There is a simple reason for this loss in generality: to do justice to the general framework of kernels and feature spaces – and to be mathematically rigorous – one requires tools and results from *reproducing kernel Hilbert spaces*. Readers familiar with this topic will see the connection in the exposition that follows. However, by being slightly less

Table 12.1 *Examples of Feature Kernels*

Polynomial kernels	$k(\mathbf{X}_i,\mathbf{X}_j) = \langle \mathbf{X}_i,\mathbf{X}_j \rangle^m$	$m > 0$
Exponential kernels	$k(\mathbf{X}_i,\mathbf{X}_j) = \exp\left(-\left\|\mathbf{X}_i - \mathbf{X}_j\right\|^\beta/a\right)$	$\beta \geq 1$ and $a > 0$
Gaussian kernels	$k(\mathbf{X}_i,\mathbf{X}_j) = \exp\left(-\left\|\mathbf{X}_i - \mathbf{X}_j\right\|^2/a\right)$	$a > 0$
Cosine kernels	$k(\mathbf{X}_i,\mathbf{X}_j) = \cos\left(\mathbf{X}_i,\mathbf{X}_j\right)$	
Similarity kernels	$k(\mathbf{X}_i,\mathbf{X}_j) = \rho(\mathbf{X}_i,\mathbf{X}_j)$	ρ a similarity derived from a distance; see Section 5.3.3

general, I avoid having to explain intricate ideas from reproducing kernel Hilbert spaces, and I hope to make the material more accessible. For details on reproducing kernel Hilbert spaces, see Cristianini and Shawe-Taylor (2000) and Schölkopf and Smola (2002), and for a more statistical and probabilistic treatment of the topic, see Berlinet and Thomas-Agnan (2004).

Let $\mathbb{X} = \left[\mathbf{X}_1\ \mathbf{X}_2 \cdots \mathbf{X}_n\right]$ be centred d-variate data. To uncover structure in $\mathbb{X}$ with Principal Component Analysis, we project $\mathbb{X}$ onto the eigenvectors of the sample covariance matrix and obtain principal component (PC) scores which are linear in the data. To exhibit non-linear structure in data, a clever idea is to include a non-linear first step in the analysis and, in a second step, make use of the existing linear theory.

Definition 12.2 Let $\mathbb{X} = \left[\mathbf{X}_1\ \mathbf{X}_2 \cdots \mathbf{X}_n\right]$ be d-dimensional centred data. Let $\left(\mathcal{F}, \mathfrak{f}, \langle\cdot,\cdot\rangle_{\mathfrak{f}}\right)$ be a feature space for $\mathbb{X}$ as in Definition 12.1. Let $\mathfrak{f}(\mathbb{X}) = \left[\mathfrak{f}(\mathbf{X}_1)\ \mathfrak{f}(\mathbf{X}_2) \cdots \mathfrak{f}(\mathbf{X}_n)\right]$ be the centred feature data which consist of n random functions $\mathfrak{f}(\mathbf{X}_i) \in \mathcal{F}$.

Let $\mathfrak{f}(\mathbb{X})^{\mathsf{T}}$ be the **transposed feature data** which consist of n rows $\mathfrak{f}(\mathbf{X}_i)^{\mathsf{T}}$. Define the bounded linear operator $S_{\mathfrak{f}}$ on $\mathcal{F}$ by

$$S_{\mathfrak{f}} = \frac{1}{n-1}\mathfrak{f}(\mathbb{X})\mathfrak{f}(\mathbb{X})^{\mathsf{T}}. \tag{12.6}$$

If $S_{\mathfrak{f}}$ has a discrete spectrum consisting of eigenvalues $\lambda_{\mathfrak{f}1} \geq \lambda_{\mathfrak{f}2} \geq \cdots$ with at most one limit point at 0, then $S_{\mathfrak{f}}$ is called a **feature covariance operator**.

For $\ell = 1, 2\ldots$, let $\eta_{\mathfrak{f}\ell} \in \mathcal{F}$ be the ℓth **eigenfunction** corresponding to $\lambda_{\mathfrak{f}\ell}$, so $\eta_{\mathfrak{f}\ell}$ satisfies

$$S_{\mathfrak{f}}\eta_{\mathfrak{f}\ell} = \lambda_{\mathfrak{f}\ell}\eta_{\mathfrak{f}\ell} \qquad \text{and} \qquad \left\langle \eta_{\mathfrak{f}j}, \eta_{\mathfrak{f}\ell}\right\rangle_{\mathfrak{f}} = \delta_{j\ell}, \tag{12.7}$$

where $\delta_{j\ell}$ is the Kronecker delta function. The ℓth **feature score** is the $1 \times n$ random vector

$$\mathfrak{f}\mathbf{W}_{\bullet\ell} = \eta_{\mathfrak{f}\ell}^{\mathsf{T}}\mathfrak{f}(\mathbb{X}) \quad \text{where } \mathfrak{f}W_{i\ell} = \eta_{\mathfrak{f}\ell}^{\mathsf{T}}\mathfrak{f}(\mathbf{X}_i) = \left\langle \eta_{\mathfrak{f}\ell}, \mathfrak{f}(\mathbf{X}_i)\right\rangle_{\mathfrak{f}} \tag{12.8}$$

is the ith entry of the ℓth **feature score**. □

If the feature data $\mathfrak{f}(\mathbb{X})$ are not centred, we centre them and then refer to the centred feature data as $\mathfrak{f}(\mathbb{X})$. In the notation of the feature score I have included the feature map $\mathfrak{f}$ to distinguish these scores from the usual PC scores.

If $\mathcal{F}$ is finite-dimensional, then $S_{\mathfrak{f}}$ is a matrix. Because $\mathcal{F}$ is a function space, the $\mathfrak{f}(\mathbf{X}_i)$ are functions, and in the spectral decomposition of $S_{\mathfrak{f}}$, the eigenvectors are replaced by the eigenfunctions. The eigenvectors have unit norm. In analogy, the eigenfunctions also have unit norm, and their norm is induced by the inner product $\langle\cdot,\cdot\rangle_{\mathfrak{f}}$. The precise form of $\mathcal{F}$ is not important. As we shall see in Theorem 12.3, all we need to know are the values $k(\mathbf{X}_i, \mathbf{X}_j)$, for $i, j \leq n$, of the feature kernel k for $\mathfrak{f}$.

The requirement that the spectrum of $S_{\mathfrak{f}}$ be discrete allows us to work with spectral decompositions similar to those used in Principal Component Analysis. For details on bounded linear operators and their spectra, see, for example, chapter 4 of Rudin (1991).

A glance back to Principal Component Analysis shows that we have used the notation $(\widehat{\lambda}_\ell, \widehat{\boldsymbol{\eta}}_\ell)$ for the eigenvalue-eigenvector pairs of the sample covariance matrix S of $\mathbb{X}$. Because I have omitted definitions for the populations and only defined the new terms for data, I use the 'un-hatted' notation $(\lambda_{\mathfrak{f}\ell}, \boldsymbol{\eta}_{\mathfrak{f}\ell})$ for eigenvalue-eigenfunction pairs of the data.

The ℓth eigenfunction $\boldsymbol{\eta}_{\mathfrak{f}\ell}$ of $S_\mathfrak{f}$ is a function, but the ℓth feature score is a $1 \times n$ vector whose ith entry refers to the ith datum $\mathbf{X}_i$. As in Principal Component Analysis, we are most interested in the first few scores because they contain information of interest. The following theorem gives expressions for the feature scores.

Theorem 12.3 *Let $\mathbb{X} = \left[\mathbf{X}_1\ \mathbf{X}_2 \cdots \mathbf{X}_n\right]$ be d-dimensional centred data. Let $\left(\mathcal{F}, \mathfrak{f}, \langle\cdot,\cdot\rangle_\mathfrak{f}\right)$ be a feature space for $\mathbb{X}$ with associated kernel k. Let K be the $n \times n$ matrix with entries K_{ij} given by*

$$K_{ij} = \frac{k(\mathbf{X}_i, \mathbf{X}_j)}{n-1} \qquad \text{for } i, j \le n.$$

Let r be the rank of K. Let $\mathfrak{f}(\mathbb{X})$ be the centred feature data, and let $S_\mathfrak{f}$ be the feature covariance operator of (12.6) which has a discrete spectrum with at most one limit point at 0. For $\ell = 1, 2, \ldots$, let $(\lambda_{\mathfrak{f}\ell}, \eta_{\mathfrak{f}\ell})$ be the eigenvalue-eigenfunction pairs of (12.7). The following hold.

1. *The first r eigenvalues $\lambda_{\mathfrak{f}1} \ge \lambda_{\mathfrak{f}2} \ge \cdots \ge \lambda_{\mathfrak{f}r}$ of $S_\mathfrak{f}$ agree with the first r eigenvalues of K.*
2. *Let $\boldsymbol{\alpha}_\ell = [\alpha_{\ell 1}, \ldots, \alpha_{\ell n}]^\mathsf{T}$ be the ℓth eigenvector of K which corresponds to the eigenvalue $\lambda_{\mathfrak{f}\ell}$; then*

$$\eta_{\mathfrak{f}\ell} = \lambda_{\mathfrak{f}\ell}^{-1/2}\mathfrak{f}(\mathbb{X})\boldsymbol{\alpha}_\ell.$$

3. *The ℓth feature scores of a random vector $\mathbf{X}_i$ and of the data $\mathbb{X}$ are*

$$\mathfrak{f}W_{i\ell} = \eta_{\mathfrak{f}\ell}^\mathsf{T}\mathfrak{f}(\mathbf{X}_i) = \lambda_{\mathfrak{f}\ell}^{1/2}\alpha_{\ell i} \qquad \text{and} \qquad \mathfrak{f}\mathbf{W}_{\bullet\ell} = \eta_{\mathfrak{f}\ell}^\mathsf{T}\mathfrak{f}(\mathbb{X}) = \lambda_{\mathfrak{f}\ell}^{1/2}\boldsymbol{\alpha}_\ell^\mathsf{T}. \tag{12.9}$$

We call K the **kernel matrix** or the **Gram matrix**.

Remark. This theorem tells us how the feature scores of $\mathbb{X}$ are related to the eigenvalues and eigenvectors of K and implies that we *do not need to know or calculate* $\mathfrak{f}(\mathbf{X}_i)$ in order to uncover non-linear structure in $\mathbb{X}$. The relationship between the feature covariance operator and the kernel matrix is particularly relevant for HDLSS data, as we only need to calculate the $n \times n$ matrix K, its eigenvalues and eigenvectors.

The kernel matrix K is often defined without the scalar $(1-n)^{-1}$. I have included this factor in the definition of K because $S_\mathfrak{f}$ includes it. If we drop the factor $(1-n)^{-1}$ in K but define the feature covariance operator as in (12.6), then the eigenvalues of $S_\mathfrak{f}$ and K differ by the factor $(1-n)^{-1}$.

To prove the theorem, it is convenient to cast the feature data into the framework of the dual matrices $Q_\mathfrak{f}^{(d)} = \mathfrak{f}(\mathbb{X})\mathfrak{f}(\mathbb{X})^\mathsf{T}$ and $Q_{(n)}^\mathfrak{f} = \mathfrak{f}(\mathbb{X})^\mathsf{T}\mathfrak{f}(\mathbb{X})$ of Section 5.5. A proof of the theorem now follows from a combination of Proposition 3.1 in Section 3.2 and Result 5.7 in Section 5.5. The main difference in our current set-up is that the matrix $Q^{(d)}$ is replaced by the bounded linear operator $Q_\mathfrak{f}^{(d)}$ derived from $S_\mathfrak{f}$, whereas $Q_{(n)}^\mathfrak{f}$, which plays the role of $Q_{\langle n\rangle}$, remains a matrix. We explore the duality of $S_{\mathfrak{f}}$ and K in the Problems at the end of Part III. The theorem mirrors the scenario used in Multidimensional Scaling: we move between the $Q_{\langle n\rangle}$ and $Q^{\langle d\rangle}$ matrix and typically use the simpler one for calculations.

Lawrence (2005) considered the probabilistic approach of Tipping and Bishop (1999) (see Section 2.8.1) and showed its connection to Kernel Principal Component Analysis. The ideas of Lawrence (2005) are restricted to Gaussian data, but for such data, one can get stronger results or more explicit interpretations than in the general case.

Example 12.3 illustrates some non-linear kernels for HDLSS data.

Example 12.3 We return to the **illicit drug market** data of Figure 1.5 in Section 1.2.2 and work with the scaled data, as in most of the previous analyses with these data. The seventeen series are the observations, and the sixty-six months are the variables.

I calculate the features scores of the scaled data as in (12.9) of Theorem 12.3 for different kernels. I first evaluate the Gram matrix for the four types of kernels listed in Table 12.1: the polynomial kernel with $m=2$, the exponential kernel with $a=36$ and $\beta=2$, the cosine kernel and the similarity kernel, which uses the correlation coefficient ρ. From the Gram matrix I calculate the first few feature scores and display the resulting three-dimensional score plots in Figure 12.1, with PC_1 on the x-axis (pointing to the right), PC_2 on the y-axis (pointing to the left) and PC_3 on the z-axis.

The top-left panel of the figure shows the 'classical' PC score plots, which are very concentrated apart from a single outlier, series 1, *heroin possession offences*. For all kernels, this series appears as an outlier. To see the structure of the remaining series more clearly in the other three panels, I have restricted the range to the remaining sixteen series. Each panel shows the 'classical' PC scores in blue for comparison. The top-right panel shows the results

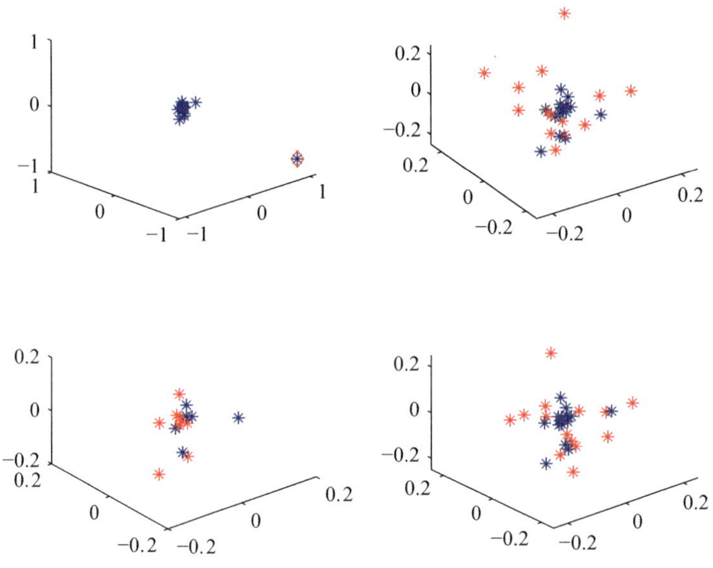

Figure 12.1 Three-dimensional score plots from Example 12.3. 'Classical' PCs in blue in all panels, with outlier series 1 in the top-left panel. In red, (*top right*) exponential kernel, (*bottom left*) cosine kernel, (*bottom right*) polynomial kernel.

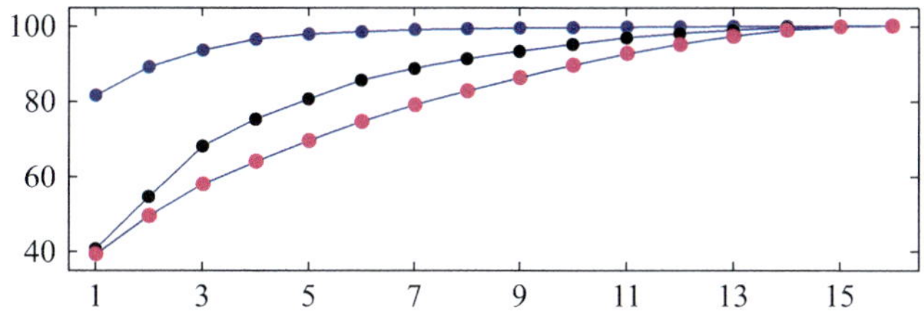

Figure 12.2 Cumulative contribution to total variance from Example 12.3. (*Top*) 'Classical' PCs, (*middle*) polynomial kernel (with *black dots*), (*bottom*) exponential kernel (with *pink dots*).

of the exponential or Gaussian kernel. As the parameter a increases, the red observations become more concentrated. For smaller values of a, the series are further spread out than shown here, and the outlier becomes less of an outlier. The bottom-left panel shows the results of the cosine kernel, which agree quite closely with the 'classical' PC scores. The outliers almost coincide, and so do many of the other series. Indeed, an 'almost agreement' appears purple, a consequence of the overlay of the colours, rather than as a blue symbol. The correlation kernel shows the same pattern, and for this reason, I have not shown its 3D score plot. The bottom-right panel shows the 3D scores of the polynomial kernel, which are more spread out than the PC scores.

The score plots of Figure 12.1 are complemented by graphs of the cumulative contribution to total variance in Figure 12.2. The top curve with blue dots shows the PC results. These are identical to the results obtained with the cosine and correlation kernels. The curve in the middle (with black dots) shows the results for the polynomial kernel, and the lowest curve shows the variance for the exponential kernel.

We note a correspondence between the 'tightness' of the observations and the variance contributions: the tighter the scores in Figure 12.1 are, the higher is the variance.

In Example 6.10 in Section 6.5.2, I calculated the PC_1 sign cluster rule (6.9) for the illicit drug market data. The membership of the series to the two clusters is shown in Table 6.9 of the example. I have applied the analogous cluster rule to the feature data, and with the exception of the exponential kernel (with $a < 36$), the partitioning is identical to that of Table 6.9. For the exponential kernel, the results depend on the choice of a; the first cluster is a proper subset of cluster 1 in Table 6.9; it grows into cluster 1 and equals it, for $a < 36$.

Overall, the first few feature scores of the non-linear kernels do not deviate much from the PC scores, and for these data, they do not exhibit interesting or different structure other than the larger spread of the scores. ■

12.2.3 Kernel Canonical Correlation Analysis

In their approach to Independent Component Analysis, Bach and Jordan (2002) first extended feature maps and kernel ideas to Canonical Correlation Analysis. We begin with their basic ideas in a form adapted to our framework. In Section 12.3, I explain Bach and Jordan's generalisation of the correlation coefficient, which is equivalent to the independence of random vectors.

In Kernel Principal Component Analysis we relate the feature covariance operator (12.6) to the kernel matrix K and take advantage of the duality between them. This duality allows computations of the feature scores from the kernel without reference to the feature map. The central building blocks in Canonical Correlation Analysis are the two matrices which relate the vectors $\mathbf{X}^{[1]}$ and $\mathbf{X}^{[2]}$: the between covariance matrix Σ_{12} and the matrix of canonical correlations C, which are defined in (3.1) and (3.3) in Section 3.2. The vectors $\mathbf{X}^{[1]}$ and $\mathbf{X}^{[2]}$ may have different dimensions, and in a kernel approach, one might want to allow distinct feature maps and kernels for the two vectors. The goal, both in 'classical' and Kernel Canonical Correlation Analysis, however, is the same: we want to find functions of the two (feature) vectors which maximise the correlation, as measured by the absolute value of the correlation coefficient 'cor'. Thus, we want vectors $\mathbf{e}_1^*$ and $\mathbf{e}_2^*$ or functions η_1^* and η_2^* such that

$$(\mathbf{e}_1^*, \mathbf{e}_2^*) = \underset{\mathbf{e}_1, \mathbf{e}_2}{\operatorname{argmax}} \left| \operatorname{cor}\left(\mathbf{e}_1^{\mathsf{T}} \mathbf{X}^{[1]}, \mathbf{e}_2^{\mathsf{T}} \mathbf{X}^{[2]} \right) \right| \tag{12.10}$$

and

$$(\eta_1^*, \eta_2^*) = \underset{\eta_1, \eta_2}{\operatorname{argmax}} \left| \operatorname{cor}\left[\eta_1^{\mathsf{T}} \mathfrak{f}_1(\mathbf{X}^{[1]}), \eta_2^{\mathsf{T}} \mathfrak{f}_2(\mathbf{X}^{[2]}) \right] \right|. \tag{12.11}$$

In the 'classical' Canonical Correlation Analysis of Chapter 3, the pairs of canonical correlation transforms $\boldsymbol{\phi}_k$ and $\boldsymbol{\psi}_k$ of (3.8) in Section 3.2 are the desired solutions to (12.10). The transforms satisfy

$$\Sigma_{12} \psi_k = \upsilon_k \Sigma_1 \phi_k \qquad \text{and} \qquad \Sigma_{12}^{\mathsf{T}} \phi_k = \upsilon_k \Sigma_2 \psi_k,$$

where υ_k is the kth singular value of $C = \Sigma_1^{-1/2} \Sigma_{12} \Sigma_2^{-1/2}$. For the sample, we replace the covariance matrices by their sample estimates and then work with $\widehat{C} = S_1^{-1/2} S_{12} S_2^{-1/2}$.

For feature spaces and feature data it is natural to define the analogue of $\widehat{C}$. This could be achieved at the level of the operators or at the level of derived scalar quantities $\eta_\nu^{\mathsf{T}} \mathfrak{f}_\ell(\mathbf{X}^{[\nu]})$ as in (12.11). We follow the simpler approach of Bach and Jordan (2002) and consider correlation coefficients based on the $\eta_\nu^{\mathsf{T}} \mathfrak{f}_\nu(\mathbf{X}^{[\nu]})$.

Definition 12.4 Consider the centred $d \times n$ data $\mathbb{X} = \begin{bmatrix} \mathbb{X}^{[1]} \\ \mathbb{X}^{[2]} \end{bmatrix}$. Put $\nu = 1, 2$. Let d_ν be the dimension of $\mathbb{X}^{[\nu]}$ and $d = d_1 + d_2$. Let $\left(\mathcal{F}_\nu, \mathfrak{f}_\nu, \langle \cdot, \cdot \rangle_{\mathfrak{f}_\nu} \right)$ be the feature space of $\mathbb{X}^{[\nu]}$. Put

$$S_{\mathfrak{f},\nu} = \frac{1}{n-1} \mathfrak{f}_\nu(\mathbb{X}^{[\nu]}) \mathfrak{f}_\nu(\mathbb{X}^{[\nu]})^{\mathsf{T}} \qquad \text{and} \qquad S_{\mathfrak{f},12} = \frac{1}{n-1} \mathfrak{f}_1(\mathbb{X}^{[1]}) \mathfrak{f}_2(\mathbb{X}^{[2]})^{\mathsf{T}}. \tag{12.12}$$

Let k_ν be the kernel associated with $\mathfrak{f}_\nu$, and define $n \times n$ matrices K_ν with entries $K_{\nu,ij}$ by

$$K_{\nu,ij} = k_\nu \left(\mathbf{X}_i^{[\nu]}, \mathbf{X}_j^{[\nu]} \right). \tag{12.13}$$

For $\eta_\nu \in \mathcal{F}_\nu$, the **feature correlation** γ is

$$\gamma(\eta_1, \eta_2) = \frac{\left\langle \eta_1, S_{\mathfrak{f},12} \eta_2 \right\rangle_{\mathfrak{f}_1}}{\left[\left\langle \eta_1, S_{\mathfrak{f},1} \eta_1 \right\rangle_{\mathfrak{f}_1} \left\langle \eta_2, S_{\mathfrak{f},2} \eta_2 \right\rangle_{\mathfrak{f}_2} \right]^{1/2}}. \tag{12.14}$$

□

Here $S_{\mathfrak{f},1}$ and $S_{\mathfrak{f},2}$ are bounded linear operators on $\mathcal{F}_1$ and $\mathcal{F}_2$, respectively, and $S_{\mathfrak{f},12}$ is a bounded linear operator from $\mathcal{F}_2$ to $\mathcal{F}_1$. If the $S_{\mathfrak{f},\nu}$ are invertible, then one can define the canonical correlation operator

$$C_{\mathfrak{f}} = S_{\mathfrak{f},1}^{-1/2} S_{\mathfrak{f},12} S_{\mathfrak{f},2}^{-1/2}.$$

For the feature data, the natural extension of the matrix $\widehat{C}$ is the operator $C_{\mathfrak{f}}$, but this definition only makes sense if the operators $S_{\mathfrak{f},\nu}$ are invertible, an assumption which may not hold. To overcome this obstacle, we use the feature correlation γ. The two quantities, $C_{\mathfrak{f}}$ and γ, are closely related: γ has a similar relationship to $C_{\mathfrak{f}}$ as Fisher's discriminant quotient $\mathfrak{q}$ has to the matrix $W^{-1}B$ of Theorem 4.6 (see Section 4.3.1). The main difference between the current setting and Fisher's is that γ is based on two functions η_1 and η_2, whereas $\mathfrak{q}$ is defined for a single vector $\mathbf{e}$. Both settings are special cases of the generalised eigenvalue problem (see Section 3.7.4).

In Kernel Principal Component Analysis we construct a matrix K from the feature covariance operator $S_{\mathfrak{f}}$. To 'kernalise' canonical correlations, we consider a kernel version of the feature correlation.

Theorem 12.5 [Bach and Jordan (2002)] *For $\nu = 1,2$, let $\mathbb{X}^{[\nu]} = \left[\mathbf{X}_1^{[\nu]}\,\mathbf{X}_2^{[\nu]}\cdots\mathbf{X}_n^{[\nu]}\right]$ be d_ν-dimensional centred data. Let $\left(\mathcal{F}, \mathfrak{f}, \langle\cdot,\cdot\rangle_{\mathfrak{f}}\right)$ be a common feature space with feature kernel k. Let K_ν be the $n\times n$ matrix with entries $K_{\nu,ij} = k(\mathbf{X}_i^{[\nu]}, \mathbf{X}_j^{[\nu]})$ as in (12.13). For $\eta_1, \eta_2 \in \mathcal{F}$, write*

$$\eta_1 = \sum_{i=1}^n \alpha_{1i}\mathfrak{f}(\mathbf{X}_i^{[1]}) + \eta_1^\perp \qquad \textit{and} \qquad \eta_2 = \sum_{i=1}^n \alpha_{2i}\mathfrak{f}(\mathbf{X}_i^{[2]}) + \eta_2^\perp, \tag{12.15}$$

where $\eta_1^\perp$ and $\eta_2^\perp$ are orthogonal to the spans of the feature data $\mathfrak{f}(\mathbb{X}^{[1]})$ and $\mathfrak{f}(\mathbb{X}^{[2]})$, respectively, and $\alpha_{\nu i} \in \mathbb{R}$, for $i \le n$. Put $\boldsymbol{\alpha}_\nu = [\alpha_{\nu 1}, \ldots, \alpha_{\nu n}]^\mathsf{T}$ and

$$\kappa(\boldsymbol{\alpha}_1, \boldsymbol{\alpha}_2) = \frac{\boldsymbol{\alpha}_1^\mathsf{T} K_1 K_2 \boldsymbol{\alpha}_2}{\left[\left(\boldsymbol{\alpha}_1^\mathsf{T} K_1 K_1 \boldsymbol{\alpha}_1\right)\left(\boldsymbol{\alpha}_2^\mathsf{T} K_2 K_2 \boldsymbol{\alpha}_2\right)\right]^{1/2}}. \tag{12.16}$$

Then, for the feature correlation γ as in (12.14),

$$\gamma(\eta_1, \eta_2) = \kappa(\boldsymbol{\alpha}_1, \boldsymbol{\alpha}_2). \tag{12.17}$$

If $\boldsymbol{\alpha}_1^*, \boldsymbol{\alpha}_2^*$ maximise κ and η_1^* and η_2^* are calculated from $\boldsymbol{\alpha}_1^*$ and $\boldsymbol{\alpha}_2^*$ as in (12.15), then η_1^* and η_2^* maximise γ.

Bach and Jordan (2002) did not explicitly state (12.16) but used it as a step towards proving (12.25) and the independence of the random vectors. If a search for the optimal $\boldsymbol{\alpha}_1$ and $\boldsymbol{\alpha}_2$ is easier than the search for the optimal η_1 and η_2, then the theorem can be used to find the *best* η_1 and η_2 from the optimal $\boldsymbol{\alpha}_1$ and $\boldsymbol{\alpha}_2$ via (12.15).

Proof Put $\langle\cdot,\cdot\rangle = \langle\cdot,\cdot\rangle_{\mathfrak{f}}$. We show that

$$\left\langle \eta_1, S_{\mathfrak{f},12}\eta_2 \right\rangle = \boldsymbol{\alpha}_1^\mathsf{T} K_1 K_2 \boldsymbol{\alpha}_2.$$

Take η_1 and η_2 as in (12.15). In the following calculations we use the kernel property (12.2).

$$\begin{aligned}\langle \eta_1, S_{\mathfrak{f},12}\eta_2 \rangle &= \frac{1}{n-1}\sum_{i=1}^{n} \eta_1^{\mathsf{T}}\mathfrak{f}(\mathbf{X}_i^{[1]})\mathfrak{f}(\mathbf{X}_i^{[2]})^{\mathsf{T}}\eta_2 \\ &= \frac{1}{n-1}\sum_{i,j,k=1}^{n} \left\langle \mathfrak{f}(\mathbf{X}_i^{[1]}), \alpha_{1j}\mathfrak{f}(\mathbf{X}_j^{[1]})\right\rangle \left\langle \mathfrak{f}(\mathbf{X}_i^{[2]}), \alpha_{2k}\mathfrak{f}(\mathbf{X}_k^{[2]})\right\rangle \\ &= \frac{1}{n-1}\sum_{i,j,k=1}^{n} \alpha_{1j}\alpha_{2k}\, k(\mathbf{X}_i^{[1]}, \mathbf{X}_j^{[1]})k(\mathbf{X}_i^{[2]}, \mathbf{X}_k^{[2]}) = \boldsymbol{\alpha}_1^{\mathsf{T}} K_1 K_2 \boldsymbol{\alpha}_2.\end{aligned}$$

Similar calculations hold for the terms in the denominators of γ and κ. ■

12.3 Kernel Independent Component Analysis

Bach and Jordan (2002) proposed a method for estimating the independent random vectors $\mathbf{S}_i$ in the independent component model $\mathbb{X} = A\mathbb{S}$ of (10.3) in Section 10.2.2. In this section I describe their approach, which offers an alternative to the methods of Chapter 10.

12.3.1 The $\mathcal{F}$-Correlation and Independence

Independence implies uncorrelatedness, but the converse does not hold in general. The relationship between these two ideas is explicit in (10.11) of Theorem 10.9 in Section 10.4.1. For Gaussian random vectors, $\mathbf{X}_1$ and $\mathbf{X}_2$, it is well known that

$$\mathrm{cor}(\mathbf{X}_1, \mathbf{X}_2) = 0 \qquad \Longrightarrow \qquad f(\mathbf{X}_1, \mathbf{X}_2) = f_1(\mathbf{X}_1) f_2(\mathbf{X}_2), \tag{12.18}$$

where cor is the correlation coefficient of $\mathbf{X}_1$ and $\mathbf{X}_2$, f is the probability density function of $(\mathbf{X}_1, \mathbf{X}_2)$ and the f_ℓ are the marginals of f corresponding to the $\mathbf{X}_\ell$. The desire to generalise the left-hand side of (12.18) – so that it becomes equivalent to independence – is the starting point in Bach and Jordan (2002). Unlike the probability density function, the correlation coefficient can be calculated easily, efficiently and consistently from data. It therefore makes an ideal candidate for checking the independence of the components of a random vector or data.

Definition 12.6 Let $\mathbb{X} = \begin{bmatrix}\mathbf{X}_1\ \mathbf{X}_2 \cdots \mathbf{X}_n\end{bmatrix}$ be d-dimensional centred data. Let $\left(\mathcal{F}, \mathfrak{f}, \langle\cdot,\cdot\rangle_{\mathfrak{f}}\right)$ be a feature space for $\mathbb{X}$ with associated kernel k. The **$\mathcal{F}$-correlation** $\rho_{\mathcal{F}}$ of the vectors $\mathbf{X}_i$, $\mathbf{X}_j$ is

$$\rho_{\mathcal{F}}(\mathbf{X}_i, \mathbf{X}_j) = \max_{g_i, g_j \in \mathcal{F}} \left| \mathrm{cor}\left[g_i(\mathbf{X}_i), g_j(\mathbf{X}_j)\right] \right|. \tag{12.19}$$

□

Definition 12.6 deviates from that of Bach and Jordan in that I define $\rho_{\mathcal{F}}$ as the maximum of the *absolute value*, whereas Bach and Jordan defined $\rho_{\mathcal{F}}$ simply as the maximum of the correlation of $g_i(\mathbf{X}_i)$ and $g_j(\mathbf{X}_j)$. In the proof of their theorem 2, however, they used the absolute value of the correlation. As in Canonical Correlation Analysis, we are primarily interested in the correlations which deviate most from zero; the sign of the correlation is not important. For this reason, I have chosen to define the $\mathcal{F}$-correlation as in (12.19).

The reproducing property (12.2) of kernels allows us to make the connection between the $\mathcal{F}$-correlation and the kernel, namely, $\langle k(\mathbf{X}_i,\cdot), g\rangle_{\mathfrak{f}} = g(\mathbf{X}_i)$ for $g \in \mathcal{F}$ and $i \leq n$.

Bach and Jordan (2002) used feature maps and feature kernels as a vehicle to prove independence. Their framework is that of Kernel Canonical Correlation Analysis, with the simplification that their feature maps $\mathfrak{f}_1$ and $\mathfrak{f}_2$ are the same. As a consequence, the associated kernels agree.

We begin with a result which shows that the $\mathcal{F}$-correlation is the right object to establish independence.

Theorem 12.7 [Bach and Jordan (2002)] *Let X_1 and X_2 be univariate random variables, and let $\mathcal{X} = \mathbb{R}$. Let k be the Gaussian kernel of Table 12.1 with $a = 2\sigma^2$ for $\sigma > 0$. Then*

$$\rho_{\mathcal{F}}(X_1, X_2) = 0 \qquad \Longleftrightarrow \qquad X_1 \text{ and } X_2 \text{ are independent.} \tag{12.20}$$

Bach and Jordan's proof of Theorem 12.7 is interesting because it exploits the reproducing property (12.2) of kernels and characterises independence in terms of characteristic functions, an idea which is also pursued in Eriksson and Koivunen (2003) (see Section 12.5.1). Bach and Jordan's proof extends to general random vectors, but the bivariate case suffices to illustrate their ideas.

Proof Independence implies a zero $\mathcal{F}$-correlation. It therefore suffices to show the converse. We use the kernel property $\langle k(X_1,\cdot), f\rangle_{\mathfrak{f}} = f(X_1)$ for $f \in \mathcal{F}$ to show that the following statements are equivalent.

$$\begin{aligned}
&\rho_{\mathcal{F}}(X_1, X_2) = 0\\
\Longleftrightarrow\ & \operatorname{cor}\left[f_1(X_1), f_2(X_2)\right] = 0 \qquad \text{for every } f_1, f_2 \in \mathcal{F},\\
\Longleftrightarrow\ & \mathbb{E}\left[f_1(X_1) f_2(X_2)\right] = \mathbb{E}\left[f_1(X_1)\right]\mathbb{E}\left[f_2(X_2)\right] \qquad \text{for every } f_1, f_2 \in \mathcal{F}.
\end{aligned} \tag{12.21}$$

Because the last statement holds for every $f \in \mathcal{F}$, we use this equality to exhibit a family of functions $f_\tau \in \mathcal{F}$ such that $\mathbb{E}\left[f_\tau(X)\right]$ converges to the characteristic function of X as $\tau \to \infty$. Consider $\chi_0 \in \mathbb{R}$ and $\tau > \sigma/\sqrt{2}$. Observe that the functions

$$f_\tau : X \longmapsto e^{-X^2/2\tau^2} e^{i\chi_0 X}$$

with Fourier transforms

$$f_\tau^{\natural} : \chi \longmapsto \sqrt{2\pi}\, e^{-\tau^2(\chi-\chi_0)^2/2}$$

belong to $\mathcal{F}$. From (12.21) it follows that for $\chi_1, \chi_2 \in \mathbb{R}$,

$$\mathbb{E}\left[e^{i(X_1\chi_1 + X_2\chi_2)} e^{-(X_1^2+X_2^2)/2\tau^2}\right] = \mathbb{E}\left[e^{iX_1\chi_1} e^{-X_1^2/2\tau^2}\right]\mathbb{E}\left[e^{iX_2\chi_2} e^{-X_2^2/2\tau^2}\right].$$

Finally, as $\tau \to \infty$,

$$\mathbb{E}\left[e^{i(X_1\chi_1 + X_2\chi_2)}\right] = \mathbb{E}\left[e^{iX_1\chi_1}\right]\mathbb{E}\left[e^{iX_2\chi_2}\right],$$

and thus the bivariate characteristic function on the left-hand side is the product of its univariate characteristic functions, and the independence of X_1 and X_2 follows from this equality. ∎

This theorem shows that independence follows from a zero $\mathcal{F}$-correlation. In the next step we look at a way of estimating $\rho_{\mathcal{F}}$ from data.

12.3.2 Estimating the $\mathcal{F}$-Correlation

Bach and Jordan (2002) used the feature correlation γ of (12.14) and its kernalisation κ to estimate $\rho_{\mathcal{F}}$ from data. To appreciate why γ is a natural candidate for estimating $\rho_{\mathcal{F}}$, we look at γ more closely.

Put $\mathbb{X} = \begin{bmatrix}\mathbb{X}^{[1]}\\ \mathbb{X}^{[2]}\end{bmatrix}$ with $\mathbb{X}^{[\nu]} = \begin{bmatrix}\mathbf{X}_1^{[\nu]}\ \mathbf{X}_2^{[\nu]} \ldots \mathbf{X}_n^{[\nu]}\end{bmatrix}$ d_ν-dimensional centred data for $\nu = 1,2$. Let $\left(\mathcal{F}, \mathfrak{f}, \langle\cdot,\cdot\rangle_{\mathfrak{f}}\right)$ be a common feature space with associated feature kernel k. Let $S_{\mathfrak{f},\nu}$ and $S_{\mathfrak{f},12}$ be the feature covariance operators of (12.12), and consider $\eta_\nu \in \mathcal{F}$. From the kernel property (12.2), it follows that

$$\left\langle \eta_\nu, S_{\mathfrak{f},\nu}\eta_\nu \right\rangle_{\mathfrak{f}} = \frac{1}{n-1}\sum_{i=1}^{n}\left[\eta_\nu(\mathbf{X}_i^{[\nu]})\right]^2 \tag{12.22}$$

and

$$\left\langle \eta_1, S_{\mathfrak{f},12}\eta_2 \right\rangle_{\mathfrak{f}} = \frac{1}{n-1}\sum_{i=1}^{n}\eta_1(\mathbf{X}_i^{[1]})\eta_2(\mathbf{X}_i^{[2]}). \tag{12.23}$$

The right-hand side of (12.22) is the sample variance of the random variables $\eta_\nu(\mathbf{X}_i^{[\nu]})$, and (12.23) is the sample covariance of the $\eta_1(\mathbf{X}_i^{[1]})$ and $\eta_2(\mathbf{X}_i^{[2]})$. The feature correlation γ of (12.14) is therefore an estimator for $\operatorname{cor}[\eta_1(\mathbf{X}_i^{[1]}), \eta_1(\mathbf{X}_i^{[1]})]$. Hence, we define an estimator $\widehat{\rho}_{\mathcal{F}}$ of the $\mathcal{F}$-correlation by

$$\widehat{\rho}_{\mathcal{F}}(\mathbb{X}) = \max_{\eta_1,\eta_2\in\mathcal{F}} |\gamma(\eta_1,\eta_2)| = \max_{\eta_1,\eta_2\in\mathcal{F}} \frac{\left|\left\langle \eta_1, S_{\mathfrak{f},12}\eta_2\right\rangle_{\mathfrak{f}}\right|}{\left[\left\langle \eta_1, S_{\mathfrak{f},1}\eta_1\right\rangle_{\mathfrak{f}}\left\langle \eta_2, S_{\mathfrak{f},2}\eta_2\right\rangle_{\mathfrak{f}}\right]^{1/2}}. \tag{12.24}$$

The following theorem is the key to the ideas of Bach and Jordan: it shows how to estimate the left-hand side of (12.20) in Theorem 12.7.

Theorem 12.8 [Bach and Jordan (2002)] *For $\nu = 1,2$, let $\mathbb{X}^{[\nu]} = \begin{bmatrix}\mathbf{X}_1^{[\nu]}\ \mathbf{X}_2^{[\nu]} \cdots \mathbf{X}_n^{[\nu]}\end{bmatrix}$ be d-dimensional centred data, and let $\mathbb{X} = \begin{bmatrix}\mathbb{X}^{[1]}\\ \mathbb{X}^{[2]}\end{bmatrix}$. Let $\left(\mathcal{F}, \mathfrak{f}, \langle\cdot,\cdot\rangle_{\mathfrak{f}}\right)$ be a common feature space for the $\mathbb{X}^{[\nu]}$ with feature kernel k. Let $K_{\nu,ij} = k(\mathbf{X}_i^{[\nu]}, \mathbf{X}_j^{[\nu]})$ be the entries of the $n\times n$ matrix K_ν. If $\eta_1,\eta_2 \in \mathcal{F}$, $\boldsymbol{\alpha}_1, \boldsymbol{\alpha}_2$ and $\kappa(\boldsymbol{\alpha}_1,\boldsymbol{\alpha}_2)$ are as in Theorem 12.5, and if $\widehat{\rho}_{\mathcal{F}}$ is given by (12.24), then*

$$\widehat{\rho}_{\mathcal{F}}(\mathbb{X}) = \max_{\boldsymbol{\alpha}_1,\boldsymbol{\alpha}_2\in\mathbb{R}^n} |\kappa(\boldsymbol{\alpha}_1,\boldsymbol{\alpha}_2)|. \tag{12.25}$$

Further, the maximisers $\boldsymbol{\alpha}_1^*$ and $\boldsymbol{\alpha}_2^*$ of (12.25) are the solutions of

$$\begin{pmatrix}\mathbf{0} & K_1K_2\\ K_2K_1 & \mathbf{0}\end{pmatrix}\begin{bmatrix}\boldsymbol{\alpha}_1\\ \boldsymbol{\alpha}_2\end{bmatrix} = \lambda\begin{pmatrix}K_1^2 & \mathbf{0}\\ \mathbf{0} & K_2^2\end{pmatrix}\begin{bmatrix}\boldsymbol{\alpha}_1\\ \boldsymbol{\alpha}_2\end{bmatrix} \quad \textit{for some } \lambda \neq 0. \tag{12.26}$$

The relationship (12.25) follows from Theorem 12.5. We recognise (12.26) as a generalised eigenvalue problem (see Section 3.7.4). To find the solution to (12.26), one could proceed as in (3.54) of Section 3.7.4. However, some caution is necessary when solving (12.26). The Gram matrices K_ν may not be invertible, but if they are, then the solution is trivial. For this reason, Bach and Jordan proposed adjusting the $\mathcal{F}$-correlation by modifying the variance terms $\langle \eta_\nu, S_{\mathfrak{f},\nu}\eta_\nu \rangle_\mathfrak{f}$ in (12.24) and the corresponding kernel matrices K_ν^2 in (12.26). The adjustment for the kernel matrices is of the form

$$K_\nu^2 \longmapsto \widetilde{K}_\nu^2 \equiv (K_\nu + c\mathbf{I})^2 \qquad \text{for some } c > 0. \tag{12.27}$$

So far we have considered two sets of random vectors and shown that two random vectors are independent if their $\mathcal{F}$-correlation is zero. To show the independence of all pairs of variables, Bach and Jordan defined a $pn \times pn$ super-kernel, for $p \leq d$, and the derived eigenvalue problem

$$\begin{pmatrix} \mathbf{0} & K_1K_2 & \cdots & K_1K_p \\ K_2K_1 & \mathbf{0} & \cdots & K_2K_p \\ \vdots & \vdots & \ddots & \vdots \\ K_pK_1 & K_pK_2 & \cdots & \mathbf{0} \end{pmatrix} \begin{bmatrix} \boldsymbol{\alpha}_1 \\ \boldsymbol{\alpha}_2 \\ \vdots \\ \boldsymbol{\alpha}_p \end{bmatrix} = \lambda \begin{pmatrix} \widetilde{K}_1^2 & \mathbf{0} & \cdots & \mathbf{0} \\ \mathbf{0} & \widetilde{K}_2^2 & \cdots & \mathbf{0} \\ \vdots & \vdots & \ddots & \vdots \\ \mathbf{0} & \mathbf{0} & \cdots & \widetilde{K}_p^2 \end{pmatrix} \begin{bmatrix} \boldsymbol{\alpha}_1 \\ \boldsymbol{\alpha}_2 \\ \vdots \\ \boldsymbol{\alpha}_p \end{bmatrix}, \tag{12.28}$$

where each $n \times n$ block $\widetilde{K}_\ell^2$ is defined by (12.27). If $p = d$, the entries $K_{\ell,ij} = k(X_{i\ell}X_{j\ell})$ of K_ℓ are obtained from the ℓth variable of $\mathbf{X}_i$ and $\mathbf{X}_j$. If we write

$$\mathcal{K}\boldsymbol{\alpha} = \lambda\mathcal{D}\boldsymbol{\alpha} \tag{12.29}$$

for (12.28), then we need to find the largest eigenvalue and corresponding eigenvector of (12.29). Bach and Jordan (2002) reformulated (12.29) and solved the related problem

$$\mathcal{C}\boldsymbol{\alpha} = \zeta\mathcal{D}\boldsymbol{\alpha} \qquad \text{where } \mathcal{C} = \mathcal{K} + \mathcal{D}. \tag{12.30}$$

I will not discuss computational aspects of solving (12.30); these are described in Bach and Jordan (2002). Algorithm 12.1 gives the main steps of Bach and Jordan's kernel canonical correlation (KCC) approach to finding independent component solutions. Their starting point is the white independent component model (10.4) of Section 10.3. For d-dimensional centred data $\mathbb{X} = [\mathbf{X}_1 \cdots \mathbf{X}_n]$ and a whitening matrix Ξ, put $\mathbb{X}^\diamond = \Xi\mathbb{X}$. Let $\mathbb{S}$ be the $d \times n$ matrix of source vectors. Let U be the orthogonal matrix of Proposition 10.5 in Section 10.3, which satisfies $\mathbb{X}^\diamond = U\mathbb{S}$. Bach and Jordan's KCC approach finds an estimate $\{B^*, \mathbb{S}^*\}$ for $\{U^\top, \mathbb{S}\}$ which satisfies

$$\mathbb{S}^* = \underset{\mathbb{S} \text{ white}}{\operatorname{argmin}}\, \widehat{\rho}_{\mathcal{F}}(\mathbb{S}) \qquad \text{subject to} \qquad \mathbb{S}^* = B^*\mathbb{X}^\diamond. \tag{12.31}$$

Algorithm 12.1 ***Kernel Independent Component Solutions***
Consider $d \times n$ data $\mathbb{X}$, and let $\mathbb{X}^\diamond$ be d-dimensional whitened data obtained from $\mathbb{X}$.

Step 1. Fix a feature kernel k – typically the Gaussian kernel of Theorem 12.7 with $\sigma = 1$. Fix M, the number of repeats.
Step 2. Let B be a candidate for $U^\top$, and put $\mathbf{S}_i = B\mathbf{X}_i^\diamond$ for $i \leq n$.
Step 3. Calculate the matrices $\mathcal{C}$ and $\mathcal{D}$ of (12.30) for the random vectors $\mathbb{S} = [\mathbf{S}_1 \cdots \mathbf{S}_n]$, and find the largest eigenvalue ζ which solves $\mathcal{C}\boldsymbol{\alpha} = \zeta\mathcal{D}\boldsymbol{\alpha}$.
Step 4. Repeat steps 2 and 3 M times. For $\ell \leq M$, let $(B_\ell, \mathbb{S}_\ell, \zeta_\ell)$ be the triples obtained in steps 2 and 3. Put $\zeta^* = \min_\ell \zeta_\ell$; then the corresponding $\{B^*, \mathbb{S}^*\}$ satisfies (12.31). ■

This algorithm finds a solution for (12.31) by selecting the white data which result in the smallest eigenvalue ζ^* of (12.29). Steps 2 and 3 are computed with the KCC version of Bach and Jordan's kernel-ica software, and their MATLAB program returns an orthogonal unmixing matrix $\widehat{B}$. Because of the random start of their iterative procedure, the resulting unmixing matrices vary. For this reason, I suggest to repeat their iterative algorithm M times as part of Algorithm 12.1, and to minimise over the ζ values in step 4. In practice, the solution vectors will not be independent because we approximate the independence criterion $\rho_{\mathcal{F}} = 0$. As in (10.7) in Section 10.3, the solutions will be as independent as possible given the specific approximation that is used.

In addition to their KCC approach, which solves (12.31), Bach and Jordan (2002) proposed a **kernel generalised variance** (KGV) approach for solving (12.31). In this second approach, they used the eigenvalues ζ_j of $\mathcal{C}\boldsymbol{\alpha} = \zeta \mathcal{D}\boldsymbol{\alpha}$ and replaced the $\mathcal{F}$-correlation by

$$\widehat{Z} = -\frac{1}{2}\sum_j \log \zeta_j. \tag{12.32}$$

For Gaussian random variables with joint probability density function f, Bach and Jordan showed that the mutual information of f agrees with $\widehat{Z}$. This fact motivated their second independence criterion. For details and properties of the kernel generalised variance approach, see section 3.4 of Bach and Jordan (2002). In Example 12.5, I calculate the independent component solutions for both independence criteria.

12.3.3 Comparison of Non-Gaussian and Kernel Independent Component Approaches

In Chapter 10 we found independent component solutions by approximating the mutual information $\mathcal{I}$ by functions of cumulants and, more specifically, by skewness and kurtosis estimators. Bach and Jordan (2002) took a different route: they started with an equivalence to independence rather than an approximation to $\mathcal{I}$. The second step in both methods is the same: estimation of the 'independence criterion' from data. Bach and Jordan's method could be superior to the cumulant-based solution methods because it only uses one approximation.

Kernel approaches have intrinsic complexities which are not inherent in the cumulant methods: the need to choose a kernel and a penalty term or tuning parameter c as in (12.27). Different kernels will result in different solutions, and these solutions may not be comparable. Bach and Jordan (2002) typically used the standard Gaussian kernel and note that the values $c = 0.02$ for $n \leq 1{,}000$ and $c = 0.002$ for $n > 1{,}000$ work well in practice.

Table 12.2 summarises the cumulant-based approaches of Sections 10.3 and 10.4 and the kernel approach of Bach and Jordan (2002). The table is based on the independent component model $\mathbf{X} = A\mathbf{S}^0$, where f is the probability density function of $\mathbf{X}$, and $\mathcal{J}$ is its the negentropy; $\mathbf{S}$ is a candidate solution for the true $\mathbf{S}^0$, and $\mathbf{S}$ is made up of the two subvectors $\mathbf{S}_1$ and $\mathbf{S}_2$. The probability density function of $\mathbf{S}$ is $\boldsymbol{\pi}$, and $\mathcal{G}$ is the skewness or kurtosis estimator. Further, $\mathcal{K}$ is the kernel matrix of (12.29) based on a sample version $\mathbb{S}$ of $\mathbf{S}$.

We compare the independent component solutions based on the non-Gaussian ideas of Chapter 10 and on the kernel approaches of this chapter for two contrasting data sets: the five-dimensional *HIV flow cytometry* data and the thirteen-dimensional *wine recognition* data. The HIV data consist of 10,000 observations, whereas the wine recognition data have

Table 12.2 *Cumulant and $\mathcal{F}$-Correlation-Based Approaches to ICA*

	Cumulant $\mathcal{G}$	$\mathcal{F}$-correlation $\rho_{\mathcal{F}}$
Aim	$\mathcal{I}(\boldsymbol{\pi}) = 0$	$\mathbf{S}_1$ and $\mathbf{S}_2$ are independent
Method	$\mathcal{I}(\boldsymbol{\pi}) \approx \mathcal{J}(f) - \mathcal{G}(\mathbf{S})$	$\mathcal{I}(\boldsymbol{\pi}) = 0 \iff \rho_{\mathcal{F}}(\mathbf{S}_1, \mathbf{S}_2) = 0$
Estimation	$\mathcal{G}(\mathbf{S})$ by sample $\widehat{\mathcal{G}}(\mathbb{S})$	$\rho_{\mathcal{F}}(\mathbf{X}_1, \mathbf{X}_2)$ using $(\mathcal{K} + \mathcal{D})\boldsymbol{\alpha} = \zeta \mathcal{D}\boldsymbol{\alpha}$
Optimisation	Maximise over $\widehat{\mathcal{G}}$	Minimise over first eigenvalue ζ

only 178 observations. Each example illustrates different aspects of independent component solutions.

Example 12.4 The collection of **HIV flow cytometry** data sets of Rossini, Wan, and Moodie (2005) comprise fourteen different subjects. The first two of these, which we previously considered, are those of an HIV^+ and HIV^- subject, respectively. In this analysis we restrict attention to the first of these, an HIV^+ data set.

Example 2.4 in Section 2.3 describes a principal component analysis of the HIV^+ data. The first eigenvector of the principal component analysis has large entries of opposite signs for variable 4, *DC8*, and variable 5, *DC4*, which increase and, respectively, decrease with the onset of HIV. The IC directions maximise criteria other than the variance, and the IC_1 directions will therefore differ from the PC_1 direction.

I calculate IC directions for the HIV^+ data based on the FastICA cumulant approximations $\mathcal{G}_3$, skewness, and $\mathcal{G}_4$, kurtosis, using Algorithm 10.1 in Section 10.6.1, and we compare these directions with the IC directions of the kernel approximations KCC and KGV of Algorithm 12.1. For each of these four approximations, I carry out 100 repetitions of the relevant algorithm.

The FastICA algorithm with the skewness criterion $\mathcal{G}_3$ finds, equally often either one only non-Gaussian direction or all five IC directions. The fifty single directions are always the same, and similarly, the most non-Gaussian out of the other 50 repetitions are also identical. The entries of this first (or only) IC direction are shown as the y-values against the variable number on the x-axis in the second panel of Figure 12.3. The black dots are the entries of the single IC direction, and the red dots are the entries of the first of five IC directions. As we can see, the difference between the two directions is very small, and the corresponding scores have absolute skewness 2.39 and 2.33. The third variable, *CD3*, has the highest weight in both IC_1 directions; this variable is also closely linked to the onset of HIV. It is interesting to note that *CD3* appears with opposite sign from *CD8* and *DC4* (variables 4 and 5) for the first ICs. The left-most panel, for comparison, shows the entries of the first eigenvector of the principal component analysis, and in this case, *CD8* and *CD4* have opposite signs.

Algorithm 10.1 with the FastICA kurtosis criterion $\mathcal{G}_4$ achieves five non-Gaussian directions in all 100 repetitions and many different 'best' IC_1 directions. The absolute kurtosis values of the IC_1 scores lie in the interval [2.6, 3.11]. In 80 of 100 repetitions, variable 3, *CD3*, had the highest absolute IC_1 entry, and for these eighty IC_1 directions, three variants occurred. These are shown in the third panel of Figure 12.3. Unlike panels 1 and 2 in the third to fifth panels I display the variable numbers 1 to 5 on the y-axis, and sort the variables of the IC_1 directions in decreasing order of absolute value of their entries. Finally I plot the sorted entries as a blue or red line showing the index on the x-axis and starting with the

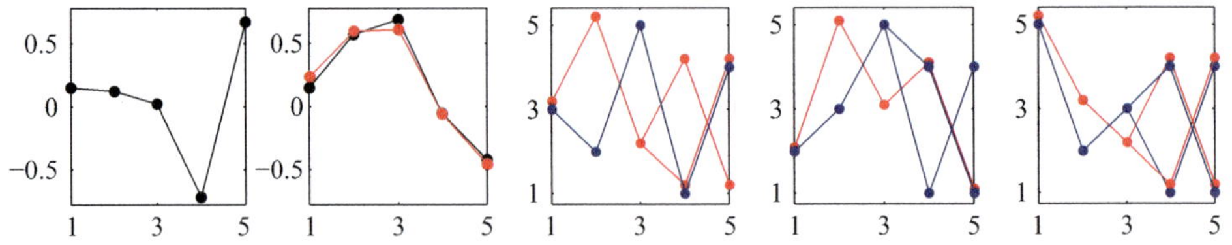

Figure 12.3 PC_1 and IC_1 directions for Example 12.4. Entries of PC_1 (*left*) and skewness IC_1 (*second panel*). Kurtosis IC_1 directions in panels 3–5, sorted by the size of the entries.

biggest one as $x = 1$. Following the blue line in the middle panel, we observe that variable 3, *CD3*, has the highest entry, variable 2, *SS*, has the second highest; and variable 4, *DC8*, has the lowest. Panels 4 and 5 show similar information for the eleven IC_1 directions whose highest entry is variable 2, *SS*, and the nine IC_1 directions with highest entry variable 5, *CD4*. Variable 1, *FS*, and variable 4, *CD8*, did not appear as 'leading' variables in the kurtosis analysis. There are three variants in the third and fourth panels each and four in the fifth panel, so there are a total ten different IC_1 directions in 100 repetitions.

In the analyses based on the skewness and the kurtosis criteria, the variables *CD8* and *CD4* have the same sign, and the sign of *CD4* and *CD8* differs from that of the other three variables. Further, variable 3, *CD3*, has the highest entry in all the skewness and most of the kurtosis IC_1 directions, which suggests a link between *CD3* and non-Gaussian structure in the data.

For the KCC and KGV approaches, I use the MATLAB code 'kernel_ica_option.m' of Bach and Jordan (2002) with their default standard Gaussian kernel. Like FastICA, KernelICA has a random start and iteratively finds the largest eigenvalue ζ in step 3 of Algorithm 12.1. In this analysis, I do not not repeat steps 2 and 3 of Algorithm 12.1; instead, we consider the first independent component directions obtained in each of 100 repetitions of steps 1–3.

Each run of Bach and Jordan's kernel software involves two iterative parts, the 'local search' and the 'polishing'. In all 100 repetitions, Bach and Jordan's KernelICA finds all five IC directions. Because their approach jointly optimises over pairs of variables, there is no obvious ranking of the IC directions, and I therefore consider the first column of the unmixing matrix for each of the 100 repetitions. I order the entries of each IC_1 direction by their absolute values and group the directions by the 'best' variable, that is, the variable that has highest absolute entry. Table 12.3 shows how often each variable occurred as 'best' in 100 repetitions for all four approaches.

The table shows that the KCC and KGV directions are almost uniformly spread across all five variables – unlike the pattern that has emerged for the non-Gaussian IC_1 directions. Figure 12.4 shows the different IC_1 directions, in separate panels for each 'best' variable, that I obtained in 100 repetitions using the same ordering and display as in the kurtosis plots of Figure 12.3. The top row of the figure refers to the KCC approach, and the bottom row to the KGV approach. As we can see, almost any possible combination is present in the top row, whereas some combinations do not appear in the KGV approach. It is difficult to draw any conclusions from this myriad of IC_1 directions.

The 100 repetitions of the KernelICA calculations took 38,896 seconds, compared with about 85 seconds for the corresponding FastICA calculations. Bach and Jordan (2002) referred to a faster implementation based on C code, which I have not used here.

Table 12.3 *Frequency of Each Largest Variable in 100 IC_1 Directions*

	FS	*SS*	*CD3*	*CD8*	*CD4*
Skewness	—	—	100	—	—
Kurtosis	—	11	80	—	9
KCC	22	14	25	22	17
KGV	20	16	27	15	22

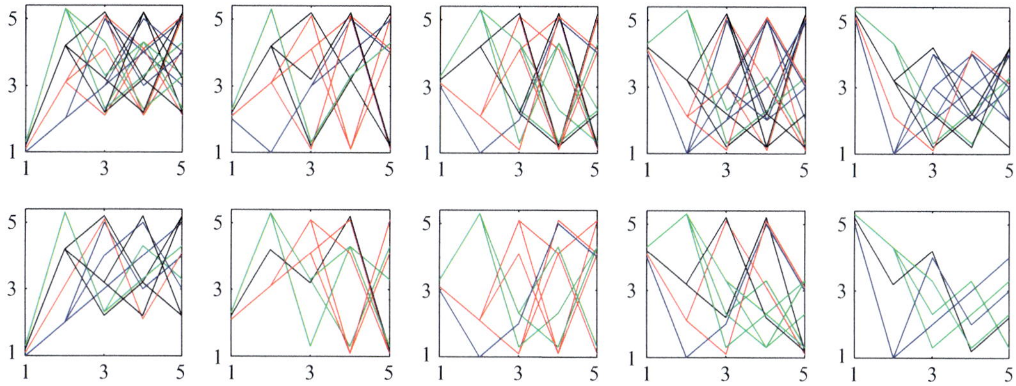

Figure 12.4 Kernel IC_1 directions for KCC in the top row and KGV in the bottom row from Example 12.4. IC_1 entries are ordered by their size, starting with the largest, and are displayed on the y-axis. Different panels have different largest entries.

For the kurtosis IC_1 directions in Figure 12.3 and the kernel IC directions in Figure 12.4, I have only shown the ranking of the variables in decreasing order rather than their actual values, and thus, many more IC directions exist than shown here, particularly for the kernel-based solutions.

For the HIV^+ data, the non-Gaussian IC_1 directions result in a small number of candidates for the most non-Gaussian direction, and the latter can be chosen from these candidates by finding that which maximises the skewness or kurtosis. No natural criterion appears to be available to choose from among the large number of possible kernel IC directions. This makes an interpretation of the kernel IC solutions difficult. ■

As we have seen in Example 12.4, the kernel IC directions are highly non-unique and seem to appear in a random fashion for these five-dimensional data. Our next example looks at a much smaller number of observations, but more variables, and focuses on different aspects of the IC solutions.

Example 12.5 The thirteen-dimensional **wine recognition** data arise from three different cultivars. In Example 4.5 in Section 4.4.2 and Example 4.7 in Section 4.5.1 we construct discriminant rules for the three classes. We now derive independent component directions for these data.

As in the previous analyses of these data, we work with the scaled data. Practical experience with Bach and Jordan's KernelICA shows that repeated runs of their code result in different values of ζ^*, different unmixing matrices and distinct scores. For the standard

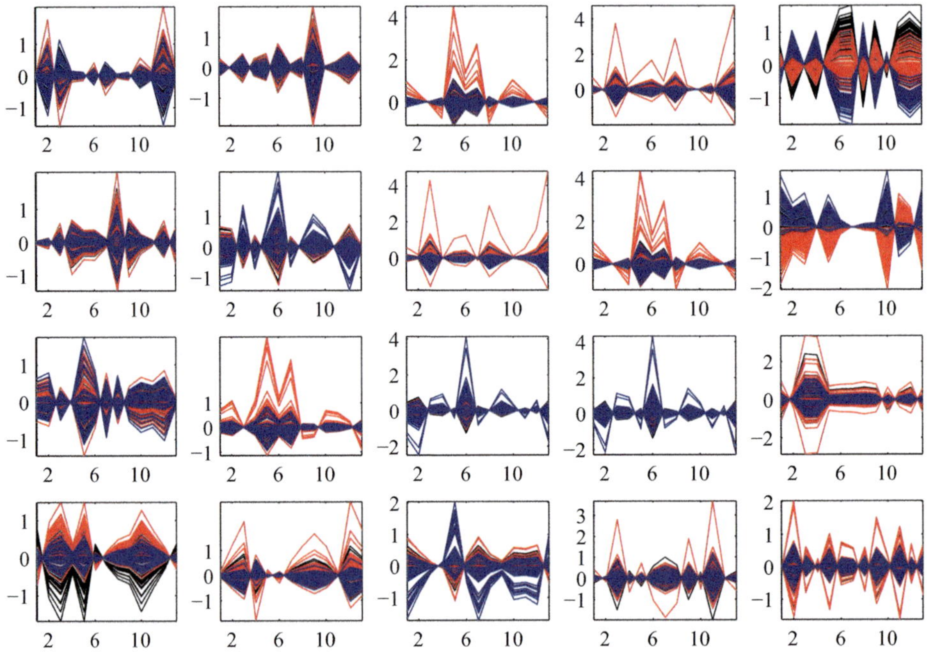

Figure 12.5 First four IC projections from Example 12.5, starting with the first projections in the top row. The three classes of observations are shown in different colours. (*Col. 1*) KCC approach; (*col. 2*) KGV approach; (*col. 3*) FastICA skewness; (*col. 4*) FastICA kurtosis; (*col. 5*) PC projections.

Gaussian kernel and $M = 100$ in step 1 of Algorithm 12.1, the columns of Figure 12.5 show the first four IC projections – starting with the first projections in the top row. The first and second columns show the projections of the KCC and KGV approaches respectively. The third and fourth columns show non-Gaussian IC projections, calculated with the FastICA skewness criteria $\mathcal{G}_3$ and kurtosis criterion $\mathcal{G}_4$ of Hyvärinen (1999). See Algorithm 10.1 in Section 10.6.1. For comparison, I include the first four PC projections in the fifth column.

In the plots of Figure 12.5, the colours black, red and blue represent the observations from the three classes – as in Figure 4.4 in Example 4.5. In many of these projection plots, the observations represented by blue and black overlap, and because blue is drawn after black here, the black lines are only visible where they differ from the others.

The KCC projections do not show any particular pattern, with the exception of the fourth projection plot: the 'red' and 'black' observations mostly have opposite signs and are relatively large, whereas the 'blue' observations are close to zero. The second to fourth KGV projections exhibit outliers. The pattern of the third KGV projection is similar to those of the first and second projections in columns 3 and 4 respectively. Indeed, the first three FastICA projections are almost the same, but the order of the first and second is interchanged. All four kurtosis-based projections in column 4 exhibit outliers, a common feature of non-Gaussian independent components scores and projections. The first PC projection appears to divide the data into the three parts, with the 'red' observations showing the smallest variability, whereas the second PC projection separates the 'red' observations from the other two groups.

Table 12.4 *Absolute Skewness and Kurtosis of First Four IC Scores*

		IC_1	IC_2	IC_3	IC_4
KCC	Skewness	0.3838	0.1482	0.5712	0.0276
	Kurtosis	2.9226	2.5324	3.2655	3.6341
KGV	Skewness	0.0703	0.7075	2.3975	0.9146
	Kurtosis	2.6155	4.1042	12.2881	6.4230
$\mathcal{G}_3$	Skewness	2.8910	2.4831	1.9992	1.5144
$\mathcal{G}_4$	Kurtosis	18.3901	16.0971	11.9537	8.1101

The IC scores obtained with FastICA are calculated efficiently, partly because the search stops when no non-Gaussian directions are found after a fixed number of iterations. The kernel IC scores take rather longer to calculate. To find the best of $M = 100$, the KCC approach took about 2,300 seconds, and the KGV approach required about 4,400 seconds, whereas the two FastICA solutions took a total of about 100 seconds.

Overall, the kernel ICA projections *look* different from the FastICA projections. Table 12.4 shows my attempt to quantify this difference: by the absolute skewness and kurtosis values of the first four scores. For the two kernel approaches, I have given both the skewness and kurtosis values, as there is no innate preference for either one of these measures. As we can see, the kernel approaches have mostly very small skewness and kurtosis, and the IC directions or scores are not ranked in decreasing order of skewness and kurtosis because the deviation from the Gaussian distribution plays no role in the $\mathcal{F}$-correlation or the kernel generalised variance approach.

In the more than 500 runs of the KCC and KGV versions of Bach and Jordan's KernelICA, no two runs produced the same results; this is only partly a consequence of the random start of the algorithm. An inspection of the KCC scores in particular reveals that most of them are almost Gaussian, and because orthogonal transformations of independent Gaussian scores result in independent Gaussians, there is a large pool of possible solutions. Further, because the kernel directions are chosen to *jointly* minimise the largest eigenvalue ζ in (12.30), there is no preferred order of the kernel IC directions, such as the variance, or measures of non-Gaussianity. A ranking of direction vectors invites an interpretation of the new directions and scores, but such an interpretation does not exist if the directions have no inherent order.

The kernel approaches find all d directions because they embrace Gaussian directions, whereas the FastICA algorithm often finds fewer than d non-Gaussian directions. Figure 12.6 summarises the deviation or the lack of deviation from the Gaussian for the different IC approaches. The top panel shows absolute skewness results, and the bottom panel shows the absolute kurtosis values, in both cases against the number of variables on the x-axis. In the calculations, I have considered the 'best' in 100 repetitions of Algorithms 12.1 and 10.1. As in Table 12.4, I calculate the absolute skewness and kurtosis for the kernel IC scores, and in the figure I show the skewness and kurtosis values of the kernel scores in decreasing order rather than in the order produced by the kernel unmixing matrices.

The skewness and kurtosis values of the cumulant-based scores obtained with FastICA and shown in black in Figure 12.6 are considerably larger than the highest skewness and kurtosis values of the kernel-based IC scores. The KGV scores, shown as blue dots, generally include some non-Gaussian directions, whereas even the most non-Gaussian KCC

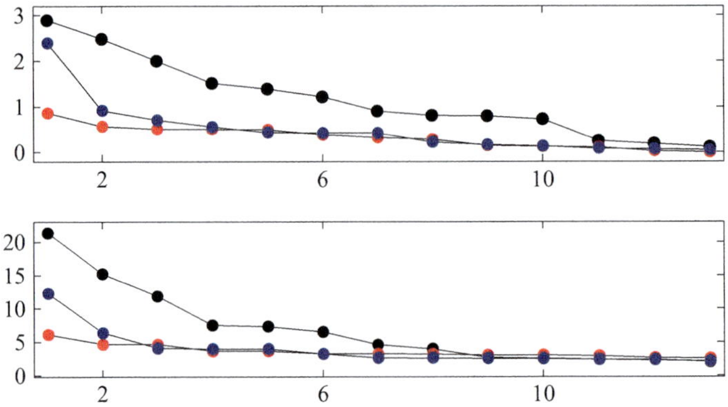

Figure 12.6 Absolute skewness (*top*) and kurtosis (*bottom*) of IC scores from Example 12.5 versus dimension. KCC (*red dots*), KGV (*blue dots*) and FastICA (*black dots*).

scores, the red dots, are almost Gaussian. The difference in the skewness and kurtosis values disappears as we consider more variables. For the FastICA scores, these almost Gaussian directions are often associated with 'noise' directions. ■

The two examples highlight an interesting difference between the FastICA directions and scores and the kernel ICA directions and scores: namely, the role that the deviation from Gaussianity plays. For the FastICA approaches, non-Gaussianity drives the process, and the almost independence of the variables appears as a consequence of the non-Gaussianity. Once we abandon the search for non-Gaussian directions, we can find as many directions as required, but we no longer have the interpretability of the results. Further, because the directions have no canonical order or ranking, it becomes more difficult to make decisions about the number of ICs which summarise the data.

Although the FastICA scores have been superior in the two examples, the kernel approaches have merit in that they have provided a new equivalence to independence. And hence they allow an innovative interpretation of independence which improves our understanding of independence and uncorrelatedness.

12.4 Independent Components from Scatter Matrices (aka Invariant Coordinate Selection)

Our next approach differs in many aspects from the kernel independent component solutions. I follow Oja, Sirkiä, and Eriksson (2006) and Tyler et al. (2009) but adapt their definitions. Both sets of authors use the acronym ICS, which stands for *independent components from scatter matrices* in Oja, Sirkiä, and Eriksson (2006) and has been named *invariant coordinate selection* in Tyler et al. (2009). Because the two approaches are closely related and discuss similar ideas and solution paths, I will treat them in the same section and move back and forth between them.

The key idea of the two approaches is to construct two scatter matrices – see Definition 12.9 – for a random vector $\mathbf{X}$ and to cleverly combine the the scatter matrices into a transformation which maps $\mathbf{X}$ onto a vector with independent components. Their approaches are reminiscent of the algebraic solutions in Principal Component Analysis, where Γ, the matrix of eigenvectors of the covariance matrix, is the key to constructing the principal components. In the methods of this section, the matrix of eigenvectors

$Q(\mathbf{X})$ – see (12.43) – of suitably chosen scatter matrices plays the role of Γ, but in this case $Q(\mathbf{X})$ is applied to the white random vector, $\mathbf{X}^{\diamond}$.

12.4.1 Scatter Matrices

Let $\mathbf{X} \sim (\mathbf{0}, \Sigma)$ and $\mathbf{T} \sim (\mathbf{0}, \mathbf{I}_{d\times d})$ be d-dimensional random vectors, and assume that $\mathbf{T}$ has independent components. In what follows, we assume that $\mathbf{X}$ and $\mathbf{T}$ satisfy the model

$$\mathbf{X} = A\mathbf{T}, \tag{12.33}$$

where A is a non-singular $d \times d$ matrix. This model agrees with the independent component model (10.1) in Section 10.2.1 when the source dimension is the same as that of $\mathbf{X}$. Unlike the notation $\mathbf{S}$ in Chapter 10, I have chosen $\mathbf{T}$ for the vector with independent components because we require different versions of the letter S for the scatter matrices that we use throughout this section.

In addition to the covariance matrix Σ, we require a second, covariance-like matrix.

Definition 12.9 Let $\mathbf{X} \sim (\boldsymbol{\mu}, \Sigma)$ be a d-dimensional random vector. Let A be a non-singular $d \times d$ matrix and $\mathbf{a} \in \mathbb{R}^d$. A matrix-valued function S, defined on $\mathbf{X}$, is called a **scatter statistic** if $\mathsf{S}(\mathbf{X})$ is a symmetric positive definite $d \times d$ matrix. We call $\mathsf{S}(\mathbf{X})$ a **scatter matrix**. A scatter statistic S may satisfy any of the following:

1. S is **affine equivariant** if for A and $\mathbf{a}$ as above,
$$\mathsf{S}(A\mathbf{X} + \mathbf{a}) = A\mathsf{S}(\mathbf{X})A^{\top}. \tag{12.34}$$
2. S is **affine proportional** if for A and $\mathbf{a}$ as above,
$$\mathsf{S}(A\mathbf{X} + \mathbf{a}) \propto A\mathsf{S}(\mathbf{X})A^{\top}. \tag{12.35}$$
3. S is **orthogonal proportional** if for $\mathbf{a}$ as above and any orthogonal $d \times d$ matrix U,
$$\mathsf{S}(U\mathbf{X} + \mathbf{a}) \propto U\mathsf{S}(\mathbf{X})U^{\top}. \tag{12.36}$$
4. S has the **independence property** if $\mathsf{S}(\mathbf{X})$ is a diagonal matrix whenever $\mathbf{X}$ has independent components.

For d-dimensional data $\mathbb{X} = \begin{bmatrix}\mathbf{X}_1 \ \mathbf{X}_2 \cdots \mathbf{X}_n\end{bmatrix}$, a **(sample) scatter statistic** S is a matrix-valued function, defined on $\mathbb{X}$, if $\mathsf{S}(\mathbb{X})$ is a symmetric positive definite $d \times d$ **scatter matrix**. □

I distinguish between a *scatter statistic*, which is a function, and a *scatter matrix*, which is the value of the scatter statistic at a random vector or data. Oja, Sirkiä, and Eriksson (2006) and Tyler et al. (2009) referred to **scatter functionals**, which include what I call a *scatter statistic*. I prefer the word 'statistic' because 'functional' has another well-defined meaning in mathematics. Tyler et al. (2009) defined their scatter functionals in terms of the distribution F of $\mathbf{X}$, and Oja, Sirkiä, and Eriksson (2006) used both forms $\mathsf{S}(F)$ and $\mathsf{S}(\mathbf{X})$.

The notion 'affine equivariant' is of interest in the model (12.33). If $\mathbf{X}$ is white and thus has the identity covariance matrix, then the more general orthogonal proportionality (12.36) becomes the property we are interested. Oja, Sirkiä, and Eriksson (2006) and Tyler et al. (2009) generalised the notion of an affine equivariant scatter statistic, but for our purpose, the preceding definitions suffice.

Probably the most common scatter statistic S is given by $\mathsf{S}(\mathbf{X}) = \Sigma$. This statistic is affine equivariant and has the independence property. Other scatter statistics are defined by

$$\mathsf{S}(\mathbf{X}) = \operatorname{var}\left[\|\mathbf{X}-\boldsymbol{\mu}\|^{p}(\mathbf{X}-\boldsymbol{\mu})\right] \qquad p=\pm 1, \tag{12.37}$$

$$\mathsf{S}(\mathbf{X}_{\text{sym}}) = \operatorname{var}\left[\left\|\mathbf{X}_{\text{sym}}\right\|^{p}(\mathbf{X}_{\text{sym}})\right] \qquad p=\pm 1, \tag{12.38}$$

$$\mathsf{S}(\mathbf{X}) = \operatorname{var}\left[\left\|\Sigma^{-1/2}(\mathbf{X}-\boldsymbol{\mu})\right\|(\mathbf{X}-\boldsymbol{\mu})\right], \tag{12.39}$$

and

$$\mathsf{S}(\mathbf{X}) = \mathbb{E}\left\{w_S\left[\mathbf{X}-\boldsymbol{\mu}_w\right]\left[\mathbf{X}-\boldsymbol{\mu}_w\right]^{\mathsf{T}}\right\},$$

where $\mathbf{X}_{\text{sym}} = \mathbf{X}_1 - \mathbf{X}_2$ in (12.38) is the *symmetrised* version of pairs of vectors $\mathbf{X}_i \sim (\boldsymbol{\mu}, \Sigma)$ for $i = 1,2$, and where the w_S in the last displayed equation are suitably chosen weights, such as $w_S = (\mathbf{X}-\boldsymbol{\mu})^{\mathsf{T}}\Sigma^{-1}(\mathbf{X}-\boldsymbol{\mu})$, and for our purpose, $\boldsymbol{\mu}_w$ is the mean or a weighted mean.

The scatter matrices (12.37) and (12.38) are proposed in Oja, Sirkiä, and Eriksson (2006), and p can take the value 1 or -1. The matrix (12.37) with $p=-1$ is a special case of the **spatial sign covariance matrix**. We recognise it as the covariance matrix of the centred direction vectors. If $p=-1$, then (12.38) is known as **Kendall's τ-matrix**. The matrices (12.38) are not affine equivariant but satisfy the orthogonal proportionality (12.36). In robust statistics, 'symmetrised' matrices such as (12.38) are preferred to (12.37).

The scatter matrix (12.39) is considered in Tyler et al. (2009). It includes the Mahalanobis distance of (5.3) in Section 5.3.1 as a weight function. A comparison of (12.37) with $p=1$ and (12.39) shows that both contain fourth-order moments, and Oja, Sirkiä, and Eriksson (2006) and Tyler et al. (2009) regarded these scatter matrices as a form of kurtosis. These 'kurtoses', however, differ from the multivariate kurtosis defined in (9.5) in Section 9.3.

12.4.2 Population Independent Components from Scatter Matrices

Let $\mathbf{X} \sim (\mathbf{0}, \Sigma)$ be a d-dimensional random vector with non-singular Σ, and assume that $\mathbf{X} = A\mathbf{T}$ as in (12.33). Oja, Sirkiä, and Eriksson (2006) and Tyler et al. (2009) used pairs of scatter statistics S_1 and S_2 and exploited the relationship between $\mathsf{S}_k(\mathbf{X})$ and $\mathsf{S}_k(\mathbf{T})$ for $k=1,2$. Both approaches take $\mathsf{S}_1(\mathbf{X}) = \Sigma$. To make their methods work, they required S_2 to satisfy the independence property and $\mathsf{S}_2(\mathbf{T})$ to be a diagonal matrix which differs from the identity matrix.

We begin with the approach of Oja, Sirkiä, and Eriksson (2006). Put $\mathsf{S}_1(\mathbf{X}) = \Sigma$. Let S_2 be another affine equivariant scatter statistic which satisfies the independence property of Definition 12.9. Oja, Sirkiä and Eriksson define a matrix-valued function M pointwise by

$$\mathsf{M} = \mathsf{S}_1^{-1/2}\mathsf{S}_2\mathsf{S}_1^{-1/2} \qquad \text{and} \qquad \mathsf{M}(\mathbf{X}) = \left[\mathsf{S}_1(\mathbf{X})\right]^{-1/2}\mathsf{S}_2(\mathbf{X})\left[\mathsf{S}_1(\mathbf{X})\right]^{-1/2}. \tag{12.40}$$

Because $\mathsf{S}_1(\mathbf{X}) = \Sigma$, it follows that

$$\mathsf{M}(\mathbf{X}) = \Sigma^{-1/2}\mathsf{S}_2(\mathbf{X})\Sigma^{-1/2}. \tag{12.41}$$

The matrix $\mathsf{M}(\mathbf{X})$ plays the role of the covariance matrix in Principal Component Analysis. It is therefore natural to consider the matrix of eigenvectors of $\mathsf{M}(\mathbf{X})$. Formally, we write the spectral decompositions of M and of the S_k as

$$\mathsf{M} = Q\Delta Q^{\mathsf{T}} \qquad \text{and} \qquad \mathsf{S}_k = \Gamma_{(k)}\Lambda_{(k)}\Gamma_{(k)}^{\mathsf{T}} \qquad \text{where } k=1,2, \tag{12.42}$$

and we interpret $\mathsf{M} = Q\Delta Q^{\mathsf{T}}$ pointwise; that is, $\mathsf{M}(\mathbf{X}) = Q(\mathbf{X})\Delta(\mathbf{X})Q(\mathbf{X})^{\mathsf{T}}$, and $Q(\mathbf{X})$ is the matrix of eigenvectors of $\mathsf{M}(\mathbf{X})$. Let $\mathbf{X}_\Sigma = \Sigma^{-1/2}\mathbf{X}$ be the sphered vector of $\mathbf{X}$. Because S_2 is affine equivariant, (12.41) implies that

$$\mathsf{S}_2(\mathbf{X}_\Sigma) = \Sigma^{-1/2}\mathsf{S}_2(\mathbf{X})\Sigma^{-1/2} = \mathsf{M}(\mathbf{X}),$$

and hence,

$$\Gamma_{(2)}(\mathbf{X}_\Sigma) = Q(\mathbf{X}) \tag{12.43}$$

by the uniqueness of the spectral decomposition. The orthogonal matrix $Q(\mathbf{X})$ is the key to obtaining independent components for $\mathbf{X}$.

Theorem 12.10 [Oja, Sirkiä, and Eriksson (2006)] *Let $\mathbf{X} \sim (\mathbf{0}, \Sigma)$ be a d-dimensional random vector with non-singular Σ, and put $\mathbf{X}_\Sigma = \Sigma^{-1/2}\mathbf{X}$. Let $\mathbf{T} \sim (\mathbf{0}, \mathbf{I}_{d\times d})$ be d-dimensional with independent components, and assume that $\mathbf{X}$ and $\mathbf{T}$ satisfy $\mathbf{X} = A\mathbf{T}$ as in (12.33). Let S_1 and S_2 be affine equivariant scatter statistics which satisfy*

1. $\mathsf{S}_1(\mathbf{T}) = \mathbf{I}_{d\times d}$ *and* $\mathsf{S}_1(\mathbf{X}) = \Sigma$.
2. $\mathsf{S}_2(\mathbf{T}) = D$ *is a diagonal matrix with d distinct and positive diagonal entries.*

Put $\mathsf{M} = \mathsf{S}_1^{-1/2}\mathsf{S}_2\mathsf{S}_1^{-1/2}$, and write $\mathsf{M} = Q\Delta Q^{\mathsf{T}}$ for the spectral decomposition. Then

$$Q(\mathbf{X})^{\mathsf{T}}\mathbf{X}_\Sigma = \mathbf{T}.$$

Remark 1. This theorem establishes an algebraic solution for the independent component model (12.33) and exhibits the orthogonal transformation $Q(\mathbf{X})$ that maps $\mathbf{X}_\Sigma$ to $\mathbf{T}$. Thus, $Q(\mathbf{X})$ plays the role of the unmixing matrix U of Proposition 10.5 in Section 10.3. The theorem invites an analogy with Principal Component Analysis: the random vector $\mathbf{X}_\Sigma$ is projected onto the direction of the eigenvectors of the positive definite matrix $\mathsf{M}(\mathbf{X})$.

The precise form of the second scatter statistic S_2 is not required in Theorem 12.10. It suffices that S_2 is affine equivariant and satisfies the independence property with distinct diagonal entries.

The proof that follows is adapted from that given in Oja, Sirkiä, and Eriksson (2006). It makes repeated use of the affine equivariance of the scatter statistics.

Proof We first consider S_1. Write $A = ULV^{\mathsf{T}}$ for the singular value decomposition of A. So $\mathbf{X} = ULV^{\mathsf{T}}\mathbf{T}$. From the affine equivariance we obtain

$$\mathsf{S}_1(\mathbf{X}) = AA^{\mathsf{T}} = UL^2U^{\mathsf{T}}, \qquad \text{and} \qquad \mathbf{X}_\Sigma = \mathsf{S}_1(\mathbf{X})^{-1/2}\mathbf{X} = UV^{\mathsf{T}}\mathbf{T}. \tag{12.44}$$

Similarly, the affine equivariance of S_2 yields

$$\mathsf{S}_2(\mathbf{X}_\Sigma) = \Sigma^{-1/2}A\mathsf{S}_2(\mathbf{T})A^{\mathsf{T}}\Sigma^{-1/2} = UV^{\mathsf{T}}DVU^{\mathsf{T}}.$$

A combination of (12.42), the uniqueness of the spectral decomposition, and (12.43) leads to

$$\Gamma_{(2)}(\mathbf{X}_\Sigma) = UV^{\mathsf{T}} = Q(\mathbf{X}).$$

The desired result follows from this last equality and (12.44). ∎

Like Oja, Sirkiä, and Eriksson (2006), Tyler et al. (2009) considered two scatter matrices, but they replaced M. Consider a d-dimensional random vector $\mathbf{X}$. Let S_1, S_2 be affine proportional scatter statistics, and assume that $\mathsf{S}_1(\mathbf{X})$ is non-singular. Tyler et al. (2009) considered the transformation N defined pointwise by

$$\mathsf{N} = \mathsf{S}_1^{-1}\mathsf{S}_2 \qquad \text{and} \qquad \mathsf{N}(\mathbf{X}) = [\mathsf{S}_1(\mathbf{X})]^{-1}\,\mathsf{S}_2(\mathbf{X}). \tag{12.45}$$

Oja, Sirkiä, and Eriksson (2006) used M of (12.40) as a vehicle for constructing an orthogonal matrix of eigenvectors, and Tyler et al. (2009) proceeded in an analogous way with their N. The problem of finding the eigenvalues and eigenvectors of $\mathsf{N}(\mathbf{X})$ is a generalised eigenvalue problems, see (3.53) in Section 3.7.4. For eigenvalue-eigenvector pairs $(\lambda, \mathbf{e})$, we have

$$\mathsf{N}(\mathbf{X})\mathbf{e} = \lambda\mathbf{e} \qquad \text{or equivalently} \qquad \mathsf{S}_2(\mathbf{X})\mathbf{e} = \lambda\mathsf{S}_1(\mathbf{X})\mathbf{e}. \tag{12.46}$$

We have met generalised eigenvalue problems in many different chapters of this book, including Chapter 4. A closer inspection of Fisher's ideas in Discriminant Analysis (see Theorem 4.6 in Section 4.3.1) shows that finding solutions to (12.45) can be regarded as an extension of Fisher's ideas. Here we require all d eigenvalue-eigenvector pairs, whereas Fisher's discriminant direction is obtained from the first eigenvalue-eigenvector pair. Tyler et al. (2009) derived properties of pairs of scatter statistics. We are primarily interested in the function N, and I therefore only summarise, in our Theorem 12.11, their theorems 5 and 6, which directly relate to independent components.

Theorem 12.11 [Tyler et al. (2009)] *Let $\mathbf{X} \sim (\mathbf{0}, \Sigma)$ be a d-dimensional random vector. Let $\mathbf{T} \sim (\mathbf{0}, \mathbf{I}_{d\times d})$ be d-dimensional with independent components, and assume that $\mathbf{X}$ and $\mathbf{T}$ satisfy $\mathbf{X} = A\mathbf{T}$ as in (12.33). Let S_1 and S_2 be affine proportional scatter statistics, and assume that $\mathsf{S}_1(\mathbf{X})$ is non-singular. For $\mathsf{N} = \mathsf{S}_1^{-1}\mathsf{S}_2$ as in (12.45), let $H(\mathbf{X})$ be the matrix of eigenvectors of $\mathsf{N}(\mathbf{X})$, and assume that the d eigenvalues of $\mathsf{N}(\mathbf{X})$ are distinct. Assume that at least one of the following holds:*

(a) *The distribution of $\mathbf{T}$ is symmetric about $\mathbf{0}$.*
(b) *The scatter statistics S_1 and S_2 are affine equivariant and satisfy the independence property of Definition 12.9.*

Then $H(\mathbf{X})^{\mathsf{T}}\mathbf{X}$ has independent components.

Version (a) of Theorem 12.11 appears to be more general than version (b): affine proportionality of the two scatter matrices suffices to construct the orthogonal matrix $H(\mathbf{X})$. An inspection of the proof in Tyler et al. (2009) shows that $\mathsf{S}_1(\mathbf{T})^{-1}\mathsf{S}_2(\mathbf{T}) \propto PDP^{\mathsf{T}}$, where P is a permutation matrix and D is the diagonal matrix of eigenvalues of $\mathsf{N}(\mathbf{X})$. These conditions are essentially equivalent to those of Oja, Sirkiä, and Eriksson (2006). The weaker assumption of affine proportionality requires the extra property that $\mathbf{T}$ has a symmetric distribution. This assumption may be interesting from a theoretical perspective, but because it cannot be checked in practice, it is less relevant for data.

Version (b) of Theorem 12.11 appears to be more general than Theorem 12.10 because it holds for any pair of scatter matrices which satisfy the assumptions of theorem. However, an inspection of their results shows that Tyler et al. (2009) exclusively referred to the matrices $\mathsf{S}_1(\mathbf{X}) = \Sigma$ and $\mathsf{S}_2(\mathbf{X}) = \mathbb{E}\left[(\mathbf{X}-\boldsymbol{\mu})^{\mathsf{T}}\,\Sigma^{-1}\,(\mathbf{X}-\boldsymbol{\mu})\,(\mathbf{X}-\boldsymbol{\mu})(\mathbf{X}-\boldsymbol{\mu})^{\mathsf{T}}\right]$ in their

Table 12.5 *Comparison of PCA, DA and ICS*

Method	PCA	DA	ICS-OSE	ICS-TCDO
Matrix	Σ	$W^{-1}B$	$\Sigma^{-1/2}\mathsf{S}_2(\mathbf{X})\Sigma^{-1/2}$	$\Sigma^{-1}\mathsf{S}_2(\mathbf{X})$
Decomposition	$\Gamma\Lambda\Gamma$	$E\Delta E^{\top}$	$Q\Delta Q^{\top}$	$H\Delta H^{\top}$
Projections	$\Gamma^{\top}\mathbf{X}$	$\eta^{\top}\mathbf{X}$	$Q(\mathbf{X})^{\top}\Sigma^{-1/2}\mathbf{X}$	$H(\mathbf{X})^{\top}\mathbf{X}$

theorems 5 and 6. It follows that version (b) of Theorem 12.11 and Theorem 12.10 are essentially equivalent. Indeed, for fixed S_1 and S_2, the following hold:

1. the eigenvalues of $\mathsf{N}(\mathbf{X})$ and $\mathsf{M}(\mathbf{X})$ agree,
2. if $\mathbf{q}$ is an eigenvector of $\mathsf{M}(\mathbf{X})$, then $\mathbf{e} = \mathsf{S}_1(\mathbf{X})^{-1/2}\mathbf{q}$ is the eigenvector of $\mathsf{N}(\mathbf{X})$ which corresponds to the same λ as $\mathbf{q}$; and
3. the same independent component solution is obtained because

$$H(\mathbf{X})^{\top}\mathbf{X} = \left[\mathsf{S}_1(\mathbf{X})^{-1/2}Q(\mathbf{X})\right]^{\top}\mathbf{X} = Q(\mathbf{X})^{\top}\mathsf{S}_1(\mathbf{X})^{-1/2}\mathbf{X} = Q(\mathbf{X})^{\top}\mathbf{X}_{\Sigma}. \tag{12.47}$$

The proof of the first two statements is deferred to the Problems at the end of Part III. It is worth noting that the matrices $\mathsf{M}(\mathbf{X})$ and $\mathsf{N}(\mathbf{X})$ are related in a similar way as the matrices $R^{[C]}$ and K of (3.22) in Section 3.5.2.

The orthogonal matrices $Q(\mathbf{X})$ and $H(\mathbf{X})$ which are obtained from the spectral decompositions of $\mathsf{M}(\mathbf{X})$ and $\mathsf{N}(\mathbf{X})$ apply to the sphered and raw data, respectively. The relationship (12.47) between $Q(\mathbf{X})$ and $H(\mathbf{X})$ is similar to that between the eigenvectors of the canonical correlation matrix C and the canonical correlation transformations of Section 3.2.

Remark 2. Theorems 12.10 and 12.11 do not make any assumptions about the Gaussianity or non-Gaussianity of $\mathbf{T}$. However, if $\mathbf{T}$ has at most one Gaussian component, then we can use Comon's result – our Theorem 10.2 in Section 10.2.1 – to infer the uniqueness of $H(\mathbf{X})^{\top}\mathbf{X}$ up to permutations of the components.

Table 12.5 summarises key quantities of Theorems 12.10 and 12.11 and compares them to similar quantities in Principal Component Analysis and Discriminant Analysis. The table highlights analogies and relationships between these four methods. In the table, 'ICS-OSE' refers to the approach by Oja, Sirkiä, and Eriksson (2006), and 'ICS-TCDO' refers to that of Tyler et al. (2009). For notational convenience, I assume that $\mathbf{X} \sim (\mathbf{0}, \Sigma)$. 'Matrix' refers to the matrix that drives the process and whose spectral decomposition is employed in finding direction vectors and projections.

For each of the methods listed in the table, the first projection is that which corresponds to the largest eigenvalue of the 'Matrix'. PCA and ICS-OSE find the largest eigenvalue of the variance and a transformed variance, whereas DA and ICS-TCDO find the maximiser of a generalised eigenvalue problem.

12.4.3 Sample Independent Components from Scatter Matrices

Let $\mathbb{X} = \left[\mathbf{X}_1\ \mathbf{X}_2 \cdots \mathbf{X}_n\right]$ be d-dimensional mean zero data with a sample covariance matrix S which is non-singular. We set $\mathsf{S}_1(\mathbb{X}) = S$. Because S_1 is fixed, the success of the methods of Oja, Sirkiä, and Eriksson (2006) and Tyler et al. (2009) will depend on the chosen expression for $\mathsf{S}_2(\mathbb{X})$.

Oja, Sirkiä, and Eriksson (2006) considered the following scatter statistics which are defined for data $\mathbb{X}$ by

$$\mathsf{S}_2(\mathbb{X}) = \begin{cases} \dfrac{1}{n-1}\sum_{i=1}^{n} \left\|\mathbf{X}_i - \overline{\mathbf{X}}\right\|^{2p} (\mathbf{X}_i - \overline{\mathbf{X}})(\mathbf{X}_i - \overline{\mathbf{X}})^{\top} & p = \pm 1, \\ \dfrac{2}{n(n-1)}\sum_{i<j}^{n} \left\|\mathbf{X}_i - \mathbf{X}_j\right\|^{2p} (\mathbf{X}_i - \mathbf{X}_j)(\mathbf{X}_i - \mathbf{X}_j)^{\top} & p = \pm 1. \end{cases} \tag{12.48}$$

The first expression is the sample version of (12.37), and the second is the sample version of the symmetrised vectors (12.38). The scatter matrices which are defined by the second expression require more computation than the weighted sample covariance matrix but should be more robust.

Put

$$\widehat{\mathsf{M}}(\mathbb{X}) = S^{-1/2}\mathsf{S}_2(\mathbb{X})S^{-1/2},$$

so $\widehat{\mathsf{M}}(\mathbb{X})$ is the sample version of (12.42) with $\mathsf{S}_2(\mathbb{X})$ chosen from (12.48). The matrix of eigenvectors $\widehat{Q}(\mathbb{X})$ is obtained from the spectral decomposition of $\widehat{\mathsf{M}}(\mathbb{X})$. Using $\widehat{Q}(\mathbb{X})$ and the sphered data $\mathbb{X}_S = S^{-1/2}\mathbb{X}$, Theorem 12.10 now leads to the transformation

$$\mathbb{X} \longmapsto \widehat{Q}(\mathbb{X})^{\top}\mathbb{X}_S = \widehat{Q}(\mathbb{X})^{\top} S^{-1/2}\mathbb{X}. \tag{12.49}$$

Tyler et al. (2009) considered the scatter matrix

$$\mathsf{S}_2(\mathbb{X}) = \frac{1}{n-1}\sum_{i=1}^{n} \left\|S^{-1/2}(\mathbf{X}_i - \overline{\mathbf{X}})\right\|^2 (\mathbf{X}_i - \overline{\mathbf{X}})(\mathbf{X}_i - \overline{\mathbf{X}})^{\top}, \tag{12.50}$$

which is the sample version of (12.39), and then defined the sample analogue of N in (12.45):

$$\widehat{\mathsf{N}}(\mathbb{X}) = S^{-1}\mathsf{S}_2(\mathbb{X}).$$

Let $\widehat{H}(\mathbb{X})$ be the matrix of eigenvectors of $\widehat{\mathsf{N}}(\mathbb{X})$. The transformation obtained from Theorem 12.11 becomes

$$\mathbb{X} \longmapsto \widehat{H}(\mathbb{X})^{\top}\mathbb{X}. \tag{12.51}$$

Despite the similarities of the two approaches, the final matrices $\widehat{Q}(\mathbb{X})^{\top}\mathbb{X}_S$ and $\widehat{H}(\mathbb{X})^{\top}\mathbb{X}$ will vary depending on the choice of the second scatter statistic.

The methods of Oja, Sirkiä, and Eriksson (2006) and Tyler et al. (2009) furnish explicit expressions, namely, $\widehat{Q}(\mathbb{X})$ and $\widehat{H}(\mathbb{X})$, for the orthogonal unmixing matrices, whereas the solutions we considered in Chapter 10 are obtained iteratively.

Theorems 12.10 and 12.11 assume that $\mathsf{S}_2(\mathbf{T})$ is diagonal whenever $\mathbf{T}$ is a random vector with independent components, and the diagonal entries of $\mathsf{S}_2(\mathbf{T})$ are distinct. Further, the independence property of S_2 yields the independence of $Q(\mathbf{X})^{\top}\mathbf{X}_{\Sigma}$ and $H(\mathbf{X})^{\top}\mathbf{X}$ for random vectors $\mathbf{X}$. For data, it is not possible to confirm the independence of the variables. The data version of $\mathsf{S}_s(\mathbf{X})$ is used in the construction of the orthogonal transformations (12.49) to (12.51), but the theorems apply to the population and may not hold exactly for the corresponding sample quantities. We will return to the connection between independent variables and the form or shape of $\mathsf{S}_2(\mathbb{T})$ in Example 12.6.

In the initial development of Independent Component Analysis, the random vector $\mathbf{X}$ and source $\mathbf{S}$ are d-dimensional. Theorems 12.10 and 12.11 use the same framework and require that $\mathsf{S}_1(\mathbf{X}) = \Sigma$ is non-singular. For data, the sample covariance matrix $\mathsf{S}_1(\mathbb{X}) = S$ will be singular when $d > n$. The ideas of the two theorems remain of interest for such cases, but some modifications are needed. If $\mathbb{X}$ are centred and $\mathsf{S}_1(\mathbb{X})$ has rank $r < d$, then one could proceed as follows:

1. For $\kappa \leq r$, calculate the κ-dimensional PC data $\mathbb{W}^{(\kappa)} = \widehat{\Gamma}_\kappa^{\mathsf{T}} \mathbb{X}$.
2. Replace $\mathbb{X}$ with $\mathbb{W}^{(\kappa)}$. Put $\mathsf{S}_1(\mathbb{W}^{(\kappa)}) = \widehat{\Lambda}_\kappa$. Consider another scatter statistic S_2, and calculate its value at $\mathbb{W}^{(\kappa)}$.
3. Apply Theorems 12.10 and 12.11 to the scatter matrices $\widehat{\Lambda}_\kappa$ and $\mathsf{S}_2(\mathbb{W}^{(\kappa)})$.

Oja, Sirkiä, and Eriksson (2006) and Tyler et al. (2009) were primarily interested in low-dimensional problems and robustness and did not consider dimension reduction issues. A first reduction of the dimension with Principal Component Analysis is natural when the dimension is large or when S is singular. A choice of the reduced dimension κ will need to be made, as is generally the case when Principal Component Analysis is employed as a dimension-reduction technique.

In the preceding analysis of the *wine recognition* data in Example 12.5, the kernel IC algorithms of Bach and Jordan (2002) resulted in different IC directions in every application of the algorithm. The calculations also reveal that most IC scores are almost Gaussian. The two approaches of this section focus on the selection of an invariant coordinate system, and the new coordinate system rotates the data to an 'independent' position, provided the second scatter matrix, $\mathsf{S}_2(\mathbb{T})$, is diagonal. We shall see that the IC directions are unique for a given choice of the second scatter statistic S_2.

Example 12.6 We continue with the **wine recognition** data and look at projections of the data onto directions obtained from scatter matrices. As in Example 12.5, we work with the scaled data $\mathbb{X}_{\text{scale}}$. We focus on the transformations which lead to the new IC scores and examine properties of these transformations and the resulting IC scores. In addition, I will comment on some invariance properties of these transformations.

Let S_1 be the scatter statistic which maps $\mathbb{X}$ into the sample covariance matrix, and put $S_1 = \mathsf{S}_1(\mathbb{X}_{\text{scale}})$. I use the four expressions of (12.48) and that of (12.50), adjusted to the scaled data, to define $\mathsf{S}_2(\mathbb{X}_{\text{scale}})$ and refer to these matrices as S_{m+}, S_{m-}, S_{s+} and S_{s-}, with m for the mean-based versions and s for the symmetrised versions of (12.48) and $+$ and $-$ for the signs of the power p. We write S_{sph} for the corresponding matrix of (12.50), which contains the sphered $\mathbf{X}_i$. I calculate the matrices $\widehat{\mathsf{M}}(\mathbb{X}_{\text{scale}})$ and $\widehat{\mathsf{N}}(\mathbb{X}_{\text{scale}})$ and their orthogonal matrices $\widehat{Q}(\mathbb{X}_{\text{scale}})$ and $\widehat{H}(\mathbb{X}_{\text{scale}})$. Put $M = \widehat{\mathsf{M}}(\mathbb{X}_{\text{scale}})$ and $N = \widehat{\mathsf{N}}(\mathbb{X}_{\text{scale}})$. When we need to distinguish between the different versions of M, we use the subscript notation I defined for the $\mathsf{S}_2(\mathbb{X}_{\text{scale}})$, so $M_{m-} = S_1^{-1/2} S_{m-} S_1^{-1/2}$. For the scatter matrices (12.48) of Oja, Sirkiä, and Eriksson (2006), the rotated 'independent' data are $\widehat{Q}(\mathbb{X}_{\text{scale}})^{\mathsf{T}} S_1^{-1/2} \mathbb{X}_{\text{scale}}$, and for the scatter matrix (12.50) of Tyler et al. (2009), the rotated 'independent' data are $\widehat{H}(\mathbb{X}_{\text{scale}})^{\mathsf{T}} \mathbb{X}_{\text{scale}}$.

The eigenvalues of the matrices M_{m+}, M_{s+} and N are more than two orders of magnitude larger than those of M_{m-} and M_{s-}. To enable a comparison between the five matrices I scale each matrix by its trace. Table 12.6 lists the traces of the five matrices and their largest

Table 12.6 *First and Last Eigenvalues* $\widehat{\lambda}_1$ *and* $\widehat{\lambda}_d$ *as a per cent of* $\text{tr}(M)$, $\text{tr}(N)$ *and* $\text{tr}(S_1)$ *for Different Scatter Matrices*

	M_{m+}	M_{m-}	M_{s+}	M_{s-}	N	S_1
$\widehat{\lambda}_1$	11.66	9.84	9.80	9.50	16.78	36.20
$\widehat{\lambda}_d$	5.60	5.37	6.32	5.13	4.55	0.80
trace	202.20	1.04	792.41	1.19	230.12	13.00

and smallest eigenvalues as percentages of the trace. The table also lists the corresponding values for the sample covariance matrix S_1.

The first eigenvalue of S_1 is relatively large – as a percentage of the trace of S_1 – compared with the relative size of the first eigenvalues of N and the four matrices M. The remaining eigenvalues of S_1 decrease quickly, which makes the first eigenvector by far the most important one. The eigenvalues of the matrices M and N are more concentrated, and any spikiness that exists in S_1 has been smoothed out in the matrices M and N.

In the IC projection plots based on $\widehat{Q}(\mathbb{X}_{\text{scale}})^{\mathsf{T}} S_1^{-1/2} \mathbb{X}_{\text{scale}}$ and $\widehat{H}(\mathbb{X}_{\text{scale}})^{\mathsf{T}} \mathbb{X}_{\text{scale}}$, the classes corresponding to black and blue lines in Figure 12.5 in Example 12.5 mostly overlap. The projection plots calculated from the scatter matrices are distinct and differ from the PC projections and IC projection plots of Example 12.5; they do not exhibit any distinctive patterns, and are therefore not shown.

Both Oja, Sirkiä, and Eriksson (2006) and Tyler et al. (2009) regarded their scatter matrices as 'a form of kurtosis'; however, because they considered a rotation of all d variables, the first new direction may not be the one with the highest kurtosis. Table 12.7 lists the kurtosis of the scores obtained from the four directions with highest absolute kurtosis together with the index of the direction. For comparison, I repeat the last line of Table 12.4, which shows the kurtosis of the first four FastICA scores.

Table 12.7 shows that the matrices M_{m+}, M_{m-} and M_{s+} have at least one direction which gives rise to a large kurtosis compared with the largest kurtosis obtained with $\mathcal{G}_4$. These large kurtoses justify the claim of Oja, Sirkiä, and Eriksson (2006) that their matrices are a form of kurtosis. The matrices M_{s-} and N do not find directions with large kurtosis for these data. The order of the directions is informed by the size of the eigenvalues of M and N rather than by the absolute value of the kurtosis of the scores, and for this reason, the highest kurtosis scores may turn out to be those derived from directions with the smallest eigenvalues, as is the case for M_{m-}.

We next turn to the independence of the scores. Theorems 12.10 and 12.11 assume the independent source model $\mathbf{X} = A\mathbf{T}$, where $\mathbf{T}$ has independent components, and further assume that the scatter statistics satisfy $\mathsf{S}_1(\mathbf{T}) = \mathbf{I}_{d\times d}$ and that $\mathsf{S}_2(\mathbf{T})$ is diagonal. If these conditions are satisfied, then the scores $\widehat{Q}(\mathbf{X})^{\mathsf{T}}\mathbf{X}_\Sigma$ and $\widehat{H}(\mathbf{X})^{\mathsf{T}}\mathbf{X}$ are independent, and $\widehat{Q}(\mathbf{X})^{\mathsf{T}}\mathbf{X}_\Sigma = \mathbf{T}$. In practice, we cannot check the independence of the scores $\widehat{Q}(\mathbb{X}_{\text{scale}})^{\mathsf{T}} S_1^{-1/2} \mathbb{X}_{\text{scale}}$ and $\widehat{H}(\mathbb{X}_{\text{scale}})^{\mathsf{T}} \mathbb{X}_{\text{scale}}$, but we can investigate the deviation of the $\mathsf{S}_\ell(\mathbb{S})$ for $\ell = 1, 2$ from diagonal matrices, where $\mathbb{S}$ refers to any of the scores $\widehat{Q}(\mathbb{X}_{\text{scale}})^{\mathsf{T}} S_1^{-1/2} \mathbb{X}_{\text{scale}}$ and $\widehat{H}(\mathbb{X}_{\text{scale}})^{\mathsf{T}} \mathbb{X}_{\text{scale}}$. This amounts to checking whether the assumptions that we can check are satisfied.

The four sets of scores $\widehat{Q}(\mathbb{X}_{\text{scale}})^{\mathsf{T}} S^{-1/2} \mathbb{X}$ are white, so they have the identity sample covariance matrix, but the sample covariance matrix of the scores $\widehat{H}(\mathbb{X}_{\text{scale}})^{\mathsf{T}} \mathbb{X}_{\text{scale}}$

Table 12.7 *Absolute Kurtosis – and Index (1–13) in Blue – of the Four IC Scores with Highest Absolute Kurtosis*

	IC_1		IC_2		IC_3		IC_4	
M_{m+}	15.47	1	9.87	3	8.92	2	3.51	4
M_{m-}	13.21	13	12.19	12	7.28	11	4.08	10
M_{s+}	16.24	2	5.93	4	5.07	1	4.59	3
M_{s-}	7.91	12	5.27	10	4.68	9	4.20	8
N	9.20	1	6.13	2	5.64	3	4.21	6
$\mathcal{G}_4$	18.39		16.10		11.95		8.11	

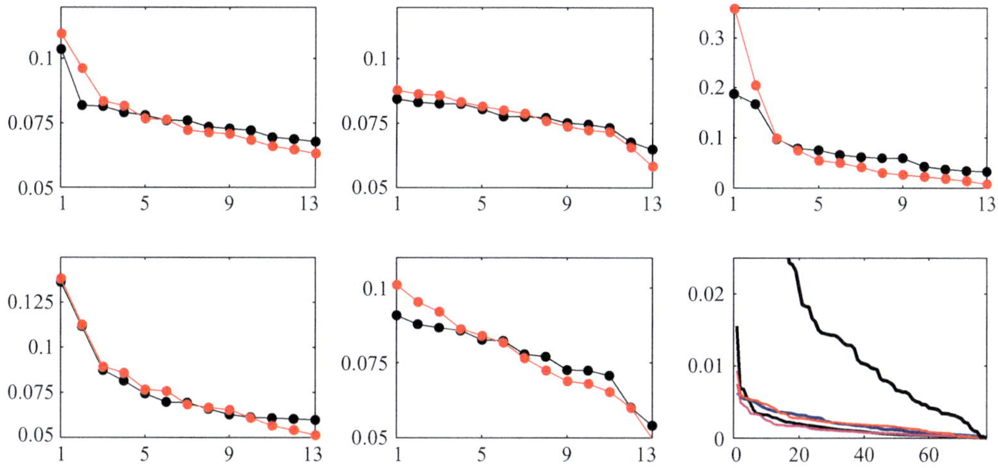

Figure 12.7 Sorted eigenvalues and diagonal entries of $\mathsf{S}_2(\mathbb{S})$ versus dimension from Example 12.6. Diagonal entries are shown in black and eigenvalues in red. From left to right: (*top row*) S_{s+}, S_{s-} and S_{sph}; (*bottom row*) S_{m+} and S_{m-}. Ordered off-diagonal entries of $\mathsf{S}_2(\mathbb{S})$ bottom right against their index.

has off-diagonal elements of the same size as the diagonal elements. The matrix $\mathsf{S}_1[\widehat{H}(\mathbb{X}_{\text{scale}})^{\mathsf{T}}\mathbb{X}_{\text{scale}}]$ is not diagonal, and hence there is no reason why the scores $\widehat{H}(\mathbb{X}_{\text{scale}})^{\mathsf{T}}\mathbb{X}_{\text{scale}}$ should be independent.

For the second scatter matrices $\mathsf{S}_2(\mathbb{S})$, I scale each matrix by its trace and then work with these scaled matrices. I calculate the diagonal entries and the eigenvalues of the scaled matrices. If $\mathsf{S}_2(\mathbb{S})$ were diagonal, then the eigenvalues and diagonal entries would agree, and all off-diagonal entries would be zero. For the scaled matrices, Figure 12.7 displays the diagonal entries and the eigenvalues, both sets sorted in decreasing order, as functions of the dimension for the matrices $\mathsf{S}_2(\mathbb{S})$. The diagonal entries are shown in black and the eigenvalues in red.

Going from left to right, the panels in the top row of Figure 12.7 refer to S_{s+}, S_{s-} and S_{sph}, and those in the bottom row refer to S_{m+} and S_{m-}. The bottom-right panel shows the size of the $d \times (d-1)/2$ off-diagonal entries, here shown in decreasing order against the indices $1, \ldots, 78$. The four overlapping curves show the entries of the $\mathsf{S}_2(\mathbb{S})$ matrices derived from M, and the isolated curve is derived similarly from N. The four plots in the left and middle panels show that the diagonal entries do not deviate much from the eigenvalues, but this is not the case for the plots in the top-right panel.

To obtain a more quantitative measure of the difference between the diagonal entries and eigenvalues, I calculate the Euclidean norm of the vector of differences between the sorted scaled diagonal entries and sorted eigenvalues for each scatter matrix $\mathsf{S}_2(\mathbb{S})$. The norms vary from 0.0102 to 0.0184 for the matrices obtained from M, whereas the norm of the corresponding difference obtained from N is 0.1882, and thus an order of magnitude bigger. The norms are an indication of the deviation from the diagonal form of each matrix. In particular, the norm calculations show that the deviation is much bigger for N than for the M matrices. There may, of course, be other and better tools for measuring the departure of a matrix from the diagonal form, including the Frobenius norm of the difference of two matrices, but I am not concerned with this issue here.

For the *wine recognition* data, the tables and graphs show that the scatter matrices of Oja, Sirkiä, and Eriksson (2006) result in scores that are a closer, and hence a better, fit to independent scores than the scatter matrix used in Tyler et al. (2009). This interpretation is confirmed by the deviation of the matrices $\mathsf{S}_2(\mathbb{S})$ from a corresponding diagonal matrix and by the deviation of $\mathsf{S}_1[\widehat{H}(\mathbb{X}_{\rm scale})^{\mathsf{T}}\mathbb{X}_{\rm scale}]$ from the identity matrix.

I conclude this example with some comments on the 'invariance part' of 'invariant coordinate selection'. The approaches of Oja, Sirkiä, and Eriksson (2006) and Tyler et al. (2009) rotate the original coordinate system by first sphering the data and then applying a specially chosen orthogonal matrix. For the *wine recognition* data, the performance of classification rules is of particular interest. In Example 4.5 in Section 4.4.2 and Example 4.7 in Section 4.5.1, I construct discriminant rules for the three classes, which result in twelve misclassified observations for Fisher's rule and zero misclassifications for the normal rule. An application of Fisher's rule and the normal linear rule to the d-dimensional scores – obtained with the scatter matrix approaches – confirms that the rules are invariant under the transformations: in each of the five transformations, I obtain the same error as for the scaled data. In the Problems at the end of Part III, we want to show that Fisher's rule and the normal rule are invariant under the transformations $\mathbf{X} \to Q(\mathbf{X})^{\mathsf{T}}\mathbf{X}_\Sigma$. ∎

The approaches based on kernels and on scatter matrices construct d-dimensional scores and thus differ from the approaches in Chapters 10 and 11, which focus on finding the 'best' directions and associated scores. Table 12.2 highlights some of the differences between the cumulant-based IC approaches of Chapter 10 and the approach based on the $\mathcal{F}$-correlation. Table 12.8 complements and extends Table 12.2 by including Projection Pursuit and the scatter matrix approaches of this section.

In the table, 'Dim reduction' refers to dimension reduction. Approaches which are suitable for dimension reduction have a 'Yes'. 'HDLSS' refers to whether a method is suitable for HDLSS data.

The methods in the table are listed in the order I describe them in the text. The two earlier methods, ICA with the $\mathcal{G}_4$ kurtosis criterion and Projection Pursuit, emphasise the non-Gaussianity of the solutions. Deviation from Gaussianity drives the process and allows a ranking of the scores. The restriction to non-Gaussian solutions limits the number of possible directions, and in the case of Projection Pursuit, we typically only search for the first direction.

The methods of this chapter abandon the search for non-Gaussian solutions. As a consequence, they find d new directions, but some or many of the resulting scores are close to the Gaussian. This shows a clear trade-off between the two opposing requirements: to find d-dimensional solutions and to find non-Gaussian solutions.

Table 12.8 *A Comparison of Approaches to Finding New or Independent Directions*

	ICA with $\mathcal{G}_4$	Projection Pursuit	Kernel-based methods	Scatter matrix methods
Sources	Indep. and non-Gauss.	Non-Gauss.	Indep.	Indep.
Criterion	Kurtosis or skewness	Kurtosis and skewness	$\mathcal{F}$-cor exact and new	Indep. prop. of scatter stat.
No. of directions	Most non-Gauss.	First	All d	All d
Dim reduction	Yes	Yes	No	No
HDLSS	Yes	Yes	No	No
Computations	Efficient iterative	(Difficult) now as ICA	Iterative slow	Efficient algebraic
Unique Solutions	Repeat of a few different ones	Many	No, all different	Unique

It is not possible to determine which method yields the 'most independent' scores. FastICA and the kernel-based methods each have their own criteria for measuring deviation from independence. The kurtosis can be calculated for *any* scores, but the criteria used in the kernel approaches are not directly applicable to the FastICA solutions. Projection Pursuit focuses on non-Gaussian criteria, and independence is not relevant. The scatter matrix approaches have algebraic solutions, which are unique, because they rely on a spectral decomposition only; the emphasis lies in a new d-dimensional coordinate selection, and an individual direction is not so important.

The various methods have different advantages, so they apply to different scenarios and data. The search for new methods continues. In the next section I outline three further ways of exploiting independence with the aim of acquainting the reader with new ideas for finding independent solutions. I will leave it to you, the reader, to experiment with these approaches and to apply them to data.

12.5 Non-Parametric Estimation of Independence Criteria

The purpose of this section is to give the reader a taste of further developments in the pursuit of interesting structure in multivariate and high-dimensional data. I restrict the discussion to a small number of approaches which show diversity and interesting statistical ideas that are supported by theoretical arguments. At the end of each approach I refer to related papers. The common theme is the independent component model $\mathbf{X} = A\mathbf{S}$ and estimation of the source $\mathbf{S}$ or its distribution $\boldsymbol{\pi}$. We look at

- the characteristic function approaches of Yeredor (2000) and Eriksson and Koivunen (2003),
- the order statistics approach of Learned-Miller and Fisher (2003), and
- the kernel density approach of Samarov and Tsybakov (2004).

12.5.1 A Characteristic Function View of Independence

Bach and Jordan (2002) proposed the $\mathcal{F}$-correlation as a means of characterising independent vectors. Unlike the probability density function and the mutual information, the $\mathcal{F}$-correlation has a natural data-based estimator and therefore has immediate appeal when searching for independent solutions. Alternatively, the independence of random vectors can be expressed in terms of the characteristic functions, and the latter have data-based estimators. In this section we consider the approaches of Yeredor (2000) and Eriksson and Koivunen (2003), who exploited the independence property of characteristic functions.

We begin with a definition and properties of characteristic functions in a multivariate context.

Definition 12.12 Let $\mathbf{X}$ be a d-dimensional random vector from a probability density function f. For $\mathbf{t} \in \mathbb{R}^d$, the **characteristic function** χ of $\mathbf{X}$ is

$$\chi(\mathbf{t}) = \mathbb{E}\left[\exp(i\mathbf{t}^{\mathsf{T}}\mathbf{X})\right]. \tag{12.52}$$

□

The characteristic function is the expected value of a function of the random vector, and we estimate this expected value by an appropriate sample average. For $\mathbb{X} = \left[\mathbf{X}_1\ \mathbf{X}_2 \cdots \mathbf{X}_n\right]$ and $\mathbf{t} \in \mathbb{R}^d$, the **sample characteristic function** $\widehat{\chi}$ is

$$\widehat{\chi}(\mathbf{t}) = \frac{1}{n}\sum_{i=1}^{n} \exp(i\mathbf{t}^{\mathsf{T}}\mathbf{X}_i). \tag{12.53}$$

The characteristic function is of interest because of its relationship to the probability density function, namely,

$$\chi(\mathbf{t}) = \int \exp(i\mathbf{t}^{\mathsf{T}}\cdot)f\ .$$

The probability density function f of a random vector $\mathbf{X}$ and the characteristic function χ of $\mathbf{X}$ are Fourier pairs, that is, there is a one-to-one correspondence between f and χ via the Fourier transform, and f can be recovered from χ by an inverse Fourier transform. In the discrete case, the integrals are replaced by multivariate sums. Characteristic functions always exist – unlike moment-generating functions. For details on properties of characteristic functions, see chapter 9 of Dudley (2002).

A random vector $\mathbf{X}$ is independent if its probability density function is the product of the marginal densities. The following proposition shows a similar characterisation of independence based on characteristic functions.

Proposition 12.13 *Let $\mathbf{X}$ be a d-dimensional random vector from a probability density function f with marginals f_j. Let χ be the characteristic function of $\mathbf{X}$. For $j \le d$, let χ_j be the characteristic function corresponding to the marginal f_j, The following statements are equivalent:*

1. *The vector $\mathbf{X}$ has independent components.*
2. *The probability density function f satisfies $f = \prod_{j=1}^{d} f_j$.*
3. *The characteristic function χ satisfies $\chi = \prod_{j=1}^{d} \chi_j$.*

A proof of this proposition can be found in Dudley (2002).

Yeredor (2000) and Eriksson and Koivunen (2003) exploited the equivalence stated in Proposition 12.13 and estimated χ by (12.53). Yeredor (2000) first proposed the use of characteristic functions as a solution to the independent component model, and Eriksson and Koivunen (2003) extended the approach and asymptotic theory of Yeredor (2000).

The mutual information measures the deviation of the joint probability density function from the product of its marginals. To measure the difference between χ and $\prod \chi_j$, Yeredor (2000) and Eriksson and Koivunen (2003) proposed the criteria Δ_1, Δ_2 and D_w of Definition 12.14.

Definition 12.14 Let $\mathbb{X} = \begin{bmatrix}\mathbf{X}_1 \ \mathbf{X}_2 \cdots \mathbf{X}_n\end{bmatrix}$ be d-dimensional data. Let χ be the characteristic function of the random vectors $\mathbf{X}_i$. The **second characteristic function** ξ of the $\mathbf{X}_i$ is defined in a neighbourhood of $\mathbf{t} = \mathbf{0}$ by

$$\xi(\mathbf{t}) = \log \chi(\mathbf{t}).$$

For $j \leq d$, let χ_j and ξ_j be the characteristic and the second characteristic function of the jth variable of the $\mathbf{X}_i$. Let $\mathbf{t} = [t_1, \ldots, t_d] \in \mathbb{R}^d$. The quantities Δ_1, Δ_2 and D_w measure the departure from independence:

$$\Delta_1(\mathbf{t}) = \chi(\mathbf{t}) - \prod_{j=1}^{d} \chi_j(t_j),$$

$$\Delta_2(\mathbf{t}) = \xi(\mathbf{t}) - \sum_{j=1}^{d} \xi_j(t_j),$$

and

$$D_w(\chi) = \int w |\Delta_1|^2,$$

where w is a non-negative weight function defined on $\mathbb{R}^d$. □

Sometimes ξ is called the **cumulant generating function** (see section 2.7.2 of Hyvärinen, Karhunen, and Oja 2001).

The main difficulty in using these 'characteristic' measures of the departure from independence is the fact that we require Δ_1 and Δ_2 in a neighbourhood of $\mathbf{t} = \mathbf{0}$. Finding enough points $\mathbf{t}_i$ close to $\mathbf{0}$ which result in good estimates for Δ_1 or Δ_2 can prove to be an arduous task. For this reason, Eriksson and Koivunen used the measure D_w. Although D_w circumvents the problem of the points in a neighbourhood of $\mathbf{0}$, it requires a judicious choice of the weight function w and a suitable truncation of a Fourier series expansion of D_w. For details regarding these choices and a description of their algorithm JECFICA, see Eriksson and Koivunen (2003).

It is interesting to note that the three measures Δ_1, Δ_2 and D_w do not, at first glance, appear to have any connection to non-Gaussian random vectors. However, for white Gaussian random vectors, $\Delta_1 = 0$, whereas for non-Gaussian white random vectors, $\Delta_1 \neq 0$.

The ideas of Eriksson and Koivunen (2003) informed the approach of Chen and Bickel (2006), who derived theoretical properties, including consistency proofs, of the unmixing matrix $B = A^{-1}$ for the independent component model $\mathbf{X} = A\mathbf{S}$. Chen and Bickel (2006) estimated the source distribution by B-splines and made use of cross-validation to choose the smoothing parameter. Starting with a consistent estimator for A^{-1} and a learning rule – see (10.26) in Section 10.5.1 – for updating the estimate B_i, they showed that the sequence B_i converges to a limit which is asymptotically efficient.

12.5.2 An Entropy Estimator Based on Order Statistics

The early approaches to Independent Component Analysis approximate the entropy by third- and fourth-order cumulants and then use the relationship between the mutual information and the entropy, that is, $\mathcal{I}(f) = \sum \mathcal{H}(f_j) - \mathcal{H}(f)$ (see (9.13) in Section 9.4), to approximate the mutual information. Section 10.4.2 describes these approximations. More recently, Learned-Miller and Fisher (2003) returned to the entropy $\mathcal{H}$ and proposed an estimator which uses ideas from order statistics.

As in Learned-Miller and Fisher (2003), we first look at a non-parametric estimator for the univariate entropy.

Let $\left[X_1\ X_2 \cdots X_n\right]$ be a random sample, and let $X_{(1)} \le X_{(2)} \le \cdots \le X_{(n)}$ be the derived ordered random variables. If the X_i are identically distributed with distribution function F and probability density function f, then

$$F(X_i) \sim U(0,1) \quad \text{the uniform distribution on } [0,1],$$

and

$$\mathbb{E}\left[F(X_{(i+1)}) - F(X_{(i)})\right] = \frac{1}{n+1} \qquad \text{for } i \le n-1. \tag{12.54}$$

For a proof of the first statement, see theorem 2.1.4 of Casella and Berger (2001). The first fact leads to (12.54), and the expectation is taken with respect to the product density $\prod f$.

Learned-Miller and Fisher defined an estimator $\widehat{f}$ of the probability density function f which is based on (12.54):

$$\widehat{f}(X) = \frac{1}{(n+1)} \frac{1}{\left[X_{(i+1)} - X_{(i)}\right]} \qquad \text{for } X_{(i)} \le X \le X_{(i+1)}. \tag{12.55}$$

From (12.55) it is but a small step to obtain an estimator for the entropy $\mathcal{H}(f) = -\int f \log f$.

Proposition 12.15 [Learned-Miller and Fisher (2003)] *Let* $\left[X_1\ X_2 \cdots X_n\right]$ *be random variables from the probability density function* f. *The entropy* $\mathcal{H}$ *of* f *is estimated from the* X_i *by*

$$\mathcal{H}(f) \approx \frac{1}{n-1} \sum_{i=1}^{n-1} \log\left\{(n+1)\left[X_{(i+1)} - X_{(i)}\right]\right\}.$$

A derivation of this approximation is straightforward but relies on the definition of $\widehat{f}$ in (12.55).

The intuitively appealing estimator of the proposition has high variance. To obtain an estimator with lower variance, Learned-Miller and Fisher chose bigger spacings between the order statistics and proposed the estimator

$$\widehat{\mathcal{H}}_m(f) = \frac{1}{n-m} \sum_{i=1}^{n-m} \log\left\{\frac{n+1}{m}\left[X_{(i+m)} - X_{(i)}\right]\right\}, \tag{12.56}$$

which they call the ***m*-spacing estimator** of the entropy. It follows from Vasicek (1976) and Beirlant et al. (1997) that this estimator is consistent as $m, n \to \infty$ and $m/n \to 0$. Typically, Learned-Miller and Fisher (2003) worked with the value $m = \sqrt{n}$.

To apply the entropy results to the independent component model $\mathbf{X} = A\mathbf{S}$, Learned-Miller and Fisher (2003) started with white signals, as in Corollary 10.10 of Section 10.4.2. The minimisation of the mutual information reduces to finding the orthogonal unmixing matrix $B = A^{-1}$ which minimises the entropy (12.56). Learned-Miller and Fisher explained their algorithm RADICAL for two-dimensional independent component problems. The complexity of the algorithm increases quickly as the dimension of the data increases. It is not clear whether their algorithm is computationally feasible for a large number of variables.

12.5.3 Kernel Density Estimation of the Unmixing Matrix

Our final approach in this chapter is based on kernels – not the feature kernels of Sections 12.2 and 12.3 but those used in density estimation which we encountered throughout Chapter 11. Kernel density estimation has enjoyed great popularity, in particular, in the analysis of univariate and bivariate data (see Scott 1992 or Wand and Jones 1995). Samarov and Tsybakov (2004) used univariate kernels when estimating the probability density function of $\mathbf{S}$ in the independent component model $\mathbf{X} = A\mathbf{S}$. Their approach allows a simultaneous estimation of the source vectors and the separating matrix. As part of their theoretical developments, they showed that their estimators converge to the true sources with the parametric rate of $n^{-1/2}$. I outline their approach and state their results. The interested reader can find the proofs in their paper.

Rather than starting with the independent component model (10.1) in Section 10.2.1, Samarov and Tsybakov (2004) used the equivalent model

$$\mathbf{S} = B\mathbf{X}, \tag{12.57}$$

where the unmixing or separating matrix $B = \begin{bmatrix}\boldsymbol{\omega}_1 \cdots \boldsymbol{\omega}_d\end{bmatrix}$ has orthogonal columns, and $\mathbf{X}$ and $\mathbf{S}$ are d-dimensional random vectors. We write Σ and f^* for the covariance matrix and the probability density function of $\mathbf{X}$. Similarly, we write D and $\pi = \prod \pi_j$ for the diagonal covariance matrix and the product probability density function of the source $\mathbf{S}$, which has independent entries. The link between $\mathbf{S}$ and $\mathbf{X}$ can be expressed in terms of their covariance matrices and their probability density functions. We have

$$D = B\Sigma B^{\mathsf{T}} \qquad \text{and} \qquad f^*(\mathbf{X}) = \det(B) \prod_{j=1}^{d} \pi_j(\boldsymbol{\omega}_j^{\mathsf{T}}\mathbf{X}). \tag{12.58}$$

The relationship (12.58) is the same as (10.23) in Section 10.5.1 but given here for the densities instead of the log-likelihoods.

Samarov and Tsybakov (2004) defined a function T on probability density functions f by

$$T(f) = \mathbb{E}\left[(\nabla f)(\nabla f)^{\mathsf{T}}\right], \tag{12.59}$$

where ∇f is the gradient of f, and the expectation is with respect to f. Introduction of the function T is crucial to the success of their method; T has the following properties, which I explain below:

1. If $f = f^*$, then
   (a) T is a function of B and thus links f^* and B, and
   (b) T and Σ satisfy a generalised eigenvalue relationship.
2. T is estimated from data by means of univariate kernel density estimates, and thus

(a) B can be estimated using 1(b), and
(b) $\mathbf{S}$ and π can be estimated using 1(a) and (12.58).

Statement 1(a) is established in the following proposition.

Proposition 12.16 [Samarov and Tsybakov (2004)] *Let* $\mathbf{X}$ *and* $\mathbf{S}$ *be* d*-dimensional random vectors which satisfy model (12.57) with* $D = B\Sigma B^{\mathsf{T}}$ *and* $f^*(\mathbf{X}) = \det(B)\prod_{j=1}^{d}\pi_j(\boldsymbol{\omega}_j^{\mathsf{T}}\mathbf{X})$. *Define* T *as in (12.59). Then*

$$T(f^*) = B^{\mathsf{T}}CB$$

is positive definite, and C *is a diagonal matrix with positive diagonal elements* c_{jj} *given by*

$$c_{jj} = [\det(B)]^2\,\mathbb{E}\left\{\prod_{k\neq j}\left[\pi_k(\boldsymbol{\omega}_k^{\mathsf{T}}\mathbf{X})\right]^2\left[\pi_j'(\boldsymbol{\omega}_k^{\mathsf{T}}\mathbf{X})\right]^2\right\},$$

where π' *is the derivative of* π.

The equalities of the proposition show that $T(f^*)$ combines information about the separating matrix B and the marginals π_j.

Statement (1b) establishes the connection between T and the covariance matrix Σ which follows from the next proposition.

Proposition 12.17 [Samarov and Tsybakov (2004)] *Let* $\mathbf{X}$ *and* $\mathbf{S}$ *be* d*-dimensional random vectors which satisfy model (12.57) with* $D = B\Sigma B^{\mathsf{T}}$ *and* $f^*(\mathbf{X}) = \det(B)\prod_{j=1}^{d}\pi_j(\boldsymbol{\omega}_j^{\mathsf{T}}\mathbf{X})$. *Define* T *as in (12.59). Put*

$$E = B^{\mathsf{T}}C^{1/2} \qquad \textit{and} \qquad \Delta = C^{1/2}DC^{1/2},$$

and write $E = \left[\mathbf{e}_1 \cdots \mathbf{e}_d\right]$. *If the diagonal elements of* Δ *are* $\delta_1, \ldots, \delta_d$, *then*

$$T(f^*) = E^{\mathsf{T}}E, \qquad E^{\mathsf{T}}[T(f^*)]^{-1}E = \mathbf{I}_{d\times d} \qquad \textit{and} \qquad E^{\mathsf{T}}\Sigma E = \Delta. \tag{12.60}$$

Further,

$$T(f^*)\Sigma\mathbf{e}_j = \delta_j\mathbf{e}_j, \tag{12.61}$$

for $j \le r$, *where* r *is the rank of* $T(f^*)\Sigma$, *and the column vectors* $\boldsymbol{\omega}_j$ *of* B *are given by*

$$\boldsymbol{\omega}_j = c_{jj}^{-1/2}\mathbf{e}_j. \tag{12.62}$$

The equalities (12.60) can be restated as a generalised eigenvalue problem (see Section 3.7.4 and Theorem 4.6 in Section 4.3.1). From properties of generalised eigenvalue problems, (12.61) follows. Proposition 12.17 also establishes explicit expressions for the columns of B which are derived from $E = B^{\mathsf{T}}C^{1/2}$.

These two propositions tell us that the task of estimating the $\boldsymbol{\omega}_j$ is accomplished if we can estimate the matrix $T(f^*)\Sigma$.

For data $\mathbb{X} = \left[\mathbf{X}_1 \cdots \mathbf{X}_n\right]$ with sample covariance matrix S, Samarov and Tsybakov (2004) estimated Σ by $S(n-1)/n$ and then defined an estimator $\widehat{T}(f^*)$ for $T(f^*)$ – without knowledge of f^* – by

$$\widehat{T}(f^*) = \frac{1}{n}\sum_{i=1}^{n}\left[\nabla\widehat{f}_{-i}^*(\mathbf{X}_i)\right]\left[\nabla\widehat{f}_{-i}^*(\mathbf{X}_i)\right]^{\mathsf{T}}, \tag{12.63}$$

where the vector $\nabla \widehat{f}^*_{-i}(\mathbf{X}_i)$ has entries

$$\frac{\partial \widehat{f}^*_{-i}}{\partial X_k}(\mathbf{X}_i) = \frac{h^{-d-1}}{n-1} \sum_{j \neq i} K_{(d)} \left(\frac{\mathbf{X}_j - \mathbf{X}_i}{h} \right),$$

for $k \leq d$. Here $K_{(d)}$ is the d-fold product of univariate kernels K_1, and $h > 0$ is the bandwidth of K_1 and also of $K_{(d)}$.

From (12.63) and (12.62), an estimator $\widehat{B} = \left[\widehat{\boldsymbol{\omega}}_1 \cdots \widehat{\boldsymbol{\omega}}_d\right]$ of B can be derived.

To estimate f^* and the π_j, Samarov and Tsybakov (2004) used a second kernel K_2 and different bandwidths h_j for $j \leq d$ and put

$$\widehat{f^*}(\mathbf{X}) = \det(\widehat{B}) \prod_{j=1}^{d} \frac{1}{n h_j} \sum_{i=1}^{n} K_2 \left(\frac{\widehat{\boldsymbol{\omega}}_j^{\mathsf{T}} \mathbf{X}_i - \widehat{\boldsymbol{\omega}}_j^{\mathsf{T}} \mathbf{X}}{h_j} \right). \tag{12.64}$$

Finally, the estimated marginal densities $\widehat{\pi}_j$ are found to be

$$\widehat{\pi}_j(\widehat{\boldsymbol{\omega}}_j^{\mathsf{T}} \mathbf{X}) = \frac{1}{n h_j} \sum_{i=1}^{n} K_2 \left(\frac{\widehat{\boldsymbol{\omega}}_j^{\mathsf{T}} \mathbf{X}_i - \widehat{\boldsymbol{\omega}}_j^{\mathsf{T}} \mathbf{X}}{h_j} \right). \tag{12.65}$$

The main result of Samarov and Tsybakov (2004) shows the asymptotic performance of the estimators $\widehat{f^*}$ of f^* and $\widehat{\pi}_j$ of π_j. The precise regularity conditions on f^* and the kernel K_2 are given in their paper.

Theorem 12.18 [Samarov and Tsybakov (2004)] *Let* $\mathbf{X}$ *and* $\mathbf{S}$ *be* d*-dimensional random vectors which satisfy model (12.57) with* $D = B \Sigma B^{\mathsf{T}}$ *and* $f^*(\mathbf{X}) = \det(B) \prod_{j=1}^{d} \pi_j(\boldsymbol{\omega}_j^{\mathsf{T}} \mathbf{X})$. *Assume that* f^* *is sufficiently smooth and satisfies a finite moment condition. Define* $\widehat{f^*}$ *and* $\widehat{\pi}_j$, *for* $j \leq d$, *as in (12.64) and (12.65), and assume that the kernel, which is used in the derivation of (12.64), satisfies suitable regularity conditions. Then the following hold as* $n \to \infty$:

1. $\left\| \widehat{\boldsymbol{\omega}}_j - \boldsymbol{\omega}_j \right\| = O_p(n^{-1/2})$ *for* $j \leq d$, *and*
2. $|\widehat{f^*}(\mathbf{X}) - f(\mathbf{X})| = O_p(n^{-s/(2s+1)})$,

where s *is related to a Lipschitz condition on the kernels.*

The result tells us that asymptotically the estimated unmixing matrix converges to the true unmixing matrix at the parametric rate of $n^{-1/2}$. Further, the density estimator $\widehat{f^*}$ converges to the true probability density function f^*, and the rate of convergence depends on properties of the kernels.

Although $\mathbf{X}$ and $\mathbf{S}$ are multivariate random vectors, the estimation of $\widehat{f^*}$ only requires univariate kernels. These theoretical results are encouraging even for moderate dimensions d, provided that $n \gg d$.

Also in 2004, Boscolo, Pan, and Roychowdhury (2004) proposed an approach to Independent Component Analysis which is based on non-parametric density estimation and the gradient of the source density. Samarov and Tsybakov (2004) focused on theoretical issues and convergence properties of their estimators, and Boscolo, Pan, and Roychowdhury (2004) were more concerned with computational rather than primarily statistical issues. Boscolo, Pan, and Roychowdhury describe a simulation study and compare a number of different ICA algorithms, including JADE and FastICA, which form the basis of Algorithm 10.1 in Section 10.6.1, the kernel ICA of Algorithm 12.1 and their own non-parametric approach.

Other non-parametric density-based approaches to Independent Component Analysis include Hastie and Tibshirani (2002), who used a penalised splines approach to estimating the marginal source densities, and Barbedor (2007), who estimated the marginal densities by wavelets.

The approaches we looked at in this chapter and the additional references at the end of each section indicate that there are many and diverse efforts which all attempt to solve the independent component problem by finding good estimators of the sources and their probability density functions. These developments differ from the early research in Independent Component Analysis and Projection Pursuit in a number of ways:

- The early methods typically approximated the mutual information or the projection index by functions of cumulants, whereas the methods discussed in this chapter explicitly estimate an independence criterion – such as the $\mathcal{F}$-correlation, the characteristic functions or the source density.
- The early methods explicitly used the strong interplay between non-Gaussianity and independence, whereas the methods of this chapter primarily pursue the independence aspect.

With the different demands data pose, it is vital to have a diversity of methods to choose from and to have methods based on a theoretical foundation. The complexity of the data, together with the aim of the analysis, may determine which method(s) one should use. For small dimensions, one may want to recover all d sources, whereas for large dimensions, a dimension-reduction aspect can be the driving force in the analysis. In addition, computational aspects may determine the types of methods one wants to use. If possible, I recommend working with more than one approach and comparing the results.

13

Feature Selection and Principal Component Analysis Revisited

Den Samen legen wir in ihre Hände! Ob Glück, ob Unglück aufgeht, lehrt das Ende (Friedrich von Schiller, *Wallensteins Tod*, 1799). *We put the seed in your hands! Whether it develops into fortune or mistfortune only the end can teach us.*

13.1 Introduction

In the beginning – in 1901 – there was Principal Component Analysis. On our journey through this book we have encountered many different methods for analysing multidimensional data, and many times on this journey, Principal Component Analysis reared its – some might say, ugly – head. About a hundred years since its birth, a renaissance of Principal Component Analysis (PCA) has led to new theoretical and practical advances for high-dimensional data and to SPCA, where ***S*** variously refers to *simple, supervised* and *sparse*. It seems appropriate, at the end of our journey, to return to where we started and take a fresh look at developments which have revitalised Principal Component Analysis. These include the availability of high-dimensional and functional data and the necessity for dimension reduction and feature selection and new and sparse ways of representing data.

Exciting developments in the analysis of high-dimensional data have been interacting with similar ones in Statistical Learning. It is not clear where *analysis of data* stops and *learning from data* starts. An essential part of both is the selection of 'important' and 'relevant' features or variables. In addition to the classical features, such as principal components, factors or the classical configurations of Multidimensional Scaling, we encountered non-Gaussian and independent component features. Interestingly, a first step towards obtaining non-Gaussian or (almost) independent features is often a Principal Component Analysis. The availability of complex high-dimensional or functional data and the demand for 'interesting' features pose new challenges for Principal Component Analysis which include

- finding, recognising and exploiting non-Gaussian features,
- making decisions or predictions in supervised and unsupervised learning based on 'relevant' variables,
- deriving sparse PC representations, and
- acquiring a better understanding of the theoretical properties and behaviour of principal components as the dimension grows.

The growing number of variables requires not only a dimension reduction but also a *selection* of important or relevant variables. What constitutes 'relevant' typically depends on the analysis we want to carry out; Sections 3.7 and 4.8 illustrate that dimension reduction and variable selection based on Principal Component Analysis do not always result in variables that are optimal for the next step of the analysis. In Section 4.8.3 we had a first look at variable ranking, which leads to a reduced number of relevant variables, provided that a suitable ranking scheme is chosen. We pursue the idea of variable ranking further in this chapter and explore its link to canonical correlations. Variable ranking is just one avenue for reducing the number of variables. Sparse representations, that is, representations with a sparse number of non-zero entries, are another; we will restrict attention to sparse representations arising from within a principal component framework in this last chapter. Sparse representations are studied in other areas, including the rapidly growing compressed sensing which was pioneered by Candès, Romberg, and Tao (2006) and Donoho (2006). Compressed sensing uses input from harmonic and functional analysis, frame theory, optimisation theory and random matrix theory (see Elad 2010).

Our last topic concerns the behaviour of principal components as the dimension grows – possibly faster than the sample size. For the classical case, $d \ll n$, we know from Theorem 2.20 in Section 2.7.1 that the sample eigenvectors and eigenvalues are consistent estimators of the population quantities. These consistency properties do not hold in general when $d > n$. Johnstone (2001) introduced 'spiked covariance' models for sequences of data, indexed by the dimension, and Jung and Marron (2009) explored the relationship between the rate of growth of d and the largest eigenvalue λ_1 of the covariance matrix. In Theorem 2.25 in Section 2.7.2 we learned that the first sample eigenvector is consistent, provided that λ_1 grows at a faster rate than d. In this chapter we examine the behaviour of the eigenvectors for different growth rates of the dimension d and the first few eigenvalues of the covariance matrix, and we also allow the sample size to grow. These results form part of the general framework for Principal Component Analysis proposed in Shen, Shen, and Marron (2012) and provide an appropriate conclusion of this chapter.

In Section 13.2 we look at the analyses of three data sets which combine dimension reduction based on Principal Component Analysis with a search for independent component features in the context of supervised learning, unsupervised learning and a test of Gaussianity. A suitable choice of the number of principal components is important for the success of these analyses. Section 13.3 details variable ranking ideas in a statistical-learning context and explores the relationship to canonical correlations. For linear regression, these ideas lead to Supervised Principal Component Analysis. For classification, there is a close connection between variable ranking schemes and Fisher's linear discriminant rule. We study the asymptotic behaviour of such discriminant rules as the dimension of the data grows and will find that d is allowed to grow faster than the sample size, but not too fast. Section 13.4 focuses on sparse PCs, that is, PCs whose non-zero weights are concentrated on a small number of variables. The section describes how ideas from regression, such as the LASSO and the elastic net, can be translated into a PC framework and looks at the SCoTLASS directions and rank one approximations of the data. The final Section 13.5 treats sequences of models, indexed by the dimension d, and explores the asymptotic behaviour of the eigenvalues, eigenvectors and principal component scores as $d \to \infty$. We find that the asymptotic behaviour depends on the growth rates of d, n and the first eigenvalue of the covariance matrix. All three are allowed to grow, but if d grows too fast, the results rapidly become inconsistent.

13.2 Independent Components and Feature Selection

A first attempt at dimension reduction is often a Principal Component Analysis. Although it is tempting to 'just do a PCA', some caution is appropriate, as a reduced number of PCs may not contain enough relevant information. We have seen instances of this in Example 3.6 in Section 3.5.1, which involves canonical correlations; in Example 3.11 in Section 3.7.3, which deals with linear regression, and in Example 4.12 in Section 4.8.3, which concerns classification. And our final Section 13.5 shows that the principal components may, in fact, contain the 'wrong' information in the sense that they are not close to the population principal components. It does not follow that Principal Component Analysis does not work, but merely that care needs to be taken.

Principal Component Analysis is only one of many dimension-reduction methods, and depending on the type of information we hope to discover in the data, a combination of Principal Component Analysis and Independent Component Analysis or Projection Pursuit may lead to more interesting or relevant features. In this section I illustrate with three examples how independent component features of the reduced principal component data can lead to interpretable results in supervised learning, unsupervised learning and a non-standard application which exploits the non-Gaussian nature of independent components. In the description and discussion of the three examples, I will move between the actual example and a description of the ideas or approaches used. I hope that this swap between the ideas and their application to data will make it easier to apply these approaches to different data.

13.2.1 Feature Selection in Supervised Learning

The first example is a classification or supervised-learning problem which relates to the detection of regions of emphysema in the human lung from textural features. The example incorporates a suitable feature selection which improves the computational efficiency of classification at no cost to accuracy. The method and algorithm I discuss apply to a large range of labelled data, but we focus on the *emphysema* data which motivated this approach.

Example 13.1 The **HRCT emphysema** data are described in Prasad, Sowmya, and Koch (2008). High-resolution computed tomography (HRCT) scans have emerged as an important tool for detection and characterisation of lung diseases such as emphysema, a chronic respiratory disorder which is closely associated with cigarette smoking. Regions of emphysema appear as low-attenuation areas in CT images, and these diffuse radiographic patterns are a challenge for radiologists during diagnosis. An HRCT scan is shown in the left panel of Figure 13.1. The darker regions are those with emphysema present. The images are taken from the paper by Prasad, Sowmya, and Koch (2008).

An HRCT scan, as in Figure 13.1, consists of 262,144 pixels on average, and for each pixel, a decision about the presence of emphysema is made by an expert radiologist, who considers small regions rather than individual pixels. The status of each pixel is derived from that of the small region to which it belongs. Such expert analyses are time-consuming, do not always result in the same assessment and often lead to overestimates of the emphysema regions. More objective, fast and data-driven methods are required to aid and complement the expert diagnosis.

For each pixel, twenty-one variables, called *textural features*, have been used to classify these data. For a detailed description of the experiment, see Prasad, Sowmya, and Koch (2008). ∎

The approach of Prasad, Sowmya, and Koch (2008) is motivated by the need to detect diseases in lung data objectively, accurately and efficiently, but their method applies more generally, and Prasad, Sowmya, and Koch apply their feature-selection ideas to the detection of other lung diseases and to nine data sets that are commonly used in supervised learning. The latter include Fisher's *iris* data and the *breast cancer* data. The nine data sets provide an appropriate testing ground for their ideas; they vary in the number of classes (with one data set consisting of twenty-six classes), the number of dimensions and the sample sizes. For details on feature selectors in machine learning, see Guyon and Elisseeff (2003).

The key idea of Prasad, Sowmya, and Koch (2008) is the determination of a suitably small number of independent components as the selected features and the use of these features as input to different discriminant rules, which include the naive Bayes rule defined by (13.1) and decision trees from the WEKA Data Mining Software (see Witten and Frank 2005).

In the analysis that follows, I will focus mostly on the naive Bayes rule for two-class problems. Consider data $\mathbb{X}$ from two classes. For $\nu = 1, 2$, let $\overline{\mathbf{X}}_\nu$ be the sample mean and S_ν the sample covariance matrix of the νth class. Let $\widehat{W} = S_1 + S_2$, as in Corollary 4.9 in Section 4.3.2, be the sum of the class sample covariance matrices. Put $\widehat{D} = \operatorname{diag}(\widehat{W})$. For $\mathbf{X}$ from $\mathbb{X}$, the **naive Bayes rule** $\mathfrak{r}_{\text{NB}}$ is defined by

$$\mathfrak{r}_{\text{NB}}(\mathbf{X}) = 1 \qquad \text{if } h(\mathbf{X}) = \left[\mathbf{X} - \frac{1}{2}(\overline{\mathbf{X}}_1 + \overline{\mathbf{X}}_2)\right]^{\mathsf{T}} \widehat{D}^{-1}(\overline{\mathbf{X}}_1 - \overline{\mathbf{X}}_2) > 0. \tag{13.1}$$

The normal rule (4.15) in Section 4.3.3 is defined for the population, whereas (13.1) is defined for the sample. Apart from this difference, and the obvious adjustment of mean and covariance matrix by their sample counterparts, the main change is the substitution of the full covariance matrix Σ with the diagonal matrix $\widehat{D} = \operatorname{diag}(\widehat{W})$. The matrix $\widehat{W}$ is not always invertible, especially for high-dimensional data. For this reason, the diagonal matrix $\widehat{D}$ replaces $\widehat{W}$. If the within-class covariance matrices are assumed to be the same, then $\widehat{D} = \operatorname{diag}(S)$. Bickel and Levina (2004) examined the performance of the naive Bayes rule and showed that it can outperform Fisher's discriminant rule when the dimension grows at a rate faster than n. I return to some of these results in Section 13.3.4.

Algorithm 13.1 ***Independent Component Features in Supervised Learning*** **(FS-ICA)**
Let $\mathbb{X} = \begin{bmatrix}\mathbf{X}_1 \cdots \mathbf{X}_n\end{bmatrix}$ be d-dimensional labelled data with sample mean $\overline{\mathbf{X}}$ and sample covariance matrix S. Write $S = \widehat{\Gamma}\widehat{\Lambda}\widehat{\Gamma}^{\mathsf{T}}$ for the spectral decomposition. Fix $K > 1$ for the number of iterations. Choose a discriminant rule $\mathfrak{r}$.

Step 1. Determine the dimension p so that the first p principal component scores explain not less than 95 per cent of total variance.
Step 2. Calculate the p-white data $\mathbb{X}(p)^\diamond = \widehat{\Lambda}_p^{-1/2}\widehat{\Gamma}_p^{\mathsf{T}}(\mathbb{X} - \overline{\mathbf{X}})$ as in (10.32) in Section 10.7.1.
Step 3. Apply Algorithm 10.1 in Section 10.6.1 to the p-white data $\mathbb{X}(p)^\diamond$ with K iterations and the FastICA skewness or kurtosis criteria.

Step 4. Calculate the derived features $\mathbb{S}^{(p)} = U^{*\mathsf{T}}\mathbb{X}(p)^{\diamond}$.

Step 5. Derive a rule $\mathfrak{r}_{p,0}$ from $\mathfrak{r}$ which is based on a training subset $\mathbb{S}_0^{(p)}$ of $\mathbb{S}^{(p)}$. Use $\mathfrak{r}_{p,0}$ to predict labels for the data in the testing subset of $\mathbb{S}^{(p)}$. ■

The value $K = 10$ was found to be adequate. The value of p which corresponds to about 95 per cent of total variance is derived experimentally in Prasad, Sowmya, and Koch (2008) and offers a compromise between classification accuracy and computational cost. Prasad, Sowmya, and Koch noted that accuracy improves substantially when the variance contributions increase from 87 to 95 per cent. But very small improvements in accuracy are achieved when the variance increases from 95 to 99 per cent, whereas computation times increase substantially.

Example 13.2 We continue with the **HRCT emphysema** data and look at the results obtained with Algorithm 13.1. I only discuss the results based on the skewness criterion of step 3 here because the kurtosis results are very similar.

For the classification step, a single partition into a training set and a testing set was derived in collaboration with the radiologist. The training set contains one-sixth of the pixels, and the remainder are used for testing. We choose p in step 2 of the algorithm, which explains at least 95 per cent of total variance. This p reduces the twenty-one variables to five and hence also to five IC features.

Prasad, Sowmya, and Koch (2008) apply three rules to the five IC features and to all twenty-one original variables and compare the results. The three rules are the naive Bayes rule, the C4.5 decision-tree learner, and the IB1 classifier from the WEKA Data Mining Software. The IB1 classifier assigns the label of the nearest training observation to an observation from the testing set. For details, see Aha and Kibler (1991). The actual choice of rule is not so important here. Prasad, Sowmya, and Koch calculate a classification error for each pixel in the testing set by comparing the predicted value with the label assigned to the pixel by the radiologist. The performance of each rule is given in Table 13.1 as a percentage of correctly classified pixels. The table shows that the performances are comparable; for the IB1 rule, the performance based on the 5 IC features is slightly superior to that based on all features.

Figure 13.1 shows the regions of emphysema, in blue, which are determined with the naive Bayes rule. The middle panel shows the classification results obtained with all twenty-one variables, and the right panel shows the corresponding results for the five IC features.

Medical experts examined these results and preferred the regions found with the five IC features, as they contain much fewer false positives.

Prasad, Sowmya, and Koch (2008) noted that classification based on IC features is generally more accurate than that obtained with the same number of PCs, which indicates that the choice of features affects the classification accuracy. ■

Algorithm 13.1 is a generalisation of Algorithm 4.2 in Section 4.8.2 in that it derives features $\mathbb{S}^{(p)} = U^{*\mathsf{T}}\widehat{\Lambda}_p^{-1/2}\mathbb{W}^{(p)}$ from the PC data $\mathbb{W}^{(p)}$ and then uses $\mathbb{S}^{(p)}$ as the input to a

Table 13.1 *Percentage of Correctly Classified Emphysema Data*

	Naïve Bayes	C4.5	IB1
5 IC features	84.63	82.33	82.71
All 21 variables	87.13	82.37	82.11

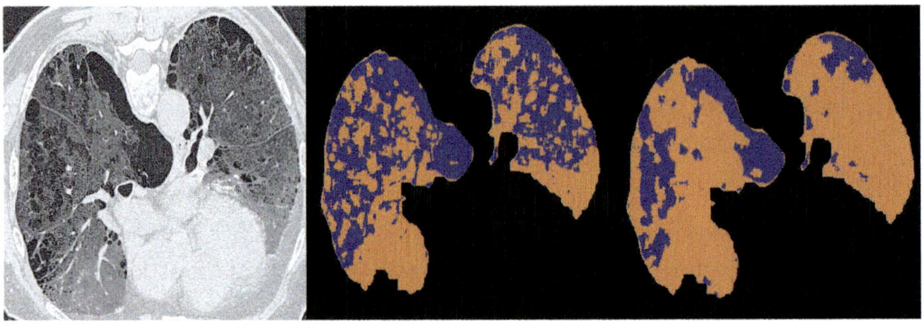

Figure 13.1 Emphysema data from Example 13.1: HRCT scan (*left*) and classification of diseased regions using a naive Bayes rule based on all features (*middle*) and based on five IC features (*right*). Regions of emphysema shown in blue. (*Source*: Prasad, Sowmya, and Koch 2008.)

discriminant rule. It may not be possible to determine theoretically whether these IC features lead to a better classification than the PC data from which they are derived. Prasad, Sowmya, and Koch (2008) observed that for the data sets they considered, the classification error based on $\mathbb{S}^{(p)}$ generally was smaller than that based on the corresponding number of PC features $\mathbb{W}^{(p)}$. These results indicate that IC features may lead to a more accurate classification than the PC data of the same dimension.

Table 13.1 shows that classification based on suitably chosen features can reduce the number of variables to less than 25 per cent essentially without loss of accuracy but with a substantial reduction in computing time. Further, the three rules seem to perform similarly. This observation is confirmed in their analysis of other data sets.

The method of Prasad, Sowmya, and Koch is to achieve a more efficient classification rather than obtaining a better overall accuracy. The latter is a task worth pursuing, but for the large number of pixels, an objective and efficient method may be more desirable, provided that the accuracy is acceptable.

13.2.2 Best Features and Unsupervised Decisions

Our second example belongs to the area of Cluster Analysis or Unsupervised Learning. The aim is to partition the Australian *illicit drug market* data into two groups using independent component features. In Example 6.10 in Section 6.5.2, I applied the PC_1 sign cluster rule to these data, and the resulting partitioning is shown in Table 6.9. Now we want to define and apply an IC_1 cluster rule. The cluster arrangement we obtain will differ from that shown in Table 6.9. There is no 'right' answer; both cluster arrangements are valuable as they offer different insights into the structure of the data and hence result in a more comprehensive understanding of the drug market.

We follow Gilmour and Koch (2006), who proposed using the most non-Gaussian direction in the data combined with a sign cluster rule. From Example 10.8 in Section 10.8, we know that the independent component scores vary with the dimension p of the p-white data. A 'good' choice of p is therefore essential.

Algorithm 13.2 ***Sign Cluster Rule Based on the First Independent Component***
Let $\mathbb{X} = \left[\mathbf{X}_1 \cdots \mathbf{X}_n\right]$ be d-dimensional data with sample mean $\overline{\mathbf{X}}$ and sample covariance matrix S. Write $S = \widehat{\Gamma}\widehat{\Lambda}\widehat{\Gamma}^{\mathsf{T}}$ for the spectral decomposition. Fix $K > 1$ for the number of iterations. Fix $\rho = 3$ for the skewness criterion or $\rho = 4$ for the kurtosis criterion.

Step 1. Use (10.40) in Section 10.8 to determine an appropriate dimension p^*_ρ for $\mathbb{X}$, and put $p = p^*_\rho$.
Step 2. Calculate the p-white data $\mathbb{X}(p)^\diamond = \widehat{\Lambda}_p^{-1/2}\widehat{\Gamma}_p^{\mathsf{T}}(\mathbb{X} - \overline{\mathbf{X}})$ as in (10.32) in Section 10.7.1.
Step 3. For $m = 1, \ldots, K$, calculate the first vector $\boldsymbol{\upsilon}_{1,m}$ of the unmixing matrix $\widehat{U}_m$ for $\mathbb{X}(p)^\diamond$ using Algorithm 10.1 in Section 10.6.1 with the FastICA skewness criterion if $\rho = 3$ or the kurtosis criterion if $\rho = 4$, and put

$$\widetilde{\boldsymbol{\upsilon}}_1 = \underset{\boldsymbol{\upsilon}_{1,m}}{\operatorname{argmax}} \left| b(\boldsymbol{\upsilon}_{1,m}^{\mathsf{T}}\mathbb{X}(p)^\diamond)\right|,$$

where b is the sample skewness b_3 of (9.8) in Section 9.3 or the sample kurtosis b_4 of (9.10) as appropriate.
Step 4. For $i = 1, \ldots, n$, let $\mathbf{X}(p)_i^\diamond$ be the p-white observation corresponding to $\mathbf{X}_i$. Define a sign rule $\mathfrak{r}$ on $\mathbb{X}$ by

$$\mathfrak{r}(\mathbf{X}_i) = \begin{cases} 1 & \text{if } \widetilde{\boldsymbol{\upsilon}}_1^{\mathsf{T}}\mathbf{X}(p)_i^\diamond \geq 0, \\ 2 & \text{if } \widetilde{\boldsymbol{\upsilon}}_1^{\mathsf{T}}\mathbf{X}(p)_i^\diamond < 0. \end{cases} \tag{13.2}$$

Step 5. Partition $\mathbb{X}$ into the clusters

$$C_1 = \{\mathbf{X}_i \, : \, \mathfrak{r}(\mathbf{X}_i) = 1\} \qquad \text{and} \qquad C_2 = \{\mathbf{X}_i \, : \, \mathfrak{r}(\mathbf{X}_i) = 2\}.$$

■

As in Algorithm 13.1, Gilmour and Koch (2006) note that $K = 10$ is an adequate number of iterations.

The rule (13.2) is similar to the PC_1 sign cluster rule (6.9) in Section 6.5.2 in that both are defined as projections of the data onto a single direction and thus are linear combinations of the data. However, because the two directions maximise different criteria, the clusters will differ.

Example 13.3 We continue with the **illicit drug market** data described in Gilmour and Koch (2006). These data consist of seventeen different series measured over sixty-six months. As we have seen in previous analyses of these data, one can get interesting results whether we regard the seventeen series or the sixty-six months as the observations. The two ways of analysing these data are closely related via the duality of $\mathbb{X}$ and $\mathbb{X}^{\mathsf{T}}$ (see Section 5.5). The emphasis in each approach, however, differs.

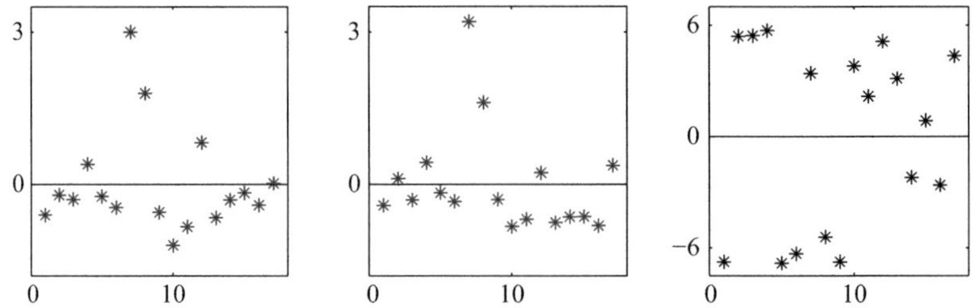

Figure 13.2 IC_1 sign rule for the illicit drug market data: cluster membership based on three PCs (*left*), six PCs (*middle*) and for comparison the PC_1 sign rule (*right*).

The high-dimension low sample size (HDLSS) set-up as in Example 6.10 in Section 6.5.2 is appropriate for this analysis, so the sixty-six months are the variables, and the seventeen series are the observations which we want to partition into two groups.

Gilmour and Koch (2006) applied Algorithm 13.2 to the scaled data. The dimension selector (10.40) in Section 10.8 yields $p^* = 3$ for both skewness and kurtosis (see Table 10.5 in Section 10.8). The skewness and kurtosis criteria result in the same cluster arrangements: the direct and indirect measures of the drug market (see Table 3.2 in Section 3.3).

For $p^* = 3$, the IC_1 scores and the resulting partitioning of the series are shown in the left subplot of Figure 13.2, with the observation or series number on the x-axis and the scores $\widetilde{\boldsymbol{v}}_1^{\mathsf{T}}\mathbf{X}(p)_i^{\diamond}$ on the y-axis. The two clusters are separated by the line $y = 0$. The series belonging to the indirect measures have non-negative $\widetilde{\boldsymbol{v}}_1^{\mathsf{T}}\mathbf{X}(p)^{\diamond}$ values. For comparison, I show two other partitions of these data. The middle panel shows the IC_1 partitioning obtained for $p = 6$, the authors' initial choice of p. The difference between the two cluster arrangements – shown in the left and middle panels in Figure 13.2 – is the allocation of series 2, *amphetamine possession offences*, which has jumped from the direct – with $\widetilde{\boldsymbol{v}}_1^{\mathsf{T}}\mathbf{X}(3)_2^{\diamond} < 0$ – to the indirect measures – with $\widetilde{\boldsymbol{v}}_1^{\mathsf{T}}\mathbf{X}(6)_2^{\diamond} > 0$ – when p changes from 3 to 6. The right subplot shows the cluster allocation arising from the PC_1 sign cluster rule (6.9) in Section 6.5.2. It is the same plot as that shown in the right panel of Figure 6.11 in the same section. The cluster allocation obtained from the PC_1 sign rule differs considerably from the other two.

Gilmour and Koch (2006) comment that series 2, *amphetamine possession offences*, fits better with the other amphetamine series, thus indicating a preference for the partitioning with the dimension selector (10.40) and $p^* = 3$ over the less well justified partition with $p = 6$. Gilmour and Koch (2006) refer to the twelve series of the larger cluster (arising from $p^* = 3$) as the 'direct measures' of the drug market because these series measure the most direct effects of the drugs.

The split into direct and indirect measures contrasts the split obtained with the PC_1 sign cluster rule, which focuses on heroin, as I commented in Example 6.10. The two ways of clustering the data, based on the PC_1 and IC_1 rules, respectively, provide different insight into the illicit drug market. If we are specifically interested in the drug heroin, then the PC_1 sign rule gives good information, whereas for a development of policies, the split into direct and indirect measures of the market is more useful. There is no right or wrong way to split these data; instead, the various analyses lead to a more complete picture and better interpretation of the different forces at work. ■

13.2.3 Test of Gaussianity

Our third example illustrates the use of independent component scores in a hypothesis test. In this case we want to test whether a random sample could come from a multivariate Gaussian distribution. Koch, Marron, and Chen (2005) proposed this test for a data *bank of* thirty-six *kidneys*. An initial principal component analysis of the data did not exhibit non-Gaussian structure in the data. Because of the small sample size, Koch, Marron, and Chen (2005) analysed the data further, as knowledge of the distribution of the data is required for the generation of simulated kidney shapes.

Independent component scores, which maximise a skewness or kurtosis criterion, are as independent and as non-Gaussian as possible given the particular approximation to the mutual information that is used (see Sections 10.4 and 10.6). IC_1 scores, calculated as in Algorithm 10.1 in Section 10.6.1, and their absolute skewness or kurtosis are therefore an appropriate tool for assessing the deviation from the Gaussian.

Example 13.4 We met the **data bank of kidneys** in Example 10.6 in Section 10.7.1. The data consist of thirty-six healthy-looking kidneys which are examined as part of a larger investigation. Coronal views of four of these kidneys and a schematic view of the fiducial points which characterise the shape of a kidney are shown in Figures 10.9 and 10.10 in Section 10.7.1.

The long-term aim of the analysis of kidneys is to generate a large number of synthetic medical images and shapes for segmentation performance characterisation. Generation of such synthetic shapes requires knowledge of the underlying distribution of the data bank of kidneys.

Example 10.6 explains the reduction of the 264 variables to seven principal components and shows, in Figure 10.9, the two most non-Gaussian IC_1 candidates in the form of smoothed histograms of the IC_1 scores. The histograms differ considerable from the Gaussian.

Visual impressions may be deceptive, but a hypothesis test provides information about the strength of the deviation from the Gaussian distribution. ■

The key idea is to use IC_1 scores and to define an appropriate test statistic which measures the deviation from the Gaussian distribution. Let $\mathbf{S}_{\bullet 1}$ be the first independent component score of the data, and let F_{S} be the distribution of $\mathbf{S}_{\bullet 1}$. We wish to test the hypotheses

$$H_0 : F_{\mathrm{S}} = \mathcal{N}(0,1) \qquad \text{versus} \qquad H_1 : F_{\mathrm{S}} \neq \mathcal{N}(0,1).$$

Suitable statistics for this test are

$$T_\rho = |\beta_\rho(\mathbf{S}_{\bullet 1})| \qquad \text{for } \rho = 3 \text{ or } 4,$$

where β_3 is the skewness of (9.4) in Section 9.3, and β_4 is the kurtosis of (9.6). Distributional properties of these statistics are not easy to derive. For this reason, Koch, Marron, and Chen (2005) simulate the distribution of T_ρ under the null hypothesis using the following algorithm.

Algorithm 13.3 ***An IC_1-Based Test of Gaussianity***

Let $\mathbb{X} = \begin{bmatrix}\mathbf{X}_1 \cdots \mathbf{X}_n\end{bmatrix}$ be $d \times n$ data with sample mean $\overline{\mathbf{X}}$ and sample covariance matrix S. Write $S = \widehat{\Gamma}\widehat{\Lambda}\widehat{\Gamma}^{\mathsf{T}}$ for the spectral decomposition. Fix $M > 0$ for the number of simulations and $K > 0$ for the number of iterations. Put $\rho = 3$ for the skewness criterion or $\rho = 4$ for the kurtosis criterion. Fix $p < d$, and calculate the p-white data $\mathbb{X}(p)^\diamond = \widehat{\Lambda}_p^{-1/2}\widehat{\Gamma}_p^{\mathsf{T}}(\mathbb{X} - \overline{\mathbf{X}})$.

Step 1. Let $\mathbf{S}^*$ be the first independent component score of $\mathbb{X}(p)^\diamond$, calculated with the FastICA options, step 1(a) or (b) of Algorithm 10.1 in Section 10.6.1. Put

$$t_\rho = \left|b_\rho(\mathbf{S}^*)\right| = \max_{\varrho=1,\ldots,K} \left|b_\rho(\mathbf{S}_\varrho)\right|, \tag{13.3}$$

where b_ρ is the sample skewness or kurtosis, as appropriate, and $\mathbf{S}_\varrho$ is the IC_1 score of $\mathbb{X}(p)^\diamond$, obtained in the ϱth iteration.

Step 2. For $j = 1, \ldots, M$, generate $\mathcal{Z}_j = \left[\mathbf{Z}_{j,1} \cdots \mathbf{Z}_{j,n}\right]$ with $\mathbf{Z}_{j,i} \sim \mathcal{N}(\mathbf{0}, \mathbf{I}_{p\times p})$. As in step 1, calculate the first independent component score $\mathbf{S}_j^{\mathcal{N}}$ of $\mathcal{Z}_j$, and put

$$\widehat{T}_{j,\rho} = \left|b_\rho(\mathbf{S}_j^{\mathcal{N}})\right|.$$

Step 3. Put $\mathcal{T}_\rho = \{\widehat{T}_{j,\rho} \,:\, j \le M\}$. Use $\mathcal{T}_\rho$ to calculate a p-value for the observed value t_ρ of (13.3), and make a decision. ∎

The test of Gaussianity depends on the dimension p of the white data. For the *kidney* data, there exists a natural choice for p which results from the experimental set-up. In other cases, p as in (10.40) in Section 10.8 provides a suitable value because it selects the relatively most non-Gaussian dimension.

Example 13.5 In Example 10.6 in Section 10.7.1, there are six IC_1 candidates for the **data bank of kidneys** after reduction to the seven-dimensional white data $\mathbb{X}(7)^\diamond$. The IC_1 corresponding to (13.3) arises from the kidney with observation number 32. This best IC_1 is displayed in the top-left panel of Figure 10.11 in Section 10.7.1 and has an absolute skewness value of 4.65.

The data bank of kidneys consist of $n = 36$ samples. Koch, Marron, and Chen (2005) used $K = 10$ and $M = 1{,}000$ for their test of Gaussianity. They generated multivariate normal data $\mathcal{Z}_\ell = \left[\mathbf{Z}_{\ell,1} \cdots \mathbf{Z}_{\ell,36}\right]$ with $\mathbf{Z}_{\ell,i} \sim \mathcal{N}(\mathbf{0}, \mathbf{I}_{7\times 7})$ for $i \le 36$ and $\ell \le M$. For the testing, they used the skewness and kurtosis criteria. Because the results based on the two criteria are almost identical, I only present the skewness results.

For each simulated data set $\mathcal{Z}_\ell$, the values $\widehat{T}_{\ell,3}$ are calculated as in step 2 of Algorithm 13.3. A smoothed histogram of the $\widehat{T}_{\ell,3}$ is shown in Figure 13.3. The histogram has a mode just above 1.5. The values of the $\widehat{T}_{\ell,3}$ are displayed at a random height and appear in the figure as the green point cloud. The red vertical line at 4.65 in the figure marks the maximum absolute skewness obtained from the kidney data. The p-value for $t_3 = 4.65$ is zero.

Although the skewness of Gaussian random vectors is zero, the absolute sample skewness over ten runs is positive. These positive values show that Independent Component Analysis finds some non-Gaussian directions even in Gaussian data. It is worth noting the green points at $x = 0$: for a total of sixteen of the 1,000 generated data sets, the maximum skewness is zero. For these sixteen cases, the FastICA algorithm failed to converge, that is, could not find a non-Gaussian solution, and the value was therefore set to zero.

The histogram and the p-value for $t_3 = 4.65$ show convincingly that the data are non-Gaussian. ∎

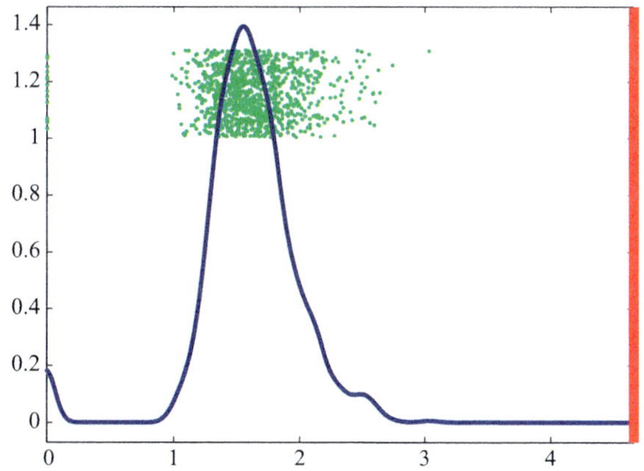

Figure 13.3 Smoothed histogram of the absolute skewness values $\widehat{T}_{j,3}$ and t_3 (red line) for the bank of kidneys.

Koch, Marron, and Chen (2005) used the result of the test to construct a transformation of the *kidney* data which makes the transformed data approximately normal and hence more appropriate for later synthetic generation of kidney shapes. Our interest is the test of non-Gaussianity based on the IC_1 scores. This test clearly demonstrate the non-Gaussian distribution of the data.

The three examples and analyses of Section 13.2 illustrate that non-Gaussian features may contain valuable information not available in the PC scores. In the last two examples, only the IC_1 score is used in the subsequent analysis. The independence of the components is not relevant in any of the three examples; instead, the non-Gaussian directions and scores in the data give rise to new information and interpretations. In each of the three examples, the choice p of the reduced PC data is important in two ways:

- to reduce the dimension and complexity of the data, and
- to reduce the dimension 'optimally'.

In the next section we consider ranking of variables as a way of reducing the dimension of the data and also derive criteria for an optimal dimension.

13.3 Variable Ranking and Statistical Learning

So far we have constructed and selected features based on Principal Component Analysis, Multidimensional Scaling and various versions of Independent Component Analysis, including Projection Pursuit. These are not the only ways one can derive *interesting* lower-dimensional subsets or features. In Section 4.8.3, I introduced the idea of variable ranking as a permutation of the variables according to some rule, with the aim of choosing the first few and best variables, where 'best' refers to the smallest classification error in Section 4.8.3.

We revisit variable ranking in the context of regression and classification and start with a regression framework. Bair et al. (2006) proposed variable ranking for univariate regression responses based on the ranking vector $\mathbf{s}$ of (4.52) in Section 4.8.3. They used the phrase 'Supervised Principal Component Analysis' for their linear regression method which integrates variable ranking and principal components into prediction of univariate regression responses.

Their variable ranking is informed by the responses, and hence the inclusion of 'supervised' in the name. Their method paved the way for further developments, which include

- a generalisation of Supervised Principal Component Analysis which
  - integrates the correlation between all variables into the ranking,
  - takes into account multivariate regression responses, and
  - selects the number of principal components in a data-driven way.
- an adaptation to Discriminant Analysis which
  - modifies the canonical correlation approach to labelled data,
  - combines variable ranking with Fisher's discriminant rule for HDLSS data, and
  - determines the best number of features.

In Section 13.3.1, I describe the variable ranking of Koch and Naito (2010), and in Section 13.3.2, I compare their approach to predicting linear regression responses with the supervised principal component approach of Bair et al. (2006). Section 13.3.3 shows how to adapt the variable ranking of Koch and Naito (2010) to a classification framework, and Section 13.3.4 explores properties of ranking vectors and discriminant rules for high-dimensional data. The results of Section 13.3.4 are based on Bickel and Levina (2004), Fan and Fan (2008) and Tamatani, Koch, and Naito (2012).

13.3.1 Variable Ranking with the Canonical Correlation Matrix C

In variable ranking for regression or discrimination, the inclusion of the responses or labels is essential in the selection of the 'most relevant' predictors. Section 4.8.3 introduced the ideas and defined, in (4.52), the ranking vector $\mathbf{s}$ of Bair et al. (2006). Because of this choice of letter, I refer to their ranking scheme the **s-ranking**. The appeal of this ranking scheme is its simplicity, but this simplicity is also a disadvantage because the $\mathbf{s}$-ranking does not take into account the correlation between the variables of $\mathbf{X}$.

Section 3.7 gave an interpretation of the canonical correlation matrix C of two random vectors $\mathbf{X}$ and $\mathbf{Y}$ as a natural generalisation of the correlation coefficient, and in Section 3.7.2 we considered how C can be used in Canonical Correlation Regression. This relationship is the key to the variable-ranking scheme of Koch and Naito (2010).

Variable Ranking of the Population Vector X. Consider d- and q-dimensional random vectors $\mathbf{X}$ and $\mathbf{Y}$ with means $\boldsymbol{\mu}_X$ and $\boldsymbol{\mu}_Y$, and assume that their covariance matrices Σ_X and Σ_Y have full rank. Recall from (3.35) in Section 3.7 that the canonical correlations matrix C links the random vectors $\mathbf{X}$ and $\mathbf{Y}$ via their sphered versions $\mathbf{X}_\Sigma$ and $\mathbf{Y}_\Sigma$, namely,

$$\mathbf{Y}_\Sigma = C^{\mathsf{T}}\mathbf{X}_\Sigma. \tag{13.4}$$

The $d \times q$ matrix C contains information about each pair of variables from $\mathbf{X}$ and $\mathbf{Y}$ but does not lead directly to a ranking of the variables of $\mathbf{X}$. The strongest relationship between $\mathbf{X}$ and $\mathbf{Y}$ is expressed by the first singular value υ_1 of C and the pair of left and right eigenvectors $\mathbf{p}_1$ and $\mathbf{q}_1$ which satisfy

$$C\mathbf{q}_1 = \upsilon_1\mathbf{p}_1 \tag{13.5}$$

(see (3.5) in Section 3.2), and the random variables $\mathbf{p}_1^{\mathsf{T}}\mathbf{X}_\Sigma$ and $\mathbf{q}_1^{\mathsf{T}}\mathbf{Y}_\Sigma$ have the strongest correlation – in absolute value – among all linear combinations $\mathbf{a}^{\mathsf{T}}\mathbf{X}_\Sigma$ and $\mathbf{b}^{\mathsf{T}}\mathbf{Y}_\Sigma$. For this reason, the left eigenvector $\mathbf{p}_1$ is a candidate for a **ranking vector** in the sense of Definition 4.23 in Section 4.8.3. We could use $\mathbf{p}_1$ to rank the sphered vector $\mathbf{X}_\Sigma$, but typically we want to rank the variables of $\mathbf{X}$ and thus make use of (13.4) to obtain

$$\mathbf{Y} - \boldsymbol{\mu}_Y = \Sigma_Y^{1/2} C^\top \Sigma_X^{-1/2} (\mathbf{X} - \boldsymbol{\mu}_X) = G^\top (\mathbf{X} - \boldsymbol{\mu}_X),$$

where $G = \Sigma_X^{-1/2} C \Sigma_Y^{1/2} = \Sigma_X^{-1} \Sigma_{XY}$. If $\mathbf{p}_1$ and $\mathbf{q}_1$ are a pair of vectors satisfying (13.5), then the first left and right canonical correlation transforms $\boldsymbol{\varphi}_1$ and $\boldsymbol{\psi}_1$ of (3.8) in Section 3.2 satisfy

$$G\boldsymbol{\psi}_1 = \upsilon_1 \boldsymbol{\varphi}_1 \qquad \text{and} \qquad \boldsymbol{\varphi}_1^\top (\mathbf{X} - \boldsymbol{\mu}_X) = \mathbf{p}_1^\top \Sigma_X^{-1/2} (\mathbf{X} - \boldsymbol{\mu}_X). \tag{13.6}$$

This last equality motivates the use of $\boldsymbol{\varphi}_1$ as a ranking vector for $\mathbf{X}$. This $\boldsymbol{\varphi}_1$ is not a unit vector, but for ranking the variables of $\mathbf{X}$, the length of $\boldsymbol{\varphi}_1$ does not matter, as we are only interested in the relative size of the entries of $\boldsymbol{\varphi}_1$.

Variable Ranking of the Data $\mathbb{X}$. For data $\mathbb{X}$ and $\mathbb{Y}$, the derivation of the ranking vector is similar to the population case if the sample covariance matrices have full rank. Assume that S_X and S_Y are non-singular. Let $\widehat{C}$ be the sample canonical correlation matrix, and let $\widehat{\mathbf{q}}_1$ be the first right eigenvectors of $\widehat{C}$. Then

$$\widehat{\boldsymbol{\varphi}}_1 = \frac{1}{\widehat{\upsilon}_1} S_X^{-1/2} \widehat{C} \widehat{\mathbf{q}}_1 \tag{13.7}$$

is a ranking vector for $\mathbb{X}$. It is the first sample canonical correlation transform and is a sample version of (13.6).

If S_X does not have full rank, then the right-hand side of (13.7) cannot be calculated. Following Koch and Naito (2010), we assume that $\mathbb{X}$ and $\mathbb{Y}$ are centred. Let $r \leq \min(d,n)$ be the rank of $\mathbb{X}$ and, as in (5.18) in Section 5.5, write

$$\mathbb{X} = \widehat{\Gamma} \widehat{\Delta} \widehat{L}^\top$$

for the singular value decomposition of $\mathbb{X}$, where $\widehat{\Delta} = \widehat{\Lambda}^{1/2}$ is an $r \times r$ matrix, and $\widehat{\Lambda}$ is the diagonal matrix obtained from the spectral decomposition of $(n-1)S_X$, $\widehat{\Gamma}$ is of size $d \times r$, and $\widehat{L}$ is of size $n \times r$, and both $\widehat{\Gamma}$ and $\widehat{L}$ are r-orthogonal. From the definition of $\widehat{C}$, it follows that

$$\widehat{C} = \widehat{\Gamma} \widehat{L}^\top \mathbb{Y}^\top (\mathbb{Y}\mathbb{Y}^\top)^{-1/2}, \tag{13.8}$$

and combining (13.8) with (13.7), we obtain the ranking vector

$$\widehat{\mathbf{b}}_1 \equiv \widehat{\boldsymbol{\varphi}}_1 = \frac{1}{\widehat{\upsilon}_1} \widehat{\Gamma} \widehat{\Delta}^{-1} \widehat{L}^\top \mathbb{Y}^\top (\mathbb{Y}\mathbb{Y}^\top)^{-1/2} \widehat{\mathbf{q}}_1. \tag{13.9}$$

Koch and Naito (2010) also considered

$$\widehat{\mathbf{b}}_2 = \frac{1}{\widehat{\upsilon}_1} \widehat{\Gamma} \widehat{L}^\top \mathbb{Y}^\top (\mathbb{Y}\mathbb{Y}^\top)^{-1/2} \widehat{\mathbf{q}}_1 \tag{13.10}$$

as a ranking vector. We recognise that $\widehat{\mathbf{b}}_2 = \widehat{\mathbf{p}}_1$, the sample version of the left eigenvector $\mathbf{p}_1$ of C. Although $\mathbf{p}_1$ is a ranking vector for $\mathbf{X}_\Sigma$, Koch and Naito observed that $\widehat{\mathbf{b}}_2$ is suitable as a ranking vector for sphered and raw data, and they suggested using both $\widehat{\mathbf{b}}_1$ and $\widehat{\mathbf{b}}_2$. For notational convenience, I will refer to ranking with $\widehat{\mathbf{b}}_1$ or $\widehat{\mathbf{b}}_2$ simply as **b-ranking** and only refer to $\widehat{\mathbf{b}}_1$ and $\widehat{\mathbf{b}}_2$ individually when necessary.

We derive (13.8) to (13.10) in the Problems at the end of Part III. Note that (13.8) to (13.10) do not depend on whether $d < n$ or $d > n$, so they apply to any data $\mathbb{X}$. The factor $1/\widehat{\upsilon}_1$ is not important in ranking and can be omitted in the **b**-ranking schemes.

Different ranking vectors will lead to differently ranked data, and similarly, the optimal m^* of step 4 in Algorithm 4.3 in Section 4.8.3 will depend on the ranking method. The **s**-ranking scheme is related to ranking based on the t-test statistic which uses the ranking vector **d** of (4.50) in Section 4.8.3.

A comparison between the ranking vectors **s** of Bair et al. (2006) and **b** of Koch and Naito (2010) shows the computational simplicity of the **s**-ranking. Koch and Naito point out that for a univariate setting, $\widehat{\mathbf{b}}_1$ is an unbiased estimator for the univariate slope regression coefficient β. The relationship with the regression coefficients is the reason for the notation: $\widehat{\mathbf{b}}_1$ is an estimator for $\boldsymbol{\beta}$. Further, the ranking vectors **b** apply to multivariate responses $\mathbb{Y}$, whereas ranking with the simpler **s**-ranking does not extend beyond univariate responses.

We compare the **s**- and **b**-ranking in a regression setting at the end of Section 13.3.2.

13.3.2 Prediction with a Selected Number of Principal Components

The *supervised* part of Supervised Principal Component Analysis has been accomplished in the preceding section, and we now follow Koch and Naito (2010) in their regression prediction for high-dimensional data. As in Section 4.8.3, 'p-ranked data' refers to the first p rows of the ranked data and hence to data of size $p \times n$.

Algorithm 13.4 summarises the main steps of the approach of Koch and Naito (2010). We explore differences between their approach and that of Bair et al. (2006), and I illustrate the performance of both methods with an example from Koch and Naito (2010), who considered the high-dimensional *breast tumour* data of Example 2.15 in Section 2.6.2.

Algorithm 13.4 ***Prediction with a Selected Number of Principal Components***

Let $\mathbb{X}$ be d-dimensional centred data, and let $\mathbb{Y}$ be the q-dimensional regression responses. Let r be the rank of $\mathbb{X}$. Let $\mathcal{E}$ be an error measure.

Step 1. Let **b** be a ranking vector, and let $\mathbb{X}_{\mathrm{B}}$ be the d-dimensional data ranked with **b**.

Step 2. For $2 \leq p \leq r$, consider the p-ranked data $\mathbb{X}_p$, the first p rows of $\mathbb{X}_{\mathrm{B}}$, as in (4.49) in Section 4.8.3, and let

$$S_p = \frac{1}{n-1}\mathbb{X}_p\mathbb{X}_p^{\mathsf{T}} = \widehat{\Gamma}_p\widehat{\Lambda}_p\widehat{\Gamma}_p^{\mathsf{T}}$$

be the sample covariance matrix of $\mathbb{X}_p$, given also in its spectral decomposition.

Step 3. Let

$$\mathbb{X}_p^{\diamond} = \widehat{\Lambda}_p^{-1/2}\widehat{\Gamma}_p^{\mathsf{T}}\mathbb{X}_p$$

be the sphered p-dimensional PC data. For $\nu \leq p$, let $\widehat{I}(\nu)$ be the kurtosis dimension selector so $\rho = 4$ in (10.39) in Section 10.8. Define the best dimension p^* as in

(10.40) in Section 10.8:

$$p^* = \underset{2\le \nu\le p}{\operatorname{argmax}} \widehat{I}(\nu). \tag{13.11}$$

Let $\widehat{\Gamma}_{p*}$ be the first p^* columns of $\widehat{\Gamma}_p$, and put $\mathbb{W}_{p*} = \widehat{\Gamma}_{p*}^{\mathsf{T}} \mathbb{X}_p$.

Step 4. Let $\widehat{\Lambda}_{p*}$ be the diagonal $p^* \times p^*$ matrix which contains the first p^* eigenvalues of $\widehat{\Lambda}_p$. The estimators $\widehat{B}_p$ for the matrix of regression coefficients B and $\widehat{\mathbb{Y}}_p$ for $\mathbb{Y}$ are

$$\widehat{B}_p = \frac{1}{n-1}\widehat{\Lambda}_{p*}^{-1}\mathbb{W}_{p*}\mathbb{Y}^{\mathsf{T}} \qquad \text{and} \qquad \widehat{\mathbb{Y}}_p = \frac{1}{n-1}\mathbb{Y}\mathbb{W}_{p*}^{\diamond\,\mathsf{T}}\mathbb{W}_{p*}^{\diamond}, \tag{13.12}$$

with $\mathbb{W}_{p*}^{\diamond} = \widehat{\Lambda}_{p*}^{-1/2}\mathbb{W}_{p*}$. Use $\mathcal{E}$ to calculate the error arising from $\mathbb{Y}$ and $\widehat{\mathbb{Y}}_p$, and call it $\mathcal{E}_p$.

Step 5. Put $p = p+1$, and repeat steps 2 to 5.

Step 6. Find $\mathfrak{p}$ such that $\mathfrak{p} = \operatorname{argmin}_{2\le p\le r} \mathcal{E}_p$, then $\mathfrak{p}^*$, calculated as in (13.11) for $p = \mathfrak{p}$, is the optimal number of principal components for the $\mathfrak{p}$-ranked data $\mathbb{X}_{\mathfrak{p}}$. ■

The optimal number of variables or dimensions $\mathfrak{p}^*$ is data-driven, as in Algorithm 4.3 in Section 4.8.3, and depends on the ranking vector. Typically, we consider the ranking vectors $\widehat{\mathbf{b}}_1$ of (13.9), $\widehat{\mathbf{b}}_2$ of (13.10), or $\mathbf{s}$ of (4.52) in Section 4.8.3, but others can be used, too. Once the variables are ranked, the ranking vector $\mathbf{b}$ plays no further role. For this reason, I have omitted reference to the ranked data $\mathbb{X}_{\mathrm{B}}$ and only refer to the p-ranked data in the definition of S_p, $\mathbb{X}_p^{\diamond}$ and so on in steps 3 to 6.

Remarks on Algorithm 13.4

Step 3. Koch and Naito observed that fewer than p combinations of the original variables typically result in better prediction. Further, the best kurtosis dimension (see (10.40)) is greater than the best skewness dimension and typically yields better results.

It is natural to consider p^* independent components as the predictors instead of p^* PCs. Koch and Naito (2010) point out that there is no advantage in using ICs as predictors. To see why, consider the sphered PC data $\mathbb{W}_{p*}^{\diamond}$ of (13.12), and let E be an orthogonal matrix such that $E\mathbb{W}_{p*}^{\diamond}$ are as independent as possible. Put $\widetilde{\mathbb{X}} = E\mathbb{W}_{p*}^{\diamond}$. From the orthogonality of E and (13.14), it follows that

$$\widehat{B}_{\mathrm{IC}} = \left[E\mathbb{W}_{p*}^{\diamond}\mathbb{W}_{p*}^{\diamond\,\mathsf{T}}E^{\mathsf{T}}\right]^{-1} E\mathbb{W}_{p*}^{\diamond}\mathbb{Y}^{\mathsf{T}} = \frac{1}{n-1}E\mathbb{W}_{p*}^{\diamond}\mathbb{Y}^{\mathsf{T}}.$$

The last identity leads to

$$\widehat{\mathbb{Y}}^{\mathrm{IC}} = \frac{1}{n-1}\mathbb{Y}\mathbb{W}_{p*}^{\diamond\,\mathsf{T}}E^{\mathsf{T}}E\mathbb{W}_{p*}^{\diamond} = \frac{1}{n-1}\mathbb{Y}\mathbb{W}_{p*}^{\diamond\,\mathsf{T}}\mathbb{W}_{p*}^{\diamond}$$

because E is an orthogonal $p^* \times p^*$ matrix. A comparison with (13.12) shows that the two expressions agree. Thus, the IC responses are identical to those obtained from p^* PCs but require computation of the matrix E for no benefit.

Step 4. The algorithm uses the least-squares expressions (3.40) in Section 3.7.2 for $\widehat{B}_p$ and $\widehat{\mathbb{Y}}_p$ with $\widetilde{\mathbb{X}} = \mathbb{W}_{p*}$. It follows that

$$\mathbb{W}_{p*}\mathbb{W}_{p*}^{\mathsf{T}} = (n-1)\widehat{\Lambda}_{p*} \qquad \text{and hence} \qquad \widehat{B}_p = \frac{1}{n-1}\widehat{\Lambda}_{p*}^{-1}\mathbb{W}_{p*}\mathbb{Y}^{\mathsf{T}}. \tag{13.13}$$

We may interpret the matrix $\mathbb{W}_{p^*}^{\diamond\mathsf{T}}\mathbb{W}_{p^*}^{\diamond}$ in (13.12) as the $Q_{\langle n\rangle}$-matrix of (5.17) in Section 5.5. The corresponding dual $Q^{\langle d\rangle}$-matrix is

$$\mathbb{W}_{p^*}^{\diamond}\mathbb{W}_{p^*}^{\diamond\,\mathsf{T}} = (n-1)\mathbf{I}_{p^*\times p^*}. \tag{13.14}$$

For $p_1 < p_2$, the p_1^* columns of $\mathbb{X}_{p_1}^{\diamond}$ are not in general contained in the p_2^* columns of $\mathbb{X}_{p_2}^{\diamond}$, and consequently, for each p, different predictors are used in the estimation of the regression coefficients.

For univariate responses $\mathbf{Y}$ and $p^* = 1$, (13.12) reduces to the estimates obtained in Bair et al. (2006) – apart from the different ranking – namely,

$$\widehat{\mathbf{Y}}^{\mathrm{PC1}} = \frac{1}{(n-1)\widehat{\lambda}_1}\mathbf{Y}\mathbb{X}_{S,p}^{\mathsf{T}}\widehat{\boldsymbol{\eta}}_1\widehat{\boldsymbol{\eta}}_1^{\mathsf{T}}\mathbb{X}_{S,p}, \tag{13.15}$$

where $\mathbb{X}_{S,p}$ are the p-dimensional data, ranked with the $\mathbf{s}$-ranking of Bair et al. (2006), and $(\widehat{\lambda}_1, \widehat{\boldsymbol{\eta}}_1)$ is the first eigenvalue-eigenvector pair of the sample covariance matrix of $\mathbb{X}_{S,p}$.

Step 6. Possible error measures for regression are described in (9.17) and (9.18) in Section 9.5. As in Algorithm 4.3 in Section 4.8.3, the best $\mathfrak{p}$ is that which minimises the error $\mathcal{E}_p$. Once we have determined the optimal PC dimension $\mathfrak{p}$, we find the IC dimension selector $\mathfrak{p}^*$ of step 3 with $p = \mathfrak{p}$, and this integer $\mathfrak{p}^*$ is the number of components used in the optimal prediction.

We conclude this section with an example from Koch and Naito (2010), in which they compared four PC-based prediction approaches including that of Bair et al. (2006).

Example 13.6 We continue with the **breast tumour gene expression** data which I introduced in Example 2.15 in Section 2.6.2. The data consist of seventy-eight patients, the observations, and 24,481 genes, which are the variables. A subset of the data, consisting of 4,751 genes, is available separately. In Example 2.15 we considered this subset only. The data have two types of responses: binary labels, which show whether a patient survived five years to metastasis, and actual survival times in months. In the analysis that follows, I use the actual survival times as the univariate regression responses and all 24,481 variables.

Following Koch and Naito (2010), we use three different ranking schemes in step 1 of Algorithm 13.4: ranking with $\widehat{\mathbf{b}}_1$ of (13.9), $\widehat{\mathbf{b}}_2$ of (13.10) and $\mathbf{s}$ of (4.52) in Section 4.8.3. The third ranking vector is that of Bair et al. (2006).

Bair et al. (2006) used PC_1 instead of calculating the reduced dimension p^* in step 3 of Algorithm 13.4. To study the effect of this choice, I consider their ranking scheme both with p^* as in (13.11) and with PC_1 only. I refer to the four analyses with the following names

1. KN-b1: ranking with $\widehat{\mathbf{b}}_1$,
2. KN-b2: ranking with $\widehat{\mathbf{b}}_2$,
3. BHPT-p*: ranking with $\mathbf{s}$ of (4.52), and
4. BHPT-1: ranking with $\mathbf{s}$ and using PC_1 instead of p^* in step 3.

Table 13.2 *Percentage of Best Common Genes for Each Ranking*

Top	1,000	200	50	1,000	200	50
		$\widehat{\mathbf{b}}_1$			$\widehat{\mathbf{b}}_2$	
$\widehat{\mathbf{b}}_2$	58.6	55	46	—	—	—
$\mathbf{s}$	17.7	9.5	8	33.8	15	14

Table 13.3 *Errors of Estimated Survival Times and Optimal Number of Variables for the Breast Tumour Data*

	KN-b1	KN-b2	BHPT-p*	BHPT-1
$\mathfrak{p}$	50	40	35	20
$\mathfrak{p}^*$	11	12	13	—
$\mathcal{E}(\mathfrak{p})$	13.84	12.23	17.80	20.56

The three different ranking schemes result in different permutations of the variables. The degree of consensus among the three schemes is summarised in Table 13.2 as percentages for the top 1,000, 200 and 50 genes that are found with each of the three ranking schemes. For example, 46 (per cent), refers to twenty-three of the fifty top variables on which the $\widehat{\mathbf{b}}_1$ and the $\widehat{\mathbf{b}}_2$ rankings agree. The table shows that the consensus between the $\widehat{\mathbf{b}}_1$ and $\widehat{\mathbf{b}}_2$ rankings is about 50 per cent, whereas the $\mathbf{s}$-ranking selects different genes from the other two ranking schemes; the agreement between $\widehat{\mathbf{b}}_1$ and $\mathbf{s}$, in particular, is very low. The table shows the degree of consensus but cannot make any assertions about the interest the chosen genes might have for the biologist.

The four analyses start with a principal component analysis, which reduces the number of possible principal components to seventy-eight, the rank of the sample covariance matrix. Koch and Naito (2010) calculated $\widehat{\mathbf{Y}}$ as in (13.12) for the first three analyses and as in (13.15) for BHPT-1 and then used the error measure $\mathcal{E} = (1/n)\sum_i |\widehat{Y}_i - Y_i|$ for each dimension p. The values $\mathfrak{p}$, $\mathfrak{p}^*$ and the error at $\mathfrak{p}$ are shown in Table 13.3. All four analyses result in $\mathfrak{p} \leq 50$ for step 4 of the algorithm, and the best dimension $\mathfrak{p}^*$ is much smaller than $\mathfrak{p}$. For BHPT-1, I have omitted the value $\mathfrak{p}^*$ in column four of the table because it is not relevant, as PC_1 is chosen for each $\mathfrak{p}$ and, by default, $\mathfrak{p}^* = 1$.

Figure 13.4 shows performance results of all four analyses with the error $\mathcal{E}(p)$ on the y-axis, and dimension p on the x-axis. Figure 13.4 shows that BHPT-1 (*black*) is initially comparable with KN-b1 (*red*) and KN-b2 (*blue*) until it reaches its minimum error at $\mathfrak{p} = 20$, and for $p > 20$, the error increases again. In contrast, KN-b1 and KN-b2 continue to decrease for $p > 20$. KN-b1 (*red*) shows a sharp improvement in performance at about $p = 20$ but then reaches its minimum error for $\mathfrak{p} = 50$. The analysis BHPT-p* performs better than the other three for small values of p, namely, up to about $p = 20$, but then flattens out, apart from a dip at $p = 35$, where it reaches its minimum error. Overall, KN-b1 and KN-b2 have lower errors and a smaller number of variables than the other two, and the use of PC_1 only in BHPT-1 results in a poorer performance.

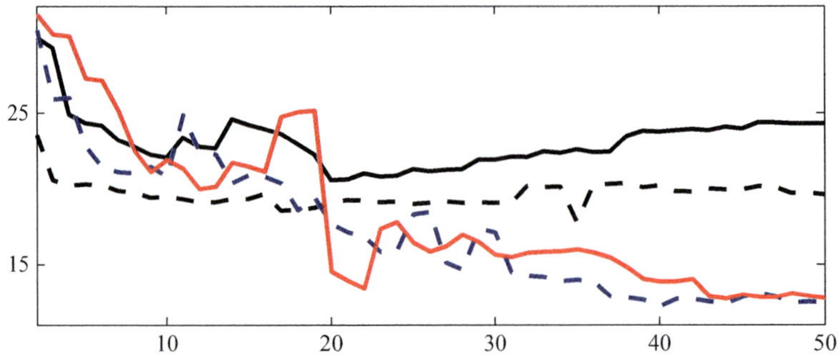

Figure 13.4 Error versus dimension: KN-b1 (*red*), KN -b2 (*blue*), BHPT-p* (*black dashed*) and BHPT-1 (*black*) from Example 13.6.

The three analyses, which calculate p^* as in step 3 of Algorithm 13.4, have similar 'best' $\mathfrak{p}^*$ values, but their optimal values $\mathfrak{p}$ and their minimum errors differ: KN-b1 has the smallest value for $\mathfrak{p}^*$, so it results in the most parsimonious model, whereas KN-b2 has the smallest error, and both perform better than the corresponding analysis based on the **s**-ranking.

The analyses highlight the compromise between computational efficiency and accuracy. Clearly, BHPT-1 is the easiest to compute, but KN-b1 or KN-b2 give higher accuracy. ■

For the HDLSS *breast tumour* data, the ranking with the canonical correlation matrix $\widehat{C}$ rather than the marginal ranking of individual genes has led to smaller errors. The example also shows that a better performance is achieved when more than one predictor is used in the estimation of the responses, that is, when the p-ranked data are summarised in more than the first PC.

13.3.3 Variable Ranking for Discriminant Analysis Based on C

In Section 13.3.1 the canonical correlation matrix C is the vehicle for constructing a ranking vector for regression data with multivariate predictors and responses. In Discriminant Analysis, the continuous response variables are replaced by discrete labels, and strictly speaking, the idea of the correlation between the predictors and responses can no longer be captured by the correlation coefficient. We now explore how we can modify the canonical correlation matrix in order to construct a ranking vector for the predictor variables in a Discriminant Analysis framework.

Section 4.2 introduced scalar-valued and vector-valued labels; so far we mostly used the scalar-valued labels with values $1,\ldots,\kappa$ for data belonging to κ classes. As we shall see, vector-valued labels, as in (4.3), are the right objects for variable ranking, and an adaptation of the canonical correlation matrix to such labels leads to an easily interpretable ranking vector. I begin with the general case of κ classes and then describe the ideas and theoretical developments of Tamatani, Koch, and Naito (2012), who considered the case of two classes and normal random vectors. For an extension of their approach and results to $\kappa \geq 2$ classes, see Tamatani, Naito, and Koch (2013).

For classification into κ classes, the label $\mathbf{Y}$ of a datum $\mathbf{X}$ is given by (4.3) in Section 4.2. Centering the vector-valued labels does not make sense; instead, the **DA(-adjusted)**

canonical correlation matrix $C_{\rm DA}$ is

$$C_{\rm DA} = \Sigma^{-1/2}\mathbb{E}\left[(\mathbf{X}-\boldsymbol{\mu})\mathbf{Y}^{\mathsf{T}}\right]\left[\mathbb{E}(\mathbf{Y}\mathbf{Y}^{\mathsf{T}})\right]^{-1/2}, \tag{13.16}$$

where $\mathbf{X} \sim (\boldsymbol{\mu}, \Sigma)$, and Σ is assumed to be invertible. Let $\mathbb{X} = \left[\mathbf{X}_1 \cdots \mathbf{X}_n\right]$ be $d \times n$ data from κ classes. Let $(\overline{\mathbf{X}}, S)$ be the sample mean and sample covariance matrix of $\mathbb{X}$, and assume that S is invertible. Let $\mathbb{Y}$ be the $\kappa \times n$ matrix of labels. The **sample DA(-adjusted) canonical correlation matrix** $\widehat{C}_{\rm DA}$ is

$$\begin{aligned}\widehat{C}_{\rm DA} &= \frac{1}{\sqrt{n-1}} S^{-1/2}\left[(\mathbb{X}-\overline{\mathbf{X}})\mathbb{Y}^{\mathsf{T}}\right](\mathbb{Y}\mathbb{Y}^{\mathsf{T}})^{-1/2}\\ &= \left[(\mathbb{X}-\overline{\mathbf{X}})(\mathbb{X}-\overline{\mathbf{X}})^{\mathsf{T}}\right]^{-1/2}\left[(\mathbb{X}-\overline{\mathbf{X}})\mathbb{Y}^{\mathsf{T}}\right](\mathbb{Y}\mathbb{Y}^{\mathsf{T}})^{-1/2}.\end{aligned} \tag{13.17}$$

In Section 13.3.1, we focused on the first left eigenvector of $\widehat{C}$. The first left eigenvector of $\widehat{C}_{\rm DA}$ is now the object of interest.

Theorem 13.1 *Let $\mathbb{X}$ be $d \times n$ labelled data from κ classes such that n_ν random vectors are from the νth class and $n = \sum n_\nu$. Let $\overline{\mathbf{X}}_\nu$ be the sample mean of the νth class. Let $(\overline{\mathbf{X}}, S)$ be the sample mean and sample covariance matrix of $\mathbb{X}$, and assume that S is invertible. Let $\mathbb{Y}$ be the $\kappa \times n$ matrix of vector-valued labels. Consider $\widehat{C}_{DA}$ of (13.17). Then*

$$\mathbb{Y}\mathbb{Y}^{\mathsf{T}} = \begin{bmatrix} n_1 & 0 & 0 & 0\\ 0 & n_2 & 0 & 0\\ & & \ddots & \\ 0 & 0 & 0 & n_\kappa\end{bmatrix}, \tag{13.18}$$

and

$$\widehat{R}^{[1]}_{DA} \equiv \widehat{C}_{DA}\widehat{C}^{\mathsf{T}}_{DA} = \frac{1}{n-1} S^{-1/2} B_{\bar{X}} S^{-1/2},$$

with

$$B_{\bar{X}} \equiv \left[(\mathbb{X}-\overline{\mathbf{X}})\mathbb{Y}^{\mathsf{T}}\right]\left(\mathbb{Y}\mathbb{Y}^{\mathsf{T}}\right)^{-1}\left[\mathbb{Y}(\mathbb{X}-\overline{\mathbf{X}})^{\mathsf{T}}\right] = \left[\sum_{\nu=1}^{\kappa} n_\nu(\overline{\mathbf{X}}_\nu-\overline{\mathbf{X}})(\overline{\mathbf{X}}_\nu-\overline{\mathbf{X}})^{\mathsf{T}}\right]. \tag{13.19}$$

Further, if $\widehat{\mathbf{p}}_{DA}$ is the eigenvector of $\widehat{R}^{[1]}_{DA}$ which corresponds to the largest eigenvalue $\widehat{\upsilon}^2$ of $\widehat{R}^{[1]}_{DA}$, and $\widehat{\mathbf{b}}_{DA} = S^{-1/2}\widehat{\mathbf{p}}_{DA}$, then

$$B_{\bar{X}}\widehat{\mathbf{b}}_{DA} = \widehat{\upsilon}^2 S\widehat{\mathbf{b}}_{DA} \qquad \textit{or equivalently} \qquad S^{-1}B_{\bar{X}}\widehat{\mathbf{b}}_{DA} = \widehat{\upsilon}^2\widehat{\mathbf{b}}_{DA}. \tag{13.20}$$

I use the notation $\widehat{R}^{[1]}_{\rm DA}$ in the theorem in analogy with (3.12) in Section 3.3. A proof of the theorem is deferred to the Problems at the end of Part III.

Because the matrix $C_{\rm DA}$ differs from a canonical correlation matrix, I have listed explicitly the terms and expressions in the theorem that lead to interesting interpretations:

1. The matrix (13.18) is a diagonal $\kappa \times \kappa$ matrix whose non-zero entries are the number of observations in each class. We can interpret $\mathbb{Y}\mathbb{Y}^{\mathsf{T}}/(n-1)$ as an 'uncentred sample covariance matrix' whose non-zero entries are the proportions in each class, and thus $\mathbb{Y}\mathbb{Y}^{\mathsf{T}}$ is natural, as it summarises relevant information.

2. The matrix $B_{\bar{X}}$ of (13.19) is essentially the between-class sample covariance matrix $\widehat{B}$ of Corollary 4.9 in Section 4.3.2; however, the subscript $_{\bar{X}}$ in $B_{\bar{X}}$ reminds us that the average of all observations is used in $B_{\bar{X}}$, rather than the class averages which we encounter in $\widehat{B}$.
3. The equalities (13.20) remind us of Fisher's generalised eigenvalue problem (4.8) in Section 4.3.1. The covariance matrix S is used in (13.20), whereas (4.8) is based on the sum of the within-class covariance matrices $\widehat{W}$. The two variance terms are closely related.
4. Formally, $\widehat{\mathbf{b}}_{\mathrm{DA}}$ is the first left canonical correlation transform of $\widehat{C}_{\mathrm{DA}}$, and we thus may interpret it as a ranking vector, in analogy with $\widehat{\mathbf{b}}_1$ of (13.9) in Section 13.3.1. In (13.20), $\widehat{\mathbf{b}}_{\mathrm{DA}}$ plays the role of Fisher's direction vector, called $\widehat{\eta}$ in Corollary 4.9 in Section 4.3.2. Thus, in a Discriminant Analysis setting, $\widehat{\mathbf{b}}_{\mathrm{DA}}$ plays the dual role of ranking vector and direction vector for Fisher's discriminant rule.

In addition to $\widehat{\mathbf{b}}_{\mathrm{DA}}$, we use $\widehat{\mathbf{p}}_{\mathrm{DA}} = S^{1/2}\widehat{\mathbf{b}}_{\mathrm{DA}}$ as a ranking vector, similar to the two ranking vectors of the regression framework in Section 13.3.1.

For the remainder of this section we consider two-class problems, the setting that is used in Bickel and Levina (2004), Fan and Fan (2008), and Tamatani, Koch, and Naito (2012). In this case, the population C_{DA}, the sample $\widehat{C}_{\mathrm{DA}}$ and their eigenvectors have simpler expressions. Corollary 13.2 gives these expressions for the population.

Corollary 13.2 [Tamatani, Koch, and Naito (2012)] *Let $\mathbf{X}$ be a d-dimensional random vector which belongs to class C_1 with probability π and to class C_2 with probability $1-\pi$. Assume that the two classes have different class means $\boldsymbol{\mu}_1 \neq \boldsymbol{\mu}_2$ but share a common within-class covariance matrix Σ. Assume further that Σ is invertible. Let $\mathbf{Y}$ be the vector-valued label of $\mathbf{X}$. Then C_{DA} as in (13.16) and its first left eigenvector $\mathbf{p}_{DA}$ are*

$$C_{DA} = \sqrt{\pi(1-\pi)}\,\Sigma^{-1/2}(\boldsymbol{\mu}_1 - \boldsymbol{\mu}_2)\begin{bmatrix}\sqrt{1-\pi} & -\sqrt{\pi}\end{bmatrix}$$

and

$$\mathbf{p}_{DA} = \frac{\Sigma^{-1/2}(\boldsymbol{\mu}_1 - \boldsymbol{\mu}_2)}{\left\|\Sigma^{-1/2}(\boldsymbol{\mu}_1 - \boldsymbol{\mu}_2)\right\|}.$$

The expression for $\mathbf{p}_{\mathrm{DA}}$ follows from the uniqueness of the eigenvector decomposition because

$$C_{\mathrm{DA}}C_{\mathrm{DA}}^{\mathsf{T}} = \pi(1-\pi)\Sigma^{-1/2}(\boldsymbol{\mu}_1 - \boldsymbol{\mu}_2)(\boldsymbol{\mu}_1 - \boldsymbol{\mu}_2)^{\mathsf{T}}\Sigma^{-1/2},$$

and $\mathbf{b}_{\mathrm{DA}} = c\Sigma^{-1/2}\mathbf{p}_{\mathrm{DA}} = \Sigma^{-1}(\boldsymbol{\mu}_1 - \boldsymbol{\mu}_2)$, where $c = \left\|\Sigma^{-1/2}(\boldsymbol{\mu}_1 - \boldsymbol{\mu}_2)\right\|$. Furthermore, the linear decision function h_β of (4.17) in Section 4.3.3 with $\boldsymbol{\beta} = \mathbf{b}_{\mathrm{DA}}$ results in

$$h(\mathbf{X}) = \left[\mathbf{X} - \frac{1}{2}(\boldsymbol{\mu}_1 + \boldsymbol{\mu}_2)\right]^{\mathsf{T}}\mathbf{b}_{\mathrm{DA}} = \left[\mathbf{X} - \frac{1}{2}(\boldsymbol{\mu}_1 + \boldsymbol{\mu}_2)\right]^{\mathsf{T}}\Sigma^{-1}(\boldsymbol{\mu}_1 - \boldsymbol{\mu}_2). \tag{13.21}$$

Because $\mathbf{b}_{\mathrm{DA}}$ has norm one, we may interpret it as the discriminant direction in Fisher's rule or, equivalently, as the direction vector in the normal linear rule.

For two-class problems and data, the ranking vector $\widehat{\mathbf{p}}_{\mathrm{DA}}$ of Theorem 13.1 reduces to

$$\widehat{\mathbf{p}}_{\mathrm{DA}} = S^{-1/2}(\overline{\mathbf{X}}_1 - \overline{\mathbf{X}}_2)/\left\|S^{-1/2}(\overline{\mathbf{X}}_1 - \overline{\mathbf{X}}_2)\right\|.$$

Table 13.4 *C-Matrix and Related Quantities for Two-Class Problems Based on the Diagonal* $D = \Sigma_{diag}$ *of the Common Covariance Matrix*

	Population	Data
C	$\sqrt{\pi(1-\pi)}\, D^{-1/2}(\boldsymbol{\mu}_1 - \boldsymbol{\mu}_2)\begin{bmatrix}\sqrt{1-\pi}\\ -\sqrt{\pi}\end{bmatrix}$	$\left(\sqrt{n_1 n_2}/n\right)\widehat{D}^{-1/2}(\overline{\mathbf{X}}_1 - \overline{\mathbf{X}}_2)\begin{bmatrix}\sqrt{n_2/n}\\ -\sqrt{n_1/n}\end{bmatrix}$
$\mathbf{p}$	$c_p^{-1} D^{-1/2}(\boldsymbol{\mu}_1 - \boldsymbol{\mu}_2)$	$c_{\widehat{p}}^{-1}\widehat{D}^{-1/2}(\overline{\mathbf{X}}_1 - \overline{\mathbf{X}}_2)$
scalar c	$c_p = \left\|D^{-1/2}(\boldsymbol{\mu}_1 - \boldsymbol{\mu}_2)\right\|$	$c_{\widehat{p}} = \left\|\widehat{D}^{-1/2}(\overline{\mathbf{X}}_1 - \overline{\mathbf{X}}_2)\right\|$
$\mathbf{b}$	$D^{-1}(\boldsymbol{\mu}_1 - \boldsymbol{\mu}_2)$	$\widehat{D}^{-1}(\overline{\mathbf{X}}_1 - \overline{\mathbf{X}}_2)$
υ	$\pi(1-\pi)c_p^2$	$(n_1 n_2/n^2)c_{\widehat{p}}^2$
$h(\mathbf{X})$	$\left[\mathbf{X} - \frac{1}{2}(\boldsymbol{\mu}_1 + \boldsymbol{\mu}_2)\right]^{\mathsf{T}} D^{-1}(\boldsymbol{\mu}_1 - \boldsymbol{\mu}_2)$	$\left[\mathbf{X} - \frac{1}{2}(\overline{\mathbf{X}}_1 + \overline{\mathbf{X}}_2)\right]^{\mathsf{T}} \widehat{D}^{-1}(\overline{\mathbf{X}}_1 - \overline{\mathbf{X}}_2).$

This vector $\widehat{\mathbf{p}}_{\rm DA}$ is used in the variable ranking scheme of Example 4.12 in Section 4.8.3, called $\mathbf{b}$ in the example. For the data of Example 4.12, ranking with $\widetilde{\mathbf{b}} = \widehat{\mathbf{p}}_{\rm DA}$ performs better than that obtained with $\widehat{\mathbf{b}}_{\rm DA}$. For this reason, I do not show or discuss $\widehat{\mathbf{b}}_{\rm DA}$ in Example 4.12.

So far we have assumed that both Σ and S are invertible. If the dimension increases, Σ may continue to be invertible, but S will be singular when $d > n$; that is, the dimension exceeds the number of observations. In this case, we cannot apply (13.21) to data because S is not invertible. In the regression setting of Section 13.3.1, we expressed $\widehat{C}$ in terms of the r-orthogonal matrices $\widehat{\Gamma}$ and $\widehat{L}$ of $\mathbb{X}$. Here we pursue a different path: we replace the singular matrix S in (13.17) with its diagonal $S_{\rm diag}$, which remains invertible, unless the variance term $s_j^2 = 0$ for some $j \leq d$.

Corollary 13.2 assumes that the two classes differ in their true means and share the same covariance matrix. This assumption is reasonable for many data sets and makes theoretical developments more tractable, especially if we simultaneously replace the covariance matrix Σ with its diagonal $D = \Sigma_{\rm diag}$.

Table 13.4 summarises the C-matrix and related quantities for the population and the sample case when the common covariance matrix has been replaced by its diagonal. For the population, the quantities follow from Corollary 13.2. If the sample within-class covariances are approximately the same for both classes, we replace Σ with $\widehat{\Sigma}$, where $\widehat{\Sigma}$ is the pooled covariance matrix or $\widehat{\Sigma} = (S_1 + S_2)/2$, and S_ν is the sample within-class covariance matrix of the νth class. We put $\widehat{D} = \widehat{\Sigma}_{\rm diag}$, the diagonal matrix derived from $\widehat{\Sigma}$. For data, we further replace $\boldsymbol{\mu}_\nu$ with $\overline{\mathbf{X}}_\nu$, the sample mean over the n_ν observations from the νth class, and put $n = n_1 + n_2$, and finally, we replace the probabilities with the ratios in each class. The ranking vector $\mathbf{b}$ is given the form which conveys the relevant information. The quantity υ in the table refers to the first singular value of the relevant matrix C.

The decision function for the data, shown in the last row of the table, is equivalent to that of the naive Bayes rule (13.1) in Section 13.2.1 as they only differ by the factor 2, which arises because $\widehat{W} = 2\widehat{\Sigma}$. This factor does not affect classification, and it is common to refer to both functions h as **naive Bayes** decision functions. I refer to the C-matrix in the table as the **naive (Bayes) canonical correlation matrix** and denote it by $C_{\rm NB}$ for the population and by $\widehat{C}_{\rm NB}$ for data. I use similar notation for the vectors $\mathbf{p}$ and $\mathbf{b}$ for the remainder of this and the next section, so the $\mathbf{p}$ for the data becomes $\widehat{\mathbf{p}}_{\rm NB}$.

Relationships to Other Ranking Vectors and Decision Functions

1. Note that $c_{\widehat{p}}\widehat{\mathbf{p}}_{\mathrm{NB}} = \mathbf{d}$, the ranking vector (4.50) in Section 4.8.3, which forms the basis for the classical t-tests. The naive canonical correlation approach based on $\widehat{C}_{\mathrm{NB}}$ therefore provides a natural justification for the ranking vector $\mathbf{d}$.
2. If we replace Σ with D and $\widehat{\Sigma}$ with $\widehat{D}$, we lose the correlation between the variables. This loss is equivalent to assuming that the variables are uncorrelated. For HDLSS data, this loss is incurred because $\widehat{\Sigma}$ is singular. However, if $\widehat{\Sigma}^{-1}$ exists, I recommend applying both ranking vectors $\widehat{\mathbf{b}}_{\mathrm{DA}}$ and $\widehat{\mathbf{b}}_{\mathrm{NB}}$ or $\widehat{\mathbf{p}}_{\mathrm{DA}}$ and $\widehat{\mathbf{p}}_{\mathrm{NB}}$ and the corresponding decision functions.
3. Example 4.12 in Section 4.8.3 showed the classification performances based on ranking with $\mathbf{d} \propto \widehat{\mathbf{p}}_{\mathrm{NB}}$, and with $\mathbf{b} \propto \widehat{\mathbf{p}}_{\mathrm{DA}}$. We glean from Figure 4.10 and Table 4.9 that for these data, $\widehat{\mathbf{p}}_{\mathrm{NB}}$ leads to a slightly smaller misclassification – eleven misclassified observations compared to twelve – however, $\widehat{\mathbf{p}}_{\mathrm{DA}}$ results in a more parsimonious model because it requires fifteen variables compared with the twenty-one variables required for the best classification with $\widehat{\mathbf{p}}_{\mathrm{NB}}$.

For HDLSS data, the approach based on the diagonal matrix $\widehat{D}$ has some advantages: as we shall see in Section 13.3.4, the eigenvector $\widehat{\mathbf{p}}_{\mathrm{NB}}$ is HDLSS-consistent provided that d does not grow too fast, and error bounds for the probability of misclassification can be derived for the naive Bayes rule h in Table 13.4.

A ranking scheme, especially for high-dimensional data, is essential in classification, and we also require a 'stopping' criterion for the number of selected variables. Algorithm 13.5 in Section 13.3.4 enlists a criterion of Fan and Fan (2008) that estimates the optimal number of variables. Instead of Fan and Fan's criterion, one might want to consider a range of dimensions, as we have done in step 6 of Algorithm 13.4, and then choose the dimension which minimises the error.

13.3.4 Properties of the Ranking Vectors of the Naive $\widehat{C}$ When d Grows

We consider a sequence of models from two classes C_1 and C_2: the data $\mathbb{X}_d = \left[\mathbf{X}_1 \cdots \mathbf{X}_n\right]$ are indexed by the dimension d, where $\log d = o(n)$ and $n = o(d)$ as $d, n \to \infty$. Using the notation of (2.22) in Section 2.7.2, the data satisfy (13.22) and (13.23):

$$\mathbf{X}_i = \boldsymbol{\mu}_\nu + \boldsymbol{\epsilon}_i \qquad \text{for } i \leq n, \tag{13.22}$$

where

$$\boldsymbol{\mu}_\nu = \begin{cases} \boldsymbol{\mu}_1 & \text{if } \mathbf{X}_i \text{ belongs to class } C_1, \\ \boldsymbol{\mu}_2 & \text{if } \mathbf{X}_i \text{ belongs to class } C_2, \end{cases}$$

and

$$\boldsymbol{\epsilon}_i \sim \mathcal{N}(\mathbf{0}_d, \Sigma_d) \text{ and } \boldsymbol{\epsilon}_i \text{ are independent for } i \leq n.$$

Let σ_j^2 be the diagonal elements of Σ_d, for $j \leq d$. The entries ϵ_{ij} of $\boldsymbol{\epsilon}_i$ satisfy Cramér's condition: there exist constants K_1, K_2, M_1 and M_2 such that

$$\mathbb{E}(|\epsilon_{ij}|^m) \leq \frac{K_1 m!}{2} M_1^{m-2}, \qquad \mathbb{E}(|e_{ij}^2 - \sigma_j^2|^m) \leq \frac{K_2 m!}{2} M_2^{m-2} \qquad \text{for } m = 1, 2, \ldots \tag{13.23}$$

I have stated Cramér's condition (13.23), as these moment assumptions on the $\boldsymbol{\epsilon}_i$ are required in the proofs of the theorems of this section. Similarly, statements [A] and [B] that follow detail notation about $\mathbb{X}$ which is implicitly assumed in Theorems 13.3 to 13.5.

[A] Let π be the probability that a random vector $\mathbf{X}_i$ belongs to $\mathcal{C}_1$, and let $1-\pi$ be the probability that the random vector $\mathbf{X}_i$ belongs to $\mathcal{C}_2$.
[B] Let n_ν be the number of observations from class $\mathcal{C}_\nu$ so that $n = n_1 + n_2$.

It will often be convenient to omit the subscript d (as in $\mathbb{X}_d$ or Σ_d) in this section, but the reader should keep in mind that we work with sequences of data and covariance matrices. Because both d and n grow, we can regard the asymptotic results in terms of d or n. To emphasise the HDLSS framework, I state the limiting results as $d \to \infty$.

For HDLSS data, pseudo-inverses or generalised inverses of the sample covariance matrix S yield approximations to Fisher's discriminant rule. Practical experience with such approximate rules in a machine-learning context and in the analysis of microarray data and comparisons with the naive Bayes rule show that the naive Bayes rule results in better classification (see Domingos and Pazzani 1997 and Dudoit, Fridlyand, and Speed 2002). These empirical observations are the starting point in Bickel and Levina (2004), who showed that for HDLSS data, Fisher's rule performs poorly in a minimax sense, and the naive Bayes rule outperforms the former under fairly broad conditions.

Bickel and Levina (2004) were the first to put error calculations for the naive Bayes rule on a theoretical foundation. Their approach did not include any variable selection, a step that is essential in practice in many HDLSS applications. Fan and Fan (2008) demonstrated that almost all linear discriminant rules perform as poorly as random guessing in the absence of variable selection. I will come back to this point after Theorem 13.5.

I combine the approaches of Fan and Fan (2008) and Tamatani, Koch, and Naito (2012). Both papers consider theoretical as well as practical issues and rely on the results of Bickel and Levina (2004). In their *features annealed independence rules* (FAIR), Fan and Fan integrated variable selection into the naive Bayes framework of Bickel and Levina. Tamatani, Koch, and Naito (2012) compared variable ranking based on $\widehat{\mathbf{p}}_{\rm NB}$ and $\widehat{\mathbf{b}}_{\rm NB}$ and observed that $\widehat{\mathbf{p}}_{\rm NB}$ is essentially equivalent to variable selection with FAIR.

I begin with the asymptotic behaviour of the vectors $\widehat{\mathbf{p}}_{\rm NB}$ and $\widehat{\mathbf{b}}_{\rm NB}$ and then present probabilities of misclassification for the naive Bayes rule.

Put

$$D = \Sigma_{\rm diag} \qquad \text{and} \qquad R = D^{-1/2}\Sigma D^{-1/2}, \tag{13.24}$$

so R is the matrix of correlation coefficients which we previously considered in Theorem 2.17 in Section 2.6.1. Let λ_R be the largest eigenvalue of R. For the asymptotic calculations, we consider the parameter space Θ, called Γ in Fan and Fan (2008), and the spaces Θ^* and $\Theta^\#$, called Γ^* and Γ^{**} in Tamatani, Koch, and Naito (2012). The three parameter spaces differ subtly in the conditions on the parameters θ:

$$\begin{aligned}
\Theta &= \left\{\theta = (\boldsymbol{\mu}_1, \boldsymbol{\mu}_2, \Sigma) \, : \, \left\| D^{-1/2}(\boldsymbol{\mu}_1 - \boldsymbol{\mu}_2)\right\|^2 \geq k_d,\, \lambda_R \leq b_0,\, \min_{j\leq d} \sigma_j^2 > 0\right\}, \\
\Theta^* &= \left\{\theta = (\boldsymbol{\mu}_1, \boldsymbol{\mu}_2, \Sigma) \, : \, \left\| D^{-1}(\boldsymbol{\mu}_1 - \boldsymbol{\mu}_2)\right\|^2 \geq k_d,\, \lambda_R \leq b_0,\, \min_{j\leq d} \sigma_j^2 > 0\right\}, \\
\Theta^\# &= \left\{\theta = (\boldsymbol{\mu}_1, \boldsymbol{\mu}_2, \Sigma) \, : \, k_d^\# \geq \left\| D^{-1}(\boldsymbol{\mu}_1 - \boldsymbol{\mu}_2)\right\|^2 \geq k_d,\, \lambda_R \leq b_0,\, \min_{j\leq d} \sigma_j^2 > 0\right\}.
\end{aligned} \tag{13.25}$$

Here b_0 is a positive constant, k_d and $k_d^\#$ are sequences of positive numbers that depend only on d and the k_d values in the three parameter spaces do not have to be the same. The condition $\left\|D^r(\boldsymbol{\mu}_1 - \boldsymbol{\mu}_2)\right\|$, with $r = -1/2, -1$, refers to the overall strength of the observations. The parameter spaces are restricted to observations which exceed some lower bound and are bounded by an upper bound in the case of $\Theta^{\#}$. The condition $\lambda_R \leq b_0$ gives a bound on the eigenvalue of R. The condition $\sigma_j^2 > 0$ ensures that all variables are random and guarantees that D is invertible.

For HDLSS data $\mathbb{X}$, which satisfy (13.22), the sample covariance matrix S is not invertible. We therefore focus on the naive Bayes approach. The population and data matrices C_{NB} and $\widehat{C}_{\text{NB}}$ and their eigenvectors are summarised in Table 13.4. Both C_{NB} and $\widehat{C}_{\text{NB}}$ are of rank one, and it therefore suffices to consider the first left eigenvectors $\mathbf{p}_{\text{NB}}$ and $\widehat{\mathbf{p}}_{\text{NB}}$. As in (2.25) in Section 2.7.2, we measure the closeness of $\mathbf{p}_{\text{NB}}$ and $\widehat{\mathbf{p}}_{\text{NB}}$ by the angle $\mathfrak{a}(\widehat{\mathbf{p}}_{\text{NB}}, \mathbf{p}_{\text{NB}})$ and say that $\widehat{\mathbf{p}}_{\text{NB}}$ is HDLSS-consistent if $\mathfrak{a}(\widehat{\mathbf{p}}_{\text{NB}}, \mathbf{p}_{\text{NB}}) \to 0$ in probability.

The ranking vectors $\mathbf{b}_{\text{NB}}$ and $\widehat{\mathbf{b}}_{\text{NB}}$ correspond to the canonical correlation transform and are not, in general, unit vectors. To measure the closeness between them, it is convenient to scale them first but still refer to them by $\mathbf{b}_{\text{NB}}$ and $\widehat{\mathbf{b}}_{\text{NB}}$. Using the notation of Table 13.4, put

$$\mathbf{b}_{\text{NB}} = c_1 D^{-1/2}\mathbf{p}_{\text{NB}} = c_2 D^{-1}(\boldsymbol{\mu}_1 - \boldsymbol{\mu}_2)$$

and

$$\widehat{\mathbf{b}}_{\text{NB}} = c_3 \widehat{D}^{-1/2}\widehat{\mathbf{p}}_{\text{NB}} = c_4 \widehat{D}^{-1}(\overline{\mathbf{X}}_1 - \overline{\mathbf{X}}_2) \tag{13.26}$$

for positive constants c_1–c_4 such that $\|\mathbf{b}_{\text{NB}}\| = \left\|\widehat{\mathbf{b}}_{\text{NB}}\right\| = 1$.

Theorem 13.3 [Tamatani, Koch, and Naito (2012)] *Let $\mathbb{X}_d$ be a sequence of $d \times n$ data which satisfy (13.22) and (13.23). Let C_{NB} and $\widehat{C}_{NB}$ be as in Table 13.4. Let $\upsilon > 0$ be the singular value and $\mathbf{p}_{NB}$ the left eigenvector of C_{NB} which corresponds to υ. Let $\widehat{\upsilon}$ and $\widehat{\mathbf{p}}_{NB}$ be the analogous quantities for $\widehat{C}_{NB}$. Let $\mathbf{b}_{NB}$ and $\widehat{\mathbf{b}}_{NB}$ be defined as in (13.26). Consider Θ of (13.25), Θ^* of (13.25) and $\Theta^{\#}$ of (13.25). Assume that* $\log d = o(n)$ *and* $n = o(d)$.

1. *If $n_2/n_1 = O(1)$ and $\left\|D^{-1/2}(\boldsymbol{\mu}_1 - \boldsymbol{\mu}_2)\right\|^2/k_d = O(1)$, then, as $d \to \infty$, for $\theta \in \Theta$,*

$$\frac{\widehat{\upsilon}}{\upsilon} = \begin{cases} 1 + o_p(1) & \text{if } d = o(nk_d) \\ 1 + (d/n)\left\|D^{-1/2}(\boldsymbol{\mu}_1 - \boldsymbol{\mu}_2)\right\|^{-2} + o_p(1) & \text{if } d/(nk_d) \to \tau,\ \tau > 0. \end{cases}$$

2. *If $d = o(nk_d)$ and $n_2/n_1 = O(1)$, then for $\theta \in \Theta$,*

$$\mathfrak{a}(\mathbf{p}_{NB}, \widehat{\mathbf{p}}_{NB}) \xrightarrow{\ p\ } 0,$$

and for $\theta \in \Theta^*$,

$$\mathfrak{a}(\mathbf{b}_{NB}, \widehat{\mathbf{b}}_{NB}) \xrightarrow{\ p\ } 0 \qquad \text{as } d \to \infty.$$

3. *If $d/(nk_d) \to \tau$ for $\tau > 0$ and $n_1/n \to 1/\xi$ for $\xi > 1$, then for $\theta \in \Theta$,*

$$\mathfrak{a}(\mathbf{p}_{NB}, \widehat{\mathbf{p}}_{NB}) \xrightarrow{\ p\ } \arccos\left[(1 + \xi\tau)^{-1/2}\right] \qquad \text{as } d \to \infty.$$

4. *Assume that $k_d^{\#}/k_d = O(1)$ and that the diagonal entries σ_j^2 of D satisfy*

$$\frac{1}{\sigma_0} \leq \min_{j \leq d} \sigma_j^2 \leq \max_{j \leq d} \sigma_j^2 \leq \frac{1}{\sigma_0^{\#}} \quad \text{for } \sigma_0, \sigma_0^{\#} > 0.$$

If $d/(nk_d) \to \tau$, $d/(nk_d^{\#}) \to \tau^{\#}$ *and* $n_1/n \to 1/\xi$, *for some* $\tau, \tau^{\#} > 0$ *and* $\xi > 1$ *then, as* $d \to \infty$, *for* $\theta \in \Theta^{\#}$,

$$\arccos\left[(1+\xi\sigma_0^{\#}\tau^{\#})^{-1/2}\right][1+o_p(1)] < \mathfrak{a}(\mathbf{b}_{NB},\widehat{\mathbf{b}}_{NB}) < \arccos\left[(1+\xi\sigma_0\tau)^{-1/2}\right][1+o_p(1)].$$

This theorem summarises theorems 4.1 to 4.3, 5.1 and 5.2 in Tamatani, Koch, and Naito (2012), and the proofs can be found in the paper. Theorem 13.3 states precise growth rates for d, namely, $d \ll nk_d$, for convergence of the singular value $\widehat{\upsilon}$ to υ and for HDLSS-consistency of the eigenvector $\widehat{\mathbf{p}}_{\text{NB}}$ and the direction $\widehat{\mathbf{b}}_{\text{NB}}$ to occur. The vectors $\widehat{\mathbf{p}}_{\text{NB}}$ and $\widehat{\mathbf{b}}_{\text{NB}}$ are HDLSS-consistent on overlapping but not identical parameter sets Θ and Θ^*, respectively.

If d grows at a rate proportional to nk_d, then $\widehat{\upsilon}$ does not converge to the population quantity, and the angle between the vectors does not go to zero, so $\widehat{\mathbf{p}}_{\text{NB}}$ is not consistent. The behaviour of the eigenvectors in parts 2 and 3 of the theorem is similar to those of the first PC eigenvectors in Theorem 13.18. I will return to this surprising similarity after stating Theorem 13.18.

The simulations of Tamatani, Koch, and Naito (2012) include the case corresponding to part 4 of Theorem 13.3. For the parameters $\xi = 2, \tau^{\#} = \tau = 0.5, \sigma_0^{\#} = 1-\epsilon, \sigma_0 = 1+\epsilon$ and $\epsilon \ll 1$, the simulations show that the angle between $\mathbf{b}_{\text{NB}}$ and $\widehat{\mathbf{b}}_{\text{NB}}$ converges to 45 degrees.

In its second role, $\widehat{\mathbf{b}}_{\text{NB}}$ is the direction of the decision function h (see Table 13.4). We now turn to the error of the rule induced by $\widehat{\mathbf{b}}_{\text{NB}}$. In Theorem 13.5 we will see that the conditions which yield the desirable HDLSS consistency of $\widehat{\mathbf{b}}_{\text{NB}}$ reappear in the worst-case classification error.

The linear rules we consider are symmetric in the two classes, and it thus suffices to consider a random vector $\mathbf{X}$ from class C_1.

Definition 13.4 Let $\mathbf{X}$ be a random vector from $\mathbb{X}$ which satisfies (13.22), and assume that $\mathbf{X}$ belongs to class C_1. Let h be the decision function of a rule that is symmetric in the two classes. Let Θ be a set of parameters $\theta = (\boldsymbol{\mu}_1, \boldsymbol{\mu}_2, \Sigma)$, similar to or the same as one of the sets in (13.25). The **posterior error** or the **(posterior) probability of misclassification** $\mathbb{P}_{\mathrm{m}}$ of h and $\theta \in \Theta$ is

$$\mathbb{P}_{\mathrm{m}}(h,\theta) = \mathbb{P}\{h(\mathbf{X}) < 0 \mid \mathbb{X} \text{ and labels } \mathbb{Y}\},$$

and the **worst-case posterior error** or **probability of misclassification** is

$$\mathbb{P}_{\mathrm{m}}^{\Theta}(h) = \max_{\theta\in\Theta} \mathbb{P}_{\mathrm{m}}(h,\theta).$$

□

Bickel and Levina (2004) considered a slightly different set of parameters from the set Θ in (13.25) in their definition of posterior error. I present the error probabilities which Fan and Fan (2008) derived for $d \propto nk_d$ as they extend the results of Bickel and Levina.

Theorem 13.5 [Fan and Fan (2008)] *Let* $\mathbb{X}_d$ *be a sequence of* $d \times n$ *data which satisfy (13.22) and (13.23). Consider* Θ *of (13.25) and* $\theta \in \Theta$. *Let* h_{NB} *be the naive Bayes decision function for* $\mathbb{X}$, *defined as in Table 13.4. Assume that* $\log d = o(n)$ *and* $n = o(d)$. *Put* $c_p = \left\|D^{-1/2}(\boldsymbol{\mu}_1 - \boldsymbol{\mu}_2)\right\|$.

1. *If* $d = o(nk_d)$, *then*

$$\mathbb{P}_{\mathrm{m}}(h_{NB},\theta) \le 1 - \Phi\left\{\frac{1}{2}\frac{c_p}{\sqrt{\lambda_R}}[1+o_p(1)]\right\},$$

and the worst-case probability of misclassification is

$$\mathbb{P}_{\mathrm{m}}^{\Theta}(h) = 1 - \Phi\left\{\frac{1}{2}\left[\frac{k_d}{b_0}\right]^{1/2}[1+o_p(1)]\right\}. \qquad (13.27)$$

2. *If* $d/(nk_d) \to \tau$ *for some* $\tau > 0$, *then*

$$\mathbb{P}_{\mathrm{m}}(h_{NB},\theta) \le 1 - \Phi\left\{\frac{\sqrt{n_1 n_2/(dn)}\, c_p^2[1+o_p(1)] + (n_1 - n_2)\sqrt{d/(nn_1n_2)}}{2\sqrt{\lambda_R}\{1 + n_1 n_2\, c_p^2[1+o_p(1)]/(dn)\}^{1/2}}\right\},$$

where Φ *is the standard normal distribution function, and* λ_R *is the largest eigenvalue of* R *in (13.24).*

This theorem shows that the relative rate of growth of d with nk_d governs the error probability. Part 2 is the first part of theorem 1 in Fan and Fan (2008), and a proof is given in their paper. Fan and Fan include in their theorem an expression for the worst-case error when $d = o(nk_d)$, which appears to be derived directly from the probability in part 2. Tamatani, Koch, and Naito (2012) obtained the upper bound in part 1 of Theorem 13.5 under the assumption $d = o(nk_d)$. For d in this regime, the worst-case error (13.27) is smaller than the worst-case error derived in theorem 1 of Fan and Fan.

Theorem 13.3 covers two regimes: $d = o(nk_d)$ and $d/(nk_d) \to \tau$, for $\tau > 0$. The former leads to the HDLSS consistency of $\widehat{\mathbf{p}}_{\mathrm{NB}}$ and $\widehat{\mathbf{b}}_{\mathrm{NB}}$. It is interesting to observe that for $d = o(nk_d)$, we obtain HDLSS consistency of $\widehat{\mathbf{b}}_{\mathrm{NB}}$ as well as a tighter bound for the error probability in Theorem 13.5 than for the second regime $d/(nk_d) \to \tau$.

If $k_d = 0$ and n_1 and n_2 are about the same, then $c_p = \left\|D^{-1/2}(\boldsymbol{\mu}_1 - \boldsymbol{\mu}_2)\right\|^2 \ge 0$, and part 1 of Theorem 13.5 yields $\mathbb{P}_{\mathrm{m}}^{\Theta}(h) \to 1 - \Phi(0) = 1/2$; that is, the naive Bayes rule may not be any better than random guessing. Theorem 2 of Fan and Fan (2008) has details. Fan and Fan avoided the potentially poor performance of the naive Bayes rule by selecting salient variables and working with those variables only. These variables are obtained by ranking and by deciding how many of the ranked variables to use. Corollary 13.6 has the details, but we first require some notation.

Consider $\mathbb{X}$ and the sample eigenvector $\widehat{\mathbf{p}}_{\mathrm{NB}}$ of $\widehat{C}_{\mathrm{NB}}$. Rank $\mathbb{X}$ using $\widehat{\mathbf{p}}_{\mathrm{NB}}$. For $p \le d$, let $\mathbb{X}_p$ be the p-ranked data, that is, the first p rows of the ranked data. We restrict the decision function h to the p-ranked data in the natural way: write $\overline{\mathbf{X}}_{p,\nu}$ for the sample mean of the νth class of $\mathbb{X}_p$ and $\widehat{D}_p$ for the diagonal matrix of the sample covariance matrix of $\mathbb{X}_p$. For $\mathbf{X}_p$ from $\mathbb{X}_p$, put

$$h_{\mathrm{NB},p}(\mathbf{X}_p) = \left[\mathbf{X}_p - \frac{1}{2}(\overline{\mathbf{X}}_{p,1} + \overline{\mathbf{X}}_{p,2})\right]^{\mathsf{T}} \widehat{D}_p^{-1}(\overline{\mathbf{X}}_{p,1} - \overline{\mathbf{X}}_{p,2}). \qquad (13.28)$$

Fan and Fan derived error probabilities for the truncated decision functions $h_{\mathrm{NB},p}$ in their theorem 4. For convenience of notation, they assumed that their data have the identity covariance matrix. I state their result for more general data $\mathbb{X}$ which satisfy (13.22) and (13.23).

Corollary 13.6 [Fan and Fan (2008)] *Assume that the data $\mathbb{X}$ are as in Theorem 13.5 and that the rows of $\mathbb{X}$ have been ranked with the eigenvector $\widehat{\mathbf{p}}_{NB}$ of $\widehat{C}_{NB}$. For $j \le d$, let $\mu_j^- = \mu_{1j} - \mu_{2j}$ be the difference of the jth class means, and let σ_j^2 be the variance of the jth variable of $\mathbb{X}$. Let $p = p_n \le d$ be a sequence of integers, and let $h_{NB,p}$ be the truncated decision function for the p-ranked data $\mathbb{X}_p$. If*

$$\frac{n}{\sqrt{p}} \sum_{j \le p} (\mu_j^- / \sigma_j)^2 \to \infty \qquad \text{as } p \to \infty,$$

then for $\theta \in \Theta$,

$$\mathbb{P}_{\mathrm{m}}(h_{NB,p}, \theta) = 1 - \Phi\left\{ \frac{\sum_{j \le p} (\mu_j^- / \sigma_j)^2 [1 + o_p(1)] + p(n_1 - n_2)/(n_1 n_2)}{2\sqrt{\lambda_R}\{\sum_{j \le p} (\mu_j^- / \sigma_j)^2 [1 + o_p(1)] + np/(n_1 n_2)\}^{1/2}} \right\},$$

and the optimal choice of the number p^+ of variables is

$$p^+ = \underset{p \le d}{\operatorname{argmax}} \frac{\left[\sum_{j \le p} (\mu_j^- / \sigma_j)^2 + p(n_1 - n_2)/(n_1 n_2)\right]^2}{\lambda_{R_p} \sum_{j \le p} (\mu_j^- / \sigma_j)^2 + np/(n_1 n_2)}, \tag{13.29}$$

where R_p is the correlation matrix of the p-ranked data, and λ_{R_p} is its largest eigenvalue.

In their *features annealed independence rules* (FAIR), Fan and Fan (2008) ranked the data with a vector which is equivalent to $\widehat{\mathbf{p}}_{\mathrm{NB}}$ and referred to the ranked variables as 'features'. They determined the optimal number of features $\widehat{p}^+$ by estimating p^+ of (13.29) directly from the data; in particular, they replaced μ_j^- and σ_j in (13.29) with the sample means and the pooled variances, respectively. Tamatani, Koch, and Naito (2012) employed the estimator of (13.29) in their ranked naive Bayes algorithm, here given as Algorithm 13.5.

Algorithm 13.5 ***Naive Bayes Rule for Ranked Data***

Let $\mathbb{X}$ be d-dimensional labelled data from two classes with n_ν observations in the νth class and $n = n_1 + n_2$. Let $\widehat{C}_{\mathrm{NB}}$ be the naive canonical correlation matrix of $\mathbb{X}$. Let $\widetilde{\mathbf{b}}$ be one of the vectors $\widehat{\mathbf{p}}_{\mathrm{NB}}$ or $\widehat{\mathbf{b}}_{\mathrm{NB}}$.

Step 1. Rank $\mathbb{X}$ with $\widetilde{\mathbf{b}}$, and write $\mathbb{X}$ for the ranked data.

Step 2. For $j \le d$, find $m_j^- = \overline{X}_{1j} - \overline{X}_{2j}$, the difference of the jth sample means of the two classes, and s_j^2, the pooled sample variance of the jth variable of $\mathbb{X}$.

Step 3. Calculate

$$\widehat{p}^+ = \underset{p \le d}{\operatorname{argmax}} \frac{\left[\sum_{j \le p} (m_j^- / s_j)^2 + p(n_1 - n_2)/(n_1 n_2)\right]^2}{\lambda_{\widehat{R}_p} \sum_{j \le p} (m_j^- / s_j)^2 + np/(n_1 n_2)}, \tag{13.30}$$

where $\widehat{R}_p$ is the sample correlation matrix of the p-ranked data, and $\lambda_{\widehat{R}_p}$ is its largest eigenvalue.

Step 4. Define a naive Bayes rule based on the decision function h_{NB,p^+} as in (13.28). ■

I conclude this section with an HDLSS example given in Tamatani, Koch, and Naito (2012), in which they compared the FAIR-based variable selection with that based on $\widehat{\mathbf{b}}_{\mathrm{NB}}$. For an extension of the approaches described in this section to a general κ-class setting with ranking vector chosen from the pool of $\kappa - 1$ eigenvectors, and selection of the number of features, see Tamatani, Naito, and Koch (2013).

Example 13.7 The **lung cancer** data contain 181 samples from two classes, malignant pleural mesothelioma (MPM) and adenocarcinoma (ADCA), and a total of 12,553 genes, the variables. There are thirty-one samples from the class MPM and 150 samples from the class ADCA. Gordon et al. (2002), who previously analysed these data, chose a training set of sixteen samples from each of the two classes, and used the remaining 149 samples for testing. Tamatani, Koch, and Naito (2012) used the same training and testing samples in their classification with the naive Bayes rule.

The aim of the classification with the naive Bayes rule is to assess the performance of the two ranking vectors $\widehat{\mathbf{p}}_{\mathrm{NB}}$ and $\widehat{\mathbf{b}}_{\mathrm{NB}}$. Although the optimal number of variables is determined by (13.30), the two ranking vectors result in different values for $\widehat{p}^{+}$.

The naive Bayes rule with the $\widehat{\mathbf{p}}_{\mathrm{NB}}$ ranking is equivalent to FAIR of Fan and Fan (2008). FAIR selects fourteen genes, has a zero training error and misclassifies eight observations of the testing set. Tamatani, Koch, and Naito (2012) referred to the naive Bayes rule with $\widehat{\mathbf{b}}_{\mathrm{NB}}$ ranking as *naïve canonical correlation* (NACC). The NACC rule selects only seven genes and results in the same training and testing error as FAIR. Only two of the genes, numbers 2,039 and 11,368, are common in the variable selection. A list of the selected genes of both approaches is given in Table 8.3 of Tamatani, Koch, and Naito (2012). It might be of interest to discuss the selected genes with medical experts, but this point is not addressed in their analysis.

The results show that we obtain quite different rankings with the two vectors. Because the NACC approach results in a smaller number of features and hence in a more parsimonious model, it is preferable. ■

For the *lung cancer* data, ranking with $\widehat{\mathbf{b}}_{\mathrm{NB}}$ results in a model with fewer features, so it is preferable because the misclassification on the testing set is the same for both ranking schemes. For the *breast cancer* data of Example 4.12 in Section 4.8.3, however, ranking with $\widehat{\mathbf{p}}_{\mathrm{DA}}$ performs better than ranking with $\widehat{\mathbf{b}}_{\mathrm{DA}}$, where $\widehat{\mathbf{p}}_{\mathrm{DA}}$ and $\widehat{\mathbf{b}}_{\mathrm{DA}}$ are the ranking vectors obtained from $\widehat{C}_{\mathrm{DA}}$ (see Theorem 13.1). In yet another analysis, namely, in the analysis of the *breast tumour gene expression* data in Example 13.6, the ranking vectors $\widehat{\mathbf{b}}_1$ and $\widehat{\mathbf{b}}_2$ yield comparable results, with $\widehat{\mathbf{b}}_2$ resulting in twelve selected variables and an error of 12.23 compared with the eleven selected variables and an error of 13.84 for $\widehat{\mathbf{b}}_1$. The ranking vector $\widehat{\mathbf{b}}_1$ of the regression setting corresponds to $\widehat{\mathbf{b}}_{\mathrm{NB}}$, and $\widehat{\mathbf{b}}_2$ corresponds to $\widehat{\mathbf{p}}_{\mathrm{NB}}$. These comparisons show that we should use more than one ranking vector in any analysis and compare the results.

My final remark in this section relates to the choice of the optimal number of variables. In Example 4.12 in Section 4.8.3 and Example 13.6, the optimal number of variables is that which minimises a given error criterion, and no distributional assumptions are required. In Example 13.7, we have used (13.29) and its estimate (13.30). The advantage of working with (13.30) is that we only have to apply the rule once, namely, with $\widehat{p}^{+}$ variables. On the other hand, (13.29) is based on a normal model, and both classes share a common covariance matrix. In practice, the data often deviate from these assumptions, and then the choice (13.30) may not be optimal. In the Problems at the end of Part III we return to Example 13.7

and find the optimal number of ranked variables in a way similar to that of Algorithm 4.3 in Section 4.8.3.

13.4 Sparse Principal Component Analysis

In Supervised Principal Component Analysis, we reduced the number of variables by ranking all variables and selecting the 'best' variables. An alternative is to reduce the number of 'active' variables for each principal component, for example, by allocating zero weights to some or many variables and thus 'deactivating' or discounting those variables. This is the underlying idea in Sparse Principal Component Analysis, to which we now turn.

The eigenvectors in Principal Component Analysis have non-zero entries for all variables. This fact makes an interpretation of the principal components difficult as the number of variables increases. Consider the *breast tumour gene expression* data of Example 2.15 in Section 2.6.2. For each of the 4,751 variables, the gene expressions, we obtain non-zero weights for each eigenvector. For $\widehat{\eta}_1$, all entries are well below 0.08 in absolute value, and therefore, no variable is singled out.

If we reduce the number of non-zero entries of the eigenvectors $\widehat{\eta}$, we lose some special properties of principal components, including the maximality of the variance contributions and the uncorrelatedness of the PCs. In Factor Analysis, rotations have been a popular tool for addressing the problem of non-zero entries, but the rotated components have other drawbacks, as pointed out in Jolliffe (1989, 1995). Informal approaches to reducing the number of non-zero entries in the eigenvectors have often just 'zeroed' the smaller entries, but without due care, such approaches can give misleading results (see Cadima and Jolliffe 1995).

Closely related to the number of non-zero entries of a direction vector, say, $\boldsymbol{\gamma}$, which induces a score $\boldsymbol{\gamma}^{\mathsf{T}}\mathbb{X}$, is the **(degree of) sparsity of a score**, that is, the number of non-zero components of $\boldsymbol{\gamma}$. We call a PC score $\boldsymbol{\gamma}^{\mathsf{T}}\mathbb{X}$ **sparse** if $\boldsymbol{\gamma}$ has a small number of non-zero entries. There is no precise number or ratio which characterises sparsity, but the idea is to sufficiently reduce the number of non-zero entries.

Since the mid-1990s, more systematic ways of controlling the eigenvectors in a Principal Component Analysis have emerged. I group these developments loosely into

1. simple and nonnegative PCs,
2. the LASSO, elastic net and sparse PCs,
3. the SVD approach to sparse PCs, and
4. theoretical developments for sparse PCs.

I will look at items 2 and 3 of this list in this section and only mention briefly the early developments (item 1) as they are mostly superseded. I describe some theoretical developments in Section 13.5 in the context of models which are indexed by the dimension.

13.4.1 The Lasso, SCoTLASS Directions and Sparse Principal Components

Early attempts at simplifying the eigenvectors of the covariance matrix S of data $\mathbb{X}$ are described in Vines (2000) and references therein. These authors replaced the eigenvectors with weight vectors with entries 0 and 1 or -1, 0 and 1. Rousson and Gasser (2004) replaced the first few eigenvectors with vectors with entries 0 and 1 but then allowed the remaining and less important vectors to remain close to the eigenvectors from which they originated. Such approaches might be appealing but may be too simplistic and are often difficult to compute. Zass and Shashua (2007) forced all entries to be non-negative. This adjustment typically results in relatively small contributions to variance of the modified PCs and is only of interest when there exist reasons for non-negative entries.

I will not further comment on these simple PCs but move on to the ideas of Jolliffe, Trendafilov, and Uddin (2003), who called their method 'simplified component technique–LASSO (SCoTLASS). The LASSO part refers to the inclusion of an ℓ_1 penalty function which Tibshirani (1996) proposed for a linear regression framework. The key idea of Tibshirani's LASSO is the following: consider predictor variables $\mathbb{X}$ and continuous univariate regression responses $\mathbf{Y}$; then, for $t > 0$, the **LASSO estimator** $\widehat{\boldsymbol{\beta}}_{\text{LASSO}}$ is the minimiser of

$$\left\| \mathbf{Y} - \boldsymbol{\beta}^{\mathsf{T}}\mathbb{X} \right\|_2^2 \qquad \text{subject to} \qquad \|\boldsymbol{\beta}\|_1 = \sum_{k=1}^{d} |\beta_k| \le t. \tag{13.31}$$

The parameter t of the ℓ_1 constraint controls the number of non-zero entries β_k, and the LASSO estimator results in fewer than d non-zero entries. For high-dimensional data, the parameter t is related to the sparsity of the estimator $\boldsymbol{\beta}$.

To integrate the ideas of the LASSO estimator into Principal Component Analysis, Jolliffe, Trendafilov, and Uddin (2003) considered centred $d \times n$ data $\mathbb{X}$. Let $\mathbb{W}^{(k)}$ be the $k \times n$ principal component data, and let $\mathbf{W}_{\bullet j}$ be the jth row of $\mathbb{W}^{(k)}$. Then

$$\mathbf{W}_{\bullet j} = \widehat{\boldsymbol{\eta}}_j^{\mathsf{T}}\mathbb{X} \tag{13.32}$$

and

$$\mathbb{X} \approx \widehat{\Gamma}_k \widehat{\Gamma}_k^{\mathsf{T}}\mathbb{X} = \widehat{\Gamma}_k \mathbb{W}^{(k)}. \tag{13.33}$$

We begin with (13.32) and in Section 13.4.2 return to (13.33), the approximation given in Corollary 2.14 in Section 2.5.2. The expression (13.32) is simply the jth PC score. If we re-interpret the $\mathbf{W}_{\bullet j}$ in a regression framework and think of it as univariate responses in linear regression, then (13.32) is equivalent to

$$\widehat{\boldsymbol{\eta}}_j = \underset{\|\boldsymbol{\gamma}\|_2 = 1}{\operatorname{argmin}} \left\| \mathbf{W}_{\bullet j} - \boldsymbol{\gamma}^{\mathsf{T}}\mathbb{X} \right\|_2 .$$

To restrict the number of non-zero entries in candidate solutions $\boldsymbol{\gamma}$, Jolliffe, Trendafilov, and Uddin (2003) solved, for $t > 0$, the constrained problem

$$\widehat{\boldsymbol{\gamma}} = \underset{\|\boldsymbol{\gamma}\|_2 = 1}{\operatorname{argmin}} \left\| \mathbf{W}_{\bullet j} - \boldsymbol{\gamma}^{\mathsf{T}}\mathbb{X} \right\|_2^2 \qquad \text{subject to} \qquad \|\boldsymbol{\gamma}\|_1 \le t. \tag{13.34}$$

The similarity between (13.34) and (13.31) is the key to the approach of Jolliffe, Trendafilov, and Uddin (2003). The number of non-zero entries depends on the choice of t because $\widehat{\boldsymbol{\gamma}}$ is a unit vector.

In a linear regression setting, (13.31) is all that is needed. In a principal component framework, we require solutions for each of the d 'responses' $\mathbf{W}_{\bullet 1}, \ldots, \mathbf{W}_{\bullet d}$, and we are interested in the relationship between the solutions: the eigenvectors $\widehat{\boldsymbol{\eta}}_j$ in Principal Component Analysis are orthogonal and maximise the variance of $\widehat{\boldsymbol{\eta}}_j^{\mathsf{T}} S \widehat{\boldsymbol{\eta}}_j$. For solutions $\boldsymbol{\gamma}$ of (13.34) that are more general than the eigenvectors of S, some of the properties of eigenvectors will no longer hold. Jolliffe, Trendafilov, and Uddin generalised eigenvectors to the SCoTLASS directions which are defined in Definition 13.7.

Definition 13.7 Let $\mathbb{X}$ be centred data of size $d \times n$ as in (2.18) in Section 2.6. Let S be the covariance matrix of $\mathbb{X}$, and let r be the rank of S. For $t \ge 1$, the **SCoTLASS directions** are d-dimensional unit vectors $\boldsymbol{\gamma}_1 \cdots \boldsymbol{\gamma}_r$ which maximise $\boldsymbol{\gamma}_j^{\mathsf{T}} S \boldsymbol{\gamma}_j$ subject to

1. $\boldsymbol{\gamma}_1^{\mathsf{T}} S \boldsymbol{\gamma}_1 > \boldsymbol{\gamma}_2^{\mathsf{T}} S \boldsymbol{\gamma}_2 > \cdots > \boldsymbol{\gamma}_r^{\mathsf{T}} S \boldsymbol{\gamma}_r$,
2. $\boldsymbol{\gamma}_j^{\mathsf{T}} \boldsymbol{\gamma}_k = \delta_{jk}$ for $j,k \le r$ and δ the Kronecker delta function, and
3. $\left\|\boldsymbol{\gamma}_j\right\|_1 \le t$, for $j \le r$. □

Jolliffe, Trendafilov, and Uddin (2003) typically worked with the scaled data, whose covariance matrix is the matrix of sample correlation coefficients R by Corollary 2.18 in Section 2.6.1. Here I will use S generically for the sample covariance matrix of the data. For the scaled data, $S = R$.

The second condition in Definition 13.7 enforces orthogonality of the SCoTLASS directions, and conditions 1 and 2 together imply, by Theorem 2.10 in Section 2.5.2, that for $j \le r$,

$$\boldsymbol{\gamma}_j^{\mathsf{T}} S \boldsymbol{\gamma}_j \le \widehat{\boldsymbol{\eta}}_j^{\mathsf{T}} S \widehat{\boldsymbol{\eta}}_j.$$

This last inequality shows that the contribution to variance of the SCoTLASS directions is smaller than that of the corresponding eigenvectors.

Because of (13.34), we interpret the jth SCoTLASS direction as an approximation to the eigenvector $\widehat{\boldsymbol{\eta}}_j$ of S. The discrepancy between $\widehat{\boldsymbol{\eta}}_j$ and $\boldsymbol{\gamma}_j$ depends on t, the tuning parameter which controls the number of non-zero entries of the $\boldsymbol{\gamma}_j$. Different values of t characterise the solutions:

- if $t \ge \sqrt{r}$, then $\boldsymbol{\gamma}_j = \widehat{\boldsymbol{\eta}}_j$, for $j \le r$,
- if $1 \le t \le \sqrt{r}$, then the number of non-zero entries in the $\boldsymbol{\gamma}_j$ decreases with t,
- if $t = 1$, then exactly one entry $\gamma_{jk} \ne 0$, for each $j \le r$, and
- if $t < 1$, then no solution exists.

Jolliffe, Trendafilov, and Uddin (2003) illustrated the geometry of the SCoTLASS directions for two dimensions in their figure 1. Their figure shows that the PC solution, an ellipse, shrinks to the square with vertices $(0, \pm 1)$ and $(\pm 1, 0)$ as t decreases to 1.

The SCoTLASS directions are a means for summarising data, similar to the eigenvectors $\widehat{\boldsymbol{\eta}}$, and it is therefore natural to construct the analogues of the PC scores from the SCoTLASS directions.

Definition 13.8 Let $\mathbb{X}$ be d-dimensional centred data with covariance matrix S of rank r. For $t \ge 1$ let $\boldsymbol{\gamma}_1 \cdots \boldsymbol{\gamma}_r$ be unit vectors satisfying conditions 1 to 3 of Definition 13.7. For $k \le r$, put

$$\mathbf{M}_{\bullet k} = \boldsymbol{\gamma}_k^{\mathsf{T}} \mathbb{X},$$

and call $\mathbf{M}_{\bullet k}$ the kth **sparse (SCoTLASS) principal component score**. □

For brevity, I refer to the $\mathbf{M}_{\bullet k}$ as the **sparse PCs**, and I will be more precise only if a distinction between different sparse scores is required. Each $\mathbf{M}_{\bullet k}$ is a row vector of size $1 \times n$. When t is large enough, $\mathbf{M}_{\bullet k} = \mathbf{W}_{\bullet k}$, but for smaller values of t, only some of the variables of $\mathbb{X}$ contribute to $\mathbf{M}_{\bullet k}$, whereas all entries of $\mathbb{X}$ contribute to $\mathbf{W}_{\bullet k}$.

Jolliffe, Trendafilov, and Uddin (2003) experimented with a range of values $1 < t < \sqrt{r}$ and illustrated their method with the thirteen-dimensional *pitprops* data. I report parts of their example.

Example 13.8 The **pitprops** data of Jeffers (1967) consist of fourteen variables which were measured for 180 pitprops cut from Corsican pine timber. Jeffers predicted the strength of

Table 13.5 *Variables of the Pitprops Data from Example 13.8*

Variable no.	Variable name
1	Top diameter of the prop in inches
2	Length of the prop in inches
3	Moisture content of the prop, expressed as a percentage of the dry weight
4	Specific gravity of the timber at the time of the test
5	Oven-dry specific gravity of the timber
6	Number of annual rings at the top of the prop
7	Number of annual rings at the base of the prop
8	Maximum bow in inches
9	Distance of the point of maximum bow from the top of the prop in inches
10	Number of knot whorls
11	Length of clear prop from the top of the prop in inches
12	Average number of knots per whorl
13	Average diameter of the knots in inches

Table 13.6 *Weights of the First Two PCs and Sparse PCs for the Pitprops Data*

			$t = 2.25$		$t = 1.75$		ζ_1	
	PC_1	PC_2	sPC_1	sPC_2	sPC_1	sPC_2	sPC_1	sPC_2
X_1	0.404	0.212	0.558	0.085	0.664		−0.477	
X_2	0.406	0.180	0.580	0.031	0.683	−0.001	−0.476	
X_3	0.125	0.546		0.647		0.641		0.785
X_4	0.173	0.468		0.654		0.701		0.620
X_5	0.057	−0.138					0.177	
X_6	0.284	−0.002	0.001	0.208		0.293		
X_7	0.400	−0.185	0.266		0.001	0.107	−0.250	
X_8	0.294	−0.198	0.104	−0.098	0.001		−0.344	−0.021
X_9	0.357	0.010	0.372		0.283		−0.416	
X_{10}	0.379	−0.252	0.364	−0.154	0.113		−0.400	
X_{11}	−0.008	0.187		0.099				
X_{12}	−0.115	0.348		0.241		0.001		0.013
X_{13}	−0.112	0.304		0.026				
% var	*32.4*	*18.2*	*26.7*	*17.2*	*19.6*	*16.0*	*28.0*	*14.4*

pitprops from thirteen variables which are listed in Table 13.5. The first six PCs contribute just over 87 per cent of total variance, and Jolliffe, Trendafilov, and Uddin (2003) compared these six eigenvectors with the SCoTLASS directions.

We look at two specific values of t and compare the first two SCoTLASS directions with the corresponding eigenvectors. Columns 2 to 7 of Table 13.6 are excerpts from tables 2, 4 and 5 in Jolliffe, Trendafilov, and Uddin (2003). The last two columns of Table 13.6 are taken from table 3 of Zou, Hastie, and Tibshirani (2006). I will return to the last two columns in Example 13.9.

The columns 'PC_1' and 'PC_2' in Table 13.6 contain the entries of the eigenvectors $\widehat{\eta}_1$ and $\widehat{\eta}_2$ of S. The non-zero entries of the SCoTLASS directions $\boldsymbol{\gamma}_1$ and $\boldsymbol{\gamma}_2$ are given for $t = 2.25$ and $t = 1.75$ in the next four columns. Blank spaces in the table imply zero weights. A comparison of the entries of $\widehat{\eta}_1$ and the two $\boldsymbol{\gamma}_1$ shows that the sign of the weights does not change, and the order of the weights, induced by the absolute value of the entries, remains the same. Weights with an absolute value above 0.4 have increased, weights well above 0.2 and below 0.4 in absolute value have decreased, and all weights closer to zero than 0.2 have become zero. For $t = 1.75$, the weight for X_6 has disappeared, whereas for $t = 2.25$, it has the value 0.001, which is also considerably smaller than the PC weight of 0.287. A similar behaviour can be observed for PC_2 with slightly different cut-off values. The sparse PCs have become simpler to interpret because there are fewer non-zero entries. The last two columns of the table refer to Algorithm 13.6, which I describe in the next section.

The last line of Table 13.6 gives the contributions to variance for $\widehat{\boldsymbol{\eta}}_k^{\mathsf{T}} S \widehat{\boldsymbol{\eta}}_k$ and the $\boldsymbol{\gamma}_k^{\mathsf{T}} S \boldsymbol{\gamma}_k$; as t decreases, the weight vectors become sparser, and the variance contributions decrease. ■

A calculation of the sparse principal components of the pitprops data reveals that these components are correlated, unlike the PC scores. The increase in correlation and the decrease in their contribution to variance are the compromises we make in exchange for the simpler expressions in the sparse PCs.

13.4.2 Elastic Nets and Sparse Principal Components

The preceding section explored how Jolliffe, Trendafilov, and Uddin (2003) integrated the LASSO estimator into a PC framework. A generalisation of the LASSO is the elastic net of Zou and Hastie (2005), which combines ideas of the ridge and the LASSO estimator. Zou, Hastie, and Tibshirani (2006) explained how the elastic net ideas fit into a PC setting, analogues to the way Jolliffe, Trendafilov, and Uddin (2003) integrated the LASSO in their approach.

I start with the basic idea of elastic nets and then explain how Zou, Hastie, and Tibshirani (2006) included them in a PC setting. As we shall see, the elastic net solutions also result in sparse PCs, but these solutions differ from those obtained with the SCoTLASS directions.

Following Zou and Hastie (2005), let $\mathbf{Y}$ be univariate responses in a linear regression setting with d-dimensional predictor variables $\mathbb{X}$. For scalars $\zeta_1, \zeta_2 > 0$, the **elastic net estimator**

$$\widehat{\boldsymbol{\beta}}_{\mathrm{ENET}} = (1+\zeta_2)\left[\underset{\boldsymbol{\beta}}{\operatorname{argmin}} \left\|\mathbf{Y} - \boldsymbol{\beta}^{\mathsf{T}}\mathbb{X}\right\|_2^2 + \zeta_2 \|\boldsymbol{\beta}\|_2^2 + \zeta_1 \|\boldsymbol{\beta}\|_1\right]. \qquad (13.35)$$

The term $\zeta_2 \|\boldsymbol{\beta}\|_2^2$ is related to the ridge estimator – see (2.39) in Section 2.8.2. Without this ℓ_2 term, $\widehat{\boldsymbol{\beta}}_{\mathrm{ENET}}$ reduces to the LASSO estimator $\widehat{\boldsymbol{\beta}}_{\mathrm{LASSO}}$ of (13.31).

Let $\mathbb{X} = \begin{bmatrix}\mathbf{X}_1 \cdots \mathbf{X}_n\end{bmatrix}$ be d-dimensional centred data with sample covariance matrix S and r the rank of S. The LASSO PC problem (13.34) is stated as a contrained problem, but for $\zeta_1 > 0$ and $j \le r$, it can be equivalently expressed as

$$\widehat{\boldsymbol{\gamma}}_{\mathrm{LASSO}} = \underset{\{\boldsymbol{\gamma}\,:\,\|\boldsymbol{\gamma}\|_2=1\}}{\operatorname{argmin}} \left[\left\|\mathbf{W}_{\bullet j} - \boldsymbol{\gamma}^{\mathsf{T}}\mathbb{X}\right\|_2^2 + \zeta_1 \|\boldsymbol{\gamma}\|_1\right].$$

Similarly, the elastic net estimator in a PC setting requires scalars $\zeta_1, \zeta_2 > 0$ and, for $j \le r$, leads to the estimator

$$\widehat{\boldsymbol{\gamma}}_{\mathrm{ENET}} = \underset{\{\boldsymbol{\gamma}\,:\,\|\boldsymbol{\gamma}\|_2=1\}}{\operatorname{argmin}} \left[\left\|\mathbf{W}_{\bullet j} - \boldsymbol{\gamma}^{\mathsf{T}}\mathbb{X}\right\|_2^2 + \zeta_2 \|\boldsymbol{\gamma}\|_2^2 + \zeta_1 \|\boldsymbol{\gamma}\|_1\right]. \qquad (13.36)$$

For notational convenience, I have ignored the fact that different tuning parameters ζ_1 and ζ_2 may be required as j varies. To solve the r minimisation problems (13.36) jointly, there are a number of ways which all capture the idea of a constrained optimisation. Let $k \le r$, and let A and B be $d \times k$ matrices. We may want to solve problems described generically in the following way

$$\underset{A}{\operatorname{argmin}} \left[\mathbb{W}^{(k)} - A^{\mathsf{T}}\mathbb{X}\right] + \text{constraints on } A,$$

$$\underset{A}{\operatorname{argmin}} \left[\mathbb{X} - AA^{\mathsf{T}}\mathbb{X}\right] + \text{constraints on } A, \tag{13.37}$$

$$\underset{A,B}{\operatorname{argmin}} \left[\mathbb{X} - AB^{\mathsf{T}}\mathbb{X}\right] + \text{constraints on } A \text{ and } B. \tag{13.38}$$

I deliberately listed the three statements without reference to particular norms; we will work out suitable norms later. The first statement is a natural generalisation of (13.34) and treats the principal component data as the multivariate responses in a linear regression setting. The second and third statements are related to the approximation (13.33) but make use of it in different ways. A comparison of (13.37) and (13.33) tells us that the matrix A replaces the k-orthogonal matrix $\widehat{\Gamma}_k$, and one therefore wants to find the optimal $d \times k$ matrix A. In (13.38) we go one step further and allow a substitution of $\widehat{\Gamma}_k$ and $\widehat{\Gamma}_k^{\mathsf{T}}$ by different matrices A and B^{T}.

Zou, Hastie, and Tibshirani (2006) solved (13.38) iteratively using separate updating steps for the matrices A and B. To appreciate their approach to solving (13.38), we examine the pieces, beginning with their theorem 3, which only refers to an ℓ_2 constraint.

Theorem 13.9 [Zou, Hastie, and Tibshirani (2006)] *Let $\mathbb{X} = \left[\mathbf{X}_1 \cdots \mathbf{X}_n\right]$ be d-dimensional centred data with sample covariance matrix S. Let r be the rank of S, and for $j \le r$, let $\widehat{\boldsymbol{\eta}}_j$ be the eigenvectors of S. Consider $\zeta_2 > 0$. Fix $k \le r$. Let A and B be $d \times k$ matrices. Put*

$$(\widehat{A}, \widehat{B}) = \underset{(A,B)}{\operatorname{argmin}} \left[\left\|\mathbb{X} - AB^{\mathsf{T}}\mathbb{X}\right\|_{Frob}^2 + \zeta_2 \|B\|_{Frob}^2\right] \quad \textit{subject to} \quad A^{\mathsf{T}}A = \mathbf{I}_{k\times k}, \tag{13.39}$$

where $\|\cdot\|_{Frob}$ is the Frobenius norm of (5.1) in Section 5.2. Then the column $\widehat{\mathbf{b}}_j$ of $\widehat{B}$ is a scalar multiple of $\widehat{\boldsymbol{\eta}}_j$, for $j = 1, \ldots, k$.

A proof of this theorem is given in Zou, Hastie, and Tibshirani (2006). The theorem states that the minimiser $\widehat{B}$ is essentially the matrix of eigenvectors $\widehat{\Gamma}_k$ provided that $A^{\mathsf{T}}A = \mathbf{I}_{k\times k}$. The matrix $A = \widehat{\Gamma}_k$ is a candidate, but $\widehat{\Gamma}_k E$ is also a candidate for A for any orthogonal $k \times k$ matrix E.

The following definition includes ℓ_1 and ℓ_2 constraints for statement (13.38).

Definition 13.10 Let $\mathbb{X}$ be d-dimensional centred data with covariance matrix S of rank r. Let $k \le r$. Let A and B be $d \times k$ matrices. Consider scalars $\zeta_2, \zeta_{1j} > 0$, with $j \le k$. Let $(\widehat{A}, \widehat{B})$ be the minimiser of the **sparse (elastic net) principal component criterion**

$$\underset{(A,B)}{\operatorname{argmin}} \left[\left\|\mathbb{X} - AB^{\mathsf{T}}\mathbb{X}\right\|_{\text{Frob}}^2 + \zeta_2 \|B\|_{\text{Frob}}^2 + \sum_{j=1}^{k} \zeta_{1j} \left\|\mathbf{b}_j\right\|_1\right] \quad \text{subject to} \quad A^{\mathsf{T}}A = \mathbf{I}_{k\times k}. \tag{13.40}$$

□

A single tuning parameter ζ_2 suffices for the ℓ_2 constraint, whereas the LASSO requires different ζ_{1j}. Let $A = [\mathbf{a}_1 \cdots \mathbf{a}_k]$ and $B = [\mathbf{b}_1 \cdots \mathbf{b}_k]$; then, as in Zou, Hastie, and Tibshirani (2006), we alternate between separate optimisation steps for A and B:

- *Fix A*. For each column $\mathbf{a}_j$ of A, define the $1 \times n$ projection $\mathbf{a}_j^{\mathsf{T}}\mathbb{X}$. For $j \leq r$, scalars $\zeta_2 > 0$ and $\zeta_{1j} > 0$, find

$$\widehat{\mathbf{b}}_j = \underset{\mathbf{b}_j}{\operatorname{argmin}} \left[\left\| \mathbf{a}_j^{\mathsf{T}}\mathbb{X} - \mathbf{b}_j^{\mathsf{T}}\mathbb{X} \right\|_2^2 + \zeta_2 \left\| \mathbf{b}_j \right\|_2^2 + \zeta_{1j} \left\| \mathbf{b}_j \right\|_1 \right]. \tag{13.41}$$

- *Fix B*. If UDV^{T} is the singular value decomposition of the $k \times d$ matrix $B^{\mathsf{T}}\mathbb{X}\mathbb{X}^{\mathsf{T}}$, then the minimiser

$$\widehat{A} = \underset{A}{\operatorname{argmin}} \left\| \mathbb{X} - AB^{\mathsf{T}}\mathbb{X} \right\|_{\text{Frob}}^2 \qquad \text{is given by} \qquad \widehat{A} = VU^{\mathsf{T}}. \tag{13.42}$$

If B is fixed, then (13.40) reduces to the simpler problem $\operatorname{argmin}_A \left\| \mathbb{X} - AB^{\mathsf{T}}\mathbb{X} \right\|_{\text{Frob}}^2$, and the orthogonal matrix $\widehat{A} = VU^{\mathsf{T}}$, which minimises this Frobenius norm, is obtained from Theorem 8.13 in Section 8.5.2.

A solution to (13.40) is not usually available in closed form. Algorithm 13.6 is based on algorithm 1 of Zou, Hastie, and Tibshirani (2006), which takes advantage of the two steps (13.41) and (13.42).

Algorithm 13.6 ***Sparse Principal Components Based on the Elastic Net***
Let $\mathbb{X}$ be d-dimensional centred data with covariance matrix S of rank r. Let $\widehat{\Gamma} = \left[\widehat{\boldsymbol{\eta}}_1 \cdots \widehat{\boldsymbol{\eta}}_r\right]$ be the matrix of eigenvectors of S. Fix $k \leq r$. Let $A = \left[\mathbf{a}_1 \cdots \mathbf{a}_k\right]$ and $B = \left[\mathbf{b}_1 \cdots \mathbf{b}_k\right]$ be $k \times d$ matrices. Consider $j \leq k$. Initialise A by putting $\mathbf{a}_j = \widehat{\boldsymbol{\eta}}_j$.

Step 1. For scalars $\zeta_2 > 0$ and $\zeta_{1j} > 0$, determine

$$\widehat{\mathbf{b}}_j = \underset{\mathbf{b}_j}{\operatorname{argmin}} \left[\left\| \mathbf{a}_j^{\mathsf{T}}\mathbb{X} - \mathbf{b}_j^{\mathsf{T}}\mathbb{X} \right\|_2^2 + \zeta_2 \left\| \mathbf{b}_j \right\|_2^2 + \zeta_{1j} \left\| \mathbf{b}_j \right\|_1 \right],$$

and put $B = \left[\widehat{\mathbf{b}}_1 \cdots \widehat{\mathbf{b}}_k\right]$.

Step 2. Calculate the singular value decomposition of $B^{\mathsf{T}}\mathbb{X}\mathbb{X}^{\mathsf{T}}$, and denote it by UDV^{T}. Update A by putting $A = VU^{\mathsf{T}}$.

Step 3. Repeat steps 1 and 2 until the vectors $\widehat{\mathbf{b}}_j$ converge.

Step 4. Put $\boldsymbol{\gamma}_j = \widehat{\mathbf{b}}_j / \left\| \widehat{\mathbf{b}}_j \right\|_2$, and call $\boldsymbol{\gamma}_j^{\mathsf{T}}\mathbb{X}$ the jth **sparse (elastic net) principal component score**. ■

Unlike the SCoTLASS directions $\boldsymbol{\gamma}_j$ of (13.34) and Definition 13.7, the directions $\boldsymbol{\gamma}_j$ of step 4 are not in general orthogonal. Although the scores of Definition 13.8 and those constructed in Algorithm 13.6 differ, I refer to both scores simply as *sparse principal component scores*. Should a distinction be necessary, then I will make clear which sparse PCs are the appropriate ones.

Algorithm 13.6 uses the scalars $\zeta_2 > 0$ and $\zeta_{1j} > 0$ and the number of variables $k \leq r$. Zou, Hastie, and Tibshirani (2006) reported that the scores do not appear to change much as the tuning parameters ζ vary, and for this reason, they suggested using a small positive value or even $\zeta_2 = 0$ when $n > d$. As in Zou and Hastie (2005), Zou, Hastie, and Tibshirani (2006) suggested picking values $\zeta_{1j} > 0$ that yield a good compromise between sufficient contributions to variance and sparsity. They do not comment on the choice of k, but in general, only the first few scores are used.

Example 13.9 We continue with the **pitprops** data and look at some of the results of Zou, Hastie, and Tibshirani (2006), who compared the weights obtained with Algorithm 13.6 with those obtained in Jolliffe, Trendafilov, and Uddin (2003). Zou, Hastie, and Tibshirani (2006) take $k = 6$, the value which was chosen in the original analysis, but I only report the weights of their first two sparse PCs.

Zou, Hastie, and Tibshirani (2006) put $\zeta_2 = 0$, so the ℓ_2 constraint disappears. This allows a more natural comparison with the results of Jolliffe, Trendafilov, and Uddin (2003). The main difference between these two approaches now reduces to the different implementations of the LASSO constraint. Jolliffe, Trendafilov, and Uddin (2003) used the same t for all variables, whereas Zou, Hastie, and Tibshirani (2006) used the vector $\boldsymbol{\zeta}_1 = (0.06, 0.16, 0.1, 0.5, 0.5, 0.5)$ for the six directions they are interested in, so $\zeta_{12} = 0.16$ applies to the second direction.

The weights of the directions $\boldsymbol{\gamma}_j$, with $j = 1, 2$, are obtained in step 4 of Algorithm 13.6. These weights are shown in the last two columns of Table 13.6 in Example 13.8 and are displayed on the y-axis in Figure 13.5, against the variable number on the x-axis. The top plot shows the weights for the first direction and the bottom plot those for the second direction. Black is used for the eigenvectors $\widehat{\eta}$ and blue for the SCoTLASS directions; the solid line corresponds to $t = 2.25$ and the dotted line to $t = 1.75$. The red line shows the weights corresponding to the elastic net for $\boldsymbol{\zeta}_1$ given at the end of the preceding paragraph. Table 13.6 shows that the signs of the 'elastic net' weights are mostly negative, whereas the $\widehat{\boldsymbol{\eta}}_1$ weights are mostly positive. For easier interpretation, I have inverted the sign of the elastic net weights for the first direction in Figure 13.5.

A comparison of the weights of the first directions shows that the large weights obtained with $\boldsymbol{\zeta}_1$ are closer to the entries of $\widehat{\boldsymbol{\eta}}_1$ than the corresponding weights obtained with the SCoTLASS directions, where 'large' refers to the absolute value. The zero entries among the various sparse weights mostly agree with one notable exception: for variable 5, the sign of the elastic net weight is opposite that of the corresponding eigenvector weight. A closer inspection reveals that the PC weight for variable 5 is very small (0.057) and has become zero in the SCoTLASS directions, yet the elastic net weight for variable 5 is non-zero,

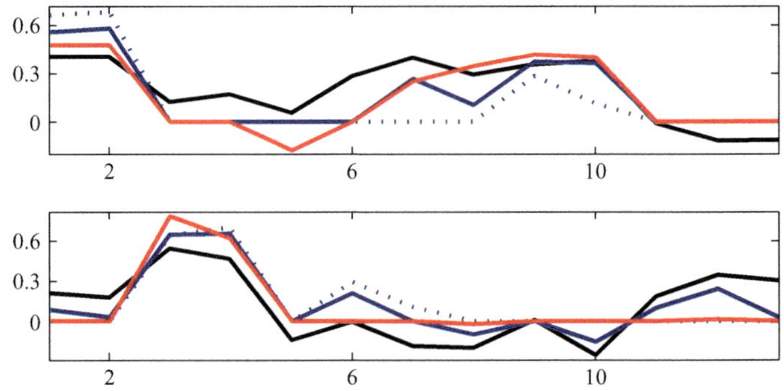

Figure 13.5 Weights of sparse PC scores for Example 13.9 on the y-axis against variable number on the x-axis. (*Top*) PC_1 weights; (*bottom*) PC_2 weights: $\widehat{\boldsymbol{\eta}}$ (*black*), t = 2.25 (*blue*), t = 1.75 (*dotted*), $\boldsymbol{\zeta}_1$ (*red*). See also Table 13.6.

although the larger PC_1 weights of variables 3, 4, 6, 12 and 13 have all been set to zero with the elastic net. A glance back to the last row of Table 13.6 shows that ζ_{11} results in a higher variance contribution of the first sparse score than the sparse PCs obtained from the SCoTLASS directions.

For the second direction vector, the value ζ_{12} and the second elastic net PC score result in a lower variance contribution than the corresponding SCoTLASS scores. Further, ζ_{12} leads to a sparser PC than the second SCoTLASS directions.

There is a trade-off between the contribution to variance and the degree of sparsity of the directions. As the non-zero entries become more sparse, the contribution to variance decreases. The different values of the entries of $\boldsymbol{\zeta}_1$ provide more flexibility than the fixed value t which is used in the SCoTLASS method, but a different choice for the ζ_{1j} requires more computational effort. ■

For high-dimensional data, the computational cost arising from step 1 of Algorithm 13.6 can be large. To reduce this effort, Zou, Hastie, and Tibshirani (2006) pointed out that a 'thrifty solution emerges' if $\zeta_2 \to \infty$ in (13.40). To see this, first note that the elastic net estimator $\widehat{\beta}_{\mathrm{ENET}}$ of (13.35) contains the factor $(1+\zeta_2)$. If we put $B' = B/(1+\lambda_2)$, and use B' instead of B in (13.40), then the equality

$$\left\|\mathbb{X} - AB^{\mathsf{T}}\mathbb{X}\right\|_{\mathrm{Frob}}^2 = \mathrm{tr}(\mathbb{X}\mathbb{X}^{\mathsf{T}}) - 2\,\mathrm{tr}(\mathbb{X}\mathbb{X}^{\mathsf{T}}BA^{\mathsf{T}}) - \mathrm{tr}(AB^{\mathsf{T}}\mathbb{X}\mathbb{X}^{\mathsf{T}}BA^{\mathsf{T}}),$$

with the rescaled B' instead of B, leads to

$$\begin{aligned}
&\left\|\mathbb{X} - AB'^{\mathsf{T}}\mathbb{X}\right\|_{\mathrm{Frob}}^2 + \zeta_2 \left\|B'\right\|_{\mathrm{Frob}}^2 + \sum_{j=1}^{k} \zeta_{1j} \left\|\mathbf{b}'_j\right\|_1 \\
&= \mathrm{tr}(\mathbb{X}\mathbb{X}^{\mathsf{T}}) - \frac{2}{1+\zeta_2}\,\mathrm{tr}(\mathbb{X}\mathbb{X}^{\mathsf{T}}BA^{\mathsf{T}}) - \frac{1}{(1+\zeta_2)^2}\,\mathrm{tr}(AB^{\mathsf{T}}\mathbb{X}\mathbb{X}^{\mathsf{T}}BA^{\mathsf{T}}) \\
&\quad + \frac{\zeta_2}{(1+\zeta_2)^2}\left\|B\right\|_{\mathrm{Frob}}^2 + \frac{1}{1+\zeta_2}\sum_{j=1}^{k} \zeta_{1j}\left\|\mathbf{b}_j\right\|_1 \\
&\longrightarrow \mathrm{tr}(\mathbb{X}\mathbb{X}^{\mathsf{T}}) + \frac{1}{1+\zeta_2}\left[-2\,\mathrm{tr}(\mathbb{X}\mathbb{X}^{\mathsf{T}}BA^{\mathsf{T}}) + \left\|B\right\|_{\mathrm{Frob}}^2 + \sum_{j=1}^{k}\zeta_{1j}\left\|\mathbf{b}_j\right\|_1\right] \qquad \text{as } \zeta_2 \to \infty.
\end{aligned}$$

The first term, $\mathrm{tr}(\mathbb{X}\mathbb{X}^{\mathsf{T}})$, does not affect the optimisation, so it can be dropped. Using the notation in Definition 13.10, and assuming that $d \gg n$ and $k > 0$, the **simplified sparse principal component criterion** which corresponds to $\zeta_2 \to \infty$ becomes

$$\underset{(A,B)}{\mathrm{argmin}}\left[-2\,\mathrm{tr}(\mathbb{X}\mathbb{X}^{\mathsf{T}}BA^{\mathsf{T}}) + \left\|B\right\|_{\mathrm{Frob}}^2 + \sum_{j=1}^{k}\zeta_{1j}\left\|\mathbf{b}_j\right\|_1\right] \quad \text{subject to } A^{\mathsf{T}}A = \mathbf{I}_{k\times k}. \tag{13.43}$$

This simplified criterion can be used in Algorithm 13.6 instead of (13.40) by substituting step 1 with step 1a:

Step 1a. For scalars $\zeta_{1j} > 0$, determine

$$\widehat{\mathbf{b}}_j = \underset{\mathbf{b}_j}{\mathrm{argmin}}\left[-2\mathbf{b}_j^{\mathsf{T}}\mathbb{X}\mathbb{X}^{\mathsf{T}}\mathbf{a}_j + \left\|\mathbf{b}_j\right\|_2^2 + \zeta_{1j}\left\|\mathbf{b}_j\right\|_1\right].$$

In addition to using step 1a, Zou, Hastie, and Tibshirani (2006) determined the entries of each $\widehat{\mathbf{b}}_j$ by thresholding. I will return to thresholding in connection with the approach of Shen and Huang (2008) in Section 13.4.3.

Zou, Hastie, and Tibshirani (2006) illustrated the performance of the simplified sparse principal component criterion (13.43) with the microarray data of Ramaswamy et al. (2001), which consist of $d = 16{,}063$ genes and $n = 144$ samples. Zou, Hastie, and Tibshirani (2006) apply the simplified criterion (13.43) to these data and conclude that about 2.5 per cent, that is, about 400 of the genes, contribute with non-zero weights to the first sparse principal component score, which explains about 40 per cent of the total variance compared to a variance contribution of 46 per cent from the first PC score. Thus the degree of sparsity is high without much loss of variance.

Remark. As we have seen, letting ζ_2 increase results in a simpler criterion for high-dimensional data. In the *pitprops* data of Example 13.9, $\zeta_2 = 0$, but for these data, $d \ll n$. We see yet again that different regimes of d and n need to be treated differently.

13.4.3 Rank One Approximations and Sparse Principal Components

The ideas of Jolliffe, Trendafilov, and Uddin (2003) and Zou, Hastie, and Tibshirani (2006) have been applied and further explored by many researchers. Rather than attempting (and failing) to do justice to this rapidly growing field of research, I have chosen one specific topic in Sparse Principal Component Analysis, namely, finding the best rank one approximations to the data, and I will focus on two papers which consider this topic: the 'regularised singular value decomposition' of Shen and Huang (2008) and the 'penalized matrix decomposition (PMD)' of Witten, Tibshirani, and Hastie (2009). Both papers contain theoretical developments of their ideas and go beyond the sparse rank one approximations on which I focus.

A starting point for both approaches is the singular value decomposition of $\mathbb{X}$. Let $\mathbb{X}$ be centred $d \times n$ data, and write $\mathbb{X} = \widehat{\Gamma}\widehat{D}\widehat{L}^{\top}$ for its singular value decomposition. Then $\widehat{D}^2/(n-1) = \widehat{\Lambda}$, the diagonal matrix of eigenvalues of the sample covariance matrix S of $\mathbb{X}$. Let r be the rank of S. For $k \le r$, the left and right eigenvectors $\widehat{\Gamma}_k = \begin{bmatrix}\widehat{\boldsymbol{\eta}}_1 \cdots \widehat{\boldsymbol{\eta}}_k\end{bmatrix}$ and $\widehat{L}_k = \begin{bmatrix}\widehat{\boldsymbol{\nu}}_1 \cdots \widehat{\boldsymbol{\nu}}_k\end{bmatrix}$ satisfy (5.19) in Section 5.5. The $n \times k$ matrix $\widehat{L}_k$ is k-orthogonal, and the k-dimensional PC data are given by $\mathbb{W}^{(k)} = \widehat{D}_k \widehat{L}_k^{\top}$.

Although the emphasis in this section is on rank one approximations, we start with the more general rank k approximations.

Definition 13.11 Let $\mathbb{X} = \begin{bmatrix}\mathbf{X}_1 \cdots \mathbf{X}_n\end{bmatrix}$ be centred d-dimensional data with covariance matrix S of rank r. Let $\mathbb{X} = \widehat{\Gamma}\widehat{D}\widehat{L}^{\top}$ be the singular value decomposition of $\mathbb{X}$. For $k \le r$, the **rank k approximation** to $\mathbb{X}$ is the $d \times n$ matrix $\widehat{\Gamma}_k \widehat{D}_k \widehat{L}_k^{\top}$. □

The columns of the rank k approximation to $\mathbb{X}$ are derived from the vectors $\widehat{\mathbf{P}}_{ij} = \widehat{\boldsymbol{\eta}}_j \widehat{\boldsymbol{\eta}}_j^{\top} \mathbf{X}_i$ (see Corollary 2.14 in Section 2.5.2), and the ith column is $\widetilde{\mathbf{X}}_i^k = \sum_{j=1}^k \widehat{\mathbf{P}}_{ij}$. Using the trace norm of Definition 2.11 in Section 2.5.2, the error between $\mathbf{X}_i$ and $\widetilde{\mathbf{X}}_i^k$ is, again by Corollary 2.14,

$$\left\| \mathbf{X}_i - \widetilde{\mathbf{X}}_i^k \right\|_{tr}^2 = \sum_{m>k} \widehat{\lambda}_m = \frac{1}{n-1} \sum_{m>k} \widehat{d}_m^2,$$

where $\widehat{\lambda}_m$ is the mth eigenvalue of S, and $\widehat{d}_m$ is the mth singular value of $\mathbb{X}$.

Torokhti and Friedland (2009) used low-rank approximations in order to develop a 'generic Principal Component Analysis'. They were not concerned with sparsity, and I will therefore not describe their ideas here.

For rank one approximations to the $d \times n$ data $\mathbb{X}$, the associated **rank one optimisation** problem is the following: from the sets of vectors $\boldsymbol{\gamma} \in \mathbb{R}^d$ and unit vectors $\boldsymbol{\varphi} \in \mathbb{R}^n$, find the vectors $(\boldsymbol{\gamma}^*, \boldsymbol{\varphi}^*)$ such that

$$(\boldsymbol{\gamma}^*, \boldsymbol{\varphi}^*) = \underset{(\boldsymbol{\gamma},\boldsymbol{\varphi})}{\operatorname{argmin}} \left\| \mathbb{X} - \boldsymbol{\gamma}\boldsymbol{\varphi}^{\mathsf{T}} \right\|_{\text{Frob}}^2. \tag{13.44}$$

A solution is given by the first pair of left and right eigenvectors $(\widehat{\boldsymbol{\eta}}_1, \widehat{\boldsymbol{\nu}}_1)$ and the first singular value $\widehat{d}_1$ of $\mathbb{X}$:

$$\boldsymbol{\gamma}^* = \widehat{d}_1 \widehat{\boldsymbol{\eta}}_1 \qquad \text{and} \qquad \boldsymbol{\varphi}^* = \widehat{\boldsymbol{\nu}}_1. \tag{13.45}$$

The solution pair $(\widehat{d}_1 \widehat{\boldsymbol{\eta}}_1, \widehat{\boldsymbol{\nu}}_1)$ yields the rank one approximation $\widehat{d}_1 \widehat{\boldsymbol{\eta}}_1 \widehat{\boldsymbol{\nu}}_1^{\mathsf{T}}$ to $\mathbb{X}$.

We could construct a rank two approximation in a similar way but instead consider the difference

$$\mathbb{X}^{(1)} \equiv \mathbb{X} - \widehat{d}_1 \widehat{\boldsymbol{\eta}}_1 \widehat{\boldsymbol{\nu}}_1^{\mathsf{T}}$$

and then solve the rank one optimisation problem (13.44) for $\mathbb{X}^{(1)}$. The solution is the pair $(\widehat{d}_2 \widehat{\boldsymbol{\eta}}_2, \widehat{\boldsymbol{\nu}}_2)$. Similarly, for $1 < k \le r$, put

$$\mathbb{X}^{(k)} = \mathbb{X}^{(k-1)} - \widehat{d}_k \widehat{\boldsymbol{\eta}}_k \widehat{\boldsymbol{\nu}}_k^{\mathsf{T}}, \tag{13.46}$$

and then the rank one optimisation problem for $\mathbb{X}^{(k)}$ results in the solution $(\widehat{d}_{k+1} \widehat{\boldsymbol{\eta}}_{k+1}, \widehat{\boldsymbol{\nu}}_{k+1})$.

After k iterations, the stepwise approach of (13.46) results in a rank k approximation to $\mathbb{X}$. Compared with a single rank k approximation, the k-step iterative approach has the advantage that at each step it allows the inclusion of constraints, which admit sparse solutions. This is the approach adopted in Shen and Huang (2008) and Witten, Tibshirani, and Hastie (2009). I describe Algorithm 13.7 in a generic form which applies to both their approaches. Table 13.7 shows the similarities and differences between their ideas. The algorithm minimises penalised criteria of the form

$$\left\| \mathbb{X} - \boldsymbol{\gamma}\boldsymbol{\varphi}^{\mathsf{T}} \right\|_{\text{Frob}}^2 + P_\delta(\boldsymbol{\gamma}), \tag{13.47}$$

where $P_\delta(\boldsymbol{\gamma})$ is a penalty function, $\delta \ge 0$ is a tuning parameter, and $\boldsymbol{\gamma} \in \mathbb{R}^d$. Typically, P_δ is a soft thresholding of $\boldsymbol{\gamma}$, an ℓ_1 or a LASSO constraint. The thresholding is achieved component-wise and, for $t \in \mathbb{R}$, is defined by

$$h_\delta(t) = \operatorname{sgn}(t)(|t| - \delta)_+. \tag{13.48}$$

For vectors $\mathbf{t} \in \mathbb{R}^d$, the vector-valued thresholding is defined by $h_\delta(\mathbf{t}) = \left[h_\delta(t_1) \cdots h_\delta(t_d)\right]^{\mathsf{T}}$.

Algorithm 13.7 ***Sparse Principal Components from Rank One Approximations***

Let $\mathbb{X}$ be d-dimensional centred data with covariance matrix S of rank r. Let $\mathbb{X} = \widehat{\Gamma}\widehat{D}\widehat{L}$ be the singular value decomposition of $\mathbb{X}$ with singular values $\widehat{d}_j$. Let $\widehat{\boldsymbol{\eta}}_j$ be the column vectors of $\widehat{\Gamma}$ and $\widehat{\boldsymbol{v}}_j$ the column vectors of $\widehat{L}$. Let $\boldsymbol{\gamma} \in \mathbb{R}^d$. Let $\boldsymbol{\varphi} \in \mathbb{R}^n$ be a unit vector. Let P_δ be one of the penalty functions which are listed in (13.47). Put $k=0$ and $\mathbb{X}^{(0)} = \mathbb{X}$.

Step 1. Determine the minimiser $(\boldsymbol{\gamma}^*, \boldsymbol{\varphi}^*)$ of the penalised problem

$$(\boldsymbol{\gamma}^*, \boldsymbol{\varphi}^*) = \underset{(\boldsymbol{\gamma}, \boldsymbol{\varphi})}{\operatorname{argmin}} \left\| \mathbb{X}^{(k)} - \boldsymbol{\gamma}\boldsymbol{\varphi}^{\mathsf{T}} \right\|_{\text{Frob}}^2 + P_\delta(\boldsymbol{\gamma}).$$

Step 2. Put $\boldsymbol{\gamma}_{k+1} = \boldsymbol{\gamma}^*$ and $\boldsymbol{\varphi}_{k+1} = \boldsymbol{\varphi}^*$. Define

$$\mathbb{X}^{(k+1)} = \mathbb{X}^{(k)} - \boldsymbol{\gamma}_{k+1}\boldsymbol{\varphi}_{k+1}^{\mathsf{T}},$$

and put $k = k+1$.

Step 3. Repeat steps 1 and 2 for $k \le r$.

Step 4. For $1 \le k \le r$, define sparse principal component scores by $\boldsymbol{\gamma}_k^{\mathsf{T}}\mathbb{X}$. ■

In practice, Shen and Huang (2008) and Witten, Tibshirani, and Hastie (2009) commonly applied the soft thresholding (13.48) in step 1 by repeating the two steps

$$\boldsymbol{\gamma}_1 \leftarrow h_\delta(\mathbb{X}\boldsymbol{\varphi}) \qquad \text{or} \qquad \boldsymbol{\gamma}_2 \leftarrow \frac{h_\delta(\mathbb{X}\boldsymbol{\varphi})}{\|h_\delta(\mathbb{X}\boldsymbol{\varphi})\|_2}$$

and

$$\boldsymbol{\varphi} \leftarrow \frac{\mathbb{X}^{\mathsf{T}}\boldsymbol{\gamma}}{\left\|\mathbb{X}^{\mathsf{T}}\boldsymbol{\gamma}\right\|_2} \qquad \text{for } \boldsymbol{\gamma} = \boldsymbol{\gamma}_1 \text{ or } \boldsymbol{\gamma} = \boldsymbol{\gamma}_2. \tag{13.49}$$

The iteration, which is part of step 1, stops when the pair of vectors $(\boldsymbol{\gamma}, \boldsymbol{\varphi})$ converges. The vectors $\boldsymbol{\gamma}_{k+1}$ and $\boldsymbol{\varphi}_{k+1}$ of step 2 are the final pair obtained from this iteration.

The approaches of Shen and Huang (2008) and Witten, Tibshirani, and Hastie (2009) differ partly because the authors present their ideas with variations. In Table 13.7, I write 'SH' for the approach of Shen and Huang and refer to their penalty P_δ with soft thresholding, the second option in (13.47). I write 'WTH' for the approach of Witten, Tibshirani, and Hastie and will focus on the 'penalized matrix decomposition PMD (L_1, L_1) criterion', their 'new method for sparse PCA'. Shen and Huang referred to two other penalty functions and included hard thresholding. Witten, Tibshirani, and Hastie also referred to a LASSO constraint for $\boldsymbol{\varphi}$. In addition, both papers included discussions and algorithms regarding cross-validation-based choices of the tuning parameters, which I omit.

The table shows that the ideas of the soft thresholding approach of Shen and Huang (2008) and the main PMD criterion of Witten, Tibshirani, and Hastie (2009) are very similar, although the computational aspects might differ. Witten, Tibshirani, and Hastie (2009) have a collection of algorithms for different combinations of constraints, including one for data with missing values and an extension to Canonical Correlation data.

One of the advantages of working with soft threshold constraints is the increase in variance of the sparse PC scores compared with the sparse scores based on the LASSO or the elastic net. For the *pitprops* data, however, the soft thresholding of Shen and Huang (2008) leads to results that are very similar to those obtained with the LASSO and elastic net. Because of this similarity, I will not present these results but refer the reader to their paper.

Table 13.7 *Details for Algorithm 13.7 Using the Notation of Algorithm 13.7 and (13.49)*

	SH	WTH
Parameters	$\theta > 0$	$c \geq 0, \Delta > 0$
Constraints	$P_\gamma(\boldsymbol{\gamma}) = 2\theta \, \|\boldsymbol{\gamma}\|_1$	$P(\boldsymbol{\gamma}) = \|\boldsymbol{\gamma}\|_1 \leq c$
Thresholding	$h_\theta(t) = \text{sign}(t)(\|t\| - \theta)_+$	$h_\Delta(t) = \text{sign}(t)(\|t\| - \Delta)_+$
Input to (13.49)	$\boldsymbol{\varphi} = \widehat{v}$ of (13.45)	$\boldsymbol{\gamma}$ with $\|\boldsymbol{\gamma}\|_2 = 1$
$\boldsymbol{\gamma}$ in (13.49)	$\boldsymbol{\gamma}_1$ from (13.49)	$\boldsymbol{\gamma}_2$ from (13.49)
Sparse direction	$\boldsymbol{\gamma}_0 = \boldsymbol{\gamma} / \|\boldsymbol{\gamma}\|_2$	$\boldsymbol{\gamma}$
Sparse PC	$\boldsymbol{\gamma}_0^{\top}\mathbb{X}$	$\boldsymbol{\gamma}^{\top}\mathbb{X} / \left\|\boldsymbol{\gamma}^{\top}\mathbb{X}\right\|_2$
Cross-validation	To determine the degree of sparsity of $\boldsymbol{\gamma}_0$	To determine constants c and Δ

The research described in Section 13.4 has stimulated many developments and extensions which focus on constructing sparse PC directions, and include asymptotic developments and consistency results for such settings. (See Leng and Wang (2009), Lee, Lee, and Park (2012), Shen, Shen, and Marron (2013), Ma (2013) and references therein.) I will not describe the theoretical developments of these recent papers here, but only mention that consistency of the sparse vectors can happen even if the non-sparse general setting leads to strongly inconsistent eigenvectors.

13.5 (In)Consistency of Principal Components as the Dimension Grows

In this final section I focus on the question: Do the estimated principal components converge or get closer to the true principal components as the dimension grows? As we shall see, there is no simple answer. I describe the ideas of Johnstone and Lu (2009) and Jung and Marron (2009) and the more recent extensions of these results by Jung, Sen, and Marron (2012) and Shen, Shen, and Marron (2012) and Shen et al. (2012), which highlight the challenges we encounter for high-dimensional data and which provide answers for a range of scenarios.

Section 2.7.1 examined the asymptotic behaviour of the sample eigenvalues and eigenvectors for fixed d, and Theorem 2.20 told us that for multivariate normal data, the eigenvalues and eigenvectors are consistent estimators of the respective population parameters and are asymptotically normal. As the dimension grows, these properties no longer hold – even for normal data. Theorem 2.23 states that, asymptotically, the first eigenvalue does not have a normal distribution when d grows with n. Theorem 2.25 gives a positive result for HDLSS data: it states conditions under which the first eigenvector converges to the population eigenvector. We pursue the behaviour of the first eigenvector further under a wider range of conditions.

13.5.1 (In)Consistency for Single-Component Models

Following Johnstone and Lu (2009), we explore the question of consistency of the sample eigenvectors when d is comparable to n and both are large so when $\mathbf{n} \succeq \mathbf{d}$ or $\mathbf{n} \preceq \mathbf{d}$ in the sense of Definition 2.22 of Section 2.7.2. We consider the sequence of single-component models described in Johnstone and Lu (2009), which are indexed by d or n:

$$\mathbb{X} = \boldsymbol{\rho}\boldsymbol{v}^{\top} + \sigma E, \qquad (13.50)$$

where $\mathbb{X}$ are the $d \times n$ data, $\boldsymbol{\rho}$ is a d-dimensional single component, $\boldsymbol{v}$ is a $1 \times n$ vector of random effects with independent entries $v_i \sim \mathcal{N}(0,1)$ and the $d \times n$ noise matrix E consists of n independent normal vectors with mean zero and covariance matrix $\mathbf{I}_{d\times d}$.

In their paper, Johnstone and Lu (2009) indexed (13.50) by n and regarded d as a function of n. Instead, one could shift the emphasis to d and consider asymptotic results as d grows. This indexation is natural in Section 13.5.2 because n remains fixed in the approach of Jung and Marron (2009).

For the model (13.50), we want to estimate the single component $\boldsymbol{\rho}$. Because $\boldsymbol{\rho}$ is not a unit vector, we put $\boldsymbol{\rho}_0 = \mathrm{dir}(\boldsymbol{\rho})$ as in (9.1) in Section 9.1. Using the framework of Principal Component Analysis, the sample covariance matrix S has $r \le \min(d,n)$ non-zero eigenvalues. Let $\widehat{\boldsymbol{\rho}}$ be the eigenvector of S corresponding to the largest eigenvalue. As in (2.25) in Section 2.7.2 and Theorem 13.3, we measure the closeness of $\widehat{\boldsymbol{\rho}}$ and $\boldsymbol{\rho}_0$ by the angle $\mathfrak{a}(\widehat{\boldsymbol{\rho}}, \boldsymbol{\rho}_0)$ between the two vectors or, equivalently, by the cosine of the angle $\cos[\mathfrak{a}(\widehat{\boldsymbol{\rho}}, \boldsymbol{\rho}_0)] = \widehat{\boldsymbol{\rho}}^{\mathsf{T}} \boldsymbol{\rho}_0$.

Recall from Definition 2.24 in Section 2.7.2 that $\widehat{\boldsymbol{\rho}}$ consistently estimates $\boldsymbol{\rho}_0$ if $\mathfrak{a}(\widehat{\boldsymbol{\rho}}, \boldsymbol{\rho}_0) \xrightarrow{p} 0$. We want to examine the asymptotic behaviour of $\cos[\mathfrak{a}(\widehat{\boldsymbol{\rho}}, \boldsymbol{\rho}_0)]$ for the sequence of models (13.50) and regard $d, \mathbb{X}$ and $\boldsymbol{\rho}$ as functions of n. The following theorem holds.

Theorem 13.12 [Johnstone and Lu (2009)] *Let $\mathbb{X} = \boldsymbol{\rho}\boldsymbol{v}^{\mathsf{T}} + \sigma E$ be a sequence of models which satisfy (13.50), and regard $\mathbb{X}$, the dimension d and $\boldsymbol{\rho}$ as functions of n. Put the component $\boldsymbol{\rho}_0 = \mathrm{dir}(\boldsymbol{\rho})$, and let $\widehat{\boldsymbol{\rho}}$ be the eigenvector of the covariance matrix S of $\mathbb{X}$ corresponding to the largest eigenvalue. Assume that, as $n \to \infty$,*

$$d/n \to c \qquad \text{and} \qquad (\|\boldsymbol{\rho}\|/\sigma)^2 \to \omega,$$

for $\omega > 0$ and some $c \ge 0$. Then

$$\cos[\mathfrak{a}(\widehat{\boldsymbol{\rho}}, \boldsymbol{\rho}_0)] \xrightarrow{a.s.} [(\omega^2 - c)_+]/(\omega^2 + c\omega) \quad \text{as } n \to \infty,$$

where the convergence is almost sure convergence.

Remark. Note that $[(\omega^2 - c)_+]/(\omega^2 + c\omega) < 1$ if and only if $c > 0$, so

$$\widehat{\boldsymbol{\rho}} \text{ is a consistent estimator of } \boldsymbol{\rho}_0 \iff d/n \to 0,$$

that is, if d grows more slowly than n. Further, if $\omega^2 < c$ and if

$$\lim_{n\to\infty} \frac{d}{n}\frac{\sigma^4}{\rho^4} \ge 1,$$

then $\widehat{\boldsymbol{\rho}}$ ultimately contains no information about $\boldsymbol{\rho}$. The latter is a consequence of the asymptotic orthogonality of the vectors involved.

A first rigorous proof of the (in)consistency of model (13.50) is given in Lu (2002), and related results are proved in Paul (2007) and Nadler (2008). Paul (2007) extended (13.50) to models with spiked covariance matrices (see Definition 2.24 in Section 2.7.2). In Paul's model, the $d \times n$ centred data $\mathbb{X}$ are given by

$$\mathbb{X} = \sum_{j=1}^{m} \boldsymbol{\rho}_j \boldsymbol{v}_j^{\mathsf{T}} + \sigma E,$$

where the unknown vectors $\boldsymbol{\rho}_j$ are mutually orthogonal with norms decreasing for increasing j, and the entries $v_{ij} \sim \mathcal{N}(0,1)$ are independent for $j \le m$, $i \le n$, and $m < d$. The noise matrix E is the same as in (13.50). Paul showed that Theorem 13.12 generalises to the estimator $\widehat{\boldsymbol{\rho}}_1$ of the first vector $\boldsymbol{\rho}_1$; that is, in his multicomponent model, $\widehat{\boldsymbol{\rho}}_1$ is consistent if and only if $d/n \to 0$.

Given this inconsistency result for large dimensions, the question arises: How should one deal with high-dimensional data? Unlike the ideas for sparse principal components put forward in Sections 13.4.1 and 13.4.2, which enforce sparsity *during* Principal Component Analysis, Johnstone and Lu (2009) proposed reducing the dimension in a separate step *before* Principal Component Analysis, and they observed that dimension reduction prior to Principal Component Analysis can only be successful if the population principal components are concentrated in a small number of dimensions. Johnstone and Lu (2009) try to capture the idea of concentration by representing $\mathbb{X}$ and $\boldsymbol{\rho}$ in a basis in which $\boldsymbol{\rho}$ has a sparse representation. Let $\mathcal{B} = \{\mathbf{e}_1 \cdots \mathbf{e}_d\}$ be such a basis, and put

$$\boldsymbol{\rho} = \sum_{j=1}^{d} \varrho_j \,\mathbf{e}_j \qquad \text{and} \qquad \mathbf{X}_i = \sum_{j=1}^{d} \xi_{ij} \mathbf{e}_j \qquad (i \le n), \tag{13.51}$$

for suitable coefficients ϱ_j and ξ_{ij}. Order the coefficients ϱ_j in decreasing order of absolute value, and write $|\varrho_{(1)}| \ge |\varrho_{(2)}| \ge \cdots$. If $\boldsymbol{\rho}$ is concentrated with respect to the basis $\mathcal{B}$, then for $\delta > 0$, there is a small integer k such that

$$\Big| \, \|\boldsymbol{\rho}\|_2^2 - \sum_{j=1}^{k} \varrho_{(j)}^2 \Big| < \delta.$$

A small δ implies that the magnitudes of the $|\varrho_{(j)}|$ decrease rapidly. Possible candidates for a sparse representation are wavelets (see Donoho and Johnstone 1994); curvelets, which are used in compressive sensing (see Starck, Candès, and Donoho 2002) or methods obtained from dictionaries, such as the K-SVD approach of Aharon, Elad, and Bruckstein (2006).

Algorithm 13.8, which is described in Johnstone and Lu (2009), incorporates the concentration idea into a variable selection prior to Principal Component Analysis, and results in sparse principal components.

Algorithm 13.8 ***Sparse Principal Components Based on Variable Selection***
Let $\mathbb{X} = \begin{bmatrix}\mathbf{X}_i \cdots \mathbf{X}_n\end{bmatrix}$ be $d \times n$ data. Let $\mathcal{B} = \{\mathbf{e}_1, \ldots, \mathbf{e}_d\}$ be an orthonormal basis for $\mathbb{X}$ and $\boldsymbol{\rho}$.

Step 1. Represent each $\mathbf{X}_i$ in the basis $\mathcal{B}$: $\mathbf{X}_i = \sum_{j=1}^{d} \xi_{ij} \mathbf{e}_j$, and put $\boldsymbol{\xi}_{\bullet j} = \begin{bmatrix}\xi_{1j} \cdots \xi_{nj}\end{bmatrix}$. Calculate the variances $\widehat{\sigma}_j^2 = \operatorname{var}(\boldsymbol{\xi}_{\bullet j})$, for $j \le d$.

Step 2. Fix $\kappa > 0$. Let I be the set of indices j which correspond to the κ largest variances $\widehat{\sigma}_j^2$.

Step 3. Let $\widetilde{\mathbb{X}}^I$ be the $\kappa \times n$ matrix with rows $\boldsymbol{\xi}_{\bullet j}$ and $j \in I$. Calculate the $\kappa \times \kappa$ sample covariance matrix S^I of $\widetilde{\mathbb{X}}^I$ and the eigenvectors $\widehat{\boldsymbol{\rho}}_1^I, \ldots, \widehat{\boldsymbol{\rho}}_\kappa^I$ of S^I.

Step 4. Threshold the eigenvectors $\widehat{\boldsymbol{\rho}}_j^I = [\widehat{\rho}_{j1}^I, \ldots, \widehat{\rho}_{j\kappa}^I]^{\mathsf{T}}$. For $j \le \kappa$, fix $\delta_j > 0$, and put

$$\widetilde{\rho}_{jk}^I = \begin{cases} \widehat{\rho}_{jk}^I & \text{if } |\widehat{\rho}_{jk}^I| \ge \delta_j, \\ 0 & \text{if } |\widehat{\rho}_{jk}^I| < \delta. \end{cases}$$

Step 5. Reconstruct eigenvectors $\widehat{\boldsymbol{\eta}}_j$ for $\mathbb{X}$ in the original domain by putting

$$\widehat{\boldsymbol{\eta}}_j = \sum_{\ell \in I} \widetilde{\rho}^{I}_{j\ell} \mathbf{e}_\ell \qquad \text{for } j \leq \kappa. \tag{13.52}$$

Use the eigenvectors $\widehat{\boldsymbol{\eta}}_j$ to obtain sparse principal components for $\mathbb{X}$. ■

Unlike the variable ranking of Sections 4.8.3 and 13.3.1, Johnstone and Lu (2009) selected their variables from the transformed data $\boldsymbol{\xi}_{\bullet j}$ and $j \leq d$. As a result, the variable selection will depend on the choice of the basis $\mathcal{B}$. If the basis consists of the eigenvectors $\boldsymbol{\eta}_j$ of the covariance matrix Σ of the $\mathbf{X}_i$, then $\boldsymbol{\xi}_{\bullet j} = \boldsymbol{\eta}_j^{\mathsf{T}} \mathbb{X}$, and the ordering of the variances reduces to that of the eigenvalues. In general, $\boldsymbol{\rho}$ will not be concentrated in this basis, and other bases are therefore preferable.

The index set I selects dimensions j such that the coefficients $\boldsymbol{\xi}_{\bullet j}$ have large variances. Implicit in this choice of the index set is the assumption that for the model (13.50), the components j with large values ϱ_j as in (13.51) have large variances $\widehat{\sigma}_j^2$.

Johnstone and Lu (2009) used a data-driven choice for the cut-off of the index set which is based on an argument relating to the upper percentile of the χ_n^2 distribution. The thresholding in step 4 is based on practical experience and yields a further filtering of the noise, with considerable freedom in choosing the threshold parameter.

For the single-component model $\mathbb{X} = \boldsymbol{\rho}\boldsymbol{\nu}^{\mathsf{T}} + \sigma E$ satisfying (13.50), Johnstone and Lu (2009) proposed an explicit form for choosing the cut-off value κ of step 2 in Algorithm 13.8. Fix $\alpha > \sqrt{12}$, put $\alpha_n = \alpha\,[\log(\max\{d,n\})/n]^{1/2}$, and replace step 2 by

Step 2a. Define the subset of variables

$$I = \{ j \,:\, \widehat{\sigma}_j^2 \geq \sigma^2(1+\alpha_n) \}, \tag{13.53}$$

and put $\kappa = |I|$, the number of indices in I.

The seemingly arbitrary value $\alpha > \sqrt{12}$ allows the authors to prove the consistency result in their theorem 2 – our next theorem.

Theorem 13.13 [Johnstone and Lu (2009)] *Let $\mathbb{X} = \boldsymbol{\rho}\boldsymbol{\nu}^{\mathsf{T}} + \sigma E$ be a sequence of models which satisfy (13.50), and regard $\mathbb{X}$, the dimension d and $\boldsymbol{\rho}$ as functions of n. Let $\mathcal{B}$ be a basis for $\mathbb{X}$, and identify $\mathbb{X}$ with the coefficients ξ_{ij} and $\boldsymbol{\rho}$ with the coefficients ϱ_j as in (13.51). Assume that $\boldsymbol{\rho}$ and the dimension d satisfy*

$$\frac{\log[\max(d,n)]}{n} \to 0 \qquad \textit{and} \qquad \|\boldsymbol{\rho}\| \to \varsigma \qquad \textit{as } n \to \infty,$$

for $\varsigma > 0$. Assume that the ordered coefficients $|\varrho_{(1)}| \geq |\varrho_{(2)}| \geq \cdots \geq |\varrho_{(d)}|$ satisfy for some $0 < q < 2$ and $c > 0$

$$|\varrho_{(\nu)}| \leq c\nu^{-1/q} \qquad \textit{for } \nu = 1, 2, \ldots.$$

Use subset I of (13.53) to calculate the eigenvector $\widehat{\boldsymbol{\rho}}_1^{I}$ of S^I as in step 3 of Algorithm 13.8. If $\widehat{\boldsymbol{\rho}}$ is the first eigenvector calculated as in (13.52), then $\widehat{\boldsymbol{\rho}}$ is a consistent estimator of $\boldsymbol{\rho}_0 = \mathrm{dir}(\boldsymbol{\rho})$, and

$$\mathfrak{a}(\widehat{\boldsymbol{\rho}}, \boldsymbol{\rho}_0) \xrightarrow{a.s.} 0 \qquad \textit{as } n \to \infty.$$

A proof of this theorem is given in their paper. The theorem relies on a fixed value α which determines the index set I. Their applications make use of Algorithm 13.8 but they also allow other choices of the index set. They use a wavelet basis for their functional data with localised features, but other bases could be employed instead. For wavelets the sparse way, see Mallat (2009).

The theoretical developments and practical results of Johnstone and Lu (2009) showed the renewed interest in Principal Component Analysis and, more specifically, the desire to estimate the eigenvectors correctly for data sets with very large dimensions. The sparse principal components obtained with Algorithm 13.8 will differ from the sparse principal components described in Sections 13.4.1 and 13.4.2 and the theoretical results of Shen, Shen, and Marron (2013) and Ma (2013), because each method solves a different problem. The fact that each of these methods (and others) uses the name sparse PCs is a reflection of the serious endeavours in revitalising Principal Component Analysis and in adapting it to the needs of high-dimensional data.

13.5.2 Behaviour of the Sample Eigenvalues, Eigenvectors and Principal Component Scores

Theorem 2.25 of Section 2.7.2 states the HDLSS consistency of the first sample eigenvector for data whose first eigenvalue is much larger than the other eigenvalues. This result is based on proposition 1 of Jung and Marron (2009). In this section I present more of their ideas and extensions of their results which reveal the bigger picture with its intricate relationships. I split the main result of Jung and Marron (2009), their theorem 2, into two parts which focus on the different types of convergence that arise in their covariance models. The paper by Jung, Sen, and Marron (2012) analysed the behaviour of the sample eigenvalues and eigenvectors for the boundary case $\alpha = 1$ that was not covered in Jung and Marron (2009). As this case fits naturally into the framework of Jung and Marron (2009), I will combine the results of both papers in Theorems 13.15 and 13.16.

In Section 2.7.2 we encountered spiked covariance matrices, that is, covariance matrices with a small number of large eigenvalues. Suitable measures for assessing the spikiness of covariance matrices are the ε_k of (2.26) in Section 2.7.2. Covariance matrices that are characterised by the ε_k are more general than spiked covariance matrices in that they allow the eigenvalues to decrease at different rates as the dimension increases. Following Jung and Marron (2009), we consider this larger class of models. As mentioned prior to Definition 2.24 in Section 2.7.2, Jung and Marron scaled their sample covariance matrix with $(1/n)$, as do Jung, Sen, and Marron (2012) and Shen et al. (2012). The scaling differs from ours but does not affect the results.

Sample eigenvectors of high-dimensional data can converge or diverge in different ways, and new concepts are needed to describe their behaviour.

Definition 13.14 For fixed n and $d = n+1, n+2, \ldots$, let $\mathbb{X}_d \sim (\mathbf{0}_d, \Sigma_d)$ be a sequence of $d \times n$ data with sample covariance matrix S_d. Let

$$\Sigma_d = \Gamma_d \Lambda_d \Gamma_d \qquad \text{and} \qquad S_d = \widehat{\Gamma}_d \widehat{\Lambda}_d \widehat{\Gamma}_d$$

be the spectral decompositions of Σ_d and S_d, respectively, with eigenvalues λ_j and $\widehat{\lambda}_j$ and corresponding eigenvectors $\boldsymbol{\eta}_j$ and $\widehat{\boldsymbol{\eta}}_j$. The eigenvector $\widehat{\boldsymbol{\eta}}_k$ of S_d is **strongly inconsistent** with its population counterpart $\boldsymbol{\eta}_k$ if the angle $\mathfrak{a}$ between the vectors satisfies

$$\mathfrak{a}(\boldsymbol{\eta}_k, \widehat{\boldsymbol{\eta}}_k) \stackrel{p}{\longrightarrow} \frac{\pi}{2} \qquad \text{or, equivalently, if} \qquad |\boldsymbol{\eta}_k^{\mathsf{T}} \widehat{\boldsymbol{\eta}}_k| \stackrel{p}{\longrightarrow} 0 \qquad \text{as } d \to \infty.$$

The eigenvector $\widehat{\boldsymbol{\eta}}_k$ is **subspace consistent** if there is a set $J = \{\iota, \iota+1, \ldots, \iota+\ell\}$ such that $1 \le \iota \le \iota+\ell \le d$ and

$$\mathfrak{a}(\widehat{\boldsymbol{\eta}}_k, \operatorname{span}\{\boldsymbol{\eta}_j,\ j \in J\}) \stackrel{p}{\longrightarrow} 0 \qquad \text{as } d \to \infty,$$

where the angle is the smallest angle between $\widehat{\boldsymbol{\eta}}_k$ and linear combinations of the $\boldsymbol{\eta}_j$, with $j \in J$. □

To avoid too many subscripts and make the notation unwieldy, I leave out the subscript d in the eigenvalues and eigenvectors.

The strong inconsistency is a phenomenon that can arise in high-dimensional settings, for example, when the variation in the data obscures the underlying structure of Σ_d with increasing d. Subspace consistency, on the other hand, is typically a consequence of two or more eigenvalues being sufficiently close that the correspondence between the eigenvectors and their sample counterparts may no longer be unique. In this case, $\widehat{\boldsymbol{\eta}}_k$ still may converge, but only to a linear combination of the population eigenvectors which belong to the corresponding subset of eigenvalues. To illustrate these ideas, I present Example 4.2 of Jung and Marron (2009).

Example 13.10 [Jung and Marron (2009)] For $\iota = 1, 2$, let $F_{\iota,d}$ be symmetric $d \times d$ matrices with diagonal entries 1 and off-diagonal entries ρ_ι which satisfy $0 < \rho_2 \le \rho_1 < 1$. Put

$$F_{2d} = \begin{pmatrix} F_{1,d} & \mathbf{0} \\ \mathbf{0} & F_{2,d} \end{pmatrix} \qquad \text{and} \qquad \Sigma_{2d} = F_{2d} F_{2d}^{\mathsf{T}}.$$

The first two eigenvalues of Σ_{2d} are

$$\lambda_1 = (d\rho_1 + 1 - \rho_1)^2 \qquad \text{and} \qquad \lambda_2 = (d\rho_2 + 1 - \rho_2)^2,$$

and the $2d \times 2d$ matrix Σ_{2d} satisfies the ε_3 condition $\varepsilon_3(d) \gg 1/d$ (see (2.26) in Section 2.7.2). The matrix F_{2d} is an extension of the matrix F_d described in (2.28) which satisfies the ε_2 condition.

For a fixed n and $d = n+1, n+2, \ldots$, consider $2d \times n$ data $\mathbb{X}_{2d} \sim \mathcal{N}(\mathbf{0}_{2d}, \Sigma_{2d})$.

The relationship between the ρ_ι governs the consistency behaviour of the first two sample eigenvectors. Let $\boldsymbol{\eta}_\iota$ and $\widehat{\boldsymbol{\eta}}_\iota$ be the first two population and sample eigenvectors. Although the eigenvectors have $2d$ entries, it is convenient to think of them as indexed by d. In the following we use the notation $\rho_1 \sim \rho_2$ to mean that the two numbers are about the same. There are four cases to consider, and in each case we compare the growth rates of the ρ_ι to $d^{-1/2}$:

1. If $\rho_1 \gg \rho_2 \gg d^{-1/2}$, then, for $\iota = 1, 2$, $\mathfrak{a}(\boldsymbol{\eta}_\iota, \widehat{\boldsymbol{\eta}}_\iota) \stackrel{p}{\longrightarrow} 0$, as $d \to \infty$.
2. If $\rho_1 \sim \rho_2 \gg d^{-1/2}$, then, for $\iota = 1, 2$, $\mathfrak{a}(\widehat{\boldsymbol{\eta}}_\iota, \operatorname{span}\{\boldsymbol{\eta}_1, \boldsymbol{\eta}_2\}) \stackrel{p}{\longrightarrow} 0$, as $d \to \infty$.
3. If $\rho_1 \gg d^{-1/2} \gg \rho_2$, then, as $d \to \infty$,

$$\mathfrak{a}(\boldsymbol{\eta}_\iota, \widehat{\boldsymbol{\eta}}_\iota) \stackrel{p}{\longrightarrow} \begin{cases} 0 & \text{if } \iota = 1, \\ \pi/2 & \text{if } \iota = 2. \end{cases}$$

4. If $d^{-1/2} \gg \rho_1 \gg \rho_2$, then, for $\iota = 1,2$, $\mathfrak{a}(\boldsymbol{\eta}_\iota, \widehat{\boldsymbol{\eta}}_\iota) \xrightarrow{p} \pi/2$, as $d \to \infty$.

The four cases show that if the ρ_ι are sufficiently different and decrease sufficiently slowly, then both eigenvectors are HDLSS consistent. The type of convergence changes to subspace consistency when the ρ_ι and hence also the first two eigenvalues, are about the same size. As soon as one of the ρ_ι decreases faster than $d^{-1/2}$, the sample eigenvector becomes strongly inconsistent. This illustrates that there are essentially two behaviour modes: the angle between the eigenvectors decreases to zero or the angle diverges maximally. ■

Instead of defining the four cases of the example by the ρ_ι one could consider growth rates of λ_ι. Jung and Marron (2009) indicated how this can be done, but because the eigenvalues are defined from the ρ_ι the approach used in the example is more natural.

Equipped with the ideas of subspace consistency and strong inconsistency, we return to Theorem 2.25 and inspect its assumptions: The ε_2 assumption is tied to the model of a single large eigenvalue. In their theorem 1, Jung and Marron show that for such a single large eigenvalue the sample eigenvalues behave as if they are from a scaled covariance matrix, and the matrix $\mathbb{X}_d^\mathsf{T}\mathbb{X}_d$ converges to a multiple of the identity as d grows. The assumption that λ_1/d^α converges for $\alpha > 1$ is of particular interest; in Theorem 2.25, λ_1 grows at a faster rate than d because $\alpha > 1$. As we shall see in Theorem 13.15, the rate of growth of λ_1 turns out to be the key factor which governs the behaviour of the first eigenvector.

Theorem 13.15 refers to the ρ-mixing property which is defined in (2.29) in Section 2.7.2.

Theorem 13.15 [Jung and Marron (2009), Jung, Sen, and Marron (2012)] *For fixed n and $d = n+1, n+2, \ldots$, let $\mathbb{X}_d \sim (\mathbf{0}_d, \Sigma_d)$ be a sequence of $d \times n$ data with Σ_d, S_d and their eigenvalues and eigenvectors as in Definition 13.14. Put $\mathbb{W}_{\Lambda,d} = \Lambda_d^{-1/2}\Gamma_d^\mathsf{T}\mathbb{X}_d$, and assume that $\mathbb{W}_{\Lambda,d}$ have uniformly bounded fourth moments and are ρ-mixing for some permutation of the rows. Let $\alpha > 0$. Assume that*

$$\frac{\lambda_1}{d^\alpha} = \sigma^2 \qquad \text{for } \sigma > 0$$

and

$$\lambda_k = \tau^2 \qquad \text{for } \tau > 0 \text{ and } 1 < k \le n.$$

Put $\delta = \tau^2/\sigma^2$. If the Σ_d satisfy the ε_2 condition $\varepsilon_2(d) \gg 1/d$, and if $\sum_{k>1} \lambda_k = O(d)$, then the following hold.

1. *The first eigenvector $\widehat{\boldsymbol{\eta}}_1$ of S_d satisfies*

$$\mathfrak{a}(\widehat{\boldsymbol{\eta}}_1, \boldsymbol{\eta}_1) \longrightarrow \begin{cases} 0 & \text{in prob. for } \alpha > 1, \\ \arccos\left[(1+\delta/\chi)^{-1/2}\right] & \text{in dist. for } \alpha = 1, \\ \pi/2 & \text{in prob. for } \alpha \in (0,1). \end{cases}$$

as $d \to \infty$, where χ is a random variable. For $1 < k \le n$, the remaining eigenvectors satisfy

$$\mathfrak{a}(\widehat{\boldsymbol{\eta}}_k, \boldsymbol{\eta}_k) \xrightarrow{p} \pi/2 \qquad \text{for all } \alpha > 0, \text{ as } d \to \infty.$$

2. *The first eigenvalue* $\widehat{\lambda}_1$ *of* S_d *satisfies*

$$\frac{n}{\sigma^2} \frac{\widehat{\lambda}_1}{\max(d^\alpha, d)} \stackrel{D}{\longrightarrow} \begin{cases} \chi & \text{for } \alpha > 1, \\ \chi + \delta & \text{for } \alpha = 1, \\ \delta & \text{for } \alpha \in (0,1), \end{cases}$$

as $d \to \infty$*, where* χ *is a random variable. For* $1 < k \leq n$*, the remaining eigenvalues satisfy*

$$\frac{n\widehat{\lambda}_k}{d} \stackrel{p}{\longrightarrow} \tau^2 \qquad \text{for all } \alpha > 0, \text{ as } d \to \infty.$$

3. *If the* $\mathbb{X}_d$ *are also normal, then the random variable* $n\widehat{\lambda}_1/\lambda_1$ *converges in distribution to a* χ^2_n *random variable as* $d \to \infty$.

As mentioned in the comments following Theorem 2.25, the $\mathbb{W}_{\Lambda,d}$ are obtained from the data by sphering with the population quantities $\Lambda_d^{-1/2}\Gamma_d^{\mathsf{T}}$. In their proposition 1, Jung and Marron (2009) proved the claims of Theorem 13.15 which relate to the strict inequalities $\alpha > 1$ and $\alpha < 1$. Jung, Sen, and Marron (2012) extended these results to the boundary case $\alpha = 1$ and thereby completed the picture for HDLSS data with one large eigenvalue. The theorem tells us that for covariance matrices with a single large eigenvalue, the behaviour of the first sample eigenvector depends on the growth rate of the first eigenvalue λ_1 with d: the first sample eigenvector is either HDLSS consistent or strongly inconsistent when $\alpha \neq 1$, and it converges to a random variable when $\alpha = 1$. The second and subsequent eigenvectors of S_d are strongly inconsistent; this follows because the remaining eigenvalues remain sufficiently small as d grows.

The consistency results ($\alpha > 1$) stated in the theorem are repeats of the corresponding statements in Theorem 2.25 in Section 2.7.2. Theorem 13.15 allows the greater range of values $\alpha > 0$, and the rate of growth of λ_1 therefore ranges from slower to faster than d and includes equality.

The single large spike result of Theorem 13.15 has two natural extensions to κ large spikes of the covariance matrix Σ_d: the eigenvalues have different growth rates, or they have the same growth rate but different constants. As we shall see, the two cases lead to different asymptotic results. For $j \leq d$, let λ_j be the eigenvalues of Σ_d, listed in decreasing order as usual. Fix $\kappa < d$. For $k > \kappa$, assume that $\lambda_k = \tau^2$ for $\tau > 0$, as in Theorem 13.15. The first κ eigenvalues take one of the following forms:

1. Assume that there is a sequence $\alpha_1 > \alpha_2 > \cdots > \alpha_\kappa > 1$ and constants $\sigma_j > 0$ such that

$$\lambda_j = \sigma_j^2 d^{\alpha_j} \qquad \text{for } j \leq \kappa. \tag{13.54}$$

2. Let $\alpha > 1$. Assume that there is a sequence $\sigma_1 > \sigma_2 > \cdots > \sigma_\kappa > 1$ such that

$$\lambda_j = \sigma_j^2 d^{\alpha} \qquad \text{for } j \leq \kappa. \tag{13.55}$$

If the eigenvalues grow as in (13.54), then each of the first κ sample eigenvalues and eigenvectors converges as described in Theorem 13.15. This result is shown in proposition 2 of Jung and Marron (2009). The situation, however, becomes more complex for eigenvalues as in (13.55). We look at this case in Theorem 13.16.

The following theorem refers to the Wishart distribution, which is defined in (1.14) in Section 1.4.1.

Theorem 13.16 [Jung and Marron (2009), Jung, Sen, and Marron (2012)] *For fixed n and $d = n+1, n+2, \ldots$, let $\mathbb{X}_d \sim (\mathbf{0}_d, \Sigma_d)$ be a sequence of $d \times n$ data, with Σ_d, S_d and their eigenvalues and eigenvectors as in Definition 13.14. Let $\mathbb{W}_{\Lambda,d} = \Lambda_d^{-1/2}\Gamma_d^{\mathsf{T}}\mathbb{X}_d$ satisfy the assumptions of Theorem 13.15. Let $\kappa < n$ and $\alpha > 0$. Assume that there is a $\tau > 0$ and a sequence $\sigma_1 > \sigma_2 > \cdots > \sigma_\kappa > 1$ such that*

$$\lambda_j = \sigma_j^2 d^\alpha \qquad \text{for } j \le \kappa \qquad \text{and} \qquad \lambda_k = \tau^2 \qquad \text{for } \kappa < k \le n.$$

If the Σ_d satisfy the $\varepsilon_{\kappa+1}$ condition $\varepsilon_{\kappa+1}(d) \gg 1/d$ and $\sum_{k>\kappa} \lambda_k = O(d)$, then, for $j \le \kappa$, the following hold:

1. *The jth eigenvector $\widehat{\boldsymbol{\eta}}_j$ and linear combinations of the eigenvectors $\boldsymbol{\eta}_1, \ldots, \boldsymbol{\eta}_\kappa$ satisfy*

$$\mathfrak{a}(\widehat{\boldsymbol{\eta}}_j, \mathit{span}\{\boldsymbol{\eta}_\iota,\ \iota = 1,\ldots,\kappa\}) \longrightarrow \begin{cases} 0 & \text{in prob. for } \alpha > 1,\\ \arccos\left[(1+\tau^2/\omega_j)^{-1/2}\right] & \text{in dist. for } \alpha = 1, \end{cases}$$

   as $d \to \infty$, where ω_j is the jth eigenvalue of a random matrix of size $\kappa \times \kappa$.
2. *The jth eigenvalue $\widehat{\lambda}_j$ satisfies*

$$\frac{n\widehat{\lambda}_j}{d^\alpha} \xrightarrow{D} \begin{cases} \omega_j & \text{for } \alpha > 1,\\ \omega_j + \tau^2 & \text{for } \alpha = 1, \end{cases}$$

   as $d \to \infty$, where ω_j is as in part 1.
3. *If the $\mathbb{X}_d$ are also normal, then, as $d \to \infty$, the random variable ω_j of part 2 is the jth eigenvalue of a matrix that has a Wishart distribution with n degrees of freedom and diagonal covariance matrix whose jth diagonal entry is σ_j^2.*

Theorem 13.16 is a combination of results: Jung and Marron (2009) dealt with the case $\alpha > 1$, and Jung, Sen, and Marron (2012) considered the case $\alpha = 1$. Proofs of the statements in Theorem 13.16 and details relating to the limiting random variables can be found in these papers. Starting with $\alpha > 1$, part 1 of Theorem 13.16 is proposition 3 of Jung and Marron (2009), part 2 is their lemma 1 and part 3 corresponds to their corollary 3. The proof of their proposition 2 follows from the proof of their theorem 2 which generalises the proposition. For the $\alpha = 1$ results, part 1 is part of theorem 3 of Jung, Sen, and Marron (2012), part 2 comes from their theorem 2 and part 3 follows from the comments following their theorem 2.

I have only stated results for the first κ eigenvalues and eigenvectors. For the sample eigenvectors $\widehat{\boldsymbol{\eta}}_k$ with $\kappa < k \le n$, the corresponding eigenvalues λ_k are small, and the eigenvectors are strongly inconsistent, similar to the eigenvectors in part 1 of Theorem 13.15.

Theorem 13.16 states that the sample eigenvectors are no longer HDLSS consistent in the sense that they converge to a single population eigenvector. The best one can achieve, even when $\alpha > 1$, is that each sample eigenvector converges to a linear combination of eigenvectors. For the boundary case $\alpha = 1$, the angle between $\widehat{\boldsymbol{\eta}}_j$ and a best linear combination of the $\boldsymbol{\eta}_j$ converges to a random variable. The eigenvalues still converge to separate random variables, but the distribution of these random variables becomes more complex. If the data are normal, then the random variables are the eigenvalues of a random matrix that has a Wishart distribution.

So far we have considered the behaviour of the eigenvalues and eigenvectors of the covariance matrix when the dimension grows. We have seen that the sample eigenvectors are

HDLSS consistent if the first κ eigenvalues of Σ_d grow at a faster rate than d. It is natural to assume that the corresponding sample PC scores converge to their population counterparts. Shen et al. (2012) observed that this is not the case; the sample PC scores may fail to converge to the population quantities even if the corresponding eigenvectors are consistent.

The following theorem describes the behaviour of the PC scores. It uses the the notation of Definition 13.14 as well as the following: let $\mathbb{X}$ be $d \times n$ data, and for $j \leq d$, let

$$\mathbf{V}_{\bullet j} = \begin{bmatrix} V_{1j} \cdots V_{1j} \end{bmatrix} \quad \text{with } V_{ij} = \lambda_j^{-1/2} \boldsymbol{\eta}_j^{\mathsf{T}} \mathbf{X}_i$$

and

$$\widehat{\mathbf{V}}_{\bullet j} = \begin{bmatrix} \widehat{V}_{1j} \dots \widehat{V}_{nj} \end{bmatrix} \quad \text{with } \widehat{V}_{ij} = \widehat{\lambda}_j^{-1/2} \widehat{\boldsymbol{\eta}}_j^{\mathsf{T}} \mathbf{X}_i \tag{13.56}$$

be the jth sphered principal component scores of $\mathbb{X}$. The row vector $\mathbf{V}_{\bullet j}$ is based on the eigenvalues and eigenvectors of Σ, and the score $\widehat{\mathbf{V}}_{\bullet j}$ is the usual sphered jth sample principal component score.

For eigenvalues λ_j, λ_k of Σ, we explore relationships between pairs of eigenvalues, as d or n grow. In the second case, n grows, while d grows in the first and third case. Write

$\lambda_k \gg \lambda_j$ if $\lim_{d\to\infty} \lambda_j/\lambda_k = 0$,

$\lambda_k \succeq \lambda_j$ if $\limsup_{n\to\infty} \lambda_j/\lambda_k < 1$, and

$\lambda_k \sim \lambda_j$ if the upper and lower limits of λ_j/λ_k are bounded above and below as d grows.

Theorem 13.17 [Shen et al. (2012)] *Let $\mathbb{X}_d \sim \mathcal{N}(\mathbf{0}_d, \Sigma_d)$ be a sequence of multivariate normal data indexed by d. Let Σ_d and S_d be the covariance and sample covariance matrices of $\mathbb{X}$, and let $(\lambda_j, \boldsymbol{\eta}_j)$ and $(\widehat{\lambda}_j, \widehat{\boldsymbol{\eta}}_j)$ be the eigenvalue-eigenvector pairs of Σ_d and S_d, respectively.*

1. *For fixed n, let $\mathbb{X}_d$ be a sequence of size $d \times n$ with $d = n+1, n+2, \ldots$. For $\kappa < n$, assume that the eigenvalues λ_j of Σ_d satisfy*

$$\lambda_1 \gg \cdots \gg \lambda_\kappa \gg \lambda_{\kappa+1} \sim \cdots \sim \lambda_d \sim 1.$$

If $d = o(\lambda_\kappa)$ as $d \to \infty$, then the scores V_{ij} and $\widehat{V}_{ij}$, derived from $\mathbb{X}_d$ as in (13.56), satisfy

$$\left| \frac{\widehat{V}_{ij}}{V_{ij}} \right| \xrightarrow{p} \zeta_j \qquad \text{for } i \leq n,\ j \leq \kappa,$$

where $\zeta_j = (n/\chi_j)^{1/2}$ is a random variable and χ_j has a χ_n^2 distribution.

2. *Let $n \to \infty$ and $d = O(n)$. For $\kappa < n$, assume that the eigenvalues λ_j of Σ_d satisfy*

$$\lambda_1 \succeq \cdots \succeq \lambda_\kappa \gg \lambda_{\kappa+1} \sim \cdots \sim \lambda_d \sim 1.$$

If $d = o(\lambda_\kappa)$ as $n \to \infty$, then the scores V_{ij} and $\widehat{V}_{ij}$, derived from $\mathbb{X}_d$ as in (13.56), satisfy

$$\left| \frac{\widehat{V}_{ij}}{V_{ij}} \right| \xrightarrow{a.s.} 1 \qquad \text{for } i \leq n,\ j \leq \kappa,$$

where the convergence is the almost sure convergence.

Theorem 13.17 deals with eigenvalues which satisfy (13.54), so the first κ sample eigenvectors are HDLSS consistent. Despite this behaviour, the first κ sample scores cannot be used to consistently estimate the population scores when n is fixed and d grows. This behaviour is a consequence of a lack of data – recall that $n \ll d$. Surprisingly, the jth ratio of sample score and population score converges to the *same* random variable for each observation $\mathbf{X}_i$. As Shen et al. (2012) remarked, these findings suggest that the score plots still can be used to explore structure in the high-dimensional data.

Part 2 of Theorem 13.17 describes a setting which remedies the lack of data: now n increases, and as a consequence, the consistency of the scores can be established. The setting of the second part of the theorem leads the way back to the more general setting of Shen, Shen, and Marron (2012). I give a brief overview of their framework and results in the final section.

13.5.3 Towards a General Asymptotic Framework for Principal Component Analysis

The (in)consistency results of Sections 13.5.1 and 13.5.2 highlight the delicate balance that exists between the sample size and dimension and between the shape of the covariance matrix and the dimension. The theorems of these sections show how the consistency of principal component directions and scores of high-dimensional data depend on these relationships. We learned the following:

1. Sample eigenvectors are typically either consistent or strongly inconsistent.
2. Spiked covariance matrices with a few eigenvalues that grow sufficiently fast and at sufficiently different rates are candidates for producing a few consistent sample eigenvectors.
3. Consistency of the first sample eigenvector is not sufficient to guarantee that the PC scores are consistent.

Johnstone and Lu (2009) and Jung and Marron (2009) considered different models with distinct growth rates of d and n. The approach of Johnstone and Lu (2009) emphasised the connection with sparse PCs and proposed the use of sparse bases, whereas Jung and Marron (2009) focused more specifically on the size of the eigenvalues for a fixed sample size. A combination of their ideas could lead to a natural generalisation by letting the number of large eigenvalues inform the choice of basis for a sparse representation of sample eigenvectors and sparse principal components.

The research of Johnstone and Lu (2009) and Jung and Marron (2009) is just the beginning, and their results have since been applied and generalised in many directions. Shen, Shen, and Marron (2013) and Ma (2013) focussed on consistency for sparse settings. Paul (2007), Nadler (2008), Lee, Zou, and Wright (2010) and Benaych-Georges and Nadakuditi (2012) proved results for the random matrix domain $\mathbf{n}{\succeq}\mathbf{d}$ and $\mathbf{n}{\preceq}\mathbf{d}$ – see Definition 2.22 in Section 2.7.2. More closely related to and also extending Jung and Marron (2009) are the results of Jung, Sen, and Marron (2012) and Shen et al. (2012) which I described in the preceding section. This emerging body of research clearly acknowledges the need for new methods and the role Principal Component Analysis plays in the analysis of high-dimensional data, HDLSS data and functional data.

Johnstone and Lu (2009) focused on the pair sample size and dimension and kept the size of the largest eigenvalue of Σ constant. In their case it does not matter whether d is a function of n or n is regarded as a function of d because the key information is the growth rate of d with n. In contrast, Jung and Marron (2009), Jung, Sen, and Marron (2012) and Shen et al. (2012) chose the HDLSS framework: n remains fixed, and the eigenvalues of Σ become functions of the dimension.

The three quantities sample size, dimension and the size of the eigenvalues of Σ are closely connected. The key idea of Shen, Shen, and Marron (2012) is to treat the sample size and the eigenvalues simultaneously as function of d and break with the '*as* $n \to \infty$' tradition. This insight leads to an elegant general framework for developing asymptotics in Principal Component Analysis. The framework of Shen, Shen, and Marron (2012) covers different regimes, as well as transitions between the regimes, and includes the settings of Sections 13.5.1 and 13.5.2 as special cases. I will give a brief outline of their ideas in the remainder of this section and start with some notation.

Let $\mathbb{X}_d = \left[\mathbf{X}_1 \cdots \mathbf{X}_n\right]$ be a sequence of Gaussian data indexed by the dimension d such that $\mathbf{X}_i \sim \mathcal{N}(\mathbf{0}_d \Sigma_d)$, for $i \le n$. Let $\lambda_1, \ldots, \lambda_d$ be the eigenvalues of Σ_d. Call $\alpha \ge 0$ the **spike index** and $\gamma \ge 0$ the **sample index**, and assume that

$$n \sim d^{\gamma}, \qquad \lambda_j \sim d^{\alpha}, \qquad \text{and} \qquad \lambda_{\kappa+\ell} \sim 1, \tag{13.57}$$

for some $1 \le \kappa \le d,\ j \le \kappa$ and $\ell \ge 1$.

In our next and last theorem, the rate of $d/(n\lambda_1)$ will be the key to the behaviour of the eigenvalues and eigenvectors of S_d. For an interpretation of the results, it will be convenient to express this ratio in terms of the sample and spike indices

$$\frac{d}{n\lambda_1} \sim \frac{d}{d^{\gamma+\alpha}}.$$

As $d \to \infty$, there are three distinct cases:

$$\gamma + \alpha \quad \begin{cases} > 1 & \text{so} \quad \dfrac{d}{n\lambda_1} \to 0, \\ = 1 & \text{so} \quad \dfrac{d}{n\lambda_1} \to c, \\ < 1 & \text{so} \quad \dfrac{d}{n\lambda_1} \to \infty. \end{cases} \tag{13.58}$$

Although not used explicitly in the form (13.58), we recognise the three cases in the following theorem which further splits the first case into two parts.

Theorem 13.18 [Shen, Shen, and Marron (2012)] *Let $\mathbb{X}_d \sim \mathcal{N}(\mathbf{0}_d, \Sigma_d)$ be a sequence of $d \times n$ data. Let $\kappa = 1$. Assume that the eigenvalues λ_j of Σ_d satisfy (13.57). Let S_d be the sample covariance matrix, and let $\widehat{\lambda}_j$ be the non-zero eigenvalues of S_d, with $j \ge 1$. Let $\boldsymbol{\eta}_j$ and $\widehat{\boldsymbol{\eta}}_j$ be the eigenvectors of Σ_d and S_d. If both $d, n \to \infty$, then the following hold.*

1. *If $d/(n\lambda_1) \to 0$, then the first eigenvector is consistent, the remaining ones are subspace consistent and*

$$\frac{\widehat{\lambda}_1}{\lambda_1} \xrightarrow{a.s.} 1 \qquad \text{and} \qquad \left|\widehat{\boldsymbol{\eta}}_1^{\mathsf{T}}\boldsymbol{\eta}_1\right| = 1 + O\left(\left[\frac{d}{n\lambda_1}\right]^{1/2}\right) \quad a.s.,$$

$$\widehat{\lambda}_j \overset{a.s.}{\sim} \frac{d}{n}, \qquad \widehat{\lambda}_k = O\left(\frac{d}{n}\right) \quad a.s.$$

and

$$\mathfrak{a}(\widehat{\boldsymbol{\eta}}_\ell, span\{\boldsymbol{\eta}_\iota,\ \iota \geq 2\}) = O\left(\left[\frac{d}{n\lambda_1}\right]^{1/2}\right) \quad a.s.,$$

where $j \in J = \{2 \cdots \min(n, d-1)\}$, $k > 1$, $k \notin J$ and $\ell > 1$.

2. *If $d/(n\lambda_1) \to 0$ and $d/n \to \infty$, then the first eigenvector is consistent, the remaining ones are strongly inconsistent and*

$$\frac{\widehat{\lambda}_1}{\lambda_1} \xrightarrow{a.s.} 1 \qquad \text{and} \qquad \left|\widehat{\boldsymbol{\eta}}_1^{\mathsf{T}}\boldsymbol{\eta}_1\right| = 1 + O\left(\left[\frac{d}{n\lambda_1}\right]^{1/2}\right) \quad a.s.,$$

$$\widehat{\lambda}_j \overset{a.s.}{\sim} \frac{d}{n} \qquad \text{and} \qquad \left|\widehat{\boldsymbol{\eta}}_j^{\mathsf{T}}\boldsymbol{\eta}_j\right| = O\left(\left[\frac{n}{d}\right]^{1/2}\right) \quad a.s. \quad \text{for } j > 1.$$

3. *If $d/(n\lambda_1) \to c \in (0, \infty)$, then the first eigenvector is inconsistent, all others are strongly inconsistent and*

$$\frac{\widehat{\lambda}_1}{\lambda_1} \xrightarrow{a.s.} 1 + c \qquad \text{and} \qquad \left|\widehat{\boldsymbol{\eta}}_1^{\mathsf{T}}\boldsymbol{\eta}_1\right| = (1+c)^{-1/2} + o(1) \ a.s.,$$

$$\widehat{\lambda}_j \xrightarrow{a.s.} \frac{d}{n} \qquad \text{and} \qquad \left|\widehat{\boldsymbol{\eta}}_j^{\mathsf{T}}\boldsymbol{\eta}_j\right| = O\left(\left[\frac{n}{d}\right]^{1/2}\right) \quad a.s. \quad \text{for } j > 1.$$

4. *If $d/(n\lambda_1) \to \infty$, then all eigenvectors are strongly inconsistent and*

$$\widehat{\lambda}_j \overset{a.s.}{\sim} \frac{d}{n} \qquad \text{and} \qquad \left|\widehat{\boldsymbol{\eta}}_j^{\mathsf{T}}\boldsymbol{\eta}_j\right| = O\left(\left[\frac{n\lambda_j}{d}\right]^{1/2}\right) \quad a.s. \quad \text{for } j \geq 1.$$

Theorem 13.18 covers theorems 3.1 and 3.3 of Shen, Shen, and Marron (2012), and proofs can be found in their paper. Shen, Shen, and Marron (2012) presented a number of theorems for the different regimes. I have only quoted results relating to the single component spike, so $\kappa = 1$ in (13.57). Theorem 13.18 provides insight into the asymptotic playing field as d, n and λ_1 grow at different rates. Shen, Shen, and Marron (2012) also considered different scenarios for multi-spike models, so $\kappa > 1$. For results relating to $\kappa > 1$, I refer the reader to their theorems 4.1–4.4.

I conclude this final section with some comments and remarks on the properties stated in the theorem and a figure. The figure summarises schematically the asymptotic framework of Shen Shen, and Marron (2012) both for the single-spike and multi-spike case. It is reproduced from Shen, Shen, and Marron (2012) with the permission of the authors.

Remark 1. Parts 1 and 2 of Theorem 13.18 tell us that the first eigenvalue converges to the population eigenvalue, and the angle between the first sample and population eigenvectors goes to zero. Thus, $\widehat{\boldsymbol{\eta}}_1$ is a consistent estimator of $\boldsymbol{\eta}_1$ provided that $n \to \infty$, and d grows at a rate slower than that of $n\lambda_1$. This setting corresponds to the case $\gamma + \alpha > 1$. The rate tells us how fast the angle between the first eigenvectors converges to 0 degrees.

Because $\lambda_j \sim 1$ for $j > 1$, $d/(n\lambda_j) = d/n$ in part 2 and, by assumption, $d/n \to \infty$. It follows that all but the first eigenvector are strongly inconsistent, so the angle between the first eigenvectors converges to 90 degrees.

Remark 2. In part 3, the growth rate of d is the same as that of $n\lambda_1$, so $\alpha + \gamma = 1$. For this boundary case, which generalises the case $\alpha = 1$ in Theorems 13.15 and 13.16, the first eigenvalue no longer converges to its population quantity, and the eigenvector fails to be consistent. This result is surprising and interesting. Here the angle goes to a constant between 0 and 90 degrees. In contrast to the strong inconsistency of the second and later eigenvectors, Shen Shen, and Marron (2012) referred to the behaviour of the first eigenvector as **inconsistent**.

Remark 3. Part 4 deals with the case $\gamma + \alpha < 1$; that is, d grows faster than $n\lambda_1$. Now all eigenvalues diverge, and the angle of the sample and population eigenvectors converges to 90 degrees, so all eigenvectors are strongly inconsistent.

Remark 4. The statements in Theorem 13.18 describe the behaviour of the first eigenvalue and that of the angle between the first population and sample eigenvectors with different growth rates of d, n and λ_1. In Theorem 13.3 we met similar asymptotic relationships between the ranking vectors $\widehat{\mathbf{p}}_{\mathrm{NB}}$ and $\mathbf{p}_{\mathrm{NB}}$ in a two-class discrimination problem based on ideas from Canonical Correlation Analysis. In Theorem 13.3, the strength of the observations, which is characterised by the sequence k_d of (13.25), takes the place of the eigenvalue λ_1 in Theorem 13.18. Theorem 13.3 is less general than Theorem 13.18 in that it can only deal with one singular value and one pair of eigenvectors. However, the similarity in the asymptotic behaviour between the two settings is striking.

Allowing for the change from k_d to λ_1, a comparison of parts 2 and 3 of Theorem 13.3 with parts 1 to 3 of Theorem 13.18 highlights the following similarities:

1. Consider the growth rates $d/(nk_d) \to 0$ and $d/(n\lambda_1) \to 0$, then the first singular value and, respectively, the first eigenvalue converge to the population quantity, and the angle between the first eigenvectors converges to zero, so the sample eigenvector is HDLSS consistent.
2. Consider the growth rates $d/(nk_d) \to c > 0$ and $d/(n\lambda_1) \to c > 0$, then the angle between the first eigenvectors converges to a non-zero angle, so the sample eigenvector is inconsistent.

It remains to discuss Figure 13.6 of Shen Shen, and Marron (2012), which succinctly summarises a general framework for Principal Component Analysis and covers the classical case, the random matrix case and the HDLSS setting. The left panel refers to the single spike setting, $\kappa = 1$, our Theorem 13.18. The right panel refers to the multi-spike setting, $\kappa > 1$, that we have looked at in Theorem 13.16 and which is covered in more detail in theorems 4.1–4.4 of Shen, Shen, and Marron (2012) and references therein. The figure shows the regions of consistency and strong inconsistency of the PC eigenvectors as functions of the spike index α and the sample index γ of (13.57), with α on the x-axis and γ on the

y-axis. More specifically, the figure panels characterise the eigenvectors which correspond to the large eigenvalues, such as $\lambda_1, \ldots, \lambda_\kappa$ in Theorem 13.16.

The white triangular regions with $0 \leq \alpha + \gamma < 1$ refer to the regimes of strong inconsistency. For these regimes, d grows much faster than $n\lambda_1$. The sample eigenvectors no longer contain useful information about their population quantities. The regions marked in grey correspond to $\alpha + \gamma > 1$, that is, regimes where the first (few) eigenvectors are consistent.

In the left panel there is a solid line which separates the consistent regime from the strongly inconsistent regime. As we have seen in Theorems 13.15, 13.16 and 13.18, the angle between the first (few) eigenvectors converges to a fixed non-zero angle, so the eigenvectors are inconsistent. The separating line includes the special cases ($\alpha = 0, \gamma = 1$) of Johnstone and Lu (2009), our Theorem 13.12, and ($\alpha = 1, \gamma = 0$) of Jung, Sen, and Marron (2012), which I report as part of Theorem 13.15 for the single-spike case.

In the multi-spike case, the interpretation of the figure is more intricate. For $\kappa > 1$ large eigenvalues, subspace consistency of the eigenvectors can occur when $\gamma = 0$, so when n is fixed as in Theorem 13.16. The line where subspace consistency can occur is marked in red in the right panel of the figure. The figure also tells us that we require $\gamma > 0$, as well as $\alpha + \gamma > 1$, for consistency. The case $\gamma > 0$ implies that n grows as in Theorem 13.18.

The classical case, d fixed, is included in both panels in the grey consistency regimes. The random matrix cases and the HDLSS scenario cover the white and grey areas, and do not fit completely into one or the other of the two regimes. The two regimes depend on the size of the large eigenvalues as well as n and d, while the random matrix cases and the HDLSS scenario are characterised by the interplay of n and d only. Indeed, the inclusion of the growth rate of the eigenvalues and the simultaneous treatment of the three key quantities dimension, sample size and size of the eigenvalues have enabled Shen Shen, and Marron (2012) to present a coherent and general framework for Principal Component Analysis.

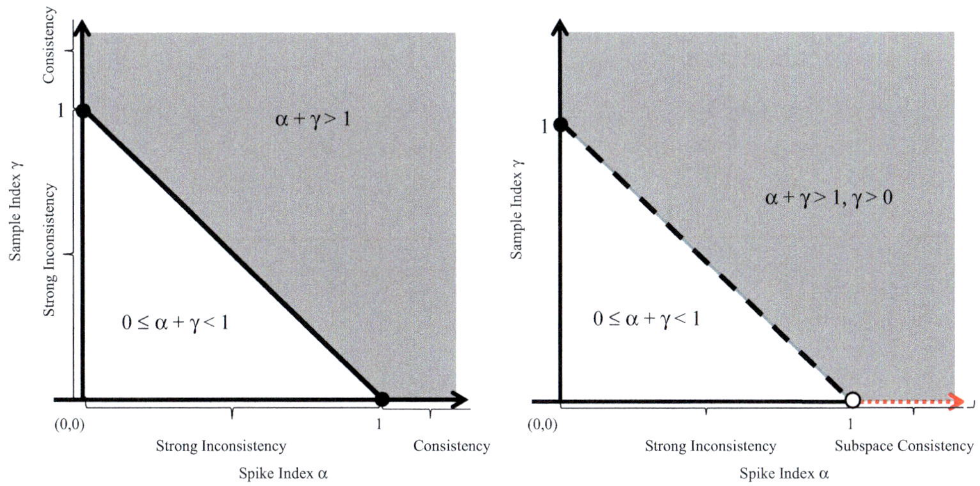

Figure 13.6 (In)Consistency of the PC eigenvectors as the dimension d, the sample size ($n = d^\gamma$) and the size of the eigenvalues ($\lambda_j = d^\alpha$) vary. Left panel refers to one large eigenvalue; right panel refers to multiple large eigenvalues. Grey regions correspond to the regime of consistent eigenvectors, and the white regions show the regime of strongly inconsistent eigenvectors. (Reprinted from Shen, Shen, and Marron (2012) with the permission of the authors.)

Problems for Part III

Independent Component Analysis

1. Give a proof of Theorem 10.2 with emphasis on the form of the matrix E. *Hint:* You may use the proof of Comon (1994).
2. Consider Fisher's *iris* data. Leave out the 'red' species *Setosa*, and repeat the analyses of Example 10.2 for the remaining two species. Compare the results of the new analysis with those of Example 10.2.
3. Give a detailed proof of Proposition 10.8 stating precisely which results are used and how.
4. Explain why it is advantageous to sphere a random vector before finding the independent components but not prior to a Principal Component Analysis. Illustrate with the *Swiss bank notes* data.
5. Let $\mathbf{X}$ and $\mathbf{Y}$ be d-dimensional random vectors with probability density functions f and g and marginals f_j and g_j. Let A be an invertible transformation, and define a function τ by $\mathbf{Z} \longmapsto A\mathbf{Z}$, where $\mathbf{Z}$ is $\mathbf{X}$ or $\mathbf{Y}$. Let f_τ and g_τ be the probability density functions of the transformed random vectors $\tau(\mathbf{X})$ and $\tau(\mathbf{Y})$. Prove the following statements.

   (a) $\mathcal{I}(f) = \mathcal{K}(f, \prod f_j)$.
   (b) $\mathcal{J}(f_\tau) = \mathcal{J}(f)$, so $\mathcal{J}$ is invariant under τ.
   (c) $\mathcal{K}(f_\tau, g_\tau) = \mathcal{K}(f, g)$, so $\mathcal{K}$ is invariant under τ.
   (d) Assume further that $g = \prod g_j$, then $\mathcal{K}(f, g) = \mathcal{I}(f) + \sum_j \mathcal{K}(f_j, g_j)$.

6. The three data sets $\mathbb{X}_1$, $\mathbb{X}_2$ and $\mathbb{X}_3$ of Figure 9.1 in Section 9.1 are generated from the following mixtures of normals. $\mathbf{X}_1 \sim \mathcal{N}(\mathbf{0}, \Sigma_1)$, $\mathbf{X}_2 \sim \{0.6 \times \mathcal{N}(\mathbf{0}, \Sigma_1) + 0.4 \times \mathcal{N}([5, 0, 0]^{\mathsf{T}}, \Sigma_2)\}$ and $\mathbf{X}_3 \sim \{0.4 \times \mathcal{N}(\mathbf{0}, \Sigma_1) + 0.35 \times \mathcal{N}([-5, 0, 0]^{\mathsf{T}}, \Sigma_2) + 0.25 \times \mathcal{N}([0, -4, 3]^{\mathsf{T}}, \Sigma_3)\}$, where

$$\Sigma_1 = \begin{pmatrix} 2.4 & -0.5 & 0 \\ -0.5 & 1 & 0 \\ 0 & 0 & 1 \end{pmatrix} \qquad \Sigma_2 = \begin{pmatrix} 2.4 & -0.5 & 0.3 \\ -0.5 & 1 & -0.4 \\ 0.3 & -0.4 & 1.5 \end{pmatrix} \qquad \Sigma_3 = 0.75 \times \Sigma_2.$$

   (a) Generate 2,000 random vectors from each of the three distributions, and determine the three-dimensional sources for these data sets with the skewness approximation $\mathcal{G}_3$ of Algorithm 10.1. Plot the three-dimensional data and the three direction vectors $\boldsymbol{\omega}_k$, where $k \leq 3$. Calculate the skewness of each of the source components and compare.
   (b) Repeat part (a) 100 times, and display the direction vectors $\boldsymbol{\omega}_k$ in an appropriate plot.

(c) For the data of part (a), calculate the direction vectors $\widetilde{\mathbf{X}}$ of (9.2) in Section 9.1. Repeat the calculations of part (b) for the direction vectors of these mixture distributions. Compare the results of parts (b) and (c), and comment.

7. Give a proof of part 1 of Theorem 10.9.
8. Let π be the probability density function of a bivariate source $\mathbf{S}$ with identical marginals. For the marginals, consider the four distributions of Example 10.5. Let A be an invertible 2×2 matrix, and put $\mathbf{X} = A\mathbf{S}$. Let f be the probability density function of $\mathbf{X}$.
   (a) Evaluate the four terms in (10.11), based on f, for the different distributions π.
   (b) Repeat part (a) for the bivariate white signal $\mathbf{X}^{\diamond}$ and its probability density function $f_{\diamond}$.
   (c) Simulate $\mathbb{S}$ from the source distributions, estimate the probability density function of $\mathbb{X} = A\mathbb{S}$ and repeat part (a) for the density estimates $\widehat{f}$ of the signals $\mathbb{X}$. Compare the results.
9. Give a detailed proof of Corollary 10.10 stating precisely which results are used and how.
10. Consider the *athletes* data of Example 8.3 in Section 8.2.2. Carry out the following analyses separately for the raw and scaled data.
    (a) Calculate the first three principal component vectors and the first three independent component vectors with the skewness and kurtosis approximations $\mathcal{G}_3$ and $\mathcal{G}_4$.
    (b) Display the PC and IC scores in three-dimensional scatterplots, and also show the scores in separate plots.
    (c) Compare the PC and IC results, and comment.
    (d) Explain briefly the difference in the results obtained from the raw and scaled data, and comment on the appropriateness of either.
11. For $k = 1, 2$, let $\mathcal{N}(\mu_k, \sigma_k^2)$ be univariate normal distributions such that $\mu_1 < \mu_2$, and let ϕ_k be the corresponding probability density functions. For $0 < \alpha < 1$, put $f = \alpha\phi_1 + (1-\alpha)\phi_2$.
    (a) Determine the mean and variance of f, and find the Gaussian probability density functions f_G which has the same mean and variance as f.
    (b) Calculate the negentropy $\mathcal{J}(f)$ and the Kullback-Leibler divergence $\mathcal{K}(f, f_G)$ of (10.12) for $\mu_1 = 0$ fixed and a range of values for μ_2. Comment on the results.
12. Consider the *breast cancer* data which we first analysed in Example 2.6 in Section 2.4.1. Calculate the first three ICs for the raw and scaled data, and interpret these scores using graphical displays or otherwise. Next, consider the p-white data for $p \geq 3$, and find the first three ICs. Discuss the changes in the ICs as the dimension p changes.
13. Consider the assumption of part 2 of Theorem 10.12. Show that for $\mathbf{T} = \mathbf{X}_0$
    (a) $\mathcal{D}_2(\mathbf{T}, \mathbf{S}) = 0$, and
    (b) $\mathcal{D}_4(\mathbf{T}, \mathbf{S}) = \mathcal{D}_4^{\mathbf{I}}(\mathbf{T}, \mathbf{S}) + c$ with $\mathcal{D}_4^{\mathbf{I}}(\mathbf{T}, \mathbf{S})$ as in the theorem.
14. Consider the *breast cancer* data of Problem 12, and call it $\mathbb{X}$. Let $\mathbb{X}_k$, with $k = 1, 2, 3$, be the following subsets of the data $\mathbb{X}$: the data $\mathbb{X}_1$ consist of the first 200 observations of $\mathbb{X}$, $\mathbb{X}_2$ consist of the last 200 observations of $\mathbb{X}$ and the $\mathbb{X}_3$ consist of the first 400 observations of $\mathbb{X}$. Apply the dimension selector $\widehat{I}_\rho$ of (10.39) to the raw and scaled versions of the three subsets using both the skewness and kurtosis versions of $\widehat{I}_\rho$. Compare your results with those obtained in Table 10.5 and with the results obtained in Problem 12.

Projection Pursuit

15. For the PCA and DA frameworks, give expressions for the projection indices $\widehat{Q}$ of the sample $\mathbb{X} = \begin{bmatrix}\mathbf{X}_1 & \cdots & \mathbf{X}_n\end{bmatrix}$. Find the maximum that is obtained by the index in each of the two cases, and explain what the index maximises in each case.
16. Give a proof of
    (a) Proposition 11.3, and
    (b) expression (11.9) for the probability density function f_Θ.
17. For the four distributions in Example 11.2, calculate numerical values for Q_U.
18. Discuss how one can define a projection index in a Canonical Correlation framework. Give an expression for a suitable index, find its maximum and give an interpretation of the index. Point out one of the main differences between an index for a PC setting and a CC setting.
19. Derive (11.8) to (11.10) for univariate and bivariate probability density functions f.
20. Show that $\theta_1(X_1)$ in (11.23) is standard normal, and using (11.23), show that $\mathbf{X}$ can be expressed in the form (11.22).
21. List the main steps in the proof of Theorem 11.6. *Hint:* Show that

$$\int_{-1}^{1}\left(f_R - \frac{1}{2}\right)^2 = \int_{-1}^{1} f_R^2 - \frac{1}{2} = \int_{-1}^{1}\sum_{j=1}^{\infty}\alpha_j p_j(R) f_R(R)\, dR - \frac{1}{2},$$

    where the coefficients

$$\alpha_j = \int_{-1}^{1} p_j(R) f_R(R)\, dR.$$

22. Derive a result analogous to Theorem 11.6 for the sample index $\widehat{Q}_D$, and sketch its proof.

Kernel and More Independent Component Solutions

23. For $z \in \mathbb{R}$, define a feature map for random variables x by $\mathfrak{f}_z(x) = a\exp[-\beta(z-x)^2]$, for some $a, \beta > 0$. If y is another random variable, define the kernel k associated with $\mathfrak{f}$ by

$$k(x, y) = \int_{\mathbb{R}} \mathfrak{f}_z(x)\mathfrak{f}_z(y)\, dz.$$

    Give an explicit expression for k.
24. For the polynomial kernel of Table 12.1 and $m = 2$, give an explicit expression of the feature map $\mathfrak{f}$, and derive expressions for the feature covariance operator $S_\mathfrak{f}$ and the feature scores.
25. Give an explicit proof of Theorem 12.3. First explain how one casts the feature data into the framework of the $Q_{\langle n\rangle}$ and $Q^{\langle d\rangle}$ matrices, and then show explicitly how this set-up is used to give the desired results.
26. For the *athletes* data of Example 8.3 in Section 8.2.2, regard the variables of Table 8.2 without the variable *Sex* as the observations.
    (a) Use the exponential kernel with $\beta = 1,2$ and a range of values for a, and calculate the feature scores.
    (b) Repeat part (a) for the polynomial kernels with $m = 2,3$.

(c) Display the features scores in two- and three-dimensional score plots. Compare these score plots with the configurations in Figure 8.2 in Section 8.2.2 and with score plots of the PC scores. Comment on the results.

27. Describe a scenario where the ideas of Kernel Canonical Correlation may be more appropriate than Canonical Correlation Analysis, and explain why and in which way they are 'better'. Explain how you would interpret $\kappa(\boldsymbol{\alpha}_1, \boldsymbol{\alpha}_2)$ of (12.16) in Theorem 12.5.
28. Consider the *athletes* data of Problem 26, but now treat the 201 athletes as the observations.
    (a) For the Gaussian kernel with $\sigma = 1$, find the kernel independent solutions with the KCC and KGV approaches.
    (b) Calculate the independent solutions with the FastICA skewness and kurtosis criteria, and compare the results of the four sets of solutions visually. Comment on the results.
29. Explain the main idea of the kernel generalised variance approach of Bach and Jordan (2002), and point out similarities and differences to the method of Theorem 12.8.
30. (a) Show that the eigenvalues of $M(\mathbf{X})$ and $N(\mathbf{X})$ as in Theorems 12.10 and 12.11 agree and that $\mathbf{q}$ is an eigenvector of $M(\mathbf{X})$ which corresponds to the eigenvalue λ if and only if $\mathbf{e} = \mathsf{S}_1(\mathbf{X})^{-1/2}\mathbf{q}$ is the corresponding eigenvector of $N(\mathbf{X})$.
    (b) In the notation of Theorem 12.10, suppose that the affine equivariant scatter statistic S_1 is replaced by an affine equivariant statistic $\widetilde{\mathsf{S}}_1$ which satisfies $\widetilde{\mathsf{S}}_1(\mathbf{T}) = \mathbf{I}_{d\times d}$ and $\widetilde{\mathsf{S}}_1(\mathbf{X}) \neq c\Sigma$, for any scalars c. How does the conclusion of Theorem 12.10 change? Prove your assertion.
31. Consider the *athletes* data of Problem 26, and treat the 201 athletes as the observations. Let S_1 be the scatter statistic of Theorem 12.10. Define S_2 by the first symmetrised matrix in (12.48) for the approach of Oja, Sirkiä, and Eriksson (2006) and by (12.50) for the approach of Tyler et al. (2009). Calculate the independent component solutions for both approaches, and compare the results.
32. Consider a random vector $\mathbf{X}$ which belongs to one of κ classes. For $\ell = 1, 2$, let S_ℓ be scatter statistics which satisfy the assumptions of Theorem 12.10, and define $\mathbf{T}_0 = Q(\mathbf{X})^{\mathsf{T}}\mathbf{X}_\Sigma$ using the notation of Theorem 12.10.
    (a) Let $\mathfrak{r}_F$ be Fisher's rule for classifying $\mathbf{X}$. Define Fisher's rule for $\mathbf{T}_0$, and show that the two rules result in the same classification.
    (b) Repeat part (a) for the normal linear rule under the assumption that the classes have the same covariance matrix Σ.
33. (a) Show that part 3 of Proposition 12.13 is equivalent to the other two parts.
    (b) Show that the second statement of (12.54) holds when f is the product density.
    (c) Sketch a proof of Proposition 12.15.
34. (a) Describe the spline-based approach of Hastie and Tibshirani (2002).
    (b) Explain the wavelet approach of Barbedor (2007).
    (c) Compare the two approaches to the non-parametric approach of Section 12.5.3.

Feature Selection and Principal Component Analysis Revisited

35. Apply Algorithm 13.1 to the *breast cancer* data of Example 4.12 in Section 4.8.3. Apply Fisher's discriminant rule, the naive Bayes rule and the IB1 classifier to the derived features of step 5 of the algorithm, and use 10 per cent cross-validation to determine the training and testing sets.

(a) Use all variables instead of choosing a subset of variables as in step 1 of the algorithm.
(b) Use 95 per cent of variance as in step 1 of the algorithm to find the first p principal components.
(c) Use Table 10.5 in Section 10.8 for the number of features in step 1 of the algorithm.
(d) Interpret your results and compare them with the results of the analyses in Example 4.12.

36. Apply Algorithm 13.2 to the *breast cancer* data of Example 4.12 in Section 4.8.3, and compare the two clusters with the partitionings obtained from the PC_1 and PC_2 sign cluster rules and the labels.
37. Assume that $\mathbf{X}$ and $\mathbf{Y}$ are mean zero random vectors and have non-singular covariance matrices Σ_X and Σ_Y respectively. Let C be the matrix of canonical correlations of $\mathbf{X}$ and $\mathbf{Y}$, and put $\widetilde{C} = \Sigma_X^{-1/2} C\, \Sigma_Y^{1/2}$. Let $\boldsymbol{\varphi}_1$ and $\boldsymbol{\psi}_1$ be the first left and right canonical correlation transforms.

(a) Show that $\widetilde{C}\boldsymbol{\psi}_1 = \upsilon_1 \boldsymbol{\varphi}_1$ for some scalar υ_1.
(b) Consider data $\mathbb{X}$ and $\mathbb{Y}$ which have non-singular covariance matrices. Let $\widehat{C}$ be the sample canonical correlation matrix. Prove (13.8).
(c) For data $\mathbb{X}$ and $\mathbb{Y}$ as in part (b), derive the expression for $\widehat{\boldsymbol{\varphi}}_1$ and $\widehat{\mathbf{p}}_1$ given in (13.9) and (13.10).

38. Give a proof of Theorem 13.1.
39. Consider the matrix $\widehat{C}_{\mathrm{NB}}$ in Table 13.4. Calculate $\widehat{C}_{\mathrm{NB}}^{\mathsf{T}}\widehat{C}_{\mathrm{NB}}$, and find its rank and the first eigenvalue and eigenvector. Explain how you can use the eigenvalue-eigenvector pair to calculate the first left eigenvector of $\widehat{C}_{\mathrm{NB}}$.
40. Consider the *lung cancer* data of Example 13.7. Rank the data with $\widehat{\mathbf{p}}_{\mathrm{NB}}$ and $\widehat{\mathbf{b}}_{\mathrm{NB}}$. As in Algorithm 4.3 in Section 4.8.3, use the naive Bayes rule on $p = 2, 3, \ldots$ ranked variables, and calculate the classification error. Find the optimal number of ranked variables as in Algorithm 4.3 for all data, and repeat for training and testing data. How do these optimal numbers of variables compare with those obtained in Example 13.7?
41. Let $\mathbb{X}$ be d-dimensional centred data with sample covariance matrix S. Let r be the rank of S. For $1 \le j \le r$, let $\mathbf{W}_{\bullet j}$ be the jth PC scores. Let $\zeta_2 > 0$, and put

$$\widehat{\boldsymbol{\gamma}}_0 = \underset{\boldsymbol{\gamma}}{\operatorname{argmin}} \left\| \mathbf{W}_{\bullet j} - \boldsymbol{\gamma}^{\mathsf{T}}\mathbb{X} \right\|_2^2 + \zeta_2 \left\| \boldsymbol{\gamma} \right\|_2^2 .$$

Show that $\widehat{\boldsymbol{\gamma}}_0 = c\widehat{\boldsymbol{\eta}}_j$ for some constant $c > 0$, where $\widehat{\boldsymbol{\eta}}_j$ is the jth eigenvector of S. Find the value of c.
42. For the *athletes* data of Problem 26, calculate the PC scores, the sparse SCoTLASS PC scores and the sparse elastic net PC scores with $\zeta_2 = 0$. Use the values of t and $\boldsymbol{\zeta}_1$ as in Table 13.6, and produce a table similar to Table 13.6 for the athletes data.
43. Let $\mathbb{X}$ be a $d \times n$ matrix of centred observations, and write $\mathbb{X} = \widehat{\Gamma}\widehat{D}\widehat{L}^{\mathsf{T}}$ for its singular value decomposition.

(a) Show that (13.44) holds.
(b) Let $k \le r$, the rank of the covariance matrix S of $\mathbb{X}$. Define the matrix $\mathbb{X}^{(k)}$ as in (13.46), and prove that $(\widehat{d}_{k+1}\widehat{\boldsymbol{\eta}}_{k+1}, \widehat{\boldsymbol{\nu}}_{k+1})$ is its best rank one solution with respect to the Frobenius norm.

44. Determine the differences between Algorithms 13.6 and 13.7, and comment on these differences. Also consider and comment on computational aspects which might affect the solutions.
45. Explain the extension to Canonical Correlation data of the method of Witten, Tibshirani, and Hastie (2009), and describe a scenario where this method would be useful in practice.

Bibliography

Abramowitz, M., and I. A. Stegun (1965). *Handbook of Mathematical Functions*. New York: Dover.

Aeberhard, S., D. Coomans and O. de Vel (1992). Comparison of classifiers in high dimensional settings. Tech. Rep. No. 92-02, Dept. of Computer Science and Dept. of Mathematics and Statistics, James Cook University of North Queensland. Data sets collected by Forina et al. and available at:www.kernel-machines.com/.

Aebersold, R., and M. Mann (2003). Mass spectrometry-based proteomics. *Nature 422*, 198–207.

Aha, D., and D. Kibler (1991). Instance-based learning algorithms. *Machine Learning 6*, 37–66.

Aharon, M., M. Elad and A. Bruckstein (2006). K-SVD: An algorithm for designing overcomplete dictionaries for sparse representation. *IEEE Trans. on Signal Processing 54*, 4311–4322.

Ahn, J., and J. S. Marron (2010). The maximal data piling direction for discrimination. *Biometrika 97*, 254–259.

Ahn, J., J. S. Marron, K. M. Mueller and Y.-Y. Chi (2007). The high-dimension low-sample-size geometric representation holds under mild conditions. *Biometrika 94*, 760–766.

Amari, S.-I. (2002). Independent component analysis (ICA) and method of estimating functions. *IEICE Trans. Fundamentals E 85A*(3), 540–547.

Amari, S.-I., and J.-F. Cardoso (1997). Blind source separation—Semiparametric statistical approach. *IEEE Trans. on Signal Processing 45*, 2692–2700.

Amemiya, Y., and T. W. Anderson (1990). Asymptotic chi-square tests for a large class of factor analysis models. *Ann. Stat. 18*, 1453–1463.

Anderson, J. C., and D. W. Gerbing (1988). Structural equation modeling in practice: A review and recommended two-step approach. *Psychol. Bull. 103*, 411–423.

Anderson, T. W. (1963). Asymptotic theory for principal component analysis. *Ann. Math. Stat. 34*, 122–148.

Anderson, T. W. (2003). *Introduction to Multivariate Statistical Analysis* (3rd ed.). Hoboken, NJ: Wiley.

Anderson, T. W., and Y. Amemiya (1988). The asymptotic normal distribution of estimators in factor analysis under general conditions. *Ann. Stat. 16*, 759–771.

Anderson, T. W., and H. Rubin (1956). Statistical inference in factor analysis. In *Third Berkeley Symposium on Mathematical Statistics and Probability 5*, pp. 111–150. Berkely: University California Press.

Attias, H. (1999). Independent factor analysis. *Neural Comp. 11*, 803–851.

Bach, F. R., and M. I. Jordan (2002). Kernel independent component analysis. *J. Machine Learning Res. 3*, 1–48.

Baik, J., and J. W. Silverstein (2006). Eigenvalues of large sample covariance matrices of spiked population models. *J. Multivar. Anal. 97*, 1382–1408.

Bair, E., T. Hastie, D. Paul and R. Tibshirani (2006). Prediction by supervised principal components. *J. Am. Stat. Assoc. 101*(473), 119–137.

Barbedor, P. (2009). Independent component analysis by wavelets. *Test 18*(1), 136–155.

Bartlett, M. S. (1938). Further aspects of the theory of multiple regression. *Proc. Cambridge Philos. Soc. 34*, 33–40.

Bartlett, M. S. (1939). A note on tests of significance in multivariate analysis. *Proc. Cambridge Philos. Soc. 35*, 180–185.

Beirlant, J., E. J. Dudewicz, L. Györfi and E. van der Meulen (1997). Nonparametric entropy estimation: An overview. *Int. J. Math. Stat. Sci. 6*, 17–39.

Bell, A. J., and T. J. Sejnowski (1995). An information-maximization approach to blind separation and blind deconvolution. *Neural Comput. 7*, 1129–1159.

Benaych-Georges, F., and R. Nadakuditi (2012). The eigenvalues and eignvectors of finite, low rank perturbations of large random matrices. *Adv. Math. 227*, 494–521.

Berger, J. O. (1993). *Statistical Decision Theory and Bayesian Analysis*. New York: Springer-Verlag.

Berlinet, A., and C. Thomas-Agnan (2004). *Reproducing Kernel Hilbert Spaces in Probability and Statistics*. Boston: Kluwer Academic Publishers.

Bickel, P. J., and E. Levina (2004). Some theory for Fisher's linear discriminant function, 'naïve Bayes', and some alternatives when there are many more variables than observations. *Bernoulli 10*, 989–1010.

Blake, C., and C. Merz (1998). UCI repository of machine learning databases. Data sets available at: www.kernel-machines.com/.

Borg, I., and P. J. F. Groenen (2005). *Modern Multidimensional Scaling: Theory and Applications* (2nd ed.). New York: Springer.

Borga, M., H. Knutsson and T. Landelius (1997). Learning canonical correlations. In *Proceedings of the 10th Scandinavian Conference on Image Nanlysis*, Lappeenranta, Finland.

Borga, M., T. Landelius and H. Knutsson (1997). A unified approach to PCA, PLS, MLR and CCA. Technical report, Linköping University, Sweden.

Boscolo, R., H. Pan and V. P. Roychowdhury (2004). Indpendent component analysis based on nonparametric density estimation. *IEEE Trans. Neural Networks 15*(1), 55–65.

Breiman, L. (1996). Bagging predictors. *Machine Learning 26*, 123–140.

Breiman, L. (2001). Random forests. *Machine Learning 45*, 5–32.

Breiman, L., J. Friedman, J. Stone and R. A. Olshen (1998). *Classification and Regression Trees*. Boca Raton, FL: CRC Press.

Cadima, J., and I. T. Jolliffe (1995). Loadings and correlations in the interpretation of principle components. *J. App. Stat. 22*, 203–214.

Calinski, R. B., and J. Harabasz (1974). A dendrite method for cluster analysis. *Commun. Stat. 3*, 1–27.

Candès, E. J., J. Romberg and T. Tao (2006). Robust undertainty principles: Exact signal reconstruction from highly imcomplete frequency information. *IEEE Trans. Inform. Theory 52*, 489–509.

Cao, X.-R., and R.-W. Liu (1996). General approach to blind source separation. *IEEE Trans. Signal Processing 44*, 562–571.

Cardoso, J.-F. (1998). Blind source separation: Statistical principles. *Proc. IEEE 86*(10), 2009–2025.

Cardoso, J.-F. (1999). High-order contrasts for independent component analysis. *Neural Comput. 11*(1), 157–192.

Cardoso, J.-F. (2003). Dependence, correlation and Gaussianity in independent component analysis. *J. Machine Learning Res. 4*, 1177–1203.

Carroll, J. D., and J. J. Chang (1970). Analysis of individual differences in multidimensional scaling via a n-way generalization of 'Eckart-Young' decomposition. *Psychometrika 35*, 283–319.

Casella, G., and R. L. Berger (2001). *Statistical Inference*. Pacific Grove, CA: Wadsworth & Brooks/Cole Advanced Books and Software.

Chaudhuri, P., and J. S. Marron (1999). Sizer for exploration of structures in curves. *J. Am. Stat. Assoc. 94*, 807–823.

Chaudhuri, P., and J. S. Marron (2000). Scale space view of curve estimation. *Ann. Stat. 28*, 408–428.

Chen, A., and P. J. Bickel (2006). Efficient independent component analysis. *Ann. Stat. 34*, 2825–2855.

Chen, J. Z., S. M. Pizer, E. L. Chaney and S. Joshi (2002). Medical image synthesis via Monte Carlo simulation. In T. Dohi and R. Kikinis (eds.), *Medical Image Computing and Computer Assisted Intervention (MICCAI)*. Berlin: Springer. pp. 347–354.

Chen, L., and A. Buja (2009). Local multidimensional scaling for nonlinear dimension reduction, graph drawing and proximity analysis. *J. Am. Stat. Assoc. 104*, 209–219.

Choi, S., A. Cichocki, H.-M. Park and S.-Y. Lee (2005). Blind source separation and independent component analysis: A review. *Neural Inform. Processing 6*, 1–57.

Comon, P. (1994). Independent component analysis, A new concept? *Signal Processing 36*, 287–314.

Cook, D., A. Buja and J. Cabrera (1993). Projection pursuit indices based on expansions with orthonormal functions. *J. Comput. Graph. Stat. 2*, 225–250.

Cook, D., and D. Swayne (2007). *Interactive and Dynamic Graphics for Data Analysis*. NewYork: Springer.

Cook, R. D. (1998). *Regression Graphics: Ideas for Studying Regressions through Graphics*. NewYork: Wiley.

Cook, R. D., and S. Weisberg (1999). *Applied Statistics Including Computing and Graphics*. New York: Wiley.

Cook, R. D., and X. Yin (2001). Dimension reduction and visualization in discriminant analysis (with discussion). *Aust. NZ J. Stat. 43*, 147–199.

Cormack, R. M. (1971). A review of classification (with discussion). *J. R. Stat. Soc. A 134*, 321–367.

Cover, T. M., and P. Hart (1967). Nearest neighbor pattern classification. *Proc. IEEE Trans. Inform. Theory IT-11*, 21–27.

Cover, T. M., and J. A. Thomas (2006). *Elements of Information Theory* (2nd ed.). Hoboken, NJ: John Wiley.

Cox, D. R., and D. V. Hinkley (1974). *Theoretical Statistics*. London: Chapman and Hall.

Cox, T. F., and M. A. A. Cox (2001). *Multidimensional Scaling* (2nd ed.). London: Chapman and Hall.

Cristianini, N., and J. Shawe-Taylor (2000). *An Introduction to Support Vector Machines*. Cambridge University Press.

Davies, C., P. Corena and M. Thomas (2012). South Australian grapevine data. CSIRO Plant Industry, Glen Osmond, Australia, personal communication.

Davies, P. I., and N. J. Higham (2000). Numerically stable generation of correlation matrices and their factors. *BIT 40*, 640–651.

Davies, P. M., and A. P. M. Coxon (1982). *Key Texts in Multidimensional Scaling*. London: Heinemann Educational Books.

De Bie, T., N. Cristianini and R. Rosipal (2005). Eigenproblems in pattern recognition. In E. Bayro-Corrochano (ed.), *Handbook of Geometric Computing: Applications in Pattern Recognition, Computer Vision, Neuralcomputing, and Robotics*, pp. 129–170. New York: Springer.

de Silva, V., and J. B. Tenenbaum (2004). Sparse multidimensional scaling using landmark points. Technical report, Standford University.

Devroye, L., L. Györfi and G. Lugosi (1996). *A Probabilistic Theory of Pattern Recognition. Applications of Mathematics*. New York: Springer.

Diaconis, P., and D. Freedman (1984). Asymptotics of graphical projection pursuit. *Ann. Stat. 12*, 793–815.

Domeniconi, C., J. Peng and D. Gunopulos (2002). Locally adaptive metric nearest-neighbor classification. *IEEE Trans. Pattern Anal. Machine Intell. PAMI-24*, 1281–1285.

Domingos, P., and M. Pazzani (1997). On the optimality of the simple Bayesian classifier under zero-one loss. *Machine Learning 29*, 103–130.

Donoho, D. L. (2000). Nature vs. math: Interpreting independent component analysis in light of recent work in harmonic analysis. In *Proceedings International Workshop on Independent Component Analysis and Blind Signal Separation (ICA2000)*, Helsinki, Finland, pp. 459–470.

Donoho, D. L. (2006). Compressed sensing. *IEEE Trans. Inform. Theory 52*, 1289–1306.

Donoho, D. L., and I. M. Johnstone (1994). Ideal denoising in an orthonormal basis chosen from a library of bases. *Comp. Rendus Acad. Sci. A 319*, 1317–1322.

Dryden, I. L., and K. V. Mardia (1998). *The Statistical Analysis of Shape*. New York: Wiley.

Dudley, R. M. (2002). *Real Analysis and Probability*. Cambridge University Press.

Dudoit, S., J. Fridlyand and T. P. Speed (2002). Comparisons of discrimination methods for the classification of tumors using gene expression data. *J. Am. Stat. Assoc. 97*, 77–87.

Duong, T., A. Cowling, I. Koch and M. P. Wand (2008). Feature significance for multivariate kernel density estimation. *Comput. Stat. Data Anal. 52*, 4225–4242.

Duong, T., and M. L. Hazelton (2005). Cross-validation bandwidth matrices for multivariate kernel density estimation. *Scand. J. Stat. 32*, 485–506.

Elad, M. (2010). *Sparse and Redundant Representations: From Theory to Applications in Signal and Image Processing*. New York: Springer.

Eriksson, J., and V. Koivunen (2003). Characteristic-function based independent component analysis. *Signal Processing 83*, 2195–2208.

Eslava, G., and F. H. C. Marriott (1994). Some criteria for projection pursuit. *Stat. Comput. 4*, 13–20.

Fan, J., and Y. Fan (2008). High-dimensional classification using features annealed independence rules. *Ann. Stat. 36*, 2605–2637.

Figueiredo, M. A. T., and A. K. Jain (2002). Unsupervised learning of finite mixture models. *IEEE Trans. Pattern Anal. Machine Intell. PAMI-24*, 381–396.

Fisher, R. A. (1936). The use of multiple measurements in taxonomic problems. *Ann. Eugenics 7*, 179–188.

Fix, E., and J. Hodges (1951). Discriminatory analysis, nonparametric discrimination: Consistency properties. Technical report, Randolph Field, TX, USAF School of Aviation Medicine.

Fix, E., and J. Hodges (1952). Discriminatory analysis: Small sample performance. Technical report, Randolph Field, TX, USAF School of Aviation Medicine.

Flury, B., and H. Riedwyl (1988). *Multivariate Statistics: A Practical Approach*. Cambridge University Press. Data set available at: www-math.univ-fcomte.fr/mismod/userguide/node131.html.

Fraley, C., and A. Raftery (2002). Model-based clustering, discriminant ananlysis, and density estimation. *J. Am. Stat. Assoc. 97*, 611–631.

Friedman, J. H. (1987). Exploratory projection pursuit. *J. Am. Stat. Assoc. 82*, 249–266.

Friedman, J. H. (1989). Regularized discriminant analysis. *J. Am. Stat. Assoc. 84*, 165–175.

Friedman, J. H. (1991). Multivariate adaptive regression splines. *Ann. Stat. 19*, 1–67.

Friedman, J. H., and W. Stuetzle (1981). Projection pursuit regression. *J. Am. Stat. Assoc. 76*, 817–823.

Friedman, J. H., W. Stuetzle and A. Schroeder (1984). Projection pursuit density estimation. *J. Am. Stat. Assoc. 79*, 599–608.

Friedman, J. H., and J. W. Tukey (1974). A projection pursuit algorithm for exploratory data analysis. *IEEE Trans. Comput. C-23*, 881–890.

Gentle, J. E. (2007). *Matrix Algebra*. New York: Springer.

Gilmour, S., and I. Koch (2006). Understanding illicit drug markets with independent component analysis. Technical report, University of New South Wales.

Gilmour, S., I. Koch, L. Degenhardt and C. Day (2006). Identification and quantification of change in Australian illicit drug markets. *BMC Public Health 6*, 200–209.

Givan, A. L. (2001). *Flow Cytometry: First Principles* (2nd ed.). New York: Wiley-Liss.

Gokcay, E., and J. C. Principe (2002). Information theoretic clustering. *IEEE Trans. Pattern Anal. Machine Intell. PAMI-24*, 158–171.

Gordon, G. J., R. V. Jensen, L. Hsiao, S. R. Gullans, J. E. Blumenstock, S. Ramaswamy, W. G. Richards, D. J. Sugarbaker and R. Bueno (2002). Translation of microarray data into clinically relevant cancer diagnostic tests using gene expression ratios in lung cancer and mesothelioma. *Cancer Res. 62*, 4963–4967.

Gower, J. C. (1966). Some distance properties of latent root and vector methods used in multivariate analysis. *Biometrika 53*, 325–338.

Gower, J. C. (1968). Adding a point to vector diagrams in multivariate analysis. *Biometrika 55*, 582–585.

Gower, J. C. (1971). Statistical methods of comparing different multivariate analyses of the same data. In F. R. Hodson, D. Kendall, and P. Tautu (eds.), *Mathematics in the Archeological and Historical Sciences*, pp. 138–149. Edinburgh University Press.

Gower, J. C., and W. J. Krzanowski (1999). Analysis of distance for structured multivariate data and extensions to multivariate analysis of variance. *Appl. Stat. 48*, 505–519.

Graef, J., and I. Spence (1979). Using distance information in the design of large multidimensional scaling experiments. *Psychol. Bull. 86*, 60–66.

Greenacre, M. J. (1984). *Theory and Applications of Correspondence Analysis*. New York: Academic Press.

Greenacre, M. J. (2007). *Correspondence Analysis in Practice* (2nd ed.). London: Chapman and Hall/CRC Press.

Gustafsson, J. O. R. (2011). MALDI imaging mass spectrometry and its application to human disease. Ph.D. thesis, University of Adelaide.

Gustafsson, J. O. R., M. K. Oehler, A. Ruszkiewicz, S. R. McColl and P. Hoffmann (2011). MALDI imaging mass spectrometry (MALDI-IMS): Application of spatial proteomics for ovarian cancer classification and diagnosis. *Int. J. Mol. Sci. 12*, 773–794.

Guyon, I., and A. Elisseeff (2003). An introduction to variable and feature selection. *J. Machine Learning Res. 3*, 1157–1182.

Hall, P. (1988). Estimating the direction in which a data set is most interesting. *Prob. Theory Relat. Fields 80*, 51–77.

Hall, P. (1989a). On projection pursuit regression. *Ann. Stat. 17*, 573–588.

Hall, P. (1989b). Polynomial projection pursuit. *Ann. Stat. 17*, 589–605.
Hall, P., and K.-C. Li (1993). On almost linearity of low dimensional projections from high dimensional data. *Ann. Stat. 21*, 867–889.
Hall, P., J. S. Marron and A. Neeman (2005). Geometric representation of high dimension low sample size data. *J. R. Stat. Soc. B (JRSS-B) 67*, 427–444.
Hand, D. J. (2006). Classifier technology and the illusion of progress. *Stat. Sci. 21*, 1–14.
Harrison, D., and D. L. Rubinfeld (1978). Hedonic prices and the demand for clean air. *J. Environ. Econ. Manage. 5*, 81–102. (http://lib.stat.cmu.edu/datasets/boston).
Hartigan, J. (1975). *Clustering Algorithms*. New York: Wiley.
Hartigan, J. A. (1967). Representation of similarity matrices by trees. *J. Am. Stat. Assoc. 62*, 1140–1158.
Harville, D. A. (1997). *Matrix Algebra from a Statistician's Perspective*. New York: Springer.
Hastie, T., and R. Tibshirani (1996). Discriminant adaptive nearest neighbor classification. *IEEE Trans. Pattern Anal. Machine Intell. PAMI-18*, 607–616.
Hastie, T., and R. Tibshirani (2002). Independent component analysis through product density estimation. In *Proceedings of Neural Information Processing Systems*, pp. 649–656.
Hastie, T., R. Tibshirani and J. Friedman (2001). *The Elements of Statistical Learning – Data Mining, Inference, and Prediction*. New York: Springer.
Helland, I. S. (1988). On the structure of partial least squares regression. *Commun. Stat. Simul. Comput. 17*, 581–607.
Helland, I. S. (1990). Partial least squares regression and statistical models. *Scand. J. Stat. 17*, 97–114.
Hérault, J., and B. Ans (1984). Circuits neuronaux à synapses modifiables: décodage de messages composites par apprentissage non supervisé. *Comp. Rendus Acad. Sci. 299*, 525–528.
Hérault, J., C. Jutten and B. Ans (1985). Détection de grandeurs primitives dans un message composite par une architecture de calcul neuromimétique en apprentissage non supervisé. In *Actes de Xème colloque GRETSI*, pp. 1017–1022, Nice, France.
Hinneburg, A., C. C. Aggarwal and D. A. Keim (2000). What is the nearest neighbor in high dimensional spaces? In *Proceedings of the 26th International Conference on Very Large Data Bases*, Cairo, Egypt, pp. 506–515.
Hotelling, H. (1933). Analysis of a complex of statistical variables into principal components. *J. Educ. Psych. 24*, 417–441 and 498–520.
Hotelling, H. (1935). The most predictable criterion. *J. Exp. Psychol. 26*, 139–142.
Hotelling, H. (1936). Relations between two sets of variates. *Biometrika 28*, 321–377.
Huber, P. J. (1985). Projection pursuit. *Ann. Stat. 13*, 435–475.
Hyvärinen, A. (1999). Fast and robust fixed-point algorithm for independent component analysis. *IEEE Trans. Neural Networks 10*, 626–634.
Hyvärinen, A., J. Karhunen and E. Oja (2001). *Independent Component Analysis*. New York: Wiley.
ICA Central (1999). available at://www.tsi.enst.fr/icacentral/.
Inselberg, A. (1985). The plane with parallel coordinates. *Visual Computer 1*, 69–91.
Izenman, A. J. (2008). *Modern Multivariate Statistical Techniques*. New York: Springer.
Jeffers, J. (1967). Two case studies in the application of principal components. *Appl. Stat. 16*, 225–236.
Jing, J., I. Koch and K. Naito (2012). Polynomial histograms for multivariate density and mode estimation. *Scand. J. Stat. 39*, 75–96.
John, S. (1971). Some optimal multivariate tests. *Biometrika 58*, 123–127.
John, S. (1972). The distribution of a statistic used for testing sphericity of normal distributions. *Biometrika 59*, 169–173.
Johnstone, I. M. (2001). On the distribution of the largest principal component. *Ann. Stat. 29*, 295–327.
Johnstone, I. M., and A. Y. Lu (2009). On consistency and sparsity for principal components analysis in high dimensions. *J. Am. Stat. Assoc. 104*, 682–693.
Jolliffe, I. T. (1989). Rotation of ill-defined principal components. *Appl. Stat. 38*, 139–147.
Jolliffe, I. T. (1995). Rotation of principal components: Choice of normalization constraints. *J. Appl. Stat. 22*, 29–35.
Jolliffe, I. T., N. T. Trendafilov and M. Uddin (2003). A modified principal component technique based on the LASSO. *J. Comput. Graph. Stat. 12*, 531–547.

Jones, M. C. (1983). The projection pursuit algorithm for exploratory data analysis. Ph.D. thesis, University of Bath.

Jones, M. C., and R. Sibson (1987). What is projection pursuit? *J. R. Stat. Soc. A (JRSS-A) 150*, 1–36.

Jöreskog, K. G. (1973). A general method for estimating a linear structural equation system. In A. S. Goldberger and O. D. Duncan (eds.), *Structural Equation Models in the Social Sciences*, pp. 85–112. San Francisco: Jossey-Bass.

Jung, S., and J. S. Marron (2009). PCA consistency in high dimension low sample size context. *Ann. Stat. 37*, 4104–4130.

Jung, S., A. Sen, and J. S. Marron (2012). Boundary behavior in high dimension, low sample size asymptotics of pca. *J. Multivar. Anal. 109*, 190–203.

Kaiser, H. F. (1958). The varimax criterion for analytic rotation in factor analysis. *Psychometrika 23*, 187–200.

Kendall, M., A. Stuart and J. Ord (1983). *The Advanced Theory of Statistics*, Vol. 3. London: Charles Griffin & Co.

Klemm, M., J. Haueisen and G. Ivanova (2009). Independent component analysis: Comparison of algorithms for the inverstigation of surface electrical brain activity. *Med. Biol. Eng. Comput. 47*, 413–423.

Koch, I., J. S. Marron and J. Chen (2005). Independent component analysis and simulation of non-Gaussian populations of kidneys. Technical Report.

Koch, I., and K. Naito (2007). Dimension selection for feature selection and dimension reduction with principal and independent component analysis. *Neural Comput. 19*, 513–545.

Koch, I., and K. Naito (2010). Prediction of multivariate responses with a selected number of principal components. *Comput. Stat. Data Anal. 54*, 1791–1807.

Kruskal, J. B. (1964a). Multidimensional scaling by optimizing goodness-of-fit to a nonmetric hypothesis. *Psychometrika 29*, 1–27.

Kruskal, J. B. (1964b). Nonmetric multidimensional scaling: A numerical method. *Psychometrika 29*, 115–129.

Kruskal, J. B. (1969). Toward a practical method which helps uncover the structure of a set of multivariate observations by fining the linear transformation which optimizes a new 'index of condensation'. In R. C. Milton and J. A. Nelder (eds.), *Statistical Computation*, pp. 427–440. New York: Academic Press.

Kruskal, J. B. (1972). Linear transformation of multivariate data to reveal clustering. In R. N. Shepard, A. K. Rommey and S. B. Nerlove (eds.), *Multidimensional Scaling: Theory and Applications in the Behavioural Sciences*, Vol. I, pp. 179–191. London: Seminar Press.

Kruskal, J. B., and M. Wish (1978). *Multidimensional Scaling*. Beverly Hills, CA: Sage Publications.

Krzanowski, W. J., and Y. T. Lai (1988). A criterion for determining the number of groups in a data set using sum of squares clustering. *Biometrics 44*, 23–34.

Kshirsagar, A. M. (1972). *Multivariate Analysis*. New York: Marcell Dekker.

Kullback, S. (1968). Probability densities with given marginals. *Ann. Math. Stat. 39*, 1236–1243.

Lawley, D. N. (1940). The estimation of factor loadings by the method of maximum likelihood. *Proc. R. Soc. Edinburgh A 60*, 64–82.

Lawley, D. N. (1953). A modified method of estimation in factor analysis and some large sample results. In *Uppsala Symposium on Psychlogical Factor Analysis*, Vol. 17(19). Uppsala, Sweden: Almqvist and Wiksell, pp. 34–42.

Lawley, D. N., and A. E. Maxwell (1971). *Factor Analysis as a Statistical Method*. New York: Elsevier.

Lawrence, N. (2005). Probabilistic non-linear principal component analysis with gaussian process latent variable models. *J. Machine Learning Res. 6*, 1783–1816.

Learned-Miller, E. G., and J. W. Fisher (2003). ICA using spacings estimates of entropy. *J. Machine Learning Res. 4*, 1271–1295.

Lee, J. A., and M. Verleysen (2007). *Nonlinear Dimensionality Reduction*. New York: Springer.

Lee, S., F. Zou and F. A. Wright (2010). Convergence and prediction of principal component scores in high-dimensional settings. *Ann. Stat. 38*, 3605–3629.

Lee, T.-W. (1998). *Independent Component Analysis Theory and Applications*. Boston: Academic Publishers, Kluwer.

Lee, T.-W., M. Girolami, A. J. Bell and T. J. Sejnowski (2000). A unifying information-theoretic framework for independen component analysis. *Comput. Math. Appl. 39*, 1–21.

Lee, T.-W., M. Girolami and T. J. Sejnowski (1999). Independent component analysis using an extended infomax algorithm for mixed subgaussian and super-Gaussian sources. *Neural Comput. 11*, 417–441.

Lee, Y. K., E. R. Lee, and B. U. Park (2012). Principal component analysis in very high-dimensional spaces. *Stat. Sinica 22*, 933–956.

Lemieux, C., I. Cloutier and J.-F. Tanguay (2008). Estrogen-induced gene expression in bone marrow c-kit+ stem cells and stromal cells: Identification of specific biological processes involved in the functional organization of the stem cell niche. *Stem Cells Dev. 17*, 1153–1164.

Leng, C. and H. Wang (2009). On general adaptive sparse principal component analysis. *J. of Computational and Graphical Statistics 18*, 201–215.

Li, K.-C. (1992). On principal Hessian directions for data visualization and dimension reduction: Another application of Stein's lemma. *J. Am. Stat. Assoc. 87*, 1025–1039.

Lu, A. Y. (2002). Sparse principal component analysis for functional data. Ph.D. thesis, Dept. of Statistics, Stanford University.

Ma, Z. (2013). Sparse principal component analysis and iterative thresholding. *Ann. Stat. 41*, 772–801.

Malkovich, J. F., and A. A. Afifi (1973). On tests for multivariate normality. *J. Am. Stat. Assoc. 68*, 176–179.

Mallat, S. (2009). *A Wavelet Tour of Signal Processing the Sparse Way* (3d ed.). New York: Academic Press.

Mammen, E., J. S. Marron and N. I. Fisher (1991). Some asymptotics for multimodality tests based on kernel density estimates. *Prob. Theory Relat. Fields 91*, 115–132.

Marčenko, V. A., and L. A. Pastur (1967). Distribution of eigenvalues of some sets of random matrices. *Math. USSR-Sb 1*, 507–536.

Mardia, K. V., J. Kent and J. Bibby (1992). *Multivariate Analysis*. London: Academic Press.

Marron, J. S. (2008). Matlab software. pcaSM.m and curvdatSM.m available at: www.stat.unc.edu/postscript/papers/marron/Matlab7Software/General/.

Marron, J. S., M. J. Todd and J. Ahn (2007). Distance-weighted discrimination. *J. Am. Stat. Assoc. 102(480)*, 1267–1271.

McCullagh, P. (1987). *Tensor Methods in Statistics*. London: Chapman and Hall.

McCullagh, P., and J. Kolassa (2009). Cumulants. *Scholarpedia 4*, 4699.

McCullagh, P., and J. A. Nelder (1989). *Generalized Linear Models* (2nd ed.), Vol. 37 of *Monographs on Statistics and Applied Probability*. London: Chapman and Hall.

McLachlan, G., and K. Basford (1988). *Mixture Models: Inference and Application to Clustering*. New York: Marcel Dekker.

McLachlan, G., and D. Peel (2000). *Finite Mixture Models*. New York: Wiley.

Meulman, J. J. (1992). The integration of multidimensional scaling and multivariate analysis with optimal transformations. *Psychometrika 57*, 530–565.

Meulman, J. J. (1993). Principal coordinates analysis with optimal transformation of the variables – minimising the sum of squares of the smallest eigenvalues. *Br. J. Math. Stat. Psychol. 46*, 287–300.

Meulman, J. J. (1996). Fitting a distance model to homogeneous subsets of variables: Points of view analysis of categorical data. *J. Classification 13*, 249–266.

Miller, A. (2002). *Subset Selection in Regression* (2nd ed.), Vol. 95 of *Monographs on Statistics and Applied Probability*. London: Chapman and Hall.

Milligan, G. W., and M. C. Cooper (1985). An examination of procedures for determining the number of clusters in a data set. *Psychometrika 50*, 159–179.

Minka, T. P. (2000). Automatic choice of dimensionality for PCA. Tech Report 514, MIT. Available at ftp://whitechapel.media.mit.edu/pub/tech-reports/.

Minotte, M. C. (1997). Nonparametric testing of the existence of modes. *Ann. Stat. 25*, 1646–1660.

Nadler, B. (2008). Finite sample approximation results for principal component analysis: A matrix perturbation approach. *Ann. Stat. 36*, 2791–2817.

Nason, G. (1995). Three-dimensional projection pursuit. *Appl. Stat. 44*, 411–430.

Ogasawara, H. (2000). Some relationships between factors and components. *Psychometrika 65*, 167–185.

Oja, H., S. Sirkiä and J. Eriksson (2006). Scatter matrices and independent component analysis. *Aust. J. Stat. 35*, 175–189.

Partridge, E. (1982). *Origins, A Short Etymological Dictionary of Modern English* (4th ed.). London: Routledge and Kegan Paul.

Paul, D. (2007). Asymptotics of sample eigenstructure for a large dimensional spiked covariance model. *Stat. Sinica 17*, 1617–1642.

Pearson, K. (1901). On lines and planes of closest fit to systems of points in space. *Philos. Mag. 2*, 559–572.

Prasad, M. N., A. Sowmya, and I. Koch (2008). Designing relevant features for continuous data sets using ICA. *Int. J. Comput. Intell. Appl. (IJCIA) 7*, 447–468.

Pryce, J. D. (1973). *Basic Methods of Linear Functional Analysis*. London: Hutchinson.

Qiu, X., and L. Wu (2006). Nearest neighbor discriminant analysis. *Int. J. Pattern Recog. Artif. Intell. 20*, 1245–1259.

Quist, M., and G. Yona (2004). Distributional scaling: An algorithm for structure-preserving embedding of metric and nonmetric spaces. *J. Machine Learning Res. 5*, 399–420.

R Development Core Team (2005). R: A language and environment for statistical computing. R Foundation for Statistical Computing, Vienna, Austria.

Rai, C. S., and Y. Singh (2004). Source distribution models for blind source separation. *Neurocomputing 57*, 501–505.

Ramaswamy, S., P. Tamayo, R. Rifkin, S. Mukheriee, C. Yeang, M. Angelo, C. Ladd, M. Reich, E. Latulippe, J. Mesirov, T. Poggio, W. Gerald, M. Loda, E. Lander and T. Golub (2001). Multiclass cancer diagnosis using tumor gene expression signature. *Proc. Nat. Aca. Sci. 98*, 15149–15154.

Ramos, E., and D. Donoho (1983). Statlib datasets archive: Cars. Available at: http://lib.stat.cmu.edu/datasets/.

Ramsay, J. O. (1982). Some statistical approaches to multidimensional scaling data. *J. R. Stat. Soc. A (JRSS-A) 145*, 285–312.

Rao, C. (1955). Estimation and tests of significance in factor analysis. *Psychometrika 20*, 93–111.

Richardson, M. W. (1938). Multidimensional psychophysics. *Psychol. Bull. 35*, 650=660.

Ripley, B. D. (1996). *Pattern Recognition and Neural Networks*. Cambridge University Press.

Rosipal, R., and L. J. Trejo (2001). Kernel partial least squares regression in reproducing kernel hilbert spaces. *J. Machine Learning Res. 2*, 97–123.

Rossini, A., J. Wan and Z. Moodie (2005). Rflowcyt: Statistical tools and data structures for analytic flow cytometry. R package, version 1. available at:: http://cran.r-project.org/web/packages/.

Rousson, V., and T. Gasser (2004). Simple component analysis. *J. R. Stat. Soc. C (JRSS-C) 53*, 539–555.

Roweis, S., and Z. Ghahramani (1999). A unifying review of linear gaussian models. *Neural Comput. 11*, 305–345.

Roweis, S. T., and L. K. Saul (2000). Nonlinear dimensionality reduction by local linear embedding. *Science 290*, 2323–2326.

Rudin, W. (1991). *Functional Analysis* (2nd ed.). New York: McGraw-Hill.

Sagae, M., D. W. Scott and N. Kusano (2006). A multivariate polynomial histogram by the method of local moments. In *Proceedings of the 8th Workshop on Nonparametric Statistical Analysis and Related Area*, Tokyo, pp. 14–33 (in Japanese).

Samarov, A., and A. Tsybakov (2004). Nonparametric independent component analysis. *Bernoulli 10*, 565–582.

Sammon, J. W. (1969). A nonlinear mapping for data structure analysis. *IEEE Trans. Computers 18*, 401–409.

Schneeweiss, H., and H. Mathes (1995). Factor analysis and principal components. *J. Multivar. Anal. 55*, 105–124.

Schoenberg, I. J. (1935). ‘Remarks to Maurice Fréchet’s article ‘Sur la définition axiomatique d’une classe d’espaces distanciés vectoriellement applicable sur l’espaces de Hilbert.’. *Ann. Math. 38*, 724–732.

Schölkopf, B., and A. Smola (2002). *Learning with Kernels. Support Vector Machines, Regularization, Optimization and Beyond*. Cambridge, MA: MIT Press.

Schölkopf, B., A. Smola and K.-R. Müller (1998). Nonlinear component analysis as a kernel eigenvalue problem. *Neural Comput. 10*, 1299–1319.

Schott, J. R. (1996). *Matrix Analysis for Statistics*. New York: Wiley.

Scott, D. W. (1992). *Multivariate Density Estimation: Theory, Practice, and Visualization*. New York: Wiley.

Searle, S. R. (1982). *Matrix Algebra Useful for Statistics*. New York: John Wiley.
Serfling, R. J. (1980). *Approximation Theorems of Mathematical Statistics*. New York: Wiley.
Shen, D., H. Shen, and J. S. Marron (2012). A general framework for consistency of principal component analysis. arXiv:1211.2671.
Shen, D., H. Shen, and J. S. Marron (2013). Consistency of sparse PCA in high dimension, low sample size. *J. Multivar. Anal. 115*, 317–333.
Shen, D., H. Shen, H. Zhu, and J. S. Marron (2012). High dimensional principal component scores and data visualization. arXiv:1211.2679.
Shen, H., and J. Huang (2008). Sparse principal component analysis via regularized low rank matrix approximation. *J. Multivar. Anal. 99*, 1015–1034.
Shepard, R. N. (1962a). The analysis of proximities: Multidimensional scaling with an unknown distance function I. *Psychometrika 27*, 125–140.
Shepard, R. N. (1962b). The analysis of proximities: Multidimensional scaling with an unknown distance function II. *Psychometrika 27*, 219–246.
Short, R. D., and K. Fukunaga (1981). Optimal distance measure for nearest neighbour classification. *IEEE Trans. Inform. Theory IT-27*, 622–627.
Silverman, B. W. (1981). Using kernel density estimates to investigate multimodality. *J. R. Stat. Soc. B (JRSS-B) 43*, 97–99.
Silverman, B. W. (1986). *Density Estimation for Statistics and Data Analysis*, Vol. 26 of *Monographs on Statistics and Applied Probability*. London: Chapman and Hall.
Starck, J.-L., E. J. Candès and D. L. Donoho (2002). The curvelet transform for image denoising. *IEEE Trans. Image Processing 11*, 670–684.
Strang, G. (2005). *Linear Algebra and Its Applications* (4th ed.). New York: Academic Press.
Tamatani, M., I. Koch and K. Naito (2012). Pattern recognition based on canonical correlations in a high dimension low sample size context. *J. Multivar. Anal. 111*, 350–367.
Tamatani, M., K. Naito, and I. Koch (2013). Multi-class discriminant function based on canonical correlation in high dimension low sample size. preprint.
Tenenbaum, J. B., V. de Silva and J. C. Langford (2000). A global geometric framework for nonlinear dimensionality reduction. *Science 290*, 2319–2323.
Tibshirani, R. (1996). Regression shrinkage and selection via the lasso. *J. R. Stat. Soc. B (JRSS-B) 58*, 267–288.
Tibshirani, R., and G. Walther (2005). Cluster validation by prediction strength. *J. Comput. Graph. Stat. 14*, 511–528.
Tibshirani, R., G. Walther and T. Hastie (2001). Estimating the number of clusters in a dataset via the gap statistic. *J. R. Stat. Soc. B (JRSS-B) 63*, 411–423.
Tipping, M. E., and C. M. Bishop (1999). Probabilistic principal component analysis. *J. R. Stat. Soc. B (JRSS-B) 61*, 611–622.
Torgerson, W. S. (1952). Multidimensional scaling: 1. Theory and method. *Psychometrika 17*, 401–419.
Torgerson, W. S. (1958). *Theory and Method of Scaling*. New York: Wiley.
Torokhti, A., and S. Friedland (2009). Towards theory of generic principal component analysis. *J. Multivar. Anal. 100*, 661–669.
Tracy, C. A., and H. Widom (1996). On orthogonal and symplectic matrix ensembles. *Commun. Math. Phys. 177*, 727–754.
Tracy, C. A., and H. Widom (2000). The distribution of the largest eigenvalue in the Gaussian ensembles. In J. van Diejen and L. Vinet (eds.), *Cologero-Moser-Sutherland Models*, pp. 461–472. New York: Springer.
Trosset, M. W. (1997). Computing distances between convex sets and subsets of the positive semidefinite matrices. Technical Rep. 97-3, Rice University.
Trosset, M. W. (1998). A new formulation of the nonmetric strain problem in multidimensional scaling. *J. Classification 15*, 15–35.
Tucker, L. R., and S. Messick (1963). An individual differences model for multidimensional scaling. *Psychometrika 28*, 333–367.
Tyler, D. E., F. Critchley, L. Dümgen and H. Oja (2009). Invariant coordinate selection. *J. R. Stat. Soc. B (JRSS-B) 71*, 549–592.

van't Veer, L. J., H. Dai, M. J. van de Vijver, Y. D. He, A. A. M. Hart, M. Mao, H. L. Peterse, K. van der Kooy, M. J. Marton, A. T. Witteveen, G. J. Schreiber, R. M. Kerkhoven, C. Roberts, P. S. Linsley, R. Bernards and S. H. Friend (2002). Gene expression profiling predicts clinical outcome of breast cancer. *Nature 415*, 530–536.

Vapnik, V. (1995). *The Nature of Statistical Learning Theory*. New York: Springer-Verlag.

Vapnik, V. (1998). *Statistical Learning Theory*. New York: Wiley.

Vapnik, V., and A. Chervonenkis (1979). *Theorie der Zeichenerkennung*. Berlin: Akademie-Verlag (German translation from the original Russian, published in 1974).

Vasicek, O. (1976). A test for normality based on sample entropy. *J. R. Stat. Soc. B (JRSS-B) 38*, 54–59.

Venables, W. N., and B. D. Ripley (2002). *Modern Applied Statistics with S* (4th ed.). New York: Springer.

Vines, S. K. (2000). Simple principal components. *Appl. Stat. 49*, 441–451.

Vlassis, N., and Y. Motomura (2001). Efficient source adaptivity in independent component analysis. *IEEE Trans. on Neural Networks 12*, 559–565.

von Storch, H., and F. W. Zwiers (1999). *Statistical Analysis of Climate Research*. Cambridge University Press.

Wand, M. P., and M. C. Jones (1995). *Kernel Smoothing*. London: Chapman and Hall.

Wegman, E. (1992). The grand tour in k-dimensions. In *Computing Science and Statistics*. New York: Springer, pp. 127–136.

Williams, R. H., D. W. Zimmerman, B. D. Zumbo and D. Ross (2003). Charles Spearman: British behavioral scientist. *Human Nature Rev. 3*, 114–118.

Winther, O., and K. B. Petersen (2007). Bayesian independent component analysis: Variational methods and non-negative decompositions. *Digital Signal Processing 17*, 858–872.

Witten, D. M., and R. Tibshirani (2010). A framework for feature selection in clustering. *J. Am. Stat. Assoc. 105*, 713–726.

Witten, D. M., R. Tibshirani, and T. Hastie (2009). A penalized matrix decomposition, with applications to sparse principal components and canonical correlation analysis. *Biostatistics 10*, 515–534.

Witten, I. H., and E. Frank (2005). *Data Mining: Practical Machine Learning Tools and Techniques* (2nd ed.). San Francisco: Morgan Kaufmann.

Wold, H. (1966). Estimation of principal components and related models by iterative least squares. In P. R. Krishnaiah (ed.), *Multivariate Analysis*, pp. 391–420. New York: Academic Press.

Xu, R., and D. Wunsch II (2005). Survey of clustering algorithms. *IEEE Trans. Neural Networks 16*, 645–678.

Yeredor, A. (2000). Blind source separation via the second characteristic function. *Signal Processing 80*, 897–902.

Young, G., and A. S. Householder (1938). Discussion of a set of points in terms of their mutual distances. *Psychometrika 3*, 19–22.

Zass, R., and A. Shashua (2007). Nonnegative sparse PCA. In *Advances in Neural Information Processing Systems (NIPS-2006)*, Vol. 19, B. Schölkopf, J. Platt, and T. Hofmann, eds., p. 1561. Cambridge, MA: MIT Press.

Zaunders, J., J. Jing, A. D. Kelleher and I. Koch (2012). Computationally efficient analysis of complex flow cytometry data using second order polynomial histograms. Technical Rep., St Vincent's Centre for Applied Medical Research, St Vincent's Hospital, Australia, Sydney.

Zou, H., and T. Hastie (2005). Regularization and variable selection with the elastic net. *J. R. Stat. Soc. B (JRSS-B) 67*, 301–320.

Zou, H., T. Hastie and R. Tibshirani (2006). Sparse principal component analysis. *J. Comput. Graph. Stat. 15*, 265–286.

Author Index

Subject Index

Data Index

abalone ($d = 8, n = 4,177$)
2 PCA, 46, 64, 68, 166
3 CCA, 112
6 CA, 191
7 FA, 239
assessment marks ($d = 6, n = 23$)
8 MDS, 276
athletes ($d = 12, n = 202$)
8 MDS, 255, 260
11 PP, 370

Boston housing ($d = 14, n = 506$)
3 CCA, 87
breast cancer ($d = 30, n = 569$)
2 PCA, 29, 32, 43, 46, 65, 166
4 DA, 137, 151, 162
6 CA, 196, 209
7 FA, 239
breast tumour ($d = 4,751, 24,481, n = 78$)
2 PCA, 50
8 MDS, 270
13 FS-PCA, 436

car ($d = 5, n = 392$)
3 CCA, 75, 79
7 FA, 228, 234
cereal ($d = 11, n = 77$)
8 MDS, 264

Dow Jones returns ($d = 30, n = 2,528$)
2 PCA, 29, 32, 166
6 CA, 211
7 FA, 232

exam grades ($d = 5, n = 120$)
7 FA, 238, 244

HIV flow cytometry ($d = 5, n = 10,000$)
1 MDD, 5
2 PCA, 24, 41, 165
6 CA, 220
12 K&MICA, 397

HRCT emphysema ($d = 21, n = 262,144$)
13 FS-PCA, 423, 425

illicit drug market ($d = 66, n = 17$)
1 MDD, 7
2 PCA, 49
3 CCA, 80, 84, 89, 98, 106
6 CA, 210
7 FA, 230
8 MDS, 259, 277
10 ICA, 329, 343
12 K&MICA, 388
13 FS-PCA, 427
income ($d = 9, n = 1,000$)
2 PCA, 166
3 CCA, 95, 102
iris ($d = 4, n = 150$)
1 MDD, 5, 6
4 DA, 124, 150
6 CA, 187, 193
10 ICA, 327

(data bank of) kidneys ($d = 264, n = 36$)
10 ICA, 336
13 FS-PCA, 429, 430

lung cancer ($d = 12,553, n = 181$)
13 FS-PCA, 448

ovarian cancer proteomics ($d = 1,331, n = 14,053$)
2 PCA, 52
6 CA, 213
8 MDS, 278, 281

PBMC flow cytometry ($d = 10, n = 709,086$)
6 CA, 202
pitprops ($d = 14, n = 180$)
13 FS-PCA, 451, 456

simulated data ($d = 2 - 50, n = 100$–$10,000$)
2 PCA, 24, 35, 165

4 DA, 125, 128, 130, 141, 145
6 CA, 195, 208
9 NG, 296
10 ICA, 342, 346
11 PP, 367
sound tracks ($d = 2, n = 24,000$)
10 ICA, 308, 329
South Australian grapevine
($d = 19, n = 2,062$)
6 CA, 204

Swiss bank notes ($d = 6, n = 200$)
2 PCA, 28, 30

ten cities ($n = 10$)
8 MDS, 250, 253, 267

wine recognition ($d = 13, n = 178$)
2 PCA, 31
4 DA, 134, 139
12 K&MICA, 399, 409